Advances in Intelligent Systems and Computing

Volume 517

Series editor

Janusz Kacprzyk, Polish Academy of Sciences, Warsaw, Poland
e-mail: kacprzyk@ibspan.waw.pl

About this Series

The series "Advances in Intelligent Systems and Computing" contains publications on theory, applications, and design methods of Intelligent Systems and Intelligent Computing. Virtually all disciplines such as engineering, natural sciences, computer and information science, ICT, economics, business, e-commerce, environment, healthcare, life science are covered. The list of topics spans all the areas of modern intelligent systems and computing.

The publications within "Advances in Intelligent Systems and Computing" are primarily textbooks and proceedings of important conferences, symposia and congresses. They cover significant recent developments in the field, both of a foundational and applicable character. An important characteristic feature of the series is the short publication time and world-wide distribution. This permits a rapid and broad dissemination of research results.

More information about this series at http://www.springer.com/series/11156

Subhransu Sekhar Dash · K. Vijayakumar
Bijaya Ketan Panigrahi · Swagatam Das
Editors

Artificial Intelligence and Evolutionary Computations in Engineering Systems

Proceedings of ICAIECES 2016

Editors
Subhransu Sekhar Dash
Department of Electrical and Electronics Engineering
SRM University
Chennai, Tamil Nadu
India

Bijaya Ketan Panigrahi
Department of Electrical and Electronics Engineering
Indian Institute of Technology Delhi
New Delhi
India

K. Vijayakumar
Department of Electrical and Electronics Engineering
SRM University
Chennai, Tamil Nadu
India

Swagatam Das
Electronics and Communication Sciences Unit
Indian Statistical Institute
Kolkata, West Bengal
India

ISSN 2194-5357 ISSN 2194-5365 (electronic)
Advances in Intelligent Systems and Computing
ISBN 978-981-10-3173-1 ISBN 978-981-10-3174-8 (eBook)
DOI 10.1007/978-981-10-3174-8

Library of Congress Control Number: 2016958719

Printed on acid-free paper

This Springer imprint is published by Springer Nature
The registered company is Springer Nature Singapore Pte Ltd.
The registered company address is: 152 Beach Road, #21-01/04 Gateway East, Singapore 189721, Singapore

Preface

The volume contains the papers presented at the International Conference on Artificial Intelligence and Evolutionary Computations in Engineering Systems (ICAIECES) held during 19–21 May, 2016 at SRM University, Chennai, India. For ICAIECES 2016, we have received 316 papers from various countries across the globe. After a rigorous peer-review process, 70 full-length articles were accepted for oral presentation at the conference. This corresponds to an acceptance rate of 23.73% and is intended for maintaining the high standards of the conference proceedings. The papers included in this AISC volume cover a wide range of topics on evolutionary programming and evolution strategies such as genetic algorithms, artificial intelligence system, differential evolution, particle swarm optimization, ant colony optimization, bacterial foraging optimization algorithm, harmony search algorithm, shuffled frog leaping algorithm, artificial bees and fireflies algorithm, optimization techniques and their applications for solving problems in the area of engineering and technology. In the conference, separate sessions were arranged for delivering the keynote address by eminent professionals from various academic institutions and industries. The various keynote speakers during the conference are Padmasree Y.S. Rajan, Vice President, Forum For Global Knowledge Sharing, Bangalore; Dr. Lipo Wang, Professor, NTU, Singapore; Dr. Ponnambalam S.G., Professor, Monash University, Malaysia; Dr. Swagatam Das, Electronics and Communication Sciences Unit, Indian Statistical Institute, Kolkata; Dr. K. Shanti Swarup, Professor, IIT Madras; Dr. P. Somasundaram, Associate Professor, Anna University, Chennai; and Dr. D. Devaraj, Dean, Academic, Kalasalingam University, Krishnankoil. All these lectures generated a great interest among the participants in paying more attention in these emerging areas.

We take this opportunity to thank the authors of all the papers for their hard work, adherence to the deadlines, and for suitably incorporating the changes suggested by the reviewers. The quality of a refereed volume depends mainly on the expertise and dedication of the reviewers. We are indebted to the Program Committee members for their guidance and coordination in organizing the review process.

We would also like to thank our sponsors for providing all the support and financial assistance. We are indebted to the Chancellor, Vice Chancellor, Advisors, Pro-Vice Chancellor, Registrar, Director (E&T), Faculty members, and Administrative Personnel of SRM University, Kattankulathur, Chennai for supporting our cause and encouraging us to organize the conference on a grand scale. We would also like to thank all the participants for their interest and enthusiastic involvement. Finally, we would like to thank all the volunteers for their tireless efforts in meeting the deadlines and arranging every detail meticulously for the smooth conduct of the conference. We hope the readers of these proceedings find the papers useful, inspiring, and enjoyable.

Chennai, India — Shubhransu Sekhar Dash
Chennai, India — K. Vijayakumar
New Delhi, India — Bijaya Ketan Panigrahi
Kolkata, India — Swagatam Das
May 2016

About the Book

The volume is a collection of high-quality peer-reviewed research papers presented in the International Conference on Artificial Intelligence and Evolutionary Computation in Engineering Systems (ICAIECES 2016) held at SRM University, Chennai, Tamilnadu, India. This conference is an international forum for industry professionals and researchers to deliberate and state their research findings, discuss the latest advancements, and explore the future directions in the emerging areas of engineering and technology. The book presents original work and novel ideas, information, techniques, and applications in the field of communication, computing, and power technologies.

Contents

About the Editors

Dr. Subhransu Sekhar Dash is presently Professor in the Department of Electrical and Electronics Engineering, SRM University, Chennai, India. He received his Ph.D. degree from College of Engineering, Guindy, Anna University, Chennai, India. He has more than 19 years of research and teaching experience. His research areas include power electronics and drives, modeling of facts controller, power quality, power system stability and smart grid. He is a Visiting Professor at Francois Rabelais University, POLYTECH, France and visiting research scholar at University of Wisconsin, Millwauke, USA.

Dr. K. Vijayakumar is presently HOD and Professor in the Department of Electrical and Electronics Engineering, SRM University, Chennai, India. He received his Ph.D. degree from SRM University, Chennai, India. He has more than 19 years of research and teaching experience. His research areas are modeling, control, and operation of power system, optimization of power system, flexible AC transmission system, and power quality.

Dr. Bijaya Ketan Panigrahi is Associate Professor of Electrical and Electronics Engineering Department in Indian Institute of Technology, Delhi, India. He received his Ph.D. degree from Sambalpur University. He is a chief editor in the International Journal of Power and Energy Conversion. His areas of interests include power quality, facts devices, power system protection, and AI application to power system.

Dr. Swagatam Das received the B.E. Tel.E., M.E. Tel.E (Control Engineering specialization), and Ph.D. degrees all from Jadavpur University, India, in 2003, 2005, and 2009, respectively. He is currently serving as Assistant Professor at the Electronics and Communication Sciences Unit of the Indian Statistical Institute, Kolkata, India. His research interests include evolutionary computing, pattern recognition, multi-agent systems, and wireless communication. He has published one research monograph, one edited volume, and more than 200 research articles in peer-reviewed journals and international conferences.

Social Media Sentiment Polarity Analysis: A Novel Approach to Promote Business Performance and Consumer Decision-Making

Arya Valsan, C.T. Sreepriya and L. Nitha

Abstract In order to have a clear understanding of the market structure as well as the customer trends toward various products, there is a need for every company to collect, monitor, and analyze the user data generated online. In this paper, the online reviews of products from two leading camera manufacturers have been utilized to analyze the user trends. After preprocessing the data, sentiment analysis techniques have been employed to mine the textual content of customers' opinion and classify them into different polarities according to the theoretical conceptualization of service and performance. The sentiment analysis results using Support Vector Machine provide a high level of accuracy in encapsulating and measuring the sentiments of customers toward the products and services as compared to the other text mining strategies. Further, a competitive analysis technique based on K-means clustering has been implemented to examine the most frequent word which is discussed by the customer. The combination of these two methods provides benefit not only to obtain the best classification but also to help the user focus on the most relevant categories that meet his/her interest.

Keywords Competitive intelligence analysis · K-means algorithm · Preprocessing · Sentiment analysis · Support vector machine

A. Valsan (✉) · C.T. Sreepriya · L. Nitha
Department of Computer Science and IT, Amrita School of Arts and Sciences, Kochi Amrita Vishwa Vidyapeetham (Amrita University), Kochi, India
e-mail: maloottyarya@gmail.com

C.T. Sreepriya
e-mail: sreepriya.lakshmi09@gmail.com

L. Nitha
e-mail: nitha.leeladharan@gmail.com

S.S. Dash et al. (eds.), *Artificial Intelligence and Evolutionary Computations in Engineering Systems*, Advances in Intelligent Systems and Computing 517,
DOI 10.1007/978-981-10-3174-8_1

1 Introduction

Data mining has become a necessary task in the world of artificial intelligence, where human interpretation of large volume data has become time-consuming and troublesome. It becomes always difficult for a customer to interpret the user-generated content in the social media since it belongs to same target industry. Especially when the user-generated reviews are larger in volume, the companies need to examine the customer-generated opinions about their products and services. They need to track opinions about their competitor's product and services through social media analytics. This evokes the necessity of a classification and clustering algorithm to perform the aforementioned tasks. This paper put forward a technique based on data mining in order to identify the social media sentiments toward the products and there by enhance business performance. The proposed system explores the capability of a classifier known as support vector machine (SVM), for performing the sentiment analysis.

As the first step, data preprocessing step is needed to ensure that raw data are transformed into usable format. In the second step, SVM performs the classification task by separating the examples of different categories by training data set and predicting the sentiments based on the trained data set. After classification, the probability of individual data set is retrieved and based on that the accuracy of classification is analyzed. In the third step, K-means algorithm is used. K-means is the most frequent text clustering algorithm in the data mining and statistics. After preprocessing the data set, we retrieve words and its number of occurrence. The result is applied as the input to the K-means algorithm which in turn forms different clusters of industry specific words. In the current study, we obtained an extensive data source of online camera reviews. The data allow us to get a better understanding of the trends and market structure. We employ the sentiment analysis technique to mine the textual content of customer's opinion and classify them into different polarity according to the theoretical conceptualization of service and performance. The sentiment analysis results give high level of accuracy in encapsulating and measuring sentiments of customer toward the products and services as compared to other text mining tools.

2 Literature Review

He et al. [1] approach presents social media analytics framework with sentiment benchmarks for gathering industry specific marketing intelligence. With this approach, a new social media analytics with sentiment benchmarks can be developed to further improve the marketing intelligence and point out the areas where business have to be improved as well as the area where they are leading. For this analysis, they have used customer opinions obtained from social media sites, such as Face book, Twitter, Blogs, etc. and for the purpose of mining the data, they have utilized supervised as well as unsupervised text mining strategies. Both these classification methods exhibit advantages and disadvantages. Unsupervised method can be executed independently but there is no guarantee that this method will generate meaningful results. On the other hand, supervised method needs manual effort and intelligence; hence it demands a large workforce. Therefore, they have combined both these methods as a hybrid approach.

Hu et al. [2] reveals the fact that the buyers look for user reviews posted on the seller's web sites while they are shopping online. The user reviews help the buyers in their decision-making process. The authors have mined and summarized the user reviews. They have used a new summarization method which is different from the traditional approach because they have mined only the features of the product from the reviews posted in sites in order to determine whether the opinion sentence is positive or negative.

Yu et al. [3] mentions that SVMs have several challenging methods for classification and regression analysis because of their stable mathematical principles, which are not provided by other approaches. For a large set of data, SVM is not suitable for pattern recognition and machine learning because of the complexity in training such huge data sets. SVM's mathematical principles convey two important features such as margin maximization and nonlinear transformation of the feature space using the kernel trick. In this paper, the margin maximization has been carried out using quadratic programming problem. The existing SVM is not feasible for large data sets and such data sets are not frequently accessed because of the highly expensive I/O operations. Therefore, a new approach to this problem has been formulated in this paper.

A novel learning approach has been presented by Durgesh et al. [4]. The authors have applied SVM on different types of data sets, which have more than two classes. The paper mentions about SVM as a very powerful learning method derived from statistical learning. In this work, they have found out the support vectors from the trained data set and have presented the results obtained by applying different kernel functions on the entire data set. They have experimented with a number of parameters which have a direct influence on the result. These parameters include selection of a kernel function, standard deviation of Gaussian kernel, number of training data set, etc.

He et al. [5] deals with the study of applying text mining methods in a large set of unstructured data. The authors point out the necessity for companies to monitor the customer-generated data posted on their sites as well as the customer reviews of the competitor's sites also. This will help the companies to increase their competitive advantages and effectively assess the competitive business environment. The final results reveal the importance of social media competitive analysis and the intensity of text mining methods that can be used as a powerful mining technique to extract business values from a large amount of data available on the social media sites.

Mostafa [6] discusses about the influence of social media sites in product purchasing. He has used a number of tweets collected from social media sites to analyze customers' sentiments toward a well-known brand of products. He had used a predefined lexicon which includes a number of seed adjectives needed for the analysis process. The final output demonstrates the positive user sentiments toward popular brands.

Hotho et al. [7] conveys that a large amount of unstructured data cannot be simply used for further processing. Therefore, they have used some preprocessing methods to identify the useful patterns. In this paper, a description of the analysis of text preprocessing, classification, clustering, etc. has been provided. The authors describe text as a challenging method on information retrieval, machine learning, and statistics. Initially, they have furnished a brief overview of the methods and later on, these methods have been related with data mining. In the next stage, the different approaches of task analysis like preprocessing, classification, clustering, information extraction, and visualization have been described.

Abraham et al. [8] explores the text mining method applied in social media, which is used by the person having interest in different models of vehicles. They posted their opinions in the online discussion forum, but this forum was not enough to identify and categorize the defects of the vehicles. In this paper, they explore a new method and decision support system to identify and prioritize the defects of vehicles in an excellent manner.

An accurate clustering method, which performs well without any pre-assignment of exact number of clusters, has been detailed in the paper by Zalik [9]. K-means algorithm has been found to perform this task accurately. The author has used minimization of the cost function for achieving this task. K-means algorithm includes two main steps. In the first step, initial clustering is performed by the preprocessing method and minimum one seed point is assigned to each cluster, while the cost function is minimized in the second step. The exact number of clusters is identified and the remaining seed points are assigned near the actual clusters when the cost function converges to a global minimum. Excellent performance has been achieved by the proposed method.

Jain [10] discusses about the clustering analysis using K-means algorithm. Cluster analysis deals with the grouping of objects based on some characteristics or similarity, without the use of any class labels. It is entirely different from classification because of the absence of category information. The clustering provides a structure in the data. K-means is the simple algorithm for cluster analysis, which requires three user-defined parameters such as number of clusters, initial clusters as well as the distance metrics. In this paper, he has given a good overview of the cluster analysis and a summarized view of the popular clustering methods.

3 Methodology

In our proposed method we combine both supervised and unsupervised approach to produce a hybrid method. Hybrid approach combines both the top-down method (supervised) and bottom-up method (unsupervised) for classifying text into industry specific categories for user's interest. In top-down approach, we use SVM for

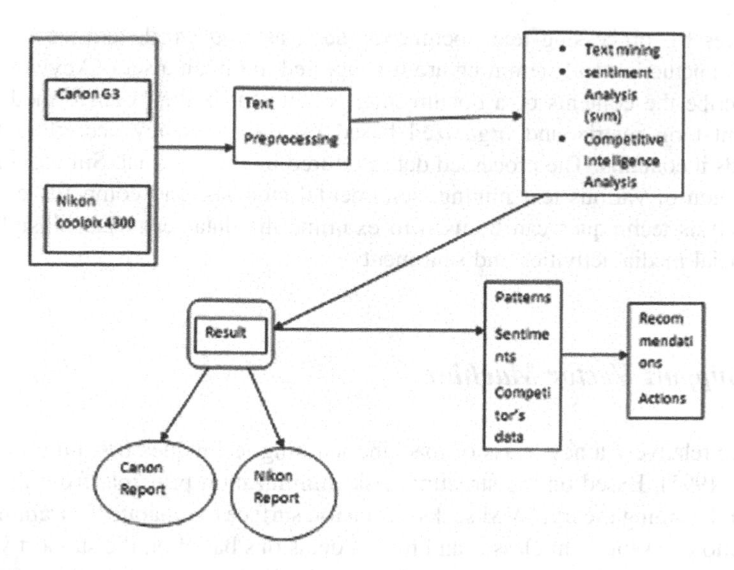

Fig. 1 Block diagram of proposed system

categorizing user's sentiments. In this paper, we use reviews of digital camera collected from review sites. The text messages are first classified into three sentiment categories (positive, neutral,. and negative). The top-down method specifically filters out the comment texts that do not match a user's interest. The bottom-up method filters analysis comment text using text analysis methods, such as clustering. Here we use K-means for cluster Analysis. The bottom-up method can be automatically processed, mine, and cluster text into the specified categories. By combining these two methods, we obtain the best classification and thereby help the user focus on the most relevant categories that meet his/her interest.

The step-by-step process of the system is shown in the Fig. 1.

3.1 Data Collection

In this step, we select a few leading companies in their target industry like Canon and Nikon. We monitor selected social media site (face book, twitter, and blogs) and collect user-generated data (such us likes, share, comments) posted by the users.

3.2 Preprocessing

In this step, we need to transform the text documents into vectors before classification since the input to SVM and K-means are in vector form. The common procedures for processing text documents, such as stop word, remove numbers; remove punctuation and stemming are first applied to obtain a set of keywords that can describe the contents of a document. A document is then transformed into a document term matrix and organized based on the frequency according to the keywords it contains. The processed data is stored in excel format. Subsequently, a combination of various text mining, sentimental analysis, and competitive intelligent analysis techniques can be used to examine the data set to gain insight into users social media activities and sentiments.

3.3 Support Vector Machine

SVM are relatively a new class of machine learning techniques first introduced by Vapnik (1995). Based on the structural risk minimization principle from the computational learning theory, SVM seeks a decision surface to separate the training data points into two sentiment classes and makes decisions based on the support vectors that are selected as the only effective elements in the training set. In this paper, we use linear SVM due to its popularity and high performance in text categorization. Here, we have test data set and training data set as the input data to SVM. Then it find the hyper plane that separate the test data set with maximum margin based on the training data as illustrated in Fig. 2. During the preprocessing step, review data set is converted into document term matrix (x_i) that contain the significant features representing the data set. In this, $w\,x_i + b > 0$ represent the positive data which is above the hyper plane, $w\,x_i + b < 0$ represent the negative data which is below the hyper plane and $w\,x_i + b = 0$ represent neutral data where w is the weight vector where $w = \{w_1, w_2, \ldots w_n\}$, x_i is the input and b is the scalar known as bias. The result is a hyper plane that has the largest distance to x_i from both sides. The classification task can then be formulated by discovering which side of the hyper plane a test sample fall into.

SVMs are supervised learning method that analyzes the data used for classification and regression analysis. With a set of trained data, each test data is marked as positive, neutral or negative categories. Therefore, a new data is assigned to any of these two categories using SVM training algorithms. The data is more closest to the hyper plane is called support vectors.

3.4 K-Means

K-means algorithm is a quantitative unsupervised learning method which partitions a group of data points into a small number of clusters. Suppose, we have n data points X_i, where $i = 1, 2, \ldots n$ that have to be partitioned into k clusters. The goal of the K-means algorithm is to partition the data point into k clusters such that all data points in each cluster are similar to each other and data points in different clusters

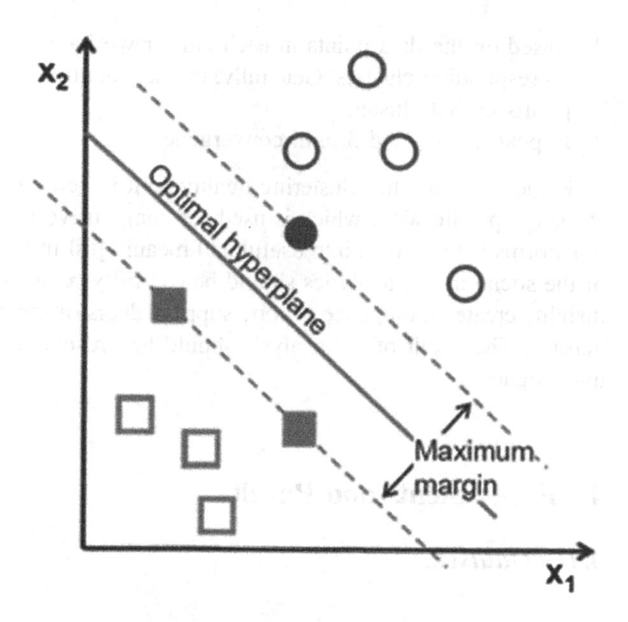

Fig. 2 An illustration of the SVM method

are different as possible. To calculate similarity and dissimilarity we use Euclidean distance method. In K-means algorithm, the main concept is the centroid, which is the center of each cluster. In Fig. 3, each cluster has a centroid, which is the most prototype of the cluster. The K-means algorithm is as follows:

1. Randomly select *k* data points and assign them as initial centroids.
2. For each point, find the nearby centroid and assign the data point to the cluster collaborate with the nearby centroid.

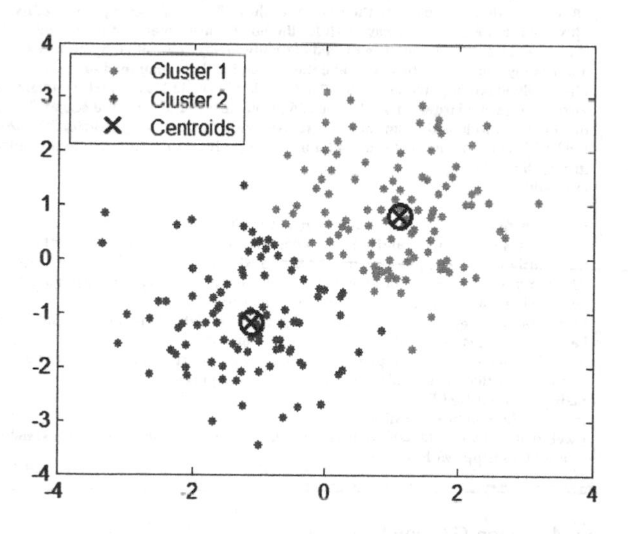

Fig. 3 An illustration of the K-means

3. Based on the data points in each cluster we have to update the centroid of the corresponding clusters. Generally, the new centroid is the average of all the data points in that cluster.
4. Repeats step 2 and 3 until convergence.

K-means assure the clustering quality so it is being used to make the clusters of industry specific word which is used for competitive intelligence analysis and al it transforms the raw data into useful and meaningful information. Finally, the results of the social media analytics should be carefully reviewed and then used to derive insight, create intelligence report, support decision-making, or make recommendations. The result of the analysis should be presented in a way that the user can understand.

4 Experiments and Results

4.1 Datasets

To evaluate the performance of our proposed system which is implemented in R, we use the customer reviews of two digital cameras: Canon G3 and Nikon Coolpix 4300. The reviews were collected from different social media sites. A sample review dataset is given in Fig. 4.

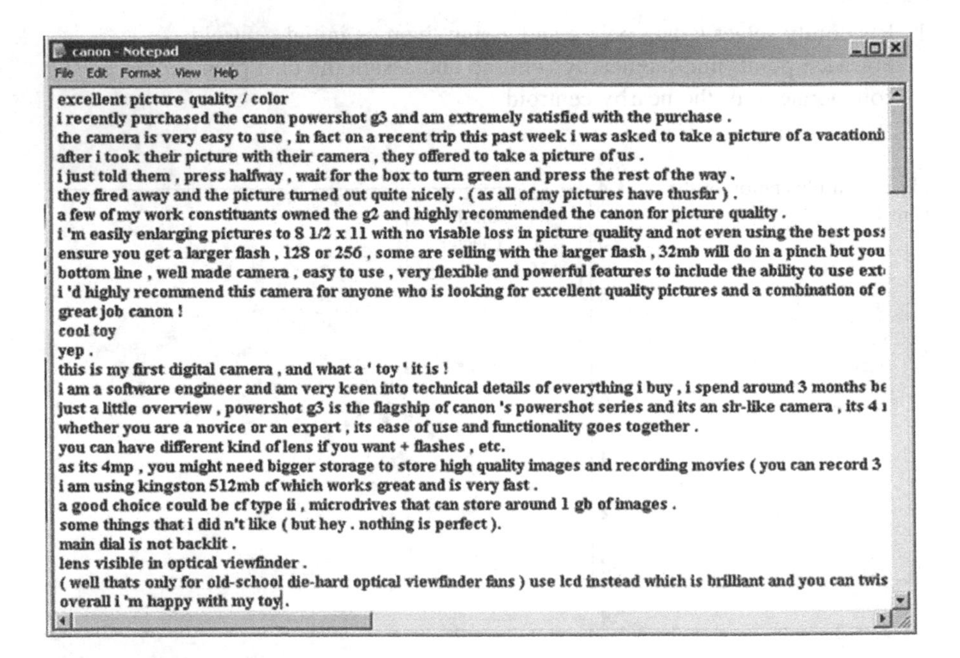

Fig. 4 Canon G3 sample dataset

4.2 *Results and Discussion*

In the preprocessing step, we use the methods and packages like stop word, stemming, etc. and "tm, SnowballC" to create the document term matrix which is our test dataset. A human reader manually reads the document term matrix, selects the industry specific words and assigns the sentiments such as 1 for positive data, −1 for negative data and 0 for neutral data which are the training dataset. Classify the trained dataset in terms of test dataset using the package "RTextTools" and "e1071" in SVM. The results show (Figs. 5 and 6) the S.V.M label of classification

Fig. 5 SVM result of Canon G3

	A	B	C
1		SVM_LABEL	SVM_PROB
2	1	1	0.911238092
3	2	1	0.867348395
4	3	1	0.859104999
5	4	1	0.892992637
6	5	1	0.850535103
7	6	1	0.925823519
8	7	1	0.854747602
9	8	1	0.843409595
10	9	1	0.846019391
11	10	1	0.88915061
12	11	1	0.848640624
13	12	1	0.866400446
14	13	1	0.845063524
15	14	1	0.901139055
16	15	1	0.882778214
17	16	1	0.842688775
18	17	1	0.8446834
19	18	1	0.845229501
20	19	1	0.843104752
21	20	1	0.842899446
22	21	1	0.845998317
23	22	1	0.865499826
24	23	1	0.844553031
25	24	1	[illegible]

resnik1

Fig. 6 SVM result of Nikon Coolpix 4300

	A	B	C
1		SVM_LABEL	SVM_PROB
2	1	1	0.421134366
3	2	1	0.834632107
4	3	1	0.920908045
5	4	0	0.556818322
6	5	0	0.49282204
7	6	1	0.799779992
8	7	1	0.888257881
9	8	1	0.936673703
10	9	1	0.832922286
11	10	1	0.839644547
12	11	1	0.99444487
13	12	1	0.812881929
14	13	1	0.810368857
15	14	1	0.885152636
16	15	1	0.856583825
17	16	1	0.815037348
18	17	1	0.912602851
19	18	1	0.851090646
20	19	1	0.829976357
21	20	1	0.948508855
22	21	1	0.894871185
23	22	1	0.950367711
24	23	-1	0.616656119

and probability of classification which depicts the sentiments of customers toward the product. Through the result we can analyze the individual report of each product.

We propose K-means for Competitive Intelligence Analysis. The result of K-means is shown in Figs. 7 and 8 here cluster 1 contain the most frequently discussed words by the customer and cluster 2 is the second most discussing words by the customer and so on. User have the privilege to specify the number of clusters based on their requirements here we choose the number of clusters = 4. Packages are "cluster," and "factoextra."

Here we got the individual result of two products so a user have to identify the most discussed words by the customer is visualized above.

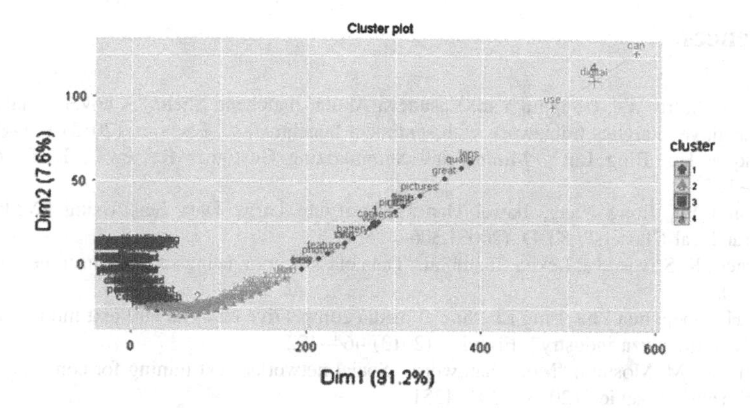

Fig. 7 K-means result of Canon G3

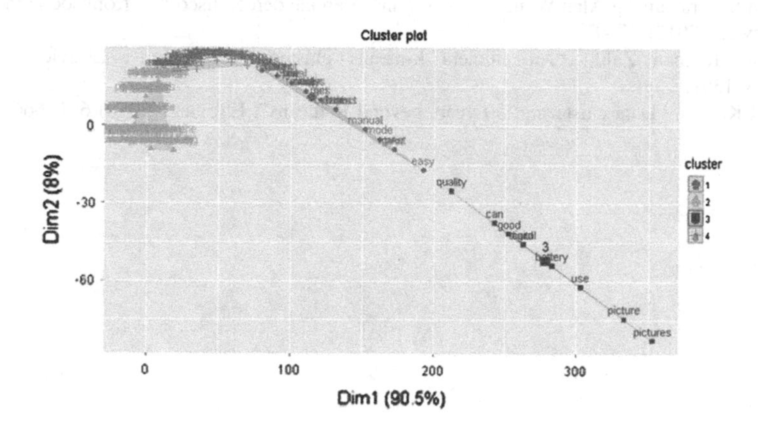

Fig. 8 K-means result of Nikon coolpix 4300

5 Conclusion and Future Work

In the system, we implement a hybrid approach which is a combination of supervised and unsupervised methods to analyze the performance of two leading companies and developed a competitive intelligence report that will helps the companies as well as the customers for their decision-making. We use text mining, sentiment analysis, and competitive intelligence analysis. SVM is used for sentiment analysis and we propose K-means for competitive intelligent analysis. Here we visualize the result which helps to interpret the output more efficiently. In future work we are planning to analysis the real-time data using cloud set up from the social media sites.

References

1. Wu He, Harris Wu, Gongjun Yan, Vasudeva Akula, Jiancheng Shen: "A novel social media competitive analytics framework with sentiment benchmarks". Elsevier. (2015) 801–812.
2. Minqing Hu, Bing Liu.: "Mining and Summarizing Customer Reviews". KDD. (2004) 168–177.
3. Hwanjo Yu, Jiong Yang, Jiawei Han: "Classifying Large Data Sets Using SVMs with Hierarchical Clusters". KDD. (2003) 306–315.
4. Durgesh K. Srivastava, Lekha Bhambhu: "Data classification using support machine". JATIT. (2010).
5. Wu He, Shenghua Zha, Ling Li: "Social media competitive analysis and text mining: A case study in the pizza industry". Elsevier. (2013) 464–472.
6. Mohamed M. Mostafa: "More than words: Social networks' text mining for consumer brand sentiments". Elsevier. (2013) 4241–4251.
7. Andreas Hotho, Andreas Nürnberger, Gerhard Paaß: "A Brief Survey of Text Mining". (2005).
8. Alan S Abrahams, g Alan Wang, Weiguo Fan: "Vehicle defect discovery from social media". Elsevier. (2012) 87–97.
9. Krista Rizman Zalik: "An efficient k-means clustering algorithm". Elsevier. (2008) 1385–1391.
10. Anil K. Jain. "Data clustering: 50 years beyond K-means". Elsevier. (2010) 651–666.

Detection of Incongruent Firewall Rules and Flow Rules in SDN

Nandita Pallavi, A.S. Anisha and V. Leena

Abstract The networking is the backbone that supports the vast area of Information Technology. SDN is the new road that takes the conventional networking to greater heights. SDN is going to aid all future innovations and developments in the field of networking. SDN stands for Software Defined Networking, this separates the network into two planes namely data plane and control plane. A data plane is the abstraction of all the hardware side of the network and the control plane is the central unit that acts like a brain controlling the entire network. This dual architecture thus helps to maintain a network that is centralized, highly scalable, flexible etc. The programmability of the network opens the window of scope for greater innovations and developments. SDN can gracefully accommodate technology shifts. At the same time SDN posses certain security issues that need to be addressed. As a widely flourishing and developing networking method, these security issues need to be tackled. In this paper we are trying to address the security issue of rewriting flow entries in switches. We propose an algorithm for the detection of incongruence between firewall rules and flow rules and thus we overcome the threat caused by modification of flow entries. The proposed system is for Open Flow based Firewalls. The system is intended to boost the security capabilities of SDN, thereby minimizing some of the security challenges in SDN.

Keywords SDN · Firewall · Open flow · HSA · Firewall rules · Flow rules · Duo lock

N. Pallavi (✉) · A.S. Anisha · V. Leena
Department of CS & IT, Amrita School of Arts and Sciences Kochi,
Amrita Vishwa Vidyapeetham, Kochi, India
e-mail: pallavinandita@gmail.com

A.S. Anisha
e-mail: anishaambili007@gmail.com

V. Leena
e-mail: vleena@gmail.com

S.S. Dash et al. (eds.), *Artificial Intelligence and Evolutionary Computations in Engineering Systems*, Advances in Intelligent Systems and Computing 517,
DOI 10.1007/978-981-10-3174-8_2

1 Introduction

Software Defined Networking is an emerging architecture that is cost-effective, dynamic, manageable, and adaptable [1]. All these features make it ideal for the high-bandwidth, dynamic nature of today's applications [2]. Using this technology a third party can introduce a new service or customize network behavior by writing simple software. Coming to the security aspect in SDN [3, 4], though some level of security [5] is provided by the SDN switches, it does not provide enough protection from all the things that can go wrong. This is where the importance of a firewall in SDN is reinstituted. The firewall filters all the incoming and outgoing traffic in the network based on prioritized firewall rules [6]. A firewall constitutes of a collection of rules that either allow or deny traffic. The flow tables are present in the switches and the entries in flow tables mark the path the traffic can traverse. There may be contradiction between the flow rules and flow table entries [7]. This contradiction is the violation that we are handling in this paper. The packet entering the switch passes through flow tables. A flow is a sequence of packets that matches a specific entry in a flow table.

2 Related Work

Hu et al. [1] has presented a comprehensive framework, FLOWGUARD, has proposed a firewall policy violation detection and resolution mechanism in dynamic OpenFlow networks. Jarschel [8] establish SDN as a widely adopted technology beyond laboratories and suggest that insular deployments requires a compass to navigate the multitude of ideas and concepts that make up SDN today [8]. Contribution represents an important step towards such an instrument. It gives a thorough definition of SDN and its interfaces as well as a list of its key attributes. Mininet [9] an instant virtual platform http://mininet.org/ helps to understand the implementation level details of SDN. Hu et al. [7] proposed that the source and destination addresses of firewall rules and flow entries are first represented by binary vector. Then, conflicts between firewall rules and flow rules are checked through comparing the shifted flow space and deny firewall authorization space. PeymanKazemian et al. [10] has developed a general and protocol-agnostic framework, called *Header Space Analysis* (HSA). Their formalism allows us to statically check network specifications and configurations to identify an important class of failures such as *Reachability* Failures, *Forwarding Loops* and *Traffic Isolation and Leakage* problems. Wang et al. [11] introduce a systematic solution for conflict detection and resolution in OpenFlow-based firewalls through checking flow space and firewall authorization space. This approach can check the conflicts between the firewall rules and flow policies based on the entire flow paths within an OpenFlow network. PeymanKazemian et al. [12] introduces a real time policy checking tool called NetPlumber.

3 Methodology

3.1 Algorithm 1: Matrix Mapper

Step 1: Convert the firewall rules into a matrix format.
Step 2: Convert the flow table rule into a matrix format.
Step 3: Create a transitive adjacency matrix based on the flow table matrix.
Step 4: By comparing firewall table and flow table matrices, derive a resultant matrix and also create a port matrix.

Based on the topology given in Fig. 1, matrices are designed here. The firewall matrix consists of values 0, 1 and 2 and it is denoted by the symbol FW. Row and column denotes the source and destination respectively. Each value in the cell of the matrix corresponds to the state of traffic between the source and destination. The state of traffic implies whether there is traffic between the corresponding source host and destination host [10, 12]. The value 0 means traffic denial, 1 means traffic is allowed, and 2 denotes partial allow. The firewall rules are analyzed from least priority to high priority to set values in the matrix. If allow or deny between two hosts is not specified then it is assumed to be allow. Table 2 is an FW matrix designed from the firewall table given in Table 1.

The Flow table matrix contains two values 0 and 1 where 0 corresponds to absence of flow and 1 corresponds to the presence of flow. It is represented by the symbol FT.

From matrix FT, transitive adjacency matrix TA is constructed. The significance and procedure of creating matrix TA from a given matrix is explained in Sect. 3.3. The corresponding TA matrix of FT shown in Table 3 is given in Table 4.

The matrix FW and TA are compared. From these two matrices resultant matrix *R* is built. Whenever a new rule is inserted or an existing rule is dropped in FW or an existing rule is modified, the *R* matrix is built. For values of matrices with values 1 or 0, an XNOR operation is performed and the result is recorded in matrix *R*. The value 0 in *R* means violation. That is the case when an action is specified in firewall and a contradicting action is specified in flow table (Table 5).

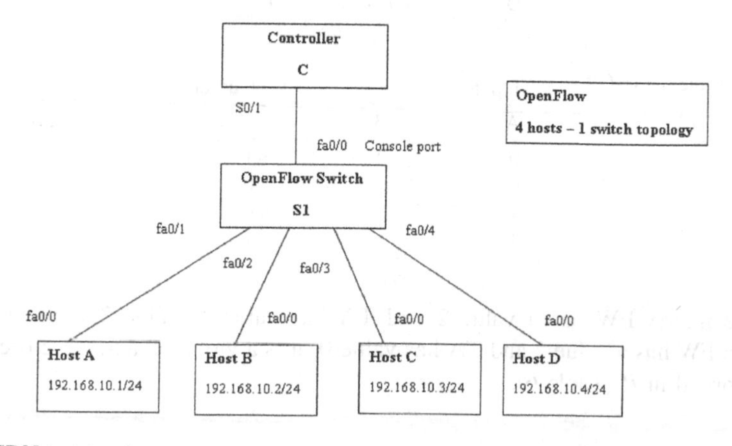

Fig. 1 SDN test topology

Table 1 Firewall table

Order	Protocol	SrcIP	*	DestIP	Dest port	Action
1	tcp	192.168.10.1	*	192.168.10.2	*	Allow
2	tcp	192.168.10.2	*	192.168.10.4	22, 25	Deny
3	tcp	192.168.10.3	*	192.168.10.2	*	Deny
4	tcp	192.168.10.4	*	192.168.10.2	*	Deny
5	tcp	192.168.10.4	*	192.168.10.3	*	Allow
6	tcp	192.168.10.3	*	192.168.10.4	*	Allow
7	tcp	192.168.10.2	*	192.168.10.1	22, 23	Allow
8	tcp	192.168.10.3	*	192.168.10.1	*	Deny
9	tcp	192.168.10.2	*	192.168.10.3	*	Deny
10	tcp	192.168.10.1	*	192.168.10.4	*	Deny

Table 2 Firewall matrix: FW

	A	B	C	D
A	1	1	1	0
B	2	1	0	2
C	0	0	1	1
D	1	0	1	1

Table 3 Flow table matrix

	A	B	C	D
A	1	1	1	0
B	0	1	0	1
C	0	1	1	0
D	0	1	0	1

Table 4 Transitive adjacency matrix

	A	B	C	D
A	1	1	1	1
B	0	1	0	1
C	0	1	1	1
D	0	1	0	1

Table 5 Table of XNOR operations

Inputs		Output
A	B	$Y = \overline{A \oplus B} = AB + \overline{A}\,\overline{B}$
0	0	1
0	1	0
1	0	0
1	1	1

If the matrix FW has a value 2 and TA has value 1, value 2 is recorded in *R*. When FW has a value 2 and TA has value 0, it is a violation and therefore value 0 is recorded in *R* (Table 6).

Table 6 Resultant matrix *R*

	A	B	C	D
A	1	1	1	0
B	0	1	1	2
C	1	0	1	1
D	0	0	0	1

Table 7 Port matrix of source B

	20	21	22	23	50	53
A	0	0	1	1	0	0

The matrix *R* helps us to detect all the incongruencies in cases other than those with partial allow (value 2), for the remaining entries (with value 2) there is a port specification associated with it. Therefore a next level of matrix needs to be constructed.

If $R_{ij} = 2$ in *R*, get the corresponding hosts. For all these hosts, create a port set for each host. With each cell with value 2, create a destination-port matrix for the corresponding source. The row contains the destination host names and column contains the partially allowed ports. An example for port matrix is shown in Table 7.

This helps to record the violation between the firewall rules and flow rules in SDN. The prioritizing errors can be detected using this strategy.

(1) A value 2 in FW matrix cannot be overwritten by 1 or 0, that is a priority violation.
(2) If specific ports of host are allowed or denied, the rest of the ports are filled with the negated value of either allow or deny.

3.2 Description of Duo Lock Algorithm

As the name suggests the algorithm is governed by two locks. The outer lock: lock_1 and the inner lock: lock_2. Whenever there is a modification in a flow entry or a firewall rule is added, deleted or modified, the control is passed through this lock. The entry is updated only when the lock is opened. If the lock is closed a violation is detected. The flow entry is updated when a mandatory overwriting is tried or changes need to be brought in the flow table based on firewall rule addition, modification or deletion. Lock_1 helps to detect and handle complete violations. In case of partial allow or deny rules in firewall, the second lock: lock_2 is used. If port wise violation does not happen then the lock is opened and flow entry is updated, otherwise if the port is closed, flow entry does not get updated and violation is reported.

In order to enable the proper execution of the concept, we introduce a transitive adjacency matrix in the locks. If the hosts A and C are denied to interact, the intruder must not be allowed to bypass this rule by an indirect interaction. Example is an indirect interaction from A to C by A to B to C. Lock_1 operates using transitive adjacency matrix and Lock_2 operates using port matrix.

3.3 Transitive Adjacency Matrix (TA)

A transitive adjacency matrix is the matrix that points whether there is a direct edge or indirect edge in a matrix. If edge (a, b) and (b, c) exists then another edge (a, c) is assumed to be existing. The transitive adjacency matrix is based on transitive law and adjacency matrix. The transitive law in mathematics and logic states that "If a**R**b and b**R**c, then a**R**c," where "**R**" is a particular relation (e.g., "… is equal to …"), a, b, c are variables (terms that may be replaced with objects) and the result of replacing a, b, and c with objects is always a true sentence (Fig. 2).

A transitive adjacency matrix is a matrix with all indirect connections also marked as an edge. This matrix is XNORed with FW matrix, if 1 then there is violation and the lock_1 is locked (Table 8).

For converting an adjacency matrix to transitive adjacency matrix a set of operations are proposed here.

For each row, r in matrix with value 1{

Find value of column with value 1 and assign column value to k

For each row of value k, retrieve column with value 1 and assign to l.

Assign TA[r][l] to 1.

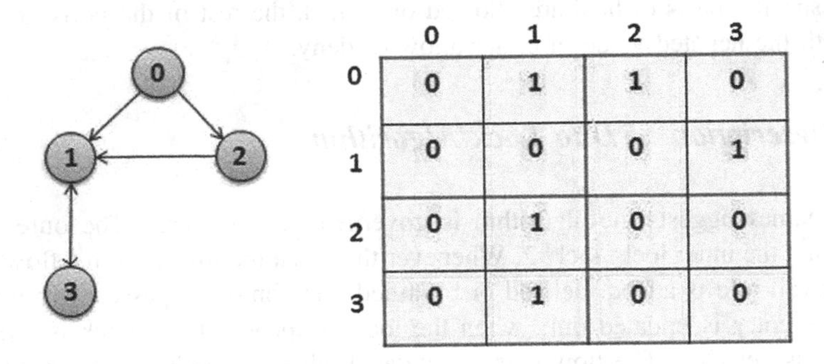

Fig. 2 Adjacency matrix representation of a directed graph

Table 8 Transitive adjacency matrix of the adjacency matrix given in Fig. 2

	0	1	2	3
0	0	1	1	1
1	0	1	0	1
2	0	1	0	1
3	0	1	0	1

3.4 Algorithm 2: Duo Lock

```
Lock1
//a prevention is better than cure approach
Matrix R (Resultant matrix of the comparison between FW and TA) checked,
If(value is 1)
        //no violation at all
        Exit
If(value is 0)
        //lock closed//violation detection
        Set lock _1 to 0

        return closed
Else
    //Lock opened
    Set lock_1 to 1

    return open

Lock 2
//cure approach
Matrix model Port matrix checked, on value 1 lock opened
Based on source destination variables corresponding port matrix is analyzed
If port deny
    //close lock_2
    Set lock_2 to 0
    return closed
Else
    //lock opened
    Set lock_2 to 1
     return open

Duo Lock:

Lock1{
    If (closed){
            Violation detection
    }
    Else {
            Flow entry updation

                    lock2{
                    If (closed){
                            Violation detection
                            Exit
                    }
                    Else{
                            //no violation
                            Flow entry updated if needed
                            Exit
                    }
            }
    }
}
```

The proposed methodology was experimented in mininet [9] using pox controller [13, 14] based on the topology given in Fig. 1. From the experiment it was clear that rules can be efficiently represented, stored and manipulated effectively using matrices. In terms of memory access also matrices are highly efficient here. As 2 check points are introduced no conflict can be bypassed. So the proposed procedure effectively detects the conflicts in a faster way. Therefore flow tables can be updated reflecting the variation proficiently.

4 Conclusion

In this paper we present a method of detection of incongruent firewall rules and flow rules in SDN. Our methods are based on the matrix model. Firewall matrix, flow table matrix, port matrix and transitive adjacency matrix is the matrix models used in this paper. Using these matrix models we have created two algorithms to detect the violations between flow table entries and flow rules. These algorithms provide an efficient way to detect violations and prevent or cure their occurrence. A method to effectively prioritize the firewall rules depending on modification, insertion or deletion of firewall rules have also been proposed in this paper. The matrix model of transitive adjacency matrix is a matrix that has been formulated according to the needs of our method, and for this we have clubbed the transitive law and adjacency matrix.

References

1. Hongxin Hu, Wonkyu Han, Gail-JoonAhn and Ziming Zhao "FLOWGUARD: Building Robust Firewalls for Software-Defined Networks" Clemson University Arizona State University.
2. Wolfgang Braun and Michael Menth "Software-Defined Networking Using OpenFlow: Protocols, Applications and Architectural Design Choices" Department of Computer Science, University of Tuebingen, Sand 13, Tuebingen 72076, Germany.
3. Phillip Porras, Steven Cheung, MartinFong, Keith Skinner, and VinodYegneswaran "Securing the Software-Defined Network Control Layer".
4. Seunghyeon Lee, Chanhee Lee, Hyeonseong Jo, Jinwoo Kim, Seungsoo Lee, Jaehyun Nam, Taejune Park, Changhoon Yoon, Yeonkeun Kim, Heedo Kang, and Seungwon Shin "A Playground for Software-Defined Networking Security" GSIS, School of Computing, KAIST.
5. Jérôme François, LautaroDolberg, Olivier Festor, Thomas EngelSnT "Network Security through Software Defined Networking: a Survey" - University of Luxembourg.
6. Michelle Suh, SaeHyong Park, Byungjoon Lee, Sunhee Yang "Building Firewall over the Software-Defined Network Controller" SDN Research Section, ETRI (Electronics and Telecommunications Research Institute), Korea.
7. Hongxin Hu, Wonkyu Han, Gail-JoonAhn, and ZimingZhao "Towards a reliable SDN firewall" Clemson University Arizona State University.
8. Michael Jarschel, Thomas Zinner, Tobias Hobfeld, Phuoc Tran-Gia. "Interfaces, attributes and use cases—a compass for SDN".

9. Mininet, an instant virtual platform http://mininet.org/.
10. PeymanKazemian, Nick McKeown, George Varghese "Header Space Analysis: Static Checking For Networks" Stanford University, UCSD and Yahoo! Research.
11. Juan Wang, Yang Wang, Hongxin Hu, Qingxin Sun, He Shi, and LangjieZeng. "Towards a Security-Enhanced Firewall Application for Openflow Network".
12. PeymanKazemian, Michael Chang, HongyiZeng, George Varghese, Nick McKeown, Scott Whyte "Real Time Network Policy Checking using Header Space Analysis".
13. Pooja, Manu Sood "SDN and Mininet: Some Basic Concepts" Department of Computer Science, Himachal Pradesh University, Shimla.
14. Sukhveer Kaur1, Japinder Singh2 and Navtej Singh Ghumman "Network Programmability Using POX Controller" 3 1,2,3 Department of Computer Science and Engineering, SBS State Technical Campus, Ferozepur, India.

An Attainment Upgrade in Audio Steganography

K. Anupriya, Revathy R. Nair and V. Leena

Abstract With the advancement of information technology, there is a visible change in economical, commercial, and technological ways. Privacy of information still remains the main concern. Audio steganography is one of the most popular techniques for shielding of information. To raise the surveillance, we focus on Audio steganography along with noise-free environment. For ensuring this, hiding in silence interval are considered because most of the current audio steganography embedding methods are improper for noise-free environment. This paper proposing an attainment upgrade in audio steganography with using RSA algorithm of three distinct prime numbers as well as hiding text in silence speech signals with high hiding capacity employing summate silence interval in speech signal. Our simulation results show that this approach maintains noise-free environment, security, and achieves a higher hiding capacity.

Keywords Steganography · Cryptography · RSA · Hiding in silence interval

1 Introduction

Internet is a global system of computer networks. So with advancement of these come up with many security problems, exceptionally in data security. Data security means defend data from the undesirable movement of illegal end users. Cryptography and steganography are the two techniques used to solve these problems. Cryptography is the branch and trade of remodeling information to make

K. Anupriya (✉) · R.R. Nair · V. Leena
Department of CS & IT, Amrita School of Arts and Sciences Kochi,
Amrita Vishwa Vidyapeetham, Kochi, India
e-mail: anupriyak92@gmail.com

S.S. Dash et al. (eds.), *Artificial Intelligence and Evolutionary Computations in Engineering Systems*, Advances in Intelligent Systems and Computing 517,
DOI 10.1007/978-981-10-3174-8_3

them sheltered and resistant to intrusion. Encryption and decryption are two important methods used under cryptography. Steganography is termed as shielded or obscure writing [1]. It is forming of shielded through uncertainty. Audio steganography is technique used to remodeling covert information by reshape audio signals in an undetectable manner [1–3].

General principles of audio steganography adopted in previous research are as follows.

In audio steganography system, to support cryptography, normally RSA algorithm is used. RSA is an unbalanced cryptographic algorithm that contains key generation (public and private), key distribution, encryption, and decryption [4, 5]. In this, prime numbers must be undisclosed and anyone will get a public key to encipher the message where any case of cracking RSA encipher known as RSA problem. Variety of methods are used for embedding text into an audio file such as LSB [6], MSB, phase coding, parity coding [6], hiding in silence interval. As our intention is to get a noise-free environment, in our proposed work, we focused on the Hiding in silence interval [2, 7]. Hiding in silence interval is a smooth and an adequate embedding method has been employed to accomplish silence interval in speech signal. Alteration in silence interval can lead to deceitful data elicitation where the initial and final interval added to the speech signal simply avoided.

- RSA algorithm with two prime number [8]
  Advantages of RSA:
  - Very agile, very simple encryption
  - Easier to perceive [8].

  Disadvantages of RSA:
  - Extremely passive key generation
  - Less covert
  - Slightly insecure in data transfer [9].
- Hiding in silence interval
  Advantages
  - Noise-free environment

  Disadvantages:
  - Low data hiding capacity [2]

Since more drawbacks are found in the existing system, we provide an efficient method with the help of a new technique.

2 Related Work

Vivek Choudhary and Praveen [9] proposed a method for progressing RSA algorithm with three prime numbers. It gives tenacity of large prime numbers. Its build upon three variables and it is generating new variables by eliminating common modulus n. Here they generated RSA algorithm with the help of three phases: key generation, encryption, and decryption.

- Generate three prime numbers such as a, b, c.
- Computing common modulus with help of three prime numbers $(a-1)(b-1)(c-1)$
- Introduce new variable '*d*'
- The '*d*' variable used for Encrypt and decrypt the message.

It has two central advantages compared with existing RSA algorithm.

- It is impossible for breaking three prime numbers.
- It makes more powerful by adding new variable and common modulus (n).

Abdul Kather and Vimal [10] proposed a new suppressing method where private data concealed in the silence part are identified by collaborative non-voice detection (CNVD) algorithm in which sample values from speech signal is deteriorate to hide covert data aim at providing emotive transparency with data hiding scope. They embed covert information into non-voice part by proposed embedding algorithm and extract secret information from non-voice part by proposed extraction algorithm. Data embedding capacity in non-voice frame of speech signals are anticipated and imperceptibility of stego files is appraised. So data hiding capacity is achieved with CNVD Algorithm.

Kamred Udham Singh [7] provided data hiding scheme in temporal domain are: low-bit encoding, echo hiding, parity coding, and hiding in silence interval approach come up with simple and adequate embedding technique. Initially determine the silence intervals of the audio and their respective length, these values are decreased by a value where $0 < z < 2^{\text{nbits}}$ and n bits is the number of bits needed to represent a value from the data to hide. Small silence intervals are left interrupted since they usually arise in continuous sentences and changing them might alter the quality of the speech. This technique has a good perceptual transparency. Modifications in silence intervals length will lead to incorrect data extraction. Besides robustness and security conventional least significant bit (LSB) technique, its alteration provide a very clear and simple way for data hiding but at a low data hiding quantity [7].

Nikita Atul Malhotra, Nikunj Tahilramani reviewed on speech and audio steganography techniques in temporal, transform, and coded domains. They discussed on various steganographic techniques which in turn classified into various domains like transform domain, temporal domain, and coded domains. They examine on classification of steganographic techniques. Thus, temporal domain apprise intends to escalate the hiding quantity [11].

3 Problem Definition

Hiding in silence interval provides noise-free environment. But to increase data hiding capacity, we need to compromise it with enhancing the silence interval which can be done by summating the length of silence interval of audio without interrupting the original signals as well as to enhance preservation, we imply reformed RSA with three distinct prime numbers.

4 Proposed System

In our audio steganographic system, we are using two techniques that are reformed RSA (three prime numbers) to strengthen security and enhanced hiding in silence intervals to increase the hiding capacity. RSA has been used for securing the secret messages. Due to the problem encountered with the RSA was of two distinct prime numbers, propose a reformed RSA technique with three prime numbers and whenever an extra bit (secret message) is added to original signal, it is obvious that original signal changing, so in order to eliminate such noise rate, we employing a techniques that Hiding in silence Intervals. But this technique supports low data rate capacity [12]. For this reason, a method is introduced to increase high data rate capacity in hiding silence interval.

4.1 Enhanced Hiding in Silence Interval

A main drawback with hiding in silence interval is less data hiding quantity [13]. So in order to prevail over this problem, we propose an enhancement in hiding in silence interval.

Algorithm:
Embedding module:
Step 1: Encrypt the message with public key
Step 2: Calculate the total length of silence interval of speech signals.
Step 3: Compare calculated length of silence interval with total cipher message size.
Step 4: If message size is larger than the calculated length,
Summate the silence interval,
Embed the data into silence interval.
Else
Embed the data into silence interval.
Step 5: At the end side, extract the original message

Summate silence interval:

When embedding cipher message to silence interval. It is important to envisage certain aspects: First, initial and final parts of speech signals are neglected even it is a silence part. Second, one will be unable to summate all the silence part within the speech signals. Third, even though it is possible to summate the silence interval, it must be noted that no human auditory system can be easily detected. As the quality of speech signal is a significant factor for summating them.

4.2 RSA Algorithm with Three Prime Numbers

RSA is asymmetric cryptography algorithm invented by Ron Rivest, Adi Shamir, and Len Adleman contribute a stable and secure system that can be employed for both public key encipher and digital signatures. In RSA, key generation, encipher, and decipher are the three steps carried out. Public key, private key, and common modulus are elected for key generation process. The message will be enciphered employing a public key, which will be got by anyone but can be deciphered employing the private key only get to the end user. Normally in RSA, two prime numbers are elected to generate modulus n but through n, there is a chance for detecting the two prime numbers. So, in this paper, RSA with three prime numbers are elected to spawn the common modulus n and with the help of three prime numbers [9] and n is used to spawn a new variable, which is proving for encipher and decipher of text data.

The procedure is as follows:

(1) Key generation

First, it prefers three prime numbers which is a, b, c. Then it appraises common modulus (or n) and Euler's totient function n (or Euler's phi function) computing by $(a-1)\,(b-1)\,(c-1)$. Then find out new variable d with help of following conditions

$$\text{If}\quad a > b\ \text{THEN}\quad n - a < d < n$$
$$\text{If}\quad a < b\ \text{THEN}\quad n - b < d < n$$

(d must be co-prime to n).

Then find out public key and private key employing d variable.

A general formula to find d: $p_k * s_k \text{mod}(d) = 1$
S_k is found by this equation: $s_k * p_k = 1 * \text{mod}(d)$

This is more secure than RSA algorithm as the new variable d can be detected only by knowing the two prime numbers a and b, which can be detected only through n, but as n is not conveyed in public key, so it is crucial to get the value of d, hence the enciphered message cannot be appreciable easily [9].

(2) Encryption

Encryption formula: $c = m^{p_k}(d)$ where p_k is public key exponent, m is message, c is cipher.

(3) Decryption

First, at the receiver side it evaluate private key (s_k) with help of current variable (d) where it computing by help of new method. Then it decrypt with private key.

Decryption formula: $m = \left[c_k^s \bmod (d)\right]^{1/2}$

Architecture:

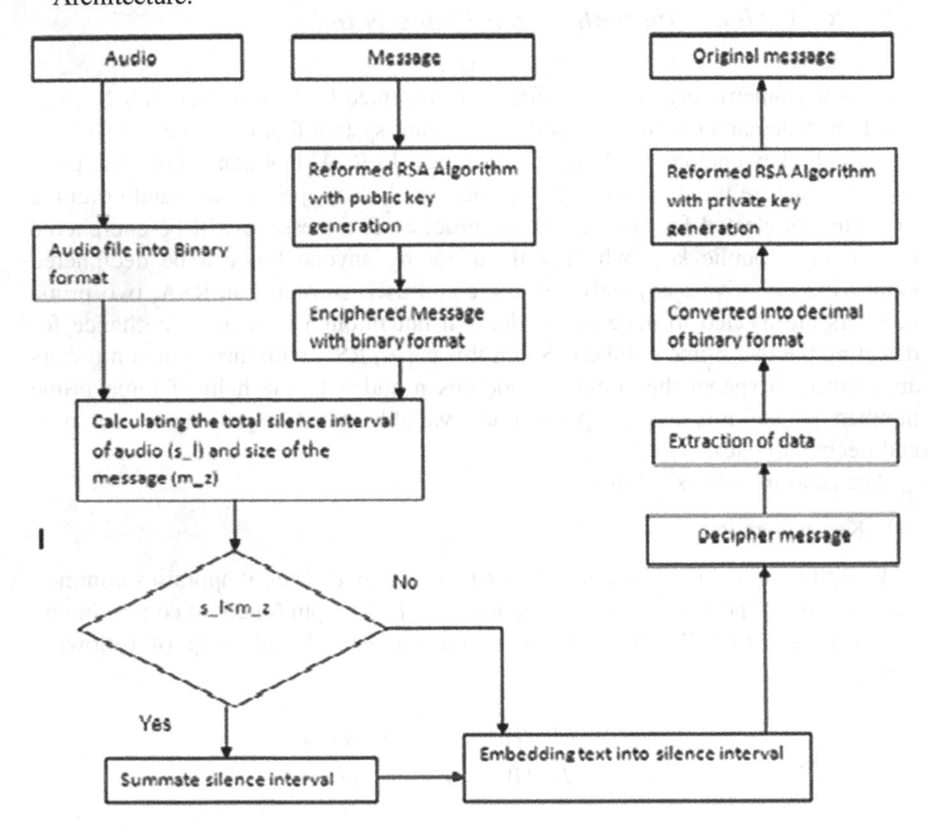

Advantage of Proposed Method

- The vitality of increasing the length of silence interval in audio, it could increase the data hiding capacity.
- It can provide noise-free environment without having any disruption.
- The intensity of reformed RSA, It could raise the surveillance.
- Eradicate of common modulus n by spawning current variable.
- It provides more surveillance on data transmission.

5 Experimental Results

5.1 Reformed RSA

The introduced system carries out with Java. The information first encrypted with reformed RSA under the cryptography. Then we first calculate the total length of silence interval of speech signals, and then calculated length of silence interval is compared with total cipher message size. If size is larger than calculated length, it move on to next process that is summate the silence interval which done manually and embedded the data into silence part of audio else simply embedded the data into silence part. At the end side, our system extracts the original message.

Table 1 shows the encipher and decipher time of reformed RSA and original RSA

Table 1 Comparison of reformed RSA and original RSA [14] in which the reformed RSA has fewer encipher time but the decipher time is higher than the aboriginal one

Algorithm	Encipher time (s)	Decipher time (s)
RSA	0.00988112	0.00086667
Reformed RSA	0.00266617	0.00112452

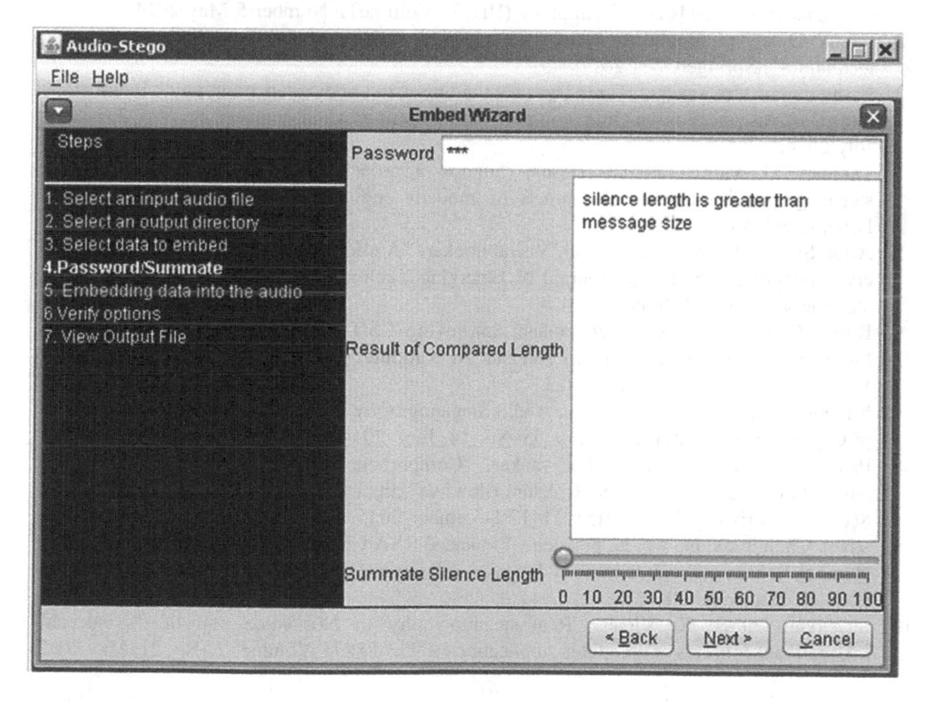

Fig. 1 Depicts the way of summating the silence interval

5.2 *Enhance Hiding in Silence Interval*

Depending on the speech signals, capacity for hiding the message can be increased using summate silence interval (Fig. 1).

6 Conclusion and Future Work

The paper proposed to improve surveillance on covert data using reformed RSA with three prime numbers and a new variable d. The paper also concentrated on providing a noise-free environment with high hiding capacity by improvising the hiding in silence interval method with the help of summating the silence interval length. From our estimated results, in order to summate the silence interval, we focused only on speech signals. So in our future work we will be trying to use all kind of audio.

References

1. Jithu Vimal, "Literature Review on Audio Steganographic Techniques", International Journal of Engineering Trends and Technology (IJETT) Volume11 Number 5-May 2014.
2. Fatiha Djaebbar, Beghdad Ayad, "comparative study of digital audio steganography techniques", SpringerOpen 2012.
3. Nishu gupta, Mrs.Shailja, "Three layer data hiding using audio steganography", International journal of Advanced researching computer and communication engineering, Vol. 3, Issue 7, July 2014.
4. Tanmaiy G. varma, Zohaib Hasan, "unique approach for data hiding using Audio steganography", International journals of modern engineering research(IJMER), Vol. 3, Issue. 4, Jul-Aug. 2013.
5. Arfan Shaikh, kirankumar Solanki, Vishal uttekar, "Audio steganography and security using cryptography", International Journal of Emerging Technology and Advanced Engineering, Volume 4, Issue 2, February 2014.
6. Burate D. j, M. R Dixit, "Performance improving LSB Audio steganography technique", International journal of advanced research in computer science and management studies, Volume 1, Issue 4, September 2013.
7. Kamred Udham Singh, "Survey on Audio Steganography Approaches", International Journal of Computer Applications, Volume 95–No. 14, June 2014.
8. Priyadarshini Patil Prashant Narayankar, "Comprehensive Evaluation of Cryptographic Algorithms: DES, 3DES, AES, RSA and Blowfish", International Conference on Information Security & Privacy (ICISP2015), 11–12 December 2015, Nagpur, INDIA.
9. Vivek Choudhary and Mr. N. praveen, "Enhanced RSA Cryptosystem Based On Three Prime Numbers", International Journal of Innovative Science, Engineering & Technology. Vol. 1 Issue 10, December 2014.
10. S. Abdul kather, K. Vimal, "Real steganography in Non-voice part of the speech", International Journals of computer Applications (0975-8887). Volume 46–No. 21, May 2012.

11. Nikita Atul Malhotra, Nikunj Tahilramani, "Survey on Speech and Audiosteganography Techniques in Temporal, Transform and Coded Domains", International Journal of Advanced Research in Computer Science and Software Engineering-March 2014.
12. Youssef, "A Two Intermediates- Audio Steganographic Technique", Journal of Emerging Trends in Computing and Information Sciences (CIS), ISSN: 2079-8407, Vol. 3, No. 11, November 2012.
13. Premalatha P, Amritha p, "Optimally Locating for Hiding Information in Audio Signal", International Journal of Computer Applications (0975–8887) March 2013.
14. M. Sundari, P. B. Revathi, "Secured Communication using Digital Watermarking with encrypted text hidden in an image", SpringerLink August 2015.

A Comparative Study Between Hopfield Neural Network and A* Path Planning Algorithms for Mobile Robot

Suhit Atul Kodgule, Arka Das and Arockia Doss Arockia Selvakumar

Abstract Path planning is an important aspect of any mobile robot navigation to find a hazard-free path and an optimal path. Currently, the A* algorithm is considered to be one of the prominent algorithms for path planning in a known environment. However, with the rise of neural networks and machine learning, newer promising algorithms are emerging in this domain. Our work compares one such algorithm namely the Hopfield neural network-based path planning algorithm with A* in a static environment. Both the Hopfield network and the A* algorithm were implemented while minimizing the total run times of the programs. For this, both the algorithms were run in MATLAB environment and a set of mazes were then executed and their run times were compared. Based on the study, the A* algorithm fared better and the Hopfield network showed promising results with scope for further reduction in its run time.

Keywords Mobile robot · Path planning · Hopfield network · A* algorithm · MATLAB

1 Introduction

Path Planning is an important aspect for field robotics and to allow the robot to traverse an optimally safe and accurate path. The major classification of path planning is divided into path planning in static environment [1, 2] and path planning in dynamic environment [3]. Apart from mobile robots, path planning can also

---

S.A. Kodgule · A. Das
School of Electrical and Electronics Engineering, VIT University, Chennai, India
e-mail: kodgule.suhitatul2013@vit.ac.in

A. Das
e-mail: arka.das2013@vit.ac.in

A.D. Arockia Selvakumar (✉)
School of Mechanical and Building Sciences, VIT University, Chennai, India
e-mail: arockia.selvakumar@vit.ac.in

S.S. Dash et al. (eds.), *Artificial Intelligence and Evolutionary Computations in Engineering Systems*, Advances in Intelligent Systems and Computing 517,
DOI 10.1007/978-981-10-3174-8_4

be applied to robotic arms [4], underwater robots [5], and autonomous vehicles [6]. Lagoudakis [7] proposes a path planning scheme using Hopfield neural network, based on which a part of our work is presented. While the path planning problem has been addressed in neural networks domain using feedforward networks, this technique lacks an internal representation of the environment and thus needs to compensate this using different methodologies. Hopfield network overcomes this barrier by generating a Hopfield surface that directly maps to the robot's environment. This is the reason why an in-depth study of the Hopfield network path planning was deemed necessary. This algorithm is not limited to mobile robots but can also be used for various other agents, such as underwater robots, software agents, and robotic arms.

Panrasee et al. proposed a modified Hopfield Network where the weight matrices are calculated by using the distance between the target and each neuron, their findings showed that it gave better results compared to Lagoudakis' work. However, as the weight matrices are not symmetric, the Hopfield energy function is not guaranteed to decrease. This may lead to a greater convergence times which is not profitable for our work. As a result, we have concentrated on meeting the *Lyapunov* function criteria to achieve a faster convergence time [8]. Kroumov et al. [9] proposes a potential field method based path planning. It overcomes the local minima problem characteristic of potential field functions and guarantees a near-optimal path for a differential drive robot.

Mahadevi et al. proposes a memory-based A* algorithm, an application of machine learning where the paths are precalculated and stored. First, the robot explores the maze and creates a map in its memory and then when it is fed a source and destination point, it recalls the path traversed while learning up the maze to get a reasonable optimized path [10]. Ankit Bhadoria et al. [11] proposes an optimized angular A* algorithm for neighbor node evaluation where the angle keeps on changing to the optimum angle between the robot's present position and the target position. Nadia Adnan et al. proposed a modified genetic algorithm method for robot path planning. The algorithm reads the already established maze map to create a reasonably optimum path using distinguished algorithm, fitness function and crossover [12]. Mohanraj et al. [13] presented a simple ant colony and ant colony meta-heuristic-based path finding algorithm, where the behavior of ant colonies in nature is implemented and their procedure of foraging food with the help of pheromones is replicated to map out the shortest path between the colony and food source.

In this study, the Hopfield neural work and A* path planning algorithms for mobile robot under static environment are studied in detail and a comparative study between Hopfield, A* and Ant Colony algorithms is carried out to find the mobile robot optimum path in a shorter time. Further complete description is mentioned in following detailed sections. Problem definition, Hopfield neural network, and A* algorithms are discussed in Sect. 2, Simulation using MATLAB is explained in Sect. 3, results and concluding remarks are discussed in Sects. 4 and 5, respectively.

2 Problem Definition

The robot environment is considered as a grid-based system where each element corresponds to a value in a matrix $n \times m$. The location of the robot, the obstacles and the target are the corresponding ordered pairs (i, j) of the matrix. Given a matrix A, the algorithms' aim is to find the shortest path from the robot's initial position to the target position while avoiding the obstacles. The position of the target and obstacles are static.

2.1 *Hopfield Network*

The Hopfield Neural Network is a fully connected recurrent neural network consisting of n neurons. Each of the given neurons takes input from all the other neurons. The connections between neurons i and j are weighted as w_{ij} and are symmetric in nature ($w_{ij} = w_{ji}$). There are no self-coupling terms, meaning $w_{ii} = 0$. Each of the weighted input to neuron i is summed with a bias unit, the importance of which is explained in Sect. 2.1.1, as show in Eq. 1.

$$x_i = \sum_{j=1}^{n} y_j w_{ij} + \theta_i, \quad \vec{X}(t) = W * \vec{Y}(t) + \vec{\theta} \tag{1}$$

where x_i is the input of neuron i, y_j is the output from neuron j and θ_i is the bias unit and $\vec{X}, \vec{Y}, \vec{\theta}$ are the discrete input, output, and bias vectors, respectively. The output of a neuron is given by the binary McCullogh-Pitts model:

$$y_i(t+1) = \varphi(x_i(t)) = \varphi\left(\sum_{j=1}^{n} y_j(t) w_{ij} + \theta_i\right) \tag{2}$$

where φ is an activation function, specifically the hyperbolic tangent function given by:

$$\mathrm{Tan\ h}(x) = \frac{\mathrm{e}^{x} - \mathrm{e}^{-x}}{\mathrm{e}^{x} + \mathrm{e}^{-x}} \tag{3}$$

whose values lie in the range of [0, 1].

The Hopfield Network has energy given by:

$$E(\vec{y}) = -\frac{1}{2}\sum_{i=1}^{n}\sum_{j=1}^{n} y_i y_j w_{ij} - \sum_{i=1}^{n} y_i \theta_i \tag{4}$$

The given energy function decreases monotonically after every evolution of the network till it reaches a stable equilibrium condition. This property of the Hopfield Network is used to stop the training of the network parameters.

2.1.1 Path Planning

Consider a configuration space C (C_1, C_2, C_3 … C_n), where each C_i corresponds to a neuron of the Hopfield Network. Each of the units can either be maximally activated or minimally activated or possess an activation (0, 1). Each obstacle, target, and robot's position corresponds to one unit of C. By common notation the target will be maximally activated. To achieve this, the bias unit θ_i are given the following values.

$$\theta_i = \left\{ \begin{array}{c} +\infty \text{ if } C_i \text{ is a target} \\ 0 \quad \text{otherwise} \end{array} \right\} \tag{5}$$

The weight matrix W is an $n^2 \times n^2$ matrix where each element (i, j) corresponds to the a function of the Euclidean distance between neuron i and j namely,

$$f(x) = \mathrm{e}^{-\propto x^2} (\propto \text{is a real positive number}) \tag{6}$$

Now, let us say one unit C_t is taken as the target unit. Correspondingly, the target neuron will be maximally activated and the rest of the neurons will be activated according to W such that the neuron closer to C_t has higher activation value compared to a neuron farther away. Once the Hopfield surface has been created, the obstacle units C_o are minimally excited by assigning a value of 0. The Hopfield surfaces with just a target and with target and obstacles are shown in Figs. 1 and 2, respectively. The surface is scaled to make the differences more apparent.

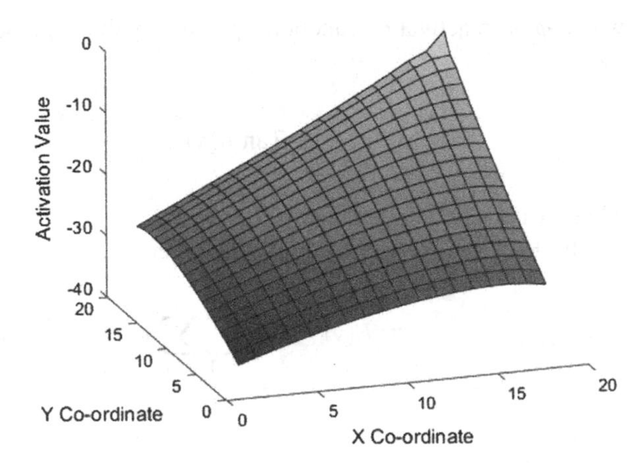

Fig. 1 Hopfield surface without obstacles

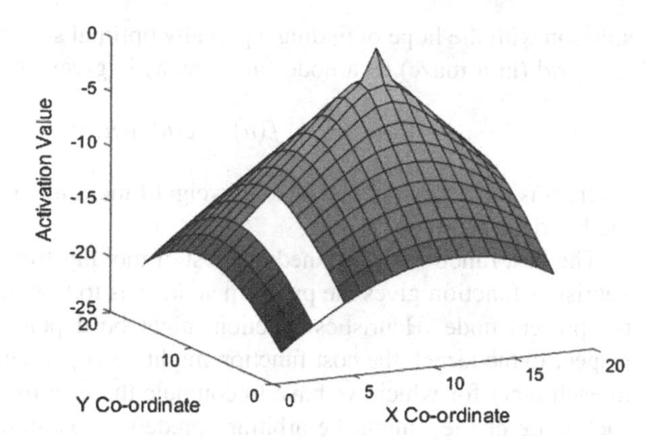

Fig. 2 Hopfield surface with obstacles

To construct a path, the Hopfield surface is used to determine the next step in the path. In case of static system, the Hopfield surface remains the same for every iteration while for the dynamic environment it differs corresponding to the position of target and obstacles. Given below is the algorithm for path construction:

1. Start
2. Assign bias unit of target a value of $+\infty$.
3. Produce the weight matrix W for a given map.
4. Iterate Output values of the Hopfield neurons until the global energy minima is reached.
5. Assign obstacle units a value of 0.
6. Calculate the gradients of robot's adjoining units using the Hopfield surface and the equation:

$$\text{gradient}(i,j) = \frac{y(j) - y(i)}{\rho(i,j)} \tag{7}$$

7. Select the unit with maximum gradient as the robot's next unit.
8. If target is reached, go to step 9.
   Else
9. Stop.

2.2 *A* Algorithm*

A* algorithm is a widely used path finding algorithm to find the shortest distance through a maze or a graph to reach the target from the source by avoiding all the obstacles. It is a heuristic search algorithm and applied greedy technique to select the best possible next node with the hope of finding the shortest distance to the target node. Greedy technique follows the method of choosing a locally optimal solution with the hope of finding a globally optimal solution. The total cost (weight) to a grid (in a maze) or a node (in a graph) is given by Eq. 8.

$$f(n) = c(n) + h(n) \tag{8}$$

where n is the nth node, f is the total weight function, c is the cost function, and h is the heuristic function.

The cost function determined the cost of moving from one node to the other and heuristics function gives the program an idea as to how far the target might be from the present node. Heuristics function might be dependent on node position with respect to the target, the cost function might be dependent on node position respect to each other for which we have to compute the cost of each path and heuristics of each node or they might be arbitrary predefined values. As the algorithm tries to find the best path, the cost of reaching a particular cell from target is equal to the sum of cost of moving from parent cell to that particular cell and cost of moving to the parent cell from target.

2.2.1 Path Finding

A* algorithm always tries to minimize the total weight function out of the all the possible reachable node with the idea to reach the target using the least cost path. If the total weight of two or more cells are same, then by applying greedy method we choose the cell with the least heuristic value with the idea that since it is closest to the target cell, it would give rise to the least cost path.

In Fig. 3, a 9 × 4 grid system is taken with '*s*' as the source and '*t*' as the target. Cost function and heuristics function is based on Pythagoras Theorem where moving in straight line incurs a cost and heuristics of 10 and moving diagonally incurs a cost and heuristics of 14.14. So initially for the first iteration.

From Table 1, we see that A and W have least equal total weight. We choose W because its heuristic value is lesser. So W will be the next chosen node.

Fig. 3 Example 9 × 4 system

1				T
2				
3				
4				
5				
6	X	A	W	
7	B	S	C	
8	Y	D	Z	
9				

Table 1 A* costs, and heuristics for Fig. 3

Grid	Cost	Heuristics	Total weight
A	10	58.28	68.28
B	10	72.42	82.42
C	10	64.14	74.14
D	10	78.28	88.28
W	14.14	54.14	68.28
X	14.14	62.42	76.56
Y	14.14	82.42	96.56
Z	14.14	74.14	88.28

2.2.2 Algorithm for Static Maze

(1) Calculate the heuristics of each cell in the maze if they are not predefined and store it in a 2D array.
(2) Create a 2D array to store the maze map.
(3) Create a 2D array to store the cost of reaching a particular grid from source grid, initialize each grid to a high value except source grid.
(4) Create a 2D array to store the address of reachable grids which stores the source grid address initially.
(5) Go for an iterative procedure till target is reached where, from the list of reachable grids, the most suitable grid is selected based on the total weight. If total weight is same in more than one grid, the one with least heuristic value is chosen. This is the node to be visited next.
(6) Update the list of reachable grid during each iteration by marking the visited grids in the maze map and taking them out of the list and putting in the newly reachable nodes from the presently visited node. A grid once visited should not be visited again.
(7) Update the cost of each reachable cell if the cost of reaching the cell from the presently visited cell is lesser than the minimum cost of reaching the cell in the previous iterations.

3 Simulation

To test the Hopfield algorithm, the simulation was carried out in MATLAB on four different mazes in static condition. These mazes have been previously utilized by [13]. We have kept the environment constant so that we can easily compare our simulation results with theirs and have a comparative study between Hopfield path planning and A* path planning proposed in this paper. The simulation process of one maze from static has been explained and the same procedure is followed to simulate the other mazes too.

3.1 A* Algorithm

In the static maze, A* algorithm has been utilized to solve the minimum path planning problem. The explanation to how the maze shown in Fig. 4b was solved is as follows. The same procedure has been utilized to solve the Figs. 4a, c of static mazes as well.

1. First the algorithm takes an input of the map in term is 0s and 1s. 0s denote obstacle and 1s denote paths that can be taken from movement. Total number of rows and columns are set to 21. The start node is given as (1, 1) and end node is given as (21, 21).
2. Here, we have taken the distance between two adjacent cells is 10 units and by Pythagoras Theorem, the distance between a grid and its corner grids are 14.14 units.

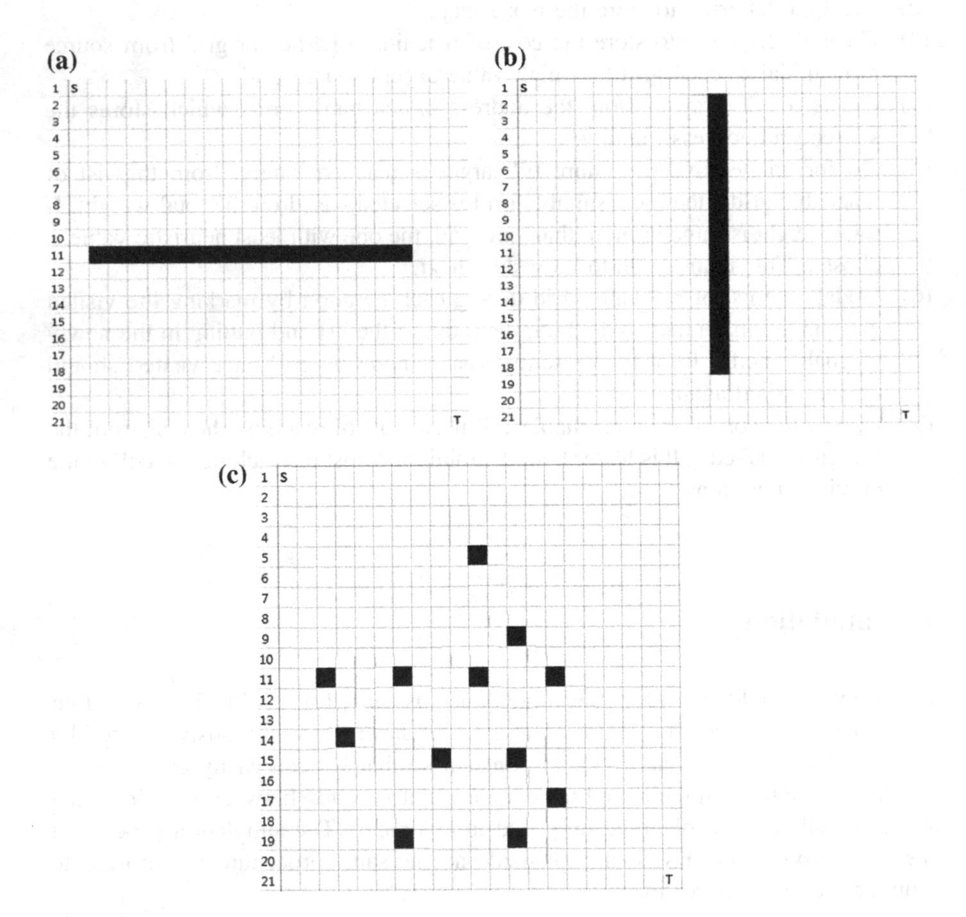

Fig. 4 Simulated mazes for A* and Hopfield

3. The heuristics function chosen here was the minimum distance between the target and a particular grid given by the formula

$$\begin{aligned} \text{Heu}(i,j) = \sqrt{2a^2} * \min(|r-i|,|c-j|) + a * \max(|r-i|,|c-j|) \\ - \min(|r-i|,|c-j|) \end{aligned} \quad (9)$$

   where Heu*(i, j)* is the heuristic function, r is the row value of target grid c is the column value of target grid, I is the row value of a particular grid whose heuristic value is being calculated, *j* is the column value of a particular grid whose heuristic value is being calculated.
4. Now we go for an iterative procedure by initially taking the source row and column, we calculate the cost of traveling to the grids surrounding it. We then calculate the total weight function by the formula,

$$f(i,j) = c(i,j) + h(i,j) \quad (10)$$

   where *i* is the row value of a particular grid, *j* is the column value of a particular grid, *f* is the total weight function, *c* is the cost function and *h* is the heuristic function.
5. We choose the local optimum solution by choosing the grid with least total weight or if more than one cell has least total weight we chose the one with the least heuristic value. The new grids surrounding the present selected grid are now taken into the list of reachable grids and the present selected grid (the source grid) is taken out of the list. The present selected grid has now been marked as visited and we will not revisit it.
   This iterative procedure continues and the local minimum grid is chosen during each iteration until the target grid is reached to construct the shortest path.

3.2 Hopfield Neural Network

The Hopfield surface was generated just once and stored in a matrix form. Every row and column index of an element corresponds to the x and y coordinates of a configurational unit of the surface. The elements of the matrix that will correspond to an obstacle are assigned a value of zero. For example, if the element of 2nd row and 5th column of a 21 × 21 matrix is an obstacle,

$$A(2,5) := 0 \text{ or } A(26) := 0$$

Since the positions are mapped to a matrix in MATLAB, an important point to note is that the coordinate system will begin from (1, 1) and not (0, 0). This is required as MATLAB begins matrix indexing from 1 and not 0. To define the robot's position in the map, a vector Pos has been used, where each position of the robot is stored sequentially in Pos. The initial position of the robot is fed directly to the vector Pos and the subsequent values are obtained from the path constructor. As stated previously, the target position is obtained by maximally activating the bias unit of the target's position. The following assignments are done to form the map.

$$A(11, 3:17) := 0, \quad I(21, 21) := +\infty, \quad \text{Pos}(1) := 1 \tag{11}$$

where I is a 21 × 21 matrix containing the bias values. Once the optimum values of these variables have been generated, the gradients of the robot's 8, 5, or 3 locations depending on whether the robot is at the edge, the corner or the center of the grid are calculated. The highest gradient is then selected and designated as the robot's next position. This continues till the target is reached. Similarly, all other static examples are simulated (Fig. 5).

4 Results and Discussion

The Hopfield and A* algorithms are implemented for the three mazes under static environment and the results are shown in Table 2 and the results are discussed in the following sections.

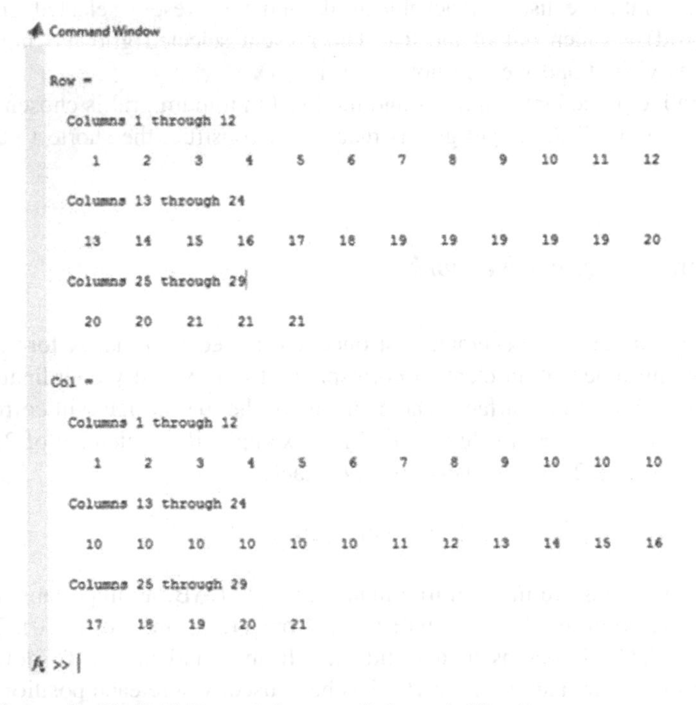

Fig. 5 Output for Hopfield Network for maze 2 shown in Fig. 4b

Table 2 Results for static maze path finding algorithms

Maze no.	Number of grids traversed for Hopfield	Run time for Hopfield (s)	Number of grids traversed for (A*)	Run time for A* (s)
1.	27	0.314	27	0.106
2.	27	0.348	27	0.099
3.	20	0.347	20	0.118

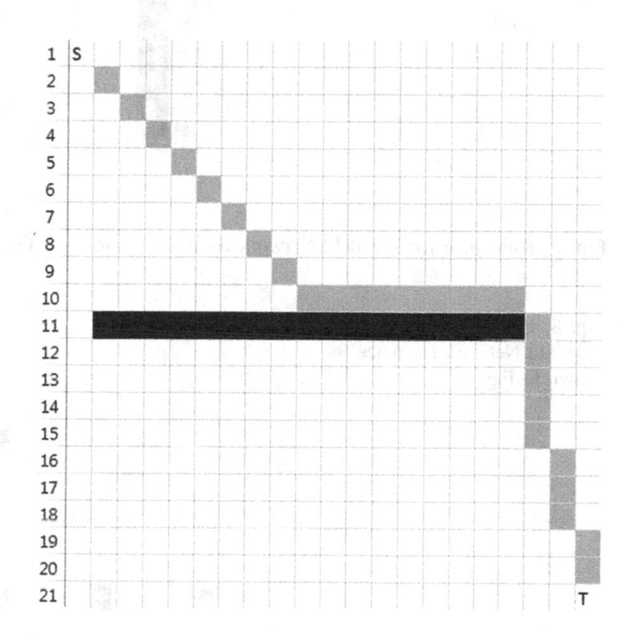

Fig. 6 Path taken for Hopfield Network for maze 1 shown in Fig. 4a

4.1 Hopfield for Static Mazes

Consider the second static maze for example. In this case, the algorithm is able to find the optimal path within 0.348 s. Here, it can be clearly seen that the path deviates from the shortest path upon encountering an obstacle and continues vertically downward till it again reaches an obstacle free zone. Majority of the time taken in this case is utilized for constructing the Hopfield matrix as it requires 100 iterations to get an optimal representation. On the other hand, a near-optimal solution can also be obtained using lesser iterations. The Hopfield Energy constraint however, needs to be satisfied in all the cases. The Hopfield Network still fares far better than the ant-colony optimization technique and the path taken results are shown in Figs. 6, 7 and 8.

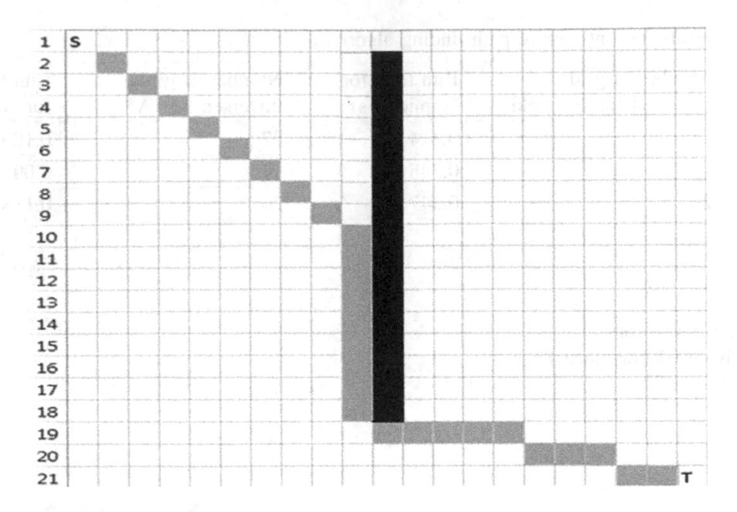

Fig. 7 Path taken for Hopfield Network for maze 2 shown in Fig. 4b

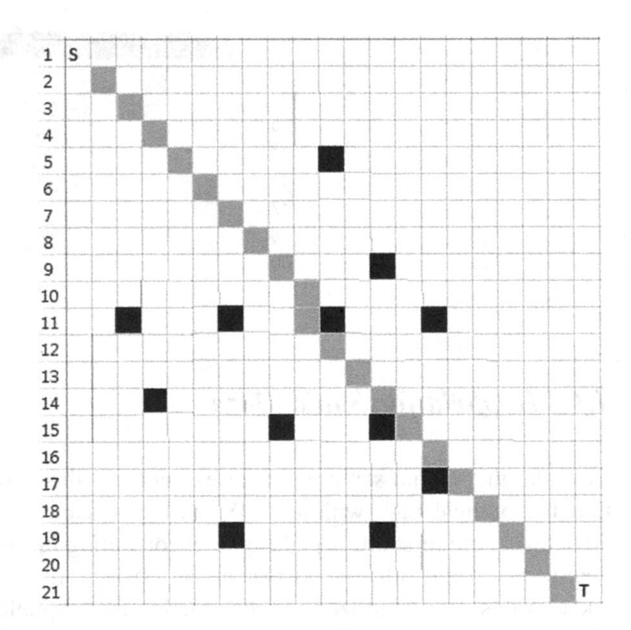

Fig. 8 Path taken for Hopfield Network for maze 3 shown in Fig. 4c

4.2 *A* for Static Mazes*

The simulation results for the static maze 2 is shown in Fig. 4b is shown in the Figs. 9 and 11. After implementing A* Algorithm, it clearly shows the path followed by the robot to reach the target grid from source grid. It is the most optimized path with the number of grids traversed in between source and destination is 27.

Fig. 9 Output for A* algorithm for static maze 2 shown in Fig. 4b

The time taken to give the output is 0.99 s. There has been considerable improvement in the path compared to the results obtained in ant-colony algorithm. The time required to run the algorithm on this particular maze has been shown in the above Table 2. We can clearly see that even the efficiency of path planning has increased. Similarly, the path taken by the robot after implementation of A* algorithm on static mazes 1 and 3 shown in Fig. 4a, c, respectively are shown in Figs. 10 and 12 respectively.

The results of A* algorithm and Hopfield Network from this work was compared with the results of simple ant colony optimization (SACO) and ant colony optimization meta-heuristics (ACO-MH) as implemented for the same static mazes and the following results were found,

- For maze 1 in Fig. 4a, for A* algorithm and Hopfield Network (100 iterations), the number of grids traversed to complete the path is 27 in both cases while in SACO it took 29 grids and ACO-MH it took 28 grids. For A* algorithm, the run time is 0.106 s. For Hopfield Network, the run time was 0.314 s (2.96 times slower than A*). For SACO the run time is 2.4 s (22.6 times slower than A* and 7.6 times slower than Hopfield). For ACO-MH, the run time is 2.37 s (22.4 times slower than A* and 7.5 times slower than Hopfield).

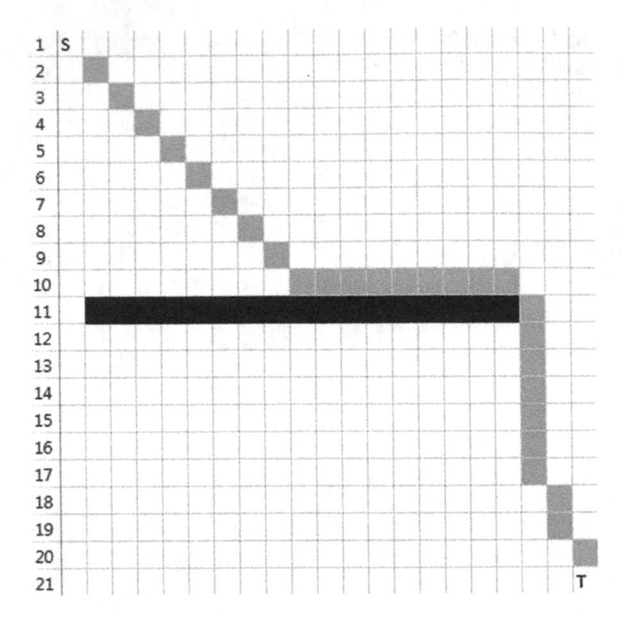

Fig. 10 Path taken for A* algorithm for maze 1 shown in Fig. 4a

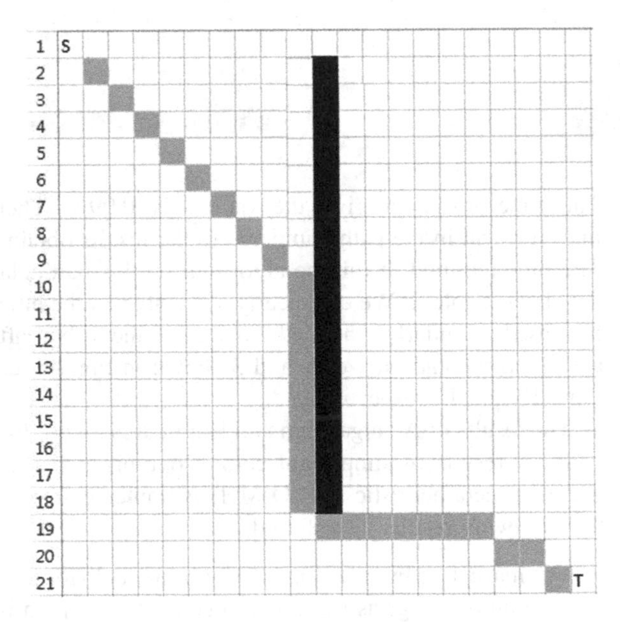

Fig. 11 Path taken for A* algorithm for maze 2 shown in Fig. 4b

- For maze 2 in Fig. 4b, for A* algorithm and Hopfield Network (100 iterations), the number of grids traversed to complete the path is 27 in both cases while in SACO it took 31 grids and ACO-MH it took 28 grids. For A* algorithm, the run time is 0.099 s. For Hopfield Network, the run time was 0.348 s. For SACO the run time is 6.57 s (66.4 times slower than A* and 18.9 times slower than

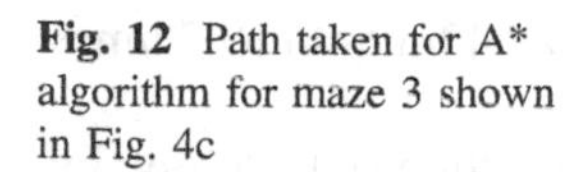

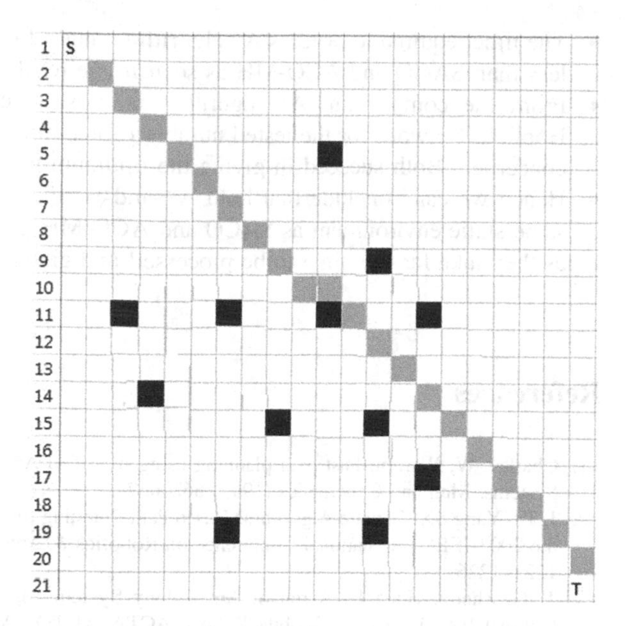

Fig. 12 Path taken for A* algorithm for maze 3 shown in Fig. 4c

Hopfield). For ACO-MH, the run time is 6.9 s (69.7 times slower than A* and 19.8 times slower than Hopfield).

- For maze 3 in Fig. 4c, for A* algorithm and Hopfield Network (100 iterations), the number of grids traversed to complete the path is 20 in both cases while in SACO it took 24 grids and ACO-MH it took 22 grids. For A* algorithm, the run time is 0.118 s. For Hopfield Network, the run time was 0.347 s. For SACO the run time is 3.67 s (31.1 times slower than A* and 10.6 times slower than Hopfield). For ACO-MH, the run time is 3.27 s (27.7 times slower than A* and 9.4 times slower than Hopfield).

5 Conclusion

Hopfield neural network and A* path planning algorithms for mobile robot under static environment has been carried out successfully. Based on the comparative study, the following concluding remarks were made,

- The A* algorithm is applied to the static mazes and it found to be work efficiently to give the global optimum path between two grid points and the processing required for finding the optimum path is very low.
- A suitable heuristic function in A* algorithm is applied to get the shortest path between any two grids in the shortest possible time. It gives a more accurate result in a much shorter time than the ant colony algorithm.

- The time required to process A* algorithm ranges from 22.4 times to 69.7 times less than SACO and ACO-MH as seen in the results.
- From the comparison, A* algorithm is most efficient followed closely by Hopfield Network for the tested number of iterations as far as time efficiency is concerned. Both succeed in giving the optimum path.
- Hence we can conclude that both A* and Hopfield Algorithms applied to the same static environment as SACO and ACO-MH generate more desired results as they take far less time to be processed and give better path planning results.

References

1. Charles. W. Warren: Fast path planning using modified A* Method. IEEE Transactions on Systems, Man and Cybernetics (1993) 662–667.
2. J. Tu, Yang SX: Genetic Algorithms Based Path Planning for a Mobile Robot. Proceedings of the 2003 IEEE International Conference on Robotics & Automation, Taipei, Taiwan (2003) 1221–1226.
3. T. G. Zheng, Huan H, Aaron S: Ant Colony System Algorithm for Real Time Globally Optimal Path Planing of Mobile Robots. ACTA AUTOMATICA SINICA vol. 33 (2007) 279–285.
4. Savsani, Poonam, R. L. Jhala, V. J. Savsani: Optimized Trajectory Planning of a Robotic Arm Using teaching learning based optimization (TLBO) and artificial bee colony (ABC) optimization techniques. *Systems Conference (SysCon) (2013) IEEE International*. IEEE (2013).
5. Garau, Bartolomé: Path planning for autonomous underwater vehicles in realistic oceanic current fields: Application to gliders in the western mediterranean sea. *Journal of Maritime Research* 6 2:5–22 (2014).
6. Chu, Keonyup, Minchae Lee, and Myoungho Sunwoo: Local path planning for off-road autonomous driving with avoidance of static obstacles. *Intelligent Transportation Systems, IEEE Transactions on* 13 4 (2012) 1599–1616.
7. M. G. Lagoudakis: Hopfield Neural Network for Dynamic Path Planning and Obstacle Avoidance. Technical Paper, University of Southwestern, Louisiana (2007).
8. Panrasee Rithipravat, Kenji Nakayama: Obstacle Avoidance by using Modified Hopfield Neural Network, ICAI.
9. Kroumov, Valeri, and Jianli Yu: Neural Network based Path Planning and Navigation of Mobile Robots. INTECH Open Access Publisher (2011).
10. Mahadevi S., K. R. Shylaja, Ravinandan M. E: Memory Based A* Algorithm for Path Planning of a Mobile Robot. International Journal of Science and Research (IJSR) (2012).
11. Ankit Bhadoria, Ritesh Kumar Singh: Optimized Angular a Star Algorithm for Global Path Search Based on Neighbor Node Evaluation. I. J. Intelligent Systems and Applications (2014).
12. Nadia Adnan Shiltagh, Lana Dalawr Jalal: Path Planning of Intelligent Mobile Robot Using Modified Genetic Algorithm. International Journal of Soft Computing and Engineering (IJSCE), Volume-3 Issue-2 (2013).
13. T. Mohanraj, S. Arunkumar, M. Raghunath, M. Anand: Mobile Robot Path Planning using Ant Colony Optimisation, International Journal of Research in Engineering and Technology, Volume: 03 Special Issue: 11 (NCAMESHE - 2014) (Jun-2014).

Vanishing Point Based Lane Departure Warning Using Template-Based Detection and Tracking of Lane Markers

Ammu M. Kumar, Philomina Simon and R. Kavitha

Abstract We propose a novel vision-based lane departure warning algorithm. The initial step of lane departure warning system is lane detection. In this paper, we present a lane detection method based on template matching. Kalman filtering is used to track the detected lanes. To make the algorithm robust to shadows, inpainting technique is done on regions with shadows. The curved lanes are handled with the help of Bezier spline fitting. Our algorithm determines the lane departure based on vanishing point margins. Experimental results show that the proposed method performs better than Caltech lane detection algorithm.

Keywords Lane detection · Lane tracking · Lane departure warning · Hough transform · Kalman filter

1 Introduction

Most of the road accidents occur due to the negligence of driver. Lane departure warning system warns the driver when the vehicle deviates from the lane without turn signal. There are three modules in a lane departure warning system: lane detection, lane tracking, and lane departure identification. A typical vision-based lane departure warning system consists of a camera placed behind the windshield of the car. The road image captured is used for lane detection. The lane detection and tracking is an important module of advanced driver assistance system. Vision-based

A.M. Kumar (✉) · P. Simon
Department of Computer Science, University of Kerala, Kariavattom, Thiruvananthapuram 695581, Kerala, India
e-mail: ammumanu24@gmail.com

P. Simon
e-mail: philomina.simon@gmail.com

R. Kavitha
Tata Elxsi Limited, ITPB Road, Whitefield, Banglore 560048, India
e-mail: kavitha.r@tataelxsi.co.in

S.S. Dash et al. (eds.), *Artificial Intelligence and Evolutionary Computations in Engineering Systems*, Advances in Intelligent Systems and Computing 517,
DOI 10.1007/978-981-10-3174-8_5

lane detection and tracking is a challenging research area in the field of computer vision. There are several challenges in lane detection such as presence of shadows, writings on road, variation in lighting conditions, and worn out lane markings. This paper proposes an approach that works well even in the presence of shadows, writings on road and neighboring vehicles. But the algorithm cannot detect lanes that are faded. The algorithm can handle curved lanes also. Section 2 describes the different methods used in the literature for lane departure warning. Section 3 discusses the proposed method. Section 4 presents the experimental results and analysis. Section 4 summarizes the paper.

2 Literature Review

The lane detection algorithms can be classified into two approaches: feature-based approach and model-based approach [1]. In feature-based approach, low-level image features such as edges are used. But model-based approach makes use of geometric parameters to model lanes. The image should be converted to inverse perspective mapping [2–4]. The edge detectors such as Canny [5–7] or Sobel [8, 9] or steerable filters [10] are used for finding the edges corresponding to lane markers. The lane marker is appeared as a white bar in dark background, template matching [11], and bar filtering [12] is used to filter the lane markers from other objects in the image. After filtering the edges, Hough transform [13–15] or polar randomized Hough transform [11] can be applied. The Hough transform can detect straight lanes but for curved lanes Catmull Rom spline fitting [16], B-spline fitting [17], RANSAC spline fitting [3, 18], hyperbola fitting [5] are used. Learning-based approaches [19], artificial neural network [20], ant colony optimization [21], Gabor filter lane matching [22], and *K*-means clustering [7, 17] are also used for lane detection. Kalman filter [13, 14], particle filter [4], and Lucas–Kanade algorithm [23] can be used for lane tracking.

3 Proposed Method

3.1 Preprocessing

The road images from a camera are the input for the vision-based lane departure warning system. Based on the intrinsic and extrinsic parameters of the camera, the image acquired will have different information contents associated to each pixel. This effect is known as perspective effect. To avoid this effect and to homogeneously distribute the information content among all the pixels, inverse perspective mapping is performed. The inverse perspective mapping produces a new image that represents the same scene as acquired from a different position [24].

The inverse perspective mapping is suitable for structured environments where the camera is mounted on a fixed position. For the lane departure warning system, the precalibrated camera is placed behind the windshield. In the actual image obtained from the camera, the lanes appear to converge at a point in the horizon due to the perspective effect. After applying the inverse perspective mapping (IPM), the top view of the road is obtained where the lanes appear to be parallel. The initial step in our algorithm is converting the input image to top view of the road image, based on flat road assumption and camera intrinsic and extrinsic parameters [3]. After converting to top view, the region of interest is reduced and processing time can be saved. In the next step, the color IPM image is converted to grayscale image using Eq. (1). The computation cost is reduced as single channel needs to be processed instead of processing the three channels in RGB image.

$$I_g(r,c) = 0.299.R(r,c) + 0.587.G(r,c) + 0.114.B(r,c), \tag{1}$$

where $I_g(r,c)$, $R(r,c)$, $G(r,c)$, $B(r,\ c)$ be the grayscale intensity, red, green, and blue channel values at (r,c) pixel position, respectively.

The grayscale image contains shadows of trees and other vehicles. An inpainting-based approach is used for handling the shadows. Digital inpainting is basically the process of reconstructing small damaged portions by filling the missing details using information from surrounding area [25]. To inpaint the shadow region, first an inpaint mask is constructed for the darker area in image corresponding to the shadows.

Inpainting is done using fast marching method (FMM) [26]. To inpaint a region, the algorithm starts with pixel P in the boundary and process the pixels inwards. Initially, FMM always processes the pixel close to the known image area. The inpainting of a pixel is determined by the values of the known image points close. The inpainting preserves the edges and other sharp details. The result of inpainting is shown in Fig. 1.

3.2 Template Matching

The shadow removed image is then filtered using normalized cross-correlation [27] with a template having similar structure as that of a lane marker. The template is shown in Fig. 2. Those pixels whose normalized cross-correlation coefficient value greater than 0.5 is kept as such and the remaining pixels are replaced with value 0. This will result in a modified grayscale image. Applying a threshold corresponding to 97.5th percentile of remaining values, a binary image is obtained [11].

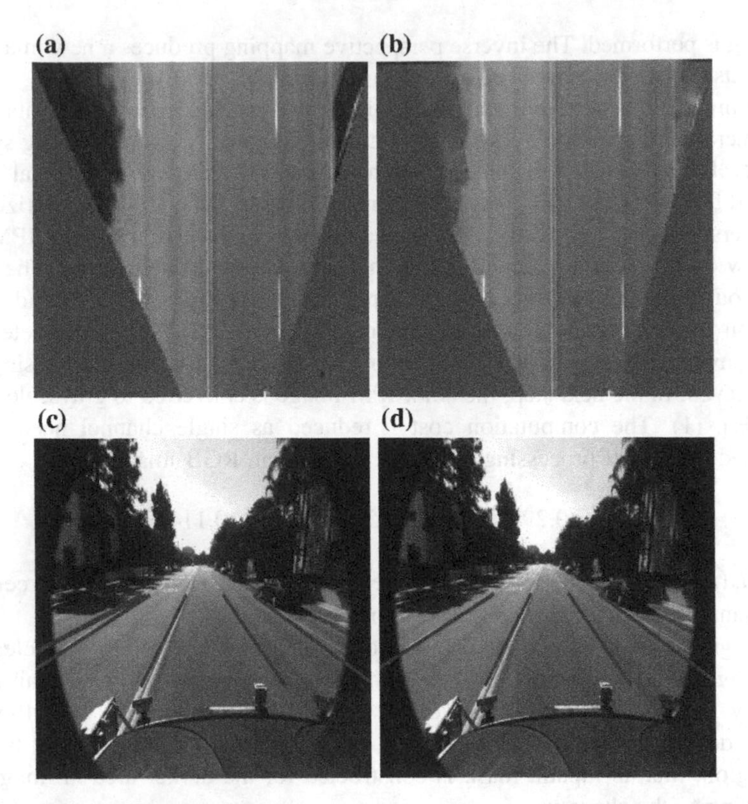

Fig. 1 **a** IPM image with shadow. **b** IPM image after inpainting (the shadow regions are now less visible). **c** Detection result without inpainting (false positive for shadow). **d** Detection result with inpainting

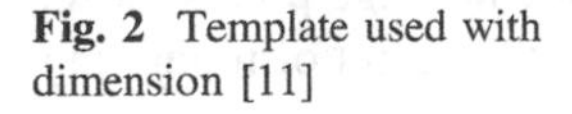

Fig. 2 Template used with dimension [11]

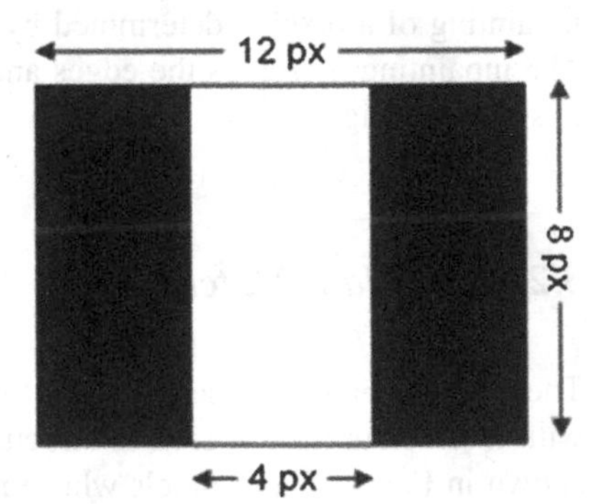

3.3 *Lane Detection*

Hough transform is applied on the binary image for detecting candidate lines representing the lane markers. The Hough transform is a parametric representation of points in the binary image. Each pixel in the binary image is transformed to Hough space by applying the given equation.

$$\rho = x\cos(\theta) + y\sin(\theta), \tag{2}$$

where ρ is the length of the perpendicular from the origin to a line passing through (x, y), and θ is the angle made by the perpendicular with the x-axis. The (ρ, θ) space is quantized into fixed sized blocks to form a 2-D accumulator array. For each (ρ, θ) pair obtained from Eq. (2), the corresponding accumulator cell is incremented [28]. Then top ten peaks in the Hough space are selected as candidate lane markers. In the inverse perspective mapping image, the lanes appear vertical and parallel, so a constraint on θ values corresponding to vertical line is applied.

To make the algorithm robust to writings on road, the lines with difference in ρ values less than a threshold with its neighboring lines are eliminated. Finally the candidate lane markers are detected. For each candidate lane marker, a rectangular region of interest enclosing the marker is considered in the binary image. Inside the region of interest for those nonzero pixels, the difference between maximum and minimum x values other than difference of x values of the start and end point of the line segment then, it is considered as a candidate for curved lane and Bezier spline fitting. Otherwise, the candidate lane marker is considered to be straight and line fitting is performed.

3.3.1 Straight Lane Detection

The line fitting is based on M-estimator. The algorithm fits a line to a set of point set by minimizing $\sum_i \rho(r_i)$, where r_i is the difference between the predicted value (based on the regression equation) and the actual, observed value of the ith point of the line and $\rho(r)$ is a distance function [29]. The distance function used in our method is given as follows.

$$\rho(r) = 2.\left(\sqrt{1 + \frac{r^2}{2}} - 1\right) \tag{3}$$

The line is fitted iteratively using weighted least squares algorithm and in each iteration, the weight is made inversely proportional to the distance function $\rho(r)$.

3.3.2 Curved Lane Detection

A third-degree Bezier spline is used to represent curved lanes. A Bezier spline is a bunch of these Bezier curves linked together. A third-degree Bezier spline fitting [3] is applied inside the bounding box for dealing with curved lanes. A third-degree

Bezier spline is defined by four control points and is a cubic polynomial and is represented using the equation:

$$P(t) = (1 - t^3)P_0 + 3t(1 - t^2)P_1 + 3t^2(1 - t)P_2 + 6t^3P_3, \quad (4)$$

where P_0, P_1, P_2, P_3 are the control points and the parameter t is proportional to cumulative sum of the euclidean distances from point P_i to the first point P_1 and the value of t lies in the interval [0, 1].

3.4 Lane Tracking

After detecting the position of bounding box in current frame, the position of bounding box in remaining frames are predicted using Kalman filter. The top-left corner and bottom-right corner of the bounding box are predicted. After that line fitting or spline fitting is applied inside the bounding box, based on the lane characteristics. To avoid the tracking to deviate from actual values, after every 2 s, lane detection is performed.

3.5 Lane Departure Detection

In the input road images, the lane markers appear to converge at a point in horizon known as vanishing point. The lane departure is identified based on the position of vanishing point vp in the image. In ideal case, the vanishing point is assumed to be almost near the center of the image. But when there is a lane departure, the vanishing point deviates from the center of image either to the right or left based on the departure direction. The departure and its direction can be determined by the deviation of vanishing point from the center of the image.

If the vanishing point moves toward the left side of the image then it is right departure otherwise left departure. To determine the departure, the difference between half width hw of the image and the x coordinate of vanishing point vp_x are compared with a threshold t. If the difference is greater than t then departure is identified. The sign of the difference is checked to determine the direction of departure. If the sign is positive, then the departure is towards right otherwise is a left side departure. The result of lane departure warning is given in Fig. 3.

$$|\mathrm{hw} - \mathrm{vp}_x| > t \quad \text{and} \quad \mathrm{hw} - \mathrm{vp}_x > 0 \quad \text{then Right departure} \quad (5)$$

$$|\mathrm{hw} - \mathrm{vp}_x| > t \quad \text{and} \quad \mathrm{hw} - \mathrm{vp}_x < 0 \quad \text{then Left departure} \quad (6)$$

Fig. 3 **a** Departing the right lane marker. **b** Departing the left lane marker

4 Results and Discussion

The algorithm is implemented in C++ using OpenCV. It is tested in Caltech Lanes dataset. The algorithm is tested on 640 × 480 images on a Intel Core i5 2.50 GHz machine. The performance evaluation metrics used are Precision, Recall and *F*-score [30]. Precision is the fraction of detected lanes markers that are actual lane markers. Recall is the fraction of actual lane markers that are detected. *F*-score is the measure that combines precision and recall and is the harmonic mean of Precision and Recall.

$$\text{Precision} = \frac{TP}{TP + FP} \tag{7}$$

$$\text{Recall} = \frac{TP}{TP + FN} \tag{8}$$

$$F\text{-score} = \frac{2\,\text{Precision} * \text{Recall}}{(\text{Precision} + \text{Recall})} \tag{9}$$

where *TP* is the number of true positives, *FP* is the number of false positives, *TN* is the number of true negatives, *FN* is the number of false negatives. Table 1 shows

Table 1 Results

Data sets	# Frame	# Actual lanes	# Detected	Precision	Recall	*F*-score	False positives per image
Cordova1	250	919	970	0.9329	0.9847	0.9581	0.26
Cordova2	406	1048	1073	0.9412	0.9637	0.9523	0.1552
Washington1	337	1274	1206	0.9668	0.9152	0.9402	0.1186
Washington2	232	931	998	0.9288	0.9957	0.9610	0.3060
Total	1224	4172	4247	0.9437	0.9607	0.9521	0.1339

the quantitative evaluation result of the lane detection algorithm tested in Caltech lane dataset [3]. The performance of the proposed algorithm is compared with Caltech lane detection [3] algorithm. The results are shown in Figs. 4, 5 and 6.

Figures 7, 8, and 9 show the quantitative comparison of the proposed method and Caltech lane detection. The precision of the proposed method is more than Caltech lane detection method. The proposed method performs better than Caltech method in terms of recall in all cases except in Washington 1 clip in the dataset. Washington 1 clip contains frames with intense illumination which causes the proposed method to miss some of the lane markers.

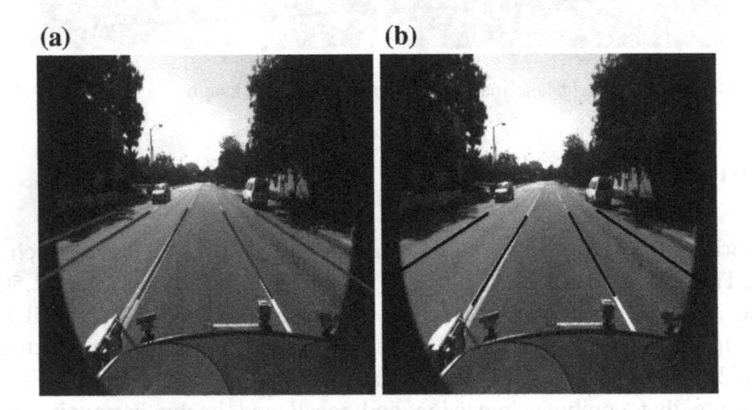

Fig. 4 Result of detection in presence of shadow. **a** Proposed method. **b** Caltech lane detection

Fig. 5 Result of detection in presence of writings on road. **a** Proposed method. **b** Caltech lane detection

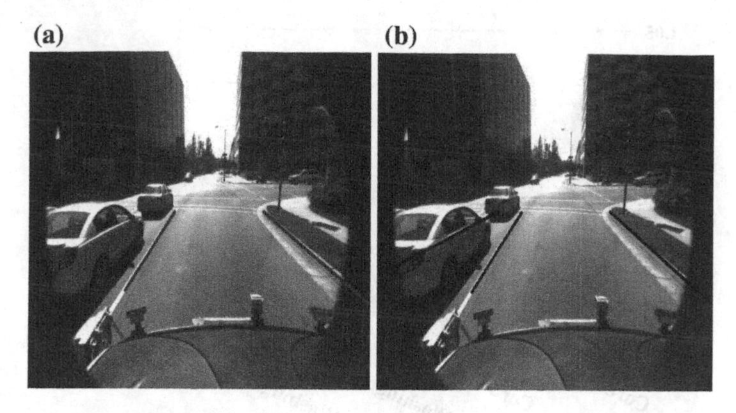

Fig. 6 Result of detection in presence of vehicles. **a** Proposed method. **b** Caltech lane detection

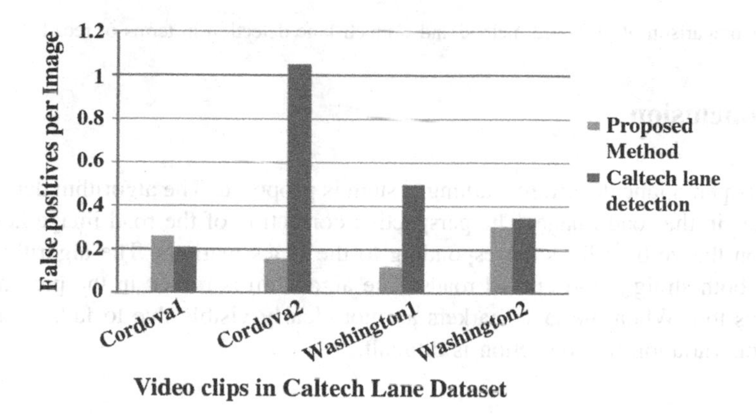

Fig. 7 Comparison of proposed method and Caltech lane detection in terms of false positives per image

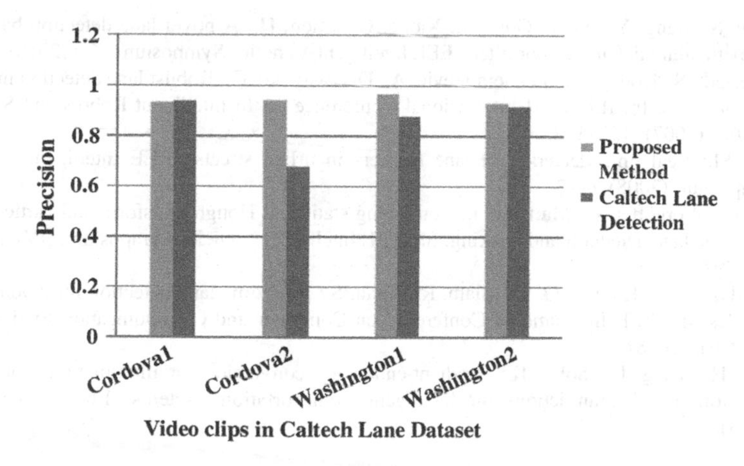

Fig. 8 Comparison of proposed method and Caltech lane detection in terms of precision

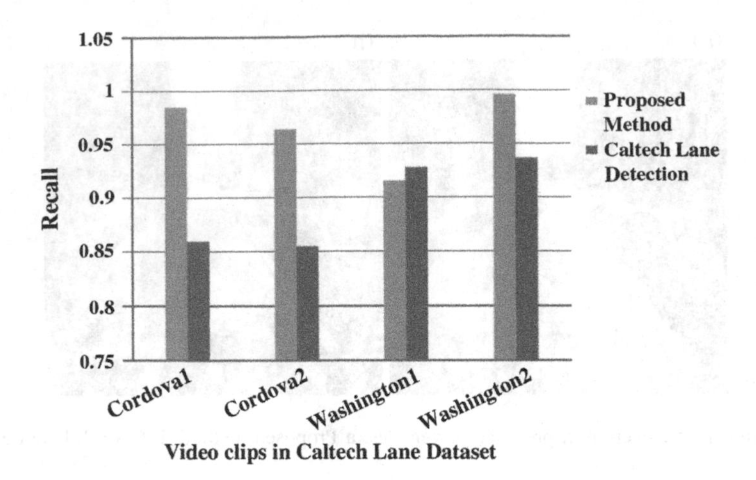

Fig. 9 Comparison of proposed method and Caltech lane detection in terms of recall

5 Conclusion

In this paper, a lane departure warning system is proposed. The algorithm detects all the lanes in the road image. The perspective correction of the road image helps to focus on the vertical lines corresponding to the lanes markers. The algorithm can handle both straight and curved roads. The algorithm is robust in the presence of shadows too. When the lane markers are not clearly visible due to fading or illumination variation, the detection is difficult.

References

1. Zhou, S., Jiang, Y., Xi, J., Gong, J., Xiong, G., Chen, H.: A novel lane detection based on geometrical model and Gabor filter. IEEE Intelligent Vehicles Symposium (IV) (2010) 59–64.
2. Sehestedt, S., Kodagoda, S., Alempijevic, A., Dissanayake, G.: Robust lane detection in urban environments. In: IEEE/RSJ International Conference on In Intelligent Robots and Systems (IROS), (2007), 123–128.
3. Aly, M.: Real time detection of lane markers in urban streets. IEEE Intelligent Vehicles Symposium (2008) 7–12.
4. Liu, G., Wörgötter, F., Markelic, I.: Combining statistical Hough transform and particle filter for robust lane detection and tracking. In: IEEE Intelligent Vehicles Symposium, (IV), (2010), 993–997.
5. Assidiq, A. A., Khalifa, O. O., Islam, R., Khan, S.: Real time lane detection for autonomous vehicles. In: IEEE International Conference on Computer and Communication Engineering, (ICCCE), (2008), 82–88.
6. Yoo, H., Yang, U., Sohn, K.: Gradient-enhancing conversion for illumination-robust lane detection. IEEE Transactions on Intelligent Transportation Systems, 14(3), 1083–1094, (2013).

7. Tran, T. T., Cho, H. M., and Cho, S. B.: A robust method for detecting lane boundary in challenging scenes. Information Technology Journal, 10(12), 2300–2307, (2011).
8. Yim, Y. U., Oh, S. Y.: Three-feature based automatic lane detection algorithm (TFALDA) for autonomous driving. IEEE Transactions on Intelligent Transportation Systems, 4(4), 219–225, (2003).
9. Tsai, S. C., Huang, B. Y., Wang, Y. H., Lin, C. W., Lin, C. T., Tseng, C. S., Wang, J. H.: Novel boundary determination algorithm for lane detection. In: IEEE International Conference on Connected Vehicles and Expo (ICCVE), (2013), 598–603.
10. Jung, H., Min, J., Kim, J.: An efficient lane detection algorithm for lane departure detection. In: IEEE Intelligent Vehicles Symposium (IV), (2013), 976–981.
11. Borkar, A., Hayes, M., Smith, M.: Polar randomized Hough transform for lane detection using loose constraints of parallel lines. In: International Conference on Acoustics, Speech and Signal Processing (ICASSP). (2011) 1037–1040.
12. Zhao, H., Teng, Z., Kim, H. H., Kang, D. J.: Annealed Particle Filter Algorithm Used for Lane Detection and Tracking. Journal of Automation and Control Engineering 1(1), (2013).
13. Li, Y., Iqbal, A., Gans, N. R.: Multiple lane boundary detection using a combination of low-level image features. In: 17th IEEE International Conference on Intelligent Transportation Systems (ITSC), (2014), 1682–1687.
14. Ozgunalp, U., Dahnoun, N.: Robust lane detection and tracking based on novel feature extraction and lane categorization. In: IEEE International Conference on Acoustics, Speech and Signal Processing (ICASSP), (2014), 8129–8133.
15. Saalfeld, A.: Topologically consistent line simplification with the Douglas Peucker algorithm. Cartography and Geographic Information Science 26(1), 7–18 (1999).
16. Guo, C., Mita, S., and McAllester, D.: Lane detection and tracking in challenging environments based on a weighted graph and integrated cues. In: IEEE/RSJ International Conference on In Intelligent Robots and Systems (IROS), (2010), 6643–6650.
17. Wang, J., Mei, T., Kong, B., Wei, H.: An approach of lane detection based on Inverse Perspective Mapping. In: 17th IEEE International Conference on Intelligent Transportation Systems (ITSC), (2014), 35–38.
18. Borkar, A., Hayes, M., Smith, M.: Robust lane detection and tracking with RANSAC and Kalman filter. In: International Conference on Image Processing. (2009) 3261–3264.
19. Gopalan, R., Hong, T., Shneier, M., and Chellappa, R.: A Learning Approach Towards Detection and Tracking of Lane Markings. IEEE Transactions on Intelligent Transportation Systems, 13(3), 1088–1098, (2012).
20. Kim, Z.: Robust lane detection and tracking in challenging scenarios. IEEE Transactions on Intelligent Transportation Systems, 9(1), 16–26, (2008).
21. Daigavane, P. M., Bajaj, P. R.: Road Lane Detection with Improved Canny Edges Using Ant Colony Optimization. In: 3rd IEEE International Conference on Emerging Trends in Engineering and Technology (ICETET) (2010),76–80.
22. Zhou, S., Jiang, Y., Xi, J., Gong, J., Xiong, G., Chen, H.: A novel lane detection based on geometrical model and Gabor filter. In: IEEE Intelligent Vehicles Symposium (IV), (2010), 59–64.
23. Bottazzi, V. S., Borges, P. V., Stantic, B., Jo, J.: Adaptive regions of interest based on HSV histograms for lane marks detection. In: Springer International Publishing on Robot Intelligence Technology and Applications 2, (2014), 677–687.
24. Bertozzi, M., Broggi, A., and Fascioli, A.: Stereo inverse perspective mapping: Theory and applications. Image and Vision Computing, 16:585590. (1998).
25. Bertalmio, M., Bertozzi, A.L., Sapiro, G.: Navier-Stokes, fluid dynamics, and image and video inpainting. In: CVPR01: Proc. IEEE Int. Conf. on Computer Vision and Pattern Recognition, Kauai, vol. I, pp. 355362. IEEE Press, New York (2001).
26. Telea, A.: An image inpainting technique based on the fast marching method. J. Graph. Tools 9(1), 23–34 (2004).
27. Lewis, J.P.: Fast Normalized Cross-Correlation. Industrial Light and Magic (1995).

28. Umbaugh, S. E: Digital image processing and analysis: human and computer vision applications with CVIPtools, 176–184. CRC, (2010).
29. Bhar, L.: Robust regression, New Delhi: IASRI, (2008), 70–78.
30. Fritsch, J., Kuhnl, T., Geiger, A.: A new performance measure and evaluation benchmark for road detection algorithms. In: 16th International IEEE Conference on Intelligent Transportation Systems (ITSC), (2013), 1693–1700.

Destination Proposal System to Forecast Tourist Arrival Using Associative Classification

S. Nair Athira, K.E. Haripriya Venugopal and L. Nitha

Abstract Tourism plays an important role on Indian economy. In this paper, we are proposing a system for finding valuable tourists and forecasting the arrival of the tourists. For analysis, we collected data through online survey and can make prediction on valuable domestic tourist arrival using classification based on association rules generated in a bidirectional approach (CARGBA). By predicting the tourist arrival in India, we can recommend valuable tourist destinations and can promote and improve the facilities in Indian tourism.

Keywords Associative classification · CARGBA algorithm · Apriori algorithm · N-gram approach

1 Introduction

Twenty-first century witnessed a tremendous growth in Indian tourism and is emerging both economically and socially. In India, one of the important sources of foreign exchange is tourism. It can provide jobs to both skilled and unskilled people. Also it popularizes national unity and international brotherhood. India is one of the popular tourist destinations in Asia. India has become the famous and favorite tourist spot in Asia. Nowadays, tourism uses all the benefits of technological advancement for advertising. Globally, tourism becomes the fast growing

S.N. Athira (✉) · K.E. Haripriya Venugopal · L. Nitha
Department Computer Science and IT Amrita School of Arts and Sciences,
Amrita Vishwa Vidyapeetham (Amrita University), Kochi, India
e-mail: athira.suresh.p@gmail.com

K.E. Haripriya Venugopal
e-mail: haripriyavenu@gmail.com

L. Nitha
e-mail: nitha.leeladharan@gmail.com

S.S. Dash et al. (eds.), *Artificial Intelligence and Evolutionary Computations in Engineering Systems*, Advances in Intelligent Systems and Computing 517,
DOI 10.1007/978-981-10-3174-8_6

and popular industry. Scope, benefit, and challenges are strongly observed and analyzed by the government, because it can affect our economic, sociocultural, environmental, and educational resources of our country. Tourism domain is a crucial application area in a recommendation system. Destination proposal system is an intermediary between user and trip advisor. We conducted an online survey through social media sites like Facebook, Google+, etc. For finding the valuable tourists, demographic features, year of visit, and favorite tourist spot as parameters. The data accessible in the internet may be heterogeneous and unstructured. For making it structured and homogeneous we do data preprocessing. In our system, we use the N-Gram based approach for the sentiment classification. Based on the preprocessed data retrieved through the online survey we use CARGBA (classification based on association rules generated in a bidirectional approach) for generating rules and building classifier. From the classifier, we generate the final rule set which can be used for forecasting tourist arrival.

2 Literature Review

Nowadays, essential part within the web atmosphere is recommender systems in the view that they represent a technique of growing consumer pride and making their seat strong in the highly competitive market of the digital industry routine [1, 2]. For decades, normal organizations have accelerated their competitiveness by the use of trade intelligence techniques supported by means of strategies like data mining. Recommender systems can be categorized depending on the kind of approach used for making ideas. Content-based methods and collaborative filtering algorithms are two essential classes and their difference lies in information used for recommending various destinations. Accordingly, suggestions are established on information involving record elements as well as on knowledge regarding the action of each user. Traditionally, there are two systems used in this types of methods. They are rule-centered systems or key phrase matching.

A recently evolved technique for web mining is sentiment classification that can function on opinions and sentiments [3]. According to [4], the main goal of sentiment classification is mining textual content of written experiences from tourist for specified destination and classifying the experiences in constructive opinions. Among N-gram model, SVM, and Naive Bayes, the studies show that N-gram and SVM produce accurate results comparable with the Naive Bayes. Machine learning algorithms can function classification very well in the sentimental polarities, if the machine learning algorithms are well trained. The accuracy level of above-mentioned algorithms will reach 80% or more of classification correctly. In our system, we use the N-Gram-based approach for the sentiment classification of online survey.

In data mining, association rule mining and classification are two predominant procedures in expertise discovery procedure [5]. Integration of these two approaches is a principal research center of attention and has many purposes in knowledge mining. Combining these approaches had created new system known as class association rule mining or associative classification process. It results higher accuracy in classification by combining these two procedures.

Association rule reveals the strong association relationships among a set of data. It extracts correlations, constantly occurring patterns, associations among sets of items in the transaction databases. Main objective of association rule mining is to search out the set of all subsets of features that regularly occur in several transactions. Classification rules are one type of conditional rules, which may be used to discover data from huge knowledge sets [6–8].

There are numerous classifications based on association approaches [7, 8, 9]. First, CBA (classification based on association) developing association rules are referred to as candidate rules, which have convinced support and confidence thresholds. To design a classifier, a short set of rules are taken from them. Second, CBA-RG (classification based on association rule generator) algorithm, data is examined numerous times. All the frequent rule items are developed through all these passes. And at the end of the pass, it determines which of the candidate rule items are absolutely frequent that can yield the CARs. Third, CBA-CB (classification based on association-classifier builder) by taking all the available subsets of the training data, it yields the best classifier evaluation. If the rules have right rule sequence with lower number of errors then that subset is selected. Fourth, an approach called CMAR (classification based on multiple association rules), here based on weighted analysis using numerous solid association rules classification is achieved. Fifth, CPAR (classification based on predictive association rule). For developing rules from the training data, it takes a greedy algorithm. For examining each rule, it uses expected accuracy and uses best *k*-rules in forecasting, then it can eliminate over fitting. Lastly, it is CARGBA (classification based on association rule generated in a bidirectional approach). CARGBA has two main parts: first part is CARGBA rule generator and second part is CARGBA classifier builder. CARGBA algorithm consists of two stages. In the first step, the rules are developed in Apriori style. They have high support and lower length. In the second step, the rules are generated more specifically. So the rules are generated in reverse Apriori fashion. In CARGBA classifier builder, it frames a classifier with important rules.

According to [10], CARGBA produce a highly accurate result when compared to decision tree classifiers. CARGBA follows bidirectional approach, which produces strong association rule. CARGBA generalize the dataset and also give exact and exceptional rules. The goal of rule development is not only drawing out knowledge, but also for better accuracy using these rules for the classification. Association rule algorithm generate rules using Apriori algorithm, some of the rules get removed if the rule have lower support count even it have high confidence. On the other hand, when we are generating rule based on reverse Apriori without support trimming, the amount of support less rules will be higher and it takes more computational time. To overcome the above-mentioned problems, a new associative classification algorithm called CARGBA is used.

3 Methodology

The step-by-step process of the system is shown in Fig. 1

3.1 *Data Collection*

In this step, we conduct online survey through social media sites like Facebook, Google+, etc., using ProProfs survey maker. This software is a knowledge management software for conducting surveys, quizzes, tests, training, and flashcards. The dataset analyzed is in comma separated values (CSV) format. A sample preview of the questionnaire is given in Fig. 2

3.2 *Preprocessing*

The data collected from the online survey may be incomplete, noisy, and inconsistent and needed to be preprocessed. During preprocessing, various steps such as data cleaning, data integration, data transformation, data reduction, and data discretization are done. Finally, after preprocessing we got the structured and homogeneous data for the further process. Here, for preprocessing N-Gram based

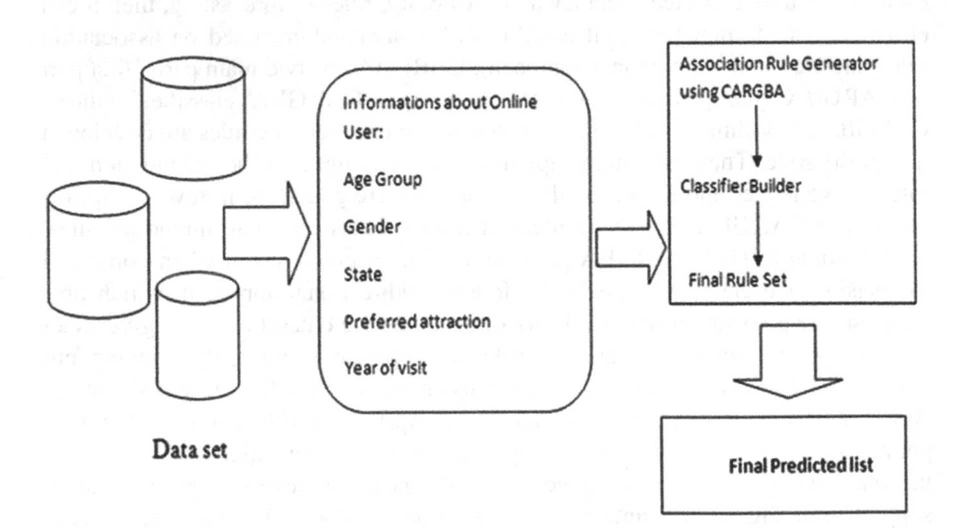

Fig. 1 Step by step process

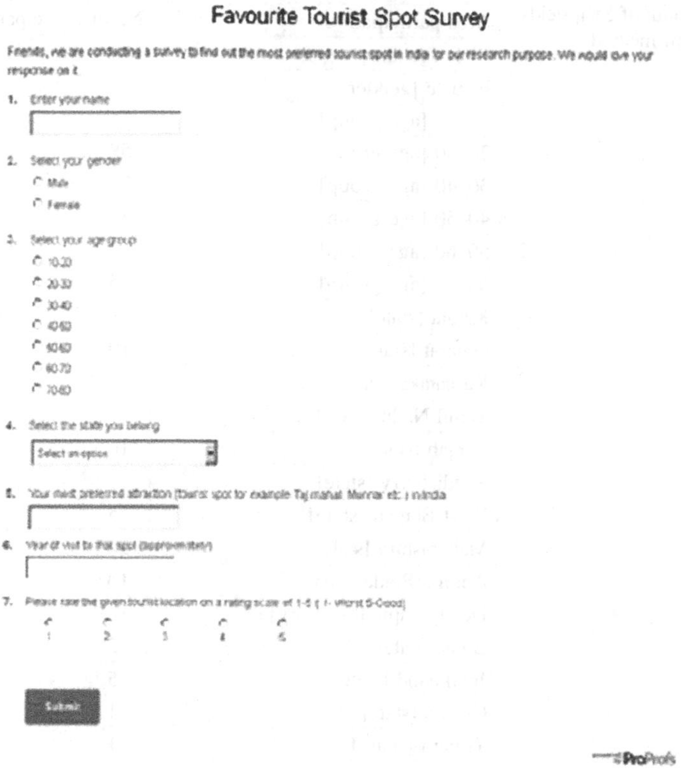

Fig. 2 Preview of online survey

approach is used. In this approach, sentiment classification is used to extract the reviews of customers. N-Gram is the best method for preprocessing sentiment polarities of reviews about travel destinations in terms of accuracy. N-gram approach is a supervised machine learning algorithm. It can reach more than 80% of classification accuracy. The data retrieved from the survey is in comma separated values file format and after the preprocessing we retrieved the count of each fields using N-gram method. It is shown in the Table 1 and the visualization shown in Fig. 3

3.3 Association Rule Generation: CARGBA

CARGBA stands for classification based on association rules generated in a bidirectional approach. It is a bidirectional rule that is used to build association rules

Table 1 Count of each fields using N-gram method

Area participation	No of participation
Male [gender]	185
Female [gender]	116
10–20 [age group]	5
20–30 [age group]	158
30–40 [age group]	76
40–50 [age group]	39
50–60 [age group]	18
60–70 [age group]	5
Kerala [state]	88
Gujarat [state]	10
Karnataka [state]	31
Tamil Nadu [state]	17
Punjab [state]	16
Pondicherry [state]	7
West Bengal [state]	15
Maharashtra [state]	26
Andhra Pradesh [state]	16
Delhi [Capital territory]	18
Bihar [state]	2
Jharkhand [state]	5
Odisha [state]	1
Tripura [state J	1
Chandigarh [state]	3
Chhattisgarh [state]	3
Rajasthan [state]	7
Madhya Pradesh [state]	3
Haryana [state]	6
Jammu And Kashmir [state]	2
Uttar Pradesh [state]	14

and it is not only simplifying the data set, but also gives exact and special rules to justify the definite behavior and irregularities in a data set.

3.3.1 CARGBA Rule Generator

The main function of CARGBA rule developing formula is to search out few rules that accept high level confidence or capable satisfactory confidence. CARGBA rule generator mainly consist two phases. These phases are given below:

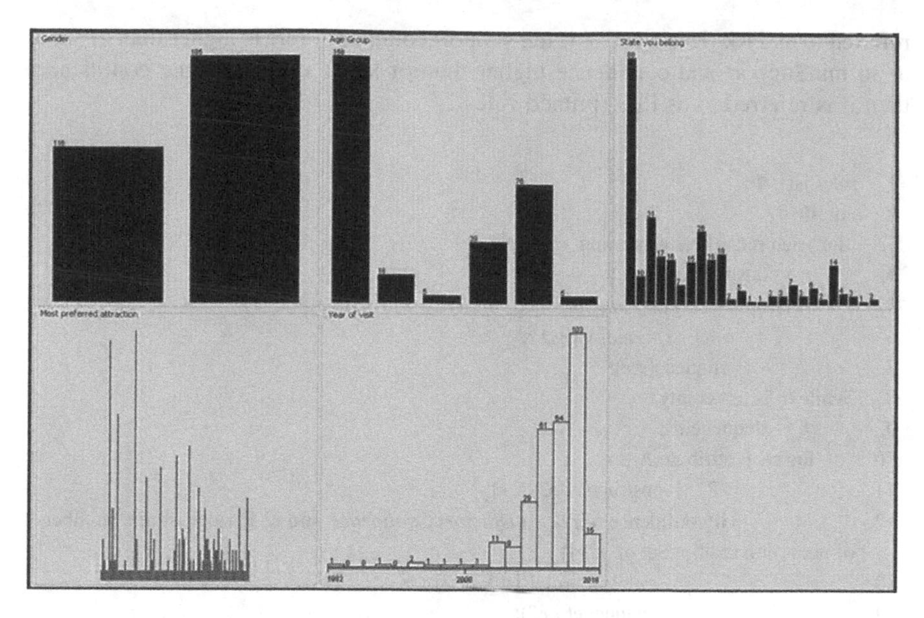

Fig. 3 Visual representation

1. $R1=\{1\text{-rules}\}$;
2. Compute *confidence* of each rule of R_1;
3. $PR_1=\{r \in R_1 |$ *confidence* of $r >=$ *SatisfactoryConfidence* **and** *support* of $r >=$ *minSupport*$\}$;
4. $R_1=\{r \in R_1 |$ *support* of $r>=$*minSupport*$\}$;
5. **for**($k=2;R_{k+1}\neq \Phi$ **and** $k\leq 1$;k++) **do**
6. $R_k=$ Apriori – Gen(R_{k-1});
7. Compute *support* **and** *confidence* of each rule of R_k
8. $PR_k=\{r \in R_k |$ *confidence* of $r>=$*SatisfactoryConfidence*$\}$;
9. $R_k=\{r \in R_k |$ *support* of $r>=$ *minSupport*$\}$;
10. $PR_{s=} \bigcup_k PR_k$;

CARGBA Rule Generator: First phase

Phase 1: All the association rules are developed in this first phase of the model 1-rules to **l**-rules that accept larger than or adequate to satisfactory confidence below support trimming wherever *k*-rule indicates a rule whose condition set has *k* things and **l** could be a constraint of the algorithmic program. For discovering association rules, Apriori algorithm is used in the first phase. The reciprocal of the algorithm is stated in the initial stage of rule generator. If the support count is less than the minSupport, then it trims the rules away at each point of rule generation.

In the initial phase of CARBA rule generator, Rk indicates the group of *k*-rules, If the confidence of *k*-rules is greater than or equal to adequate confidence, then it is referred to as PRk (Pruned Rk). If the support count each rule is higher than or same as to minSupport and confidence higher than or same as to adequate confidence, then it is referred to as PRs (pruned rules).

1. ruleList= **Φ**;
2. q= **Φ**;
3. **for** each record *rec* ∈ training examples
4. *r* = constructRule(*rec*);
5. **if** (*confidence* of r≥ *satisfactoryConfidence* and r ∉ ruleList)
6. ruleList = ruleList ∪ r;
7. enqueue(*q*,*r*);
8. **while** (*q* is not empty)
9. *r* = dequeue(*q*);
10. **for** each attribute A ∈ *r*
11. *r2* = constructRule2(A,*r*);
12. **if**(confidence of r2≥ *satisfactoryConfidence* **and** r2 ∉ ruleList **and** number of items at conditionSet of r2>l)
13. ruleList=ruleList ∪ *r2*;
14. enqueue(*q*,*r2*);

CARGBA Rule Generator: Second phase

Phase 2: Association rules of the form $(l + 1)$-rules to the n-rules are developed in this second phase, and here the total number of non-class attributes are labeled as '*n*', and the confidence of this rules might be greater than or equal to adequate or satisfactory confidence. This step is done in reverse Apriori fashion. So it is known as "Reverse Rule Generation Algorithm." Here '*q*' indicates a queue and 'rule List' carries all the developed rules. A record named 'rec' is taken from the training example for constructing a rule named '*r*' and it is constructed using a method called "construct Rule." Confidence of rule '*r*' is computed using this same method. By eliminating the characteristic A of rule '*r*' we can create a rule '*r*2' by using a method called "constructRule2." Confidence of '*r*2' is computed using this same method. Lastly, using the equation "PRs = PRs ∪ ruleList" we can combine the rules developed from phase 1 and 2.

3.3.2 CARGBA Classifier Builder

CARGBA classifier builder algorithm is described in this portion. Large number of rules are developed using the rule generator and these rules are stored in PRs. All these rules will not be used to classify test examples. Then a subset of rules from pruned rules is selected for including the dataset. Now arrange these selected rules in decreasing order of support, confidence, and rule length. Following is the algorithm for classifier builder:

1. *FinalRuleSet* = **Φ**;
2. *DataSet* = D;
3. Sort(*pr*Rs);
4. **For** each rule r ∈ PRs **do**
5. **If** *r* correctly classifies at least one training example *d* ∈ *dataset* **then**
6. Remove *d* from *dataset;*
7. Insert *r* at the end of *finalRuleSet*

CARGBA classifier builder

Here the list of rules used in the classifier lies in the finalRuleSet. PRs are sorted in the decreasing order of confidence using sort () function. From lines 4–7 training examples are classified using the rules in the finalRuleSet. All the rules in the finalRuleSet are placed only after arranging support, confidence, and rule length in the descending order. The classifier applies the primary rule in the finalRuleSet to classify a new test example. If all the rules of the classifier fail to include the test example then this will be classified to a default class.

Through Proprofs we had collected around 300 reviews of peoples from all over India. We issued the link through social media sites such as Facebook, Google+, Twitter, and Gmail, for collecting relevant and vital information. People belonging to different age group, gender, and states made their active participation through the survey. Sample data of survey is shown in Fig. 4.

Then we process the preprocessed data using the package "a rules" in R for processing. First, we perform the Apriori algorithm with a minimum support count of *five* and minimum confidence as 80%. Then we perform the reverse of Apriori algorithm. After performing Apriori, the rules with higher confidence are rejected

	A	B	C	D	E
1	Gender	Age Group	State you belong	Most preferred attraction	Year of visit
2	Female	20-30	Kerala (KL)	alappuzha	2013
3	Female	20-30	Kerala (KL)	goa	2010
4	Female	20-30	Kerala (KL)	Coorg	2009
5	Female	20-30	Kerala (KL)	GOA	2000
6	Female	20-30	Kerala (KL)	Brindavan gardens	2009
7	Female	20-30	Kerala (KL)	Munnar	2015
8	Female	20-30	Kerala (KL)	kullu manali	2011
9	Female	20-30	Kerala (KL)	Wayanad	2010
10	Female	20-30	Kerala (KL)	munnar	2008
11	Male	20-30	Kerala (KL)	Munnar	2015
12	Female	20-30	Kerala (KL)	Kashmir	2016
13	Female	20-30	Kerala (KL)	Taj Mahal	2005
14	Female	20-30	Kerala (KL)	Taj Mahal	2003
15	Female	20-30	Kerala (KL)	Humayun's tomb	2008
16	Female	20-30	Gujarat (GJ)	girnar	2010
17	Female	20-30	Kerala (KL)	Rohtang Pass	2014
18	Male	20-30	Kerala (KL)	VAGAMON	2015
19	Female	20-30	Kerala (KL)	KODAIKANAL	2009
20	Male	20-30	Kerala (KL)	Silent Valley	2013
21	Female	20-30	Kerala (KL)	iravikulam	2015
22	Female	20-30	Kerala (KL)	Hyderabad	2006
23	Male	20-30	Kerala (KL)	tajmahal	2010
24	Female	20-30	Kerala (KL)	Munnar	2011
25	Female	20-30	Kerala (KL)	Munnar ,simla	2011
26	Female	20-30	Kerala (KL)	Manali	2013
27	Male	20-30	Karnataka (KA)	Gangtok	2011
28	Male	50-60	Karnataka (KA)	Lalbagh	1982

Fig. 4 Sample data

```
projectfinalrules - Notepad
File Edit Format View Help
"4" "{State.you.belong=Kerala (KL),Year.of.visit=2015}" "{Age.Group=20-30}" 0.04318936 1 1.9050632
"6" "{Most.preferred.attraction=OOTTY}" "{Gender=Male}" 0.03986710963 1 1.62702702702703
"10" "{Age.Group=40-50,Year.of.visit=2010}" "{Gender=Male}" 0.033222591 1 1.627027027
"15" "{State.you.belong=Puducherry (PY)}" "{Age.Group=20-30}" 0.02325581 1 1.905063
"16" "{Most.preferred.attraction=Miramar}" "{Gender=Male}" 0.023255813 1 1.62702702
"22" "{Age.Group=20-30,Most.preferred.attraction=OOTTY}" "{Gender=Male}" 0.0265780 1 1.627027
"25" "{Age.Group=30-40,State.you.belong=National Capital Territory of Delhi (DL)}" "{Gender=Male}" 0.023255 1 1.627027
"26" "{Age.Group=20-30,State.you.belong=National Capital Territory of Delhi (DL)}" "{Gender=Male}" 0.02657 1 1.62702702
"35" "{Gender=Female,State.you.belong=Kerala (KL),Year.of.visit=2014}" "{Age.Group=20-30}" 0.023255 1 1.9050632
"36" "{Gender=Female,State.you.belong=Kerala (KL),Year.of.visit=2015}" "{Age.Group=20-30}" 0.0265780 1 1.905063
"37" "{Most.preferred.attraction=Kanjirakolly}" "{State.you.belong=Puducherry (PY)}" 0.0132890365448505 1 43
```

Fig. 5 Final best rule list generated using "*R*"

since minimum support count is not satisfactory. In the next step, reverse Apriori is used where the rules with satisfactory confidence with support lower than five are only retrieved and other rules are rejected. The results of Apriori and Reverse Apriori are combined together forming the rule set. Finally, the rules are arranged in decreasing order of confidence generating the final rule set which is displayed in Fig. 5. By analyzing the final rule set, the end user can make predictions, such as which categories (age, gender, etc.) of users may choose a particular location.

4 Conclusion

In our paper, we recommended a system for finding valuable tourists and forecasting the arrival of the tourists for the development of Indian tourism industry. The system put forward proposals that help the department of tourism and trip advisors to plan destinations and to attract more tourists. Advancement in tourism sector not only brings the economy to the country but also helps in the cultural development of the country.

References

1. Joel P. Lucas a, Nuno Luz b, María N. Moreno, Ricardo Anacleto b, Ana Almeida Figueiredo b,Constantino Martins b, A hybrid recommendation approach for a tourism system, Expert Systems with Applications, Volume 40, Issue 9, July 2013.
2. Yuanchun Jiang, Jennifer Shang, Yezheng Liu, Maximizing customer satisfaction through an online recommendation system: A novel associative classification model, Decision Support Systems, Volume 48, Issue 3, 2010.
3. Minqing Hu and Bing Liu Opinion Extraction and Summarization on the Web, AAAI, 2006.
4. Qiang Ye, Ziqiong Zhang, Rob Law,Sentiment classification of online reviews to travel destinations by supervised machine learning approaches, 2009.

5. Bangaru Veera Balaji, Vedula Venkateswara Rao, Improved Classification Based Association Rule Mining, International Journal of Advanced Research in Computer and Communication Engineering Vol. 2, Issue 5, 2013.
6. Maria-Luiza Antonie, Osmar R. Zaïane, An Associative Classifier based on Positive and Negative Rules. Proceedings of the 9th ACM SIGMOD Workshop on Research Issues in Data Mining and Knowledge Discovery, DMKD 2004, Paris, France, June 13, 2004.
7. D. Sasirekhal, A. Punitha, A Comprehensive Analysis on Associative Classification in Medical Datasets, Indian Journal of Science & Technology, Volume 8, Issue 33, 2015.
8. Divya Ramani, Harshita Kanani,Chirag Pandya, Ensemble of Classifiers Based on Association Rule Mining, International Journal of Advanced Research in Computer Engineering & Technology (IJARCET),Volume 2, Issue 11, 2013.
9. Divya Ramani, Chirag Pandya, Harshita Kanani, Classification method based on Association Rule Mining. IJSRD, Volume2, Issue:1, 2014.
10. Gourab Kundu, Sirajum Munir, Md. Faizul Bari, Md. Monirul Islam, and K. Murase, A Novel Algorithm for Associative Classification.
11. N. D. Solank, Mansi patel Computer Engineering Department, A Survey On Classification And Association Rule Mining, © 2014 IJIRT Volume 1 Issue 6| ISSN: 2349–6002.

Analyzing the Impact of Software Design Patterns in Data Mining Application

Nair Puja Prabhakar, Devika Rani, A.G. Hari Narayanan and M.V. Judy

Abstract Data Mining is all about analyzing the data gathered from different sources, based on different context and summarizing it into meaningful information. The belief of utilizing design pattern to improve the data mining application is relatively high. Design Pattern is a broad and general repeatable solution to a problem that occurs frequently in a software design. There are only a few number of researches illustrating their benefit. In this paper, we try to expose the relation between data mining architecture and design patterns. We have here taken a layered architecture for data mining environment and analyzed the impact of design pattern in every components of given architecture. This paper presents a survey on various design patterns used in data mining which is used to fulfill functional and non-functional requirements (quality attributes).

Keywords Data mining · Design patterns · Layered architecture · Quality attributes

1 Introduction

We are living in the information age and thus a large amount of data is gathered on daily basis. The important need is to analyze such data. Data mining is used to turn such large collection of data into meaningful knowledge. Data mining is completely

N.P. Prabhakar (✉) · D. Rani · A.G. Hari Narayanan · M.V. Judy
Department of CS and IT, Amrita School of Arts and Sciences,
Amrita Viswa Vidyapeetham, Amrita University, Kochi, India
e-mail: pujaprabhakaran@gmail.com

D. Rani
e-mail: devikahari07@gmail.com

A.G. Hari Narayanan
e-mail: hariag2002@gmail.com

M.V. Judy
e-mail: judy.nair@gmail.com

S.S. Dash et al. (eds.), *Artificial Intelligence and Evolutionary Computations in Engineering Systems*, Advances in Intelligent Systems and Computing 517,
DOI 10.1007/978-981-10-3174-8_7

an interdisciplinary area that can be explained in different ways. It is a process of examining data from many different perspective or dimensions and then summarizing the identified result. The various data mining application include marketing and fraud detection. This marketing contains customer detailing and retaining, identifying probable customer, market partitioning, and fraud detection is all about identifying fraud in credit card, intrusion detection. It can be viewed as an essential step in the process of knowledge discovery, where other steps include the following:

1. Data Cleaning
2. Data Integration
3. Data Selection
4. Data Transformation
5. Data Mining
6. Pattern evaluation
7. Knowledge presentation

In software, a design pattern is a general reiterative solution to a commonly existing problem in software design. They are not a finished design that can be applied directly to code. In many different situations, they are the fingerprint or impression for solving the same.

Commonly, a pattern has four essential elements, which are listed below:

1. The **pattern name** is a controller which can be used to describe a design problem, where it occurs and its solution in a word.
2. The **problem** describes the context where the given pattern can be applied.
3. The **solution** is used to explain the element that forms a design domain, their work flow, collaboration among each design.
4. The **consequences** are the outcome or result that we get after applying the pattern.

An efficient design but if it takes an hour to generate a solution then it is of no use to the person who needs result at that moment. Carefully developed and well-defined designs are the foundation of better engineering. Design pattern is a good option for maintainability, as it provides a new means for reusing existing designs and architecture. Understanding the present techniques through design patterns makes them more accessible to developers of new architecture. They help you to choose design option in such a way that makes a system reusable and avoid options that do not support reusability. With a genuine design pattern we can thus improve the maintainability. So we can say this design pattern describes the problem which has a chance to occur again and again in given environment, and then define the core of the solution in such a way that they can be used million times repeatedly without ever doing the same thing again. Here we have taken a layered architecture of data mining environment and analyzed the impact of design pattern in the different components of this architecture. Through this way Data mining algorithms' adaptability and maintainability can be improved.

2 Related Work

"Acceptor–Connector an Object Creational Pattern for Connecting and Initializing Communication Services". In this paper, they discuss about the acceptor–Connector pattern, when a service is initiated this design pattern decouples connection generation and service initialization from the already performed processing. To perform this decoupling here three components are used: (1) acceptors, (2) connectors, (3) service handlers. A connector powerfully produces a connection with acceptor, and a service handler is initialized in order to produce a data exchange. Where an acceptor continuously waits to get request for connection, when such request is received the connection is established and data are exchanged.

Benefits provided by such patterns are the following:

1. Reusability is achieved
2. Maintains portability
3. Extensibility is achieved

So here they have mentioned the benefits of using such pattern in distributed environment [1].

"A Pattern-Based Data Mining Approach" Patterns are very useful in software design as the well known, GOF pattern for OO design pattern. Pattern can be applied to much wider context, with the creation of such pattern in various fields it seems data mining algorithms' adaptability and maintenance can be improved. A pattern can be identified as the best one through the way they generalize and solve the actual problems. The quality and standard of a design pattern can change over time when new one gains acceptance or better solutions are available. This paper gives a brief introduction about different design pattern available in data mining. They have already identified a collection of other design patterns in the area of data mining, which are listed below:

1. Combine Voting
2. Training/retraining
3. Solution analysis

Pattern can be applied to any area of human interest. Through this pattern, we will be able to give users the possibility to model data mining, and also to better understand the data mining step [2].

"A Pattern-Based Approach for GUI modeling and Testing" In this paper, they have discussed about a pattern-based approach for GUI modeling and testing. User Interface (UI) patterns are used widely in the software design. This UI patterns exhibit commonly occurring solution to solve common UI problems. GUI is the fundamental part in software, which makes the software user friendly by offering flexibility in how user tasks are performed. They have made two observations: (1) Developers use distinct tools to benefit the creation of GUI. (2) But, it is a fact that these tools and toolkits are only able to support the beginning phases of GUI's development steps. To support the above two observation, the specific tool is used for user interface patterns [3].

"A Survey on Software Design Pattern Tools for Pattern Selection and Implementation" This paper presents a survey on various methods to select the suitable design pattern for problem in hand. It further discusses the different tools that are available to apply this design pattern into code. They ended the paper saying that in the upcoming years software design knowledge can be observed in the form of strategic and tactical patterns that solve the problem with various client/server programming, distributed analysis, software maintainability, real-time applications, embedded system, game creation, and user interface design [4].

"A Software Architecture for Data Mining Environment" The technological effectiveness, predominant the use of paradigm, are a major factor that influences the acceptance and the impartiality of such background. They have here addressed issues such as integration, where new tools can be added into environment, interoperability over numerous dimensions and the interaction with other framework available, other issue is agreement between legacy tools while choke in the environment and its accommodation with the interoperability components which is enforced [5].

"A Comprehensive Approach Toward Data Preprocessing Techniques and Association Rules". Data mining is a technique that details with big data in that preprocessing plays an important role in this mining process and huge contribution in success of project. This is also useful for association rule algorithm. In this paper, they have processed a methodology and implementation of data preprocessing and then mined the rule with prevailing association rule algorithm [6].

3 Data Mining Architecture with Design Pattern

Software architecture is a general topic which gives less concern about implementation part than to the design. It usually includes the concept necessary to describe the structure and concepts, lighting toward its evaluation. Our description is based on the fact that how this architecture explains the structure or the structure of this data mining application. The architecture that we have taken here is layered one and can be described as three layered architecture which is mentioned below:

1. User interface, which the user has interaction with, it is first thing to which user puts attention and acts as a bridge between user and business components.
2. Business components is the layer that takes care of processing tools and preprocessing part. This is basically a two group layer.

The storage component and connectors are used to store all the data that are gathered and maintained which need to be manipulated later in the journey of data analyzing (Fig. 1).

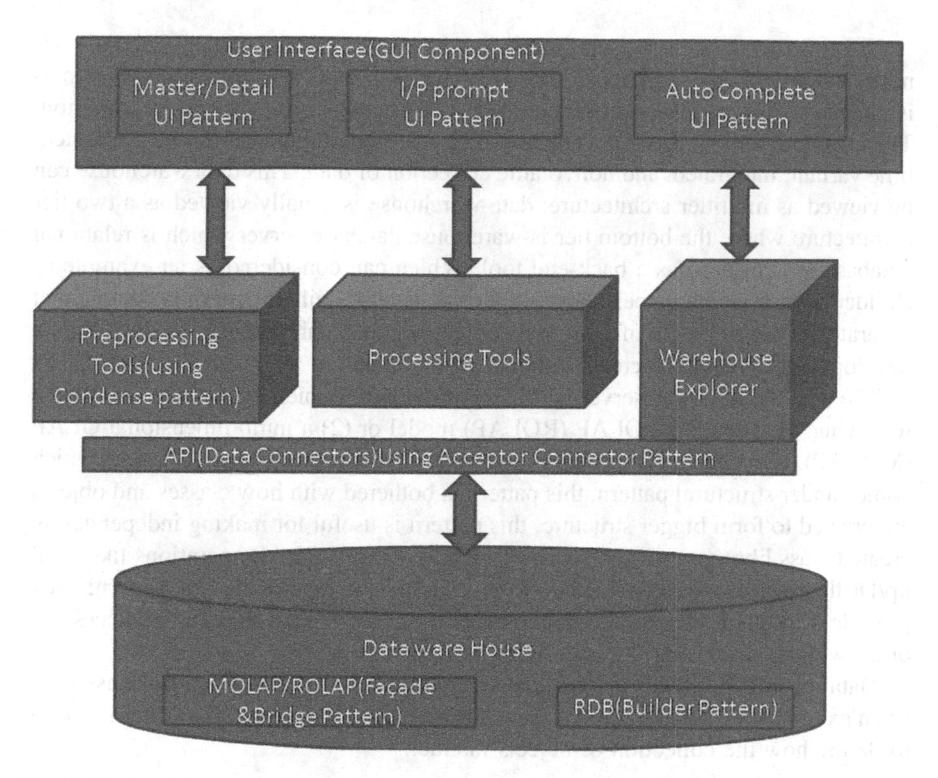

Fig. 1 Design patterns used in the layered architecture for data mining environment

User interface is a detailed graphical implementation of certain functionality, but the fact is layout may be different based on functionality, but our work is based on common behavior that is repeated over time. To get an entry, we researched on different user interface design patterns in data mining.

The Master/Detail UI Pattern is the one that divides the screen, into two areas master and detail area, based upon the selection made in the master area, the content in the detail area is displayed.

The Input Prompt UI Pattern allows the user to enter data into the computing system.

The Auto complete UI Pattern allows a word-based searching which means when a user inputs a word, a list of all possible outcomes appears and finally desired one can be selected [3].

The next component is business component which contains two parts: first is preprocessing tool and other one is processing tool. The preprocessing tool is used to ensure data quality which has various factors like accuracy, completeness, consistency. The preprocessing tools like filters use patterns like condense pattern, which is one of the most known data mining pattern [2]. This condense pattern is basically used as preprocessing pattern.

The third component present in this architecture is storage component and connectors example data warehouse. Data warehouse provides various tools to administrator for organizing the data methodically and use this data to make better decision. There are four major keywords related to data warehouse, which are subject-oriented, time variant, integrated, and nonvolatile collection of data. This data warehouse can be viewed as multitier architecture; data warehouse is usually viewed as a two-tier architecture where the bottom tier is warehouse database server which is relational database system used as a back-end tool. Which can considered as an example of Builder pattern which comes under creational pattern, builder pattern is the one that separate the development of a complex object from its embodiment so that the same development process can create different embodiment.

The next tier is OLAP server which is consistently implemented using one of the following: (1) Relational OLAP (ROLAP) model or (2) a multidimensional OLAP (MOLAP). Here we can use either of the following patterns bridge or façade, which comes under structural pattern, this pattern is bothered with how classes and objects are created to form bigger structure, this pattern is useful for making independently created class library work with each other. The various OLAP operations like Drill up/Drill down can be considered as an example of iterator design pattern; they provide a means to access the components of an aggregate object in a successive order without disclosing the underlying representation.

Data loading that takes place in data ware house or data mart can be considered as an example of mediator pattern, which is a behavioral pattern, this pattern is used to define how the collection of objects interact.

4 Impact of Design Patterns in Achieving the Quality Attributes of Data Mining Application

There are various quality attributes that has to be achieved in data mining projects. Some of the prominent quality attributes to be achieved are maintainability, reusability, performance, and fault proneness. Survey shows that the design patterns support these quality attributes and following table represents some of the quality

Maintainability	Reusability	Performance	Fault Proneness
Abstract-Factory	**Adapter**	**Facade**	**Singleton**
Composite	**Command**	**Command**	**Template**
Decorator	**Visitor**	**State**	**State**
Observer	**State**		**Adapter**
Acceptor-Connector	**Builder**		**Observer**
	Acceptor-Connector		

Fig. 2 The impact of design pattern in quality attribute in data mining

attributes and their corresponding design patterns used in the data mining and machine learning environment (Fig. 2).

5 Conclusion

Patterns are utility that converts problems in a certain situation into a solution, and are something which is meaningful to all users. There are certain benefits of using design patterns in different components of data mining architecture. From the result analysis, it is clear that by applying various design patterns in different components of data mining architecture, we can improve the quality attributes such as reusability, maintainability, extensibility, adaptability, performance, fault proneness. Thus, we can conclude that the impact of software design patterns in data mining architecture is an efficient approach to increase the quality of components.

Acknowledgments This work is supported by the DST Funded Project, (SR/CSI/81/2011) under Cognitive Science Research Initiative in the Department of Computer Science, Amrita School of Arts and Sciences, Amrita Viswa Vidyapeetham University, Kochi.

References

1. Douglas C. Schmidt, “Acceptor-Connector An Object Creational Pattern for Connecting and Initializing Communication Services”, An earlier version of this paper appeared in a chapter in the book Pattern Languages of Program Design 3.
2. Boris Delibaši, Kathrin Kirchner, Johannes Ruhland, “A Pattern Based Data Mining Approach”, Conference paper. January 2007, DOI:10.1007/978-3-540-78246-9-39.source: DBLP.
3. Rodrigo M. L. M. Moreira, Ana C. R. Paiva, Atif Memon, “A Pattern-Based Approach for GUI modeling and Testing”, 978-1-4799-2366-3/13/$31.00 ©2013 IEEE.
4. S. Saira Thabasum, Dr. U. T. Mani Sundar, “A Survey on Software Design Pattern Tools for Pattern Selection and Implementation”, International Journal of Computer Science & Communication Networks, Vol 2(4), 496–500.
5. Georges Edouard, “A Software Architecture for Date Mining Environment”, KOUAMOU National Advanced School of Engineering, Cameroon.
6. Jasdeep Singh Malik, Prachi Goyal, Mr. Akhilesh K Sharma, “A Comprehensive Approach Towards Data Preprocessing Techniques and Association Rules”.
7. Ampatzoglou A, Apostolos K, Elvira Maria A, Antonis G, Fragkiskos C, Ioannis S, “An empirical investigation on the impact of design pattern application on computer game defects.” In Proceedings of the 15th International Academic Mind Trek Conference: Envisioning Future Media Environments, pp. 214–221. ACM, 2011.
8. A.V. Krishna Prasad, Dr. S. Ramakrishna, “Software Architectures Design Patterns Mining for Security Engineering”, et al./(IJCSIT) International Journal of Computer Science and Information Technologies, Vol. 1 (5), 2010, 408–413.
9. Francesca Arcelli Fontana, Marco Zanoni, “A tool for design pattern detection and software architecture reconstruction”, Information Sciences Volume 181, Issue 7, 1 April 2011, Pages 1306–1324.

10. Gamma E, Helm R, Johnson R and Vlissides J, "Design Patterns. Elements of Reusable Object-Oriented Software".
11. Rudolf Ferenc, Arpad Beszedes, Lajos Fulop, Janos Lele, "Design Pattern Mining Enhanced by Machine Learning".
12. Ichiro Hirata, kenshero mitsulani, Toshiki yamoka, "A study on systematizing GUI design patterns for embedded system products".
13. Prechelt L, Unger B, Tichy W.F, Brossler P, Votta L.G, "A controlled experiment in maintenance comparing design patterns to simpler solutions", IEEE Transactions on Software Engineering, vol. 27, no. 12, pp. 1134–1144, Dec 2001.
14. Cline M, "The pros and cons of adopting and applying design patterns in the real world, Communications of ACM," 39(10), 47–49.
15. Vokac M, "Defect frequency and design patterns: an empirical study of industrial code", IEEE Transactions on Software Engineering, vol. 30, no. 12, pp. 904–917, Dec. 2004.
16. Gatrell M, Counsell, "Design patterns and fault-proneness a study of commercial C# software," Fifth International Conference on Research Challenges in Information Science (RCIS), vol., no., pp. 1–8.
17. Gary P. Moynihan, Abhijit Suki, Daniel J. Fonseca. "An expert system for the selection of software design patterns: Expert System Journal".
18. Afacan T, "State Design Pattern Implementation of a DSP processor: A case study of TMS5416C", 6th IEEE International Symposium on Industrial Embedded Systems (SIES), vol., no., pp. 67–70, 15–17 June 2011.

A Survey for Securing Online Payment Transaction Using Biometrics Authentication

M. Hari Priya and N. Lalithamani

Abstract In emerging lifestyle, people prefer online shopping as the most convenient way for purchasing their products. In order to provide secure transaction in e-commerce site, biometric features are incorporated in authentication. Biometric features are considered to be easier and secure because it solves the exasperation by the customer to recall the long account number and their password and it also has unique feature all over the globe. The use of biometrics authentication in online site provides more secure way of transaction than any other means [1–3]. In this work, we summarize the study of the various kinds of biometric traits that are used for online payment compared with one another.

Keywords Easy pay · Pay as you go · Payment system · Online banking · Secure transaction

1 Introduction

In day-to-day life, online payments are increasing and people prefer it for its various other reasons like quick mode of transport and convenient to purchase from any location. Before getting into detail, let us first see "why biometrics" and why not any other measures. Biometrics authentication is used for identity management. Conventional methods of identifying the person based on the knowledge, i.e., remembering the password and based on token, e.g., using Identity cards. The efficacy of the authentication mechanism can be achieved when they beat various types of attacks. Biometrics offers negative recognition and non-repudiation to resolve this attack. Negative recognition is the process in which an individual has to

M. Hari Priya (✉) · N. Lalithamani
Department of Computer Science and Engineering, Amrita School of Engineering, Amrita Vishwa Vidyapeetham, Amrita University, Coimbatore, India
e-mail: haripriyam2979@gmail.com

N. Lalithamani
e-mail: n_lalitha@cb.amrita.edu

S.S. Dash et al. (eds.), *Artificial Intelligence and Evolutionary Computations in Engineering Systems*, Advances in Intelligent Systems and Computing 517,
DOI 10.1007/978-981-10-3174-8_8

enroll in the system and receives denial of access, whereas non-repudiation is the process that whosoever accesses the resources cannot deny it later. There are three major category of biometrics.

- Physical—Hand, Fingerprint, Palm Print, Facial Recognition, Iris, Retina, and Sclera;
- Behavioral—Handwriting, voice, Signature, Gait; and
- Chemical/Biological—Perspiration and Skin Composition [4].

Figure 1 shows four main modules involved in biometric authentication process such as sensor unit, feature extraction unit, decision algorithm phase, and Database [4]. The components required for these processes are sensors, computers, and software. However, whatever the biometric feature is, now our concern is about to provide a secure online payment [5]. This concern for securing online payment transaction study is carried over in this work.

2 Prior Research

Lack of security in the e-commerce site is put forward people to do research in this domain. One of such technologies is Sidebar—Fingerprint PAY-BY-TOUCH was a privately held company which enabled consumer to pay for goods and their service with swipe of their finger on the sensor. It allowed secure access to checking, credit card and other personal information, through the unique biometric features. Apart

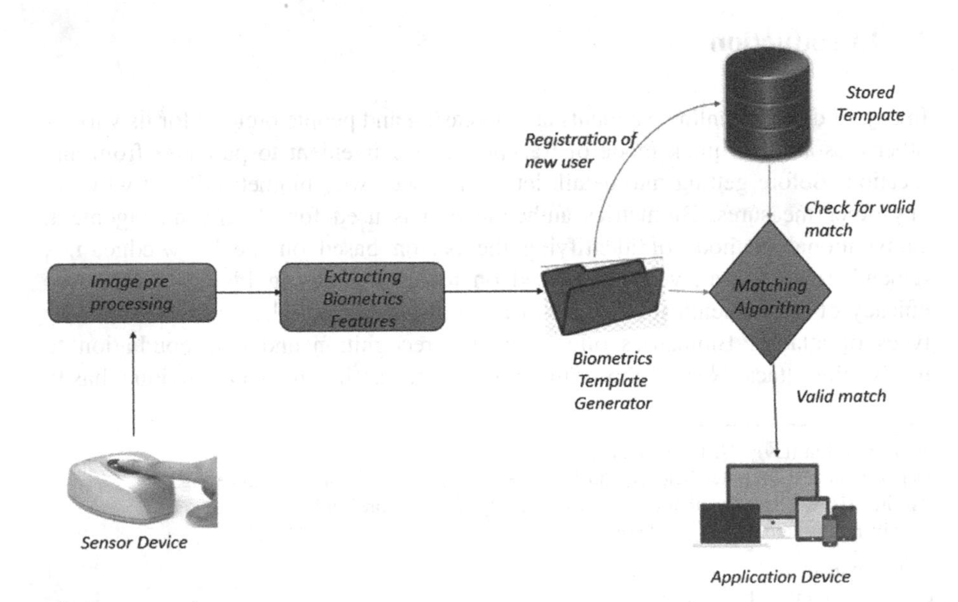

Fig. 1 Biometrics authentication process

from fingerprint, there are other biometrics traits used in the various form of security in securing the payment mode. The research from various studies says that using biometrics reduces two commonly faced problems. They are: 1. Identifying enrollment fraud for fake Identity in credit transaction over the network and 2. Increasing frauds like card trapping and hacking the account details from the card to bank [6].

There are two basic ideas of understanding to impose biometrics into this area of concern.

2.1 *Multifactor Authentication*

Incorporation of biometrics features is one of the processes involved for authentication procedure, which is used in multimodal security, i.e., along with other external key of token (such as PIN, password, and security key), biometrics is also used for authentication. This method is used in integrating the biometric features into the existing card issued by the bank. These cards can be used in ATM and various retail outsources. Biometric features are stored in a chip, which is integrated into the ATM card that contains the individual's key features, which is further saved as template format and this template is matched against the one that is available in the backend of the banking application. Once the matching is successful, then it allows for transaction.

2.2 *Single Factor Authentication (Biometrics Features as a Standalone)*

The second methodology is usage of the biometrics for authenticating an individual key component. This method is quite famous where pay-by-touch trending technology has been implemented in commercial sectors, which uses the biometric fingerprint as token for accessing all the bank account details about the customer and does payment transaction in retail shops.

These two approaches help to meet one set of goal to achieve secure payment transaction through online using either of the above method as base structure. Using this approach, it helps in financial service efficiency, reduction in online transaction frauds, reduction in payment frauds, and improve customer convenience.

Table 1 shows a clear picture of biometrics of each trait that is used in market for payment transaction [4]. From this understanding, we are going to see different way of implementation procedure for transaction process and their contribution to each methods.

Table 1 Biometrics market in transaction process

Biometric traits	Transaction authentication
Fingerprint	Dominant usage
Face	Limited usage
Iris	None
Hand geometry	None
Voice	Limited usage

3 Related Work

In this section, we are going to focus refined view of four different approaches of how biometric features are handled in different forms.

3.1 Biometrics or One-Time Password (OTP) in ATM

Figure 2 shows choosing either OTP (A) [7, 8] or biometrics (B) in ATM [9]. The ATM transaction is based on any one of the processes. If OTP is chosen, then it proceeds by generating OTP number, check for validation and proceed to transaction. If chosen as biometrics, it asks for biometric features, verifies those features of biometrics and allows them for transaction.

The steps involved in the process are:

- Inserting card and PIN.
- User has an option to choose either OTP or biometrics.

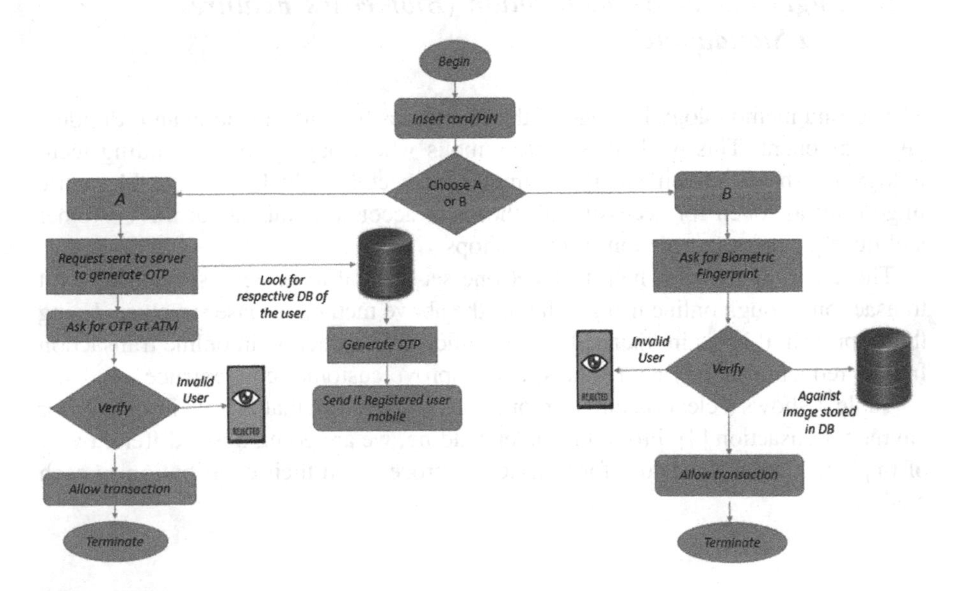

Fig. 2 Option based validation for ATM transaction

- If chosen OTP, then OTP number will be sent to the customers registered mobile.
- If biometrics is chosen, ATM with biometrics scanner will scan the customer identity that they possess.
- After verification of authenticity, the system allows for transaction.

3.2 Biometrics Face Recognition for ATM

Figure 3 exhibits face recognition implementation in ATM [10]. This technique adopts the facial recognition in three different phases such in front facing, left 90°, and right 90°. Once the verification of all three different phases is valid, then proceed for transaction.

The steps involved in the process are:

- Insert the ATM card and PIN details.
- Scan for face in three different angles (front, left angle 90°, right angle 90°).
- Once the verification for all three phases are done and matching is successful then proceed for transaction.

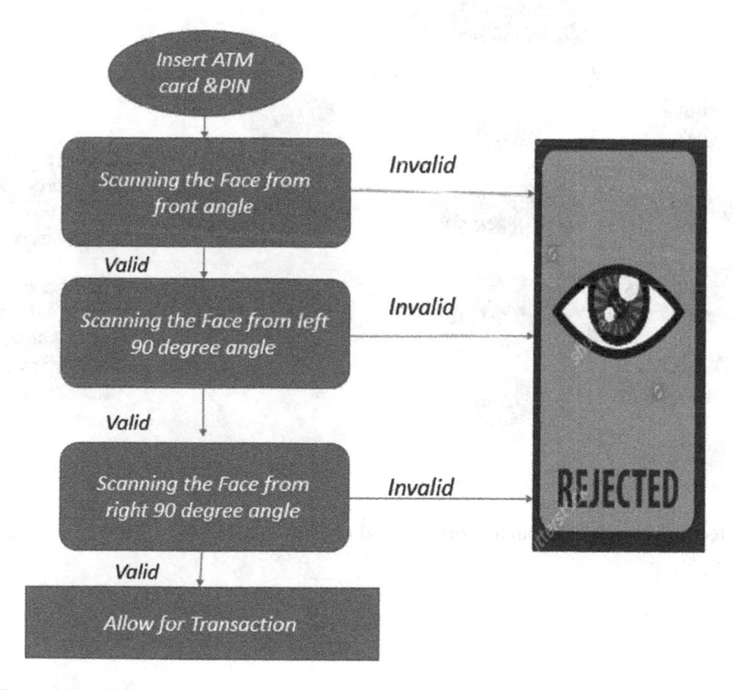

Fig. 3 Transaction using face recognition

3.3 Biometrics Fingerprint for ATM

Figure 4 shows the basic flow sequence followed for processing payment transaction in ATM using fingerprint and stored BioHASH values of the fingerprint [11–13]. This form of process transaction is based on biochip, which is present in the ATM card. The chip has biometrics information and verifies the individual based on information retrieved from biometrics database. This chip technology is used in the area of electronic banking.

The steps involved in the process are:

- The ATM card is validated.
- ATM processor retrieves the information stored in the card.
- Scan the fingerprint of the customer.
- Both the fingerprint and information from the card are send to the server.
- The server generates the BioHASH template to check the match.
- Upon successful authentication procedure, customers are allowed to perform their transaction.

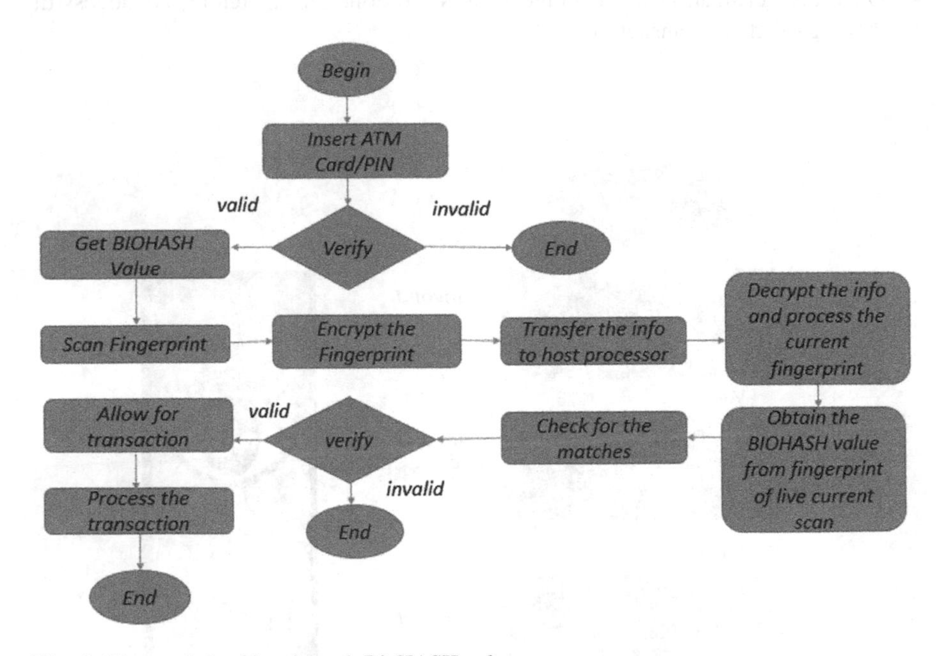

Fig. 4 Electronic banking through BioHASH value

3.4 Reliable Payment System in Nigeria

Figure 5 shows the sequence of the process that are followed in Central Bank of Nigeria [14]. Here bank has incorporated to two biometrics traits such as fingerprint and facial recognition for achieving multimodel security services [15–17].

The steps involved in the process are:

- Customer registration with biometrics feature enabled in the card.
- Verification at ATM during e-transaction.
- Provide PIN or fingerprint in stage one and followed by face recognition in stage two.
- Fusion of fingerprint and face using decision function [18–20].
- After authentication, allow for transaction

4 Proposed System for Securing Online Payment Transaction

The online payment is done using biometric traits (face and iris) as token of authentication for user convenience along with OTP generation. This work of multimodal biometrics provides high level of security.

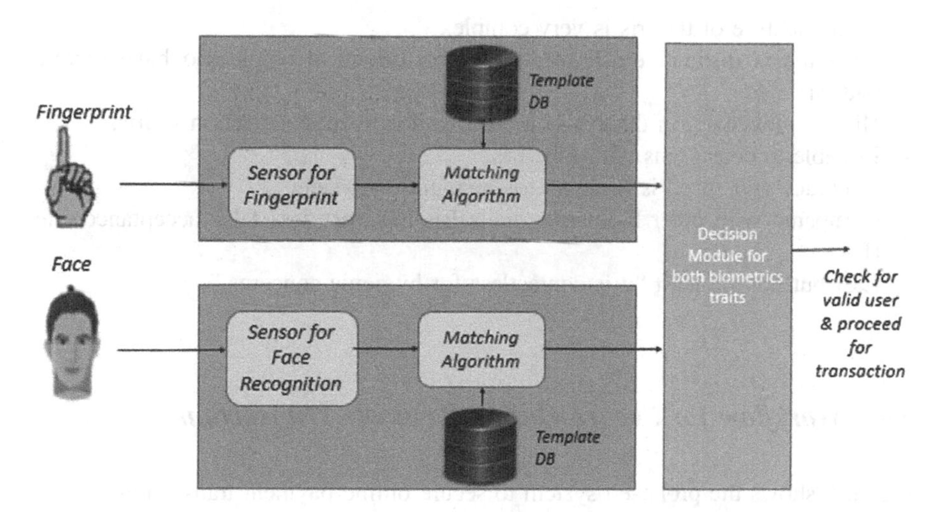

Fig. 5 Multimodal Biometrics using face and finger traits

4.1 Need for OTP

To have Multimodel security for the system, OTP (one-time password) verification takes place [7, 8]. The authenticity of client in accessing the site is validated. To avoid anomaly activity of the system scanning in the initial phase of online transaction, OTP is being used. Once the OTP reaches the client, they will use it as an authentication token and then begin the transaction process using biometric scanning of face and iris.

4.2 Usage of Face and Iris

As mentioned earlier, to have multimodal biometrics [21] which is more secure, face and iris traits are applied. Both the traits have very high level of security. These biometric traits together are unique over globe [4]. They possess user-convenient features to expose. They are contactless and hence cannot be forged.

4.3 Biometrics Iris

There are many features that make iris the more secure for biometric. Few points are highlighted below [4, 22].

- Visual texture of the iris is very complex.
- Each iris is distinctive all over the globe (identical twins also have unique pattern)
- Hippus movement in the eye is a measure for liveness detection in iris.
- Possible to detect lens with fake iris.
- Accuracy and speed is faster in iris recognition system.
- Compared with other biometrics traits Iris has very less false acceptance rate (FAR).
- Last but not the least "it is contactless for hygienic concerns."

4.4 Workflow for Secure Online Payment Transaction

Figure 6 shows the proposed system to secure online payment transaction.

- Stage 1: Initially, the user requests for transaction through online payment. Once request reaches the server, they will generate an OTP and is sent to users registered mobile phones [23, 24].

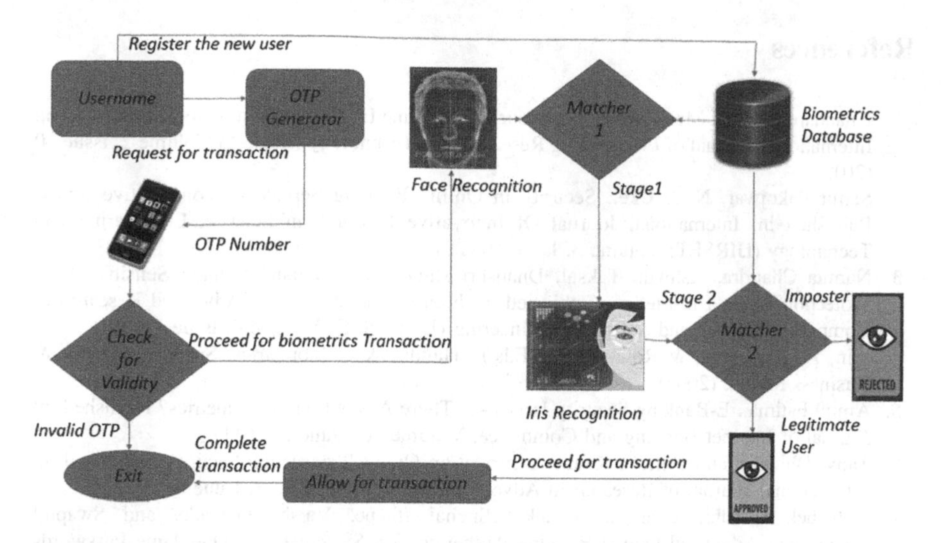

Fig. 6 Workflow to secure online payment transaction

- Stage 2: On successful validation of OTP, biometrics authentication where face and iris are used together, the system processes face recognition first and then Iris.
- Stage 3: Check for match with the database and those valid users are allowed for transaction.

5 Conclusion and Future Work

This work was done based on the motivation of securing the most recent and emerging trends in our day-to-day life with higher level of security to e-commerce site through online payment mode of transaction without any external key factors such as card, account number, PIN, passwords, etc. This work helps even the layman with basic understanding of usage of mobile phones can make quicker transaction in the most secure way of biometrics features involved that cannot be forged and transaction fraud is completely eliminated. Future work of this survey would be an implementation part of the proposed work of securing online payment transaction using biometrics authentication [25].

Acknowledgments I would like to thank my Guide, Ms. Lalithamani. N, Assistant Professor (SG), Department of Computer Science and Engineering, Amrita Vishwa Vidyapeetham, Coimbatore, for her guidance, valuable comments, and reviews that helped a lot in writing this paper. I also thank my friends for providing their valuable reviews and helpful suggestions.

References

1. Khandare, N., and Meshram, B. B.: Security of Online Electronic Transactions. Published in: International Journal of Engineering Research and Technology (ESRSA) Volume 2, Issue 10 (2013)
2. Samir Pakojwar, N. J. Uke.: Security in Online Banking Services-A comparative Study. Published in: International Journal Of Innovative Research in Science, Engineering and Technology (IJIRSET) Volume 3, Issue 10 (2014)
3. Namita Chandra, Ashwini Taksal, Dhanshri shinde and Archana Lomte.: Sensitive Data Protection Using Bio-Metircs. Published in: International Journal of Advanced Research In Computing Science and Software Engineering (IJARCSSE) Volume 4, Issue 1(2014)
4. Jain, A., Flynn, P., & Ross, A. A. (Eds.).: Handbook of biometrics. Springer Science & Business Media. (2007)
5. Amtul Fatima.: E-Banking Security Issues –Is There A Solution in Biometrics? Published in: Journal of Internet Banking and Commerce. Volume 16, Isuue 2 (2011)
6. Tanvi Dhingra, and Gurleen Kaur.: A Survey on Online Transaction Security. Published in: International Journal of Research in Advent Technology. Volume 3, Isuue 6 (2015)
7. Abhishek Gandhi, Bhagwat Salunke, Snehal Ithape, Varsha Gawade, and Swapnil Chaudhari.: Advanced Online Banking Authentication System Using One Time Passwords Embedded in Q-R Code. Published in: International Journal of Computer Science and Information Technologies (IJCSIT) Volume 5, Issue 2 (2014)
8. Sourabh Kumar Choundary, Rohini Temkar, and Nilesh Raj Bhatta.: QR Code Based Secure OTP Distribution Scheme for Authentication in Net-Banking. Published in: International Journal of Information Science and Intelligent System. Volume 2, Issue 4 (2013)
9. Mohammed Hamid Khan.: Securing ATM with OTP and Biometric, Published in: International Journal on Recent and Innovation Trends in Computing and Communication (IJRITCC), Volume 3 Issue 4 (2015)
10. Deepa Malviya.: Face Recognition Technique: Enhanced Safety Approach for ATM, Published in: International Journal of Scientific and Research Publications, Volume 4, Issue 12 (2014)
11. Kwakye, Michael Mireku, Hanan Yaro Boforo, and Eugene Louis Badzongoly.: Adoption of Biometric Fingerprint Identification as an Accessible, Secured form of ATM Transaction Authentication, Published in: International Journal of Advanced Computer Science & Applications 1, Volume 6, Issue 10 (2015)
12. Susmita Mandal.: A Review on Secured Money Transaction with Fingerprint Technique in ATM System. Published in: International Journal of Computer Science and Network. Volume 2, Isuue 4 (2013)
13. Das, Shimal, and Jhunu Debbarma.: Designing a biometric strategy (fingerprint) measure for enhancing ATM security in Indian e-banking system. Published in: International Journal of Information and Communication Technology Research (IJICT), Volume 1, Issue 5 (2011)
14. Aranuwa, F. O., and Ogunniye, G. B.: Enhanced Biometric Authentication System for Efficient and Reliable e-Payment System in Nigeria. Published in: International Journal of Applied Information systems (IJAIS), Volume 4, Issue 2 (2012)
15. Ghodke, A. P. S. S., Kolhe, H., Chaudhari, S., Deshpande, K., and Athavle, S.: ATM Transaction Security System Using Biometric Palm Print Recognition and Transaction Confirmation System. Published in: International Journal of Engineering and Computer Science (IJECS) Volume 3, Issue 4 (2014)
16. Omogbhemhe M.I, Momodu I.B.A., Sadiq F.I.: Conceptual Model for E-Banking System using Indigenous Biometric Technique. Published in: International Journal of Computer Applications. Volume 126, Issue 9 (2015)
17. Veronica S. Moertini, Asdi A. Athuri, Hery M. Kemit, and Nico Saputro.: The Development of Electronic Payment System for Universities in Indonesia: on Resolving Key Success

Factors. Published in: International Journal of Computer Science & Information Technology (IJCSIT). Volume 3, Issue 2 (2011)
18. J. Fierrez-Aguilar, J. Ortega-Garcia, D. Garcia-Romero and J. Gonzalez Rodriguez.: A Comparative evaluation of fusion strategies for multimodal biometric verfication. Published in: International Conference on Audio-and Video-Based Biometric Person Authentication. Springer Berlin Heidelberg. Volume 2688 (2003)
19. Qihui Wang, Bo Zhul, Yu Liu, Lijun Xie, and Yao Zheng.: Iris-Face Fusion and Security Analysis Based On Fisher Discriminant. Published in: International Journal on Smart Sensing and Intelligent System. Volume 8, Issue 1 (2015)
20. Dapinder Kaur and Gaganpreet Kaur.: Level of Fusion in Multimodal Biometrics: a Review. Published in: International Journal of Advanced Research in Computer Science and Software Engineering. Volume 3, Issue 2 (2013)
21. Sheetal Chaudhary and Rajender Nath.: A New Multimodal Biometric Recognition System Integrating Iris, Face and Voice. Published in: International Journal of Advanced Research in Computer Science and Software Engineering. Volume 5, Issue 4 (2015)
22. Judith Liu-Jimenez, Raul Sanchez-Reillo, and Bolen Fernandez-Saavedra.: Iris Biometrics for Embedded Systems. Volume 19, Issue 2 (2011)
23. Raina, V. K., Pandey, U. S., & Makkad, M.: A user friendly transaction model of mobile payment with reference to mobile banking in India. Published in: International Journal of Information Technology Volume 18, Issue 2 (2012)
24. SayaliShelar, PoojitaSahani, Anjali Sonsale, ShaileshBendale, and AshwiniJagdale.: An Authentication System for Online Banking Using Smart Phones. Published in: International Journal of Advanced Engineering and Global Technology (IJAEGT) Volume 3, Issue 5 (2015)
25. SavitaChoudhary.: Design of Biometric Based Transaction System using Open Source Software Development Environment. Published in: International Conference On Advances in Computer & Communication Engineering (ACCE) Volume 1, Issue 5 (2014)

NFC Logging Mechanism—Forensic Analysis of NFC Artefacts on Android Devices

Divya Lakshmanan and A.R. Nagoor Meeran

Abstract Near field communication (NFC) is a technology that facilitates communication between NFC-enabled devices by utilizing low power and radiating signals around a close proximity. The interaction is always binary, between the initiator that emits signals within a range of 4 cm and the target which enters the field of the initiator to begin the interaction. The communication may be one way (passive) or two way (active). The data shared in the session is neither encrypted not authenticated. These two factors aid in potential communication and data transfer. However, this becomes downside to this upcoming technology, when the data provided by the initiator may be subjected to data spoofing or data corruption. When the target processes such data, it could leads to unintended behaviour thereby compromising the integrity of the device. When a forensic investigator is handed a compromised device and asked to recreate the alleged crime, he relies on the presence of nonvolatile data on the device. In the Android operating system, there is no mechanism to provide the nonvolatile artefacts ensuing an NFC interaction. Therefore, any digital crime on Android devices abetted by NFC remains unsolved and the case gets deferred. The main aim of this research is to develop a logging mechanism for Android devices that will log all the interactions taking place through the NFC hardware, and the presence of these nonvolatile logs along with other volatile artefacts would benefit the forensic investigator to comprehend the exact sequence of activities that jeopardized the conventional operation of the android device.

Keywords Near field communication · Android · Forensics · Logs

D. Lakshmanan (✉) · A.R. Nagoor Meeran
SRM University, Kattankulathur, Chennai 603203, India
e-mail: divya.lakshmanan27@gmail.com

A.R. Nagoor Meeran
e-mail: nagooris@gmail.com

S.S. Dash et al. (eds.), *Artificial Intelligence and Evolutionary Computations in Engineering Systems*, Advances in Intelligent Systems and Computing 517,
DOI 10.1007/978-981-10-3174-8_9

1 Introduction

1.1 Near Field Communication

Near field communication (NFC) was introduced in Android devices from version 2.3, (Gingerbread) [1]. A typical NFC interaction begins by touching the devices together or bringing them into close proximity at a distance of 4 cm or less. An Android mobile device becomes an NFC enabled device when it has the NFC chip embedded in it and has the appropriate drivers to handle the data read from another compatible device.

At any instant, only two devices can participate in an NFC interaction. The initiator, which is a self-powered device, keeps supplying radio frequency signals at a close range. When the target enters this field provided by the initiator, it gets powered up and the two devices can communicate.

1.2 NFC Communication Modes

In the first mode, known as passive mode [2], only unidirectional data transfer takes place. The NFC enabled android device receives data from gadgets like NFC tags, NFC stickers, smart posters, etc. to behave like an electronic ID, pay for goods at a store, be an event ticket, download music, download shopping coupons, etc. In the second mode, known as active mode, there is bidirectional or unidirectional data transfer between the two devices. It is used to transfer data between both the devices, like contact information, web browser pages, YouTube links, etc.

1.3 NFC Operation Modes

The first is the 'Reader–Writer Mode' [2]. Here as a 'Reader', the device can read data from NFC tags, NFC stickers, smart posters, etc. and behave as dictated by them. As a 'Writer', the device can write to NFC tags, NFC stickers, smart posters, etc. and define their action. The second mode is 'Card Emulation Mode', where the NFC device emulates the action of a smart card. The NFC device can be used in a payment gateway to behave like a banking smartcard or as an electronic key to a hotel room. The third mode, known as 'Peer-to-Peer Mode', allows bidirectional data transfer between the interacting NFC devices. It can be put to use by activating the 'Android Beam' feature found in Android devices.

From a technology perspective, to standardize the operation and behaviour of Android devices with NFC functionality, the standards set forth by NFC Forum [3] are adopted. NFC Forum is an association that has agreed on the use of four distinct tag types across various devices. They have emerged with technical specification

Table 1 NFC Forum tag types

Type 1	Based on ISO-1443A spec Memory size: from 96 B to 2 kB Communication speed: 106 kB/s Can be read only or read/write capable Can be rewritten multiple times Not preformatted No anti-collision protection
Type 2	Based on ISO-1443A spec Memory size: from 48 B to 2 kB Communication speed: 106 kB/s Can be read only or read/write capable Can be rewritten multiple times Not preformatted Anti-collision support
Type 3	Based on ISO-18092 spec Memory size: 1 MB Communication speed: 212 kB/s or 424 kB/s Comes preconfigured from manufacturer Can be written only once Anti-collision support
Type 4	Based on ISO-14443A/B spec Memory size: up to 32 kB Communication speed: 106–424 kB/s Comes preconfigured from manufacturer Can be written only once Anti-collision support

documents that detail the memory layout, data organization and data behaviour of the four different tag types. A primer on the four tag types can be found below [2] (Table 1).

1.4 NFC Data Exchange Format (NDEF)

NDEF is the format which dictates the structure of the data exchanged in an NFC interaction. Whenever two NFC devices interact, a single NDEF message is exchanged between them. This message may have one or more records. Each record is defined by a header and payload. The header gives information about how the payload should be treated and the payload gives information about the operation to be performed by the receiving device, through a RTD (Record Type Definition) structure.

Some special cases of the record type are 'text record' (containing only textual data), 'URI record' (containing a URL or URI to be processed by receiving device), 'Smart Poster' record (a combination of text record, URI record, title record, etc.), record containing business card details, record that enables the bootstrap of technologies like WiFi, Bluetooth, etc. record that transfers small files between two applications on the interacting devices, record to exchange web page links, etc. The processing of all the aforementioned data is defined by the NFC Forum specification documents [4]. For an average Android device user, the incorporation of NFC has aided the automation of a myriad of everyday tasks.

2 Research Overview

In addition to robotizing a plethora of activities, NFC is also susceptible to inviting compromises to the Android device. An android device may harmlessly interact with another NFC enabled device with the intent of performing a simple operation. But if the data source is spoofed or corrupted by someone with unethical intentions, when such data is read and processed by the android device, the result deviates away from conventional operation [5]. For example, the URL present in a smart poster may be obfuscated to cause web browser attacks. This scenario, taking place without the knowledge of the device owner, presents a major privacy risk.

When an Android device is questionable to foul play, one of the factors to be considered is 'Was the mobile device tapped against any other electronic device or smart device to automate a task?' If the answer is affirmative to the above question, then the Near Field Communication capability of the device can be considered to have caused the compromise. When such a compromised device is handed over to a forensic investigator, with the intent of him tracking the activity of the supposed crime and finding ways to restore the regular operation of the device; the need for concrete evidence arises [6]. The existence of nonvolatile artefacts is the prime source of evidential treasure for any forensic investigator. He should be able to retrieve the nonvolatile artefacts effortlessly and utilize his competency to process the same.

Considering NFC, the developers of android did not include any mechanism that would store all details of any NFC interaction on the average android mobile, even after the device is switched off. There are no nonvolatile artefacts left behind to showcase the NFC activity that has taken place. In the event of a forensic investigation, this becomes a major problem for the forensic analyst and would put an abrupt end to any investigative process.

This research aims to provide a way to store nonvolatile artefacts ensuing an NFC interaction. This idea is implemented by a 'Logging Mechanism', which stores every interaction with the NFC hardware, i.e. the complete flow of data between the NFC controller and the Android operating system. The 'NFC Logging Mechanism' is a system application on Android that stores all required log entries in a database powered by SQLite Database technology [7].

Table 2 Hardware and software used

Hardware
1. Samsung S3-GTI9300 running Android Jelly Bean
2. NFC Forum compliant Type 1 tags
Software
1. Android Studio IDE
2. NetBeans IDE
3. Ubuntu 15.04 with Android adb installed

3 Implementation

3.1 Developing and Deploying the Android Application

In this research, the system application that performs NFC logging was developed using Android Studio IDE and installed on a test device. The developed application is called 'NFC Logs'. Following is a list of the tools used in this research (Table 2).

For the purpose of demonstration, 'NFC Logs'—the Android application that performs the logging mechanism has been tested with NFC Forum compliant Type 1 tags. Data was written into a Type 1 tag using 'TagWriter' Android application developed by NXP semiconductors, that is available on the Google Play Store [8]. To test the working of 'NFC Logs', two sets of data were tested. In the first set, textual data containing the text 'Hello,World!' was written into the tag and tapped against the device. In the second, the URL 'http://www.google.com' was written into the tag. When the second tag was tapped against the device connected to the Internet, the web browser opened and the user was redirected to the Google homepage.

To the user, 'NFC Logs' will display the unique tag id and timestamp of when the tag was read. Internally on the device, other details about the header and payload of the tag are stored as hexadecimal data in a SQLite database file located in the following path in the Android file system.

/data/data/<package_name>/databases/nfc_type1_log.db

The <package_name> parameter corresponds to 'NFC Logs' and its development platform.

3.2 Analysing the Logs Generated by 'NFC Logs'

In the next step, the logs stored in 'nfc_type1_log.db' generated by 'NFC Logs' was analysed. The file was extracted from the Android file system via the Android Debug Bridge (ADB) and running the following commands in the same sequence (Table 3).

Table 3 Commands to extract database

Commands
adb shell
cd /data/data/<application_name>/databases/
cat nfc_type1_log.db>/<internal_storage_directory>/nfc_type1_log.db
adb pull /storage/sdcard0/nfc_type1_log.db/<FORENSIC WORKSTATION>

Now the database file can be viewed using a tool called 'SQLite Browser for Ubuntu' [9]. For a chosen log entry, most of the relevant details are stored as hex data. It has to be made into human readable form to aid in analysis for a forensic investigator.

For this purpose a desktop tool 'NDEF Log Analyser' was built using NetBeans IDE, to parse the log entries. This custom made tool processes the logs generated by 'NFC Logs'. This tool allows the user to choose a database file from the operating computer. In our case, 'nfc_type_log.db' retrieved from the Android device is given as input. The tool then displays all the log entries present in the database in a table. The forensic investigator can choose one log entry and 'parse' it to convert the hex data to human readable form. Detailed information about the header and payload in the tag can be observed. The header has details like tag type, tag size and manufacturer details and the payload contains the NDEF message and its size, its mime type, etc.

3.3 *Forensic Analysis of NFC Artefacts Using 'NFC Logs' and 'NDEF Log Analyser'*

At this point we have two newly developed tools:

(1) NFC Logs—an Android application that logs all NFC Forum Type 1 tag related interaction
(2) NDEF Log Analyser—a desktop tool on Ubuntu to parse the logs generated by NFC Logs.

To show how both the tools can be used efficiently together to aid a forensic investigator, a compromise was made on the Android device. Consider this scenario. A departmental store pastes a smart poster near its entrance. A smart poster is one that has an NFC tag attached to it and some text instructing readers about the use of the tag. In this case the text informs the user that when an NFC enabled device is tapped against the tag and when he is connected to the Internet, he would be directed to the organization's website to automatically download an Android application that would apprise him about the new offers in the store. This avoids the hassle of maintaining a Google Play Store account to download Android applications. When an attacker enters the picture, he may modify the content on the NFC tag, thereby leading users to a phished web page which may contain a malicious version of the Android application to be downloaded. Unbeknownst to the user, this version may have the capability to send user data from the device to the attacker which results in privacy invasion or it may alter the regular behaviour of the device. All this occurs due to the spoofed URL content on the NFC tag. This scene was replicated for our research.

The following steps were performed for the same:

1. Install 'NFC Logs' application in the test device.
2. Set up an NFC Forum compliant Type 1 tag to be used with a smart poster. The tag would be attached to the poster and the poster informs the user about how to download the Android application when connected to the Internet. Since the tag is to be used with a smart poster, it has a smart poster record which is usually a combination of a text record and a URI record. The URI record is supposed to contain a URL to lead the user to a website to download the Android application, but it is spoofed to lead the user to a malicious webpage inorder to download the malicious Android application.
3. Setup an Apache Web Server on a Linux System. One directory in the server contains the clean files and another directory contains the malicious files, the latter supposedly set up by the attacker. The URI record in the smart poster contains the URL for the malicious files, by manipulating the URL for the actual files. This manipulation is evident when analysis is done on the URI record later.

Actual URL: http://<IP Address of server>/downloads/apps/Coupon_download.php.

Manipulated URL: http://<IP Address of server>/downloads/apps/Coupon_download.php/../../../Downloads/Apps/Coupon_download.php.

4. Tap the test device against this tag. Since any interaction with a Type 1 tag is logged by 'NFC Logs', when the device is tapped against the malicious tag, a log entry is made in the database of 'NFC Logs' application. Now the test device is led to the phished webpage and also downloads the malicious Android application, which sends data from the device back to the attacker's directory in the server.
5. The downloaded Android application is also made to alter the regular operation of the mobile device. It sets up a service that begins its execution whenever the device boots up. This service enables and disables the Bluetooth connectivity of the device automatically, every 30 s. Ultimately, whenever the user reboots his device, the Bluetooth feature will enable and disable itself, preventing any complete Bluetooth transfer to take place.

Now the test device was subjected to forensic analysis. The ACPO principles were considered while handling evidence [10]. The volatile evidence from main, events, radio and system log buffers were recovered first using ADB [11]. The evidence recovered may or may not prove recent NFC activity, depending on the timeframe between the activity and the seizure of the phone. This is because the evidence collected from the log buffers is volatile.

Since it is known that 'NFC Logs' application provides reliable, nonvolatile evidence, its database was extracted onto a computer running Ubuntu 15.04 operating system using the technique specified in Sect. 3.2. The database was processed using NDEF analyser to look out for suspicious entries. The malicious smart poster record was identified and data from that tag was scrutinized further. It was seen that the URI record was subjected to a directory traversal attack. The actual URL was overridden to refer to another path on the server, presumably the attacker's.

The device was connected to the Forensic Work Station, here the machine running Ubuntu 15.04 with ADB installed. Now nonvolatile evidence was collected from various parts of the Android file system. Since it has been found that an NFC tag with a URL has been read at a specific time, the databases of all the browsers installed in the device are collected and analyzed. This is done to find out if the user had visited the webpage specified by the URL. It was found that right after reading the tag, the user had visited the URL specified by the tag. Keeping the applications of smart poster in mind, that is, read a tag to download songs, coupons, etc. The database in the Android file system that holds details of all downloaded files was extracted. It was found that after visiting the website, an Android application package was installed.

Now that downloaded Android application was subjected to analysis. Initially its basic working was studied. The original intention of the application was to only display available offers in the store to customers. On analysis of its services and permissions, it was identified that the application was designed to start running in the background, immediately after the boot process of the device was completed. This is a strange occurrence because an application which only displays offers to the user does not have the requirement to be started after boot. Also when the permissions used by the application were analysed using a third party tool, it was found that the app required access to the Bluetooth hardware and the message inbox of the phone, both of which are practically not necessary for an app only displaying offers.

So it was concluded that the application downloaded by tapping an NFC tag, stole user data and modified the regular operation of the device. The nonvolatile data collected from the NFC Logs database, browser history database, downloads database and analysis of services and permissions used, helped to establish the timeline of operations that may have taken place which led to the compromise of the device. The falter in the operation of the device occurred due to the spoofing of data in the NFC tag. The presence of nonvolatile evidence for NFC activity, aided the forensic investigation to yield fruitful results. It is vital for all Android devices to be able to maintain nonvolatile logs of NFC activity.

When 'NFC Logs' application is incorporated into all Android devices that support NFC, a reliable source of evidence is assured for a forensic investigator. The desktop tool 'NDEF Log Analyser' can be used as a secondary tool to handle the NFC Logs.

4 Future Work

The logging mechanism can be made to log interactions with NFC Forum compliant tag types—2, 3, 4, NXP semiconductors compliant NFC tags. The Log Analyser can also be upgraded to parse all relevant log entries. Log entries related to active NFC interaction can be stored in a separate location on the Android file system.

References

1. https://en.wikipedia.org/wiki/Near_field_communication.
2. Igoe. T, Coleman. D., Jepson. B., (2014). In NFC and RFID. Beginning NFC-Near Field Communication with Arduino, Android and PhoneGap. 11–18. O'Reilly Media Inc.
3. https://en.wikipedia.org/wiki/Near_field_communication#NFC_Forum.
4. http://nfc-forum.org/our-work/specifications-and-application-documents/.
5. Coskun. V., Ozdenizci. B., Ok. K., (2012). A Survey on Near Field Communication(NFC) Technology. Wireless Personal Communication. Volume 71, Issue 3. 2259–2294.
6. Mandia. K., & Prosise. C., (2003). Evidence Handling. In Pepe. M., Bejtlich R., & Rose. C., (Eds.) Incident Response and Computer Forensics. 198–213. Osborne: The McGraw-Hill Companies.
7. https://www.sqlite.org/different.html.
8. https://play.google.com/store/apps/details?id=com.nxp.nfc.tagwriter.
9. https://apps.ubuntu.com/cat/applications/sqlitebrowser/.
10. https://www.cps.gov.uk/legal/assets/uploads/files/ACPO_guidelines_computer_evidence%5B1%5D.pdf.
11. Andriotis. P., Oikonomou G., Tryfonas T., "Forensic Analysis of Wireless Networking Evidence of Android Smartphones" in WIFS 2012.

Control of Quadrotors Using Neural Networks for Precise Landing Maneuvers

U.S. Ananthakrishnan, Nagarajan Akshay, Gayathri Manikutty and Rao R. Bhavani

Abstract Aerial and ground robots have been widely used in tandem to overcome the limitations of the individual systems, such as short run time and limited field of view. Several strategies have been proposed for this collaboration and all of them involve periodic autonomous precision landing of the aerial vehicle on the ground robot for recharging. Intelligent control systems like neural networks lend themselves naturally to precision landing applications since they offer immunity to system dynamics and adaptability to various environments. Our work describes an offline neural network backpropagation controller to provide visual servoing for the landing operation. The quadrotor control system is designed to perform precise landing on a marker platform within the specified time and distance constraints. The proposed method has been simulated and validated in a Gazebo and robot operating system simulation environment.

Keywords UAV-UGV collaboration · Quadrotor landing · Vision based navigation · Neural networks · Backpropagation algorithm · ROS architecture

U.S. Ananthakrishnan (✉) · N. Akshay · G. Manikutty · R.R. Bhavani
AMMACHI Labs, Amrita School of Engineering, Amritapuri Amrita Vishwa Vidyapeetham
Amrita University, Kollam, Kerala, India
e-mail: krishnan793@gmail.com

N. Akshay
e-mail: akshayn@am.amrita.edu

G. Manikutty
e-mail: gayathrim@am.amrita.edu

R.R. Bhavani
e-mail: bhavani@am.amrita.edu

S.S. Dash et al. (eds.), *Artificial Intelligence and Evolutionary Computations in Engineering Systems*, Advances in Intelligent Systems and Computing 517,
DOI 10.1007/978-981-10-3174-8_10

1 Introduction

Unmanned aerial vehicle (UAV) and Unmanned Ground Vehicle (UGV) collaboration has frequently been used in humanitarian operations such as disaster response, search and rescue, robotic demining, and inspection and structural assessment of critical infrastructure among others [1–3]. For indoor and cluttered environments, rotor crafts with vertical take-off and landing (VTOL) capability have been preferred. Quadrotors are one such VTOL designs with high maneuverability, mechanical simplicity, and agility. Due to the limited flight time of less than 30 min and typical load carrying capacity of 2.5 kg, quadrotors cannot be used alone for most of these operations. UGVs on the other hand have battery life in hours and a high load carrying capacity. But the limited field of view of an UGV needs to be addressed in order to use it alone in operations like search and rescue, disaster response, etc. One of the solutions for these constraints is to use the UAV in combination with UGV wherein the limited field of view of the UGV is resolved by combining the situational awareness from the UAV.

Much of the prior works on intelligent controllers have focused on fixed wing aircraft [4–6]. A comprehensive idea of different landing techniques are reviewed in [7]. There are several methods to address the problem of limited flight time of a UAV. One can use a tethered power supply [8] for the UAV or one can swap out the batteries periodically [9, 10]. The former approach limits the UAV mobility especially in areas where there are tall trees and cluttered buildings. This is the general landscape of the region where this research is being conducted. The latter would necessitate periodic landing. This motivates the need for frequent high precision landing and an autonomous landing algorithm can offset the need for a trained pilot to perform this task. Several nonlinear solutions [11, 12] have been proposed for the control of quadrotors and for the landing problem [13]. A different approach to the same problem is applied using intelligent control systems like fuzzy and neural networks. The advantage of the present approach is the degree to which the flight controllers can be adapted to various environments. Lee et al. have demonstrated through their preliminary study that image based visual servoing (IBVS) using neural networks has been successful in real time tracking and autonomous landing [14]. Hence, we have chosen a neural network (NN)-based IBVS approach in order to develop a robust autonomous solution for precise quadrotor landing. The proposed learning for the network here uses a backpropagation technique, which is an offline learning algorithm. This type of system produces a network weight that can be used either as a starting weight for an online learning NN that adapts to the environment or can be used as such for precise landing maneuvers.

Robot operating system (ROS) is a framework for robotic applications. It is a peer-to-peer, tools-based, multilingual, thin, and open source system [15]. The proposed neural network controller system has been implemented in ROS and is available publicly as a ROS package ar2landing neural on github. Further, we will discuss an analysis of the vision controlled landing maneuvers tested on a Parrot AR2 Drone quadrotor in the Gazebo simulation environment.

2 System Description

The commercially available quadrotor, AR2 Drone 2.0 by Parrot USA is chosen as the aerial quadrotor for our implementation and experiment. The AR Drone quadrotor is capable of take-off, landing, hover and can perform pitch, roll, and yaw motions. Since the quadrotor is WiFi enabled, it can be controlled remotely. Further, the system consists of one vertical (720p) and one horizontal VGA camera. To accomplish the task of detecting the landing pad on the UGV and for accurate autonomous landing of the quadrotor, a GCS (Ground Control Station) is used which communicates with the AR Drone through WiFi. The video feed from the AR Drone is processed by the GCS and proper control commands are sent back to the UAV for positioning and landing. The system for the landing operation of the Parrot AR Drone 2.0 tracks the marker placed on the UGV and performs visual servoing to land on the UGV. The various blocks in the system for this operation include a NN for servoing and training, a marker tracking system implementation (artrack alvar) and simulation using Gazebo.

The software architecture shown in Fig. 1 for the implementation consists of a ROS stack for communication with the drone via WiFi. The ROS middle ware also interfaces with the NN for servoing, and training, the tracking system implementation (artrack alvar), Gazebo and the AR SDK that provides AT commands for varying velocity and yaw, pitch, and roll on the drone. All these communications are through TCP/IP, and hence it provides greater flexibility. In order to change the implementation from simulation to a real drone, the only thing that has to be done is connect the drone to the same network.

ROS provides a framework for information management and is now becoming a de facto tool for robotics applications. Sensor and control data topics are subscribed to and from the ardrone autonomy, ar track alvar, ar2landing neural, and ar2landing

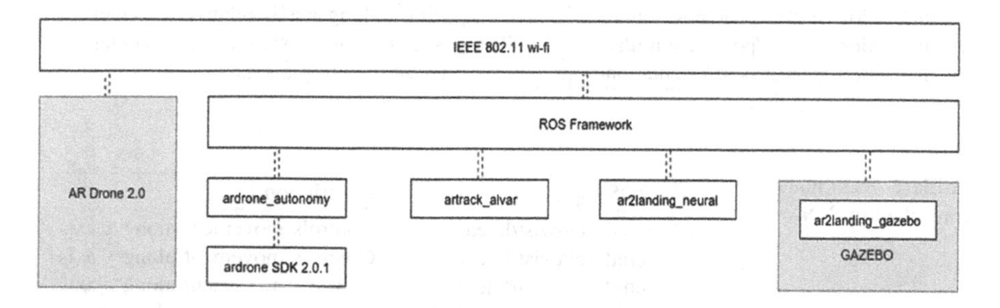

Fig. 1 System architecture illustrating ROS framework communicating with the drone, simulation environment and the controller packages

gazebo packages. ROS maintains and keeps track of all these data and also provides better debugging tools. Gazebo is a robotics simulation tool that work closely with ROS and offers the ability to accurately and efficiently simulate robots in complex indoor and outdoor environments. For this project, we used a ROS Gazebo package called tum simulator [16]. The urdf model of the AR Drone 2.0 along with necessary nodes for simulation is available in that package. A custom world is created for simulating the drone in Gazebo. In order to simulate a landing procedure, a landing platform with a tag is created and spawned in the world along with the AR Drone 2.0 urdf model.

The augmented reality (AR) tags available in artrack alvar are shown in Fig. 2. The pose estimation from artrack alvar is used as the inputs to the NN. The list of topics subscribed from artrack alvar is listed in Table 1.

The first step of any offline learning algorithms is to collect training data. The training data contains all the information in Tables 1 and 2.

The information flow of the system is shown in Fig. 3. There are four interesting outputs from ar track alvar. The three parameters correspond to position and an orientation element. This data is updated at the same rate as that of camera, which is 60 fps in this case. These four outputs are subscribed by the neural network to

Fig. 2 AR marker tags used for image based visual servoing

Table 1 ROS topics subscribed from artrack alvar

ROS topic	Description
/visualization marker/pose/position/x	Tag position along x axis relative to center
/visualization marker/pose/position/y	Tag position along y axis relative to center
/visualization marker/pose/position/z	Tag position along z axis relative to center
/visualization marker/pose/orientation/z	Tag orientation along z axis

Table 2 ROS nodes controlled by NN

ROS topic	Description
/cmd vel/twist/linear/x	Controls movement along x axis
/cmd vel/twist/linear/y	Controls movement along y axis
/cmd vel/twist/linear/z	Controls movement along z axis
/cmd vel/twist/angular/z	Controls orientation along z axis

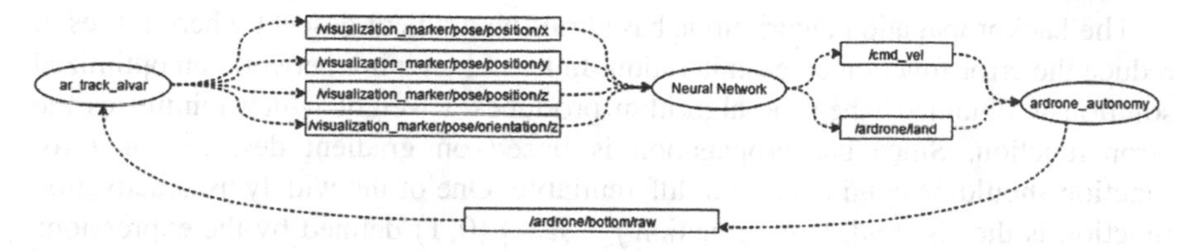

Fig. 3 Process flow

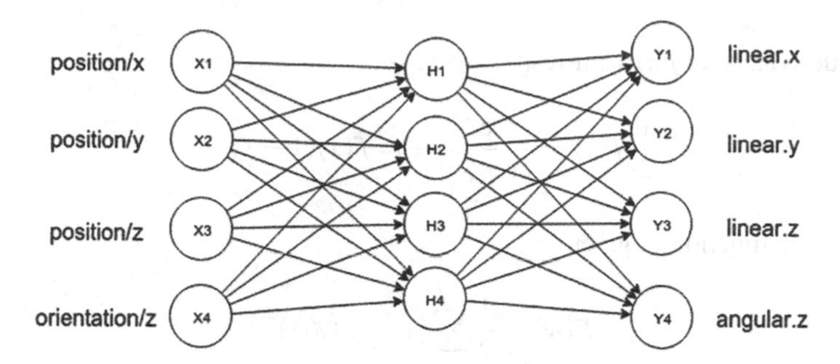

Fig. 4 Neural network layout

predict the correct control commands. These commands correspond to the linear and angular velocities of quadrotor which in turn maps to pitch, roll, and yaw motions. These topics or outputs are updated and properly communicated to the drone using the ardrone autonomy package. The ardrone autonomy package also provides the raw camera feed to the ar track alvar package.

The data for offline training is collected manually by landing the quadrotor several times. There were 1390 training data samples that correspond to around 10 manual landings. A separate routine was created for capturing the data set. These data was used to train the network such that the squared error is minimum. After several trial and checks, a NN with a configuration of (4, 4, 4), as shown in Fig. 4, is found to be sufficient. Multiple configurations were analyzed and cross checked before concluding the network layout.

The four inputs are (x, y, z, θ) where x, y, z are position of tag with respect to the AR Drone downward facing camera and θ corresponds to the orientation of drone with respect to z axis. The output of the neural network corresponds to linear x, y and z velocities which are mapped to pitch, roll, yaw, and the throttle (y_1, y_2, y_3, y_4).

$$[y_1, y_2, y_4]^{\mathrm{T}} = [x, y, z, \theta][W_{12}][W_{21}]$$

where W_{12} and W_{21} correspond to the weight matrix of layers 1 and 2, respectively. These weights are calculated during the training phase.

The backpropagation algorithm is based on the gradient descent where it tries to reduce the error function on each iteration. In neural learning network, an optimized solution is obtained when the algorithm produces a weight which minimizes the error function. Since backpropagation is based on gradient descent, the error function should be continuous and differentiable. One of the widely used activation function is the sigmoid, a real function $f_c : \Re \rightarrow (0, 1)$ defined by the expression:

$$f(x) = \frac{1}{1 + \mathrm{e}^{-x}}$$

The derivative of $f(x)$ with respect to x, is

$$\frac{\mathrm{d}}{\mathrm{d}x} f(x) = \frac{\mathrm{e}^{-x}}{(1 + \mathrm{e}^{-x})^2} = f(x)(1 - f(x)$$

The error function is given by,

$$E(w) = \frac{1}{2} \sum (y_i - f(X_j))^2$$

The backpropagation algorithm can be consolidated as,

Consider a network with M layers m = 1, 2, …, M and use V_i^m to represent the output of the ith unit in the mth layer. $V_i^0 = x_i$ is the ith input.

1. Initialize the weights with random numbers.
2. Choose $V_i^0 = x_i^p$,
   Where x_i^p is of the form (x, y, z, θ).
3. Propagate the signal forwards through the network.

$$V_i^m = f(u_i^m) = f\left(\sum w_{ij}^m V_j^{m-1}\right)$$

   for each i and m until the final outputs V_i^m is obtained.
4. Compute the deltas for the output layer

$$\delta_i^M = f'(u_i^M)(y_i^p - V_i^M)$$

5. Compute the deltas for previous layers by propagating errors backwards

$$\delta_i^{m-1} = f'(u_i^{m-1}) \sum_j w_{ij}^m \delta_j^m$$

   for $m = M, M - 1, \ldots, 2$.
6. Update the weights by

$$w_{ij}^{new} = w_{ij}^{old} + \Delta w_{ij}$$

where

$$\Delta w_{ij} = \eta \delta_i^m V_j^{m-1}$$

7. Go to Step 2 and repeat for the next input. The algorithm stops when no substantial weight changes occur on subsequent iterations.

The training phase has to be done once in order to find the tuned weights. These weights are then used for subsequent trials and checks. We also had to make sure these weights are generalized (does not have any over fitting problem). If the network weights are not generalized, the network will not be able to solve all the possible initial conditions (position and orientation). Once the AR tag is in the field of view, the neural network will take control of the quadrotor. The four inputs $X = (x, y, z, \theta)$ are acquired and the outputs $(y1, y2, y3, y4)$ that are mapped to pitch, roll, yaw, and altitude control are calculated using

$$\begin{bmatrix} y_1 \\ y_2 \\ y_3 \\ y_4 \end{bmatrix} = [X][W_{12}][W_{21}]$$

where both W_{12} and W_{21} are 4×4 matrix of the form

$$W_{xx} \begin{bmatrix} w_{11} & w_{12} & w_{13} & w_{14} \\ w_{21} & w_{22} & w_{23} & w_{24} \\ w_{31} & w_{32} & w_{33} & w_{34} \\ w_{41} & w_{42} & w_{43} & w_{44} \end{bmatrix}$$

3 Results

The neural network was implemented in ROS and found to be promising for landing a quadrotor autonomously after training with data samples (roughly 1400) that corresponds to 10 manual landing. Additionally, a neural network of four inputs and four outputs with one hidden layer with four nodes was found to have its weights converge for the landing procedure of the quadrotor (See Fig. 5). The network converges after 25,000 iterations. The paper demonstrates the success in implementation of a neural network controller along with image markers-based visual servoing. In order to test the generalized behavior of the neural network, sever trials were performed with random initial position and orientation. All of the trials were successful and the network weights were found to be generalized rather than specific to the trained data set.

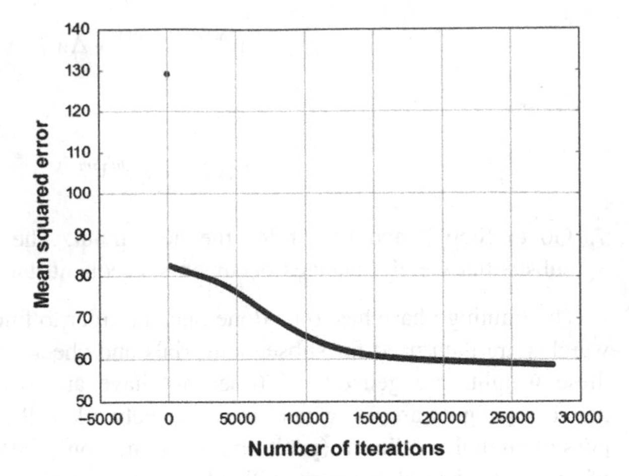

Fig. 5 Mean squared error vs. number of iterations

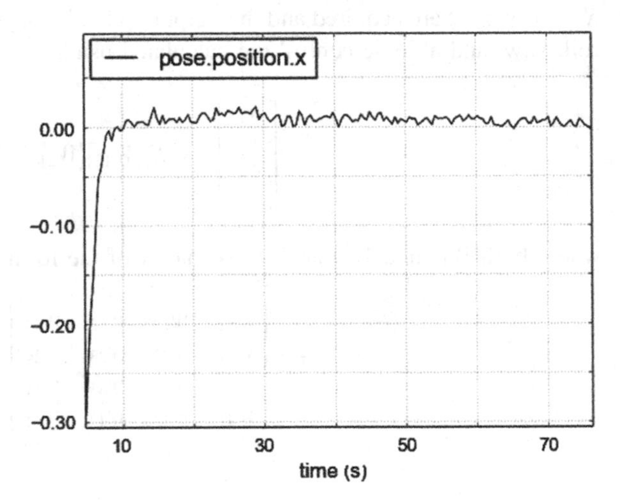

Fig. 6 Position/*x* vs. time

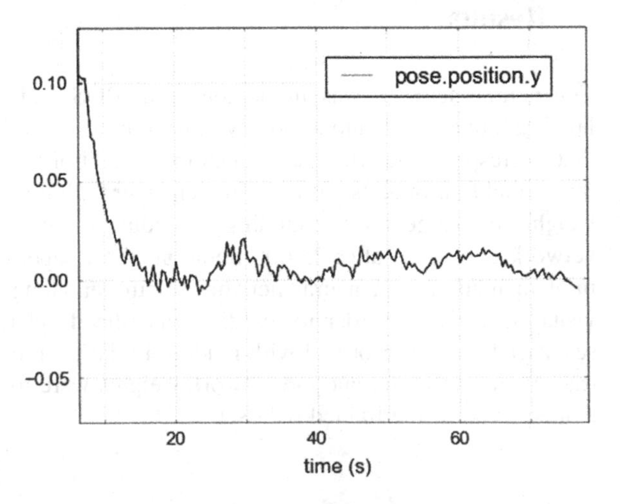

Fig. 7 Position/*y* vs. time

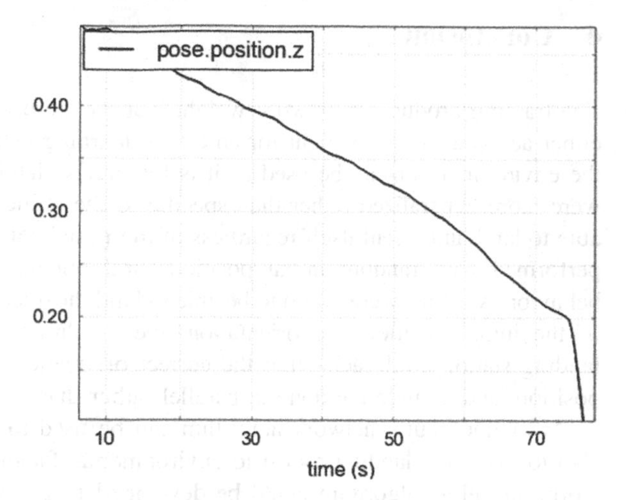

Fig. 8 Position/z vs. time

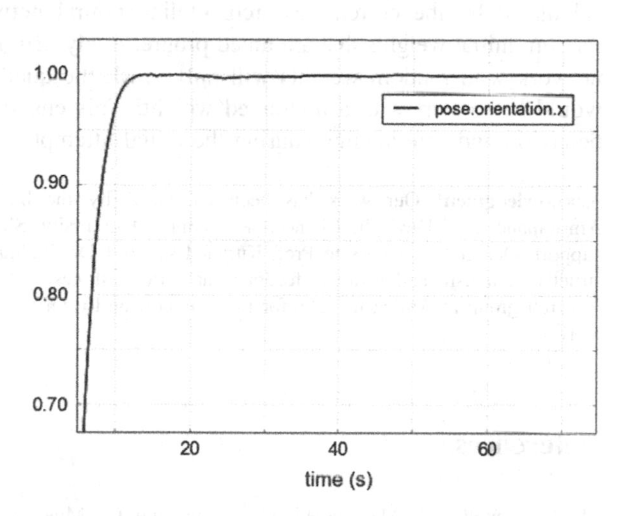

Fig. 9 Orientation/z vs. time

In order to validate the proposed approach, several trials were performed in simulation and the trajectory of each trial is plotted. Also the pose of the quadrotor for one such trial is plotted against time to show the response of system. From Figs. 6, 7, 8, 9, it can be noted that the network was able to perform the landing in 75 s over a height of 1 m with an offset of 0.3 m from center. As an example of tracking, in Fig. 6, the desired position/x is 1, which corresponds to 0°.

It can be observed that the controller efficiently reduced the position error in x to zero within 5 s in the landing maneuver. The pose correction of the system is visualized in Figs. 6, 7, 8, 9. The desired pose is (0, 0, 0, 1) that corresponds to (position/x, position/y, position/z, orientation/z). The steep response of position/z versus time in Fig. 8 after 75 s is due to the fact that the network initiated a landing command to the AR Drone.

4 Conclusion

The training produces a network weight that can be used for two purposes. It can either act as a starting weight for an online learning neural network that adapts to the environment or can be used as it is for precise landing. The network weights were more generalized rather than specific to the trained data. The quadrotor was able to land and orient itself regardless of the initial conditions. Several trials were performed with random initial positions and orientation to test its generalized behavior. All trials were found to be able to land the quadrotor correctly. Regardless of the initial position and orientation, the quadrotor positions itself above the landing station while achieving the correct orientation. Correction of orientation, position, and altitude are done in parallel rather than in a series fashion.

An online neural network algorithm can be used to increase the accuracy and also to make the landing robust to environmental factors like the wind. As future work, an online algorithm could be developed that can make use of the weights calculated by the current research. Online neural networks normally start with random initial weights that are tuned progressively. But for applications like this, a network with random weights will only crash the quadrotor. So a better practice would be to start with a pretrained weight. This ensures fast learning, predictive behavior, and safe landings during the initial attempts.

Acknowledgment Our work has been motivated by the humanitarian initiatives of Mata Amritanandamayi Devi, the Chancellor of Amrita University. She has been our inspiration and support. Our gratitude goes to Prof. Kurien Issac and Dr. Radhamani Pillay, who offered constructive criticism and valuable feedback at various stages of this project. We also thank the research group at Ammachi Labs for their support and offering us the resources that aided our work.

References

1. L. Cantelli, M. Mangiameli, C. Melita, and G. Muscato, "UAV/UGV cooperation for surveying operations in humanitarian demining," in Safety, Security, and Rescue Robotics (SSRR), 2013 IEEE International symposium on, pp. 1–6, IEEE, 2013.
2. P. Maini and P. Sujit, "On cooperation between a fuel constrained uav and a refueling ugv for large scale mapping applications," in Unmanned Aircraft Systems (ICUAS), 2015 International Conference on, pp. 1370–1377, IEEE, 2015.
3. G.-J. M. Kruijff, F. Colas, T. Svoboda, J. Van Diggelen, P. Balmer, F. Pirri, and R. Worst, "Designing intelligent robots for human-robot teaming in urban search and rescue," in AAAI Spring Symposium: Designing Intelligent Robots, 2012.
4. D. Chaturvedi, R. Chauhan, and P. K. Kalra, "Application of generalised neural network for aircraft landing control system," Soft Computing, vol. 6, no. 6, pp. 441–448, 2002.
5. H. Izadi, M. Pakmehr, and N. Sadati, "Optimal neuro-controller in longitudinal auto-landing of a commercial jet transport," in Control Applications, 2003. CCA 2003. Proceedings of 2003 IEEE Conference on, vol. 1, pp. 492–497, IEEE, 2003.

6. J.-G. Juang, L.-H. Chien, and F. Lin, "Automatic landing control system design using adaptive neural network and its hardware realization," Systems Journal, IEEE, vol. 5, no. 2, pp. 266–277, 2011.
7. A. Gautam, P. Sujit, and S. Saripalli, "A survey of autonomous landing techniques for uavs," in Unmanned Aircraft Systems (ICUAS), 2014 International Conference on, pp. 1210–1218, IEEE, 2014.
8. C. Papachristos and A. Tzes, "The power-tethered uav-ugv team: A collaborative strategy for navigation in partially-mapped environments," in Control and Au- tomation (MED), 2014 22nd Mediterranean Conference of, pp. 1153–1158, IEEE, 2014.
9. D. Lee, J. Zhou, and W. T. Lin, "Autonomous battery swapping system for quadrotor," in Unmanned Aircraft Systems (ICUAS), 2015 International Conference on, pp. 118–124, IEEE, 2015.
10. T. Toksoz, J. Redding, M. Michini, B. Michini, J. P. How, M. Vavrina, and J. Vian, "Automated battery swap and recharge to enable persistent uav missions," in AIAA Infotech@ Aerospace Conference, 2011.
11. A. Tayebi and S. McGilvray, "Attitude stabilization of a four-rotor aerial robot," in Decision and Control, 2004. CDC. 43rd IEEE Conference on, vol. 2, pp. 1216–1221, IEEE, 2004.
12. T. Luukkonen, "Modelling and control of quadrotor," Independent research project in applied mathematics, Espoo, 2011.
13. P. Benavidez, J. Lambert, A. Jaimes, and M. Jamshidi, "Landing of an ardrone 2.0 quadrotor on a mobile base using fuzzy logic," in World Automation Congress (WAC), 2014, pp. 803–812, IEEE, 2014.
14. D. Lee, T. Ryan, and H. J. Kim, "Autonomous landing of a vtol uav on a moving platform using image-based visual servoing," in Robotics and Automation (ICRA), 2012 IEEE International Conference on, pp. 971–976, IEEE, 2012.
15. M. Quigley, K. Conley, B. Gerkey, J. Faust, T. Foote, J. Leibs, R. Wheeler, and A. Y. Ng, "Ros: an open-source robot operating system," in ICRA workshop on open source software, vol. 3, p. 5, 2009.
16. H. Huang and J. Sturm, "tumsimulator ros wiki," 2014. [Online; accessed 23-December-2015; http://www.ros.org/wiki/tum_ardrone].

Superimposed Pilot Based Channel Estimation for MIMO Systems

S. Sarayu, Janaki Radhakrishnan and S. Kirthiga

Abstract Estimation of the channel has been one of the major concerns in any communication system. Although various channel estimation techniques are available for MIMO (Multiple Input Multiple Output) systems, the problem of wastage of bandwidth due to the transmission of training sequences prior to the transmission of actual data exists. Use of superimposed pilot sequence that accomplishes transmission of both the pilot and data simultaneously has been proposed as an effective alternative. The work carried out includes both simulation and real-time implementation of channel models and channel estimation techniques. Appropriate channel models chosen include shadowing model and Markov model to address large-scale fading effects, and Rician model for small-scale fading effects. Employing channel estimation techniques for the above listed channel models is the novel idea aimed for satellite applications. Also different algorithms for channel estimation like least square (LS), minimum mean square error (MMSE), and linear minimum mean square Error (LMMSE) have been compared. Furthermore, merits and demerits of superimposed and conventional pilots are analyzed by making performance comparisons. Hardware implementation is done using Universal Software Radio Peripheral (USRP).

Keywords MIMO · Channel modeling · Channel estimation · LS · MMSE · LMMSE · USRP

S. Sarayu (✉) · J. Radhakrishnan · S. Kirthiga
Department of Electronics and Communication Engineering, Amrita School of Engineering, Coimbatore, Amrita Vishwa Vidyapeetham, Amrita University, Coimbatore 641112, India
e-mail: sarayussv@gmail.com

J. Radhakrishnan
e-mail: janaki26794@gmail.com

S. Kirthiga
e-mail: s_krithiga@cb.amrita.edu

S.S. Dash et al. (eds.), *Artificial Intelligence and Evolutionary Computations in Engineering Systems*, Advances in Intelligent Systems and Computing 517,
DOI 10.1007/978-981-10-3174-8_11

1 Introduction

Channel estimation is important as it provides information about the channel characteristics. When a signal is transmitted across the channel, its nature varies due to noise and distortions. The knowledge of the channel is used to nullify the effects caused by the noisy channel to perform accurate decoding of the transmitted signal. An estimate of the channel can give the necessary characteristics required for analysis purpose. In channel estimation, initially a known sequence called pilot or training sequence is transmitted and channel matrix is estimated which is used for obtaining the data that was transmitted.

Three types of training sequences exist such as conventional pilot (CP), overlay pilot (OP), and superimposed pilot (SIP). Of the three, SIP is capable of giving higher throughput. In SIP, pilot superimposed over low power data is sent in one time slot and data alone in the next time slot. Hence, SIP is used in this paper.

Channel modeling is done initially to understand the channel nature intuitively. Channel models available for MIMO systems that support the above pilot schemes such as shadowing model, Markov model, and Rician model, have been implemented.

In our work, both simulation and real-time implementation of channel model and channel estimation techniques have been done. Performance comparisons between SIP and CP, and between various estimation algorithms is carried out. Hardware platform called USRP (Universal Software Radio Peripheral) is used for transmission and reception of a signal in real time. An elaborate discussion is presented in the following sections.

The next section summarizes the work being implemented by describing the overall system, Sect. 3 gives an insight into various channel models and channel estimation techniques suitable for the desired application, Sect. 4 discusses the conclusions that can be drawn from the results obtained using simulation and USRP. The last section concludes the work and describes the future scope and the improvements that can be made from the current work.

2 System Overview

This section deals with the description of the entire system being implemented. The block diagram of the overall system is given in Fig. 1. First part of the work is channel modeling. MIMO channel models used include shadowing model, Markov model, and Rician model that take into consideration all the channel distortions that are predominant in any satellite communication. Second part is channel estimation using different techniques and training sequences.

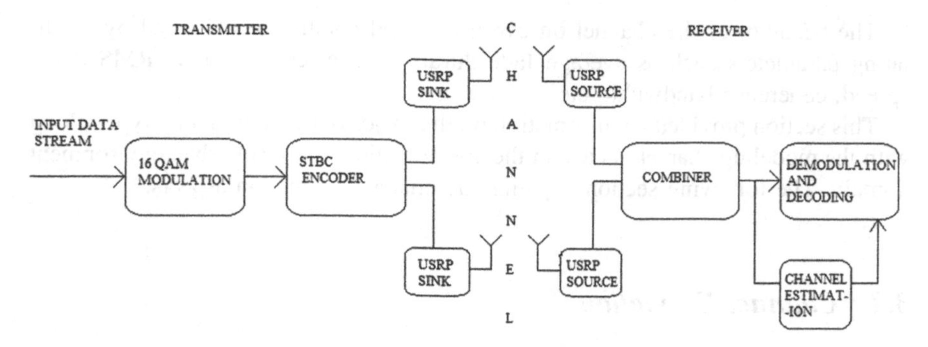

Fig. 1 MIMO transmitter and receiver

The entire work can be described in a nutshell as follows:

1. Generation of superimposed pilot (SIP) and conventional pilot (CP).
2. Performing simulation for both channel modeling and channel estimation by implementing appropriate channel models and channel estimation algorithms.
3. Estimation of channel parameters.
4. Carrying out real-time implementation using USRP.
5. Obtain results in both cases for SIP as well as CP and perform comparisons between the two to find out the most suitable one for the application being used.
6. Validating simulation with results obtained from USRP.

3 Channel Modeling and Channel Estimation

3.1 Channel Modeling

For a communication system, signal fading is often individually explained as a component of path loss, small-scale fading and large-scale fading. In order to understand these distortions, channel modeling is being implemented.

Empirical stochastic model is chosen for convenience to model the effects described above. The values after conducting experiments are taken from [1]. To model large-scale fading effects, shadowing model is used. There are two types of shadowing: high and low shadowing. To switch between high and low shadowing Markov model is used. The equations involved are taken from [1]. The small-scale fading effects are obtained from modeling the channel as a Rician fading channel. These models are understood and applied for simulation as is given from [1]. The antennas used for transmission are dual circularly polarized antenna [Left-Hand Circularly Polarized antenna (LHCP) and Right-Hand Circularly Polarized antenna (RHCP)] [1–3] as it helps to reduce the effect of Faraday rotation prominent in satellite communication.

The behavior of the channel on every deviated condition is attained by calculating parameters such as average fade duration, level crossing rate, RMS delay spread, coherence bandwidth, etc.

This section provided an information on the major types of modeling types along with the modeling characteristics on the consideration of the suburban environment domain. The following section explains the channel estimation aspects.

3.2 Channel Estimation

This section describes the SIP-based channel estimation for MIMO systems. As was discussed earlier, three types of training sequences exist. The structure of these is shown in Fig. 2.

The figure has been obtained from [4]. The generation of SIP has been done by taking into consideration the digital watermarking techniques [4, 5].

To begin with, in SIP, the there are two modes of transmission, SIP mode and the data mode. In SIP mode, the pilot is superimposed or in other words placed over the data which has low power. This mode will be transmitted in time slot T_t. The pilot in this mode has a power of σ_t^2 and the data has a power of σ_{dt}^2.

The equation in this mode is given by (1)

$$Y_t = \left(\sqrt{\frac{\sigma_t^2}{M}} X_t + \sqrt{\frac{\sigma_{dt}^2}{M}} X_{dt} \right) H + N_t \tag{1}$$

In (1), M is the number of transmitter antennas used. The total time slot T is the sum of T_t and T_d. Y_t is the receiver matrix of the size $T_t \times N$, X_t and X_{dt} is of the order $T_t \times M$ and transmitted symbol matrix, respectively.

In the data mode, power used is σ_d^2 and is transmitted in the time slot T_d. The equation in this mode is given by (2)

$$Y_d = \left(\sqrt{\frac{\sigma_d^2}{M}} X_d \right) H + N_d \tag{2}$$

where X_d is the transmitted data matrix, and Y_d is the received data matrix. Here, $T_d = T - T_t$.

In this method, first the transmission of the SIP mode takes place. From the known pilot, channel matrix H is appropriately estimated using Eq. (1). Next the data is transmitted and is decoded using the previously estimated channel matrix H.

The algorithms used for analysis are described as follows:

Minimum mean square error algorithm (MMSE): This algorithm is aimed at minimizing the error of the channel estimate. Due to this reason, it is considered to be efficient in almost all cases. The problem with this algorithm is the complexity of

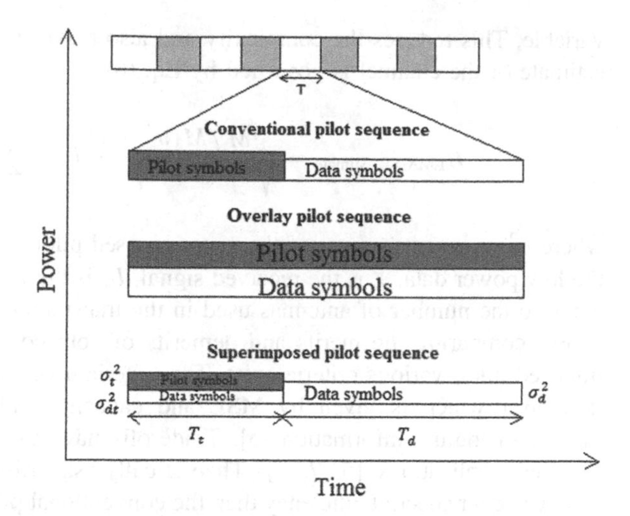

Fig. 2 Structure of the training sequences

certain systems that require huge and difficult computations [6, 7]. The channel estimate obtained by using this algorithm is given by Eq. (3)

$$H_{\text{MMSE estimate}} = \frac{\rho}{M} YX^H \left(R_H^{-1} + \frac{\rho}{M} XX^H \right)^{-1} \tag{3}$$

where ρ the signal-to-noise ratio is, M is the number of antennas used for transmission, and R_H^{-1} represents the inverse of the autocorrelation. The error of the estimation is given as in Eq. (4)

$$\text{error} = \text{tr}\left\{ \left(R_H^{-1} + \frac{\rho}{M} XX^H \right)^{-1} \right\} \tag{4}$$

Least square algorithm (LS): As the name suggests it is aimed at minimizing the square distance between the received signal and the original signal [8]. For the sake of simplicity, the noise is neglected. The least square estimate of the channel matrix is given by Eq. (5)

$$H_{\text{LS estimate}} = YX^H \left(XX^H \right)^{-1} \tag{5}$$

Here, Y represents what was received, X represents what was transmitted, and X^H gives the transpose of the complex conjugate of X.

First the channel estimate using the pilot as the input is obtained which is then used for finding the mean square error of the channel estimate. In case of the LS algorithm, the channel estimation error is proportional to the trace of $(XX^H)^{-1}$.

Linear minimum mean square error algorithm (LMMSE): Linear minimum mean square error algorithm assumes the estimate to be a linear function of another variable. This reduces the complexity and also makes the estimation optimal. The estimate of the channel is governed by Eq. (6)

$$H_{\text{LMMSE estimate}} = \sqrt{\frac{M}{\sigma_t^2}}\left(\frac{M(\sigma_{dt}^2+1)}{\sigma_t^2}I_M + \sum^2\right)^{-1}\sum Y \tag{6}$$

where σ_t^2 represents power of the superimposed pilot, σ_{dt}^2 represents the power of the low power data, Y is the received signal, I_M is the identity matrix of dimension equal to the number of antennas used in the transmitter side.

For comparing the merits and demerits of both conventional pilot and superimposed pilot, various criteria exist. The criteria used include the channel estimation error which is given by MSE and capacity achieved which is given by maximum mutual information [5]. Trade-offs have to be made according to the required applications [6, 7, 9]. Theoretically, superimposed pilot is known to achieve better spectral efficiency than the conventional pilot. Both SIP and CP have been compared both using simulation as well as real-time implementation using a hardware platform USRP. Also since different algorithms exist for performing channel estimation, effective results will be obtained only if efficient algorithm of all is chosen. So, comparisons between LS, LMMSE, and MMSE algorithms have also been performed.

The next section gives the results obtained through simulation and experimental setup.

4 Results and Discussions

Previous analysis gave us an understanding on the major modeling and estimation techniques used as part of the discussion in the project and how the channel is found out, analyzed, modeled, and estimated to avail a fine tune of the setup proposed. This section would generally deal with the results which are obtained by performing an experimental analysis and the results obtained from the same. Further on, the simulation analysis is taken to the next level of the hardware implementation to what is called as USRP. Finally, the results availed with the simulated and real-time values are compared and concluded.

4.1 Simulation Results

The transmitted sequence and received signal used for the purpose of simulation is given in Table 1.

Table 1 Sequence being transmitted

Type	Transmitted sequence	Received sequence
SIP	$\begin{bmatrix} 4.4721-4.4721i & 4.4721i \\ 4.4721i & 4.4721+4.4721i \end{bmatrix}$	$\begin{bmatrix} 0.7715-0.4990i & 0.5380+0.1375i \\ 0.2717+1.0392i & 0.6756+1.2160i \end{bmatrix}$
CP	$\begin{bmatrix} 4.7434-4.7434i & 1.5811+4.7434i \\ -1.5811+4.7434i & 4.7434+4.7434i \end{bmatrix}$	$\begin{bmatrix} 0.7239-0.5376i & 0.3335+0.1450i \\ 0.5637+1.1053i & 0.9026+1.2870i \end{bmatrix}$

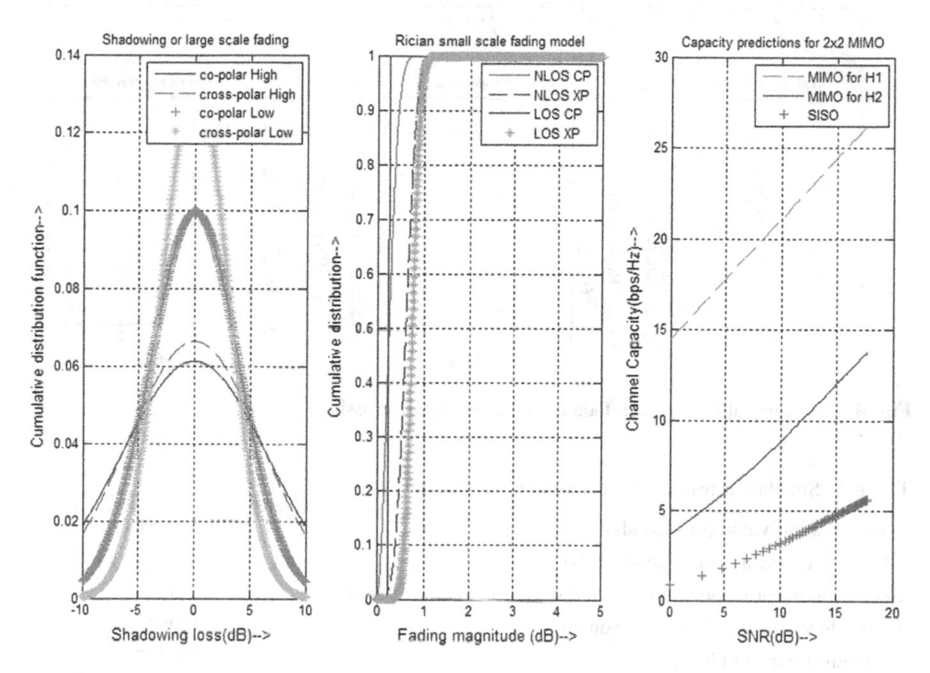

Fig. 3 Probability density functions and capacity predictions

Figure 3 shows the plots for probability density functions and capacity predictions, respectively [10]. Figure 4 shows the simulated resultant graphs used for the determination of the second-order statistics which mainly include the autocorrelation functions, level crossing rate and the average fade durations as considered for a particular communication system. The results obtained are tabulated in Table 2. The observations lead to the conclusion that the channel is a flat fading one. Figure 5 shows the performance comparisons between conventional pilot and superimposed pilot sequence. Figure 5a, b shows the throughput comparisons and Fig. 5c, d shows the comparison between channel estimation techniques for the two types of inputs used.

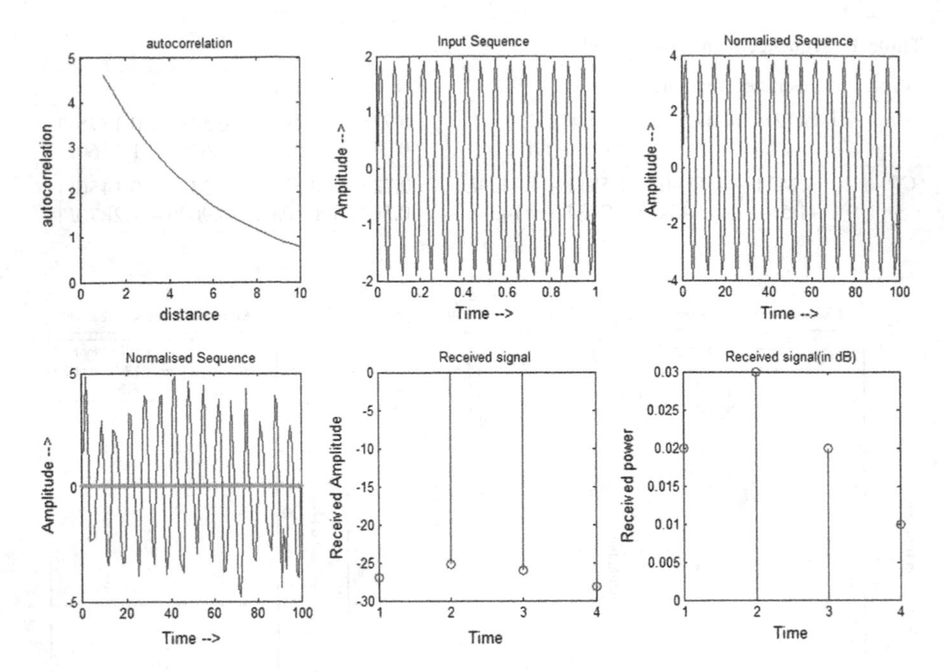

Fig. 4 Autocorrelation, average fade duration and level crossing rate

Table 2 Simulation results of second-order statistics

Fade duration value (in seconds)	10.7718
Mean excess delay (in microseconds)	1.25
Second moment of power delay profile (in microsecond square)	2.5
RMS Delay spread (in microseconds)	0.9682
Coherence bandwidth (in Hertz)	0.0207

This, being the simulated approach, analyzes, which of the two types of input has better capacity or throughput with respect to the increase has in SNR values. The capacity performance of superimposed pilot with respect to the conventional pilot is better. Also, the error rate for SIP is greater than the CP sequence type [11, 12].

4.2 Experimental Setup

The real-time implementation is carried out using USRP. The values obtained after transmission of SIP and CP through USRP in the receiver side are given in Table 3.

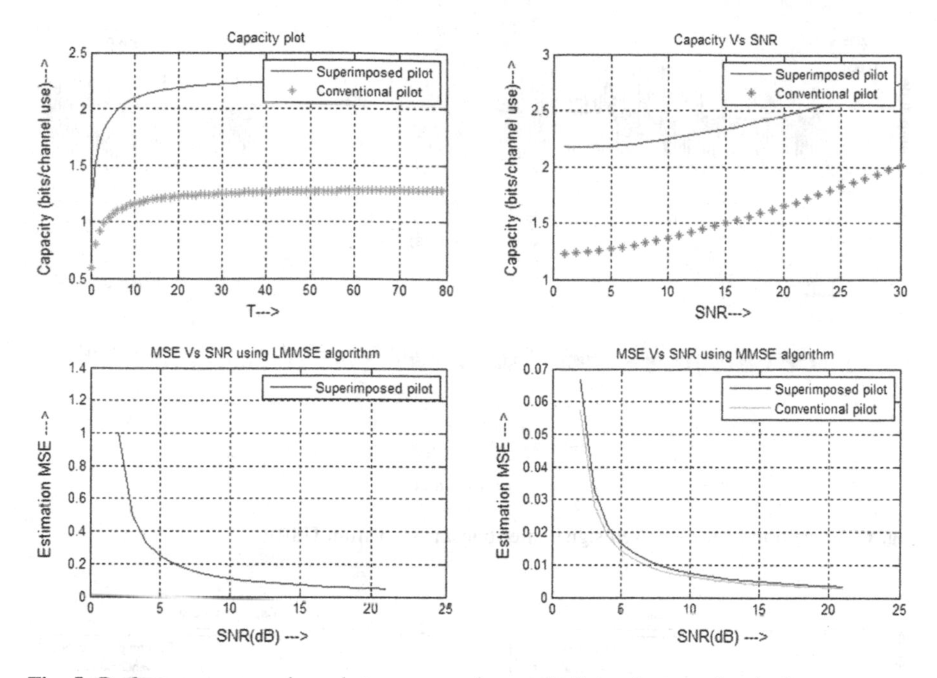

Fig. 5 Performance comparisons between superimposed pilot and conventional pilot

Table 3 Sequence being transmitted and received

Type	Transmitted sequence	Received sequence
SIP	$\begin{bmatrix} 4.4721-4.4721i & 4.4721i \\ 4.4721i & 4.4721+4.4721i \end{bmatrix}$	$\begin{bmatrix} -0.2566-0.1064i & -0.8504-0.5642i \\ -0.0004-0.0010i & -0.6983-0.4525i \end{bmatrix}$
CP	$\begin{bmatrix} 4.7434-4.7434i & 1.5811+4.7434i \\ -1.5811+4.7434i & 4.7434+4.7434i \end{bmatrix}$	$\begin{bmatrix} -0.6983-0.5196i & -0.5866-0.3466i \\ -0.6872-0.6425i & -0.9218-0.4078i \end{bmatrix}$

First, the Blocks are implemented in GNU radio. The output of the STBC from simulation is used as the input in GNU radio. It is this input that is transmitted across two antennas according to the scheme proposed by Alamouti.

The transmission is possible only due to the use of USRP sink block. Correspondingly, the reception is achieved by using USRP source block. Then the values are taken appropriately from the graph generated by the GNU radio companion for further analysis. This received signal is used in the code for computing the channel estimate and the estimation error and for further comparisons of SIP and CP. The waveform obtained as a result of transmitting through USRP is shown in Fig. 6. Since a complex number is transmitted, two plots are obtained separately, one for real part and another for complex part, each showing the amplitude of the signal being received with time.

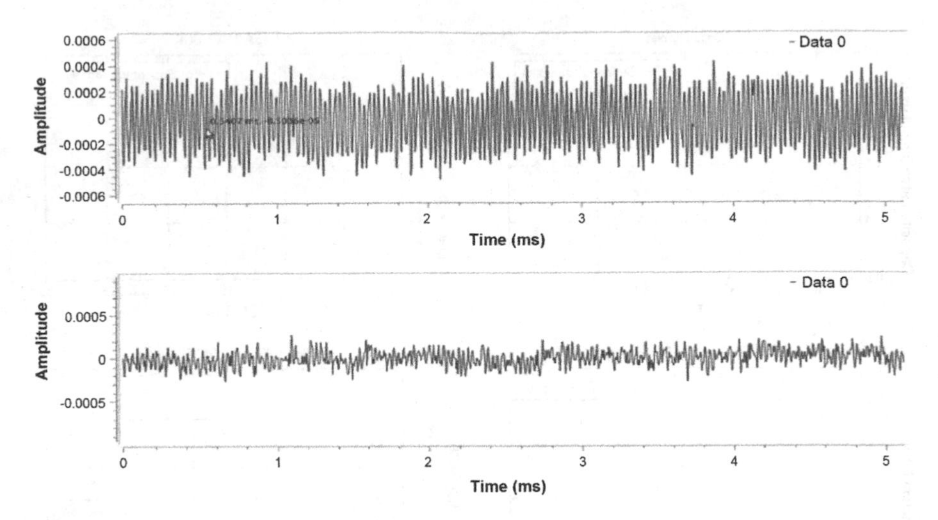

Fig. 6 Waveform of the received signal after transmission from USRP

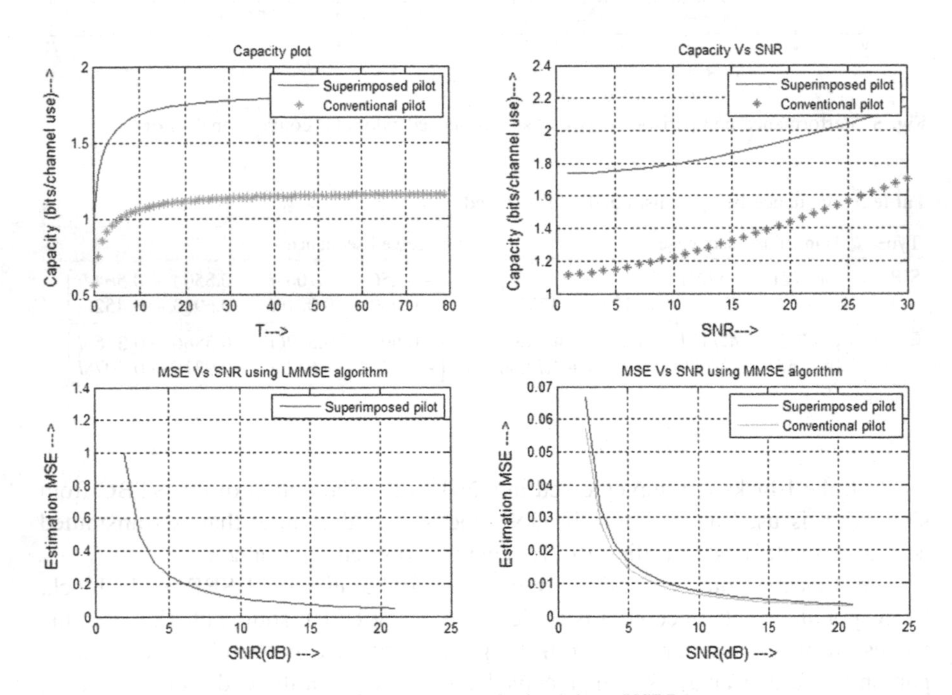

Fig. 7 Performance comparisons for the values obtained from USRP

In Fig. 6, the first plot shows the amplitude of real part versus time and the second plot shows the amplitude of the complex part versus time. The next plot is obtained for MSE versus snr from values obtained in real time.

Figure 7 shows mean square error versus SNR plots and capacity plots for both SIP and CP. The results are obtained by utilizing the values from Table 3. However, it is found that the experimental results and simulated results are nearly the same, thus validating the simulation results.

The next section concludes the entire work.

5 Conclusion

We have analyzed the performance of SIP over CP using different channel estimation algorithms. It can be found that SIP is definitely a better alternative than CP if overall system throughput is desired. On the other hand, it does have a disadvantage of producing more channel estimation error. Neglecting this anomaly, we conclude that SIP is one of the finest ways of estimating the channel. Also LMMSE algorithm is an optimal solution that can be used for the purpose of channel estimation. The given conclusions are drawn from capacity plots, and MSE plots.

Finally, the project was extended for the application in real time using USRP. It can be seen that USRP is a user-friendly hardware that makes the transmission and reception a lot easier. USRP typically can be used for any software defined radio application. It is new emerging hardware tool that has been used for academic and research purposes. Also software used for interfacing USRP with the computer is an open source software which can be easily used and altered according to our application. There are avid chances of stepping this concept to the next level of analysis where the same concept with a slight change in the algorithm used can be implemented. The work can be extended to time variant channels for which Kalman filters can be used. This defines the future scope of the project.

Acknowledgements We would like to extend special sense of gratitude to our project advisor Dr. S. Kirthiga for her wonderful and immense support throughout the completion of the project. We also thank Mr. R. Gandhiraj for helping us understand the whereabouts of USRP. We also thank our friends who assisted in every aspect of the work. We finally thank the department for giving us the necessary platform for carrying out the project successfully.

References

1. Peter R. King, Tim W.C. Brown, Argyrios Kyrgiazos and Barry G. Evans.: Empirical-Stochastic LMS-MIMO channel model Implementation and Validation. In: IEEE ANTENNAS AND PROPOGATION, vol. 60, no. 2, pp. 606–614, Feb. 2012.
2. Konstantinous P. Ziolis, Jesus Gomez-Vilardibo, Enrico Casini, and Ana. I. Perz-Neira.: Statistical Modelling of Dual-polarized MIMO Land mobile satellite channels. In: IEEE COMMUNICATIONS, vol. 58, no. 11, pp. 3077–3083, Nov. 2010.
3. Peter R King and Stavros Stavron.: Low Elevation Wideband Land Mobile Satellite MIMO channel characteristics. In: IEEE WIRELESS COMMUNICATIONS, vol. 60, no. 7, pp. 2712–2720, Jul. 2007.

4. Mikael Coldrey (Tapio) and Patrik Bohlin.: raining-Based MIMO Systems-Part I: Performance Comparison. In: IEEE SIGNAL PROCESSING, vol. 55, no. 11, pp. 5464–5476, Nov. 2007.
5. Hartung, F. and Kutter, M.: Multimedia watermarking techniques. In: IEEE TRANSACTIONS ON SIGNAL PROCESSING, vol. 87, no. 7, pp. 1079–1107, Jul. 1999.
6. Xia Liu, Shiyang Lu, Marek E. Bialkowski and Hon Tat Hui.: MMSE Channel Estimation for MIMO System with Receiver Equipped with a Circular Array Antenna. In: Proceedings of Asia-Pacific Microwave Conference, 2007.
7. Bohlin, P. and Coldrey, M.: Performance evaluation of MIMO communication systems based on superimposed pilots. In: IEEE INTERNATIONAL CONFERENCE ON ACOUSTICS, SPEECH AND SIGNAL PROCESSING, pp. 425–428, May 2004.
8. Kay, S.M.: Fundamentals of Statistical Signal Processing. In.: Englewood Cliffs, NJ: Prentice-Hall, 1993.
9. Aditya K. Jagannatham and Bhaskar D. Rao. SUPERIMPOSED PILOTS VS. CONVENTIONAL PILOTS FOR CHANNEL ESTIMATION an article paper.
10. King P.R.: Modelling and Measurement of the Land Mobile Satellite MIMO Radio Propagation Channel, University of Surrey, April 2007, a PHD thesis submission.
11. Mohammad-Ali Khalighi and Salah Bourennane.: Semi blind Single-Carrier MIMO Channel Estimation Using Overlay Pilots. In.: IEEE TRANSACTIONS ON VEHICULAR TECHNOLOGY, vol. 57, no. 3, pp. 1951–1956, May 2008.
12. Meenu S Kumar and Kirthiga S.: Review of Parametric Radio Channel Prediction Schemes for MIMO Systems. In.: ICCPCT, 2016.

Artificial Bee Colony Optimization Algorithm for Fault Section Estimation

M.A. Sobhy, A.Y. Abdelaziz, M. Ezzat, W. Elkhattam, Anamika Yadav and Bhupendra Kumar

Abstract This paper introduces an optimization technique that uses an artificial bee colony (ABC) algorithm to solve the fault section estimation (FSE) problem. FSE is introduced as an optimization problem, where the objective function includes the status of protective relays and circuit breakers. The ABC algorithm is a new population-based optimization technique inspired by behavior of the bee colony to search honey. In order to test the effectiveness of the proposed technique, two sample systems are tested under various test cases. Also the results obtained by the proposed ABC algorithm is compared with those obtained using two other methods. The results show the accuracy and high computation efficiency of the ABC algorithm. The ABC algorithm has a main advantage that it has only two parameters to be controlled. Therefore, the tuning of the proposed algorithm is easier and has a higher probability to reach the optimum solution than other competing methods.

Keywords Fault section estimation · Power systems · Artificial bee colony

1 Introduction

Communications and monitoring in power systems have been widely improved. A large number of system alarms and signals can now be detected and processed, this reduce the number of system control operators. When the fault occurs, operators would have to respond to a large number of alarm messages and the task can be very difficult and complicated [1]. Thus, there is a great need to implement

M.A. Sobhy · A.Y. Abdelaziz · M. Ezzat · W. Elkhattam
Department of Electrical Power & Machines, Faculty of Engineering,
Ain Shams University, Cairo, Egypt

A. Yadav (✉) · B. Kumar
Department of Electrical Engineering, National Institute of Technology,
Raipur, C.G, India
e-mail: ayadav.ele@nitrr.ac.in

S.S. Dash et al. (eds.), *Artificial Intelligence and Evolutionary Computations in Engineering Systems*, Advances in Intelligent Systems and Computing 517,
DOI 10.1007/978-981-10-3174-8_12

efficient fault section estimation methods on SCADA systems in order to help the operators to judge the situation before taking the action.

Fault section estimation (FSE) detects the faulty section using the current status of the relays and circuit breakers. The status of the protective devices is provided as 0–1 signals from SCADA systems.

Methods of solving FSE are divided into two main categories: logical methods, and optimization methods. Logical methods determine the fault section using logical rules, including some techniques such as expert system (ES) [2, 3], fuzzy logic (FL) [4, 5], rough set theory (RST) [6], Petri net (PN) [7, 8], artificial neural networks (ANN) [9, 10] and Bayesian network (BN) [11].

ES-based techniques are inspired by the behavior of experts to reach the optimum solution. But these methods have two drawbacks that they need knowledge maintenance and involve slow time response [1]. FL, RST, PN, and BN are used mainly to solve the problems with a great percentage of uncertainty and incomplete data, this type of techniques is helpful in fault section estimation problems when the information about the status of the relays and circuit breakers are incomplete. The results obtained by these methods are satisfactory, but these methods have drawbacks when they are used to detect the faulty section in large scale power systems.

The second category is optimization-based methods, which formulates the FSE problem as a $0-1$ integer optimization problem. In this case, FSE can then be solved using a global optimization method such as Boltzmann machine [12], genetic algorithm (GA) [13], tabu search (TS) [14], or swarm intelligence algorithms [15, 16].

In this paper, artificial bee colony (ABC) algorithm is proposed to solve the fault section estimation problem. Artificial bee colony (ABC) algorithm is a population-based algorithm, which is inspired by the behavior of bee colony in searching honey. The results show that the ABC algorithm is efficient and has high computation efficiency.

This paper is structured as follows: the mathematical model of FSE is given in Sect. 2, and then the proposed ABC algorithm is given in Sect. 3. The computational steps used to reach the solution of FSE problem are described in Sect. 4. In Sect. 5, the results are introduced. The conclusions are given is Sect. 6.

2 Mathematical Model

A power system can be divided into three types of protective zones: bus bars, transformers, and line sections [17].

In this paper, the study is based on the concepts of selective protection, where the protection schemes are main and backup protection. Main protection relays are designed to operate for faults in their protective zones as fast as possible. While the backup protection relays operate in case of failure of main protection relays [1].

According to the previous definitions, protective relays can be classified into three types of relays: main protective relays, primary backup protective relays, and remote backup protective relays. All three types of relays are installed at transformer sections and line sections, but each busbar section has one main relay only. When a fault occurs in any section, its main protective relay is activated to isolate the fault from the system. If there is any problem with the fault clearing process, the primary and remote backup relays are activated instead. The main function of protective relays and circuit breaker is to detect and clear the fault as fast as possible [17]. Now we will express the fault section estimation problem in a mathematical formula as an objective function. The objective function will depend on two main factors. The first one is the probability of fault occurrence in a certain section. The second factor is the status of the protective relays and circuit breakers. The mathematical expression of a fault section estimation problem is derived as follows [15, 16]:

$$\begin{aligned} \text{Min Obj } (P,R) = \exp\Bigg\{ & \sum_{b=1} S^{\text{bus}}_{m,b}(P,R) \\ & + \sum_{t=1} \left[S^{\text{TR}}_{m,t}(P,R) + S^{\text{TR}}_{p,t}(P,R) + S^{\text{TR}}_{s,t}(P,R) \right] \\ & + \sum_{l=1} \Big[S^{\text{line}}_{\text{send},m,l}(P,R) + S^{\text{line}}_{\text{send},p,l}(P,R) + S^{\text{line}}_{\text{send},s,l}(P,R) \\ & + S^{\text{line}}_{\text{rec},m,l}(P,R) + S^{\text{line}}_{\text{rec},p,l}(P,R) + S^{\text{line}}_{\text{rec},s,l}(P,R) \Big] \Bigg\} \end{aligned} \tag{1}$$

where Obj is the objective function, S is the protective function of a certain section. The protective function depends on two variables. The first variable is P that is a vector including the probability of a fault occurring at each section. Hence, the elements of vector P are numbers with a minimum value of "0" and maximum value of "1". The second variable is R which is a vector including the status of the relays and circuit breakers responsible for protecting bus bars, transformers, and line sections. Hence, the elements of vector R can have two values only "zero" or "one", "zero" indicates non operation while "one" indicates operation.

The suffix "m" indicates the main protective relay, the suffix "p" indicates the primary backup relay and the suffix "s" indicates the secondary (remote) backup relays.

The following relations are the mathematical formulation of S^{bus}_{m}, S^{TR}_{m}, $S^{\text{line}}_{\text{send},m}$, $S^{\text{line}}_{\text{rec},m}$ which are the functions of main protective relay for busbar, transformer, and the sending and receiving ends of line section. These functions are:

$$S^{\text{bus}}_{m,b} = \left(1 - 2R^{\text{bus}}_{m,b}\right) B_b \tag{2}$$

$$S^{\text{TR}}_{m,t} = \left(1 - 2R^{\text{TR}}_{m,t}\right) \text{TR}_t \left(1 - R^{\text{TR}}_{p,t}\right) \tag{3}$$

$$S^{\text{line}}_{\text{send},m,l} = \left(1 - 2R^{\text{line}}_{\text{send},m,l}\right)L_1\left(1 - R^{\text{line}}_{\text{send},p,l}\right) \tag{4}$$

$$S^{\text{line}}_{\text{rec},m,l} = \left(1 - 2R^{\text{line}}_{\text{rec},m,l}\right)L_1\left(1 - R^{\text{line}}_{\text{rec},p,l}\right) \tag{5}$$

While the functions of primary backup relays S^{TR}_p, $S^{\text{line}}_{\text{send},p}$, $S^{\text{line}}_{\text{rec},p}$ are expressed as

$$S^{\text{TR}}_{p,t} = \left(1 - 2R^{\text{TR}}_{p,t}\right)\text{TR}_t\left(1 - R^{\text{TR}}_{m,t}\right) \tag{6}$$

$$S^{\text{line}}_{\text{send},p,l} = \left(1 - 2R^{\text{line}}_{\text{send},p,l}\right)L_1\left(1 - R^{\text{line}}_{\text{send},m,l}\right) \tag{7}$$

$$S^{\text{line}}_{\text{rec},p,l} = \left(1 - 2R^{\text{line}}_{\text{rec},p,l}\right)L_1\left(1 - R^{\text{line}}_{\text{rec},m,l}\right) \tag{8}$$

The functions of secondary backup relays S^{TR}_s, $S^{\text{line}}_{\text{send},s}$, $S^{\text{line}}_{\text{rec},s}$ are expressed as

$$S^{\text{TR}}_{s,t} = \left(1 - 2R^{\text{TR}}_{s,t}\right) \times \left\{ - \prod_{k=1} \left[1 - P_{t,k}\left(1 - R_{m,k}\right)\left(1 - R_{p,k}\right)\text{CB}_k\right] \right\} \tag{9}$$

$$S^{\text{line}}_{\text{send},s,l} = \left(1 - 2R^{\text{line}}_{\text{send},s,l}\right) \times \left\{ 1 - \prod_{g=1} \left[1 - P_{l,g}\left(1 - R_{m,g}\right)\left(1 - R_{p,g}\right)\left(1 - \text{CB}_g\,\text{CB}_{\text{send},l}\right]\right. \right\} \tag{10}$$

$$S^{\text{line}}_{\text{rec},s,l} = \left(1 - 2R^{\text{line}}_{\text{rec},s,l}\right) \times \left\{ 1 - \prod_{g=1} \left[1 - \text{P}_{l,g}\left(1 - R_{m,g}\right)\left(1 - R_{p,g}\right)\left(1 - \text{CB}_g\right)\text{CB}_{\text{rec},l}\right] \right\} \tag{11}$$

where $P_{t,k}$ represents the probability that a fault occurs at the zone k which is around the transformer (t). The zone k is protected by the secondary backup relay $R^{\text{TR}}_{s,t}$ of the transformer (t) if both main relay $R_{m,k}$ and primary backup relay $R_{p,k}$ of zone k do not operate. CB_k is the circuit breaker for zone k. In (10) and (11), $P_{l,g}$ represents the probability that a fault occurs at the zone g near the line (l). $R_{m,g}$, $R_{p,g}$, and CB_g represent main relay, primary backup relay, and circuit breaker of zone g, and $\text{CB}_{\text{send,l}}$ and $\text{CB}_{\text{rec},l}$ are circuit breakers of sending and receiving ends of the line (l). There is no backup relay for bus bars as mentioned before.

3 Proposed Approach

The artificial bee colony (ABC) algorithm is a new population-based optimization technique, introduced in 2005 by Karaboga [18]. When this algorithm was introduced, it was used only for unconstrained optimization, and then it was developed to deal with constrained optimization problems [19]. According to the ABC algorithm the bee colony is divided into three types of bees; the employed bees, the onlooker bees and the scout bees. Initially each employed bee is initialized at one food source position. Then the employed bee searches for a better food position around its initial one. If the nectar amount of the new position is better than the previous one, then the employed bee replaces the old position by the new one in its memory. This crossover process is known as greedy selection process. The onlooker bees are in the hive and watch the dances of the employed bees to determine the fitness of all food positions. According to the information shared by the employed bees, the onlooker bees choose the food position according to the probability of the position, which depends on its fitness. Hence, the onlooker bee phase uses probabilistic selection. Similar to the employed bee, the onlooker bee search around the selected food position and if the new position is better than the old one then the new position replace the old one. If a food position is not improved for a certain predetermined number of times, then this position is abandoned. The employed bee responsible for the abandoned position becomes a scout bee, which searches for a new food position randomly. Therefore, scout bees perform the job of "exploration," whereas employed and onlooker bees perform the job of "exploitation" [21].

The ABC technique initially generates N solutions randomly. Each solution is composed of D control variables. For example we can consider the food position "i" then, $X_i = [x_{i,1}\ x_{i,2}\ x_{i,3}\ \ldots\ x_{i,D}]$. The equation that is used to generate food positions (solutions) randomly is given by (12):

$$X_{i,j} = X_{\max,j} + \text{rand}(0,1)\left(X_{\max,j} - X_{\min,j}\right) \tag{12}$$

where i represent number of food position, i = 1,2 … N and j represents the number of control parameter, j = 1, 2, …., D. $x_{\max,j}$ represents the maximum boundary value of the control parameter "j," and $x_{\min,j}$ represents the minimum boundary value of the control parameter "j". Each employed bee X_i dances at one food position. Then it applies a modification on its original position (solution) to get a new position (solution) V_i. This random modification is done using the search equation:

$$V_{i,j} = X_{i,j} + u\left(X_{i,j} - X_{k,j}\right) \tag{13}$$

where V_i represents the modified food source, X_i represents the current food source and X_k represents a randomly chosen food source. u is a uniform random number in the range [−1, 1]. k is an integer selected randomly from [1 to N] but with a condition that $k \neq i$.

Then the employed bees determine the nectar amount (fitness) of the new food source (modified solution). If the fitness of the modified food source V_i is higher than that of the original food source X_i, then V_i replaces X_i the employed bee memorizes V_i and forgets X_i. In other words, the employed bees apply a greedy selection process. After this process is done by all the employed bees, then they share this information with the onlooker bees waiting in the hive. Then the onlooker bee phase starts. The onlooker bees determine the nectar amount (fitness) of the food sources shared by the employed bees. The onlooker bees select a food source probabilistically according to Eq. (14). After selecting its food source X_i, each onlooker bee applies a modification on its original food source producing a new food source V_i by using Eq. (13). Then a greedy selection process is applied. If the fitness of the modified food source V_i is higher than that of the original food source X_i, then V_i replaces X_i the onlooker bee memorizes V_i and forgets X_i.

The onlooker bee selects its food source probabilistically using the following equation:

$$P_i = \frac{\text{fitness}_i}{\sum_{i=1}^{N} \text{fitness}_i} \tag{14}$$

where fitness_i is the fitness value of a solution i, and N is the total number of food source positions (solutions). If a food source is not improved after a certain pre-determined number of trials, then this food source is abandoned and the employed bees responsible for this source become a scout bee. The scout bee searches for a new food source randomly using Eq. (12).

4 ABC Algorithm for Fault Section Estimation

In this section, the flowchart and the solution steps are shown. The flowchart of the proposed ABC algorithm is shown in Fig. 1. The steps of the algorithm are as follows [20]:

Step 1 Enter the colony size and the maximum number of iterations
Step 2 Generate the initial solutions (food sources)
Step 3 Evaluate the fitness of each solution using the following equation:

$$\text{fitnees}_i = \frac{1}{1 + \text{obj}_i} \tag{15}$$

where obj_i is the value of the objective function when substituting with solution "i".

Step 4 Generate modified solutions in the neighbor hood of each initial solution of the employed bee's Eq. (13)

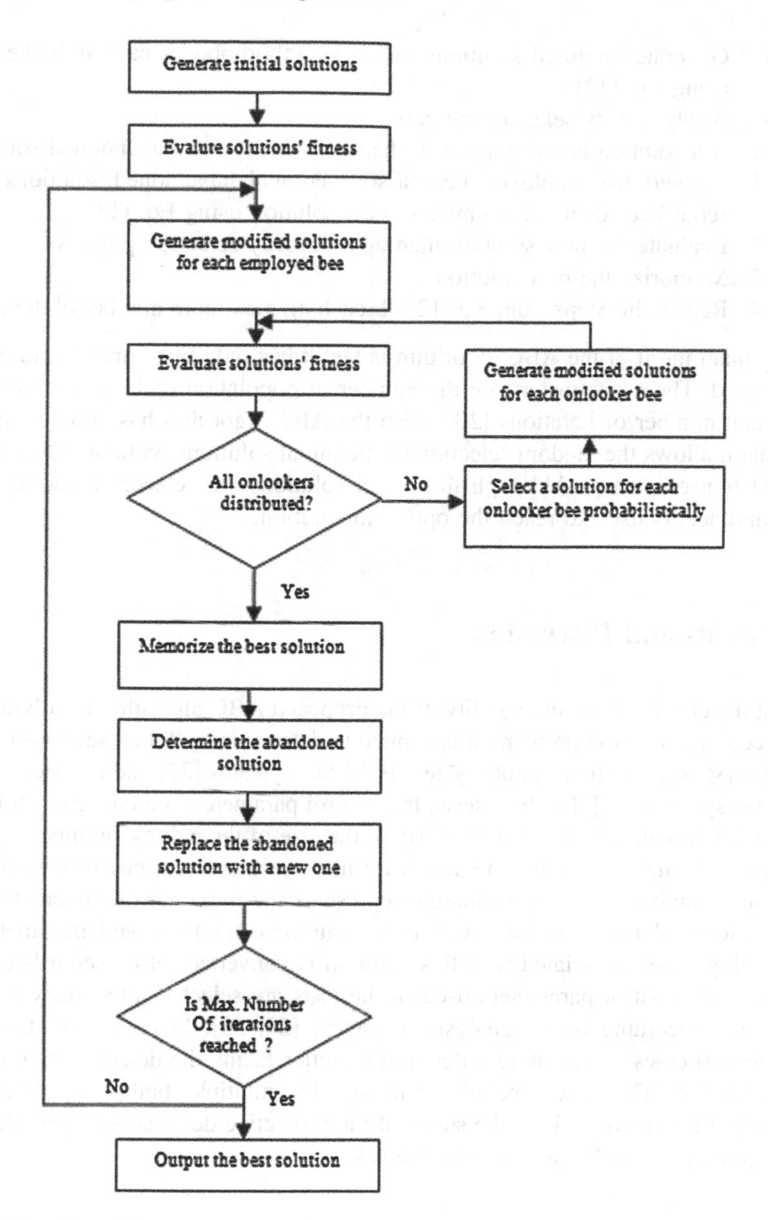

Fig. 1 Flowchart of ABC algorithm

Step 5 Apply greedy selection process
Step 6 Evaluate the probability of each solution using Eq. (14)
Step 7 Each onlooker bee selects a solution according to its probability calculated in step (6)

Step 8 Generate modified solutions in the neighborhood of each onlooker bee using Eq. (13)
Step 9 Apply greedy selection process
Step 10 If a solution is not improved, then consider it as an abandoned solution
Step 11 Convert the employed bees associated with abandoned solutions into scout bees then get completely new solution using Eq. (13)
Step 12 Evaluate the new solution then apply greedy selection process
Step 13 Memorize the best solution
Step 14 Repeat the Steps from 3 to 12 till reaching maximum number of iterations

The main merit of the ABC algorithm is that it has only two control parameters to be tuned. These parameters are the number of population (colony size) and the maximum number of iterations [20]. Also the ABC algorithm has another advantage that it allows the random selection of the initial solutions without affecting the computation efficiency. Although the initial solutions are selected randomly, the algorithm adapts itself to reach the optimum solution.

5 Results and Discussion

To test the effectiveness and validity of the proposed ABC algorithm in solving the fault section estimation problem, it is applied to determine the faulty sections of two sample systems. The first sample system is 10 bus systems [22] and the second one is 28 bus systems [23]. For both tests, the control parameters are chosen as follows the maximum number of iterations = 1000, the size of the colony (number of food sources) = 40 and the maximum number for nectar evaluating times = 6, where the maximum number of nectar evaluating represents the maximum number of times allowed for a solution to be improved. If the solution is not improved after that then the employed bee associated with this solution is converted into a scout bee. The values of the control parameters used in the tests are selected after many experiments and according to the analysis of papers [22, 23]. In each test, there are different test cases representing different difficulties facing the determination of the faulty section. The cases include examples for multiple faults, examples for incomplete information about the status of the protective devices and examples for maloperation cases of relays or circuit breakers.

5.1 Test 1

The first sample system is shown in Fig. 2. It consists of 10 sections, 14 circuit breakers, and 28 relays. The ten sections are as follows: four bus bars (B1, B2, B3, and B4), four transformers (T1, T2, T3, T4), and two line sections (L1 and L2), where B stands for busbar, T stands for transformer, and L stands for line section.

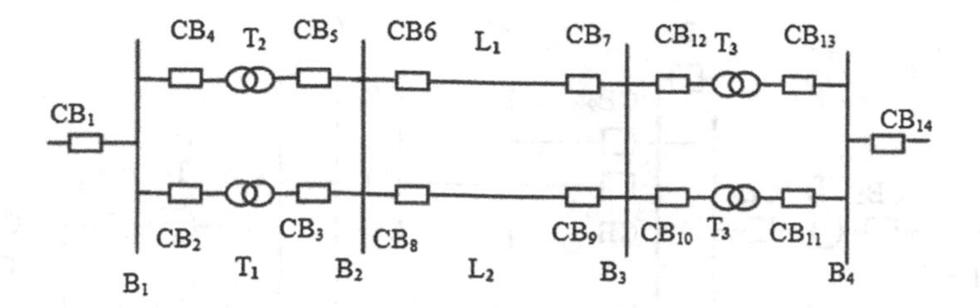

Fig. 2 Test system 1

Table 1 Test scenarios and estimation results of test 1

Test no.	Operated relays	Tripped circuit breakers	Estimation results
1	$L^{m}_{send,2}$ $L^{m}_{rec,2}$ $L^{p}_{send,2}$ T^{s}_{3} T^{s}_{4}	CB11, CB13, CB8	B_4, L_2
2	$B^{m}_{1}T^{p}_{1}T^{s}_{2}$	CB1, CB2, CB4	B_1, T_1
3	T^{p}_{4}	CB13	T_4

The relay configuration is as follows: any busbar has only one main protective relay, while each transformer and line section has main, primary backup, and secondary backup protective relays. Table 1 shows three test scenarios and their corresponding results obtained by the proposed algorithm.

In scenario 1, the data received indicates that the main protective relays $L_{send,2}$ and $L_{rec,2}$ operate; also CB8 operates, so there is a fault occurred at L_2. Meanwhile, the remote backup relays $T_{3,s}$ and $T_{4,s}$ operate to trip CB11 and CB13, so there is a fault at B4. It is obvious that there's incomplete relay information in this scenario as B_{4m} should have operated. Scenarios (2, 3) investigate multiple faults. For the three scenarios, the results obtained by the proposed algorithm are satisfactory.

5.2 *Test 2*

Figure 3 shows the second sample system with 28 sections, 40 circuit breakers, and 84 relays. The sections are divided as follows: 12 bus bars (B1, …, B12), 8 transformers (T1, …, T8), and 8 line sections (L1, …, L8).

The relay configuration is as follows: any busbar has only one main protective relay, while each transformer and line section has main, primary backup, and secondary backup protective relays.

Table 2 shows eight test cases and the corresponding results after applying a genetic algorithm [13] and the proposed algorithm. The simulation results obtained by the ABC algorithm are satisfactory in cases of incomplete information of relays, in cases of malfunction of protective devices, and in cases of multiple faults, where scenarios (from 1 to 5) investigate multiple faults. Also scenarios (6, 7, and 8) have

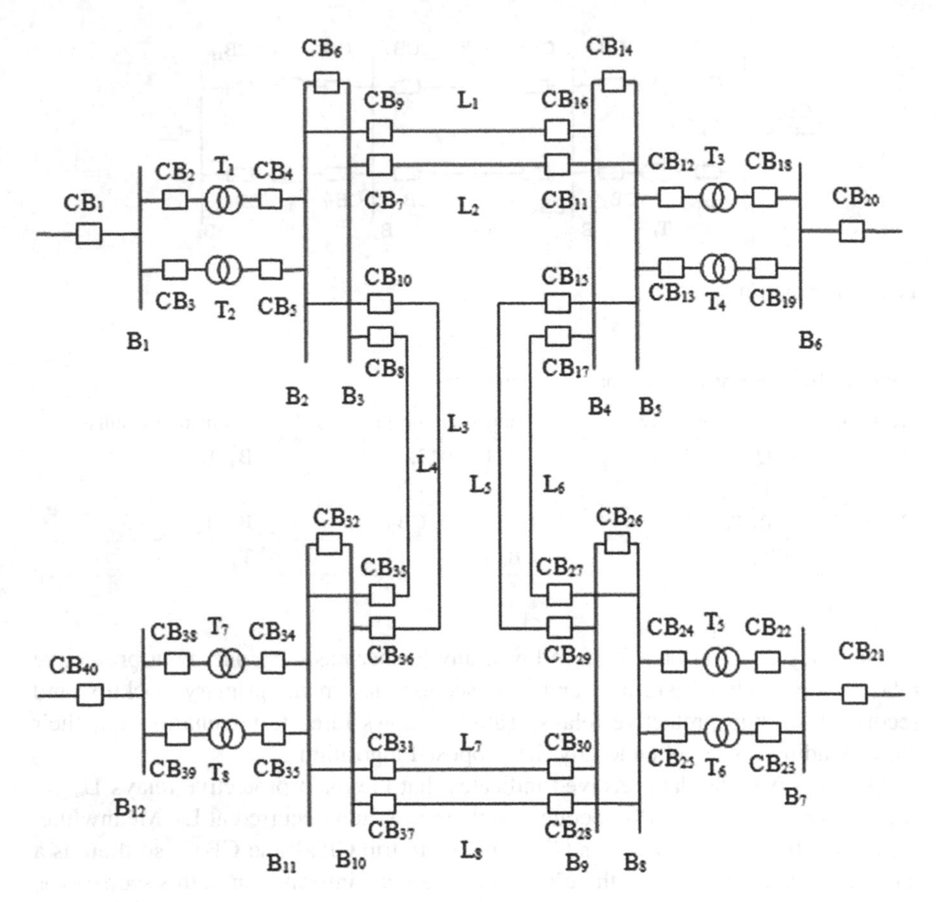

Fig. 3 Test system 2

incomplete relay information and device malfunction. For the eight scenarios, the results obtained by the proposed algorithm are acceptable. Now, we will illustrate two different scenarios in details.

For scenario 1, a possible explanation is as follows: there could be a fault on B_2, so the relay B_2^m operates, and the circuit breakers CB4, CB5, and CB10 are opened. However, the control center does not receive a signal that CB10 has operated. Simultaneously, another fault may occur on L_1, but $L^m_{send,1}$ does not operate, so CB9 is opened by the operation of $L^p_{send,1}$ or B_2^m. In scenario 8, the main relay B_{12}^m fails to operate, so the remote backup relays T_7^s and T_8^s operate to trip CB38, CB39 instead. It is obvious that GA cannot give reasonable results for scenarios (2, 4, and 5),this is because the objective function used in [15] does not include the effect of relay failure, while the objective function used in this paper consider the effect of relay failure, which help to get acceptable results.

Table 2 Test scenarios and estimation results of test 2

Test no.	Operated relays	Tripped circuit breakers	Estimation results	
			ABC	GA
1	$B_2^m\ L_{rec,1}^m\ L_{send,1}^p$	CB4, CB5, CB6, CB9, CB10, CB16	B_2, L_1	B_2, L_1
2	$T_3^p\ L_{send,7}^p\ L_{rec,7}^p$	CB12, CB18, CB30, CB31	T_3, L_7	Four solutions: (a) L_7, T_3 (b) L_7, (c) T_3 (d) No fault
3	$B_2^m B_3^m L_{send,1}^m\ L_{rec,1}^p$ $L_{rec,2}^m\ L_{send,2}^p$	CB4, CB5, CB6, CB7, CB8, CB9, CB10, CB11, CB16	B_2, B_3, L_1, L_2	B_2, B_3, L_1, L_2
4	$T_5^m\ T_6^p\ B_8^m\ B_9^m$ $L_{send,5}^m L_{rec,5}^p\ L_{send,6}^s$ $L_{send,7}^p\ L_{rec,7}^m\ L_{send,8}^s$	CB15, CB17, CB22, CB23, CB24, CB25, CB26, CB29, CB30, CB31, CB37	B_8, B_9, L_5, L_7, T_5, T_6	Two solutions: (a) B_8, B_9, L_5, L_7, T_5, T_6 (b) B_9, L_5, L_7, T_5
5	$L_{send,1}^m L_{rec,1}^p\ L_{send,2}^p$ $L_{rec,2}^p\ L_{send,7}^p\ L_{rec,7}^m$ $L_{send,8}^m L_{rec,8}^m$	CB47, CB9, CB11, CB16, CB28, CB30, CB31, CB37	L_1, L_2, L_7, L_8	Two solutions: (1) L_1, L_2, L_7, L_8 (2) L_1, L_7, L_8
6	B_2^m , $L_{rec,2}^s$, $L_{rec,4}^s$	CB4, CB5, CB9, CB10, CB11, CB35	B_2	B_2
7	T_7^p	CB34	T_7	T_7
8	$T_7^s\ T_8^s$	CB38, CB39	B_{12}	B_{12}

The faulty sections obtained using the ant colony for the eight test cases are almost similar to those obtained by the proposed ABC algorithm [16]. But the difference between the two methods is in the computation efficiency where the proposed ABC algorithm proved to be better than the ant colony method.

Table 3 shows a comparison between the computation time of the proposed method and another two methods, where method 1 stands for genetic algorithm [13], method 2 stands for ant colony method [16] and method 3 stands for the proposed ABC algorithm. In order to make a fair comparison, the number of iterations is chosen = 1000 and the population size = 40 for the three methods

Table 3 Computation time of each method in test 2

Method	Scenario	Average computation time (s)
Method 1 (GA)	From 1 to 4	5
Method 2 (Ant system)	From 1 to 8	0.4
Method 3 (ABC)	From 1 to 8	0.2

under comparison. The results show that the proposed ABC algorithm has better computation efficiency than the other techniques.

6 Conclusion

This paper proposes an optimization approach that employs an artificial bee colony (ABC) algorithm to solve the fault section estimation problem. The objective function is to determine the faulty section. Two sample systems are tested under different scenarios in order to investigate multiple faults and incomplete information about the status of the protective devices. The proposed algorithm determines the faulty sections in all scenarios with satisfactory results. The proposed algorithm is compared with two other methods, which are genetic algorithm and ant colony algorithm. The ABC algorithm show more accuracy and higher computation efficiency than the other two methods especially in case of multiple faults. The proposed algorithm has a main advantage than other techniques that it has only two parameters to be tuned which are the colony size and the maximum number of iterations. Therefore, the tuning of the two parameters towards the optimum values has a higher probability of success than other methods.

References

1. W. A. Fonseca, U. H. Bezerra, M. V.A. Nunes, Fabiola.G.N Barros, and J. A. P. Moutinho "Simultaneous Fault Section Estimation and Protective Device Failure Detection Using Percentage Values of the Protective Devices Alarms", *IEEE Transactions on Power Systems,* vol. 28, no. 1, February 2013.
2. C. Fukui and J. Kawakami, "An expert system for fault section estimation using information from protective relays and circuit breakers," *IEEE Trans. Power Delivery*, vol. PD-1, pp. 83–90, Oct. 1986.
3. Y. C. Huang, "Fault section estimation in power systems using a novel decision support system," *IEEE Trans. Power Syst.,* vol. 17, no. 2, pp. 439–444, May 2002.
4. H. J. Cho and J. K. Park, "An expert system for fault section diagnosis of power systems using fuzzy relations, "*IEEE Trans. Power System*, vol. 12, no. 1, pp. 342–348, Feb. 1997.
5. W. H. Chen, C. W. Liu, and M. S. Tsai, "On-line fault diagnosis of distribution substations using hybrid cause-effect network and fuzzy rule-based method," *IEEE Trans. Power Delivery*, vol. 15, no. 2, pp. 710–717, Apr. 2000.
6. Ching-Lai Hor, Peter A. Crossley, and Simon J. Watson, "Building knowledge for substation-based decision support using rough sets" *IEEE Trans. Power Delivery*, vol. 22, no. 3, pp. 1372–1379, July. 2007.
7. K. L. Lo et al., "Power systems fault diagnosis using Petri nets," in *Proc. Inst. Elect. Eng. ener. Transm. Distrib.*, vol. 144, May 1997, pp. 231–236.
8. Xu Luo, and Mladen Kezunovic," Implementing fuzzy reasoning petri-nets for fault section estimation" *IEEE Trans. Power Delivery,* vol. 23, pp. 676–685, April. 2008.
9. Z. E. Aygen, S. Seker, and M. Bagriyanik, "Fault section estimation in electrical power systems using artificial neural network approach," in *Proc. IEEE Trans. Dist. Conf.*, vol. 2, pp. 466–469,1999.

10. Ghendy Cardoso, Jr., Jacqueline Gisèle Rolim, and Hans Helmut Zürn. "Application of Neural-Network modules to electric power system fault section estimation". *IEEE Trans. Power Delivery,* vol. 19, pp. 1034–1038, Oct. 2004.
11. Zhu Yongli, Huo Limin, and Lu Jinling, "Bayesian Networks-based approach for power systems fault diagnosis" *IEEE Trans. Power Delivery*, vol. 21, pp. 634–638, April. 2006.
12. Oyama T. "Fault section estimation in power system using Boltzmann machine". *Proceedings of the Second International Forum on Application of Neural Networks to Power Systems* (ANNPS'93), 1993. p. 3–8.
13. F.S. Wen and Z. X. Han, "Fault section estimation in power systems using a genetic algorithm," *Electr. Power Syst. Res.*, vol. 34, no. 3, pp. 165–172, 1995.
14. F.S. Wen and C.S. Chang, "Possibilistic-diagnosis theory for fault-section estimation and state identification of unobserved protective relays using Tabu-search method". *IEE Proc Generation, Transmissions Distrib* vol. 145 no. 6, pp. 722–730, 1998.
15. Leao F, Pereira RAF, Mantovani JRS. Fault section estimation in electric power systems using an artificial immune system algorithm. *Int. J Innov Energy Syst Power* 2009; 4:14–21.
16. Chang CS, Tian L, Wen FS. A new approach to fault section estimation in power systems using ant system. *Electr Power Syst Res* 1999; 49:63 70.
17. Shyh-Jier Huang, Xian-Zong Liu, Wei-Fu Su, and Ting-Chia Ou, "Application of Enhanced Honey-Bee Mating Optimization Algorithm to Fault Section Estimation in Power Systems", *IEEE Transactions on Power Delivery,* vol. 28, no. 3, July 2013.
18. D. Karaboga, An idea based on honey bee swarm for numerical optimization, Technical Report TR06, *Computer Engineering Department, Erciyes University*, Turkey, 2005.
19. D. Karaboga and B. Basturk, *Artificial Bee Colony (ABC) Optimization Algorithm for Solving Constrained Optimization Problems*. Berlin, Germany: Springer-Verlag, 2007, vol. LNAI 4529, pp. 789–798.
20. Fahad S. Abu-Mouti, and M. E. El-Hawary, "Optimal Distributed Generation Allocation and Sizing in Distribution Systems via Artificial Bee Colony Algorithm", *IEEE Transactions on Power Delivery,* vol. 26, no. 4, October 2011.
21. Lai LL, Sichanie AG, Gwyn BJ. Comparison between evolutionary programming and a genetic algorithm for fault-section estimation. *IEE Proc: Gener Transmiss Distrib* 1998; 145:616–20.
22. Srinivasa Rao R, Narasimham SL, Ramalingaraju M. Optimization of distribution network configuration for loss reduction using artificial bee colony algorithm. *Int J Electr Power Energy Syst Eng 2008*; 1:116–22.
23. Hemamalini S, Simon SP. Artificial bee colony algorithm for economic load dispatch problem with non-smooth cost functions. *Electr Power Compon Syst 2010*; 38:786–803.

CSTS: Cuckoo Search Based Model for Text Summarization

Rasmita Rautray and Rakesh Chandra Balabantaray

Abstract Exponential growth of information in the web became infeasible for user to sieve useful information very quickly. So solution to such problem now a day is text summarization. Text summarization is the process of creating condensed version of original text by preserving important information in it. This paper presents for the first time a nature inspired cuckoo search optimization algorithm for optimal selection of sentences as summary sentence of intelligent text summarizer. The key aspects of proposed summarizer focus on content coverage and length while reducing redundant information in the summaries. To solve this optimization problem, this model uses inter-sentence relationship and sentence-to-document relationship by considering widely used similarity measure cosine similarity. The inputs for this model are taken from DUC dataset. Whereas the result is evaluated by ROUGE tool and compared with state-of-the-art approaches, in which our model in multi-document summarization have shown significant result than others.

Keywords Summarization · Content coverage · Length · Redundancy · Cuckoo search

1 Introduction

The aim of text summarization (TS) is to create a compressed version of text document preserving essential information, called summary. The main objective of summary is to express concepts in a document in less space [23]. A TS system is

R. Rautray (✉)
Department of Computer Science and Engineering, Siksha 'O' Anusandhan University, Bhubaneswar 751030, Odisha, India
e-mail: rashmitaroutray@soauniversity.ac.in

R.C. Balabantaray
Department of Computer Science, IIIT, Bhubaneswar, Odisha, India
e-mail: rakesh@iiit-bh.ac.in

S.S. Dash et al. (eds.), *Artificial Intelligence and Evolutionary Computations in Engineering Systems*, Advances in Intelligent Systems and Computing 517,
DOI 10.1007/978-981-10-3174-8_13

presumes to possess a high coverage of relevant contents and eliminate redundant information [13]. It can be classified into many ways depending on different dimensions such as source size, approach, language, purpose, and media. Based on source document size, summary extraction from a document is termed as single document summarization otherwise termed as multi document summarization. In other hand, both the single and multi-document summarization either of extractive or abstractive summarization. Selecting important sentences, paragraphs, etc. from the original document and concatenating them into shorter form is referred as extractive summarization, whereas an abstractive summarization attempts to develop an understanding of the main concepts in a document and then express those concepts in clear natural language. Abstractive summarization involves information fusion, sentence compression, and reformulation so summary of this method could be more concise compared to extractive summarization [2]. Based on language, summary depends on its input and out text. It can be categorized into three types. *Monolingual*: Here, the systems only accept documents with specific language, i.e., English and the generated summary is based on that language too. Thus, the input and output languages are same. *Multilingual:* In this case, the systems can accept documents in different languages and the users can choose the languages of the output summary. *Crosslingual:* The output and input languages differs from each other [18]. Therefore, implement point of view extractive monolingual summaries are more feasible and this paper is focus on extractive monolingual multi-document summarization.

For improvising the performance of text selection in document summarization using statistical tools, in 2000s a number of global optimization techniques were considered as effective solutions in literature [25] till date like particle swarm optimization (PSO), differential evolution (DE), genetic algorithm (GA). Unlike PSO, cuckoo search (CS) is another nature inspired algorithm has been proposed and gained a significant result in the field of optimization. In Dash et al. [11] convergence curve of various nature inspired algorithms have been shown and performance of CS for automatic generation control of thermal system problem shows significant result. Similarly in Mulani et al. [20] performance against inverse heat transfer problems is analyzed, clustering of web search results based on the cuckoo search algorithm is presented in Cobos et al. [10], also used for traffic signal controller [6] and reconfiguration of network for power loss minimization [21], finding vulnerability and mitigation techniques using CS for computer security problem in [29], achievement of minimized combinatorial test suite for software functional testing [1], hydrothermal scheduling using CS in Nguyen and Vo [22], energy conservation [12], solar radiation [26], satellite image segmentation using CS [7], etc. For the first time cuckoo search optimization algorithm is used for text summarization problem and it is being compared with state-of-the-art (recent advanced) approaches.

The structure of paper presents investigation of cuckoo search algorithm-based research work, proposed extractive summarization model framework, description of cuckoo search algorithm, simulation study, and lastly it addresses the conclusions.

2 Summarization Model

The framework of proposed summarization system involves three basic steps, such as preprocessing, processing, and summary generation is discussed in the following algorithm.

Overview of Summarization model

1. *Preprocessing*:

   1.1. *Sentence segmentation*
   1.2. *Tokenization*
   1.3. *Stop word removal*
   1.4. *Stemming.*

2. *Processing*:

   2.1. *Input representation*
   2.2. *Sentence informative score calculation*
   2.3. *Input optimization.*

3. *Summary generation*:

   3.1. *Sentence selection.*

In the preprocessing step, input to the summarizer is text document D are preprocessed by individual segmentation of sentences, i.e., $D = \{s_1, s_2, \ldots, s_n\}$, where s_i denotes ith sentence in the document, n is the number of sentences in document followed by term wise representation $T = \{t_1, t_2, \ldots, t_m\}$ represents all the distinct terms occurring in D, where m is the number of terms, removal of nonsignificant words and word stemming. Second in input representation step, the previous output is prepared for implementation of optimization problem by calculating sentence informative score. Informative score of each sentence is sum of the term frequencies of a sentence specifies the sentence weight. Importance of a sentence is highly related to a higher sentence weight value. The objective of generating summary of document sets to find subsets of sentences from the documents containing useful information, for this need to confine the data set space that covers most relevant information in the document. That can be measured by inter-sentence similarity. Inter-sentence similarity can be obtained by taking a sentence compared with other sentences using a similarity metric. Among a number of similarity matrices Cosine similarity is a popular similarity metric in the field of text summarization. For finding Cosine similarity sentences are represented as vectors. The most common model for vector representation of sentence is vector space model (VSM). Each sentence S_i in vector space model is represented in terms of weighted vector, $s_i = [w_{i1}, w_{i2}, \ldots, w_{im}]$, where w_{ik} is the weight of term t_k defined by using t_f (term frequency) present in sentence S_i. The cosine measure between two sentences $s_i = [w_{i1}, w_{i2}, \ldots, w_{im}]$ and $s_j = [w_{j1}, w_{j2}, \ldots, w_{jm}]$ is computed as:

$$\mathrm{Sim}\left(s_i, s_j\right) = \frac{\sum_{k=1}^{m} w_{ik} w_{jk}}{\sqrt{\sum_{k=1}^{m} w_{ik}^2 \sum_{k=1}^{m} w_{jk}^2}} \quad \text{where } i, j = 1, 2, \ldots, n \tag{1}$$

3 Cuckoo Search-Based Model

Cuckoo search (CS) is newly introduced and under exploration optimization algorithm. The concept of cuckoo search algorithm was inspired by the species of bird called the cuckoo. Cuckoos are fascinating birds because of their aggressive reproduction strategy and beautiful sounds, by which mature cuckoos lay their eggs in the nests of other host birds or species [19, 28]. The nest containing each egg represents a solution, and specifically each egg of a cuckoo represents a new solution. In the simplest form, each nest has one egg. The algorithm can be extended to more complicated cases in which each nest has multiple eggs representing a set of solutions. To produce a new egg in CS, Lévy flight is used as both local and global searches in the random solution domain. A Lévy flight contains successive random steps [15, 16], and is characterized by a sequence of rapid jumps elected from probability density function which has a power law tail. The step for generating a new egg can be represented by the following equation.

$$x_i^{t+1} = x_i^t + \alpha\theta \mathrm{Levy}(\lambda), \tag{2}$$

where α is step size, which should be proportional to scale of optimization problem (i.e., $\alpha > 0$), θ is entry wise move during multiplication and Levy(λ) is the Lévy distribution. For implementation of these concepts in text summarization problem, authors of [28] have employed the rules which have been discussed in the following.

Cuckoo Search is based on three idealized rules [27]:

1. One egg is laid by each cuckoo at a time in a random nest to represent a solution sets;
2. The best eggs contained in the nests will carry over to the next generation;
3. The number of available nests is fixed, and a host bird can discovered an alien egg with a probability $P_a \in (0, 1)$. If this condition satisfies, either the egg can be discarded or abandon the nest by the host, and built a new nest elsewhere.

In-addition, the overall steps involved in cuckoo search-based text summarization (CSTS) algorithm is discussed below:

Step 1 *Population initialization of host nests.*
Step 2 *Calculate sentence informative score.*
Step 3 *Evaluate fitness function to set best Gbest value.*

Step 4 *Generate new solutions using Levy flight.*
Step 5 *Calculate score of each sentence with maximum fitness.*
Step 6 *If iteration (it) at maxit then stop otherwise it = it + 1 and go to step-3.*
Step 7 *Select sentences with respect to threshold (i.e., 25, for suitable comparison of system generated summary with DUC summary) to generate summary sentences.*

4 Simulation Study

This section conduct experiments to test proposed summarization system empirically. For implementation and validation of our procedure, the suggested parameters like probability $p_a = 0.75$, $\alpha = 0.5$ and $\lambda = 0.8$ are used with maximum iteration = 200 to obtained the best result.

4.1 Problem Formulation

The objective of the TS problem is to maximize informativeness while reducing redundancy and preserving length of the generated summary. Therefore, the objective function $f(S)$ of TS problem formalize as three subfunctions, such as $f_{\mathrm{cov}}(S), f_{\mathrm{red}}(S)$ and $f_{\mathrm{len}}(S)$ to optimize summary.

$$f(S) = f_{\mathrm{cov}}(S) + f_{\mathrm{red}}(S) + f_{\mathrm{len}}(S) \tag{3}$$

$f_{\mathrm{cov}}(S)$: The content coverage of each sentence in the summary.

$$f_{\mathrm{cov}}(S) = \mathrm{Sim}(s_i, O) \quad i = 1, 2, \ldots, n, \tag{4}$$

where O = the center of the main content collection of sentences and similarity between s_i and O measures importance of the sentences.

$f_{\mathrm{red}}(S)$: The redundancy between the sentences in the summary.

$$f_{\mathrm{red}}(S) = 1 - \mathrm{Sim}(s_i, s_j) \quad i \neq j = 1, 2, \ldots, n \tag{5}$$

Higher value of $f_{\mathrm{red}}(S)$ specifies less overlap and vice versa.

$f_{\mathrm{len}}(S)$: Defines required summary length.

$$f_{\mathrm{len}}(S) = \mathrm{Sim}(s_i, s_j) \quad i \neq j = 1, 2, \ldots, n \tag{6}$$

Higher value of $f_{\mathrm{len}}(S)$ specifies higher diversity of the summary.

4.2 Dataset

The open bench mark datasets from DUC (Document Understanding Conference) were used for the evaluation of summarization result. Table 1 provides a short description of DUC data sets. For the evaluation of result, stop words of data sets were removed using provided stop list in net and the terms were stemmed using the most common stemmer in English called Porter's stemmer.

4.3 Evaluation Metric

For summary evaluation, ROUGE-1.5.5 package were used [17]. It is used as the official evaluation metric for text summarization. ROUGE includes different methods, such as ROUGE-L, ROUGE-N, ROUGE-S, ROUGE-W, and ROUGE-SU to measures the n-gram match between systems generated summaries and human summaries. Here, ROUGE-N metric compares N-grams of two summaries, and counts the number of matches:

$$\text{ROUGE} - N = \frac{\sum_{S \in \text{Summ}_{\text{ref}}} \sum_{\text{N-gram} \in S} \text{Count}_{\text{match}}(\text{N-gram})}{\sum_{S \in \text{Summ}_{\text{ref}}} \sum_{\text{N-gram} \in S} \text{Count}(\text{N-gram})} \tag{7}$$

where N stands for the length of the N-gram, count match (N-gram) is the highest number of N-grams co-occurring in candidate summary and reference-summaries. Count (N-gram) is the number of N-grams in the reference summaries.

4.4 Comparison with Different Methods

Here, we compare the performance of the proposed summarization method with other well-known query-based as well as generic-based baseline summarization methods discussed in Table 2.

Table 1 Dataset description

Data set parameters	Size (DUC2006)
Number of clusters	50
Number of documents in each clusters	25
Average no. of sent. per doc.	30.12
Maximum no. of sent. per doc.	79
Minimum no. of sent. per doc.	5
Data source	AQUAINT
Summary length (in words)	250

Table 2 Description of different summarization methods

Sl. No.	Methods	Summary type	Basic concept
1	MMR [8]	Query	Iteratively selection of a sentence with the highest similarity to the query and lowest similarity to the already selected sentences, in order to promote novelty
2	LEX [14]	Query	It focuses on four objectives namely content coverage, significance, cohesiveness and redundancy based on identified core terms and main topics
3	HybHSum [9]	Query	It is based on bi-step hybridization model: a generative model for hierarchical topic discovery and a regression model for inference
4	MCMR [3]	Query	It is an optimization based approach, which considers summarization model as an integer linear programming problem and it attempt to optimize redundancy and relevancy of the summary
5	LFIPP [5]	Generic	This approach uses both inter relationship between sentence-to-document and the sentence-to-sentence to select salient sentences from input documents and reduce redundancy in the summary
6	Centroid [24]	Generic	In this approach, the score for each sentence is calculated from a linear combination of the weights (centroid value, positional value and first sentence overlap)
7	MCLR [2]	Generic	It is considered as quadratic Boolean problem, based on the identified salient sentences and overlap information between selected sentences

4.5 Performance Analysis

This section analyses performance of our method and other state-of-the-art methods by using ROUGE-1 and ROUGE-2 metrics since they highly correlated with human judgments. ROUGE-1 measures the overlap of unigrams between the system summary and the manual summaries created by human while ROUGE-2 compares the overlap of bigrams [4].

The evaluation of methods (in Table 3) is based on content coverage, length, and nonredundancy of the sentences in the summary. With the comparison of average ROUGE values for different state-of-the-art methods, CSTS can achieve significant improvement. Here also a relative improvement $\frac{(\text{our}_{\text{\_method}}-\text{other}_{\text{\_methods}})}{\text{other}_{\text{\_methods}}} \times 100$ of methods has been shown for comparison.

Table 3 Methods ROUGE score on DUC2006 dataset

Methods	ROUGE 1	ROUGE 2	Improvement of CSTS (in %)	
			ROUGE 1	ROUGE 2
CSTS	**0.4311**	**0.1398**	–	–
MMR	0.3716	0.0757	16.01	84.67
LEX	0.4030	0.0913	6.97	53.12
HybHSum	0.4300	0.0910	0.59	53.62
MCMR	0.4184	0.0928	3.03	50.64
LFIPP	0.4209	0.0934	2.42	49.67
Centroid	0.3639	0.0726	18.46	92.56
MCLR	0.3975	0.0850	8.45	64.47

Bold value of CSTS model represents significantly better summary result in terms of ROUGE-1 and ROUGE-2 score than the state-of-the-art approaches

5 Conclusions

This work is devoted as multi-document summarization. As it is considered as optimization problem, to solve this cuckoo search algorithm is created. Cuckoo search (CS) algorithm is for the first time in text summarization to optimized summary by covering essential contents while reducing redundancy. The CS based result is compared with different existing state-of-the-art methods and performance is measured in terms of ROUGE score. This paper has found that the CSTS summary comparatively showed better than other state-of-the-art methods but this may not be valid for different datasets.

References

1. Ahmed BS, Abdulsamad TS, Potrus MY(2015) Achievement of minimized combinatorial test suite for configuration-aware software functional testing using the Cuckoo search algorithm. Information and Software Technology.66:13–29.
2. Alguliev RM, Aliguliyev RM, Hajirahimova MS(2012) GenDocSum + MCLR: Generic document summarization based on maximum coverage and less redundancy. Expert Systems with Applications. 39(16):12460–73.
3. Alguliev RM, Aliguliyev RM, Hajirahimova MS, Mehdiyev CA(2011a) MCMR: Maximum coverage and minimum redundant text summarization model. Expert Systems with Applications. 38(12):14514–22.
4. Alguliev RM, Aliguliyev RM, Isazade NR. CDDS(2013) Constraint-driven document summarization models. Expert Systems with Applications. 40(2):458–65.
5. Alguliev RM, Aliguliyev RM, Mehdiyev CA(2011b) Sentence selection for generic document summarization using an adaptive differential evolution algorithm. Swarm and Evolutionary Computation. 1(4):213–22.

6. Araghi S, Khosravi A, Creighton D(2015) Intelligent cuckoo search optimized traffic signal controllers for multi-intersection network. Expert Systems with Applications. 42(9):4422–31.
7. Bhandari AK, Singh VK, Kumar A, Singh GK(2014) Cuckoo search algorithm and wind driven optimization based study of satellite image segmentation for multilevel thresholding using Kapur's entropy. Expert Systems with Applications. 41(7):3538–60.
8. Carbonell J, Goldstein J(1998) The use of MMR, diversity-based reranking for reordering documents and producing summaries. In Proceedings of the 21st annual international ACM SIGIR conference on Research and development in information retrieval (335–336). ACM.
9. Celikyilmaz A, Hakkani-Tur D(2010) A hybrid hierarchical model for multi-document summarization. In Proceedings of the 48th Annual Meeting of the Association for Computational Linguistics (pp. 815–824). Association for Computational Linguistics.
10. Cobos C, Muñoz-Collazos H, Urbano-Muñoz R, Mendoza M, León E, Herrera-Viedma E (2014) Clustering of web search results based on the cuckoo search algorithm and balanced Bayesian information criterion. Information Sciences. 281:248–64.
11. Dash P, Saikia LC, Sinha N(2014) Comparison of performances of several Cuckoo search algorithm based 2DOF controllers in AGC of multi-area thermal system. International Journal of Electrical Power & Energy Systems. 55:429–36.
12. Dos Santos Coelho L, Klein CE, Sabat SL, Mariani VC(2014) Optimal chiller loading for energy conservation using a new differential cuckoo search approach. Energy. 75:237–43.
13. He R, Qin B, Liu T(2012) A novel approach to update summarization using evolutionary manifold-ranking and spectral clustering. Expert Systems with Applications. 39(3):2375–84.
14. Huang L, He Y, Wei F, Li W(2010) Modeling document summarization as multi-objective optimization. InIntelligent Information Technology and Security Informatics (IITSI), 382–386.
15. Kanagaraj G, Ponnambalam SG, Jawahar N(2014) Reliability-based total cost of ownership approach for supplier selection using cuckoo-inspired hybrid algorithm. The International Journal of Advanced Manufacturing Technology. 1–6.
16. Lim WC, Kanagaraj G, Ponnambalam SG(2012) Cuckoo search algorithm for optimization of sequence in pcb holes drilling process. In Emerging trends in science, engineering and technology (207–216). Springer India.
17. Lin CY, Hovy E(2003) Automatic evaluation of summaries using n-gram co-occurrence statistics. In Proceedings of the 2003 Conference of the North American Chapter of the Association for Computational Linguistics on Human Language Technology-Volume 1 (71–78). Association for Computational Linguistics.
18. Lloret E, Palomar M(2012) Text summarisation in progress: a literature review. Artificial Intelligence Review. 37(1):1–41.
19. Mohamad AB, Zain AM, Nazira Bazin NE(2014) Cuckoo search algorithm for optimization problems—a literature review and its applications. Applied Artificial Intelligence. 28(5):419–48.
20. Mulani K, Talukdar P, Das A, Alagirusamy R(2015) Performance analysis and feasibility study of ant colony optimization, particle swarm optimization and cuckoo search algorithms for inverse heat transfer problems. International Journal of Heat and Mass Transfer. 89:359–78.
21. Nguyen TT, Truong AV(2015) Distribution network reconfiguration for power loss minimization and voltage profile improvement using cuckoo search algorithm. International Journal of Electrical Power & Energy Systems. 68:233–42.
22. Nguyen TT, Vo DN(2015) Modified cuckoo search algorithm for short-term hydrothermal scheduling. International Journal of Electrical Power & Energy Systems. 65:271–81.
23. Radev DR, Hovy E, McKeown K(2002) Introduction to the special issue on summarization. Computational linguistics. 28(4):399–408.
24. Radev DR, Jing H, Styś M, Tam D(2004) Centroid-based summarization of multiple documents. Information Processing & Management. 40(6):919–38.

25. Rautray R, Balabantaray RC(2015) Comparative Study of DE and PSO over Document Summarization. In Intelligent Computing, Communication and Devices,371–377), Springer India.
26. Wang J, Jiang H, Wu Y, Dong Y(2015) Forecasting solar radiation using an optimized hybrid model by Cuckoo Search algorithm. Energy. 81627–44.
27. Yang XS, Deb S(2009) Cuckoo search via Lévy flights. In Nature & Biologically Inspired Computing, 2009. NaBIC 2009. (210–214). IEEE.
28. Yang XS, Deb S(2010) Engineering optimisation by cuckoo search. International Journal of Mathematical Modelling and Numerical Optimisation. 1(4):330–43.
29. Zineddine M(2015) Vulnerabilities and mitigation techniques toning in the cloud: A cost and vulnerabilities coverage optimization approach using Cuckoo search algorithm with Lévy flights. Computers & Security. 48:1–8.

A New 2D Shape Descriptor Generation Method for Different Craters Based on the Intensity Values

R. krishnan and Andhe Dharani

Abstract The topographical features like craters are forming in different size and shape in different categories that will give more information about the planet. The 2D shape descriptor from top view of the crater is very difficult to interpret the crater type. Here 2D shape descriptor is developed based on the intensity values of the crater image. The reverse order summation and smoothing function is applied to these intensity values, which will depict the cross-section of the crater image. This can be used for classification, 3D model development, and to retrieve other information.

Keywords Shape descriptor · Crater detection · Smoothing

1 Introduction

This paper deals with different shape descriptors for crater detection and its issues. The literature survey describes the different shape descriptors used for crater detection. The major problem is related to detection and classification of craters because of its irregularity in size and shape for various types.

In this work, the 2D shape descriptor is formed based on the intensity values of the crater image, here all the values of the image is not considering; instead of that only, central intensity values across the crater is considered. Depending on the

R. krishnan (✉)
VTU Research Centre, Department of MCA,
R. V. College of Engineering, Bangalore 560059, India
e-mail: krishnanr@rvce.edu.in

A. Dharani
Department of MCA, R. V. College of Engineering,
Bangalore 560059, India
e-mail: andhedharani@rvce.edu.in

S.S. Dash et al. (eds.), *Artificial Intelligence and Evolutionary Computations in Engineering Systems*, Advances in Intelligent Systems and Computing 517,
DOI 10.1007/978-981-10-3174-8_14

shades of the sunlight on different crater images, the intensity values are retrieved across the crater with the consideration of region 'line'. All the intensity values are retrieved where the 'line' is passed in all crater images. All four categories of the crater considered for work, are circular in shape, so if the 2D shape descriptor based on the top view of the crater, it is very difficult to distinguish; but if the consideration about central line, the intensity variation is almost depicting the structure of the crater. After applying summation and smoothing, the 2D shape descriptor is generated. These generated will compare with expected one, the crater image is detected. The 2D shape descriptor will compare with all standard descriptors of all four types of craters and best match is selected. This gives a real resemblance of the 3D object and differentiate each object in a better way compared to top view shape descriptor.

2 Literature Survey

Image feature extraction can be divided into two types: geometric and texture features. The geometric types are Haar-like features, Canny edge detector, Gabor-like primitives, and shape context descriptor explicitly record locations of edges/bars, etc. The texture features are depicted using histogram statistics [1]. Region-based descriptors Zernike moments (ZMs) and pseudo-Zernike moments were used as the feature sets. The ZM shape descriptor $f_s^I = \{\mathrm{ZM}_{mn}\}$ where order m and repetition n of the shape content along the radial and angular directions [2]. It is calculated using vector projection of I on the basis of ZM_{mn}; Due to interdependent features of f_s^I, the weighted Euclidean distance is used to find the dissimilarity between two f_s^Is of two images a and b, i.e.,

$$d_s(a,b) = \left\|(f_s^a - f_s^b)\right\|_2^{\wedge_s} \tag{1}$$

In DSR descriptor, Depth buffer descriptor, Silhouette descriptor, and Radical extent descriptor are used. To develop 2D views, the 3D object is projected perpendicular to hyper planes. A 2D FFT transform is applied to depth image to develop depth buffer images; for 3D model retrieval, here three silhouette images and six depth buffer images are extracted [3]. The ORG algorithm is a tree representation of a 3D image; here each voxel is a node and edges between nodes form the path between two voxels, these paths must have a least minimum intensity constraint. Let $\min(p_{ij})$ be the minimum intensity of each voxel in the path p between voxel i and j, and let g_{ij} be the path obtained by the path obtained by the graph traversal from node i to node j: it is guaranteed that $\min(g_{ij}) \geq \min(p_{ij})$ for every other p_{ij} [4]. A boundary is an enclosed region with coordinate pairs ($x0$, $y0$), ($x1$, $y1$), …, (xk, yk) … ($xN - 1$, $yN - 1$). These boundary coordinates are expressed as $x(k) = xk$ and $y(k) = yk$. The boundary is expressed as

$$s(k) = x(k) + jy(k) \tag{2}$$

where $k = 0, 1, 2, \ldots N - 1$; $j = \sqrt{-1}$.

The Fourier descriptors which uses Fourier transform to represent a boundary $s(k)$ in frequency domain

$$F(u) = \sum_{k=0}^{N-1} s(k)e^{-(j2\Pi uk/N)} \tag{3}$$

where $u = 0, 1, 2 \ldots, N - 1$. The complex coefficient $F(u)$ are the Fourier descriptors of the boundary $s(k)$ [5].

Other techniques for the shape descriptors are ellipticity, ratio between geodesic distances, curvature variance, salience variance, color curvature, color average value, average number of regional minima in the gradient image inside the object, perimeter, area, second, third, and fourth moments [6]. The scale invariant feature transform (SIFT) and Edge-SIFT applied in images that uses histogram of gradients and fuse them, results another shape descriptor method [7].

In symmetry-based shape descriptor, the midpoint of the geodesic path between planar curve β and its reflection β' under d_1 is symmetric in shape where

$$d_1([q_1]_1, [q_2]_1) = \min_{\gamma \in \Gamma} \mathrm{d}c(q_1, (q_2 o \gamma)\sqrt{\dot{\gamma}}) \tag{4}$$

and square root velocity function of a curve β is defined as q [8].

In contour-based shape feature extraction method, a shape is defined by $[x, y]$ coordinates of boundary coordinates. The object curve is represented in parametric equation form given as,

$$\Gamma(\mu) = [x(\mu) \quad y(\mu)] \tag{5}$$

where μ $\{1, 2, \ldots N\}$. Here x and y are related using with the parameter μ. The length of arc, centroid distance, and centroid angle are the parameters used to describe an object shape [9]. Another method shape-based transfer functions for volume visualization uses curve skeleton, decomposition of curved segments, merging of skeleton regions, incorporating curve skeleton, and volume data gives shape value to the skeleton regions [10].

3 Issues of Circle and Ellipse Shape Descriptor

The existing crater detection algorithms with circle or ellipse characterization that shows deviation from exact circle or ellipse circumference, and difficulties in setting the center point of the crater to fit circle or ellipse geometry because of the irregularity in crater shape. This intensity based gives the cross-sectional information of all four types of craters;. all four categories of craters with circle or ellipse circumference, with difference in middle portion. In simple crater, there is no feature present inside, whereas the central peak contains peak, domed crater contains dome, flat floor contains flat surface inside the crater. This can easily extracted by considering the cross-sectional view of the crater. This can be achieved by intensity values passed through the middle line of the crater image.

If the size of the crater is small, then the circle characterization shape descriptor is difficult to estimate the feature present inside crater, but these information easily can retrieve by considering the intensity values.

Consider the shape descriptor based on the shading effect of the crater which will vary depending on the sunlight. These information will vary for other crater images; as well as due to the sun directions the dark and light shading position in the crater will change, these issues are solved through this method.

4 New Methodology to Develop 2D Shape Descriptor

The steps to create a 2D shape descriptor as follows:

1. Consider region as 'line' and extract the intensity values across the crater image coordinates say (X_0, Y_0), (X_1, Y_1), (X_2, Y_2) … (X_n, Y_n) are I_0, I_1, … I_n
2. $$I_0^n = \sum_{k=0}^{n} (I_k + I_{n-k})$$
3. Plot the intensity summation values of I_0^n
4. Apply Smoothing function moving average of five point on I_0^n

$$SI_0^n = \sum_{k=0}^{n} \sum_{j=k}^{k+5} I_k$$

5. Plot the values of SI_0^n.

5 Result Analysis

Based on the shape of crater and the dark/light shade of the crater image due to sun light, the intensity graph will not depict the exact shape of the topographical object as shown in Fig. 1, region 'line' is considered here, so that other image pixel calculations are avoided.

By considering the intensity values from left to right and right to left in the region line for different craters, and the average intensity is plotted, as shown in Fig. 2.

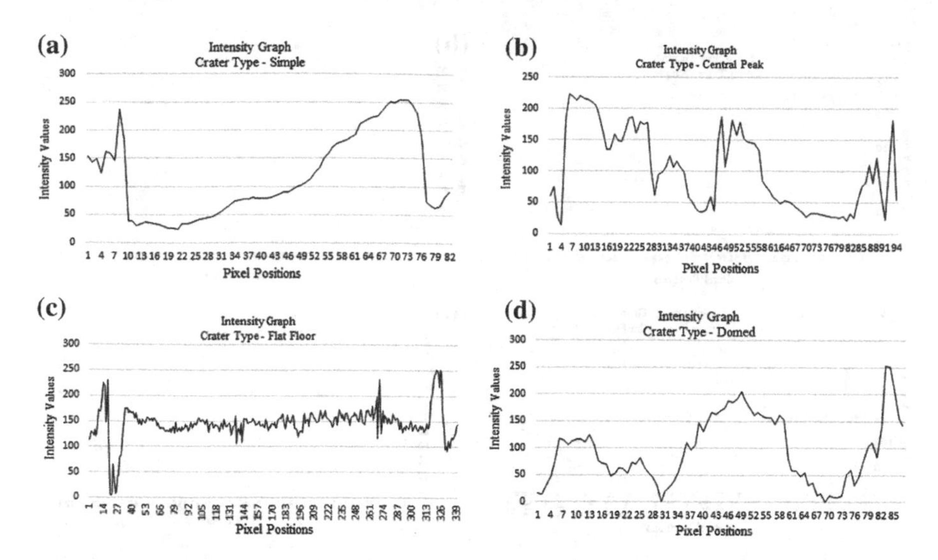

Fig. 1 Intensity graph with region line, for different types of craters **a** simple, **b** central peak, **c** flat floor and **d** domed

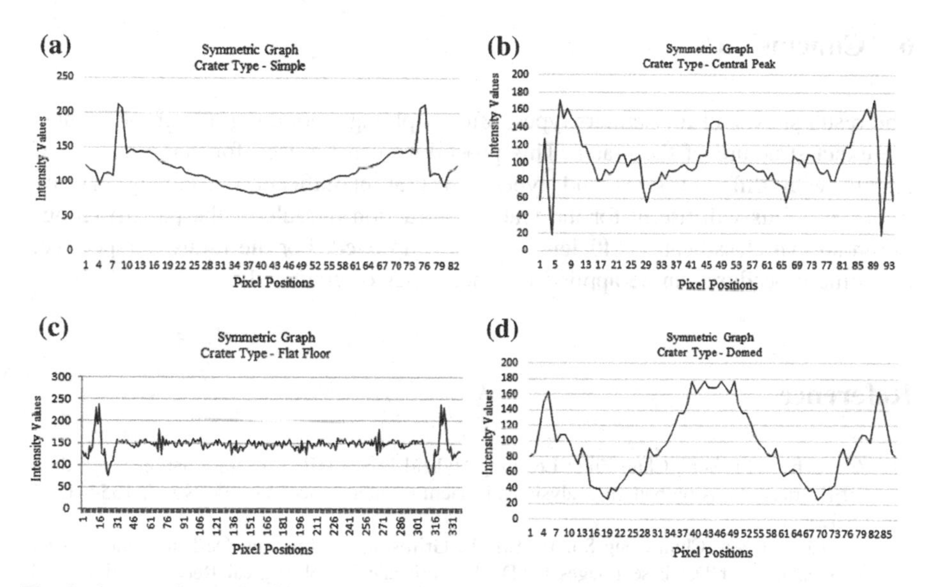

Fig. 2 Symmetric graph with region line, for different types of craters **a** simple, **b** central peak, **c** flat floor and **d** domed

The Smoothing is done by considering the average intensity values from left to right and right to left in the region line for different craters, and the average intensity is plotted, based on moving average of five points as shown in Fig. 3, which will depict the shape of the crater.

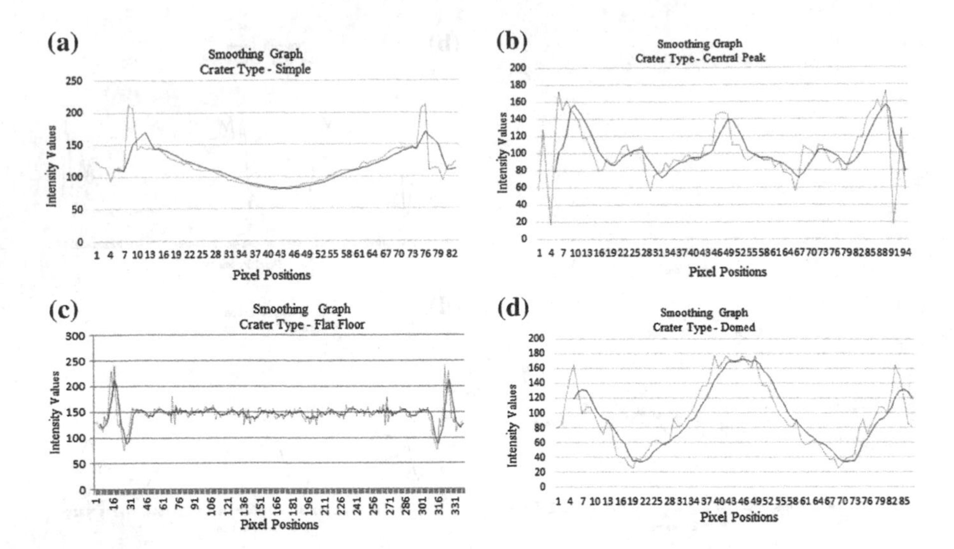

Fig. 3 Smoothing graph with region line, for different types of craters **a** simple, **b** central peak, **c** flat floor and **d** domed

6 Conclusion

The result shows all four craters types, after applying smoothing, the graph depicts the exact structure of the crater. This procedure can be used for classification of craters with different sizes and types, so that classification accuracy can be improved. This will useful for the feature extraction module of the pattern recognition so that classification efficiency can be improved. For the future perspective, the same procedure can be applied to other types of craters also.

References

1. Zhangzhang Si, Song-Chun Zhu.: Learning Hybrid Image (HIT) by Information Projection, IEEE Transactions on pattern analysis and machine intelligence, Vol 34, No. 7, 1354–1367 (2012).
2. Jiann-Jone Chen, Chun-Rong Su, W. Eric L. Grimson, Jun-Lin Liu, De-Hui Shiue: Object Segmentation of Database Images by Dual Multiscale Morphological Reconstructions and Retrieval Applications, IEEE Transactions on Image Processing, Vol 21, No. 2, 828–843 (2012).
3. Petros Daras, Apostolos Axenopoulos, Georgios Litos: 'Investigating the Effects of Multiple Factors Towards more accurate 3D Object Retrieval', IEEE Transactions on multimedia, Vol 14, No. 2, 374–388, (2012).
4. Marco Livesu, Fabio Guggeri, Riccardo Scateni: 'Reconstructing the Curve-Skeletons of 3D Shapes using the Visual Hull', IEEE Transactions on visualization and computer graphics, Vol. 18, No. 11, 1891–1901, (2012).

5. Yanjun Zhao, Saeid Belkasim, 'Multiresolution Fourier Descriptors for Multiresolution Shape Analysis', IEEE Signal Processing Letters, Vol. 19, No. 10, 692–695, (2012).
6. Celso T, N. Suzuki, Jancarlo F. Gomes, Alexandre X. Falcão, João P. Papa*, and Sumie Hoshino-Shimizu: Automatic Segmentation and Classification of Human Intestinal Parasites From Microscopy Images', IEEE Transactions on biomedical Engineering, Vol. 60, No. 3, 802–812, (2013).
7. Lingxi Xie, Qi Tian, Meng Wang, Bo Zhang, 'Spatial Pooling of Heterogeneous Features for Image Classification', IEEE Transactions on Image Processing, Vol. 23, No. 5, 1994–2008, (2014).
8. Sebastian Kurtek, Mo Shen, Hamid Laga, 'Elastic Reflection Symmetry Based Shape Descriptors', IEEE Trans, 293–300, (2014).
9. P. Arjun, T.T Mirnalinee, M. Tamilarasan, 'Compact Centroid Distance Shape Descriptor based on Object Area Normalization', IEEE International Conference on Advanced Communication Control and Computing Technologies (ICACCCT), 1650–1655, (2014).
10. Jorg-Stefan Prabni, Timo Ropinski, Jorg Mensmann, Klaus Hinrichs, 'Shape Based Transfer functions for volume visualizations', IEEE Pacific Visualization Symposium, 9–16, 2010.

A TEP-Based Approach for Optimal Thrust Direction of Lunar Soft Landing

Naveen Pragallapati and N.V.S.L. Narasimham

Abstract Determination of optimal thrust direction or steering angle for lunar soft landing trajectory is attempted in this article. The problem is complex due to the presence of system constraints and local minima. An exhaustive search of optimal thrust direction incurs high computational costs. The problem was solved as an optimum initial value estimation problem. Taboo evolutionary programming (TEP) is utilized to compute the optimal estimates. The study gives the integration of TEP technique in solving the governing nonlinear differential equations where a control parameter involved. The results are compared with available results in literature and it shows that the solution based on TEP algorithm is comparable to the counterpart. Further, sensitivity of design parameters such as terminal altitude, true anomaly, and terminal velocity over the final landing mass at the touch down is also examined.

Keywords Evolutionary programming · Soft landing · Trajectory optimization · Parameter control

1 Introduction

Exploration of lunar environments and its resources has begun in 50s. Generally, Landing on the Moon is broadly classified into two ways namely the hard landing and the soft landing. The soft landing approach has several practical uses against the hard landing approach in terms of ground experiments, surface mapping, etc. Following the literature [1–3] lunar soft landing process begins from a perilune altitude of 15 km for landing on to the surface of Moon. Further, the mass of the

N. Pragallapati (✉) · N.V.S.L.Narasimham
Department of Humanities & Mathematics, G. Narayanamma Institute of Technology & Science, Shaikpet, Hyderabad 500104, Telangana, India
e-mail: naveen@gnits.ac.in

N.V.S.L.Narasimham
e-mail: nvsl_simham@yahoo.co.in

S.S. Dash et al. (eds.), *Artificial Intelligence and Evolutionary Computations in Engineering Systems*, Advances in Intelligent Systems and Computing 517,
DOI 10.1007/978-981-10-3174-8_15

lunar module is always limited, it is extremely important that the fuel consumption is minimized. Several optimization algorithms have been proposed to solve lunar soft landing problem for the last two decades. A temporal finite element method [4] was followed to solve the optimal control problem of lunar soft landing. Several strategies for soft landing [5] from a lunar parking orbit were examined using controlled random search. Rapid trajectory optimization for a soft lunar landing using a direct collocation method was attempted in [6]. Some other researchers [6–10] employed intelligent algorithms. These intelligent algorithms are inspired by biological behavior or physical phenomena because of applying efficient stochastic searching techniques.

In 2006, a new method of global optimization technique, Taboo evolutionary programming (TEP) motivated by combining Taboo Search (TS) and Evolutionary programming (EP) was first introduced in [11]. TEP essentially combines the features of an EP, called single-point mutation [12] with TS. The results were found to be in good agreement with that of analytical results. Determination minimum energy opportunities for interplanetary trajectories using TEP were attempted [13, 14]. In these works TEP algorithm successfully identified the global optimum in terms of total energy requirements for the trajectory transfer. The results were compared with genetic algorithm and it indicates that TEP is viable alternative optimization technique while identifying the global optimum.

In the present study, Lunar module soft landing with minimal fuel consumption control strategy using TEP has been carried out. The TEP is a stochastic algorithm. A two-dimensional polar coordinate approach of lunar soft landing problem is considered in this study. The problem is transformed into a two-point boundary value problem (TPBVP) which is amenable for integration. TEP algorithm is utilized to obtain the solution of TPBVP, which subsequently results the optimal control strategy for lunar soft landing.

1.1 Model Equations

The planar motion of the lander probe [15] is represented by the variables, r, ϕ, u, v, where r is the radial distance from the center of the Moon; ϕ is the range angle or true anomaly; u is the horizontal velocity; and v is the vertical velocity. The assumptions while deriving the governing equations include: Target landing location lie in the orbital plane, gravity field of the Moon is spherical.

The equations describing the motion of the probe are

$$\dot{r} = v \tag{1}$$

$$\dot{\phi} = \frac{u}{r} \tag{2}$$

$$\dot{u} = -\frac{uv}{r} + \frac{T}{M}\cos\beta \tag{3}$$

$$\dot{v} = \frac{u^2}{r} - \frac{u}{r^2} + \frac{T}{M}\sin\beta, \tag{4}$$

where T is the thrust acting on the spacecraft, M is the instantaneous mass of the spacecraft, and β is the control angle measured clockwise (Fig. 1) from the local horizontal to the thrust direction.

For an optimal design of the soft landing trajectory, the time profile of the control variable (i.e., the thrust angle $\beta(t)$ with t being the time in seconds) is to be determined. In order to obtain the control variable profile, Pontryagin's minimum principle is used. Subsequently, the two-point boundary value problem in terms of the optimal state and the co-state vectors is obtained to compute $\beta(t)$. The resulted model equations are as follows:

$$\dot{p}_r = \frac{p_\phi u}{r^2} + \frac{p_v u^2}{r^2} - \frac{p_u uv}{r^2} - 2\frac{p_v u}{r^3} \tag{5}$$

$$\dot{p}_\phi = 0 \tag{6}$$

$$\dot{p}_u = -\frac{p_\phi}{r} - 2\frac{p_v u}{r} + \frac{p_u v}{r} \tag{7}$$

$$\dot{p}_v = -p_r + \frac{p_u u}{r}, \tag{8}$$

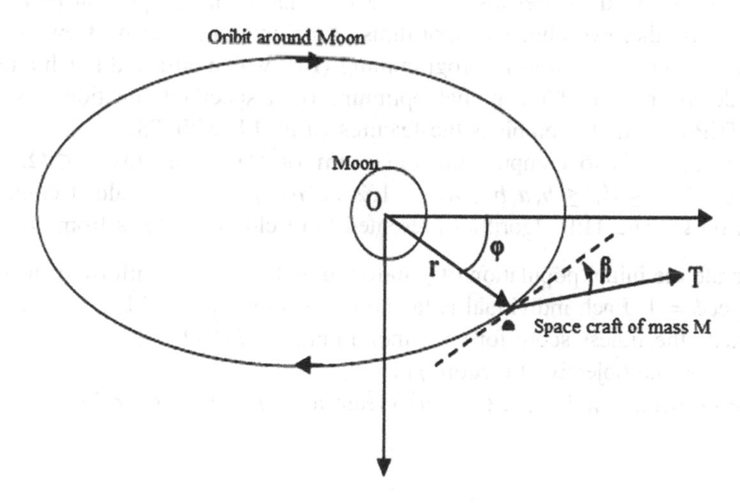

Fig. 1 Polar coordinate system of lunar soft landing

where p_r, p_ϕ, p_u, p_v be the co-state variables.

The boundary conditions are as follows:

$$\left.\begin{array}{llll} r(t_i) = r_i(\text{given}), & \phi(t_i) = \phi_i(\text{given}), & u(t_i) = u_i(\text{given}), & v(t_i) = v_i(\text{given}), \\ p_r(t_i) = ?, & p_\phi(t_i) = 0, & p_v(t_i) = ?, & p_u(t_i) = ?, \\ r(t_f) = r_f(\text{given}), & \phi(t_f) = \text{free}, & u(t_f) = u_f(\text{given}), & v(t_f) = v_f(\text{given}), \\ p_r(t_f) = \text{free}, & p_\phi(t_f) = 0, & p_v(t_f) = \text{free}, & p_u(t_f) = \text{free}, \end{array}\right\} \quad (9)$$

while the optimal thrust direction is described by equation

$$\beta = \tan^{-1}\left(\frac{p_v}{p_u}\right) \quad (10)$$

Here the suffix 'i' denotes the specified initial value and 'f' denotes the specified terminal values of the variable. Thus in order to solve optimal control problem one needs to solve TPBVP governed by the set of differential equations (1)–(8) along with the boundary conditions (9). This requires an appropriate initial values of $p_r(t_i)$, $p_v(t_i)$, and $p_u(t_i)$. Computing optimum initial values were done by Taboo Evolutionary Programming (TEP).

1.1.1 Taboo Evolutionary Programming

In 1995, Taboo (or 'Tabu' being a different spelling of the same word) search (TS), a stochastic optimization method, originally developed in [16, 17] and it was extended to continuous valued functions. Performance of TS with other methods is well studied earlier. The convergence [11] of TS for continuous function optimization is well studied. Results clearly reveal that TS technique can be a viable alternative to other evolutionary algorithms. An important branch of evolutionary algorithms (EA) is evolutionary programming (EP) which attracted much attention for the determination of the global optimum of a specified function. As noted earlier, TEP essentially combines the features of an EP with TS.

The objective is to compute the minimum of $f(x)$, such that $x \in \Omega$, where $\Omega = \{x \in R^n : a \leq x(j) \leq b, a, b \in R, i = 1, 2, \ldots n\}$, f is a real-valued continuous function on Ω. The TEP algorithm presented here closely follows from [11]:

1. Generate the initial population of μ individuals based on a uniform distribution, and set $k = 1$. Each individual is taken as a vector, $x_i\ \forall i \in \{1, 2, \ldots, \mu\}$.
2. Evaluate the fitness score for each individual, $x_i\ \forall i \in \{1, 2, \ldots, \mu\}$ of the population on the objective function, $f(x_i)$.
3. For each parent x_i, $i \in \{1, 2, \ldots, \mu\}$ create a single offspring x_i' by

$$x_i'(j_i) = x_i(j_i) + \eta \cdot N_i(0,1), \quad \eta = \eta \cdot e^{-\alpha},$$

where j_i is randomly chosen from the set $\{1, 2, \ldots, n\}$ and the other components of x_i' are equal to the corresponding x_i's. $N(0, 1)$ denotes a normally distributed one-dimensional random number with a mean of zero and a standard deviation of one. Here, the parameter $\alpha = 1.01$. The initial value of η is $\frac{b-a}{2}$ and whenever $\eta < 10^{-4}$ then η is set to its initial value.

4. Calculate the fitness of each offspring x_i', $i \in \{1, 2, \ldots, \mu\}$.
5. Perform the search using the following improved paths:
   (a) Choose an improved path as follows. For each $i \in \{1, 2, \ldots, \mu\}$ if $f(x_i') \leq f(x_i)$ then $y_i = x_i'$, $d_i = x_i' - x_i$ is an improved path. Put a pair of vectors (y_i, d_i) into the set A.
   (b) Choose r best fitness individuals from the set A as the parents with improved paths. Note that $(y_m, d_m), m = 1, 2, \ldots, r$, where y_m is an objective variable and d_m is the corresponding improved path. Set $A = \phi$, where ϕ is a null set.
   (c) Calculate fitness: for each $m \in \{1, 2, \ldots, r\}$, $y_m' = y_m + \rho d_m$, $\rho = \rho \times e^{-\alpha}$. The initial value of ρ is 1 and whenever $\rho < 10^{-6}$, then ρ is set to its initial value.
   (d) For each, $m \in \{1, 2, \ldots, r\}$ if $f(y_m') \leq f(y_m)$, then set (y_m', d_m) as a parent of the next generation with improved search paths, and put into the set A.
   (e) Record the number τ of members in set A.
6. Choose the parents for the next generation.
   (a) Perform a comparison over the union of parents x_i and offspring $x_i', \forall i \in \{1, 2, \ldots, \mu\}$ for each individual, q opponents are chosen uniformly at random from all the parents and offspring. For each comparison, if the individual's fitness is equal to or greater than the opponent's, it scores a '*win*'. Select the $\mu - \tau$ individuals out of x_i and $x_i', \forall i \in \{1, 2, \ldots, \mu\}$, which have the most *wins* to be put into the set B.
   (b) Make the individuals x, x' and y from sets B and A the parents of the next generation. Set $B = \phi$.
7. Check the Taboo status as follows:
   (a) Record the current optimal fitness, f_k^* and the current optimal solution x_k^*.
   (b) When $k > L$ and for a specified $\sigma 1$ and σ_2 (sufficiently both are small real numbers) compare the optimal fitness of the current generation with the optimal fitness of the previous L generations. Thus, if $|f_k^* - f_{k-L}^*| \leq \sigma_1$, then $f^* = f_k^*, x^* = x_k^*$. Put the pair of vectors (f^*, x^*) into the taboo table ψ. The length of taboo table ψ is l.
   (c) For any (f^*, x^*) in ψ, if the current optimal fitness f_k^* and the optimal solution x_k^* satisfy the taboo conditions $|f_k^* - f^*| \leq \sigma_1$ and $\|x_k^* - x^*\| \leq \sigma_2$,

then generate the initial population of μ individuals and set new individuals as the kth generation.

8. Terminate if k is more than maximum number of generations. Otherwise, set $k = k + 1$ and go to Step (3).

1.1.2 Methodology

As discussed in Sect. 1.1 above, the solution of TPBVP is obtained by appropriate values for the co-state variables (i.e., p_r, p_v, and p_u) at the initial time. Here in this study p_r is fixed at the initial time. The terminal boundary conditions to be achieved are handled through a function and it is set as the objective function. The objective function to meet the terminal conditions at the final instant of time is formulated as

$$F = \left[u(t_f) - u_f\right]^2 + \left[v(t_f) - v_f\right]^2 \tag{11}$$

The function F must be zero if the terminal conditions are to be met. The methodology to solve TPBVP is given in Fig. 2. For each specified initial co-state values, the differential equations (1)–(4) and (5)–(8) are numerically integrated with the optimal thrust direction described by Eq. (10). The function value is evaluated from Eq. (11). The procedure will continue until the maximum number of generations is reached. For the present study, a maximum of 200 generations with a Taboo list of size 6 from 50 simulations is implemented. The optimal parameters (from 50 simulations) where the objective function attains extreme value are computed.

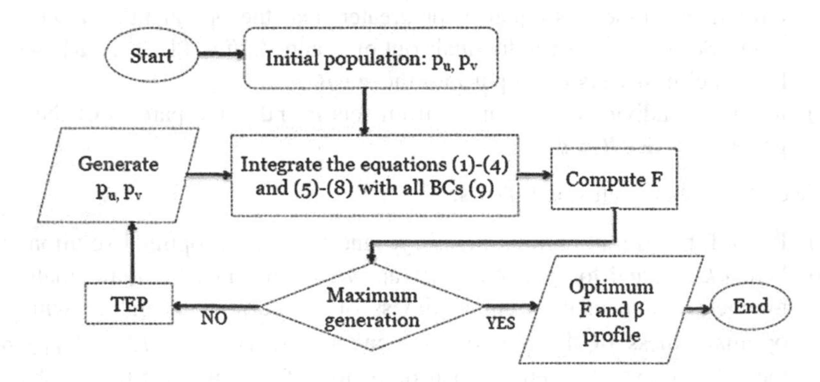

Fig. 2 Flowchart to compute optimum thrust angle direction for lunar soft landing

1.1.3 Computational Results

In the present study, the soft landing from an initial orbit has been carried out in one phase: A powered horizontal breaking phase which starts from an initial orbit, generally from the perilune of the initial orbit, and ends with zero altitude with zero horizontal velocity but a definite small vertical velocity. The formulation has been validated with the results available in the literature. An initial mass of 300 kg, 440 N thrusters with 310 s specific impulse is assumed here. Terminal vertical velocity of 5 m/s and a horizontal velocity of 0 m/s are the conditions achieved at touchdown. Table 1 gives a comparison of optimal payload mass at touch down from the lunar parking orbit. It could be seen that the methodology results compare very well with the literature results.

A detailed result for the orbit 100 × 15 km from where optimal control is exercised is given below. Initially, the space craft starts from the perilune of 100 × 15 km orbit. Terminal vertical velocity of 5 m/s and a horizontal velocity of 0 m/s are the conditions achieved at touchdown. The altitude profile of the optimal trajectory is plotted in Fig. 3. The optimal steering (or thrust angle, β) profile of the thrust direction is plotted in Fig. 4. It is clear from Table 1 for all the cases the optimum payload achieved by use of TEP is more when compared to its counterpart. Further analysis is carried out to assess the impact of the variations in

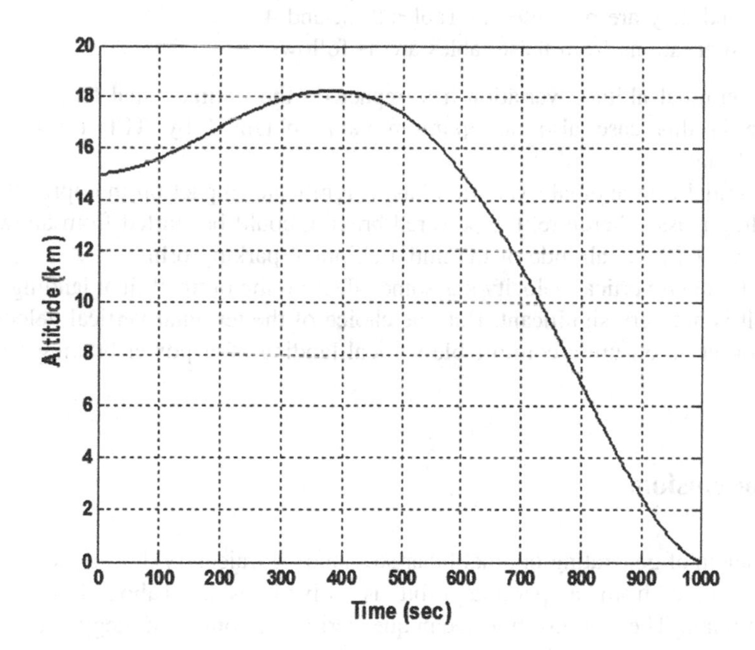

Fig. 3 Time history profile of altitude for optimal landing trajectory

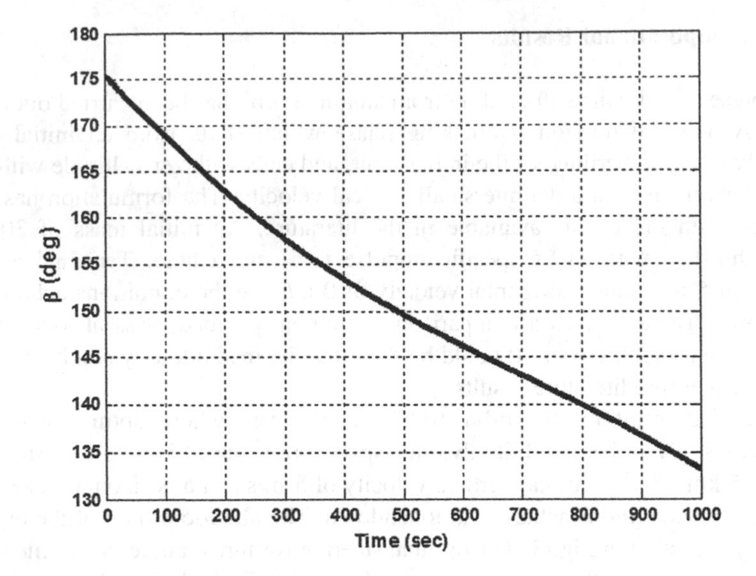

Fig. 4 Time history profile of optimal thrust direction

various parameters namely terminal altitude, initial true anomaly, terminal vertical velocity and they are presented in Tables 2, 3, and 4.

The observations from these tables are as follows:

1. The terminal altitude variation has impact on the optimal final landing mass. Here in this case also an optimum value obtained by TEP technique is significant.
2. The initial true anomaly variation has insignificant impact on the optimal final landing mass. Therefore, the powered braking could be started from anywhere near the perilune altitude of the initial elliptical parking orbit.
3. The terminal vertical velocity has some effect on the optimal final landing mass but it is not very significant. But, the choice of the terminal vertical velocity is important to ensure a comfortable vertical landing after power braking phase.

2 Conclusions

The problem of generating optimal lunar soft landing trajectory during the powered braking phase from a parking orbit is solved using Taboo Evolutionary Programming. The optimization technique derives a control strategy (i.e., thrust direction with respect to time) by estimating the co-state parameters at the beginning. The formulation is validated with those results available in literature. The

Table 1 Lunar parking orbits of different sizes and final landing masses at touchdown

Parameters	Size of lunar parking orbit (km)						
	100 × 100	100 × 50	100 × 25	100 × 20	100 × 15	100 × 10	100 × 5
Duration (s)	1037.01 (1036.99)	1016.801 (1012.06)	1008.28 (1001.14)	1006.720 (999.128)	1005.22 (997.146)	1003.762 (995.210)	1002.364 (993.340)
Landing mass (kg)	149.91 (149.766)	152.834 (152.126)	154.067 (153.144)	154.293 (153.323)	154.511 (153.498)	154.722 (153.671)	154.924 (153.845)

Numbers in parenthesis is from [5]

Table 2 Impact of terminal altitude on final mass (horizontal braking phase)

Parameters	Target altitude (km)				
	1.0	2.0	3.0	4.0	5.0
Duration (s)	984.22 (976.48)	983.47 (975.75)	982.73 (975.01)	981.99 (974.28)	981.25 (973.56)
Mass (kg)	157.550 (156.757)	157.659 (156.864)	157.766 (156.970)	157.873 (157.076)	157.979 (157.181)

Initial orbit: 100 × 15 km; Isp = 310 s; Thrust = 440 N; Initial Mass = 300 kg; Term. Vert. Velo = 0.005 km/s; Term. Horz. velo = 0 km/s; True anomaly = 0°

Table 3 Impact of initial true anomaly on final mass (horizontal braking phase)

Parameters	True anomaly			
	0.0	10.0	20.0	30.0
Duration (s)	1002.735	1002.610	1002.832	1003.433
Mass (kg)	154.870	154.888	154.856	154.769

Target Alt. = 3.0 km; initial orbit: 100 × 15 km; Isp = 310 s; Thrust = 440 N; Initial Mass = 300 kg; Term. Vert. Velo = 0.005 km/s; Term. Horz. velo = 0 km/s

Table 4 Impact of terminal vertical velocity during powered braking

Parameters	Terminal vertical velocity (m/s)					
	3	4	5	6	6.5	7
Duration (s)	1003.683	1003.208	1002.735	1002.266	1002.032	1001.799
Mass (kg)	154.733	154.802	154.870	154.938	154.972	155.006

Target Alt. = 3.0 km; initial orbit: 100 × 15 km; Isp = 310 s; Thrust = 440 N; Initial Mass = 300 kg; Term. Horz. velo = 0 km/s; True anomaly = 0.0°

results are very well compared. Optimal trajectories for powered braking phase from given altitude to touchdown is generated. Because a landing mission requires vertical landing, apart from near-zero velocities at the time of touchdown, different strategies are to be explored to identify an appropriate one.

References

1. Kawakatsu, Y., Kaneko, Y. and Takizawa, Y. (1998), Trajectory design of SELENE Lunar orbiting and Landing, AAS/AIAA Astrodynamics Conference, AAS 98–320.
2. Oono, H., Ishikawa, S., Nakajima, K., Hayashi, K. and Odaka, R. (1998), Navigation and Guidance Error Analysis of Lunar Lander Considering Orbit Determination Error, AAS Astrodynamics Conference AAS 98–319.
3. Zhou, Jingyang (2011), Optimal guidance and control in space technology, Ph.D. Curtin University, Department of Mathematics and Statistics.
4. Vasile, M. and Flobergghagen, R. (1998), Optimal Trajectories for Lunar Landing Missions, AAS Astrodynamics Conference. AAS 98–321.

5. Ramanan, R. V. and Lal, M. (2005), Analysis of optimal strategies for soft landing on the moon from lunar parking orbits, Journal of Earth System Science, 114, pp. 807–813.
6. Tu, L., Yuan, J., Luo, J., Ning, X., and Zhou, R. (2007), Lunar soft landing rapid trajectory optimization using direct collocation method and nonlinear programming, Proceedings of the 2nd International Conference on Space Information Technology, Wuhan, China.
7. Alireza Askarzadeh and Alireza Rezazadeh (2012), An Innovative Global Harmony Search Algorithm for Parameter Identification of a PEM Fuel Cell Model, IEEE Transactions on industrial electronics, 59(9), pp. 3473–3480.
8. Ruben Barros Godoy, Joao O. P. Pinto, Carlos Alberto Canesin, Ernane Antonio Alves Coelho, and Alexandra M. A. C. Pinto (2012), Differential-Evolution-Based Optimization of the Dynamic Response for Parallel Operation of Inverters With No Controller Interconnection, IEEE Transactions on industrial electronics, 59(7), pp. 2859–2866.
9. Radu-Emil Precup, Radu-Codrut David, Emil M. Petriu, Stefan Preitl, and Mircea-Bogdan Radac (2012), Fuzzy Control Systems With Reduced Parametric Sensitivity Based on Simulated Annealing, IEEE Transactions on industrial electronics, 59(8), pp. 3049–3061.
10. Songtao Chang, Yongji Wang, and Xing Wei (2013), Optimal Soft Lunar Landing Based on Differential Evolution, Proceedings - IEEE International Conference on Computer Science and Automation Engineering, 3, pp. 152–156.
11. Mingjun Ji, Jacek Klinowski (2006), Taboo evolutionary programming: a new method of global optimization, Proc. R. Soc. A 462, pp. 3613–3627.
12. Ji, M., Tang, H. & Guo, J. (2004), A single-point mutation evolutionary programming, Inf. Process. Lett. 90, pp. 293–299.
13. Mutyalarao, M., Sabarinath, A. and Xavier James Raj, M. (2011), Taboo Evolutionary Programming Approach to Optimal Transfer from Earth to Mars, SEMCCO 2011, Part II, LNCS 7077, Springer-Verlag, Berlin, pp. 122–131.
14. Mutyalarao, M., and Xavier James Raj, M., (2013), Optimal Velocity Requirements for Earth to Venus Mission Using Taboo Evolutionary Programming, SEMCCO 2013, Part I, LNCS 8297, pp. 762–772.
15. Kirk, D. E. (1970), Optimal Control Theory: An Introduction, Prentice Hall.
16. Glover, F (1989), Tabu search- part I, ORSA Journal of Computing, 1, pp. 190–206.
17. Glover, F. (1990), Tabu search-part II, ORSA Journal of Computing, 2, pp. 4–32.

Pre-filters Based Synchronous Rotating Reference Frame Phase Locked Loop (SRF PLL) Design for Distorted Grid Conditions

K. Sridharan, B. Chitti Babu, B. Naga Parvathi and P. Kartheek

Abstract In this paper presents the analysis of pre-filters based SRF PLL for distorted grid conditions. An exact detection of phase and fundamental frequency of grid current is essential for the control algorithm of grid connected power converter circuit. A control model of the SRF PLL is developed and is made on tuning the system under distorted grid conditions like harmonics and dc offsets. The SRF PLL can be completely implemented in software with and without pre-filters. The pre-filters are band pass and high pass filters these can be used to reduce harmonics present in the input and high pass filter alone can reduce the dc offset present in the input. The effects of high pass and band pass filters on dc offsets are analyzed. The superior performance of proposed pre-filter-based SRF PLL phase detection system is studied and the obtained results are compared with SRF PLL-based phase detection to confirm the feasibility of the study under different grid environment such as high-harmonic injection and dc offset. All analytical results are verified using MATLAB software.

Keywords Phase locked loop · Band pass filter · High pass filter · Total harmonic distortion

K. Sridharan
Department of Electrical Engineering, Saveetha School of Engineering, Saveetha University, Chennai, India
e-mail: srimaky@yahoo.com

B. Chitti Babu
Department of Electrical Engineering, The University of Nottingham Malaysia Campus, Semenyih, Malaysia
e-mail: bcbabunitrkl@ieee.org

B. Naga Parvathi · P. Kartheek (✉)
Department of Electrical Engineering, Saveetha School of Engineering, Chennai, India
e-mail: kc9777@gmail.com

B. Naga Parvathi
e-mail: nagaparvathi8@gmail.com

S.S. Dash et al. (eds.), *Artificial Intelligence and Evolutionary Computations in Engineering Systems*, Advances in Intelligent Systems and Computing 517,
DOI 10.1007/978-981-10-3174-8_16

1 Introduction

Phase locked loop is the control system which gives the output phase angle that matches input phase angle. Phase locked loops are widely used in grid connected power converter circuits to synchronize the grid voltage. An ideal PLL can provide the quick and accurate information of synchronization. PLL is most widely using synchronization technique for time-varying signal [1]. To ensure safe and reliable operation of power system based on new and renewable sources, usually power system operators should satisfy the grid code requirements such as grid stability, fault ride through, power quality improvement, grid synchronization, and power control, etc. [2]. In PLL, the difference between the input phase angle and the output phase angle is detected by the phase detector and passes to the loop filter and output of the loop filter drives the VCO (voltage controlled oscillator) and provides the phase angle which follows the input phase angle [3].

There are several advanced single phase and three phase PLL's proposed to bring the improvement in the performance of PLL during non-idle grid voltages and some topologies of PLL address dc offset problem. However the SRF-PLL implementation is simple over the advanced PLL's. SRF PLL will save considerable digital resources and computation time reduces for any low-end digital controller used for the implementation. So it is necessary to learn detailed design for the basic SRF-PLL when the input contains harmonics and dc offsets. The SRF- in PLL design objectives can be stated as [4–7]

(i) The response time and settling time for the SRF-PLL must be least for given worst case of dc offset input.
(ii) SRF PLL should satisfy the grid interconnection standards such as IEEE 1547-2003 must produce unit vectors

The phase and magnitude errors must be negligible when there is deviation in frequency of the grid voltage.

1.1 Stationary Reference Frame V_α and V_β

The phase angle is tracked by transferring the three phase voltages V_a, V_b, V_c to V_α, V_β two phase stationary system [5]. The three phase voltages are

$$V_a = V_m \sin\theta \tag{1}$$

$$V_b = V_m \sin(\theta - 120) \tag{2}$$

$$V_c = V_m \sin(\theta + 120) \tag{3}$$

The general αβ transformation matrix is

$$\begin{bmatrix} V_\alpha \\ V_\beta \end{bmatrix} = \begin{bmatrix} \frac{2}{3} & -\frac{1}{3} & -\frac{1}{3} \\ 0 & \frac{1}{\sqrt{3}} & -\frac{1}{\sqrt{3}} \end{bmatrix} \begin{bmatrix} V_a \\ V_b \\ V_c \end{bmatrix} \tag{4}$$

$$V_a = V_m \sin\theta \tag{5}$$

$$V_a = V_m \cos\theta \tag{6}$$

1.2 *Synchronous Rotating Reference Frame (SRF)*

By synchronizing the voltage space vector along quadrature (q) axis or direct (d) axis phase angle can be tracked. If the voltage space vector is synchronized with quadrature axis [4]. It means d axis is made zero as shown in Fig. 1.

If the voltage space vector is synchronized with q axis. Then, the transformation matrix is

$$\begin{bmatrix} V_d \\ V_q \end{bmatrix} = \begin{bmatrix} \cos\theta_e & \sin\theta_e \\ -\sin\theta_e & \cos\theta_e \end{bmatrix} \begin{bmatrix} V_\alpha \\ V_\beta \end{bmatrix} \tag{7}$$

The estimated phase angle θ_e is the integral of the estimated frequency ω_e which is the sum of PI-controller output and feed forward frequency ω_{ff}. PI regulator gain is designed, so that V_d follows the reference value $V_d = 0$. This results in an output estimated phase angle θ_e equals to input phase angle θ (Fig. 2).

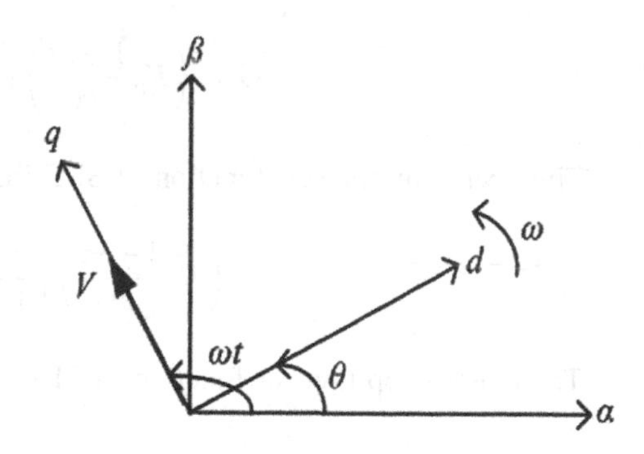

Fig. 1 Synchronous rotating reference frame

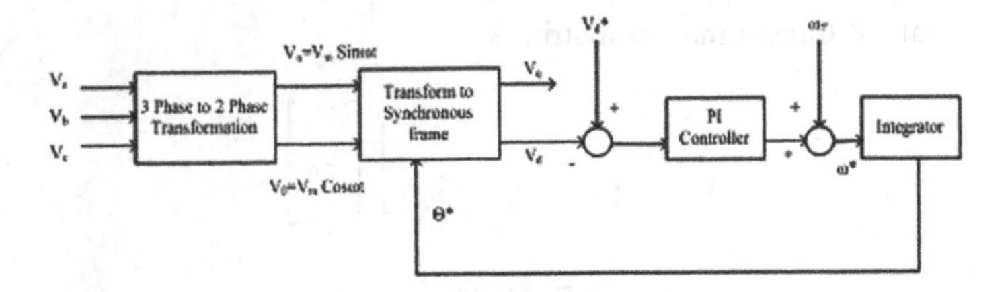

Fig. 2 Block diagram of SRF PLL

2 PI Controller

PI controller is normally used to eliminate the steady state error which results from P controller. However, in terms of the speed of the response, the overall stability of the system has a negative impact. This controller is mostly used in areas where speed of the system is not an issue. Since PI controller has no ability to estimate the future errors of the system; it cannot change or decrease the rise time and eliminate the oscillations. If applied, any amount of integration guarantees set point overshoot [8–11].

2.1 *Design of PI Regulator Gains* K_p *and* T

There are several different methods to design the gain parameters of PI regulator. The suitable method depends upon the criteria of the regulator. We have second-order system and a suitable method to use is symmetrical optimum method (SO) which is mostly used for PLL grid connecting applications.

The system transfer function is

$$G = \left(K_{\mathrm{p}}\frac{1+s\tau}{s\tau}\right)\left(\frac{1}{1+sT_s}\right) \tag{8}$$

The open-loop transfer function of SRF PLL

$$G_{\mathrm{ol}} = \left(K_{\mathrm{p}}\frac{1+s\tau}{s\tau}\right)\left(\frac{1}{1+sT_s}\right)\left(\frac{V_{\mathrm{m}}}{s}\right) \tag{9}$$

The closed-loop transfer function of SRF PLL

$$G_{\mathrm{cl}} = \frac{G_{\mathrm{ol}}}{1+G_{\mathrm{ol}}} \tag{10}$$

The relation between S domain and Z domain is

$$s = \frac{z-1}{T_s} \tag{11}$$

The transfer function of SO method is

$$G = \frac{\omega_o^2(ks + \omega_o)}{s^2(s + k\omega_0)} \tag{12}$$

By comparing the transfer function of SO method and open loop transfer function gives the following identifications:

$$\begin{aligned} \omega_c &= \frac{1}{aT_s} \\ \tau &= a^2 T_s \\ k &= \frac{1}{aV_{\mathrm{m}}s} \end{aligned} \tag{13}$$

which results in the regulator gains using SO method [5]. For the sampling time period of $T_s = 100$ us and by adjusting the normalizing factor a cross over frequency can be chosen.

For the second-order system cross over frequency and bandwidth are designed using small signal state space modeling. Higher phase margin which gives less oscillatory response, lower value of τ decreases the settling time and value of gain effects both phase margin and bandwidth (Fig. 3).

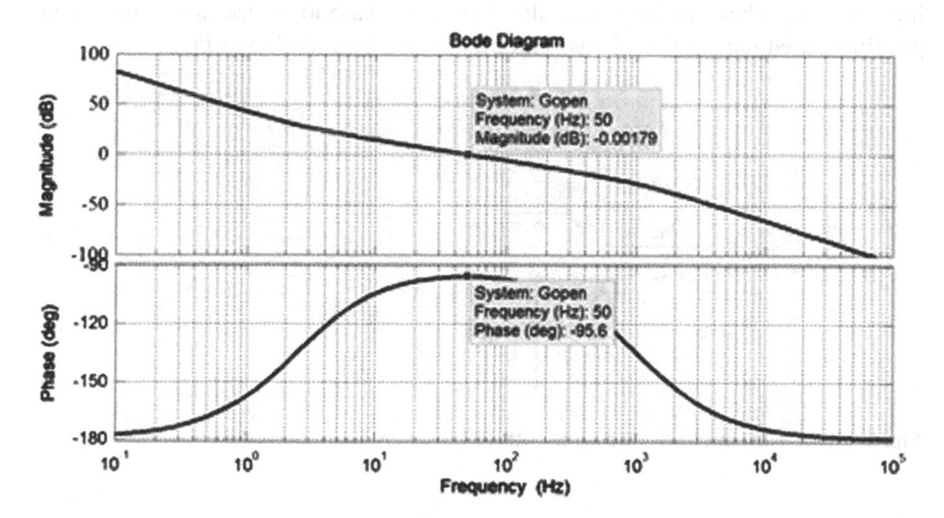

Fig. 3 Bode pot for open loop system of the SRF PLL

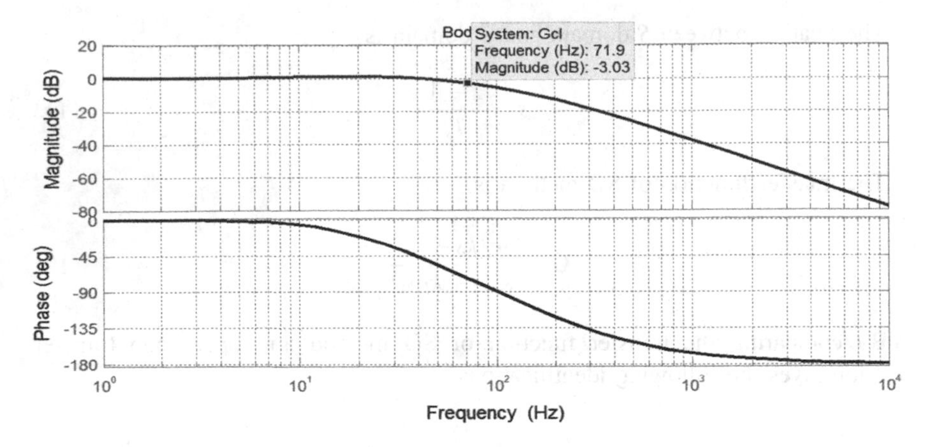

Fig. 4 Bode pot for closed-loop system of the SRF PLL

The bode plot shows that both the phase and magnitude curve is symmetric around the crossover frequency ($\omega_c = 2 * \pi * 50$). The phase margin is 84.4° at ω_c (Fig. 4).

The closed loop of the SRF PLL system have the characteristics of a low pass filter with bandwidth $\omega_B = \omega_C/0.7 = 71.9$ Hz.

3 SRF PLL with Pre-filters

Pre-filters are used to eliminate the dc offsets and to reduce harmonics present in input voltage. They can be placed after the three phase to αβ transformation. The pre-filters used here are band pass and high pass filters as shown Fig. 5.

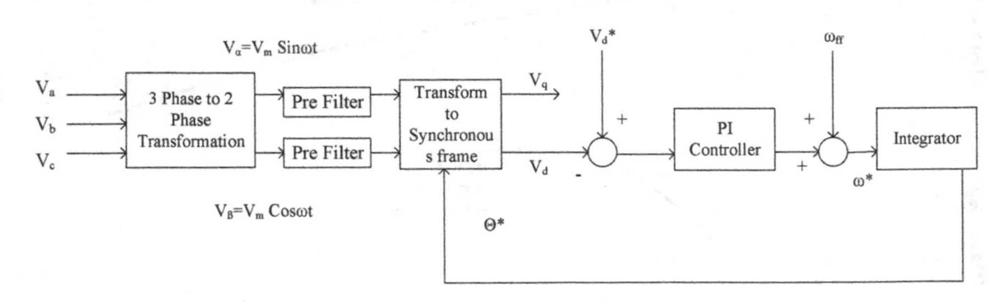

Fig. 5 Block diagram of pre-filter based SRF PLL

3.1 *Effect of Band Pass Filter*

The transfer function of band pass filter is given by

$$G_{\text{BPF}} = \frac{2K\omega_0 S}{S^2 + 2K\omega_0 S + \omega_0^2} \tag{14}$$

K can be taken as $1/\sqrt{2}$ = 1.707 and the term $\omega_0 = 2 * \text{pi} * f$ Where $f = f_0$ nominal frequency 50 Hz.

When the grid frequency is equal to ω_0, The magnitude and the phase shift given by this band pass filter will be 1 and 0 respectively, there will be no magnitude and phase errors. When the grid frequency changes from the nominal value, unwanted gains, and phase shifts will be there which results in magnitude and phase error.

The addition of the band pass filters affects the transient response time of the SRF PLL. And additional settling time due to band pass filter is lesser than using the high pass filter.

Additional advantage of using BPF is additional harmonic attenuation and has no impact when the input contains unbalance.

3.2 *Effect of High Pass Filter*

The transfer function of HPF is given by

$$G_{\text{HPF}} = \frac{s/\omega_c}{1 + s/\omega_c} \tag{15}$$

Corner frequency chosen is one decade less than nominal frequency

$$\omega_c = \frac{\omega_0}{10} \tag{16}$$

It also has the limitation of phase errors and magnitude for frequency deviations values up to 5 Hz. It is observed that using HPF magnitude and phase errors are lesser than comparing with BPF.

4 Software Implementation of SRF PLL

See Fig. 6, Table 1.

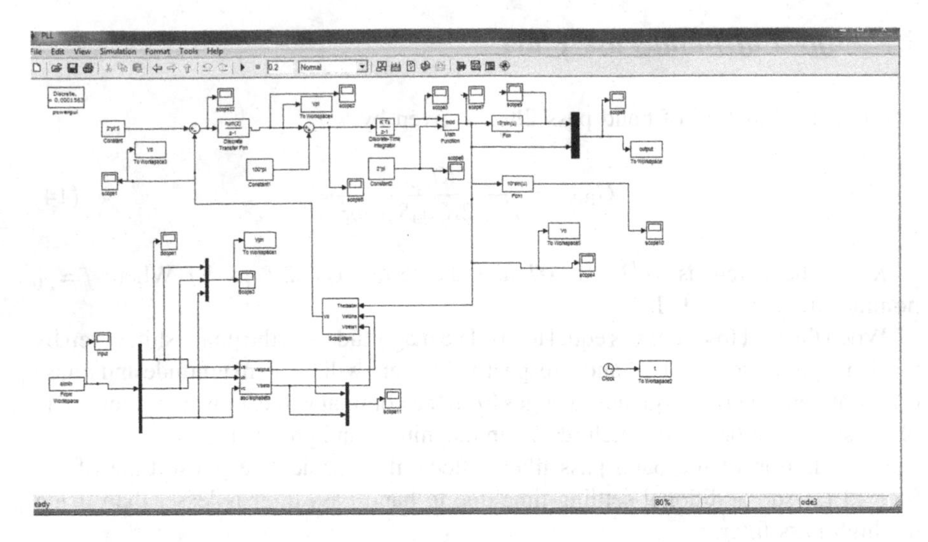

Fig. 6 Simulation of SRF PLL without pre-filters

Table 1 Design parameters for SRF PLL and pre-filter based SRF PLL

Design parameter	Value
Peak value of input voltage V_m	1 (p.u)
Fundamental supply frequency f	50 Hz
Number of sample N	128
Proportional constant K_P	0.01
Integral constant K_I	0.0026
Sampling time T_s	(1/6400)

4.1 Simulation Results for SRF PLL Without Pre-Filters

The three phase input signal contains fundamental frequency of 50 Hz(1pu), with 33% of 3rd(0.33pu), 25% of 5th(0.25pu), 17% of 7th(0.17pu), 13% of 9th(0.13pu), and 8% of 11th(0.08pu) harmonics which is shown in Fig. 7. The harmonics contain input given to SRF PLL the stationary reference voltages (V_α, V_β,) as shown in Fig. 8. The phase angle has been tracted from SRF PLL as shown in Fig. 9 and the total harmonic distortion (THD) of the harmonic content input signal is 26.79% and fundamental signal extracted by SRF PLL with the THD of the extracted signal is 11.25% as shown in Figs. 10 and 11 respectively.

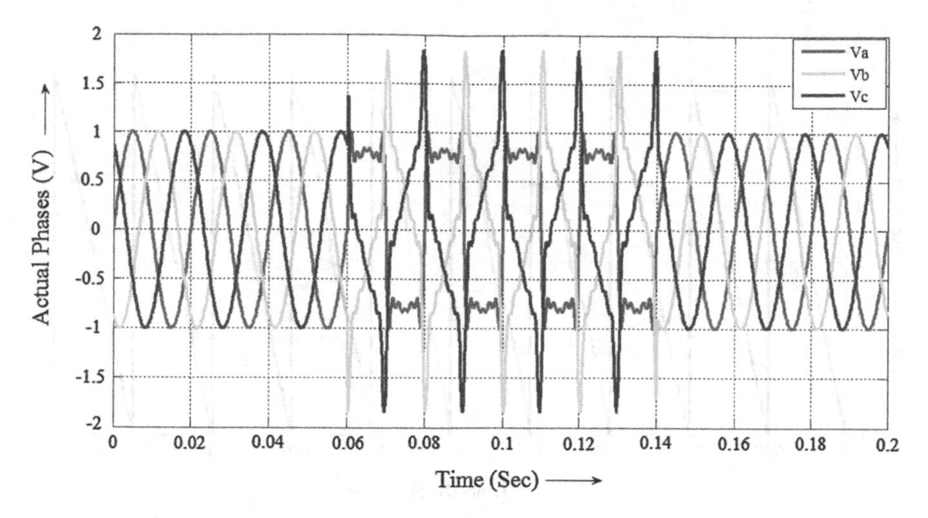

Fig. 7 Three phase input wave form with harmonics

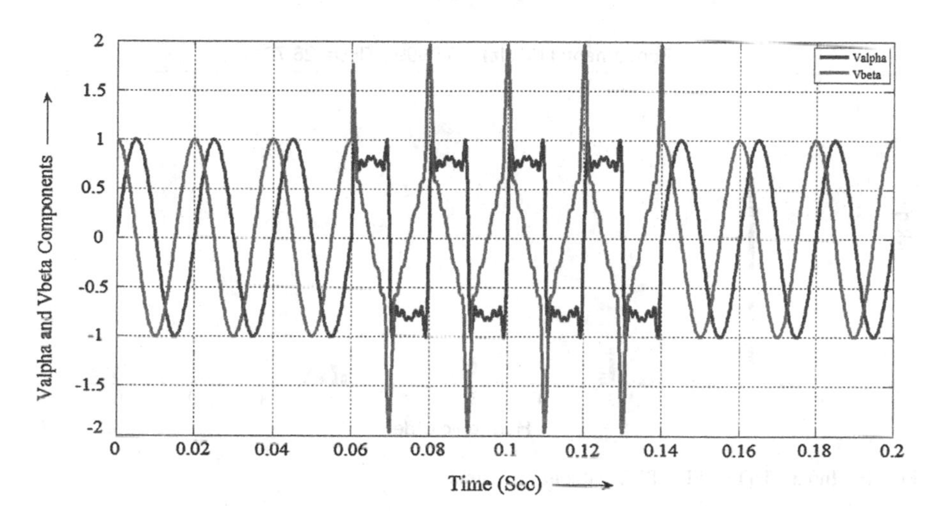

Fig. 8 V_α and V_β waveforms of SRF PLL

4.2 Simulation Results of SRF PLL with Pre-filters

4.2.1 SRF PLL with High Pass Filter

The three phase input signal contains fundamental frequency of 50 Hz(1pu), with 33% of 3rd(0.33pu), 25% of 5th(0.25pu), 17% of 7th(0.17pu), 13% of 9th(0.13pu), and 8% of 11th(0.08pu) harmonics which is shown in Fig. 7. The harmonics contain input given to SRF PLL with high pass filter the stationary reference

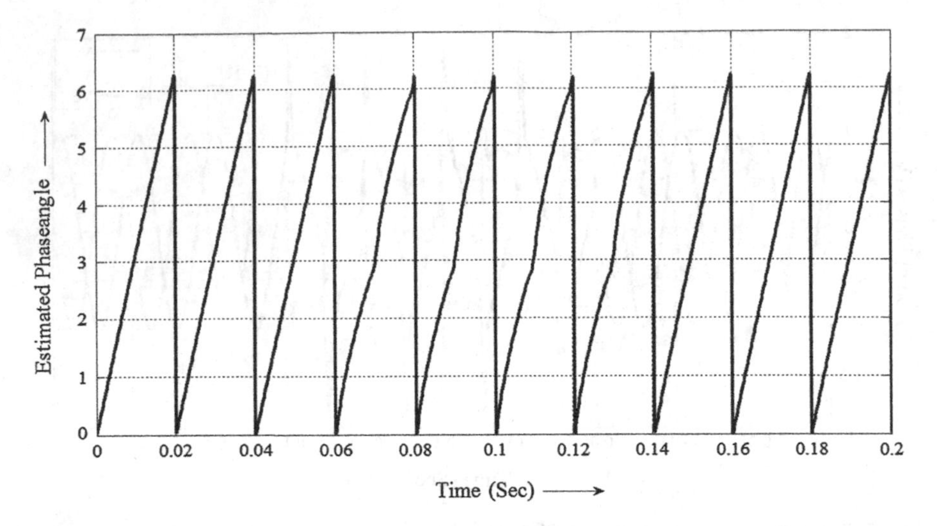

Fig. 9 Estimated phase angle of SRF PLL without pre-filters

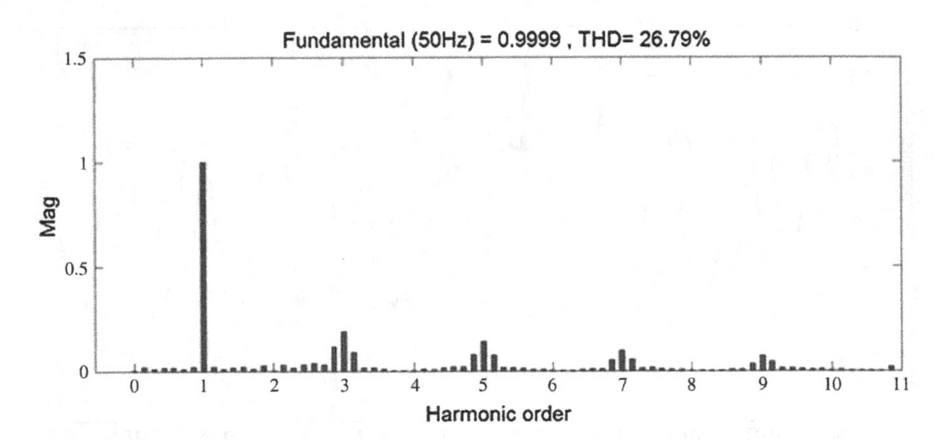

Fig. 10 Input THD of SRF PLL without pre-filters

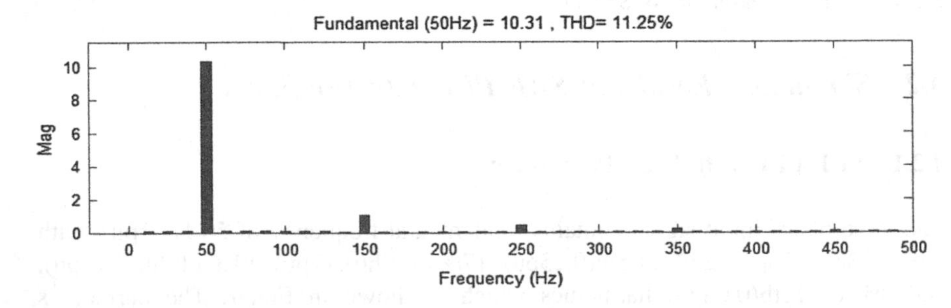

Fig. 11 Output THD of SRF PLL without pre-filters

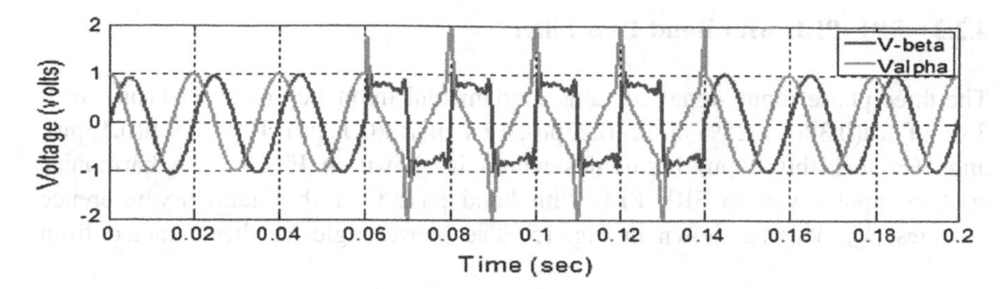

Fig. 12 *V* alpha and *V* beta waveforms of SRF PLL using high pass filter

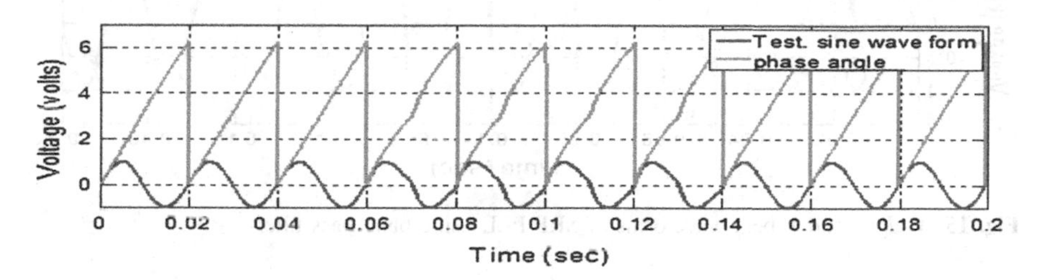

Fig. 13 Output phase angle wave forms of SRF PLL using high pass filter

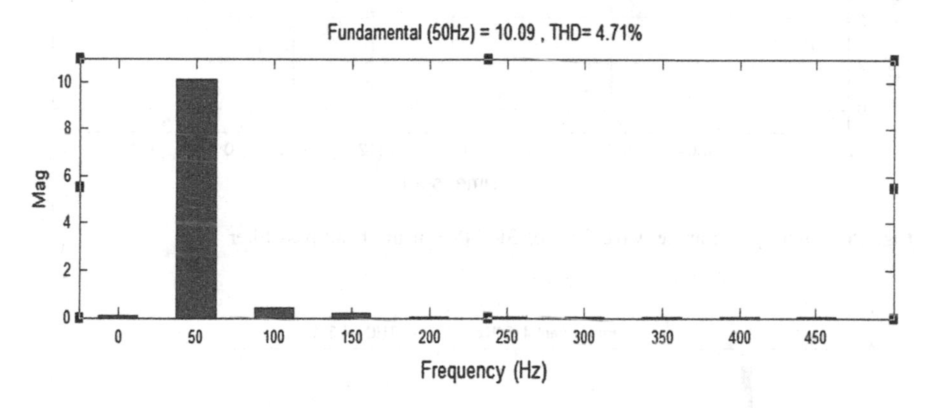

Fig. 14 Output THD of SRF PLL using high pass filter

voltages (V_α, V_β,) as shown in Fig. 12. The phase angle has been tracted from SRF PLL with high pass filter as shown in Fig. 13 and fundamental signal extracted by SRF PLL with high pass filter the THD of the extracted signal is 4.71% as shown Fig. 14.

4.2.2 SRF PLL with Band Pass Filter

The three phase input signal contains fundamental frequency of 50 Hz(1pu), with 33% of 3rd(0.33pu), 25% of 5th(0.25pu), 17% of 7th(0.17pu), 13% of 9th(0.13pu), and 8% of 11th(0.08pu) harmonics which is shown in Fig. 7. The harmonics contain input given to SRF PLL with band pass filter the stationary reference voltages (V_{α}, V_{β},) as shown in Fig. 15. The phase angle has been tracted from

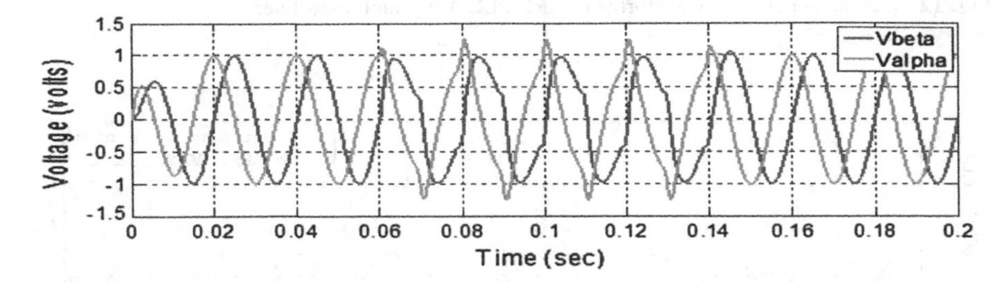

Fig. 15 *V* alpha and *V* beta waveforms of SRF PLL using band pass filter

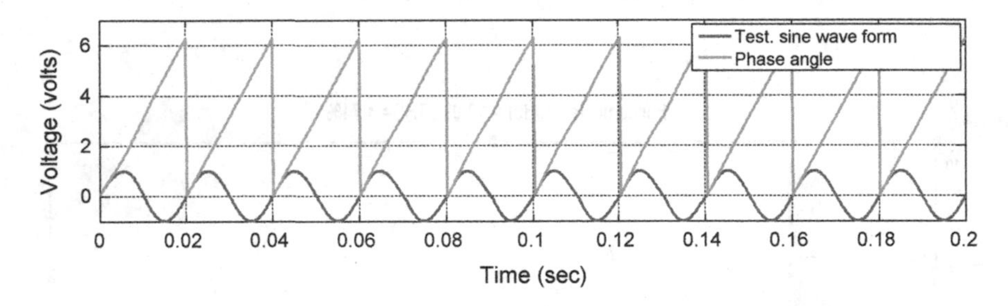

Fig. 16 Output phase angle wave form of SRF PLL using band pass filter

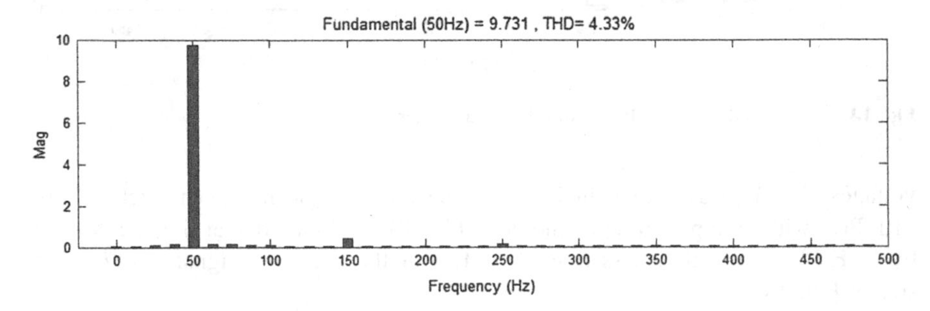

Fig. 17 Output THD of SRF PLL using band pass filter

SRF PLL with band pass filter as shown in Fig. 16 and fundamental signal extracted by SRF PLL with band pass filter the THD of the extracted signal is 4.71% as shown Fig. 17.

4.2.3 Tabulation for Input and Output THD of SRF PLL and Pre Filter Based SRF PLL

See Table 2.

Table 2 Tabulation of THD values between different pre-filters based SRF PLL and SRF PLL

Design	Input total harmonic distortion (THD) (%)	Output THD (%)
SRF PLL without pre-filters	26.79	11.65
SRF PLL using high pass filter	26.79	4.71
SRF PLL using band pass filter	26.79	4.33

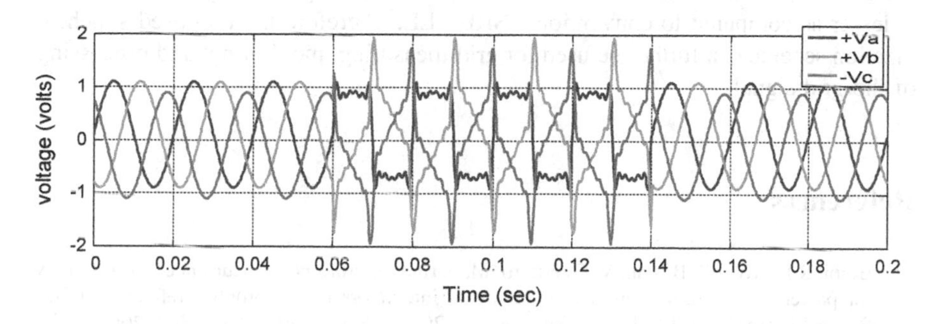

Fig. 18 Three phase input waveform with dc offsets and harmonics

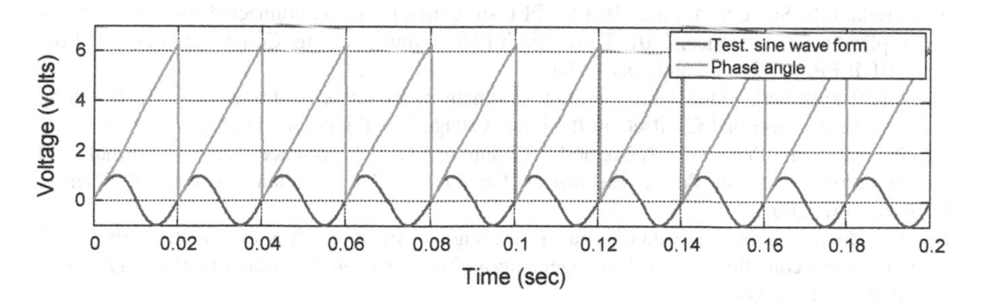

Fig. 19 Output phase angle wave from using band pass filter

4.3 Simulation Results with DC Offsets and Harmonics in Input

The three phase input signal contains fundamental frequency of 50 Hz(1pu), with 33% of 3rd(0.33pu), 25% of 5th(0.25pu), 17% of 7th(0.17pu), 13% of 9th(0.13pu), and 8% of 11th(0.08pu) harmonics with dc offset of inputs V_a, V_b and V_c have dc offsets of +0.02, +0.02 and −0.02, as shown in Fig. 18. It is given to pre filter based SRF PLL the phase angle has been extracted as shown in Fig. 19.

5 Conclusion

The proposed pre-filter-based phase detection system for synchronization of power converters is introduced in this paper. From the proposed study, one can be seen that, the pre-filters are used to extract the fundamental frequency of the phase accurately during harmonic and dc offset injection of the grid. In the proposed pre-filter-based SRF PLL when harmonic injection to the grid, the grid current THD is lesser as compared to conventional SRF PLL. Therefore the proposed synchronization scheme can further be used for grid measuring, monitoring, and processing of the grid signal.

References

1. Bojoi, R.I, Griva, G, Bostan, V, Guerriero, M, Farina, F, Profumo, F.: Current control strategy for power conditioners using sinusoidal signal integrators in synchronous reference frame. Power Electronics, IEEE Transactions on, vol. 20, no. 6, pp. 1402, 1412, Nov. 2005.
2. Tran, T.-V., Chun, T.-W., Lee, H.-H., et al.: 'PLL-based seamless transfer control between grid-connected and islanding modes in grid-connected inverters', IEEE Trans. Power Electron., 2014, 29, (10), pp. 5218–5228.
3. Arruda, L.N, Silva, S.M. Filho, B. J C.: PLL structures for utility connected systems. Industry Applications Conference, 2001. Thirty-Sixth IAS Annual Meeting. Conference Record of the 2001 IEEE, pp. 2655–2660. Oct. 2001.
4. Abhijit kulkarani, vinod john, "Design of synchronous reference frame phase locked loop with the presense of DC offsets in the input voltage", IET journals, ISSN 1755–4535.
5. Ghoshal, A and Vinod, J.: A Method To Improve PLL Performance Under Abnormal Grid Conditions. National Power Electronics Conference 2007. Indian Institute of Science, Bangalore, 2007.
6. H.-S. Song, K. Nam,: Instantaneous phase-angle estimation algorithm under unbalanced voltage-sag conditions, IEEE Proc. Generation. Transmission. Distribution. Vol. 147, no, 6, pp. 409–415, 2000.
7. R. Weidenbrug, F.P. Dawson, and R.Bonert.: New Synchronization Method for Thyristor Power Converters to Weak AC Systems. IEEE Trans. On Industrial Electronics, Vol. 40, No.05, pp. 505–511. 1993.
8. Meral, M.E.: 'Improved phase-locked loop for robust and fast tracking of three phases under unbalanced electric grid conditions', IET Gener. Transm. Distrib., 2012, 6, (2), pp. 152–160.

9. Salamah, A.M., Finney, S.J., Williams, B.W.: 'Three-phase phase-lock loop for distorted utilities', IET Electr. Power Appl., 2007, 1, pp. 937–945. Bacon, V.D., da Silva, S.A.O., Campanhol, L.B.G., et al.: 'Stability analysis and performance evaluation of a single-phase phase-locked loop algorithm using a non-autonomous adaptive filter', IET Power Electron., 2014, 7, (8), pp. 2081–2092.
10. Alfonso-Gil, J.C., Vague-Cardona, J.J., Orts-Grau, S., et al.: 'Enhanced grid fundamental positive-sequence digital synchronization structure', IEEE Trans. Power Deliv., 2013, 28, (1), pp. 226–234.
11. Elrayyah, A., Sozer, Y., Elbuluk, M.: 'Robust phase locked-loop algorithm for single-phase utility-interactive inverters', IET Power Electron., 2014, 7, (5), pp. 1064–1072.

Dynamic Performance Enhancement of Three-Phase PV Grid-Connected Systems Using Constant Power Generation (CPG)

Rajesh Kasthuri, K. Bhageeratha Reddy, J. Prasanth Ram, T. Sudhakar Babu and N. Rajasekar

Abstract Surmount demand requirement in electrical energy has given a provision of integrating renewable energy to the three-phase grid-connected system. In particular, the penetration of distributed generation system has contributed more in electrical power generation. But the system connected produces adverse effects in grid due to the unpredictable atmospheric conditions. To handle such condition, in this research article, PV system is considered under study and a new two-stage CPG-P&O MPPT method is used for interfacing PV with grid systems. In addition, a wide range of feed-in power to grid capacity at inverter level is briefly addressed. Further, the algorithm is designed to reduce the increased thermal stresses in switch and to get rid of the grid overshoot and power loss. The desired objective was acknowledged by the simulation results by ensuring delivery of optimal performance.

Keywords CPG-Constant power generation · MPPT–Maximum power point tracking · P&O-Perturb & observe

R. Kasthuri · K. Bhageeratha Reddy · J. Prasanth Ram · T. Sudhakar Babu · N. Rajasekar (✉)
Solar Energy Research Cell, School of Electrical Engineering
VIT University, Vellore, India
e-mail: natarajanrajasekar@gmail.com

R. Kasthuri
e-mail: rajesh.kasthuri@gmail.com

K. Bhageeratha Reddy
e-mail: bhagee215@gmail.com

J. Prasanth Ram
e-mail: jkprasanthram@gmail.com

T. Sudhakar Babu
e-mail: sudhakarbabu66@gmail.com

S.S. Dash et al. (eds.), *Artificial Intelligence and Evolutionary Computations in Engineering Systems*, Advances in Intelligent Systems and Computing 517,
DOI 10.1007/978-981-10-3174-8_17

1 Introduction

With increased production cost and uncertain by its availability, the conventional energy resources are decaying day by day. On the other hand, renewable energy power generation has become an important assert to solve the expected future demand. The key advantage of using renewable energy source is zero maintenance in exhaustibility, reliability, zero noise available at free of cost, empowerment of residential application using solar and wind power generation [1]. With a vision of extending these sources to grid-connected system, the authors have proposed a scholarly research in solar PV grid-connected systems.

To be specific, solar installed capacity in 2014 is 185 MWp and planed to contribute 12,500 MW to the power demand of India till 2016–2017 [2]. This shows the future reliability on solar power generation is increasing year by year. With wide scope of extending PV system to grid-connected applications. The certain factors like: (1) switching losses of inverter due to intermittency resulting in lifetime of semiconductor switches, (2) Instability in grid loading conditions under peak power, and (3) Restricted utilization of PV inverter has encouraged the researchers to introduce constant power generation (CPG) in PV grid-connected applications.

Understanding the importance of DG system, In [3] a general overview on DG involving PV, fuel cell, and wind generation systems are interfaced with grid and an interior discussion in design of controller is addressed. In [4], a closed-loop control of PV interfaced with interleaved boost converter is designed. This paper provides a two-stage conversion for a grid-connected PV system. Further, a high voltage gain is introduced to meet the grid requirement. However, these methods have not utilized the MPPT mode in PV and results with higher power loss when connected to grid. Understanding the problem, a new method of low voltage grid integration is proposed in [5]. A new regulation involving 70% of active power feed-into the grid is considered with irrespective of reactive power flow. However, the techniques discussed above has suffered from thermal stresses on switches and subjected to overloading conditions on grid.

In general, despite of the dynamic overloading condition the PV grid-connected system should be capable of operating at both reduced power mode (RPM) and MPPT mode. This is taken as an objective and a 2 kW prototype model for a PV is effectively implemented with the help of P&O algorithm [6]. Various MPPT algorithms have been implemented for extraction of maximum power from solar PV system [7–9]. Once after MPPT is introduced in PV grid-connected system, the scope for limited power point tracking is raised. In this connection, the author in [10] investigated the problem of PV under high power production levels. Hence, a novel limited power point tracking algorithm (LPPT) is successfully tested with grid-connected mode. In another approach [11] to utilize the instantaneous available PV power, a new method involving MPPT alleged with constant power generation (CPG) is proposed. This paper has given excellent results in handling thermal stresses across switches and grid overloading conditions. Further, an extended robust control in CPG mode is discussed in [12]. A single-phase system with effective 2965 W of power is produced with 80% rated power. However in practical situations, the DG systems involving wind require three-phase system to interface with grid. In view to the point, an effective PV system is constructed with 1 kW capacity. In addition to avoid too much of ripples and to elevate the system performance, an interleaved boost converter is constructed. The scheme introduced with CPG has given high resolute performance in three-phase grid-connected system. The remaining flow of the paper is organized as follows.

In Sect. 2, the modeling of PV is discussed. In Sect. 3, limitation of feed-in PV power is discussed. In Sect. 4, a brief discussion in conventional and hybrid CPG-P&O MPPT is explained. In addition, PV interfacing with grid through interleaved boost converter is experimented. In Sect. 5, the simulation result along with explanation is briefed.

2 Limiting Feed-in Power of PV Systems

The voltage levels of PV system connected to grid is unpredictable due to its change in irradiation. At the same time, it is also required to maintain grid standards under boundary limits. Hence, viability of study in limiting power of PV systems is required. In this section, the possibilities of limiting PV power production through feed-in power are analyzed.

The main objective of constant power generation (CPG) in PV systems is achieved through implementing MPPT with possible limitations. The schematic three-phase PV grid-connected system involving an interleaved boost converter is shown in Fig. 1. For the proposed system, two-stage converter strategies are adapted. This arrangement helps to have flexible control to the PV power when sudden change in PV power occurring due to change in irradiation. However, the PV system connected with large capacity will produce high frequency oscillations that may induce adverse effect in grid. Hence in these conditions, the PV is not advisable for the grid connection. Further, the control of feed-in power through constant power generation and active power control is feasible method to control grid power. Similar to active power limitation, constant power generation method is also one of the efficient methods to handle voltage amplitudes beyond boundary limits. Hence from the above discussion it is inferred that to avoid over voltages occurring at consumer side, CPG mode of power generation techniques can be used in DG system.

The CPG technique can be achieved by controlling PV power at inverter level in addition to this the CPG technique will exhibit major advantageous such as reduces system cost and it runs for long period. In this work, a case study of 1KW PV annual power generation is considered and its profile under different temperature levels and dynamic change in irradiation condition is recorded.

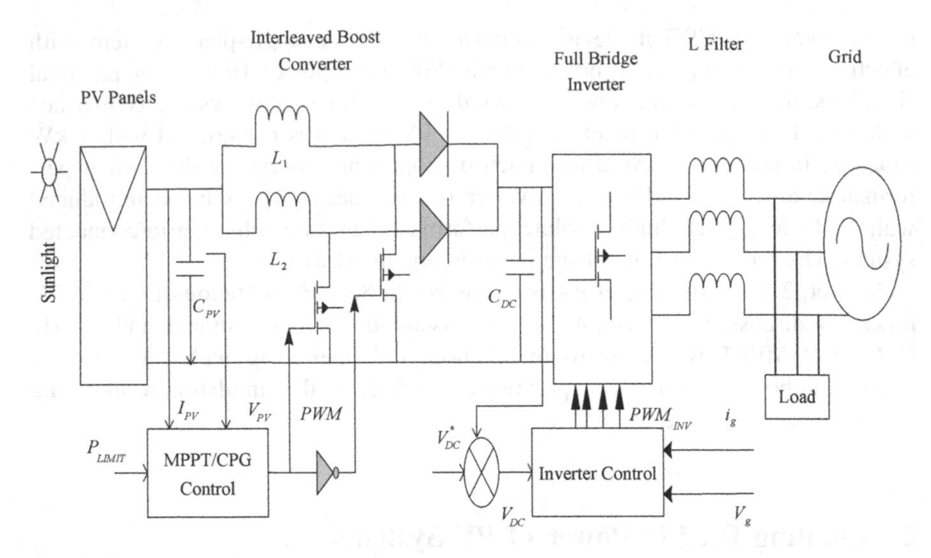

Fig. 1 Two-stage three-phase PV grid-connected system

3 CPG Operation for PV Grid-Connected System

3.1 Conventional CPG Algorithm

To come up with accurate tracking and to introduce a novelty in PV grid-connected applications, the authors in literature have followed two modes of operation (MPPT & CPG) in conventional technique. The operating principle behind hybrid P&O-CPG is explained with the help of Fig. 2.

From Fig. 2, it is seen that five feasible regions (I, II, III, IV & V) are present in power-time characteristics. The power varying due to irradiation change is clearly visible. But the grid constraints cannot be met, when power fluctuates with respect to time. So, P_{LIMIT} is used to reduce the feed-in power of PV when captured power goes out of boundary limits. With respect to P_{LIMIT}, the regions II & IV are exceeding the constraint. Hence in this regions are operated at CPG mode, while the regions I, III & V produce power ranging below P_{LIMIT}. To enhance the performance of these regions and to utilize the available maximum power, P&O MPPT is implemented.

For better understanding, the working of CPG/MPPT (P&O) algorithm and the control of CPG/MPPT method using PI controller is explained further. In this method, a novel interleaved boost converter is used to regulate the extracted maximum power from the PV. In addition, the converter helps to reduce the ripple content that appears in the output. Based on the instantaneous power available at PV panels, the PV output power can be calculated as follows:

From the discussion, it is clear that grid operation through CPG technique is one of the possible ways to integrate PV systems with reduced stresses. However, the

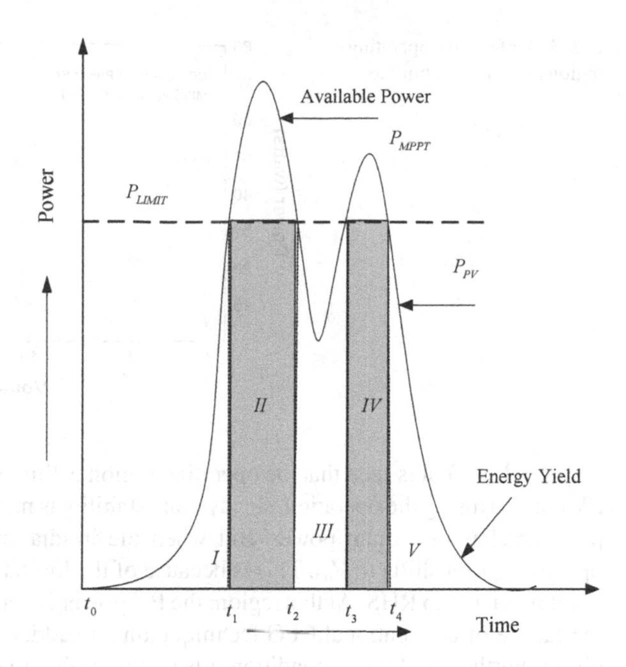

Fig. 2 Operating modes of P&O and CPG

energy attained from solar PV is not fully utilized. To accomplish aforementioned objective, CPG operation is mandatory. The operation of CPG is explained in next section.

$$P_{\text{pv}} = \begin{cases} P_{\text{MPPT}}, \text{ when, } P_{\text{pv}} \leq P_{\text{LIMIT}} \\ P_{\text{LIMIT}}, \text{ when, } P_{\text{pv}} > P_{\text{LIMIT}}. \end{cases} \quad (1)$$

where P_{pv} is the PV output power. P_{MPPT} is the maximum power available at present irradiation and P_{LIMIT} is selected based on the annual energy yield (%). The principle behind the implementation of CPG is discussed further. When the PV power output is less than the P_{LIMIT} value, the system is operated at MPPT mode. When PV output power is greater than P_{LIMIT}, i.e., when PV power exceeding the limits of grid standard, the output power is limited to P_{LIMIT}.

3.2 Limitation of CPG Operation

To explain the limitation of CPG operation, PV curve for 80 W panel at 1000 and 800 W/m^2 is plotted. To ensure PV system operating at CPG, It is very much needed to implement MPPT to the system. The researchers in the literature have followed the unique principle of implementing MPPT at inverter level to reduce cost of the system. But there are limitations observed in their proposed P&O-CPG method. The PV curve plotted in Fig. 3 explains the limitation of CPG technique.

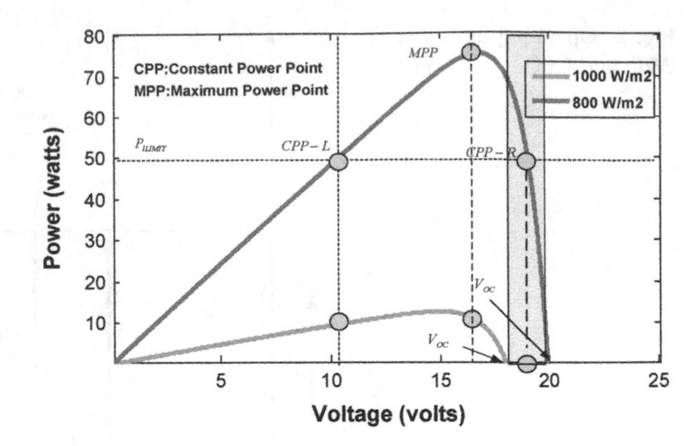

Fig. 3 Unfeasible operating regions in CPG technique

From Fig. 3, it is seen that the operating region is limited on right-hand side of the PV curve. During the operation, steady-state stability is maintained until the operating point reaches maximum power. But when the irradiation decreases suddenly, the operating point shifts to V_{oc}. This is because of the limitation constraint of operating point restricted to RHS. At this region, the PV power is zero. Hence, this experiments the failure of conventional CPG technique under sudden change in irradiation condition. Further, under this condition it is not advisable to operate PV panel.

3.3 *Proposed P&O-CPG Technique for Three-Phase Grid-Connected System*

From literature, it is well understood that two common modes are followed to extract maximum power from PV array. (1) MPPT mode: In this mode the standard P&O algorithm is used. The voltage perturbations in this algorithm make the system to reach MPP in certain iterations. It is observed that steady-state oscillations are appearing around maximum power point ($P_{pv} \leq P_{LIMIT}$). (2) CPG mode: The real task of controlling power beyond P_{LIMIT} is effectively done here. In this mode, the voltage perturbations are continuously given to reach MPP. Once the power levels go beyond P_{LIMIT}, the operating power is strictly restricted to CPP level, i.e., the voltage perturbation V_{pv}^{*} is given based on difference between present voltage ($V_{pv,n}$) and step voltage ($V_{step},$). The formula used in restricting voltage perturbation is shown in Eq. (2).

$$V_{pv}^{*} = \begin{cases} V_{MPPT}, \text{ when}, P_{pv} \leq P_{LIMIT} \\ V_{pv,n} - V_{step}, \text{ when}, \ P_{pv} > P_{LIMIT} \end{cases} \quad (2)$$

where V_{pv}^{*} is the voltage perturbation to be sent, V_{MPPT}, is the maximum power point voltage in normal P&O mode and $V_{pv,n} - V_{step}$, is the difference in present and step voltage in CPG mode.

3.4 *Problems in P&O-CPG Algorithm*

The performance of P&O-CPG algorithm is highly efficient under steady change in irradiation conditions. But in practical situations, dynamic change in irradiation changes happen due to unpredictable climatic conditions. This may raise considerable power loss and peak overshoots in the system. In particular, the PV with higher capacity when exposed to higher irradiation will induce high voltage spikes to the grid-connected system. Aforementioned problem is explained with the help of Fig. 3. Under MPPT mode, the sudden irradiation rise happens, the operating point shifts from A to B in this condition high level of voltages transients are induced because the captured output power is higher than P_{LIMIT}. On the other hand, when irradiation level reduces abruptly from point *C* to *D*, then it is very much needed to operate PV panel at MPP for the corresponding irradiation. However, power loss occurs when operating point moves from point *D* to *E*. Thus to deliver optimal performance, it is mandatory to control overshoot and power losses.

3.5 *High-Performance P&O-CPG Algorithm*

In consideration to the factors (i) augmented potential in CPG algorithm, (ii) power generation from PV can be effectively regulated, and (iii) Enriched scope of integrating PV to grid, the high-performance CPG-P&O algorithm is proposed. An enhanced attempt to reduce the drawbacks in conventional P&O-CPG algorithm is made in this section. In particular, the power loss due to fall in irradiation and overshoot in sudden raise in irradiation condition is given major concentration.

3.5.1 Reducing Overshoots

It is well known fact that the operation of grid is limited to constraint limits in terms of frequency and power. When PV with high installation capacity is interfaced to grid, the uncertain change in irradiation shifts the operating point in P–V curve (discussed in previous section) [8, 9]. In this regard, a high power is produced by violating the limitations of grid. Though the increase in power is limited to P_{LIMIT}, the panel requires some settling time to shift the operating from B to C. This will make the system to suffer from peak overshoots. To reduce this effect, the voltage perturbations in the algorithm are provided with long step size based on the irradiation condition. Hence, stress of grid at transient period is reduced. The equation to detect irradiation change is given below.

$$\mathrm{IC} = \begin{cases} 1, \text{ when}, P_{\mathrm{pv},n} - P_{\mathrm{LIMIT}} > \varepsilon_{\mathrm{inc}} \\ 0, \text{ when}, P_{\mathrm{pv},n} - P_{\mathrm{LIMIT}} \leq \varepsilon_{\mathrm{inc}} \end{cases} \tag{3}$$

where $P_{\mathrm{pv},n}$ is the available instantaneous power and $\varepsilon_{\mathrm{inc}}$ is the incremental constant. From the above equation, it can be deduced that irradiation change is detected when the difference between instantaneous power and P_{LIMIT} exceeds incremental constant ($P_{\mathrm{pv},n} - P_{\mathrm{LIMIT}} > \varepsilon_{\mathrm{inc}}$). Under this condition, the voltage perturbation with long step size is generated using the following.

$$V_{\mathrm{PV}}^{*} = V_{\mathrm{pv},n} - \left[(P_{\mathrm{pv},n} - P_{\mathrm{LIMIT}}) \frac{P_{\mathrm{LIMIT}}}{P_{\mathrm{mp}} * \gamma} \right] * V_{\mathrm{step}} \tag{4}$$

where V_{step} is the step size, $P_{\mathrm{LIMIT}}/P_{\mathrm{mp}}$ is the ratio used to control the step size, and γ is used to regulate the speed of algorithm.

3.5.2 Reducing Power Losses

Similar to the fast increase in irradiation, in a cloudy day, it is also expected to have sudden fall in irradiation levels. This will reduce the output power and the PV panel operating at maximum power point is not ensured. Since the irradiation is less, it is advisable to operate panel in maximum power point. But to shift the PV operating point to MPP will induce power loss in the grid systems. The detection of irradiation change is founded using the following equation.

$$\mathrm{IC} = \begin{cases} 1, \text{ when}, \ P_{\mathrm{pv},(n-1)} - P_{\mathrm{LIMIT}} > \varepsilon_{\mathrm{dec}} \\ 0, \text{ when}, \ P_{\mathrm{pv},(n-1)} - P_{\mathrm{LIMIT}} \leq \varepsilon_{\mathrm{dec}} \end{cases} \tag{5}$$

The irradiation change in the system is detected based on the difference between previous iteration power $P_{\mathrm{pv},(n-1)}$ and P_{LIMIT}. When this value exceeds decrement constant $\varepsilon_{\mathrm{dec}}$, the irradiation change is occurred. Further, the above equation clarifies the requirement in knowledge of previous operating mode. This can be founded using the below formula.

$$\mathrm{MOP} = \begin{cases} \mathrm{MPPT}, \text{ when}, \left|P_{\mathrm{pv},(n-1)} - P_{\mathrm{LIMIT}}\right| \geq \varepsilon_{\mathrm{ns}} \\ \mathrm{CPG}, \text{ when}, \left|P_{\mathrm{pv},(n-1)} - P_{\mathrm{LIMIT}}\right| < \varepsilon_{\mathrm{ns}} \end{cases} \tag{6}$$

The above equation determines the mode of operation based on sampling time. From (5) to (6), it can be inferred that $\varepsilon_{\mathrm{ns}}$ and $\varepsilon_{\mathrm{dec}}$ are used in detecting decrease off irradiation, where the constant $\varepsilon_{\mathrm{ns}}$ value is given as 1–2% of the rated power. During this period, the voltage to be perturbed is given based on the following. Since the panel is operated below P_{LIMIT}, the MPPT mode makes the system to reach MPP with reduced power loss.

$$V_{pv}^{*} = k * V_{oc} \quad (7)$$

V_{pv}^{*} is the voltage to be perturbed, k is the constant usually taken 70–80% of open circuit voltage V_{oc}.

4 Simulation Results and Discussion

To experiment three-phase grid-connected system, 1 kW PV panel model is designed. The modeling of PV involves 20 panels with each panel having the rating of 80 W (Kotak 80 W panel). Further, the two-stage converter involving interleaved boost converter and grid-connected inverter is interfaced with the PV system. The designed system is modeled and executed in the MATLAB platform with system configuration 4 GB RAM and i5 processor. The design specifications for interleaved boost converters are switching frequency 10Khz L1, L2 are 3 mH, respectively, DC link capacitor 100 μF, L-filter is 100 mH.

In general, for a grid-connected system, three common factors are required to maintain the stability. The parameters like phase angle, voltage, and frequency are highly sensitive to the change in supply side generation. If the system connected is not capable to meet the requirement of the grid, it is forced to disconnect from the grid, since it may induce high overshoot and considerable power loss. To avoid aforementioned problem in the design and to uphold the system with constant voltage, a DC link capacitance connected in parallel into the grid. In addition, this arrangement reduces the ripple content in the output system. Further to ensure frequency and phase angle in limited value, a phase lock loop (PLL) is constructed in grid side system.

The PV panels in practical condition may produce varying voltage due to change in irradiation condition. In our system, a typical two mode P&O-CPG is adopted to enhance the tracking ability by PV system. To test the system, a three level step change in irradiation is generated and the PV powers attained under all the irradiations are recorded. To ensure safer operation of grid, the feed-in power to grid is limited to 70–80% of the PV capacity is successfully accomplished in the designed model. The grid voltage and current levels attained in irradiation change is shown Fig. 4.

Since the irradiation being direct relation with current changes, the current value of the system is also varied. In consideration of PV connected to grid, the previous work is proposed for single-phase grid-connected system. It is observed that the settling time of the current between irradiations is high. Hence, during this period stability collapse in grid is occurred. This being the objective, the authors in this article have given enlarged concentration to reduce system disturbances using hybrid CPG technique. With this effect, the current value of the system is brought to steady state in quick succession. Also, the feed-in power is maintained under standards. This PV current under irradiation change is shown in Fig. 5.

Thus, In consideration to the factors: (i) fast irradiation change (ii) thermal overloading in the switches, and (iii) regulating PV power to grid, the possibilities in enhancing the PV performance connected to three-phase grid-connected system using CPG technique is experimented. Irrespective of the uncertain nonlinearity present in PV and unpredictable irradiation change, the designed is able to deliver

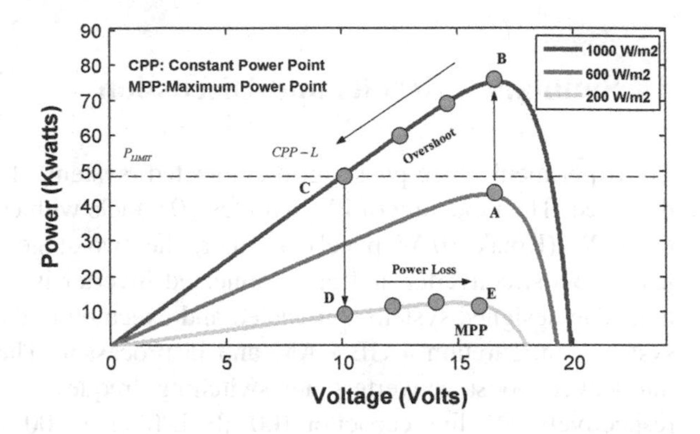

Fig. 4 Operating points of PV under dynamic change in irradiation

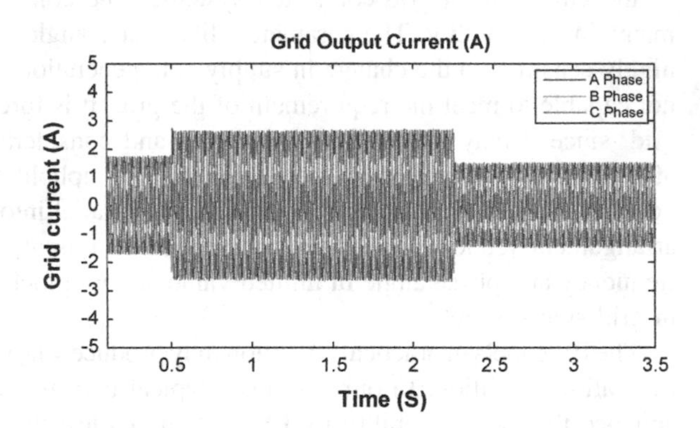

Fig. 5 Grid current during irradiation change

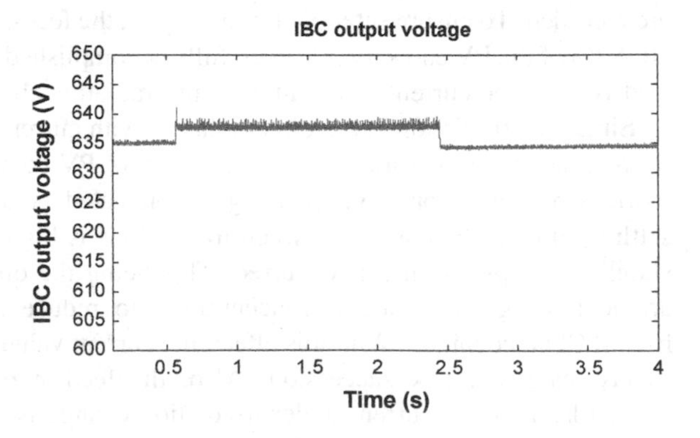

Fig. 6 IBC output voltage during irradiation changes

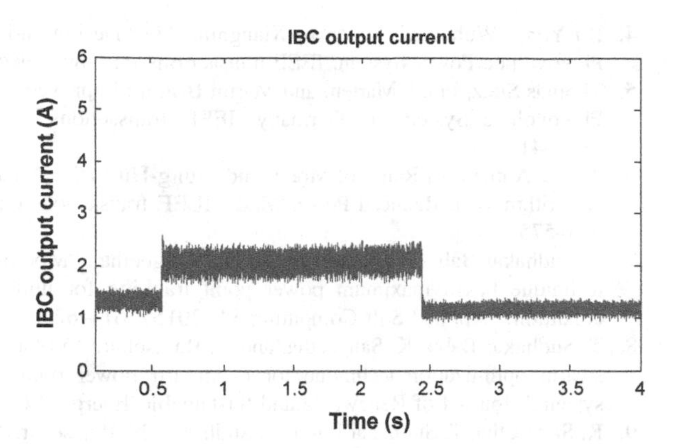

Fig. 7 IBC output current during irradiation change

optimal performance. Further, if the system capacity is increased, the stable performance can be achieved by proper parameter selection, i.e., selection of L & C values. The voltage and current levels attained by interleaved boost converter under irradiation change is shown in Figs. 6 and 7.

5 Conclusion

With the vision of integrating PV with grid, a novel approach of implementing hybrid CPG-P&O method is adapted for PV systems. The most common problem of PV like overshoots and power loss is briefly analyzed and discussed. Further, a new interleaved boost converter is preferred in DC–DC conversion to reduce the harmonic content present in the system. In order to improve the stability in system, a detailed analysis in operating point of PV under irradiation change and fluctuations in feed-in power with regulated constraints are presented. Thus, the designed system has delivered robust performance under all varying atmospheric conditions and desired stability is ensured.

References

1. Rajasekar, N., et al. "Application of modified particle swarm optimization for maximum power point tracking under partial shading condition." Energy Procedia 61 (2014): 2633–2639.
2. Deshkar, Shubhankar Niranjan, et al. "Solar PV array reconfiguration under partial shading conditions for maximum power extraction using genetic algorithm." Renewable and Sustainable Energy Reviews 43 (2015): 102–110.
3. Frede Blaabjerg, Remus Teodorescu, Marco Liserre, Adrian V. Timbus,: Overview of Control and Grid Synchronization for Distributed Power Generation Systems, ieee transactions on industrial electronics, 53 (2006) 1398–1408.

4. Bo Yang, Wuhua Li, Yi Zhao, Xiangning He.: Design and Analysis of a Grid-Connected Photovoltaic Power System. IEEE transactions on power electronics, 25 (2010) 992–999.
5. Thomas Stetz, Frank Marten, and Martin Braun,: Improved Low Voltage Grid-Integration of Photovoltaic Systems in Germany. IEEE transactions on sustainable energy, 4 (2013) 534–541.
6. Ashraf Ahmed, Li Ran, Sol Moon, and Joung-Hu Park,: A Fast PV Power Tracking Control Algorithm With Reduced Power Mode. IEEE transactions on energy conversion, 28 (2013) 565–575.
7. T. Sudhakar Babu, N. Rajasekar, and K. Sangeetha.: "Modified particle swarm optimization technique based maximum power point tracking for uniform and under partial shading condition." Applied Soft Computing 34 (2015): 613–624.
8. T. Sudhakar Babu, K. Sangeetha, and N. Rajasekar.: "Voltage band based improved particle swarm optimization technique for maximum power point tracking in solar photovoltaic system." Journal of Renewable and Sustainable Energy 8.1 (2016): 013106.
9. K. Sangeetha, T. Sudhakar Babu, N. Sudhakar, N. Rajasekar.: "Modeling, analysis and design of efficient maximum power extraction method for solar PV system." Sustainable Energy Technologies and Assessments. 15 (2016) 60–70.
10. Andoni Urtasun, Pablo Sanchis, and Luis Marroyo.: Limiting the Power Generated by a Photovoltaic System. 2013 10th International Multi-Conference on Systems, Signals & Devices (SSD) Hammamet, Tunisia, (2013).
11. Yongheng Yang, Huai Wang, Frede Blaabjerg, and Tamas Kerekes, A Hybrid Power Control Concept for PV Inverters With Reduced Thermal Loading. ieee transactions on power electronics. 29 (2014) 6271–6275.
12. Yongheng Yang, Frede Blaabjerg, Huai Wang. Constant Power Generation of Photovoltaic Systems Considering the Distributed Grid Capacity. Twenty-Ninth Annual IEEE, Applied Power Electronics Conference and Exposition (APEC), (2014).

Prediction of Mechanical Soil Properties Based on Experimental and Computational Model of a Rocker Bogie Rover

S. Nithin, B. Madhevan, Rima Ghosh, G.V.P. Bharat Kumar and N.K. Philip

Abstract Lack of knowledge on the mechanical soil properties have resulted in large inaccuracy of the rover's mobility prediction in the past. This paper deals with the prediction of mechanical properties of the soil based on the experimental and computational model of a six-wheeled rocker bogie rover. The work is divided into two parts. First, a physical model of the rover was fabricated and was made to travel on an unknown loose soil on earth. For this, a known reference value of revolutions per minute (RPM) was given to the direct current (DC) motors and the corresponding linear speed of the rover was measured. Next, a terramechanics based dynamics model was developed for a nominal value of the mechanical soil properties. The RPM needed to maintain the same linear speed as the experimental value was computed for the assumed mechanical soil properties. These soil properties were altered within a range such that the RPM obtained from the experimental and the computational results were similar to maintain the same linear velocity. The results were tested and validated for different RPM values for the predicted mechanical soil properties, which proved to be satisfactory.

Keywords Rocker bogie · Cohesion · Internal friction · Prediction · Arduino · L293D (motor driver)

S. Nithin (✉) · B. Madhevan
School of Mechanical and Building Sciences, VIT University, Chennai, India
e-mail: nithin.s.nair90@gmail.com

B. Madhevan
e-mail: madhevan.b@vit.ac.in

R. Ghosh · G.V.P. Bharat Kumar · N.K. Philip
Control Dynamics and Simulation Group, ISRO Satellite Center (ISAC), Bengaluru, India
e-mail: rimag@isac.gov.in

G.V.P. Bharat Kumar
e-mail: bharat@isac.gov.in

N.K. Philip
e-mail: philip@isac.gov.in

S.S. Dash et al. (eds.), *Artificial Intelligence and Evolutionary Computations in Engineering Systems*, Advances in Intelligent Systems and Computing 517,
DOI 10.1007/978-981-10-3174-8_18

1 Introduction

For the prediction of mobility of the rovers, the knowledge of mechanical soil properties must be known in advance. Lack of knowledge of these properties would result in inaccurate estimation of rover wheel–soil interaction parameters such as slip, torque etc. Gallina [1] has explained in his paper about the two kinds of uncertainty which the rover would face during its mobility. First, deals with the soil where the rover is tested, will be different from the one it is traveling on and second deals with the lack of knowledge of the soil properties. Wong [2] in his paper has dealt with the former uncertainty where he has predicted the performance of the rover wheels on planetary terrain from the results obtained on earth. He has predicted that sinkage and motion resistance are a function of gravity. This paper deals with the latter uncertainty where the soil property of a given terrain can be predicted based on the experimental and computational results of the rover's mobility.

The mechanical soil properties that affects the mobility of the rover includes: '*c*'—cohesion stress, 'φ'—friction angle, 'k_c' and 'k_φ'—pressure-sinkage module, *k*—soil deformation module and '*n*'—sinkage exponent. These soil properties are usually found by least square methods or using Kalman filters. Bevameter experiments were carried out where the relationship between pressure, sinkage, normal force, and torque were found [1]. Krenn [3] has used a method called soil contact method. These parameters can even be experimentally found where cohesion and internal friction angle were found by shear strength test; k_c, k_φ, and *n* by pressure-sinkage relationship [4]. Bauer [5] has found the parameters such as *n*, k_c and k_φ using a flat-plate experiment and internal friction angle φ was found from the soil slope. Nagatani [6] has found the value of soil deformation module for various slip ratios. But the cons of such experiments are that it involves the equipment setup cost and that these experiments are carried out offline. Sojourner rover had used Bevameter type of device to find internal friction angle and soil cohesion which is again an offline method [7]. But it is important for rovers to sense the changing terrain conditions and alter its control strategies to ensure safe operation. For this the rover must first sense the changing terrain and alter its terrain parameters. Lagnemma [7] in his paper has used unmanned ground vehicles (UGV) where it visually classifies the type of terrain and assumes a nominal value for internal friction and cohesion. He has explained another method to classify the terrain using a vibration-based technique [8]. Gareth [9] has explained in his paper about the relationship between the pressure and diameter of the rover wheel for a given value of soil deformation module.

In this paper, the mechanical soil properties are predicted based on the experimental and computational results of the mobility of the rover. These soil properties are altered for the range given by Shilby [10]. A six-wheeled rocker bogie rover [11] is used for the experimental and computational analysis. The study here is conducted for the terrain on earth. For the experimental analysis, the rover was fabricated and tested on an unknown terrain. Six direct current geared motors are used where the speed is controlled by an Arduino Uno microcontroller interfaced with L293D motor driver circuit. The computation was based on Wong and Reece model [12] and MATLAB was used for the simulation purpose. No sensors are used for this study and thus the wheel-soil interaction parameters such as slip, torque etc. were mathematically computed. Finally, the mechanical soil properties are altered such that the RPM needed by the motors from the experimental and computational model are similar for a given tolerance range maintaining the same linear speed.

2 Methodology

This section explains the development of the rocker bogie rover and the controller used for the DC motor control. Later, a computational terramechanics model is explained to compute the RPM needed by the DC motor.

2.1 Rocker Bogie Rover

A six-wheeled rocker bogie design is chosen as it would be able to move over obstacles maintaining constant contact with the surface. Figure 1 shows the rocker bogie model used for this study.

It consists of two rockers and two bogie arrangement. All the six wheels of the rover are independently driven by six geared DC motors. L_1 and L_2 are the link lengths of the rocker. L_3 and L_4 are the link lengths of the bogie. The rover specifications are given in Table 1.

The angle between L_1–L_2 and L_3–L_4 is 120°. The DC motors used have a gear ratio of 20. As the rover moves over an undulating terrain, the angle between L_2 and

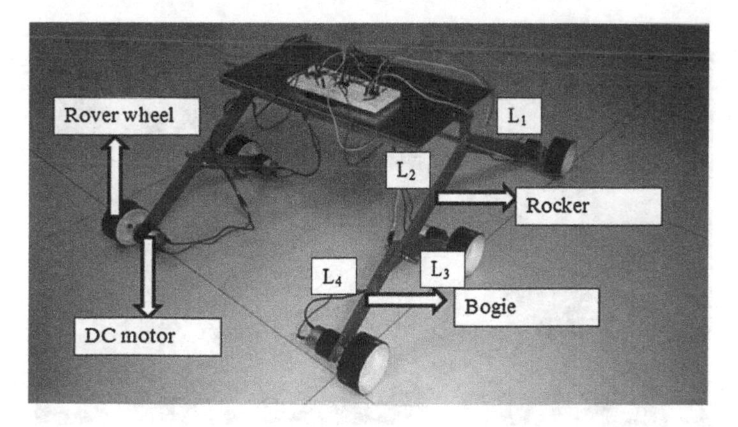

Fig. 1 Rocker bogie rover model

Table 1 Rocker bogie rover specifications

Parameter	Dimension	Unit
L_1	0.4	m
L_2	0.200	m
L_3	0.200	m
L_4	0.200	m
Wheel radius (r)	0.0375	m
Wheel width (b)	0.035	m
Rover weight	2.6	kg

L_3 changes which decides the weight acting on the rover. Figure 2 shows rover configuration as it moves over an obstacle.

2.2 *Arduino Uno and L293D*

Arduino Uno is the microcontroller used in this research to control the speed of the DC motors. It uses ATmega328P integrated circuit (IC) to control the DC motors by pulse width modulation (PWM) techniques. Function called analogWrite() is used to enter the needed speed. The program is written and uploaded into the Arduino Uno where the DC motors can be controlled for the time delay given in the program.

In order to drive the six DC motors, three L293D motor driver ICs are used. It works like H-Bridge circuit where the direction of the motors can be controlled. Each IC has four input pins and four output pins to control two DC motors at the same time. Figure 3 shows the wiring connections used in this work.

A 12 V power supply is given to the Arduino and motor driver. The PWM signal based on the duty cycle programmed to it gives the corresponding voltage output

Fig. 2 Rocker bogie rover on undulating surface

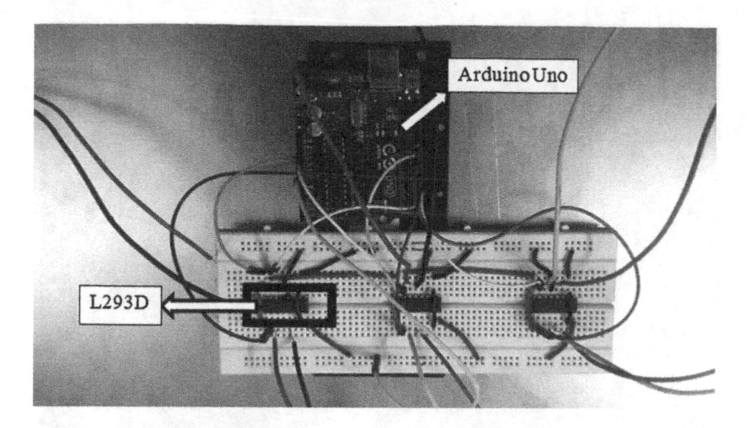

Fig. 3 Arduino Uno and L293D interface

from the Arduino Uno to the input pins of the motor driver. This corresponding voltage is used to drive the DC motors thus controlling it.

2.3 *Terrain Test Bed*

For the experimental analysis, terrain consisting of dry sand with partial clay content is used. The terrain used is 0.26 m in length and 0.085 m in width. Figure 4 shows the terrain used for this study.

For the study, a flat terrain was considered to know the behavior of the rover as it moves over the terrain. A considerable amount of sinkage was noted and the rover wheel's track print was visible when the rover is commanded to travel at 51 RPM. Figure 5 shows the image of the rover wheel, its corresponding sinkage and the track print.

2.4 *Computational Terramechanics Model*

The terramechanics model is based on Bekker's model where the RPM needed by the motor was computed based on the slip the rover wheels will experience as it traverse over the terrain. The weight acting on the rover is obtained from kinematics as per Fig. 6.

The below equation was used to compute the RPM of the motor. These equations were taken from [13, 4]

Fig. 4 Experimental terrain setup

$$z = \left[\frac{3w \cos(\text{slope})}{b(3-n)\left(\frac{k_c}{b} + k_\phi\right)\sqrt{2r}} \right]^{\frac{2}{2n+1}} \quad (1)$$

$$W = rb \left[\int_{\theta_2}^{\theta_1} \sigma(\theta) \cos\theta \, \mathrm{d}\theta + \int_{\theta_2}^{\theta_1} \tau(\theta) \sin\theta \, \mathrm{d}\theta \right] \quad (2)$$

$$\sigma(\theta) = r^n \left(\frac{k_c}{b} + k_\phi \right) [\cos\theta - \cos\theta_1]^n \quad (3)$$

$$\tau(\theta) = (c + \sigma(\theta) \tan\phi) \left[1 - \mathrm{e}^{-j(\theta)/k} \right] \quad (4)$$

$$j(\theta) = r[\theta_1 - \theta - (1-i)(\sin\theta_1 - \sin\theta)] \quad (5)$$

$$DP = rb \left[\int_{\theta_2}^{\theta_1} \tau(\theta) \cos\theta \, \mathrm{d}\theta - \int_{\theta_2}^{\theta_1} \sigma(\theta) \sin\theta \, \mathrm{d}\theta \right] \quad (6)$$

Fig. 5 Rover track print

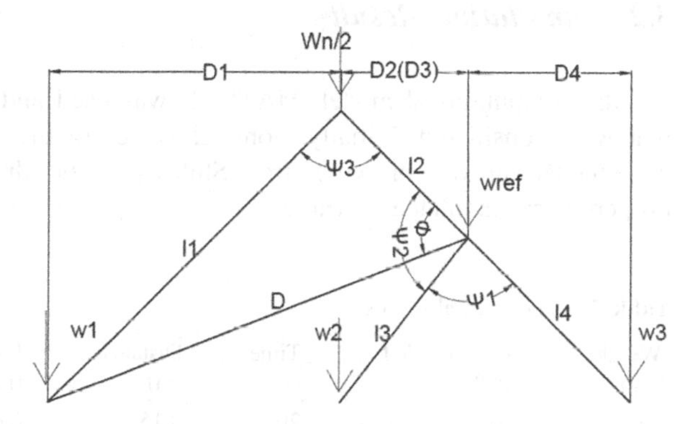

Fig. 6 Kinematics weight distribution, where, W_n is the total weight acting on the rover in Newton

$$DP = w * \sin(\text{slope}) \tag{7}$$

$$\omega = \frac{v}{(r - ir)} \quad [\text{for } (0 \leq i < 1)] \tag{8}$$

$$\text{RPM} = (\omega * \text{gr}) * \left(\frac{60}{2\pi}\right) \tag{9}$$

where 'z' is the sinkage, 'θ_1' is the entry angle, 'DP' is the drawbar full force, 'ω' is the angular velocity, 'v' is the linear velocity, 'i' is the slip ratio, and 'gr' is the gear ratio.

Entry angle is computed by iterative method where the weight obtained from kinematics must be equal to the weight in Eq. (2). Similarly, slip ratio is obtained for the value where DP in Eq. (6) is equal to the value in Eq. (7). The angular velocity and RPM is obtained for a needed linear velocity and slip ratio for the terrain.

3 Experimental and Simulation Results

3.1 Experimental Results

The rover with the specifications mentioned in Table 1 was made to move on the experimental terrain setup. Arduino Uno was used to command the motor to rotate at 51 RPM for 20 s. The results are tabulated in Table 2. It was noted that for a 51 RPM, the rover travels with a linear velocity of 0.0075 m/s.

3.2 Simulation Results

For the computational model, MATLAB was used and the results obtained from that were considered. Initially, nominal values of the mechanical soil properties were the terrain was considered from Shilby [10]. For dry sand, the mechanical soil properties are tabulated in Table 3.

Table 2 Experimental results

Weight (kg)	Commanded RPM	Time (s)	Distance (m)	Linear speed (m/s)	Sinkage (m)
2.6	51	20	0.15	0.0075	0.004

Table 3 Dry sand property from Shilby [10]

Parameter	Value	Units
n	1.1	–
k_{eq}	1,549,114.28	$\frac{N}{m^{(n+2)}}$
k	0.025	m
φ	30	deg
c	1000	$\frac{N}{m^2}$

Where k_{eq} is $\frac{k_c}{b} + k_\phi$

Table 4 Computational results for dry sand

Weight (kg)	Reference linear speed (m/s)	Time (s)	Distance (m)	Computed RPM	Sinkage (m)
2.6	0.008	20	0.16	134.47	0.0081

Table 5 Predicted mechanical soil properties

Parameter	Value	Units
n	0.8	–
k_{eq}	500,000	$\frac{N}{m^{(n+2)}}$
k	0.01	m
φ	30	deg
c	1000	$\frac{N}{m^2}$

For the values in Table 3, the results for the computational model are tabulated in Table 4 for a linear speed of 0.008 m/s.

It can be noted in Table 4 that the RPM of 134.47 is obtained instead of 51 RPM to maintain the same linear speed. Also the sinkage obtained in the computational model was double of that of the experimental values. Thus, the mechanical soil properties considered does not hold good for the soil in this study. From [10], the range of values of the soil was obtained. It was noted that the values of sinkage exponent, pressure-sinkage module, and soil deformation module were reduced for a clay surface. Since the soil considered for this study has clay content in it, the mechanical soil properties were altered. Around 20 combinations of values were performed and for the mechanical soil properties tabulated in Table 5, the computational results were similar to the experimental results.

For the soil properties as per Table 5, the corresponding simulation results are tabulated.

On comparing Table 2 with Table 6, it can be noted that to maintain the same linear speed the RPM was similar with a negligible difference of 0.64 RPM. Also the sinkage obtained was similar with a negligible difference of 0.0017 m.

3.3 Validation

In order to ensure that the predicted results are authentic, the rover was made to travel on the test terrain setup at various RPM and these results were compared with the computational results. The results from the experimental test are tabulated in Table 7 and the results for the computations test are tabulated in Table 8.

For the computational model, a tolerable limit of 0.0004 m/s is chosen. Comparing the second column of Table 7 and fifth column of Table 8, it can be seen that for the predicted mechanical soil properties the RPM needed by the motors is the similar to maintain the same linear speed. This, it can be inferred that

Table 6 Computational results for dry and clay sand

Weight (kg)	Reference linear speed (m/s)	Time (s)	Distance (m)	Computed RPM	Sinkage (m)
2.6	0.008	20	0.16	51.64	0.0057

Table 7 Experimental results for random RPM

Weight (kg)	Commanded RPM	Time (s)	Distance (m)	Linear speed (m/s)
2.6	41	20	0.111	0.0056
2.6	51	20	0.150	0.0075
2.6	63	20	0.186	0.0093
2.6	67	20	0.214	0.0107

Table 8 Computational results for corresponding linear speed from experimental data

Weight (kg)	Reference linear speed (m/s)	Time (s)	Distance (m)	Computed RPM
2.6	0.006	20	0.12	40.67
2.6	0.0079	20	0.158	50.99
2.6	0.0097	20	0.194	62.61
2.6	0.0103	20	0.206	66.49

for the predicted mechanical soil property the experimental and computational results were within an acceptable tolerance limit.

3.4 Maximum Weight Computation

Study was carried out to find the maximum load carrying capacity of the rover on a flat rigid surface and on a flat loose surface. The rover weight was measured to be 2.6 kg. The weights were added on the rover in a step of 1 kg and the results were tabulated for a flat rigid surface in Table 9 and for a flat loose surface in Table 10.

Table 9 Experimental rover weight carrying capacity on a flat rigid surface

Weight (kg)	RPM	Time (s)	Distance (m)	Linear speed (m/s)
2.6	63	20	0.25	0.0125
3.6	63	20	0.235	0.0118
4.6	63	20	0.22	0.0110
5.6	63	20	0.198	0.0099
6.6	63	20	0.188	0.0094
7.6	63	20	0.174	0.0087

Table 10 Experimental rover weight carrying capacity on a flat loose surface

Weight (kg)	RPM	Time (s)	Distance (m)	Linear speed (m/s)
2.6	63	20	0.186	0.0093
3.6	63	20	0.146	0.0073
4.6	63	20	0	0
5.6	63	20	0	0
6.6	63	20	0	0
7.6	63	20	0	0

From Table 9, it is seen that the maximum weight the rover can carry is around 7.6 kg and from Table 10, maximum weight the rover can carry is around 3.6 kg. This is because as the rover traverse over the terrain, the soil offers resistance to the wheels of the rover. On addition of weight, the load acting on the wheels will increase making it unable to overcome this resistance. In other words, the torque output from the DC motor is less when compared to the torque needed by the wheels to move over the terrain when the weight acting on the rover is greater than 3.6 kg.

4 Conclusion and Future Work

In this study, the mechanical soil properties of an unknown terrain were predicted based on the experimental and computational results of the rover mobility. Since the soil considered for this study had clay content in it, the soil deformation module and the sinkage exponent values were decreased. About 20 combinations of values were tested and the predicted values as per Table 5 were proved to be satisfactory. These results were validated for various RPM in the experimental test and compared with the simulation results. Also, the maximum weight carrying capacity of the rover was found both on a rigid and loose soil. All the tests were carried on a flat terrain and for earth soil.

This work can be further extended for an undulating surface and also for planetary soils. For the application on planetary surface, visual odometry techniques can be used to measure the distance traveled by the rover and compute its corresponding linear speed.

Acknowledgments The authors would like to thank Mr. Shamrao, Spacecraft Mechanisms Group, ISAC (ISRO) for his valuable suggestions and support throughout the research. His significant comments and remarks have guided us to obtain the needed results.

References

1. Alberto Gallina, Rainer Krenn, Bernd Schafer; "On the Treatment of Soft Soil Parameter Uncertainties in Planetary Rover Mobility Simulations", Journal of Terramechanics 63 (2016) 33–47, 2015 ISTVS. Published by Elsevier, Ltd.
2. J.Y. Wong, "Predicting the Performances of Rigid Rover Wheels on Extraterrestrial Surfaces Based on Test Results Obtained on Earth", Journal of Terramechanics 49 (2012) 49–61, 2011 ISTVS. Published by Elsevier Ltd.
3. Rainer Krenn, Gerd Hirzinger; "Simulation of Rover Locomotion on Sandy Terrain - Modeling, Verification and Validation", Conference: 10th ESA Workshop on Advanced Space Technologies for Robotics and Automation-ASTRA, 2008.
4. Genya Ishigami, Akiko Miwa, Keiji Nagatani, Kazuya Yoshida; "Terramechanics-Based Model for Steering Maneuver of Planetary Exploration Rovers on Loose Soil", Journal of Field Robotics Special Issue: Special Issue on Space Robotics, Part I Volume 24, Issue 3, pages 233–250, March 2007.
5. Robert Bauer, Winnie Leung, Tim Barfoot; "Experimental and Simulation Results of Wheel-Soil Interaction for Planetary Rovers", Intelligent Robots and Systems, 2005. (IROS 2005). 2005 IEEE/RSJ International Conference.
6. Keiji Nagatani, Ayako Ikeda, Keisuke Sato and Kazuya Yoshida; "Accurate Estimation of Drawbar Pull of Wheeled Mobile Robots Traversing Sandy Terrain Using Built-in Force Sensor Array Wheel", The 2009 IEEE/RSJ International Conference on Intelligent Robots and Systems October 11–15, 2009 St. Louis, USA.
7. Karl Iagnemma, Steven Dubowsky; "Terrain Estimation for High-Speed Rough-Terrain Autonomous Vehicle Navigation", Proceedings of the SPIE, Volume 4715, p. 256–266 (2002).
8. Karl Iagnemma, Christopher Brooks, Steven Dubowsky; "Visual, Tactile, and Vibration-Based Terrain Analysis for Planetary Rovers, Aerospace Conference, 2004. Proceedings. 2004 IEEE.
9. Gareth Meirion-Griffith, Matthew Spenko; "A New Pressure-Sinkage Model for Small, Rigid Wheels on Deformable Terrains", Proceedings of the Joint 9th Asia-Pacific ISTVS Conference and Annual Meeting of Japanese Society for Terramechanics Sapporo, Japan, September 27 to 30, 2010.
10. H. Shibly, K. Iagnemma, S. Dubowsky; "An Equivalent Soil Mechanics Formulation for Rigid Wheels in Deformable Terrain with Application to Planetary Exploration Rovers", Journal of Terramechanics 42 (2005) 1–13, Published by Elsevier Ltd.
11. Herve Hacot; "Analysis and Traction Control of a Rocker-Bogie Planetary Rover", Master of Science in Mechanical Engineering Thesis Massachusetts Institute of Technology, 1998.
12. Jo-Yung Wong and A. R. Reece; "Prediction of Rigid Wheel Performance Based on the Analysis of Soil-Wheel Stresses Part I. Performance of Driven Rigid Wheels", Journal of Terramechanics, 1967, Vol. 4, No. 1, pp. 81 to 98. Pergamon Press Ltd.
13. J.Y. Wong; "Theory of Ground Vehicles", Third edition.

Exploiting VLC Technique for Smart Home Automation Using Arduino

K.P. Swain, M.V.S.V. Prasad, G. Palai, J. Sahoo and M.N. Mohanty

Abstract The use of visible light communication (VLC) in the area of smart home automation is presented in this paper. The VLC link between two Arduino development board containing microcontroller to transfer the controlling ASCII character through LED and LDR is also discussed. The Proteus ISIS simulation software is used to realize the data transfer between two microcontrollers. Simulation result revealed that different electrical gadgets can be controlled by using VLC link. The Experiment is also carried out to show its validation.

Keywords VLC · Arduino · Proteus ISIS · Home automation

1 Introduction

Visible light communication (VLC) technology [1, 2] is one of the competing and rapidly emerging technology, where a current driven semiconductor LED and photo diode is primarily used for data communication in the visible range and offers attractive benefits, such as: larger bandwidth, high security, harmless for human health, unregulated frequency, no electromagnetic interference, etc. In this technology, LEDs are mostly used due to its fast switching action which enables its dual functionality lighting and wireless data communication. VLC conveys information by adopting mostly On-off keying and pulse position modulation in which the light intensity is modulated by LED at a very fast rate. Most of the research in this area includes the structure of the LED, minimization of BER (bit error rate), range

K.P. Swain · M.V.S.V.Prasad · G. Palai (✉) · J. Sahoo
GITA, Bhubaneswar, Odisha, India
e-mail: gpalai28@gmail.com

K.P. Swain
e-mail: kaleep.swain@gmail.com

M.N. Mohanty
I.T.E.R, Bhubaneswar, Odisha, India
e-mail: mihir.n.mohanty@gmail.com

S.S. Dash et al. (eds.), *Artificial Intelligence and Evolutionary Computations in Engineering Systems*, Advances in Intelligent Systems and Computing 517,
DOI 10.1007/978-981-10-3174-8_19

extension, etc. In some cases, organic LED and Organic photo diodes are used along with some equalization technique to reduce the BER. In order to reduce the cost and easy implementation the light intensity modulation and direct detection methods are used where light emitted by an LED is directly converted to its equivalent current by a photo diode or LDR for recovering the original data [3–5]. A microcontroller is used to control the data flow for the same. On the other hand, VLC suffers certain limitation compared to other wireless technology. The main disadvantage is the range as data rate falls drastically with increases of distance and data rate decreases when photo diode exposed to sunlight directly. Also line of sight communication is essential for this type of communication and light should be in power on position in order to send the data [6].

Though, smart home automation is already being implemented by using different technologies like Bluetooth, Wireless, Zigbee but the use of visible light communication (VLC) in the same area is hardly used. For validation, first VLC link is confirmed between two Arduino development boards and ASCII data are transferred from transmitting Arduino through the LED by modulating the signal in On-Off Keying (OOK) process and same ASCII characters is received by receiving Arduino containing a LDR which transferred the light energy to its equivalent electrical signals. The speed of data transmitting can be controlled by controlling the toggling time of LED. The VLC link is first simulated by using a Proteus ISIS simulator and data are checked in hyper terminal. Then, the same process is experimentally tested by using two Arduino development boards and data are also being checked through hyper terminal and found both transmitting and receiving data are same. After this preliminary process, a smart home automation circuit is designed where different electrical gadgets are controlled by a local computer and Arduino through visible light and controlling process is simulated by using the processing (open-source programming language) sketch which redirects the data from hardware to Proteus ISIS software environment.

2 Proposed Architecture

To realize this, we use a novel architecture by which one can carry out the automation, which is shown in Fig. 1. In Fig. 1, the main focus is given to the transportation of data between two microcontroller using two Arduino development board which is clearly mentioned in the box of the Fig. 1. This section not only simulated in Proteus ISIS software, but also physically implemented successfully.

The rest functionalities as shown in the block diagram is only simulated in Proteus ISIS by redirecting the data from Arduino to Proteus software. The appliances are assigned with two IDs one for turning it ON and other for turning OFF. These IDs are 8-bit character data which is sent with the help of a local computer to the Arduino through serial port. The Arduino then sends the data via a sender LED and the photosensitive element (LDR in this case) receives the data and triggers the state of the appliance according to the assigned ID.

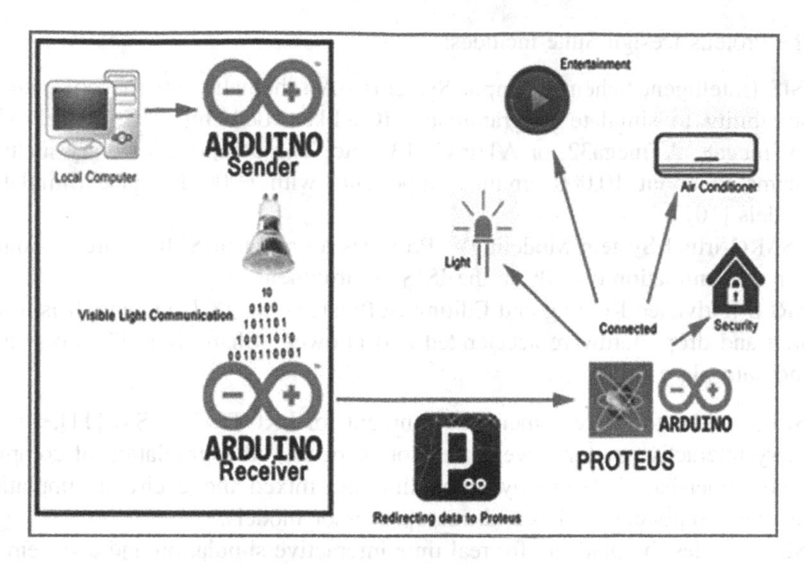

Fig. 1 Proposed architecture of smart home automation using VLC

2.1 Arduino Development Board

Arduino is an open-source physical computing [7, 8] platform based on a simple microcontroller board and a development environment that implements the processing language. It was originally meant for artists and designers to create electronic prototypes. They would be able to create these designs easily with a little knowledge of programming and electronics without going too deep into it. Electronic prototyping was traditionally only associated with engineering and engineers. Arduino microcontroller is an open-source electronics mixed hardware and software platform for the development and research. The programming language of Arduino is based on a simplified version of C/C++. Merely a few commands are required for programming the Arduino to perform in some useful application.

In this paper, we use the Arduino UNO [9] which is a microcontroller board based on the Atmega 328. It bears a 16 MHz ceramic resonator, 14 digital input/output pins (of which 6 can be used as PWM outputs), 6 analog inputs, an ICSP (in-circuit serial programming) header, a power jack, a USB connection, and a reset button. A general Arduino board contains every unit to support the microcontroller for any application; simply connect it to a computer with a USB cable or power it with an AC-to-DC adapter or battery to get started. So it became a first choice for engineers. Analog pins are capable of reading in 1024 bits and digital writing capability of 256 bit. 2.3 Proteus ISIS Simulation.

Proteus is a software package developed by Labcenter Electronics, for computer-aided design, schematic capture, microcontroller simulation, and printed circuit board (PCB) layout design.

The Proteus Design Suite includes:

- ISIS (Intelligent Schematic Input System)—A schematic capture tool with the possibility to simulate programmable ICs like Microchip PIC, Atmel AVR (ATmega8, ATmega32, or ATtiny2313), etc. The component library includes claims to about 10,000 circuit components with 6000 Prospice Simulation models [10].
- VSM (Virtual System Modeling)—Provides a graphical SPICE circuit simulation and animation directly in the ISIS environment.
- ARES (Advance Routing and Editing Software)—for PCB layouts. It is to use drag and drop, hardware accelerated and allowed "Shape Based" auto routing and auto placement.

ISIS provides the development environment for PROTEUS VSM [11], a revolutionary interactive system level simulator. It permits the simulation of complete microcontroller-based design by combining the mixed mode circuit simulation, interactive component models and microprocessor models.

ISIS provides the platform for real time interactive simulation and a system for managing the source and object code associated with each project. In summation, a number of graphical objects can be ordered on the schematic to set, frequency, conventional time, and swept variable simulation to be performed.

2.2 Processing

Processing [12] is an open-source programming environment based upon Java programming language and the programs written for this is known as 'Sketches'. Also, it supports a multi-language environment which means processing sketches can be of Java or Javascript or Python according to the choice of the programmer. Processing has an in-built and third-party libraries which support serial communication and many other hardware to software interfaces. In this work, Processing environment with Arduino board is used together and the processing sketch is used to redirect the data from hardware to Proteus environment and the executable code generated from the processing environment is very lightweight and platform independent (Fig. 2).

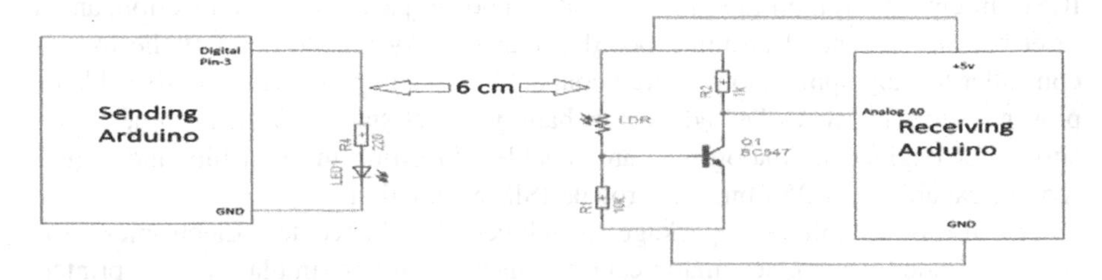

Fig. 2 Circuit diagram of data communication between two Arduino

3 Working

3.1 Transmitter Circuit

The sender circuit comprises of an Arduino Development Board (UNO) and an LED for sending the data using visible light. The anode of the LED is connected to the digital output of the Arduino (pin 3) and the cathode is connected to the Ground terminal of the Arduino. The sender Arduino encodes the ASCII characters to its equivalent binary combinations and turns the LED ON and OFF for sending data, but before sending the actual data it sends (1, 0) trigger bits for synchronizing with the receiver. The transmitting mechanism is given as follows:

- *Waiting for input:* When data are to be transmitted in the form of ASCII code (single character or multiple characters), after pressing the desired key on the keyboard, the 'ENTER' key must be pressed.
- *Storing of data*: After the ENTER key is pressed, the whole data are stored in a variable defined in the program.
- *Sending of Trigger Bits:* To synchronize the both transmitter and receiver, two trigger bits (1 & 0) are sent along with the actual data.
- *Sending the Information:* Then LED is on for 1 and off for 0 in a very fast manner according to time is set in the program to send the whole data through visible light (Fig. 3).

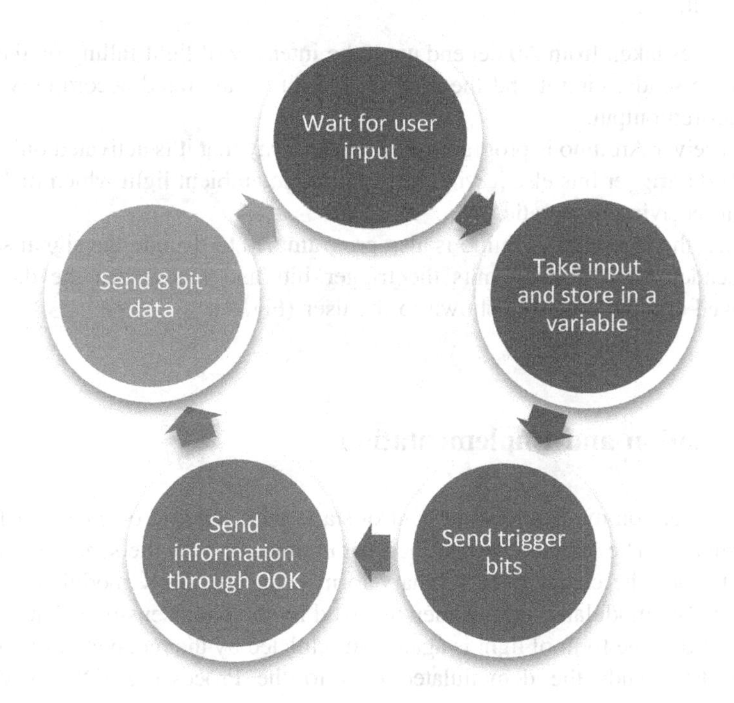

Fig. 3 Flow diagram of transmitting mechanism

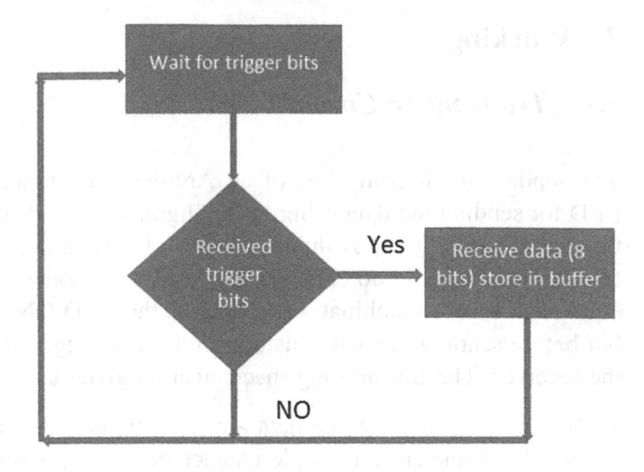

Fig. 4 Flow diagram receiver mechanism

3.2 *Receiver Circuit*

The receiver circuit comprises of a transistor BC547, LDR, 10 k resistor, and 1 k resistor for biasing of the transistor. The output of the sender circuit is taken across R2 resistor shown in the diagram and is connected to the Arduino analog Input pin A0. Receiver circuit decodes the data from the sender circuit, and shows the equivalent ASCII character for the binary combination. The receiving mechanism is given as follows:

- The values taken from A0 depend upon the intensity of light falling on the LDR from the sender circuit and the receiver circuit is calibrated accordingly to get the desired output.
- The receiver Arduino is programmed in such a way that it is activated only when it gets the trigger bits else it willll not respond to ambient light which makes the signal receiving automatic.
- Further, the receiving Arduino is also programmed to decode the signal sent by the sender, such that it omits the trigger bits and recognizes the data it is received and the output is shown to the user (Fig. 4).

4 Simulation and Implementation

The process of communication and control starts with sending of character from a local computer. The local computer, then communicates with the sending Arduino, so that it sends the character in a binary form using the OOK modulation of the LED light. The modulated light is then received by the photosensitive element. The data received in the form of light is again demodulated by the receiver Arduino. The Arduino then sends the demodulated data to the Processing IDE. A custom processing sketch is written such that it receives the data from the receiving Arduino and redirects to the Proteus Simulation environment. The emulated Arduino then listens to the data from the Processing sketch and responds to the data accordingly.

Due to absence of virtual appliances in the Proteus Environment a fan has been simulated as a simple DC motor and some other appliances as LEDs. In this circuit character 'L' has been assigned to turn on the lamp and 'K' to turn it off. Similarly 'F' to turn the fan ON and 'G' to turn the fan OFF. All these communication process is done via short range Visible light communication.

Figure 5 shows the simulation of visible light communication and the result is verified in the putty terminal. Here, putty is the freeware software which used as the replacement for hyper terminal to show data communication via serial port.

Figure 6 shows the implementation of the communication process through hardware using two Arduino board through VLC link. It can be marked the same data is received in the receiver port as transmitted in Fig. 7.

As shown in Fig. 8, the Arduino IDE serial monitor is used to send the required controlling character to control the appliances in a home and the output is shown in processing output window whereas the result is shown in Proteus Window. Here 'L' is used to control the light that can be marked by the indication in the lamp and also in the LCD.

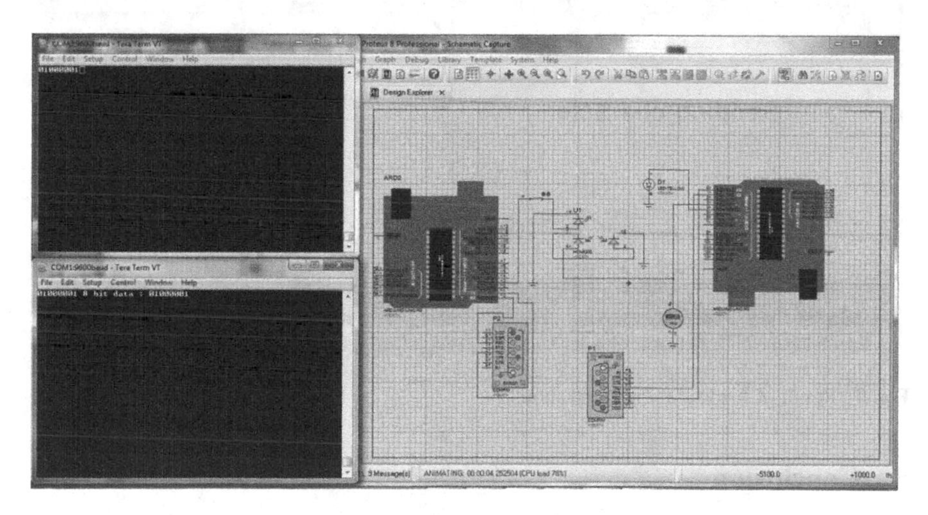

Fig. 5 Simulation diagram of communication between two Arduino with result

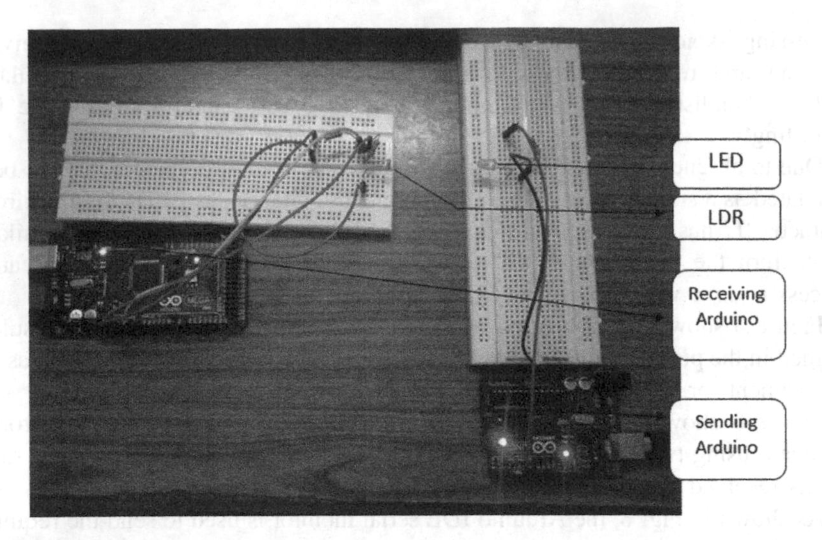

Fig. 6 Hardware implementation

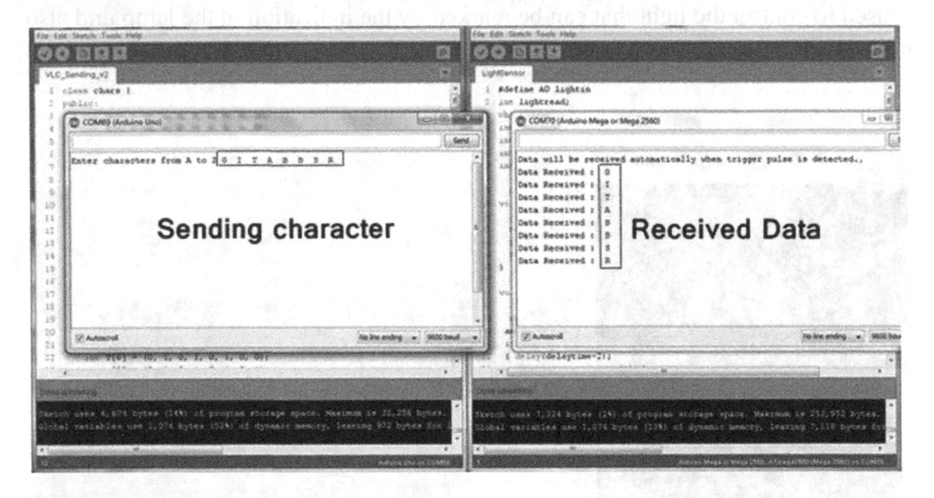

Fig. 7 Result of hardware implementation

5 Conclusion

The use of visible light communication (VLC) link in the area of smart home automation is thoroughly explored in this paper. Two Arduino development boards containing microcontroller along with LED and LDR are cogitated during the process of automation. VLC link between two Arduino boards is simulated using Proteus ISIS software and then an experiment is carried out to validate the simulation result. The present result reveals that the speed of data transmission varies

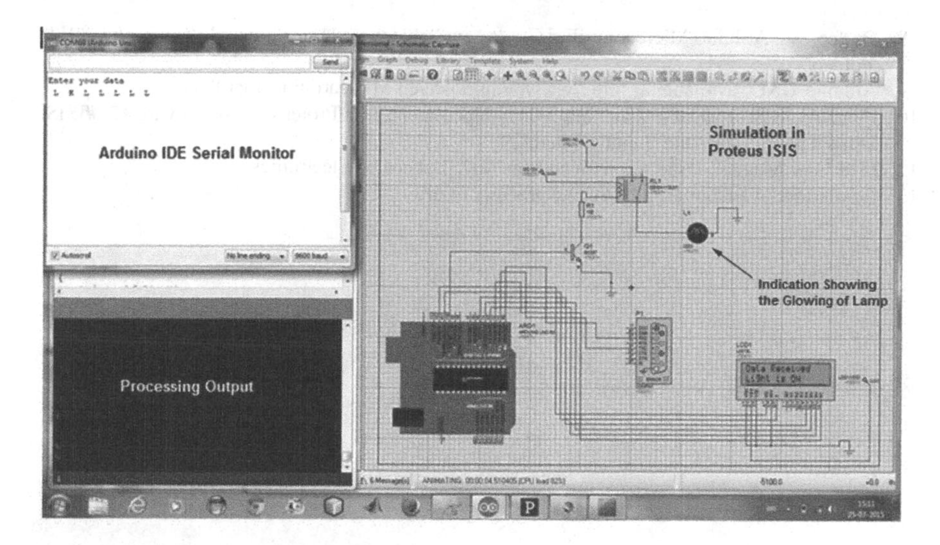

Fig. 8 Total simulation of home automation in proteus with result

with the variation of on-off time delay of LED. Simulation is also carried out to control the different electrical gadgets by transferring the ASCII characters from controlling Arduino. Finally, the software, 'processing' is employed to redirect the data into Proteus ISIS and the result infers that the accurate investigation using VLC can be used successfully for the automation of a smart home.

References

1. Haigh, P.A., Ghassemlooy, Z., Le Minh, H., Rajbhandari, S., Arca, F., Tedde, S. F., Oliver, H., Papakonstantinou, I.: Exploiting Equalization Techniques for Improving Data Rates in Organic Optoelectronic Devices for Visible Light Communications. Journal of lightwave technology, 30 19(2012), 3081–3088
2. Grobe, L., Paraskevopoulos, A., Hilt, J., Schulz, D., Lassak, F., Hartlieb, F., Kottke, C., Jungnickel, V., Langer, K.D.: High-Speed Visible Light Communication Systems. IEEE Communications Magazine, (2013) 60–66
3. Sugiyama, H., Haruyama, S., Nakagawa, M.: Experimental investigation of modulation method for visible-light communications. IEICE, Trans. Commun., E89-B 12 (2006) 3393–3400
4. Kim, H.S., Kim, D.R., Yang, S.H., Son, Y.H., Han, S.K.: An Indoor Visible Light Communication Positioning System Using a RF Carrier Allocation Technique. Journal of lightwave technology, 31 1(2013) 134–144
5. Lee, S.H., Jung, S.Y., Kwon, J.K.: Modulation and Coding for Dimmable Visible Light Communication. Optical Communication, IEEE Communications Magazine (2015) 136–143
6. Jovicic, A., Li, J., Richardson, T., Research, Q.: Visible Light Communication: Opportunities, Challenges and the Path to Market. IEEE Communications Magazine(2013) 26–32
7. Galadima, A.A.: Arduino as a learning tool. International Conference on Electronics, Computer and Computation (ICECCO) (2014) 1–4

8. Durfee, W.: Arduino Microcontroller Guide" available in www.me.umn.edu/courses/me2011/arduino/
9. Arduino UNO, available at http://www.arduino.cc/en/Main/arduinoBoardUno
10. Proteus ISIS available at http://de.wikipedia.org/wiki/Proteus_%28Software%29#ISIS.28Intelligent_Schematic_Input_System.29
11. ISIS User Manual, Issue 6.0, November 2002, Labcenter Electronics
12. Processing, available at: http://www.processing.org

Intuitionistic Hesitant Fuzzy Soft Set and Its Application in Decision Making

R.K. Mohanty and B.K. Tripathy

Abstract There are several models of uncertainty found in the literature like fuzzy set, rough set, intuitionistic fuzzy set, soft set, and hesitant fuzzy set. Also, several hybrid models have come up as a combination of these models and have been found to be more useful than the individual models. In everyday life we make many decisions. Making efficient decisions under uncertainty needs better techniques. Many such techniques have been developed in the recent past. These techniques involve soft sets and intuitionistic fuzzy sets. It is well known that intuitionistic hesitant fuzzy sets are more general than intuitionistic fuzzy sets. In this paper, we define intuitionistic hesitant fuzzy soft sets (IHFSS) and we also propose a decision making technique, which extends some of the recently developed algorithms. We also provide an application from real-life situations, which illustrates the working of the algorithm and its efficiency over the other algorithms.

Keywords Soft set · Fuzzy sets · Fuzzy soft sets · Intuitionistic fuzzy set · Hesitant sets · Intuitionistic fuzzy soft set · Decision making

1 Introduction

The notion of fuzzy sets introduced by Zadeh [1] in 1965 is one of the most fruitful models of uncertainty and has been extensively used in real-life applications. In order to bring topological flavor into the models of uncertainty and associate family of subsets of a universe to parameters, Molodtsov [2] introduced the concept of soft sets in 1999. A soft set is a parameterized family of subsets. Many operations on soft sets were introduced by Maji et al. [3, 4]. Hybrid models are obtained by suitably combining individual models of uncertainty have been found to be more

R.K. Mohanty (✉) · B.K. Tripathy
School of Computer Science and Engineering, VIT University, Vellore, Tamilnadu, India
e-mail: rknmohanty@gmail.com

B.K. Tripathy
e-mail: tripathybk@vit.ac.in

S.S. Dash et al. (eds.), *Artificial Intelligence and Evolutionary Computations in Engineering Systems*, Advances in Intelligent Systems and Computing 517,
DOI 10.1007/978-981-10-3174-8_20

efficient than their components. Several such hybrid models exist in literature. Maji et al. [5] put forward the concept of fuzzy soft set (FSS) by combining the notions of fuzzy set and soft set. Tripathy et al. [6] defined soft sets through their characteristic functions. This approach has been highly authentic and helpful in defining the basic operations like the union, intersection, and complement of soft sets. Similarly, defining membership function for FSSs will systematize many operations defined upon them as done in [7]. Many of soft set applications have been discussed by Molodtsov in [2]. An application of soft sets in decision making problems is discussed in [3]. Among several approaches, in [5], FSS and operations on it are defined. This study was further extended to the context of fuzzy soft sets by Tripathy et al. in [7], where they identified some drawbacks in [3] and took care of these drawbacks while introducing an algorithm for decision making. It has been widely known that the concept of intuitionistic fuzzy set (IFS) introduced by Atanassov [8] is a better model of uncertainty than the fuzzy set. The notion of non-membership function introduced, which does not happen to be one's complement of the membership function, introduces more generality and reality to IFS. The hesitation function generated as a consequence is what real-life situations demand. In case of fuzzy sets the hesitation component is zero. The intuitionistic fuzzy sets can only handle the incomplete information considering both the truth membership (or simply membership) and falsity membership (or non-membership) values. It does not handle the indeterminate and inconsistent information which exists in belief system.

Jiyang [9] introduced the concept of IVIFSS by combining the interval valued intuitionistic fuzzy sets (IVIFS) and soft set model. The concept of hesitant fuzzy soft sets was introduced by Sunil et al. They also discussed an application of decision making. In this paper, we introduce intuitionistic hesitant fuzzy soft sets. Here, we follow the definition of soft set due to Tripathy et al. [6] in defining IHFSS. Applications of various hybrid models are discussed in [7, 10–18]. The major contribution in this paper is introducing a decision making algorithm which uses IHFSS for decision making and we illustrate the suitability of this algorithm in real-life situations. Also, it generalizes the algorithm introduced in [5] while keeping the authenticity intact.

The concept of hesitant fuzzy sets was introduced by Torra [19]. This is an extension of fuzzy sets. It is sometimes difficult to determine the membership of an element into a set and in some circumstances this difficulty is caused by a doubt between a few different values. Some operations on hesitant fuzzy sets are defined in [20]. He also discussed an application of decision making. In this paper, we introduce the concept of IHFSS with the help of membership function.

2 Definitions and Notions

In this section, we introduce some of the definitions to be used in the paper. We assume that U is a universal set and E is a set of parameters defined over it.

Definition 2.1 A fuzzy set *A* is defined through a function μ_A called its membership function such that $\mu_A: U \to [0, 1]$.

Definition 2.2 An intuitionistic fuzzy set over *U* is associated with a pair of functions $\mu_A, \nu_A: U \to [0, 1]$ such that for any $x \in U$, $0 \le \mu_A(x) + \nu_A(x) \le 1$.

The hesitation function π_A is defined as $\pi_A(x) = 1 - \mu_A(x) - \nu_A(x)$, $\forall x \in U$.

Definition 2.3 A soft set over the soft universe (*U*, *E*) is denoted by (*F*, *E*), where

$$F: E \to P(U) \tag{1}$$

Here *P*(*U*) is the power set of *U*.

Let (*F*, *E*) be a soft set over (*U*, *E*). Then in [7] it was defined as a parametric family of characteristic functions $\chi_{(F,E)} = \{\chi^a_{(F,E)} | a \in E\}$ of (*F*, *E*) as defined below.

Definition 2.4 For any $a \in E$, we define the characteristic function $\chi^a_{(F,E)}: U \to \{0, 1\}$ such that

$$\chi^a_{(F,E)}(x) = \begin{cases} 1, & \text{if } x \in F(a); \\ 0, & \text{otherwise.} \end{cases} \tag{2}$$

Definition 2.5 A hesitant fuzzy set on *U* is defined in terms of a function that returns a subset of [0, 1] when applied to *U*, i.e.,

$$T = \{\langle x, h(x)\rangle | x \in U\} \tag{3}$$

where *h*(*x*) is a set of values in [0, 1] that denote the possible membership degrees of the element $x \in U$ to *T*.

Definition 2.6 A pair (*F*, *E*) is called a hesitant fuzzy soft set if $F: E \to \mathrm{HF}(U)$, where HF(*U*) denotes the set of all hesitant fuzzy subsets of *U*.

3 Intuitionistic Hesitant Fuzzy Sets

In this section, we introduce the notion of intuitionistic hesitant fuzzy soft sets.

Definition 3.1 A pair (*F*, *E*) is called a intuitionistic hesitant fuzzy soft set if $F: E \to \mathrm{IHF}(U)$, where IHF(*U*) denotes the set of all intuitionistic hesitant fuzzy subsets of *U*.

An IHFSS H on U is defined in terms of its membership function $\mu_H : E \rightarrow P(\text{IHFS})$, $\nu_H : E \rightarrow P(\text{IHFS})$ such that $\forall a \in E$ and $\forall x \in U$, $\mu_H^a(x), \nu_H^a(x) \in P([0, 1])$ such that $0 \leq \sup \mu_H^a(x) + \sup \nu_H^a(x) \leq 1$.

Given three IHFEs in an IHFSS is represented by h, h_1, and h_2. Then we can define union and intersection operations as follows.

Definition 3.2 For any two IHFSSs (*F*, *E*) and (*G*, *E*) over a common universe (*U*, *E*), the union of (*F*, *E*) and (*G*, *E*) is the IFSS (*H*, *E*) and $\forall a \in E$ and $\forall x \in U$, we have

$$\begin{aligned} h_1^a \cup h_2^a &= \{(\alpha_1, \alpha_2) | \alpha_1 \in h_1^a, \alpha_2 \in h_2^a\} \\ &= \left\{ \max\left\{ \mu_{\alpha_1}^a(x), \mu_{\alpha_2}^a(x) \right\}, \min\left\{ \upsilon_{\alpha_1}^a(x), \upsilon_{\alpha_2}^a(x) \right\} \right\} \end{aligned} \tag{4}$$

where $\alpha_1 \in h_1^a, \alpha_2 \in h_2^a$. h_1^a and h_2^a denote the hesitant fuzzy set.

Definition 3.3 For any two IHFSSs (*F*, *E*) and (*G*, *E*) over a common universe (*U*, *E*), the intersection of (*F*, *E*) and (*G*, *E*) is the IVIHFSS (*H*, *E*) and $\forall a \in E$ and $\forall x \in U$, we have

$$\begin{aligned} h_1^e \cap h_2^e &= \{\min(\alpha_1, \alpha_2) | \alpha_1 \in h_1^a, \alpha_2 \in h_2^a\} \\ &= \left\{ \min\left\{ \mu_{\alpha_1}^a(x), \mu_{\alpha_2}^a(x) \right\}, \max\left\{ \upsilon_{\alpha_1}^a(x), \upsilon_{\alpha_2}^a(x) \right\} \right\}. \end{aligned} \tag{5}$$

Definition 3.4 The complement of IHFSS (*F*, *E*), represented as $(F, E)^c$, is defined as

$$h^c = \left\{ \left(\nu_{(F,E)}^a, \mu_{(F,E)}^a \right) \right\}. \tag{6}$$

Definition 3.5 An IFHSS (*F*, *E*) is said to be a null IHFSS if and only if it satisfies

$$\mu_{(F,E)}^a(x) = 0 \quad \text{and} \quad \nu_{(F,E)}^a(x) = 1\,. \tag{7}$$

Definition 3.6 An IFHSS (*F*, *E*) is said to be an absolute IHFSS if and only if it satisfies

$$\begin{array}{c} \mu^e_{(F,E)}(x) = 1 \\ \text{and} \\ \nu^e_{(F,E)}(x) = 1 \end{array} . \quad (8)$$

4 Application of Intuitionistic Hesitant Fuzzy Set

In [2], Molodtsov has given several applications of soft set. In [10] the decision making example given depends on the decision of a single person. Here we discuss an application of DM in IHFSSs.

Many of researchers have tried to provide solutions for the decision making problems in lot many situations. Some of these approaches are preference ordering, utility values, preference values.

The parameters can be categorized as of two types [7].

We introduce the formula (9) to get a fuzzy value as score from an intuitionistic fuzzy value. It reduces the complexity and makes the comparison easier.

$$\text{Score} = \mu(1 + h) \quad (9)$$

The score will decrease with the increasing ν value and score will increase when either μ or h value increases. But, when μ changes, the impact will be more in comparison to the h value as both the factors of the equation depends on μ, whereas only one factor depends on h. Value of μh (2nd factor in equation) is inversely proportional to ν value. So there is no need to consider ν value again. The equation reduces to only μ value in case of fuzzy soft set, i.e., if $h = 0$.

$$\text{Normalized Score} = \frac{\sum_{K=\{o,p,n\}} (C - R_K)^2}{|K| \times C^2} \quad (10)$$

where $|K|$ is the number of approaches (e.g., optimistic, pessimistic, neutral, etc.), $|C|$ is the number of objects to choose from. R_K is the rank with respect to approach x.

Consider the case of a company that wants to select a cloud service provider from the available service providers. Before the company selects the service provider, he needs to consider the parameters of the service provider. The parameters considered for comparison are efficiency, through put, security, delay, price, and feedback. Some parameters like price inversely affect the decisions. Those parameters are called negative parameters.

Algorithm

1. Input the parameter data table by ranking according to the absolute value of parameter priorities. If the priority for any parameter has not given, then take the value as 0 by default and that column can be opt out from further computations. The boundary condition for a positive parameter is [0, 1] and for a negative parameter is [−1, 0].
2. Input the IHFSS table
3. Construct optimistic IFSS table by taking the maximum of membership values, minimum of non-membership values from IHFSS table and compute the hesitation values accordingly.
4. Construct pessimistic IFSS table by taking the minimum of membership values, maximum of non-membership values from IHFSS table, and compute the hesitation values accordingly.
5. Construct neutral IFSS table by taking the mean of the membership values and mean of the non-membership values from IHFSS table and compute the hesitation values accordingly.
6. Procedure Deci_make(IFSS table)
   - 6.1. Multiply the priority values with the corresponding parameter values to get the priority table
   - 6.2 Construct the comparison table by finding the entries as differences of each row sum in priority table with those of all other rows taking membership and hesitation values separately.
   - 6.3 Construct the decision table by taking the sums of membership values and hesitation values separately for each row in the comparison table. Compute the score for each candidate using the formula (9).
   - 6.4 Assign rankings to each candidate based upon the score obtained.
     - 6.1 If there is more than one candidate having same score than who has more score in a higher ranked parameter will get higher rank and the process will continue until each entry has a distinct rank or they are equal with respect to all parameters. In later case, the whole group of candidates who are having that same score will get same rank.
   - 6.5 Return decision table.
7. Construct the decision tables for all approaches (optimistic, pessimistic, and neutral) using the procedure Deci_make given in step 6.
8. Construct the rank matrix by taking the rank columns of all decision tables. Compute the normalized score using the formula (10).
9. Compute the final ranks for all candidates. The candidate having highest normalized score will get the highest rank and so on.

9.1 In case of more than one same normalized score, resolve the conflict by taking the scores given in the highest ranked parameter and continue the process until each entry has a distinct rank.

Let U be a set of cloud providers given by $U = \{p1, p2, p3, p4, p5, p6\}$ and E be the parameter set given by $E = \{e1, e2, e3, e4, e5, e6\}$, where $e1$, $e2$, $e3$, $e4$, $e5$ and $e6$ represents efficiency, throughput, security, delay, price, and feedback, respectively.

The parameter data table is given in the Table 1. It contains all the details about the parameters. The parameter rank is decided by comparing the absolute priority value of a parameter.

Consider an IHFSS (F, E) which shows three opinions about the quality of service by the cloud service providers as shown in Table 2.

Table 1 Parameter data table

Parameter	*e*1	*e*2	*e*3	*e*4	*e*5	*e*6
Priority	0.4	0.3	0.25	−0.3	0.1	0
Parameter rank	1	2	4	2	5	6

Table 2 IHFSS table

U	*e*1		*e*2		*e*3		*e*4		*e*5		*e*6	
	m	*n*	*m*	*n*	*m*	*n*	*m*	*n*	*m*	*n*	*m*	*n*
*p*1	0.4	0.1	1	0	0.4	0.5	0.4	0.5	0.2	0.7	0.1	0.3
	0.8	0	0.4	0.4	0.6	0	0.2	0.6	1	0	0.5	0.3
	0.7	0.2	0.1	0.7	0.5	0.3	0.7	0.3	0.2	0.4	0.3	0.3
*p*2	0.1	0.7	0.2	0.6	0.2	0.7	0.6	0.1	0	0.1	0.1	0
	0.5	0	1	0	0.5	0.3	0.1	0.5	0.2	0.2	0.4	0.2
	0.8	0.1			0.8	0	0.4	0.1	0.9	0	0.1	0.6
*p*3	0.6	0.3	0.9	0	0	0.6	1	0	0.9	0	0.2	0.6
	1	0	0	0.1	0.4	0.2	0.1	0.5	0	0.6	0.2	0.7
	0.4	0.2	0.8	0	0.9	0	0.3	0.2	0.8	0	1	0
*p*4	0.4	0	0.9	0	0.2	0.1	0.1	0.1	0.3	0.3	0	0.1
	0	0.1	0.2	0.7	0.5	0.2	0.4	0.4	0.6	0.3	0.9	0
	0.5	0.2	0.5	0.1	0	0.6	0.1	0.8	0.3	0.1	0.8	0.1
*p*5	0.8	0.1	0.3	0.6	1	0	0.6	0.2	0.5	0.5		
	0	0.8	0.4	0.3	0.5	0.5	0.5	0.3	0.6	0.1	0.7	0.2
			0.3	0.5	0.9	0	0.2	0.6	0.4	0.2	0.2	0.6
*p*6	0.1	0.7	0.8	0.1	0.4	0	0.6	0.4	0.4	0.2	1	0
	0.8	0.1	0.3	0.1	0.6	0.3	0.4	0.5	1	0	0.2	0.3
	0.4	0.6	0	0.3	0.4	0.2	0	0.6	0.7	0.1	0.6	0.4

Table 3 Optimistic IFSS table

U	e1			e2			e3			e4			e5			e6		
	m	*n*	*h*	*m*	*n*	*h*	*m*	*n*	*h*	*m*	*n*	*h*	*m*	*n*	*h*	*m*	*n*	*h*
*p*1	0.8	0.0	0.2	1.0	0.0	0.0	0.6	0.0	0.4	0.7	0.3	0.0	1.0	0.0	0.0	0.5	0.3	0.2
*p*2	0.8	0.0	0.2	1.0	0.0	0.0	0.8	0.0	0.2	0.6	0.1	0.3	0.9	0.0	0.1	0.4	0.0	0.6
*p*3	1.0	0.0	0.0	0.9	0.0	0.1	0.9	0.0	0.1	1.0	0.0	0.0	0.9	0.0	0.1	1.0	0.0	0.0
*p*4	0.5	0.0	0.5	0.9	0.0	0.1	0.5	0.1	0.4	0.4	0.1	0.5	0.6	0.1	0.3	0.9	0.0	0.1
p5	0.8	0.1	0.1	0.4	0.3	0.3	1.0	0.0	0.0	0.6	0.2	0.2	0.6	0.1	0.3	0.7	0.2	0.1
*p*6	0.8	0.1	0.1	0.8	0.1	0.1	0.6	0.0	0.4	0.6	0.4	0.0	1.0	0.0	0.0	1.0	0.0	0.0

Table 4 Pessimistic IFSS table

U	e1			e2			e3			e4			e5			e6		
	m	*n*	*h*	*m*	*n*	*h*	*m*	*n*	*h*	*m*	*n*	*h*	*m*	*n*	*h*	*m*	*n*	*h*
*p*1	0.4	0.2	0.4	0.1	0.7	0.2	0.4	0.5	0.1	0.2	0.6	0.2	0.2	0.7	0.1	0.1	0.3	0.6
*p*2	0.1	0.7	0.2	0.2	0.6	0.2	0.2	0.7	0.1	0.1	0.5	0.4	0.2	0.2	0.6	0.1	0.6	0.3
*p*3	0.4	0.3	0.3	0	0.1	0.9	0	0.6	0.4	0.1	0.5	0.4	0	0.6	0.4	0.2	0.7	0.1
*p*4	0	0.2	0.8	0.2	0.7	0.1	0	0.6	0.4	0.1	0.8	0.1	0.3	0.3	0.4	0	0.1	0.9
*p*5	0	0.8	0.2	0.3	0.6	0.1	0.5	0.5	0.0	0.2	0.6	0.2	0.4	0.5	0.1	0.2	0.6	0.2
*p*6	0.1	0.7	0.2	0	0.3	0.7	0.4	0.3	0.3	0	0.6	0.4	0.4	0.2	0.4	0.2	0.4	0.4

Optimistic IFSS table can be constructed by taking the maximum of membership values, minimum of non-membership values from IHFSS table and compute the hesitation values accordingly. Optimistic IFSS table is shown in Table 3.

Pessimistic IFSS table can be constructed by taking the minimum of membership values, maximum of non-membership values from IHFSS table and compute the hesitation values accordingly. Pessimistic IFSS table is shown in Table 4.

Neutral IFSS table can be constructed by taking the mean of the membership values and mean of the non-membership values from IHFSS table and compute the hesitation values accordingly. Neutral IFSS table is shown in Table 5.

Table 5 Neutral IFSS table

U	e1			e2			e3			e4			e5			e6		
	m	*n*	*h*	*m*	*n*	*h*	*m*	*n*	*h*	*m*	*n*	*h*	*m*	*n*	*h*	*m*	*n*	*h*
*p*1	0.633	0.100	0.267	0.500	0.367	0.133	0.500	0.267	0.233	0.433	0.467	0.100	0.467	0.367	0.167	0.300	0.300	0.400
*p*2	0.467	0.267	0.267	0.600	0.300	0.100	0.500	0.333	0.167	0.367	0.233	0.400	0.367	0.100	0.533	0.200	0.267	0.533
*p*3	0.667	0.167	0.167	0.567	0.033	0.400	0.433	0.267	0.300	0.467	0.233	0.300	0.567	0.200	0.233	0.467	0.433	0.100
*p*4	0.300	0.100	0.600	0.533	0.267	0.200	0.233	0.300	0.467	0.200	0.433	0.367	0.400	0.233	0.367	0.567	0.067	0.367
*p*5	0.400	0.450	0.150	0.333	0.467	0.200	0.800	0.167	0.033	0.433	0.367	0.200	0.500	0.267	0.233	0.450	0.400	0.150
*p*6	0.433	0.467	0.100	0.367	0.167	0.467	0.467	0.167	0.367	0.333	0.500	0.167	0.700	0.100	0.200	0.600	0.233	0.167

Table 6 Optimistic priority table

U	$e1$		$e2$		$e3$		$e4$		$e5$	
	m	h	m	h	m	h	m	h	m	h
$p1$	0.32	0.08	0.3	0	0.15	0.1	−0.21	0	0.1	0
$p2$	0.32	0.08	0.3	0	0.2	0.05	−0.18	−0.09	0.09	0.01
$p3$	0.4	0	0.27	0.03	0.225	0.025	−0.3	0	0.09	0.01
$p4$	0.2	0.2	0.27	0.03	0.125	0.1	−0.12	−0.15	0.06	0.03
$p5$	0.32	0.04	0.12	0.09	0.25	0	−0.18	−0.06	0.06	0.03
$p6$	0.32	0.04	0.24	0.03	0.15	0.1	−0.18	0	0.1	0

The priority tables can be constructed by multiplying the values in IFSS table with the respective priority values. The tables are not having non-membership value column further because the formula is dependent on membership and hesitation values only. Main idea behind this is, the change of non-membership value is always reflects in the values of either membership value or hesitation value or in both. The optimistic priority table is shown in Table 6.

In the same way, priority tables for pessimistic and neutral approach can be constructed.

Comparison tables can be constructed by taking the entries as differences of each row sum in priority table values. The optimistic comparison table is shown in Table 7.

In the same way, comparison of tables for pessimistic and neutral approach can be constructed.

Decision table can be constructed using formula (9). The optimistic decision table is shown in Table 8.

In the same way, the decision of tables for pessimistic and neutral approach can be constructed.

Table 7 Optimistic comparison table

U	P1		P2		P3		P4		P5		P6	
	m	*h*	*m*	*h*	*m*	*h*	*m*	*h*	*m*	*h*	*m*	*h*
*p*1	0.000	0.000	−0.070	0.130	−0.025	0.115	0.125	−0.030	0.090	0.080	0.030	0.010
*p*2	0.070	−0.130	0.000	0.000	0.045	−0.015	0.195	−0.160	0.160	−0.050	0.100	−0.120
*p*3	0.025	−0.115	−0.045	0.015	0.000	0.000	0.150	−0.145	0.115	−0.035	0.055	−0.105
*p*4	−0.125	0.030	−0.195	0.160	−0.150	0.145	0.000	0.000	−0.035	0.110	−0.095	0.040
*p*5	−0.090	−0.080	−0.160	0.050	−0.115	0.035	0.035	−0.110	0.000	0.000	−0.060	−0.070
*p*6	−0.030	−0.010	−0.100	0.120	−0.055	0.105	0.095	−0.040	0.060	0.070	0.000	0.000

Table 8 Optimistic decision table

	sum m	sum h	Score	Rank
$p1$	0.15	0.305	0.19575	2
$p2$	0.57	−0.475	0.29925	1
$p3$	0.3	−0.385	0.1845	3
$p4$	−0.6	0.485	−0.891	6
$p5$	−0.39	−0.175	−0.32175	5
$p6$	−0.03	0.245	−0.03735	4

Table 9 Rank matrix

	Optimistic rank	Pessimistic rank	Neutral rank	Normalized score	Final rank
$p1$	2	1	2	0.76	1
$p2$	1	3	3	0.573333	2
$p3$	3	5	1	0.466667	3
$p4$	6	6	6	0	6
$p5$	5	4	4	0.12	5
$p6$	4	2	5	0.28	4

From Table 9, it can be easily asses the quality of service provided by different service providers, which will help to take a suitable decision. Here, $p1$ is the best choice and so on.

5 Conclusion

In this article, we introduced a new definition of IHFSS, which uses the more authentic characteristic function approach for defining soft sets provided in [6]. This provides several authentic definitions of operations on IHFSS and made the proofs of properties very elegant. Earlier fuzzy soft sets were used for decision making in [4]. Their approach had many flaws. We pointed out those flaws and provided solutions to rectify them in [7]. This made decision making more efficient and realistic. Here, we proposed an algorithm for decision making using IHFSS, which uses the concept of negative parameters. Also, an application of this algorithm in solving a real-life problem is demonstrated.

References

1. L.A. Zadeh, "Fuzzy sets", Information and Control, vol. 8, 1965, pp. 338–353.
2. D. Molodtsov, "Soft Set Theory - First Results", Computers and Mathematics with Applications, vol. 37, 1999, pp. 19–31.

3. P.K. Maji, R. Biswas, A.R. Roy, "An Application of Soft Sets in a Decision Making Problem", Computers and Mathematics with Applications, vol. 44, 2002, pp. 1007–1083.
4. P.K. Maji, R. Biswas, A.R. Roy, "Soft Set Theory", Computers and Mathematics with Applications, vol. 45, 2003, pp. 555–562.
5. P.K. Maji, R. Biswas, A.R. Roy, "Fuzzy Soft Sets", Journal of Fuzzy Mathematics, vol. 9(3), 2001, pp. 589–602.
6. B.K. Tripathy and K.R. Arun, "A New Approach to Soft Sets, Soft Multisets and Their Properties", International Journal of Reasoning-based Intelligent Systems, vol. 7, no. 3/4, 2015, pp. 244–253.
7. B.K. Tripathy, Sooraj, T.R., RK Mohanty, "A new approach to fuzzy soft set and its application in decision making", Advances in intelligent systems and computing, vol. 411, 2016, pp. 305–313.
8. K. Atanassov, "Intuitionistic Fuzzy Sets", Fuzzy Set Systems, vol. 20, 1986, pp. 87–96.
9. Y. Jiang, Y. Tang, Q. Chen, H. Liu, J. Tang, "Interval-valued intuitionistic fuzzy soft sets and their properties", Computers and Mathematics with Applications, vol. 60, 2010, pp. 906–918.
10. B.K. Tripathy, RK Mohanty, Sooraj T.R., K.R. Arun, "A New Approach to Intuitionistic Fuzzy Soft Sets and its Application in Decision-Making", Advances in Intelligent Systems and Computing, vol. 439, 2016, pp. 93–100.
11. Sooraj T.R., RK Mohanty, B.K. Tripathy, "Fuzzy Soft Set Theory and its Application in Group Decision Making", Advances in Intelligent Systems and Computing, vol. 452, 2016, pp. 171–178.
12. B.K. Tripathy, Sooraj T.R., RK Mohanty, "Advances Decision Making Usisng Hybrid Soft Set Models", International Journal of Pharmacy and Technology, vol. 8(3), 2016, pp. 17694–17721.
13. B.K. Tripathy, RK Mohanty, T.R. Sooraj, "On Intuitionistic Fuzzy Soft Sets and Their Application in Decision-Making", Lecture Notes in Electrical Engineering, vol. 396, 2016, pp. 67–73.
14. B.K. Tripathy, RK Mohanty, Sooraj T.R., "On Intuitionistic Fuzzy Soft Set and its Application in Group Decision Making", In: Proceedings of the International Conference on Emerging Trends in Engineering, Technology and Science (ICETETS), Pudukkottai, 2016, pp. 1–5.
15. B.K Tripathy, RK Mohanty, Sooraj T.R., A. Tripathy, "A Modified Representation of IFSS and Its Usage in GDM", Smart Innovation, Systems and Technologies, Vol. 50, 2016, pp. 365–375.
16. RK Mohanty, Sooraj T.R., B.K. Tripathy, "An application of IVIFSS in medical diagnosis decision making", International Journal of Applied Engineering Research, vol. 10(92), pp. 85–93.
17. B.K. Tripathy, Sooraj T.R., RK Mohanty, "A New Approach to Interval-valued Fuzzy Soft Sets and its Application in Decision-Making", Advances in Intelligent Systems and Computing, Vol. 509, 2017, pp. 3–10.
18. RK Mohanty, Sooraj T.R., B.K Tripathy, "IVIFS and Decision-Making", Advances in Intelligent Systems and Computing, Vol. 468, 2017, pp. 319–330.
19. Torra V, Hesitant fuzzy sets and decision, International Journal of Intelligent Systems 25 (6), 2010, pp. 395–407.
20. Torra V, Narukawa Y, On hesitant fuzzy sets and decision, The 18th IEEE international conference on Fuzzy Systems, Jeju Island, Korea, 2009, pp. 1378–1382.

Multi-document Text Summarization Using Sentence Extraction

Ravinder Ahuja and Willson Anand

Abstract This paper presents a method for generating multi-document text summary building on single document text summaries and by combining those single document text summaries using cosine similarity. For the generation of single document text summaries features like document feature, sentence position feature, normalized sentence length feature, numerical data feature, and proper noun feature are used. Single document text summaries are combined after calculating cosine similarity between the different single document text summaries generated and from each combination, sentences with high total sentence weight are extracted to generate multi-document text summary. The average F-measure of 0.30493 on DUC 2002 dataset has been observed, which is comparable to two of five top performing multi-document text summarization systems reported on the DUC 2002 dataset.

Keywords Multi-document text summarization · Cosine similarity · Extraction of features

1 Introduction

With the increasing availability of text documents on the World Wide Web, it has become extremely difficult for the average human reader gathering information on individuals, issues, events, or topics to read each and every document with an aim to find the useful information. Therefore, automatic text summarization systems are needed to overcome this problem as they generate summary from the particular text document(s). A summary is defined as "a text that is produced from one or more texts, that conveys important information in the original text(s), and that is no longer than half of the original text(s) and usually significantly less than that" [1].

R. Ahuja (✉) · W. Anand
Jaypee Institute of Information Technology, Noida, Uttar Pradesh 201307, India
e-mail: ravinder.ahuja@jiit.ac.in

W. Anand
e-mail: willsonanand.20@gmail.com

S.S. Dash et al. (eds.), *Artificial Intelligence and Evolutionary Computations in Engineering Systems*, Advances in Intelligent Systems and Computing 517,
DOI 10.1007/978-981-10-3174-8_21

In most of the cases, a human reader is interested in gathering information from multiple text documents on a specific topic. In that situation, a summary should also be such which does not contain repeated information. The redundancy issue is more eminent in the summary generated from multiple text documents than the one generated form single text document [2]. In this paper, an approach for generating multi-document text summary is presented relying on summaries generated from single text documents and the concept of cosine similarity.

2 Literature Survey

The research on automatic text summarization has been continuing since 1950s. The most difficult task in text summarization is to recognize the more important part of the text than the less important one. Statistical techniques based on word frequency [3], sentence position [4], and key sentences and cue words [5] were among the early approaches toward text summarization. In 1990s, various machine learning methods that employed statistical techniques were proposed. These text summarization systems were based on Naïve-Bayes methods [6], hidden Markov models [7], and log linear models [8]. Other approaches toward automatic text summarization are marginal minimum relevance [9] for sentence extraction, graph-based models [10], and centroid-based summarization [11].

3 Our Approach

The proposed method for the multi-document text summarization is given in Fig. 1 followed by the explanation of each step.

3.1 Preprocessing of Text Documents

The preprocessing of text documents involves sentence splitting, tokenization, stemming, stop words removal, calculation of term frequency, and identification of nouns using part-of-speech tagger.

Sentence splitting is done to identify the end of the sentences through word-boundary markers.

Tokenization involves the breaking of text into words, symbols, etc. The words are separated from white space and punctuation characters.

Stemming is reducing words to their word stem. Porter Stemmer is used for this step.

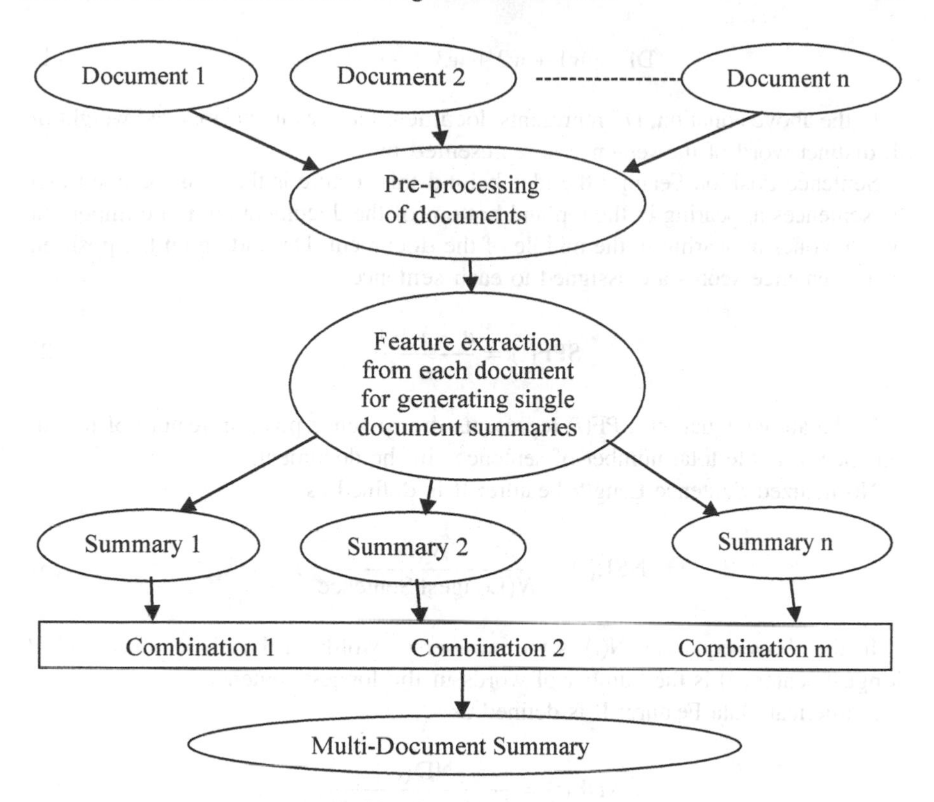

Fig. 1 Multi-document summary generation

Stop words like is, am, etc., are removed as they are not considered important.

The term frequency is calculated and normalized by dividing with the maximum term frequency of the document.

Nouns represent the name of humans, objects, actions, etc., and are identified using part-of-speech tagger.

3.2 Extraction of Features

This step aims to extract the features on the basis of which summary from each preprocessed document will be generated. The features used are document feature, sentence position feature, numerical data feature, proper noun feature, and normalized sentence length feature.

Document feature: It is the weight of the sentence with respect to the complete document. The sum of weights of distinct words appearing in the sentence is called sentence weight.

$$\mathrm{DF} = w1 + w2 + w3 + \cdots + wn\,. \tag{1}$$

In the above equation, DF represents document feature and normalized weight of *i*th distinct word of the sentence is represented by *wi*.

Sentence Position Feature: the idea behind this feature is that it is assumed that the sentences appearing in the top and bottom of the document are more important than the ones appearing in the middle of the document. Depending on the position of the sentence scores are assigned to each sentence.

$$\mathrm{SPF}(i) = \frac{n - i + 1}{n}. \tag{2}$$

In the above equation, SPF(*i*) represents the sentence position feature of the *i*th sentence, *n* is the total number of sentences in the document.

Normalized Sentence Length Feature: It is defined as

$$\mathrm{NSL}(i) = \frac{N(i)}{N(\text{Longest Sentence})}. \tag{3}$$

In the above equation, N(*i*) is the number of words in the *i*th sentence and *N* (longest sentence) is the number of words in the longest sentence.

Numerical Data Feature: It is defined as

$$\mathrm{NDF}(i) = \frac{\mathrm{ND}(i)}{\text{Sentence Length}}. \tag{4}$$

In the above equation, NDF(*i*) represents the numerical data feature of the *i*th sentence and ND(*i*) represents the total number of times numerical data occurring in sentence *i*.

Proper Noun Feature: The proper noun feature is calculated as

$$\mathrm{PNF}(i) = \frac{\text{Total proper nouns in the sentence}}{\text{Sentence Length}}. \tag{5}$$

In the above equation, PNF(*i*) represents the proper noun feature of the *i*th sentence.

3.3 Generation of Single Document Text Summaries

The single document summary is generated on the basis of total sentence weight which is computed by considering all the above explained features.

$$\mathrm{TSW}(i) = p * \mathrm{DF} + q * \mathrm{SPF} + r * \mathrm{NSL} + s * \mathrm{NDF} + t * \mathrm{PNF}. \tag{6}$$

In the above equation, TSW(i) represents the total sentence weight of the ith sentence, DF represents the document feature, SPF represents the sentence position feature, NSL represents normalized sentence length feature, NDF represents the numerical data feature, and PNF represents the proper noun feature. The constants are experimentally set as $p = 0.5$, $q = 0.2$, $r = 0.1$, $s = 0.1$, $t = 0.1$.

The total sentence weights are normalized by dividing with total sentence weight of the ith sentence by the maximum total sentence weight present in the document.

The sentences are ranked on the basis of normalized total sentence weight and top m sentences are extracted from document to generate single document text summary. The value of m depends on the percentage of summary to be generated in comparison to the complete document.

3.4 *Generation of Multi-Document Text Summary*

The summaries from single documents are combined on the basis of cosine similarity with respect to other single document text summaries generated. From each combination of most similar single document summaries the sentences with high total sentence weight are extracted. In the final multi-document text summary generated, the sentences are extracted in the same order as they appear in the original document.

$$\text{Cosine Similarity } (\mathrm{d1}, \mathrm{d2}) = \frac{\text{Dot Product } (\mathrm{d1}, \mathrm{d2})}{||\mathrm{d1}||\,||\mathrm{d2}||}. \tag{7}$$

The above equation gives the measure of Cosine similarity between two documents d1 and d2. The numerator in the equation is the product of summation of weight vector for each term occurring in document d1 and d2. The denominator is the product of square root of summation of weight vector of each term occurring in document d2 and d2. The weight vector for a term in a document is given by the multiplication of term frequency and inverse document frequency. The number of times a term occurs in a document is called term frequency whereas the measure of how much information the word provides is called inverse document frequency.

$$w(\mathrm{t}, \mathrm{d}) = (1 + \log(f(\mathrm{t}, \mathrm{d})) * \log\left(\frac{N}{n(\mathrm{t})}\right). \tag{8}$$

In the above equation, f(t, d) represents the term frequency of a term t in a document d, N is the total number of documents, and n(t) is the number of documents in which the term t occurs.

The document d1 is combined only with the document with which it has highest cosine similarity from the set of documents and so on for all documents.

After making combinations of single document summaries, the sentences with high total sentence weight are extracted from each combination to generate the multi-document text summary.

4 Dataset and Experiments

For performing experiments, the DUC 2002 dataset has been used. The DUC 2002 dataset contains 533 news articles with the gold summary of 100 words of each article for single document summary generation. It also contains 60 document sets with the gold summaries of 10 words, 50 words, 100 and 200 words for multi-document text summarization.

The single document summaries and multi-document summaries generated by the system are evaluated against the gold summaries given in DUC 2002 dataset using ROUGE 2.0.

ROUGE 2.0 is recall-oriented understudy for Gisting Evaluation 2.0. It is a set of metrics and a software package used for evaluating automatic summarization. The metrics compare an automatically produced summary against a reference or a set of references summary on parameters like average recall, average precision and F-measure.

The single document summaries generated by the system have been evaluated against 100 words single document gold summaries of DUC 2002 dataset. The average F-measure of 0.44598 for 533 documents given in DUC 2002 dataset was observed. The evaluation results of single document summaries are shown in Table 1.

In order to evaluate the multi-document summaries generated by the system, we have compared them to DUC 2002, 10 words/50 words/100 words/200 words multi-document gold summaries depending on the size of summary generated by the system. We observed the average F-Measure of 0.30493 over 60 document sets of DUC 2002 dataset. Table 2 shows the result for evaluation of multi-document summaries.

Table 1 Evaluation results of single document text summaries

Average recall	0.43649
Average precision	0.45589
Average F-measure	0.44598

Table 2 Evaluation results for multi-document text summary

Average recall	0.32548
Average precision	0.28685
Average F-measure	0.30493

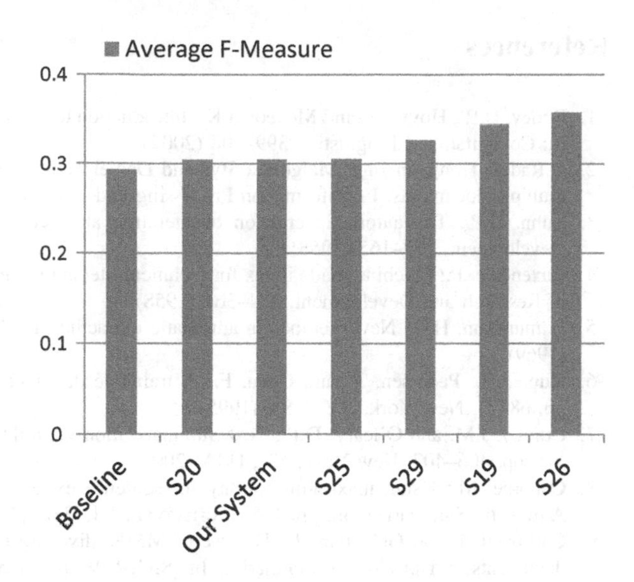

Fig. 2 Five best performing systems along with our approach versus average F-measure

5 Results

In evaluation of single document text summaries, the average F-measure of 0.44598 was observed. In evaluation of multi-document text summaries, the average F-measure of 0.30493 was observed. The five best performing multi-document text summarization systems on the dataset of DUC 2002 along with our method against average F-measure are shown in Fig. 2. The average F-measure generated by our approach is comparable to average F-measure of fourth and fifth best performing systems.

6 Conclusion and Future Work

This paper presents and evaluates an approach for generating multi-document text summaries from single document text summaries. For the generation of single document summaries, features like document feature, sentence position feature, numerical data feature, proper noun feature, and normalized sentence length feature are used. Single document summaries generated by extracting the sentences based on total sentence weight are clustered using cosine similarity. From the clusters formed, the sentences with high total sentence weight are extracted to generate the multi-document text summary. In future, we would like to evaluate our system on the further datasets released by DUC like DUC 2004, DUC 2005, etc.

References

1. Radev, D.R., Hovy, E., and Mckeown K.: Introduction to the special issue on summarization. In: Computational Linguistic., 399–408 (2002).
2. D.Radev, Hongyan Jing, Malgortza Stys and Daniel Tam: Centroid-based summarization of multiple documents. In: Information Processing and Management, 919–938 (2004).
3. Luhn, H.P.: The automatic creation of literature abstracts. In: IBM Journal of Research Development, 159–165 (1958).
4. Baxendale, P.: Machine-made index for technical literature – an experiment. In: IBM Journal of Research and Development, 354–361 (1958).
5. Edmundson, H.P.: New methods in automatic extracting. In: Journal of the ACM, 264–285 (1969).
6. Kupiec, J., Pedersen, J., and Chen, F.: A trainable document summarizer. In: SIGIR '95, pp. 68–73, New York, NY, USA (1995).
7. Conroy, J.M. and O'leary, D.P.: Text Summarization via hidden markov models. In: SIGIR '01, pp. 406–407, New York, NY, USA (2001).
8. Osborne, M.: Using maximum entropy for sentence extraction. In: ACL'02 Workshop on Automatic Summarization, pp. 1–8, Morristown, NJ, USA (2002).
9. Carbonell, J. and Goldstein, J.: The use of MMR, diversity based reranking for reordering documents and producing summaries. In: SIGIR'98, pp. 335–336, New York, NY, USA (1998).
10. Erkan, G. R.: LexRank: Graph-based Lexical Centrality as Salience in Text Summarization. In: Journal of Artificial Intelligence Research, 457–479 (2004).
11. Radev, D.R., Jing, H., and Budzikowska, M.: Centroid-based summarization of multiple documents: sentence extraction, utility based evaluation, and user studies. In: NAACL-ANLP 2000 Workshop on Automatic summarization, pp. 21–30, Morristown, NJ, USA (2000).

Tourism Recommendation Using Social Media Profiles

S. Kavitha, Vijay Jobi and Sridhar Rajeswari

Abstract The need to analyse social media content is on the rise with more people using it for sharing their day-to-day activities. This has prompted multiple studies in the area of social data mining. One innovative approach is to analyse tweets posted by the user in twitter, and generates a structure that uniquely captures all the information related to the user in a single location. This project collects metadata from twitter users, analyses their tweets and gets a set of associated probabilities for a set of topics generated from a Latent Dirichlet Allocation model. The collected information is mined and is stored in a unique data structure consisting of various layers that represent a user. This structure is stored using a hash function, which facilitates the easy storage and retrieval of the structure. A destination tree is built based on the data mined from the tourist websites of states in India, the leaves of the tree representing a tourist destination in India and each having its associated set of probabilities for a set of topics generated from the Latent Dirichlet Allocation model. Finally, the user node built is used to recommend tourist destinations in India by comparing the user node with the leaves of the tourist destination tree.

Keywords Social media mining · Twitter · Recommendation system

S. Kavitha (✉) · Vijay Jobi · Sridhar Rajeswari
Department of Computer Science and Engineering,
Anna University, Chennai, Tamil Nadu, India
e-mail: kavithasivakumar91@gmail.com

Vijay Jobi
e-mail: jobivijay2010@gmail.com

Sridhar Rajeswari
e-mail: rajisridhar@gmail.com

S.S. Dash et al. (eds.), *Artificial Intelligence and Evolutionary Computations in Engineering Systems*, Advances in Intelligent Systems and Computing 517,
DOI 10.1007/978-981-10-3174-8_22

1 Introduction

The emergence of social media services such as Facebook, Twitter, etc. in the last decade has allowed more and more users to participate in online social activities such as posting updates or tweets, uploading photos, etc. This high popularity has forked an unprecedented rate of data production which in turn has led to complexities in mining data [1].

Data obtained from social networking websites is of great value to businesses in identifying products and recommending them to users of similar interest. One simple way which is employed in mining user's reviews about products is by means of sentiment analysis [2]. Gathering and building profiles of users based on what they posted in social media is an innovative way to use the available data [3]. Tweets and posts of a user in twitter and Facebook can be mined extensively to build the topic of interest of users. Additionally, user metadata can also be mined from the profiles of the user which provides the basic information of each user and is used to recommend tourist destinations.

This chapter is organised as follows: Sect. 2 discusses some existing work on recommendation systems and social media mining, Sect. 3 gives a detailed design of the proposed system, Sect. 4 discusses experimental results, and Sect. 5 provides conclusion and possible extensions.

2 Literature Survey

2.1 Recommendation Systems

The three main types of recommendation techniques are collaborative filtering, content-based, and hybrid methods [4]. Collaborative filtering helps people to make choices based on the opinions of the other people who share similar interests. Content-based recommendation techniques recommend items that are similar to those previously preferred by the user. Hybrid methods combine both techniques together. Fuzzy preference tree-based recommender [5] is a hybrid method which uses both user reviews and item-to-item similarity to recommend products to users. The proposed system uses certain ideas from the fuzzy recommender and applies it in a business to consumer environment to recommend tourist destinations.

2.2 Social Media Mining

Social media has been mined in the past for identifying a particular trend or to gather information regarding a topic of interest. Student's informal conversations on social media were mined for understanding their learning experiences [6].

A two-step analysis framework that focuses on ascertaining user opinion of cancer treatment was proposed [7] which made use of data in forums.

In this study, "Constructing consumer profiles from social media data" [3], the authors had proposed an innovative method for constructing user profiles from tweets of the user. They used AQL to mine Boolean attributes from the user's tweets. The problem here lies in writing the set of rules that can mine a particular attribute. A user profiling and stream filtering approach based on the topical categorisation of users' posted URLs is discussed [8]. But, in this approach, only the URLs posted by the user are taken into consideration. Latent Dirichlet Allocation was evaluated for the application of topic modelling for the purpose of supporting a greater understanding of trends in social media [9]. In the study the author had mined a specific stream of twitter messages and had not done it in a per-user basis.

Even though social media has been mined extensively for various reasons, user profiling from social media is a relatively new area of research. Main drawbacks of the existing work include mining data only for a particular set of predefined features. Alternative approach would be using topic models to classify the set of posts into various categories. Though work has been done in this area, it has not been put to use to make useful recommendations to users. Furthermore, the results obtained from the classification are not aggregated and stored in a manner where retrieval is easier. This study proposes an innovative way to store the user information retrieved from social media and uses the mined results to make recommendations of tourist destinations in India.

3 System Architecture

In order to build a user node, meta data and tweets posted by each user have to be collected so that the probabilities associated with each topic obtained from the Latent Dirichlet Allocation (LDA) model can be calculated. The LDA model is built from the reviews scraped from trip advisor and data about tourist destinations manually collected. Such a node built is stored using a computed hash value. This node is then used to recommend tourist destinations for the user. Figure 1 illustrates the complete block diagram.

3.1 Random Sampling and Data Collection

Random user ids are obtained using twitter API to query tweets that contain the word 'the'. Once the duplicates are removed, the user ids obtained are used to fetch tweets of users.

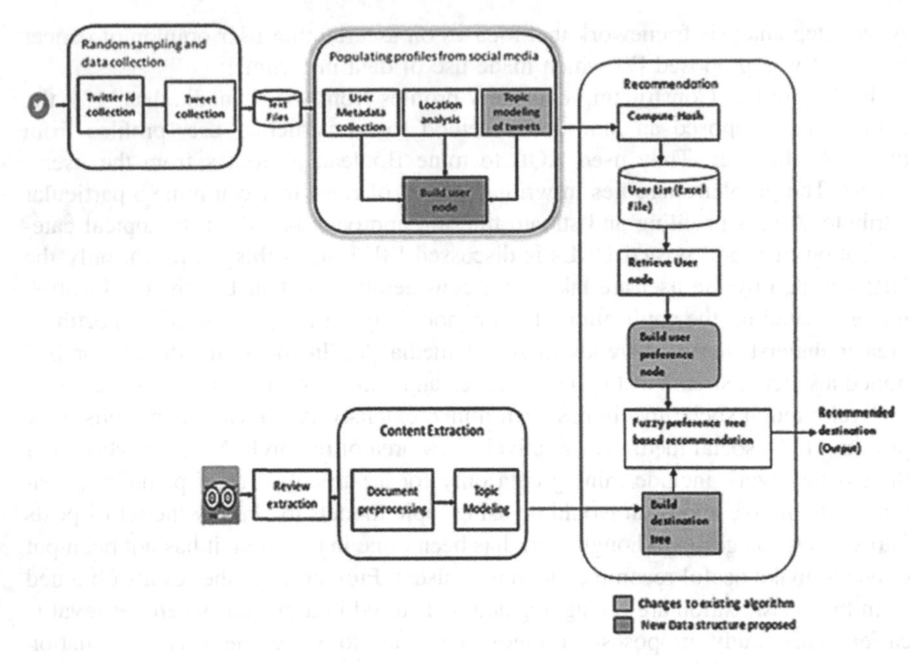

Fig. 1 Basic block diagram

3.2 Populating Profiles from Social Media

Populating profiles involve three sub-modules [10, 11] , i.e. user metadata collection, location analysis and topic modelling of tweets.

User metadata collection is done using the twitter user object. The random user ids obtained are used. Location analysis is done by getting places tagged in the tweets and location attribute set by the user. As part of topic modelling, at first the tweets are pre-processed, i.e. stop words, URLs, Hash tags are removed and stored in text files.

From the pre-processed tweets a set of attribute vectors are computed for each document by tokenising the strings and representing each document as a set of words and their respective counts. Attribute vectors form the input for the textual graph builder. The graph builder removes infrequent tokens and assigns a vertex for the remaining tokens. Edges are drawn and weights are assigned between vertices based on the co-occurrence frequency. Once the graph is built, keywords are extracted by calculating closeness centrality for each vertex. The top N vertices with the highest closeness centrality are chosen.

Latent Dirichlet Allocation model built using scraped reviews from trip advisor and tourist destination information is used to obtain probabilities for the set of tweets of each user. The set of keywords obtained from the tweets using the twitter keyword graph are given as test data for the LDA model. The probabilities hence obtained are populated into user's node. The process of topic modelling is illustrated in Fig. 2.

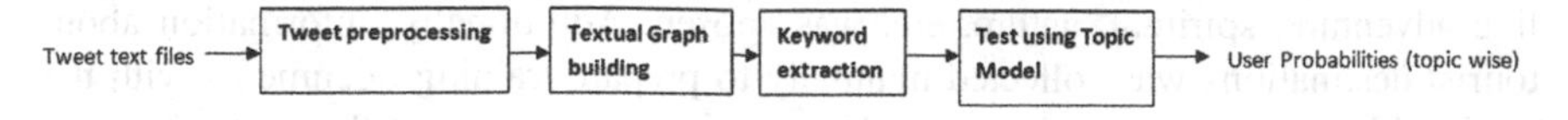

Fig. 2 Topic modelling

Fig. 3 User node

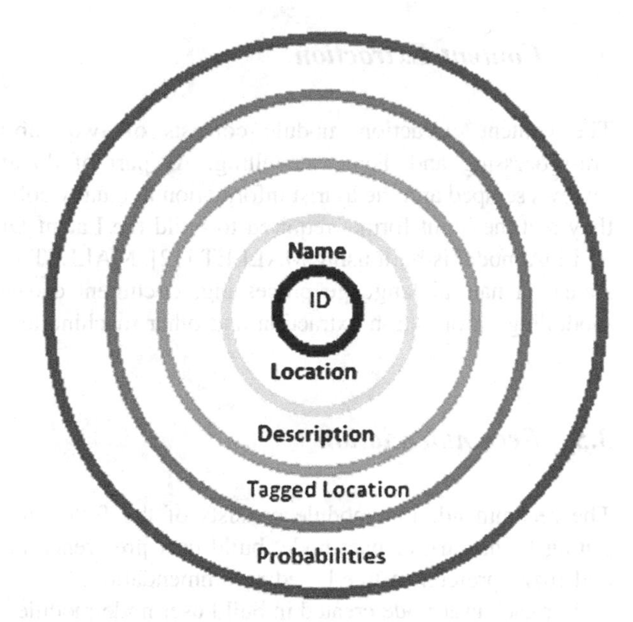

Each user is represented by the user node, which is built from the data obtained from the previous steps. User meta data, tagged locations and results from topic modelling of tweets are populated into a user node. The node is a two-dimensional structure which has a spiral representation. Each layer of the node is independently accessible and each layer represents data mined from either the user's tweets or his/her twitter account. The layers could be expanded to accommodate additional attributes in future. An illustrative representation is shown in Fig. 3. Such a structure enables one to collate all the information mined about a user at a single place.

User id is the basis by which a user is uniquely identified. So, the first layer is populated with the user id. Name, location and description are meta data obtained from the users. Tagged locations are obtained from tweets if the user has enabled geo tagging. Probabilities are obtained from the LDA model built.

3.3 Review Extraction

A web scraper is used to retrieve reviews written by users for tourist locations in India. A sample of about 25–30 reviews for each destination under varying topics like adventure, spiritual, nature, etc. was chosen. Additionally, information about tourist destinations was collected manually to prepare training documents with the perfect blend of user experiences and tourist information about the destination.

3.4 Content Extraction

The content extraction module consists of two sub-modules, i.e. Document pre-processing and Topic modelling. As part of document pre-processing, the reviews scraped and the tourist information manually collected are formatted so that they suit the input format required to build the Latent Dirichlet Allocation model.

LDA model is built using MALLET [12]. MALLET is a Java-based package for statistical natural language processing, document classification, clustering, topic modelling, information extraction and other machine learning applications to text.

3.5 Recommendation

The recommendation module consists of the following sub–modules, viz. computing hash, retrieve user node, build user preference node, build destination tree and fuzzy preference tree-based recommendation.

For each user node created in build user node module, a hash value is calculated for the user id based on the following hash function:

$$h(s) = \sum_{i=0}^{n-1} s[i] \times 31^{n-i-1}; \tag{1}$$

where n is the length of the string and $s[i]$ refers to the character at ith position of the string. This value is again converted into its short value and mod() function is applied (to convert the range to a positive one) to obtain the hash value for the user id string. This hash value is used to index into a database and correspondingly user data is filled at the corresponding fields. In case of collision, chaining is used as the method of collision resolution. Each user node is retrieved from the database by computing hash for the user id using Eq. (1), converting it into short value and taking mod().

Once the node is retrieved from the database, each node is parsed to obtain the probabilities, which is again fed as input into for the user preference node.

A user preference node is a node that contains the probabilities of each topic with respect to the tweets of each user. This node is similar to the destination node created while building the tree.

Destination tree is the key component in the recommendation engine. The database having probabilities assigned to each state is used to create a destination

Fig. 4 Destination tree

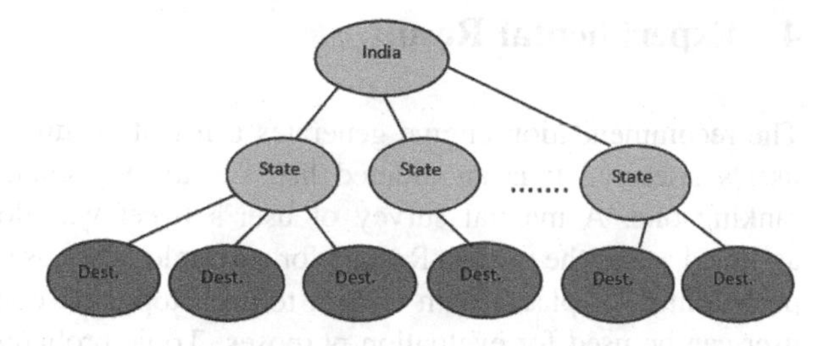

tree. The destination tree contains two types of nodes, a leaf node and an intermediate node. An Intermediate node contains just name attribute. A leaf node inherits from an intermediate node; it contains a probability field in addition to the fields inherited from the intermediate node. The tree is built iteratively with each state representing an intermediate node and each destination representing a leaf node. The destination tree hence built will be an m-ary tree. Figure 4 shows a diagrammatic representation of destination tree built; intermediate nodes are coloured blue while the leaves are coloured red. User preference node and destination tree built are compared to recommend places to users. Algorithm 1 is used to generate recommendation list for top three topics of the user:

Algorithm 1: Recommender

Input: User preference node, destination tree

Output: Recommendation list

Compare the probabilities and obtain the top three topics for user from the user preference node

Recursively iterate the tree and obtain the top three places for each topic by comparing the probabilities.

Output the lists obtained for each topic

For the top destination in each list

Obtain the state of the destination

Obtain the top three destinations of the state with respect to the user's interesting topics

end for

The recommender algorithm generates a list of top three places for each topic of interest of the user. Furthermore, it even lists out top three places to visit in each state based on the topic of interests of the user.

4 Experimental Result

The recommendation engine generates a list of destinations in India based on the user's interests. It is an ordered list, i.e. the top element would be the highest ranking one. A manual survey of user's tweet was done and topics of interest assigned as per the tweets. Results for a sample of six users are listed here. Since the probabilities of places with respect to each topic are constant, the interest of each user can be used for evaluation purposes. Topic probabilities were generated for a total of fifty eight destinations encompassed under fifteen Indian states. Metadata and thousand tweets of users were collected for a set of hundred users.

Accuracy is the proportion of true results (both true positives and true negatives) among the total number of cases examined. It is calculated using Eq. (2).

$$\text{Accuracy} = \frac{\{\text{Expected}\} \cap \{\text{Obtained}\}}{\text{Expected}} \times 100 \tag{2}$$

The results obtained are given in Table 1.

In most cases, the accuracy values yield a value of about 66.67%. Topic 1 is repeated in all the results, and this is due to the fact that Topic 1 is quite generic topic. Accuracy achieved is not 100% because the expected topics were not obtained directly from the user, and a human classifier was used to identify topic of interest of users instead:

$$\text{recall} = \frac{\text{true positives}}{\text{true positives} + \text{false negatives}} \tag{3}$$

$$\text{precision} = \frac{\text{true positives}}{\text{true positives} + \text{false positives}}. \tag{4}$$

Recall gives the fraction of preferred item that is recommended, whereas precision gives the fraction of recommended items that is actually preferred by the user. Recall is calculated using Eq. (3). Equation (4) is used to obtain precision (Figs. 5 and 6).

On an average the system gives a precision of 0.67. A high value of precision is preferred because it means that every place recommended is of interest to the user. A high value of recall means that all the user's places of interest were retrieved and

Table 1 Accuracy

User id	Expected	Obtained	Accuracy
14159285	{5, 2, 1}	{1, 3, 5}	66.67
43126513	{1, 5, 6}	{1, 6, 2}	66.67
147973367	{1, 2, 4}	{6, 1, 4}	66.67
244748979	{1, 2, 6}	{1, 6, 2}	100
24862770	{6, 1, 2}	{1, 6, 3}	66.67
383483409	{2, 1, 5}	{1, 6, 3}	33.33

Fig. 5 Precision

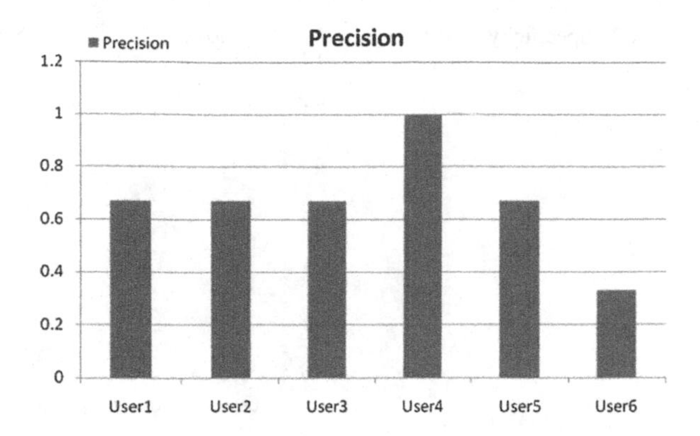

Fig. 6 Recall

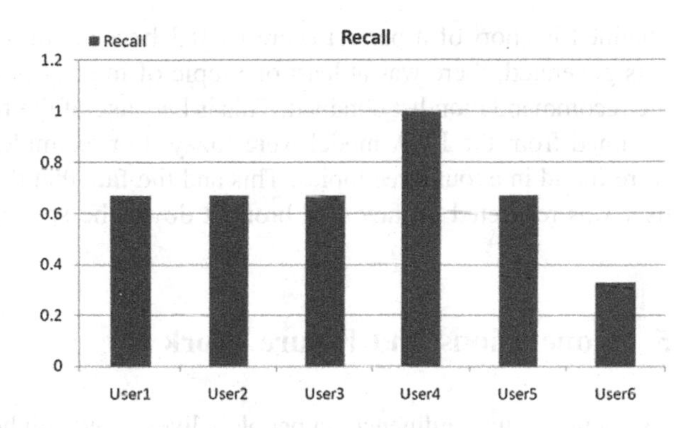

recommended. From the graph, we can infer that on an average the system gives a recall of 0.67. The precision and recall obtained are not 100% because of few reasons, viz. use of human classifier to find topic of interest of user, tweets that are not very topic specific, and limit in the number of interesting topics considered.

Specificity or true negative rate relates to the recommendation system's ability to correctly detect the topics that are not of interest for the user. Specificity is obtained using Eq. (5):

$$\text{Specificity} = \frac{\text{true negative}}{\text{true negative} + \text{false positive}}. \tag{5}$$

The values computed are plotted in Fig.7.

A high value of specificity is preferred, because it implies that topics recommended by the system are of high interest for the user. Almost all users have specificity greater than 0.7, and this is because both in the expected and obtained results almost all topics were the same and topics which were of low interests to the user were not recommended by the recommendation system. The specificity

Fig. 7 Specificity

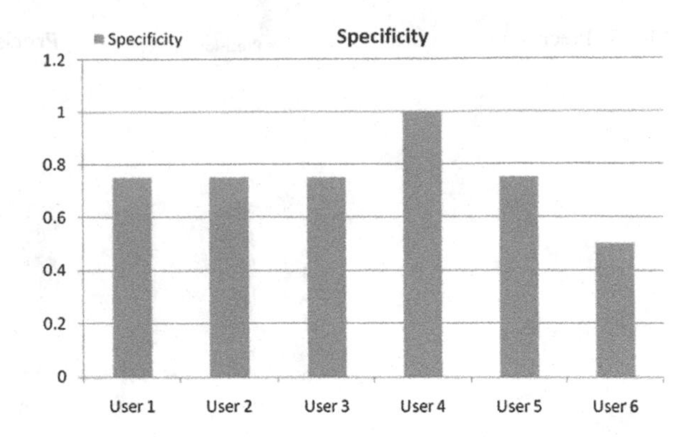

obtained is short of a perfect score by 0.3 because in almost all recommendation lists generated, there was at least one topic of interest of the user which was not in the recommendation list obtained. This is because of the fact that many of the topics obtained from the LDA model were fuzzy. For example, words relating to nature were found in about three topics. This and the fact that the topic of interest for each user was restricted to three had brought down the specificity by a bit.

5 Conclusions and Future Work

As social media's influence on people's lives grows higher and higher, the reviews posted on them gain a whole new level of importance. This paper presents an innovative way to represent a user's information retrieved from social media site.

The tweets of the user are processed and probabilities are calculated for each of the seven topics obtained from the LDA model. These probabilities along with the metadata information of user are populated in 2-D user structure with layers. This populated structure is stored by computing a hash value. Hashing enables in faster storage and retrieval of the structure. Simultaneously, a destination tree is built from the information scraped from popular tourist websites. The user node and the destination tree are compared to obtain tourist destination recommendation lists for user.

Possible extensions would be to mine data from multiple social networking sites like Facebook, Flickr, Twitter, etc. Additionally, photos and links posted can also be analysed to further enhance the profile built for each user. Another extension to the proposed work can be to construct links between user nodes so that recommendation lists can be shared among friends.

References

1. Kleinberg, Jon M, "Challenges in mining social network data: processes, privacy, and paradoxes", 1 In Proceedings of the 13th ACM SIGKDD international conference on Knowledge discovery and data mining, pp. 4–5, 2007.
2. Neri, Federico, Carlo Aliprandi, Federico Capeci, Montserrat Cuadros, and Tomas By, "Sentiment analysis on social media", In Proceedings of the 2012 International Conference on Advances in Social Networks Analysis and Mining, pp. 919–926, 2012.
3. Hernandez, Mauricio, Kirsten Hildrum, Paril Jain, Rohit Wagle, Bogdan Alexe, Ram Krishnamurthy, Ioana Roxana Stanoi, and Chitra Venkatramani, "Constructing consumer profiles from social media data", in proceedings of IEEE International Conference on Big Data, pp. 710–716, 2013.
4. G. Adomavicius and A. Tuzhilin, "Toward the next generation of recommender systems: A survey of the state-of-the-art and possible extensions," IEEE Transactions on Knowledge Data Eng., vol. 17, no. 6, pp. 734–749, 2005.
5. Wu, Dianshuang, Guangquan Zhang, and Jie Lu. "A fuzzy preference tree-based recommender system for personalized business-to-business e-services. "IEEE Transactions on Fuzzy Systems, vol.23, no. 1 pp. 29–43, 2015.
6. Chen, Xin, Mihaela Vorvoreanu, and Krishna PC Madhavan. "Mining social media data for understanding students' learning experiences.", IEEE Transactions on Learning Technologies, pp. 246–259, 2014.
7. Akay, Altug, Andrei Dragomir, and Bjorn-Erik Erlandsson. "Network-Based Modelling and Intelligent Data Mining of Social Media for Improving Care." IEEE Journal of Biomedical and Health Informatics, 210–218, 2015.
8. Garcia Esparza, Sandra, Michael P. O'Mahony, and Barry Smyth. "Catstream: categorising tweets for user profiling and stream filtering", In Proceedings of the International conference on Intelligent user interfaces ACM, pp 25–36, 2013.
9. Ostrowski, David Alfred. "Using latent dirichlet allocation for topic modelling in twitter", In proceedings of IEEE International Conference on Semantic Computing (ICSC), pp. 493–497, 2015.
10. Weng, Jianshu, et al. "Twitterrank: finding topic-sensitive influential twitterers", In Proceedings of the Third ACM international conference on Web search and data mining. ACM, pp. 261–270, 2010.
11. Abilhoa,Willyan D., and Leandro N. de Castro. "A keyword extraction method from twitter messages represented as graphs", In proceedings of Applied Mathematics and Computation conference, pp. 308–325, 2014.
12. McCallum, Andrew Kachites. "MALLET: A Machine Learning for Language Toolkit", http://mallet.cs.umass.edu, 2002.

Demand Response Program Based Load Management for an Islanded Smart Microgrid

K. Gokuleshvar, S. Anand, S. Viknesh Babu and D. Prasanna Vadana

Abstract There has been an ever-growing demand for electrical energy causing supply–demand imbalance in the power system. To overcome this imbalance, a reliable, efficient, less centralised, and more interactive energy management system (EMS) has to be realized. One of the methods to bridge the supply–demand gap using EMS is through Load Management. The evolution of EMS allows loads to respond to the demand (Demand Response) and assist customers to make informed decisions about their energy consumption, adjusting both the timing and quantity of their electricity use. This paper deals with the development of Fuzzy logic-based load management scheme using Demand Response Programs in smart microgrids. The developed system is tested on a smart micro-grid simulator (SMGS) operated in islanded mode installed in the Renewable Energy Laboratory in Amrita Vishwa Vidyapeetham, Coimbatore.

Keywords Energy management system · Load management · Demand response

1 Introduction

Microgrids show great promise as a means of integrating distributed energy resources (DERs) into the public grid distribution system and providing uninterrupted, high-quality power to local loads [1]. Microgrids are modern, locally controlled, small-scale version of the existing centralised energy grid comprising various distributed generators, storage devices and loads [2]. They are designed to operate either in grid connected or islanding mode. Microgrids are provided with

K. Gokuleshvar (✉) · S. Anand · S. Viknesh Babu · D. Prasanna Vadana
Department of Electrical and Electronics Engineering, Amrita School of Engineering, Amrita Vishwa Vidyapeetham, Amrita University, Coimbatore, India
e-mail: gokuleshvar@gmail.com

D. Prasanna Vadana
e-mail: d_prasanna@cb.amrita.edu

S.S. Dash et al. (eds.), *Artificial Intelligence and Evolutionary Computations in Engineering Systems*, Advances in Intelligent Systems and Computing 517,
DOI 10.1007/978-981-10-3174-8_23

intelligence to island from the main grid intentionally or in case of fault [3, 4]. The micro-grid with real-time measurement and communication facilities becomes a smart microgrid (SMG) [5].

Normally, public power grids operate based on the demand and hence, is load following in nature. But, the supply side and demand side are uncontrolled in a microgrid [6]. In case of a mismatch between power generated and demand, fluctuations are observed in the microgrid frequency and voltage which results in unstable operation or even leads to microgrid blackout [7]. The frequency imbalance in grid connected mode of SMG is handled by a central governor system present in the legacy grid. On the contrary, the Islanded SMG lacks such a feature and hence is very sensitive to frequency. Thus, microgrids demand the presence of an energy management system (EMS) which mimics the operation of the centralised governor to handle the frequency imbalance [8]. EMS must be capable to monitor the status of the microgrid at every instant and take necessary steps to maintain the frequency with the permissible limit of 49.9–50.05 Hz [9] as per Indian standards.

Different methods for autonomous energy management in microgrids were proposed. The authors of [10] discuss automated demand side management including online energy management, peak load reduction, and priority-based isolation of local controllable loads. In [11], economic dispatching and operation optimization in a laboratory model of a microgrid is addressed. A central control scheme is proposed which is responsible for communication with energy control centers, manage distributed generation units and storage modules to address the issue of energy imbalance in microgrids.

To handle the supply–demand mismatch, a Dynamic EMS is developed in [12] which is capable of monitoring the status of SMG at regular intervals and take necessary corrective measures by controlling the charge–discharge transactions for the energy storage systems in real time. In islanded mode, the frequency of the SMG is very sensitive which makes its operation more critical and challenging. Apart from charge–discharge transactions related to storage elements, Load Management (LM) becomes essential in Islanding mode to manage the frequency imbalance. LM is a type of demand response program (DRP) that allows utilities to reduce demand for electricity during peak hours by direct intervention of loads.

The conventional LM techniques suggest shedding a fixed amount of load if frequency crosses a certain threshold. However, this technique fails to achieve the optimal load shedding. This is due to the fact that conventional techniques shed loads without estimating the actual power imbalance. This may lead to either over-shedding, resulting in power quality problems, or under-shedding, resulting in total power system tripping [13, 14]. Time priority LM based on load profile is discussed in [15], which is optimized than the conventional LM techniques.

Some of the LM techniques like peak clipping, valley filling, and load shifting are discussed in [16]. The authors of [17] discuss implementation methods like Time of Use (ToU), Direct Load Control (DLC), and distribution loss reduction.

DRPs control electricity usage by the loads from their normal consumption patterns when electric grid eventualities threaten supply–demand balance or market conditions occur that raise electricity costs [18]. When customer participates in a DRP, there are three ways in which their use of electricity can be changed [19]

- Reducing their energy consumption through load curtailing strategies;
- Moving energy consumption to a different time period;
- Using onsite standby generated energy, thus limiting the dependence on SMG.

Thus, DRPs pave the way for utilizing the consumer as a demand side resource. Over the years, manual DRP that involves labor intensive approach has evolved to semi-automated DR where a pre-programmed strategy is initiated by a person via a central control system. However, as the system becomes complex with multiple communications, there is a need for a Fully Automated DRP [20]. This paper proposes a Regional Load Management System (RLMS) to perform DRP based LM for a Laboratory model of SMG in Islanding mode.

2 Description of the System

2.1 System Overview

The proposed RLMS aims to limit the frequency excursions in Islanding mode of SMG caused due to the mismatch in supply–demand by performing LM. The RLMS is designed to operate in a region consisting of different type of loads as shown in Fig. 1, viz. Domestic (residential), Industrial (factories, workshops), Commercial (restaurants, theaters, shopping malls), and Essential (hospitals, data centers). These loads are linked to the RLMS wirelessly through Zig Bee (IEEE 802.15.4), forming a Regional Area Network (RAN).

2.2 Solar Linked Demand Response Program

The solar-linked DRP is suggested in this work. Solar-linked DRP demands the presence of solar panels, local storage with inverter, control circuitry, and a Local Controller (LC) linked with the RLMS. Customers enrolled in this DRP establish an agreement that permits the utility to maintain direct control over their loads. Upon receiving signal from the RLMS in case of a contingency, the LC energizes the premises by the local storage or by the SMG. Eventually, customers enrolled in solar linked DRP enjoy the benefit of uninterrupted power supply even when disconnected from the SMG.

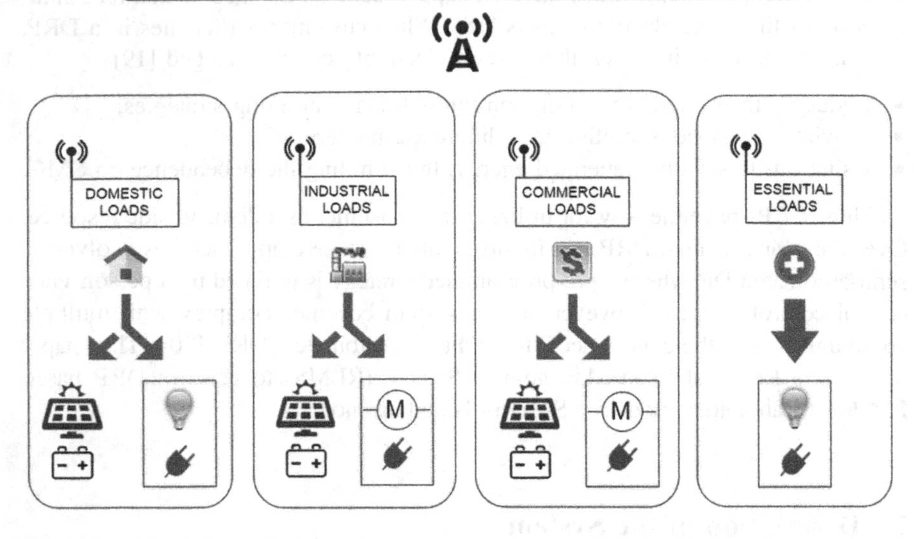

Fig. 1 Local loads in the RAN

2.3 *Load Distribution in RAN*

The RAN under consideration is designed to have a total load capacity of 500 W connected to the SMG in Islanded mode, comprising of 12 load groups out of which six groups of consumers (240 W) are enrolled to solar linked DRP. Each load group is assumed to be a cluster of 10 units, each of equal load capacity. The detailed distribution of load groups is as shown in Table 1.

Table 1 Distribution of load groups in RAN

S.no.	Load group	Capacity (W)	Classification (W)	
1	Domestic	160	DRP	15
			DRP	25
			DRP	40
			Non-DRP	40
			Non-DRP	40
2	Industrial	200	DRP	40
			DRP	60
			Non-DRP	40
			Non-DRP	60
3	Commercial	100	DRP	60
			Non-DRP	40
4	Essential	40	Non-DRP	40

3 Load Management Scheme

3.1 Priority-Based Load Management

Frequency mismatch can be handled by varying the loads connected to the Islanded SMG. When frequency is less than its permissible lower limit, loads have to be excluded from the SMG and when more than its permissible upper limit, loads have to be included. However, the exact quantity of load to be excluded/included to the SMG (in watts) is unknown.

Based on relative consumer importance, a priority based LM scheme is employed as shown in Fig. 2, in which loads are varied sequentially to limit the frequency within permissible limits. Essential loads are given the highest priority and hence shutdown at last. However, this process is time consuming and unfair to consumers at the bottom of the hierarchy. The exact quantity of load to be included/excluded to/from the SMG can be determined by trial and error, as verified through experimentation, in which human reasoning played a very significant role.

3.2 Fuzzy Logic-Based Load Management

Evidently, this scenario demands the presence of a tool based on human reasoning, which is capable of identifying the exact quantity of load to be varied and balancing system frequency in the shortest possible time. Fuzzy logic is such a tool, which resembles human reasoning the most, and is capable of making rational decisions in an environment of imprecision and uncertainty [21].

The current frequency of SMG and trend in frequency, calculated as difference between current and previous frequency, are taken as input variables to Fuzzy logic

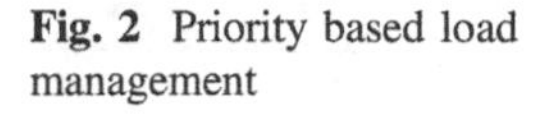
Fig. 2 Priority based load management

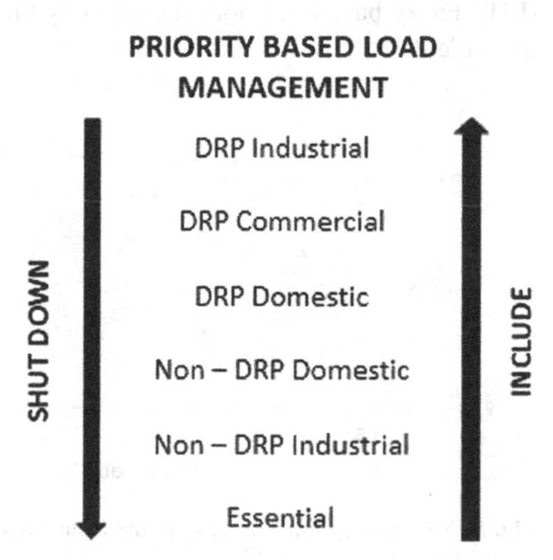

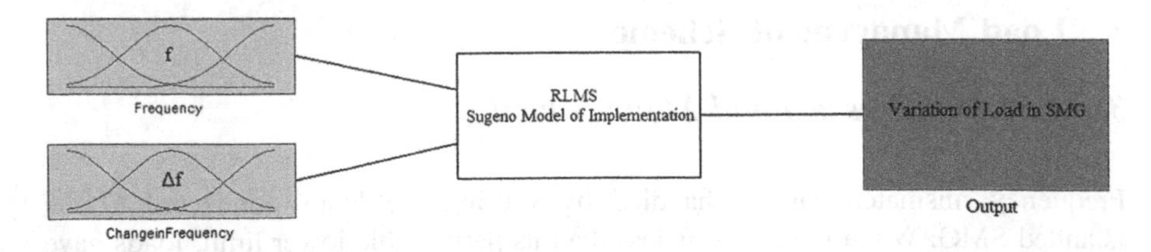

Fig. 3 Implementation of Fuzzy logic

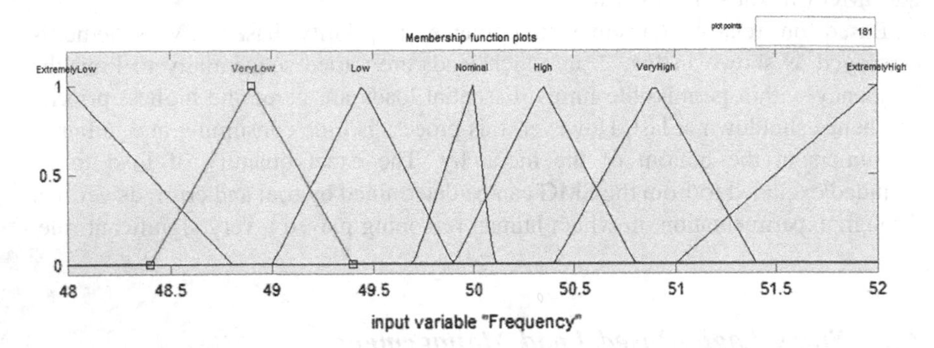

Fig. 4 Membership function plots for the input variable frequency

as shown in Fig. 3. Sugeno model of implementation is employed, as the output (quantity of load to be excluded/included) is a discrete value. There are seven membership functions for each of the input variables as shown in Figs. 4 and 5. A total of 49 rules are possible for which the decisions are assigned as shown in surface view of the rules in Fig. 6.

The load groups are divided into seven categories, viz., Extra light (EL), Very light (VL), Light (L), Medium (M), Heavy (H), Very Heavy (VH), and Extra Heavy (EH). Fuzzy-based decisions are taken by the RLMS based on the inputs as shown in Table 2.

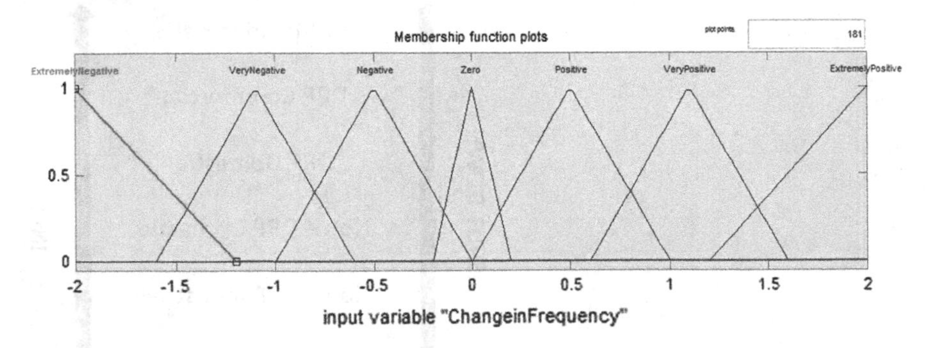

Fig. 5 Membership function plot for the input variable change in frequency

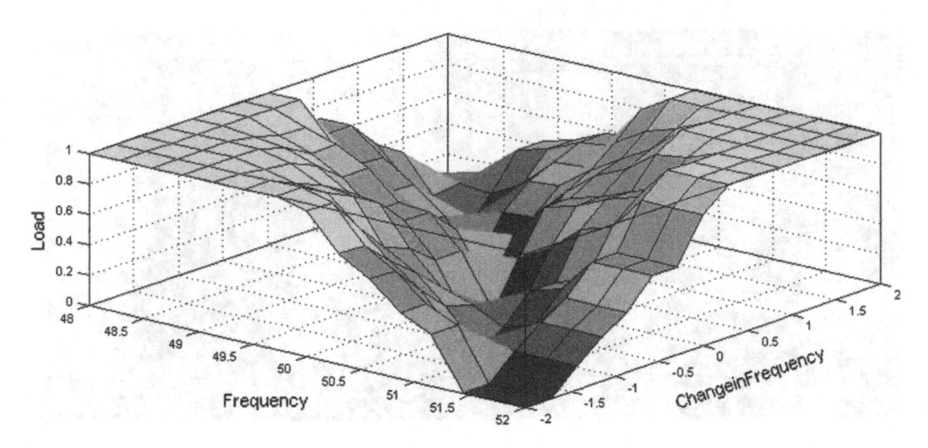

Fig. 6 Surface view of the RLMS decisions for various combinations of the input variables—frequency and change in frequency

Table 2 RLMS decisions based on Fuzzy logic

Case	Decisions	Case	Decisions
1	Exclude extra light loads	8	Include extra light loads
2	Exclude very light loads	9	Include very light loads
3	Exclude light loads	10	Include light loads
4	Exclude medium loads	11	Include medium loads
5	Exclude heavy loads	12	Include heavy loads
6	Exclude very heavy loads	13	Include very heavy loads
7	Exclude extra heavy loads	14	Include extra heavy loads

4 Hardware Implementation

4.1 Smart Microgrid Simulator (SMGS)—Test Bed for Realization of RLMS

Smart micro-grid simulator (SMGS) shown in Fig. 7, installed in Renewable Energy Laboratory at Amrita Vishwa Vidyapeetham, Coimbatore is used for validation of automated and closed loop operation of RLMS.

The SMGS is five bus system (as shown in Fig. 8) which is energized by distributed generators: (i) Micro-Hydroelectric Power Plant (MHPP), (ii) Wind Power Plant (WPP) and (iii) Solar Power Plant (SPP). Energy Storage Systems such as Battery storage with Charge–Discharge Controller (CDC) and PH (Pumped Hydro) unit with VSD (Variable Speed Drive) are connected to the SMG [22]. Table 3 shows the actual and downscaled specifications of all the components present in the SMGS.

Fig. 7 SMG simulator in renewable energy laboratory of Amrita Vishwa Vidyapeetham, Coimbatore

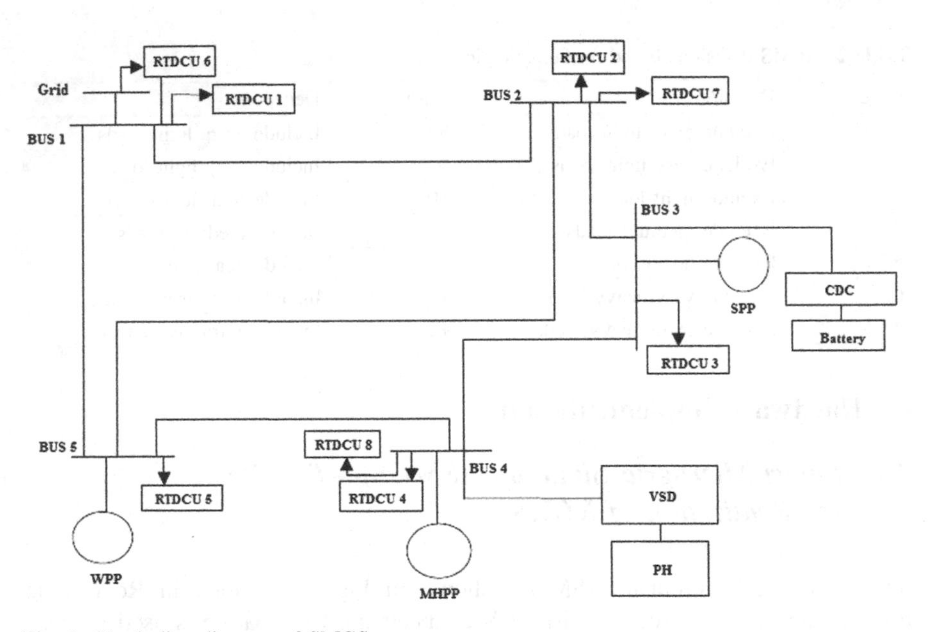

Fig. 8 Single line diagram of SMGS

In the SMGS, the real-time operational data are continuously monitored using Real-Time Data Collection Units (RTDCUs). A RTDCU is provided at every bus to measure current, voltage and frequency, and transmit the data to the RLMS wirelessly.

Table 3 Specifications of microgrid, distributed generators and energy storage systems on SMGS

S.no.	Component	Field size	Downscaled size
1	MHPP	5 MW	1 kW
2	SPP	0.625 MW	600 W
3	WPP	6 MW	1.5 kW
4	PH unit	2.5 MW	300 W
5	Battery	375 kWh	24 V, 300 Ah

4.2 *Experimental Verification of RLMS Operation*

The RAN setup connected to SMGS is shown in Fig. 9 which houses RLMS with appropriate displays. Each load group in every type (Domestic/Industrial/Commercial/Essential) is modeled as an incandescent lamp with appropriate wattage as shown in Table 1. The operating frequency of the SMGS is transmitted to the RLMS periodically (every 4 s) by the RTDCU installed in the SMG, based on which frequency trend is calculated. RLMS responds to the frequency change by varying the load connected to the Islanded SMG.

When the frequency is within permissible limits, the RLMS remains in standby mode. If the frequency deviates beyond permissible limits, the RLMS performs Fuzzy-based LM on DRP customers depending on the status of the loads. However, if the Fuzzy based LM fails to restrict the frequency, the RLMS performs priority based LM on non-DRP customers. The operation of LM scheme in RLMS is shown in Fig. 10.

Fig. 9 Hardware setup of RAN

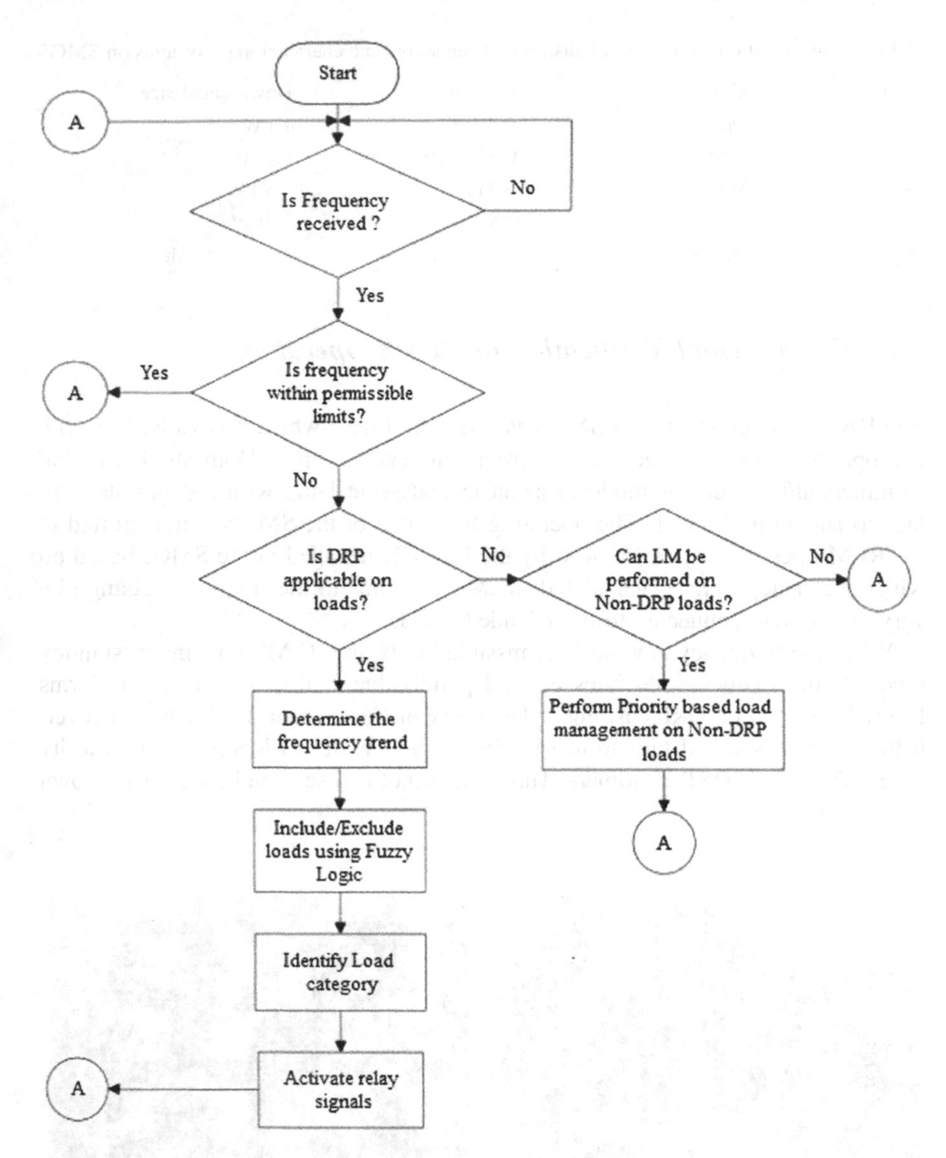

Fig. 10 Realisation of load management scheme in RLMS

5 Results and Discussions

The RLMS developed for the RAN setup is integrated with the SMGS which is operated in Islanded mode. A disturbance is introduced to the system by manually including/excluding loads, resulting in frequency excursions. The RLMS takes necessary actions based on Fuzzy logic/priority to bring back the frequency within

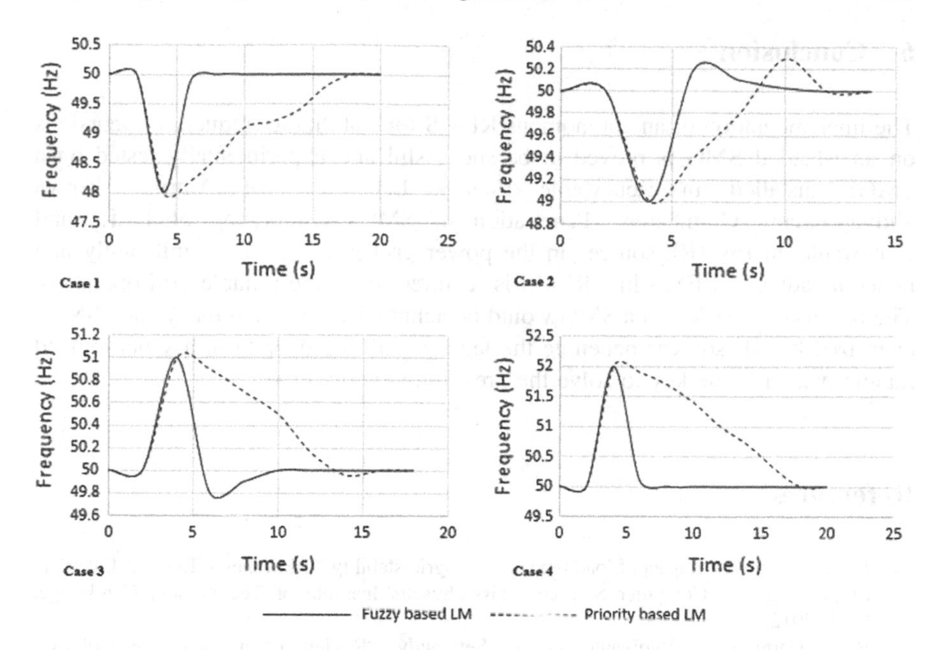

Fig. 11 Comparative analysis of Fuzzy and priority based LM schemes for an Islanded SMG

Table 4 Experimental results of LM scheme

Case	Previous frequency (Hz)	Current frequency (Hz)	Change in frequency (Hz)	Action taken (W) (include/exclude)	Resultant frequency (Hz)
1	51.33	52	1.67	**I** 240	50.1
2	52	51.5	−0.5	**I** 135	50.23
3	50.54	51	0.46	**I** 175	49.75
4	50.4	50.5	0.1	**I** 85	49.95
5	50.14	49.5	−0.64	**E** 135	50.18
6	49.49	49	−0.49	**E** 175	50.2
7	47.98	48.5	0.52	**E** 145	49.82
8	49.21	48	−1.21	**E** 240	49.9

permissible limits. Conventional priority-based LM is performed on all loads and the results are compared with Fuzzy based LM as shown in Fig. 11.

For instance, in case 1, the frequency recovery time is 16 s when priority based LM is performed. On the contrary, it is 6 s when Fuzzy based LM is performed which is far superior and an essential requirement for autonomous operation of SMG. A similar response is observed in other instances as shown in Table 4. A minimal overshoot is observed for few cases in Fuzzy based LM scheme. This is attributed to the hardware restrictions on quantity and diversity of loads in the RAN under consideration.

6 Conclusion

The implementation of an autonomous RLMS for stabilizing frequency excursions on an Islanded SMG is proved to be successful and experimentally tested on a SMGS installed in Renewable Energy Laboratory of Amrita Vishwa Vidyapeetham, Coimbatore. Penetration of SMGs comprising of DERs and Renewable energy (RE)sources in the power grid is increasing significantly and hence an advanced EMS like RLMS is required to ensure reliable grid operation. The presence of RLMS in a SMG would be acknowledged when many such SMGs energized by RE sources penetrate the legacy grid and operate in a synchronized manner which is the key to solve the prevailing energy crisis.

References

1. Jared P. Monnin, "Impact of load type on microgrid stability," M.S. thesis, Dept. of Electrical Engineering and Computer Science, Massachusetts Institute of Technology, Cambridge, USA, 2012.
2. B. S. Hartono, Y. Budiyanto and R. Setiabudy, "Review of microgrid technology," *International Conference on QiR (Quality in Research)*, Yogyakarta, 2013, pp. 127–132.
3. Z. Zhang, Y. Li and W. Chen, "The research on micro-grid mode conversion," *China International Conference on Electricity Distribution (CICED)*, Shanghai, 2012, pp. 1–7.
4. N. Hatziargyriou, H. Asano, R. Iravani and C. Marnay, "Microgrids," in *IEEE Power and Energy Magazine*, vol. 5, no. 4, pp. 78–94, July–Aug. 2007.
5. F. Berkel, D. Gorges and S. Liu, "Load-frequency control, economic dispatch and unit commitment in smart microgrids based on hierarchical model predictive control," *IEEE 52nd Annual Conference on Decision and Control (CDC)*, Firenze, 2013, pp. 2326–2333.
6. C. Chen, S. Duan, T. Cai, B. Liu and G. Hu, "Smart energy management system for optimal microgrid economic operation," in *IET Renewable Power Generation*, vol. 5, no. 3, pp. 258–267, May 2011.
7. H. Bevrani, M. R. Feizi and S. Ataee, "Robust Frequency Control in an Islanded Microgrid: H_∞ and μ-Synthesis Approaches," in *IEEE Transactions on Smart Grid*, vol. 7, no. 2, pp. 706–717, March 2016.
8. D. P. Vadana and S. K. Kottayil, "Energy-aware intelligent controller for Dynamic Energy Management on smart microgrid," *Power and Energy Systems Conference: Towards Sustainable Energy, 2014*, Bangalore, 2014, pp. 1–7.
9. *The Indian Electricity Act*, Sub-regulation (m) of Regulation 5.2, 2014.
10. Adeel Abbas Zaidi and Friederich Kupzog, "Microgrid Automation—A Self: Configuring Approach," Proc. INMIC 2008, pp. 565–570.
11. Ling Su, Jianhua Zhang and Ziping Wu. "Energy Management Strategy for Lab Microgrid," Proc. Asia-Pacific Power and Energy Conference, 2011, pp. 1–4.
12. G. K. Venayagamoorthy; R. K. Sharma; P. K. Gautam; A. Ahmadi, "Dynamic Energy Management System for a Smart Microgrid," in *IEEE Transactions on Neural Networks and Learning Systems*, vol. PP, no. 99, pp. 1–14.
13. J. A. Laghari, H. Mokhlis, A. H. A. Bakar, and H. Mohamad, "Application of computational intelligence techniques for load shedding in power systems: A review," Energy Convers. Manage., vol. 75, pp. 130–140, 2013.

14. B. Delfino, S. Massucco, A. Morini, P. Scalera, and F. Silvestro, "Implementation and comparison of different under frequency load-shedding schemes," in Proc. IEEE Power Eng. Soc. Summer Meeting, 2001, vol. 1, pp. 307–312.
15. K. U. Rao, S. H. Bhat, G. Jayaprakash, G. G. Ganeshprasad and S. N. Pillappa, "Time priority based optimal load shedding using genetic algorithm," *Communication and Computing (ARTCom 2013), Fifth International Conference on Advances in Recent Technologies in*, Bangalore, 2013, pp. 301–308.
16. G. Thomas Bellarmine P.E., "Load Management Techniques", *IEEE Southeastcon*, pp. 139–145, Nashville, TN, 2000.
17. Zahir J. Paracha and Parviz I. Doulai, "Load management techniques and Methods in electric power system," *International conference on Energy Management and Power Delivery*, Vol. 1, pp. 213–217, 1998.
18. Federal Energy Regulatory Commission, "Assessment of demand response and advanced metering" [Online]. Available: http://www.ferc.gov/legal/staff-reports/2010-dr-report.pdf, Feb. 2011.
19. Pierluigi Siano, "Demand response and smart grids—A survey", Renewable and Sustainable Energy Reviews, Vol. 30, pp. 461–478, 2014.
20. Mary Ann Piette, David Watson, Naoya Motegi, and Sila Kiliccote, "Automated Demand Response Strategies and Commissioning Commercial Building Controls," *National Conference on Building Commissioning*, pp. 1–13, 2006.
21. Timothy J. Ross, "Logic and Fuzzy Systems," in Fuzzy Logic with Engineering Applications, 3rd ed. UK: John Wiley & Sons Ltd., 2010, pp. 117–148.
22. D. PrasannaVadana, Sasi K. Kottayil, Autonomous Control of Smart Micro Grid in Islanding Mode, Procedia Technology, Volume 21, 2015, Pages 204–211.

Design of 4 × 2 Corporate Feed Microstrip Patch Antenna Using Inset Feeding Technique with Defective Ground Plane Structure

Tharian Joseph Pradeep and G. Kalaimagal

Abstract Radar and satellite communication uses X-band antennas. Also, high gain and minimal size antennas are key requirements to latest wireless and satellite communication systems. The main parameters or constraints to the design of patch antennas include length and width of the antenna and substrate thickness and type. In this paper, the proposed 4 × 2 corporate feed microstrip patch antenna which is designed to operate at 10 GHz. It was found that the 4 × 2 patch antenna delivers a higher gain and return loss. The significant reduction in mutual coupling has been experimented using Defective Ground Plane (DGS). The paper introduces design of patch feed network using the high frequency structural simulator (HFSS) and the simulation results were compared with that of 4 × 1 patch antenna. The gain, return loss, and radiation pattern are analyzed and presented.

Keywords Patch antenna · HFSS · Wireless communication · Radiation pattern · Gain · DGS

1 Introduction

The antenna is a device that can radiate or receive radio waves [1]. Antennas act as an interface for electromagnetic energy to propagating between free space and guided medium. The sudden growth in wireless communication requires antennas that can receive and transmit. Microstrip antennas are commonly used becoming increasingly useful because they can be printed straightly on a circuit board. Patch antenna are low cost have a low profile and are easily fabricated. The fringing fields around the antenna explains why the microstrip antenna radiates, the center of the

T.J. Pradeep (✉) · G. Kalaimagal
Department of Electronics and Communication, SRM University,
Chennai, Tamil Nadu, India
e-mail: tharianjosephpradeep@gmail.com

G. Kalaimagal
e-mail: Kalaimagal.g@ktr.srmuniv.ac.in

S.S. Dash et al. (eds.), *Artificial Intelligence and Evolutionary Computations in Engineering Systems*, Advances in Intelligent Systems and Computing 517,
DOI 10.1007/978-981-10-3174-8_24

half wave patch current is maximum and the current is zero at beginning of the patch. This low current value at the feed explains why the impedance is high when fed at the end. Patch antennas are most common in mobile communication and there are several methods are used for improving the antenna gain and size reduction. The change in the natural environment has no effect on the patch antenna, because it is independent of rain, snow, etc. An antenna is a device that converts radio frequency power into electromagnetic radiation.

The fringing field near the surface of the patch antenna are both in the $+y$ direction. So the fringing electric fields add up in phase on the edge of the microstrip antenna and produce the radiation of microstrip antenna that is equal current but with opposite direction on the ground plane, which cancels the radiation due to voltage distribution.

The main drawback of microstrip patch antennas is their inefficiency, narrow frequency bandwidth, and low gain. To overcome the bandwidth and gain limitations, some techniques are used such as number of antenna elements or feeding matching networks may be employed. In most application, the microstrip patch antenna is fed using either inset microstrip line or coaxial probe feed both are direct contact methods providing high efficiency [2]. This paper analyzes design of 4×1-patch and 4×2-patch antenna array with corporate feed rectangular microstrip patch array antenna with inset feed technique. In array, the effect of mutual coupling will reduce the antenna efficiency for reducing the effect of mutual coupling single slots is introduced in the ground plane.

1.1 Microstrip Patch Antenna

A Microstrip Patch antennas are fabricated by etching the desired pattern in metal trace attached to a dielectric substrate. It consists of conducting patch of any planar or nonplanar geometry on one side of a dielectric substrate with a ground plane on the other side. The most common geometrical structures for patch antenna are square, rectangular, circular, triangular, and elliptical. Rectangular geometry are separable in nature and their analysis is also simple [3]. The basic structure of patch antenna consists of substrate, ground plane, and metal trace. The input impedance of the inset fed microstrip patch antenna varies in accordance with the inset distance Y_0 and the inset width [4]. In general, the inset gap and feed line width are set at least the same as possible [5] this is done for making the resonant resistance and feedline impedance equal. The principle behind making inset cut in the patch is to match the impedance of the feed line to the patch so there is no need for any additional matching element [3]. This is done by properly selecting the location of inset (Fig. 1).

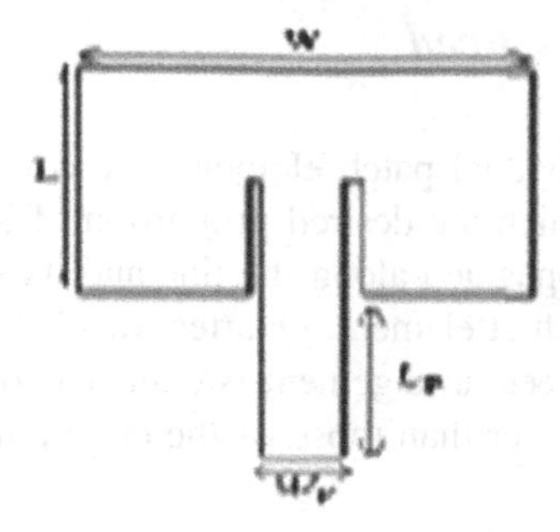

Fig. 1 Single element inset feed patch antenna

2 Methodology

The method is used to excite to radiate direct or indirect contact. There are many different methods of feeing the most common four methods are microstrip line feed, coaxial probe, corporate feed, aperture coupling, and proximity coupling [6].

2.1 Microstrip Line Feed

Microstrip line feed is common and easy to fabricate. here a conducting strip is connected straight to the edge of the Microstrip patch and they can be considered as extension of patch [7]. The conducting strip width is smaller as compared to the patch and this kind of feed arrangement has the advantage such as the feed is etched into the same substrate so that it provides a planar structure. The disadvantage of this method is that when the substrate thickness increases, surface wave spurious feed radiation increases that leads to the limitation on the bandwidth (Fig. 2)

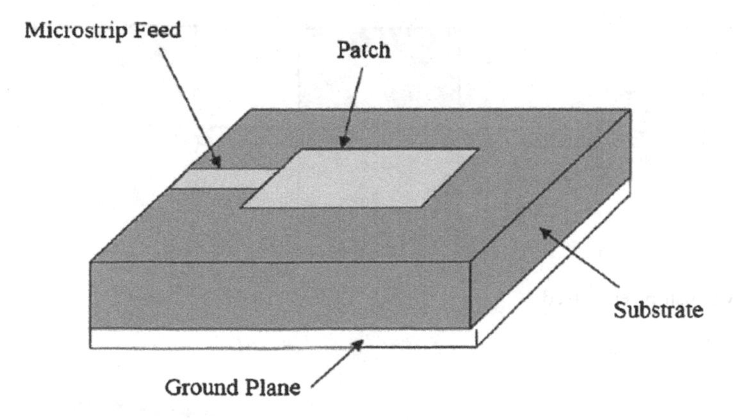

Fig. 2 Microstrip line feed

2.2 Microstrip Series Feed

In this technique, the individual patch elements are connected in a series using a transmission line from, which the desired proportion of RF energy is coupled into the individual element propagated along the line and its shown in Fig. 3. Here the input power is feed at the first element. Quarter wavelength transformer method is used in this method. The feed arrangement is compact so the line losses compared to this type of array are lower than those of the corporate feed type [8].

2.3 Microstrip Pin Feed

In this method, the coaxial connector is used for feeding the RF power to the patch. The coaxial connector has an inner conductor and an outer conductor. The inner conductor is soldered to the metallic patch element, which is drilled through the substrate plate. The outer conductor is connected to the metallic ground plane. The main advantage of this feeding technique is that the feed can be placed at any desired location. It is shown in Fig. 4 [1].

2.4 Microstrip Corporate Feed Network

A corporate feed is most widely used feeding techniques to fabricate the antenna arrays. In this case, the incident power is equally splitting and distributing to the individual antenna elements [9]. The corporate feeding technique can provide power splits of $2k$ (where $k = 2; 4; 8; 16 \ldots$). Here, in Fig. 5, the patch elements are linked together using the quarter wavelength Impedance transformers.

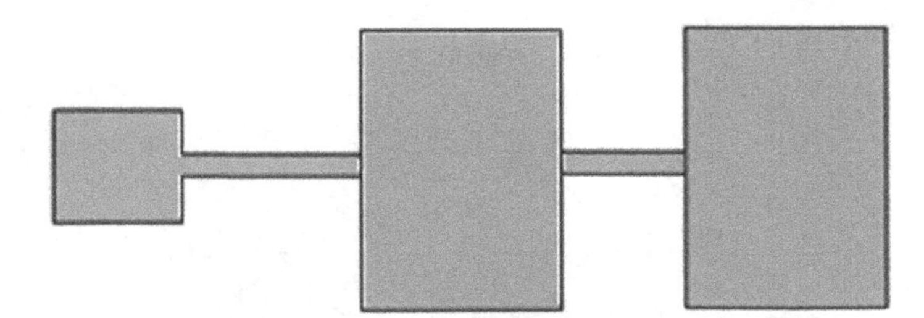

Fig. 3 Microstrip series feed

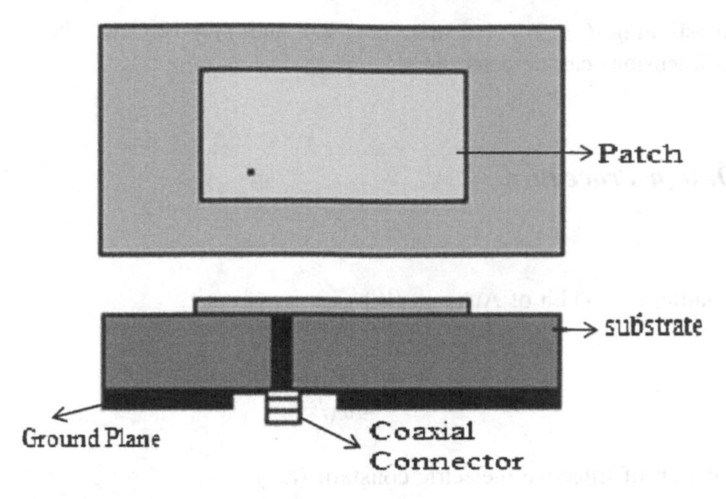

Fig. 4 Microstrip pin feed

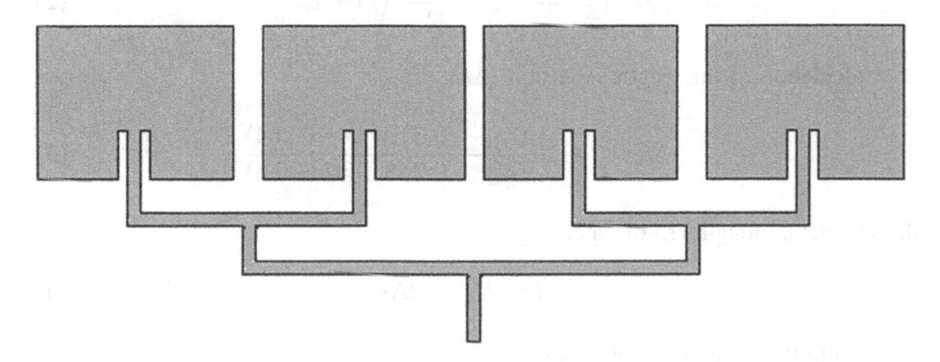

Fig. 5 4 × 1 corporate feed microstrip array antenna

3 Design Consideration for Patch Antenna

The parameters considered for the design of a rectangular microstrip patch antennae are:

1. Resonant frequency of microstrip patch antenna (f_O): The antenna has been designed for the X-band. The resonant frequency of the proposed antenna is 10 GHz.
2. Height of dielectric substrate (h): For designing the microstrip patch antenna to be used in X-band application, it is essential that the antenna should be light and compact [10]. Hence, the height of the substrate is chosen as 1.57 mm.
3. Dielectric constant of the substrate (ε_r): The dielectric material chosen for this design is FR4 substrate which has dielectric constant of 4.4.

By substituting $C = 3 \times 10^8$ m/s, $\varepsilon_r = 4.4$ and $f_0 = 10$ GHz, the values of antenna dimensions can be determined.

3.1 Design Procedure

1. Calculating the Width of Antenna (W)

$$W = \frac{\mathrm{C}}{2f_0\sqrt{\frac{\varepsilon_r + 1}{2}}} \tag{1}$$

2. Calculation of Effective dielectric constant (ε_{eff})

$$\varepsilon_{\text{reff}} = \frac{\varepsilon_r + 1}{2} + \frac{\varepsilon_r - 1}{2}\sqrt{1 + 12\frac{h}{w}} \tag{2}$$

3. Calculation of the length extension (ΔL)

$$\Delta L = 0.412h\frac{(\varepsilon_{\text{reff}} + 0.3)(w/h + 0.264)}{(\varepsilon_{\text{reff}} + 0.258)(w/h + 0.8)} \tag{3}$$

4. The actual length (L) of patch

$$L = L_{\text{eff}} - 2\Delta L \tag{4}$$

5. Length extension (ΔL) of patch

$$L_{\text{eff}} = \frac{C}{2f_0\sqrt{\varepsilon_{\text{reff}}}} \tag{5}$$

FR stands for Flame Retardant and it is very commonly available dielectric substrate in the market and very cheap (Table 1).

Table 1 Patch Dimensions

Patch parameters	Dimensions
Patch width (W)	9.13 mm
Effective dielectric constant ($\varepsilon_{\text{reff}}$)	5.6755
Effective length (L_{eff})	6.296 mm
length extension (ΔL)	0.65 mm
The actual length (L)	4.996 mm

4 Simulation Results

The simulation software used to design and simulate the rectangular patch antenna is HFSS software. 4 × 1 corporate feed antennas and 4 × 2 patch with and without slot are designed and their parameters are measured and compared.

4.1 4 × 1 Corporate Feed Patch Antenna

Figure 6 shows 4 × 1 corporate feed patch antenna with inset feed technique simulated using HFSS software. Return loss, radiation pattern, and VSWR measured using the simulator software is shown below.

(a) Return Loss
    The return loss plot for 4 × 1 corporate feed patch antenna is shown in Fig. 7. In the figure, the return loss obtained at 10.3 GHz is −15.68 dB.
(b) Radiation Pattern
    3-D radiation pattern can be obtained using the simulator software. Figure 8 shows that the maximum gain obtained is 4.10 dB.
(c) VSWR
    The plot for VSWR is shown in Fig. 9. The figure shows that the VSWR is 1.39 at 10.3 GHz.

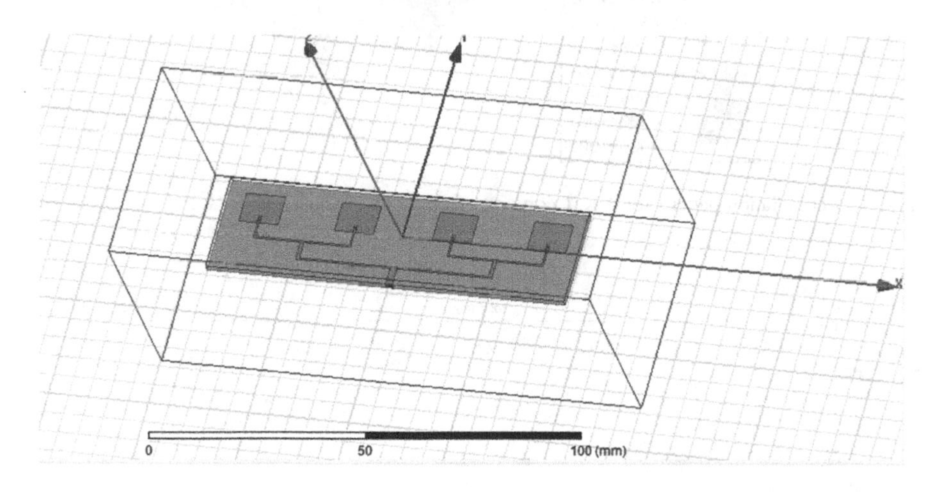

Fig. 6 4 × 1 corporate feed patch antenna

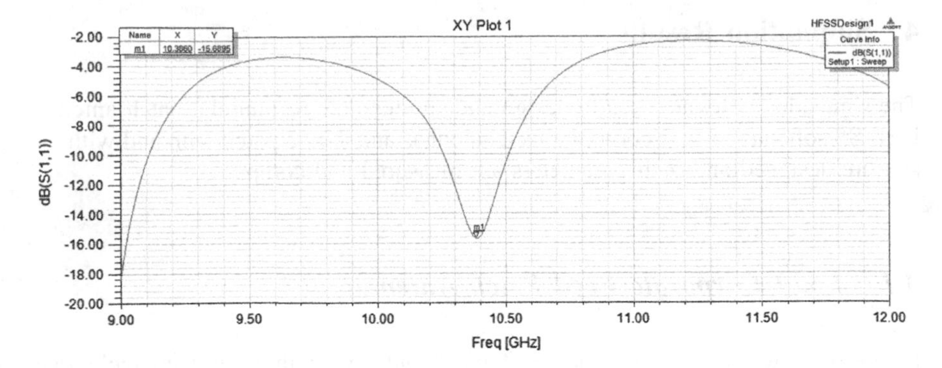

Fig. 7 Return loss of 4 × 1 corporate feed patch antenna

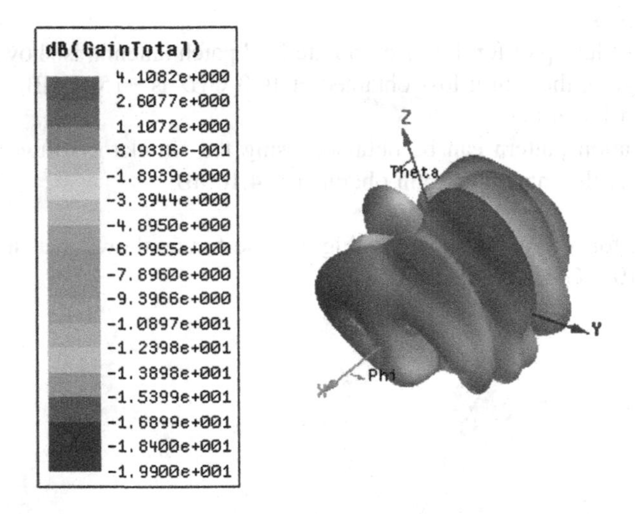

Fig. 8 3-D radiation pattern for 4 × 1 corporate feed patch antenna

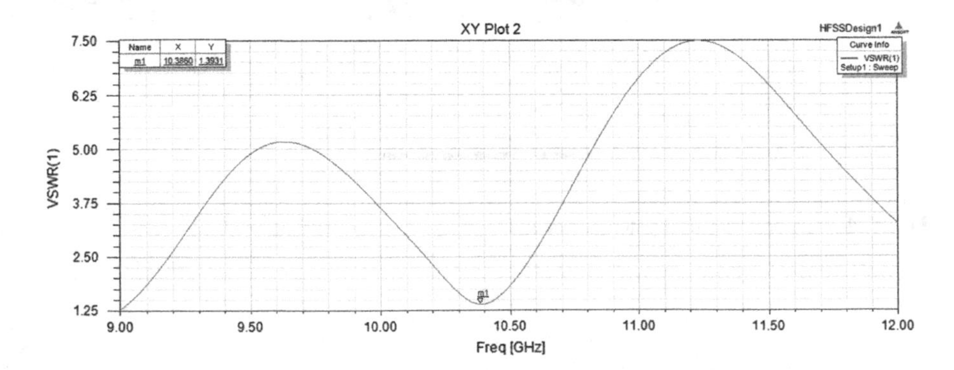

Fig. 9 VSWR for 4 × 1 corporate feed patch antenna

4.2 4 × 2 Corporate Feed Patch Antenna

Figure 10 shows 4 × 2 corporate feed patch antenna model designed using HFSS software. Return loss, radiation pattern, mutual coupling, and VSWR measured using the simulator software is shown below.

(a) Return Loss
The return loss plot for 4 × 2 corporate feed patch antenna is shown in Fig. 11. In the figure, the return loss obtained for $S(1, 1)$ 10.44 GHz is −32.33 dB and for $S(2, 2)$ 10.44 GHz is −29.56 dB.

(b) Mutual coupling
The mutual coupling between the two antenna arrays obtained using the simulation software. Figure 12 shows the mutual coupling between $S(1, 2)$ and $S(2, 1)$ obtained at 10.44 GHz is −36.58 dB

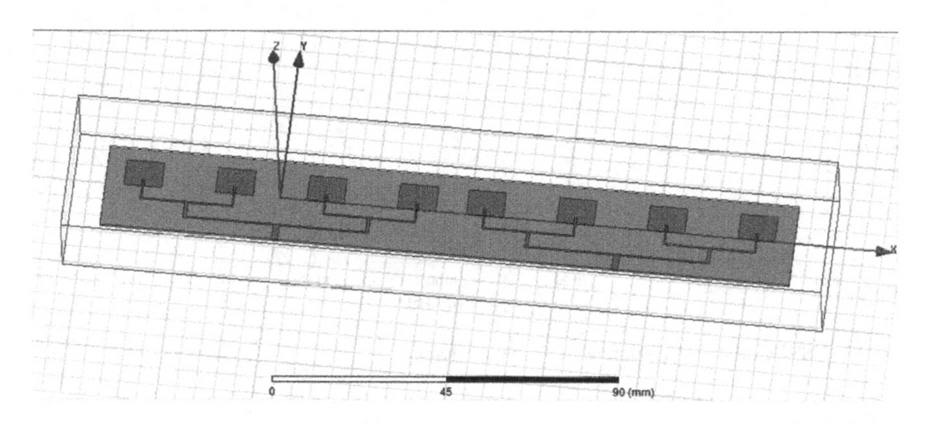

Fig. 10 4 × 2 corporate feed patch antenna

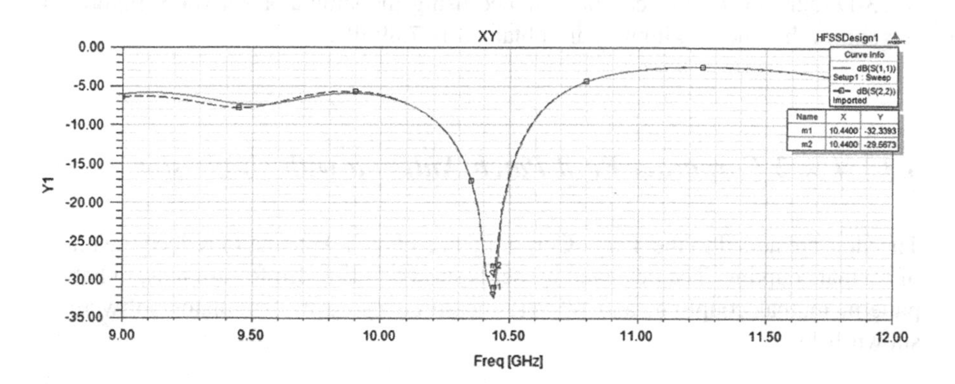

Fig. 11 Return loss of 4 × 2 corporate feed patch antenna

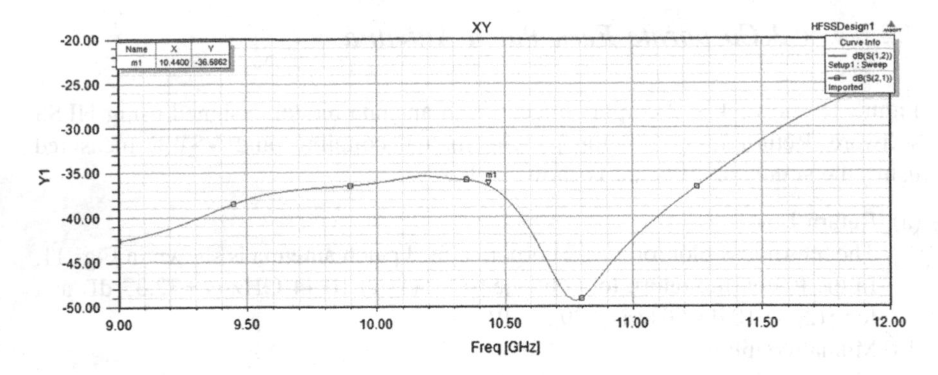

Fig. 12 Mutual coupling of 4 × 2 corporate feed patch antenna

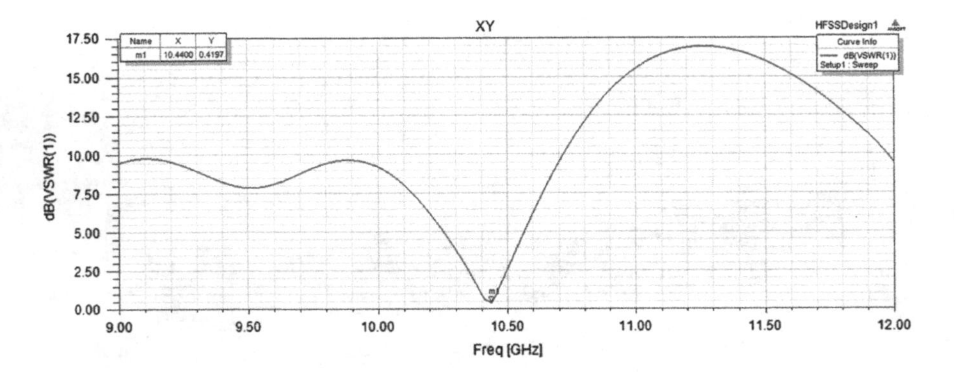

Fig. 13 VSWR of 4 × 2 corporate feed patch antenna

(c) VSWR
    The plot for VSWR is 0.41 at 10.44 GHz is shown in Fig. 13.
(d) Radiation Pattern
    3-D radiation pattern can be obtained using the simulator software. Figure 14 shows that the maximum gain obtained is 7.89 dB.

4.3 4 × 2 Corporate Feed Patch Antenna with single slot

The slots are added to the 4 × 2 Corporate Feed Patch Antenna to reduce the effect of mutual coupling. The designed model is shown in Fig. 15. Return loss, radiation pattern, mutual coupling, and VSWR measured using the simulator software is shown below.

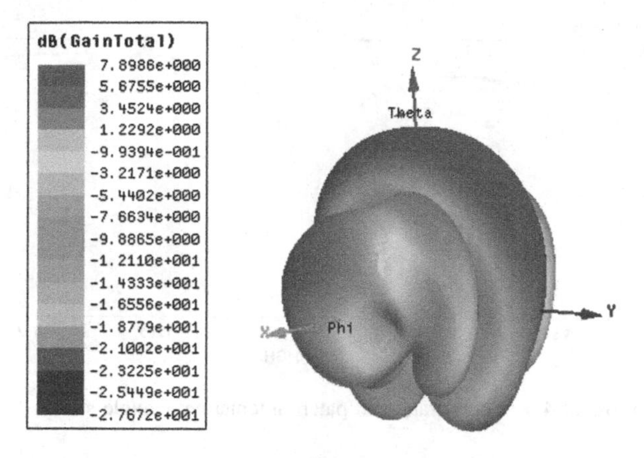

Fig. 14 Gain of 4 × 2 corporate feed patch antenna

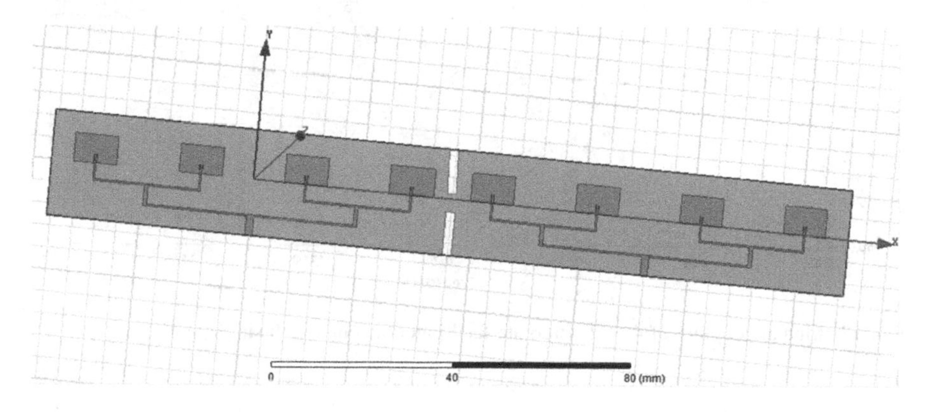

Fig. 15 4 × 2 corporate feed patch antenna with single slot

(a) Return Loss
The return loss plot for 4 × 2 corporate feed patch antenna is shown in Fig. 16. In the figure, the return loss obtained for $S(1, 1)$ 10.44 GHz is −37.80 dB and for $S(2, 2)$ 10.44 GHz is −41.25 dB.

(b) Mutual coupling
The mutual coupling between the two antenna arrays obtained using the simulation software. Figure 17 shows the mutual coupling between $S(1, 2)$ and $S(2, 1)$ obtained at 10.44 GHz is −35.89 dB.

(c) VSWR
The VSWR obtained using the simulation software is 0.22 at 10.44 GHz.

(d) Radiation Pattern
3-D radiation pattern can be obtained using the simulator software. Figure 18 shows that the maximum gain obtained is 7.82 dB.

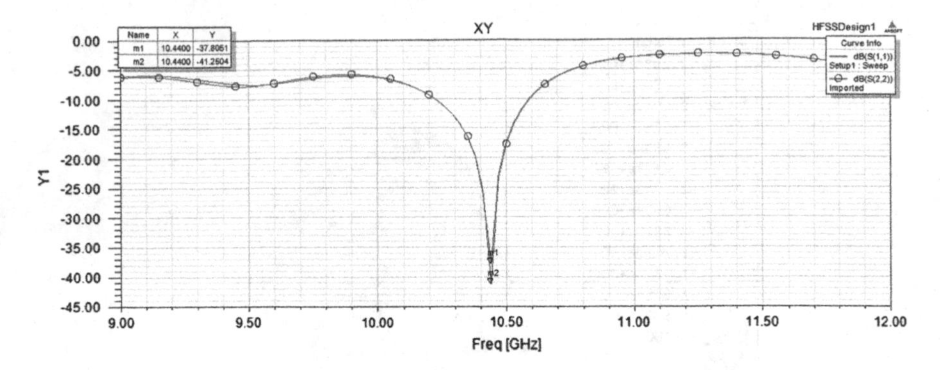

Fig. 16 Return loss of 4 × 2 corporate feed patch antenna with single slot

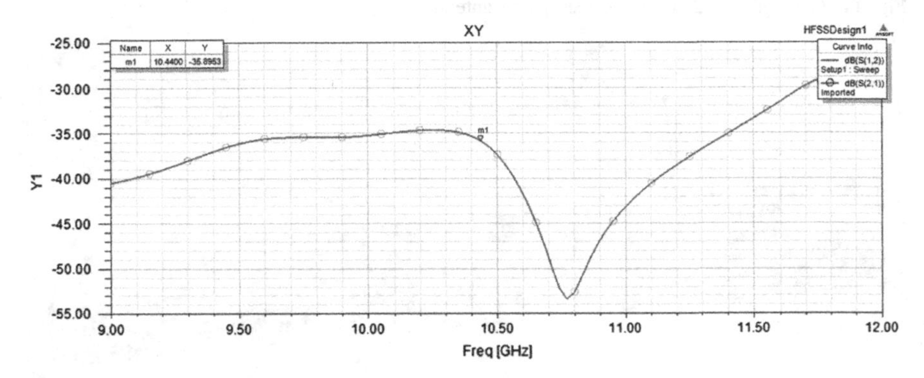

Fig. 17 Mutual coupling of 4 × 2 corporate feed patch antenna with single slot

Fig. 18 Gain of 4 × 2 corporate feed patch antenna with single slot

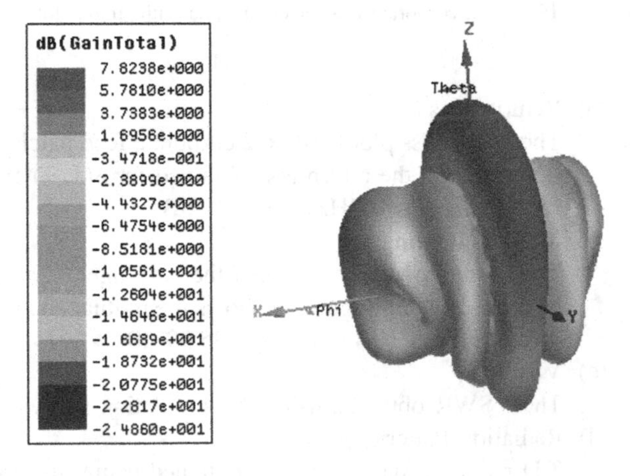

4.4 4 × 2 Corporate Feed Patch Antenna with Multiple Slot

Additional slots are added to the 4 × 2 corporate feed patch antenna to reduce the effect of mutual coupling. The designed model is shown in Fig. 19 and the simulator software is shown below.

(a) Return Loss
The return loss plot for 4 × 2 corporate feed patch antenna is shown in Fig. 16. In the figure, the return loss obtained for *S*(1, 1) 10.44 GHz is −35.50 dB and for *S*(2, 2) 10.44 GHz is −30.46 dB.
(b) Mutual coupling
The mutual coupling between the two antenna arrays obtained using the simulation software. Figure 17 shows the mutual coupling between *S*(1, 2) and *S*(2, 1) obtained at 10.44 GHz is −41.83 dB.
(c) VSWR
The VSWR obtained using the simulation software is 1.03 at 10.44 GHz.

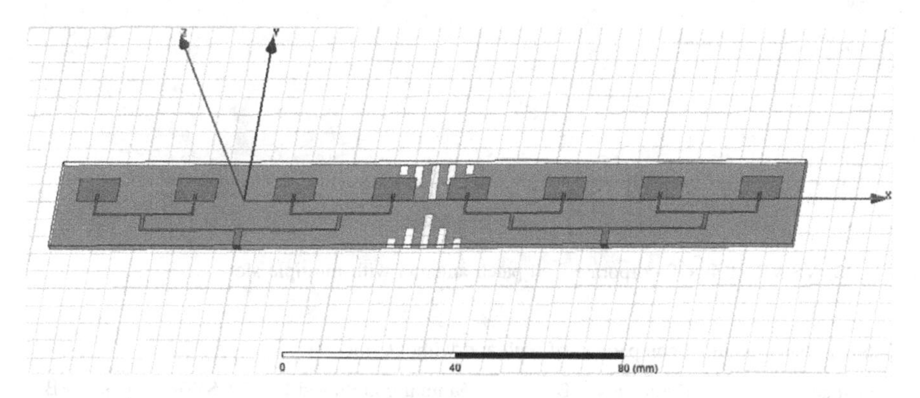

Fig. 19 4 × 2 corporate feed patch antenna with multiple slot

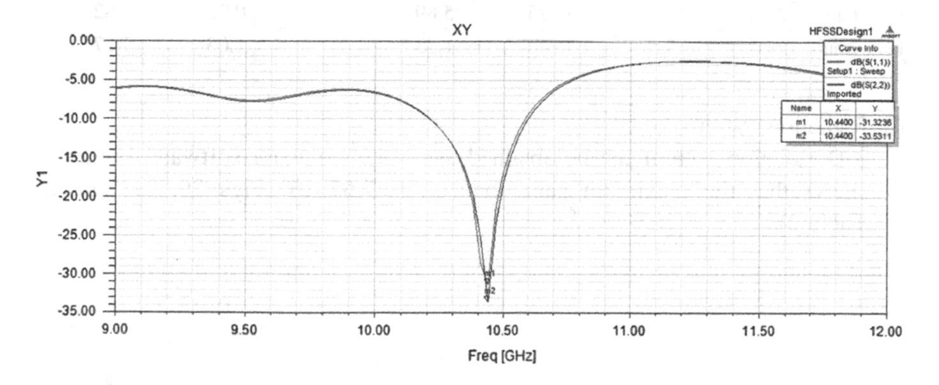

Fig. 20 Return loss of 4 × 2 corporate feed patch antenna with multiple slot

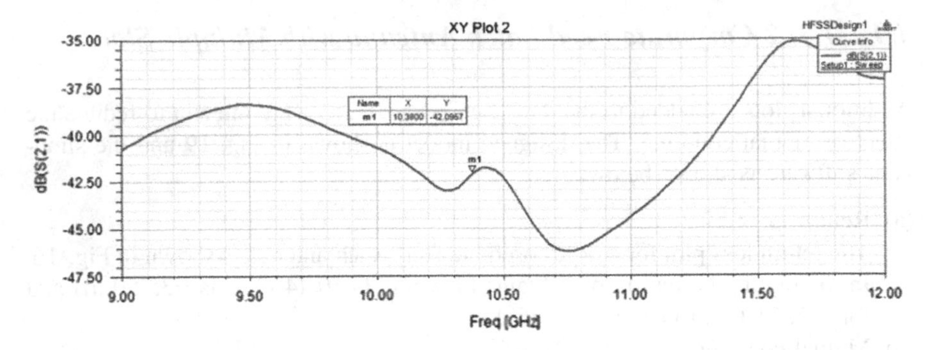

Fig. 21 Mutual coupling of 4 × 2 corporate feed patch antenna with multiple slot

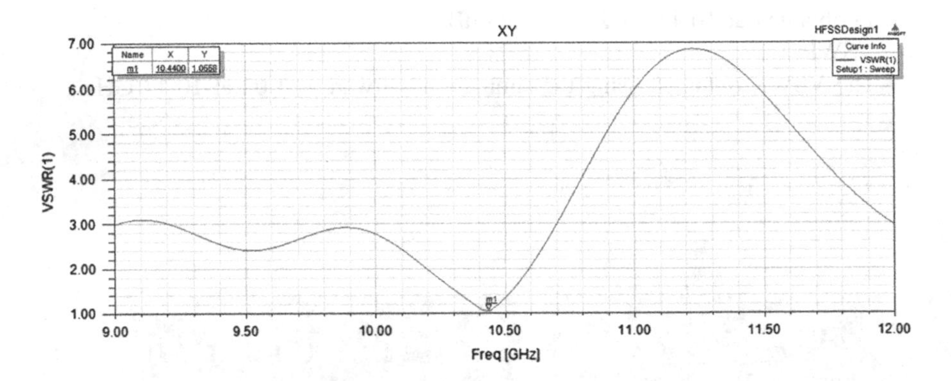

Fig. 22 VSWR of 4 × 2 corporate feed patch antenna with multiple slot

Table 2 Patch antenna comparison of with and without slot

Substrate	Return loss dB		Mutual coupling dB	VSWR	Gain dB
	S(1, 1)	*S*(2, 2)			
FR4	−32.33	−29.56	−36.56	1.39	7.89
FR4 (single slot)	−37.80	−41.25	−35.89	0.22	7.82
FR4 (multi slot)	−35.50	−30.46	−41.83	1.03	7.57

(d) Radiation Pattern
3-D radiation pattern can be obtained using the simulator software. Figure 18 shows that the maximum gain obtained is 7.57 dB (Figs. 20, 21 and 22; Table 2).

5 Conclusion

The 4 × 2 corporate feed microstrip patch antenna designed and their performance and characteristic are analyzed by using High Frequency Structure Simulator (HFSS) software for X-band application. The propose design achieves center frequency at 10.44 GHz. The results for return loss and mutual coupling for rectangular microstrip patch antenna with and without slot are analyzed and the results have been presented in this paper

References

1. Smith, T.F., Waterman, M.S.: Identification of Common Molecular subsequences. J. Mol. Biol. 147, 195–197 (1981) The Fundamentals of Patch Antenna Design and Performance, Technical Media, LLC, March 2009.
2. Rodney B Waterhouse, "Microstrip Patch Antennas, A Designers Guide".
3. Alka verma, Neelam shrivastava, "Analysis and design of Rectangular Microstrip Antenna in X Band," MIT International Journal of Electronics and Communication Engineering Vol. 1, No. 1, Jan. 2011, pp. (31–35).
4. M. Ramesh and YIP KB, Design Formula for Inset Fed Microstrip Patch Antenna, Journal of Microwaves and Optoelectronics, Vol. 3, No 3, December 2003.
5. Y.T. Lo, S.W. Lee, "Antenna Handbook", Vol 2.
6. Constantine A. Balanis, "ANTENNA THEORY"—Analysis and Design", Second Edition: Reprint 2007, John Wiley Publications.
7. Ramesh Garg, Prakash Bhatia, Inder Bahl, and Apisak Ittipiboon, "Microstrip Antenna Design Handbook" Artech House, Inc. 2001.
8. R. Garg, Microstrip antenna design handbook. Boston, Mass. [U.A.]: Artech House, 2001.
9. H. J. Visser, "Array and Phased Array Antenna Basics," ed: John Wiley & Sons.
10. Weiwei Wu, Jiaxian Yin, Naichang Yuan, "Design of an efficient X-Band Waveguide-Fed Microstrip Patch array", IEEE Transactions on Antennas and Propagation, July 2007, Vol. 55, No. 7. pp (1933–1939).

Synergistic Fibroblast Optimization

P. Subashini, T.T. Dhivyaprabha and M. Krishnaveni

Abstract Movement is the main characteristic of living species and is the major source of biologically inspired computational systems. The migration of cell organism can take the form of either movement of cells or movement within cells, and it is also capable of changing the shape as a result of reversible or irreversible contraction. This paper simulates synergistic fibroblast optimization (SFO), a multi-agent heuristic technique that models the migration and methodical behavior of the fibroblast. The proposed SFO technique exhibits the role of fibroblast in dermal wound healing process, by migrating the individual cell through the connective tissue, and synthesizes the collagen in the extracellular matrix for the new tissue formation during wound healing. Compared to the related technique particle swarm optimization (PSO), SFO produces better results in terms of both accuracy and performance. An analysis of the proposed algorithm indicates that the two most important factors contributing to SFO effectiveness are fibroblast collaborative nature and goal-oriented feature. The results suggest that SFO is a promising new optimization technique, which may be particularly applicable to find optimal maxima or minima, among the candidate solutions in the nonlinear complicated optimization problem.

Keywords Computational biology · Collagen · Extracellular matrix · Fibroblast · Heuristic · Multi-agent systems · Nature-inspired algorithms · Optimization

---

P. Subashini · T.T. Dhivyaprabha (✉) · M. Krishnaveni
Department of Computer Science, Avinashilingam Institute for Home Science and Higher Education for Women, Coimbatore 641043, Tamil Nadu, India
e-mail: ttdhivyaprabha@gmail.com

P. Subashini
e-mail: mail.p.subashini@gmail.com

M. Krishnaveni
e-mail: krishnaveni.rd@gmail.com

S.S. Dash et al. (eds.), *Artificial Intelligence and Evolutionary Computations in Engineering Systems*, Advances in Intelligent Systems and Computing 517,
DOI 10.1007/978-981-10-3174-8_25

1 Introduction

Evolutionary computation algorithms are developed, based on the characteristics of natural phenomena, including foraging behavior, evolution, cell and molecular phenomena, cognition and neurosystems, alignment phenomena in microscopes, nonbiological systems and geoscience-based techniques [1]. Since 1970s, meta-heuristic algorithms are continuously evolved to find an efficient optimization approach in order to solve complex problems effectively [2]. Optimization problems are common to which there is always a requisite to find solutions which are optimal or near-optimal. There are different techniques to estimate the intractable optimization problems and one such type is the model based on behavior of natural systems [3]. Although all the heuristic algorithms have different inspirational sources, the objective of them is to find the optimum. The aim of the proposed work is to introduce a simulated algorithm Synergistic Fibroblast Optimization (SFO) model that exhibits the behavior of the fibroblast organism. Various characteristics of fibroblast include differentiation, proliferation, inflammation, migration, reorientation, alignment, ECM synthesis, collaborative, goal-oriented, interaction, regeneration, self-adaptation and evolution, inspired to design and develop a Synergistic Fibroblast Optimization (SFO) algorithm. The goal of this algorithm is to find minima or maxima optimum in order to solve nonlinear complex problems.

Fibroblast is a kind of cellular organism which is able to migrate from one location to another in the wounded region [4]. It is an ubiquitous stromal cell which is found in most of the mammalian connective tissue and it plays a significant role in the wound healing process. When the skin gets injured, white blood cells occupy the wounded region for blood clotting, which includes various process such as inflammation, formation of tissue, angiogenesis, contraction of tissue, and tissue remodeling. The crucial part in the inflammation event is the interaction of fibroblast cells through the extracellular matrix [ECM]. Trajectories of individual fibroblast are determined as they migrate toward the wounded region under the combined influence of fibrin/collagen alignment and gradients in a paracrine chemo attractant produced by leukocytes [4, 5]. The effect of collagen alignment is influenced by a number of parameters such as fibroblast concentration, cellular speed, fibroblast sensitivity to chemoattractant concentration and chemoattractant diffusion coefficient. The interaction of cells depends on the density of the population increase to include the effect of crowding [6]. The self-organization and co-operative behavior of fibroblast to resolve injury is greatly associated with evolutionary algorithm strategy to discover the optimal solution in the problem space. The paper is structured in the following way: Sect. 1 provides a brief introduction of biological phenomena of fibroblast. Statistical model of fibroblast is discussed in Sect. 2. Section 3 describes the computational model of fibroblast. Section 4 explains the SFO implementation. The experimental results are assessed and conversed in Sect. 5, and Sect. 6 draws the conclusion of the work.

2 Statistical Model of the Fibroblast Behavior

In a mathematical model, collagen matrix is represented as continuum values $c(t, \theta)$ and the cells as discrete entities $f(t, \theta)$, respectively, where t denotes time and θ symbolizes angular orientation [7, 8]. The cell paths are denoted by $f^i(t)$, where i represents the particular fibroblast cell. The collagen matrix is denoted by $c(x, t)$, where x represents the Cartesian co-ordinates of a point in the plane. The vector c has unit length and their direction (θ) indicates the predominant orientation of the collagen matrix at the spatial location [6].

The mathematical model of cell motion and position is represented in Eqs. (1) and (2):

$$v^i(t) = (1-\rho)c\big(f^i(t)\big) + \rho\frac{f^i(t-\tau)}{\|f^i(t-\tau)\|} \tag{1}$$

$$\frac{\mathrm{d}f^i(t)}{\mathrm{d}t} \equiv f^i(t) = s\frac{v^i(t)}{\|v^i(t)\|} \tag{2}$$

where

$v^i(t)$ = velocity of ith at time t
s = cell speed
ρ = diffusion coefficient
τ = time lag
f^i = ith fibroblast cell.

The cell determines the moving direction, based on the fibrin deposition, found in the injured region [6]. It is calculated by a weighted average of their previous direction of motion, and the direction of collagen matrix at their current location is given in Eq. (3):

$$f(x,t) = \sum_{i=0}^{N} w_i(x,t)\frac{f^i(t-\tau)'}{\|f^i(t-\tau)'\|} \tag{3}$$

where τ is a time lag, N is the total number of cells, and L is the length of moving fibroblast with typical range of values of the order 100 μm. The weight function, used in Eq. (4), is defined by

$$w_i(x,t) = a_1 a_2, \quad a_j = \max\left\{1 - \frac{\left|f^i(t)_j - x_j\right|}{L}, 0\right\} \tag{4}$$

where

$x = (x_1, x_2)$ = previous and current location of ith cell at time t.
$L = 10$ μm.

The rate of change of collagen orientation in the direction of motion θ is described by Eq. (5):

$$\frac{\mathrm{d}\theta(t)}{\mathrm{d}t} = \kappa \|f\| \sin(\phi - \theta) \tag{5}$$

where

θ is the angle of collagen direction.
$\phi(x, t)$ is the angle of fibroblast located in a position x at time t.

When the difference between angular orientation of collagen deposition and cell alignment is small, the speed at which variable change during a specific interval of time is minimum. On the other hand, when the direction followed by cell formation is orthogonal, the probability of variable change attained is maximum. In addition, it is periodic, so that, when fibrin and cells are oriented either in the same direction or 180° apart, the fibrin does not change the direction [9, 6].

Interaction of fibroblasts with collagen and fibrin proteins can be performed in two ways. First, the cell reorients collagen but not fibrin and the cell moves faster in fibrin. The cells produce collagen at a constant rate and degrade fibrin at a rate proportional to the amount of collagen produced in the extracellular matrix. It degrades fibrin at a rate proportional to the amount present [6].

The equations determining collagen and fibrin proteins density are given by (6) and (7):

$$\frac{\mathrm{d}\|c\|}{\mathrm{d}t} = (p_c - d_c\|c\|) \sum_{i=1}^{M} w(f^i - x) \tag{6}$$

$$\frac{\mathrm{d}\|b\|}{\mathrm{d}t} = (-d_b\|b\|) \sum_{i=1}^{M} w(f^i - x) \tag{7}$$

where

p_c, d_c, and d_b are positive constants.
c, b = collagen and fibrin protein densities.
f^i = i^{th} fibroblast cell.
x = position of a cell.
w = mean weight of n cells.

3 Synergistic Fibroblast Optimization (SFO) Algorithm

Fibroblast-based Optimization Algorithm imitates the migration characteristics of fibroblast to replace fibrin protein with collagen matrix in the wounded region. SFO algorithm is applied to solve nonlinear complex problem, to find the optimal (minima or maxima) solution, which is problem dependent on the n-dimensional problem space. The movement of fibroblast, by the normal tissue secreted by the collagen protein to the wounded region, determines the new tissue formation. The original process for implementing SFO is given in the algorithm below:

Step 1: Initialize the population of fibroblast cells f_i, $i = 1, 2, \ldots, n$; with randomly generated position (x_i), velocity (v_i), and collagen deposition (ecm) in n-dimensional problem space. Define the parameter values of cell speed (s) and diffusion coefficient (ρ).
Step 2: Repeat
Evaluate the individual fibroblast using fitness function $F(f_i)$ in n variables for n times.
Step 3: The reorientation of cell can be performed to find optima (maxima or minima) in the evolutionary region.
Step 4: Update the velocity (v^i) and position (x^i) of a cell using the following Eqs. (8) and (9):

$$v^{i(t+1)} = v^i(t) + (1-\rho)c\left(f^i(t), t\right) + \rho * \frac{f^i(t-\tau)}{\|f^i(t-\tau)\|} \tag{8}$$

where

t = current time;
τ = time lag;
v^i = velocity of ith cell;
$\rho = 0.5$.

$$x^{i(t+1)} = x^i(t) + s * \frac{v^i(t)}{\|v^i(t)\|} \tag{9}$$

where
$s = \frac{s}{k_{ro}L}$, $k_{ro} = 10^3$ μ/min, L = cell length;

Step 5: Remodeling of collagen deposition (c_i) can be synthesized in the extracellular matrix (ecm).
Until the predetermined conditions/maximum iterations is met.

Figure 1 depicts the core idea underlying in this technique that is given a population of fibroblasts, it stochastically choses collagen for the survival of fitness

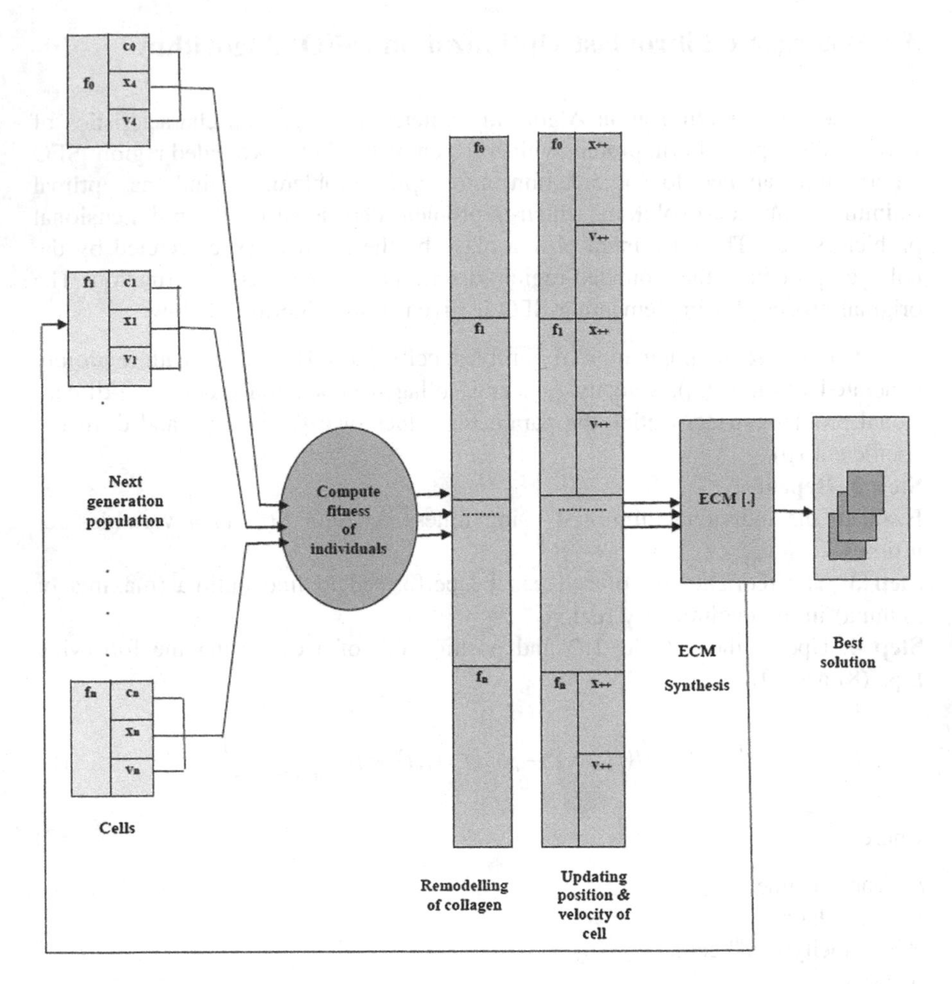

Fig. 1 Architecture of SFO algorithm

measure. Based on the fittest solution, few of the better candidate solutions are selected to seed the next generation population by applying both synthesis of extracellular matrix and remodeling of collagen orientation. This process can be repeated until a best solution is found or predefined time limit is reached.

4 SFO Implementation

To examine the efficiency of the simulated fibroblast optimization model described in this paper, an SFO algorithm was implemented as a computer program. Here, the cells are considered as a number of simple entities placed in the search space of complex problem or function, and each cell evaluates using the objective function in the search space. Each cell then determines its movement through the search space using the knowledge of previous position with some random perturbations. The next iteration takes place after all cells have been moved. Eventually, fibroblast as a whole, like the population of cell survival for wound healing, is likely to migrate close to an optimum solution by quantifying the fitness function. Each individual present in the cellular population is composed of two n-dimensional vectors, where n is the dimensionality, x_i denotes the position and v_i represent velocity of ith cell. For every iteration, of the algorithm, the group of cells that deposit collagen is evaluated as a problem solution. If that collagen deposition is better than any that has been found so far, then the co-ordinates are stored in the vector (c_i). The value of the fittest function result achieved so far is stored in a variable that can be called cbest, is compared with pre-iteration value for every time in order to yield qualitative solution. The objective of cell is that it moves to a better position and improves its acceleration by updating position (x_i) and velocity (v_i) equations to find the global optimal (minima or maxima) solution which is problem dependent. Cells moved to new positions are determined by adding (v_i) co-ordinates to (x_i), and the algorithm executes with the synthesis of collagen repository extracellular matrix (ecm).

5 Experiments and Observations

In order to investigate the effectiveness of the simulated Synergistic Fibroblast Optimization Algorithm (SFO), it is implemented using Java programming language. The characteristics of sphere test function such as continuous, unimodal, differentiable, separable and scalable, which are well suited for the natural behavior of fibroblast survival for the fitness solution are observed [10]. Figure 2 depicts the sphere function of SFO in ten dimensions; $f(x) = x_1^2 + x_2^2 + x_3^2 + x_4^2 + x_5^2 + x_6^2 + x_7^2 + x_8^2 + x_9^2 + x_{10}^2$, the mathematical representation of sphere function is defined as Eq. (10):

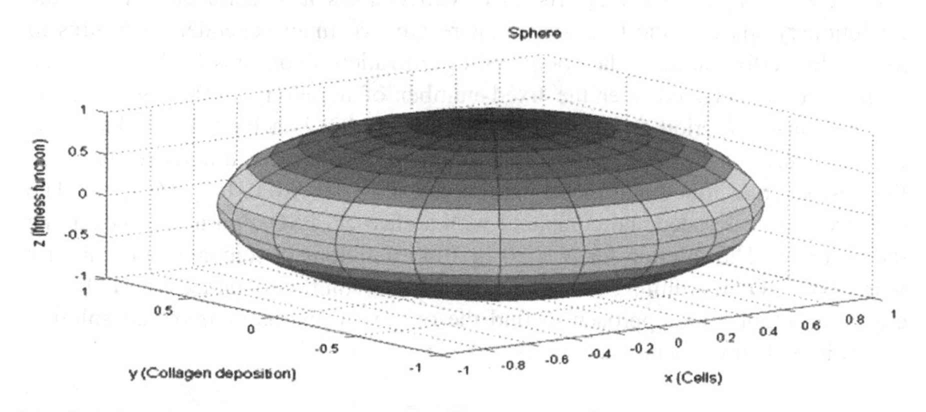

Fig. 2 Sphere function of SFO implementation

$$f(x) = \sum_{i=1}^{D} x_i^2 \tag{10}$$

Since SFO algorithm is a multi-agent optimization technique, in addition to implementing a simulated fibroblast object, it must apply a set of simulated fibroblasts along with definite control variables. The control variables consist of the population of simulated fibroblasts to use, dimensionality of the problem domain, initialization of position and velocity with random generations and parameters include cell speed and diffusion coefficient which control the cell movement and prevent from deviating in the search space. The fitness evaluation of SFO algorithm is executed, with the condition that, either maximum number of iterations or predetermined criteria is to be met. The global optimal solution, found by any fibroblast in the entire population in most situations, is returned as the best solution to the target minimization problem.

The proposed work is implemented to solve error minimization problem [11] using mean squared error metric with respect to unknown parameter $\hat{\theta}$. The mathematical equation of Mean Squared Error [MSE] [12] is represented as Eq. (11):

$$\mathrm{MSE} = E\left[\hat{\theta} - \theta\right]^2 \tag{11}$$

The experimental results of SFO and classical PSO algorithm, based on MSE metric, are given in Table 1. The two primary independent variables are the amount of collagen deposition found in the extracellular matrix (*N*) and the maximum number of iterations (*t*). It is inferred that when the number of candidate solution gets increased in the problem space, the probability of finding the optimal solution is ultimately improved in SFO than the classical PSO algorithm. It reveals that the interaction among the cells is steadily improving when the function is optimized. When the number of iterations and ecm population size gets increased, SFO algorithm gets initially stuck to give worse result while changing the environment. But the self-adaptation characteristics of cell enables it to learn and fit into the evolutionary space in the later stage. There are two main dependent measures to assess the performance of the compared optimization algorithms [13]. First is the optimal result obtained after the fixed number of iterations (1000). Second, the predetermined criterion set to get optimal value (0.02) is achieved by 5000 iterations. It describes the error as measured by the sum of squared differences between the best solution found and the optimal solution of {0.0, 0.0, 0.0, 0.0} [11]. The proposed work is tested and compared with a classical PSO algorithm, to examine the accuracy. The solution for $N = 20$ in 5000 iterations is comparable or slightly worse for SFO algorithm. The results suggest that solution obtained by SFO is an effective optimization approach to find the optimal (minima or maxima) solution which is problem dependent.

Table 1 Comparative analysis of error minimization problem

Population Size	1000 iterations				5000 iterations			
	Classical PSO	MSE	Proposed SFO	MSE	Classical PSO	MSE	Proposed SFO	MSE
$N = 10$	0.09	**0.0081**	0.02	**0.0004**	0.05	**0.0025**	0.02	**0.0004**
$N = 20$	0.07	**0.0049**	0.03	**0.0009**	0.01	**0.0001**	0.06	**0.0036**
$N = 30$	0.09	**0.0081**	0.02	**0.0004**	0.26	**0.0676**	0.07	**0.0049**
$N = 40$	0.25	**0.0625**	0.03	**0.0009**	0.27	**0.0729**	0.08	**0.0064**
$N = 50$	0.4	**0.16**	0.06	**0.0036**	0.11	**0.0121**	0.02	**0.0004**
$N = 60$	0.59	**0.3481**	0.09	**0.0081**	0.37	**0.1369**	0.13	**0.0169**
$N = 70$	0.16	**0.0256**	0.18	**0.0324**	0.45	**0.2025**	0.04	**0.0016**
$N = 80$	0.17	**0.0289**	0.07	**0.0049**	0.71	**0.5041**	0.07	**0.0049**
$N = 90$	0.14	**0.0196**	0.09	**0.0081**	0.85	**0.7225**	0.03	**0.0009**
$N = 100$	0.37	**0.1369**	0.23	**0.0529**	0.88	**0.7744**	0.06	**0.0036**

The experimental results of MSE obtained by classical PSO and the proposed SFO algorithm shown in Table 1 illustrated that the stochastic process of SFO algorithm has the capability to solve error minimization problem and it produces promising results than classical PSO algorithm. When the population size has been increased in the evolutionary space, the interaction among the swarm is improved that enables to achieve optimum solution. It is confirmed that the self-adaptation behaviour of SFO can resolve the diverse sort of optimization problem

6 Conclusions

Many nature-inspired computing algorithms already exist to solve nonlinear complicated problems. But the complexity of the problem is dynamically increased, due to many factors, such as fitness function is noisy, design variables and parameters alter against environmental change, the solutions obtained by the optimization problem modifies over time. In this paper, a novel synergistic fibroblast optimization (SFO) is introduced to which the results and experimental data is limited in scope. In order to determine which features of SFO are the most significant in producing its apparent effectiveness, different combinations of components of the algorithm were disabled, and the algorithm was executed against the sphere function. The results suggested that both collaborative nature and self-adaptation characteristic are key factors in SFO accuracy when compared to PSO. It is also identified that SFO appeared most sensitive when the collagen density gets increased in extracellular matrix (ecm) that replicate in the ignition of the process. The proposed work suggests that SFO is a multi-agent technique pertinent to distribute data in a natural way, and it may be particularly well suited to analyze and solve potential complex problems. Areas of future research in SFO include the investigation factors such as incorporating the apoptosis process and cell regeneration characteristics of fibroblast cellular organisms.

References

1. Hongwei Mo. Evolutionary computation as a direction in nature-inspired computing. Ubiquity, an acm publication. 2012; 1–9. doi:http://dx.doi.org/10.1145/2390009.2390011.
2. Carlos Cruz, Juan Gonzalez R, David Pelta A. Optimization in dynamic environments: a survey on problems, methods and measures. Springer. 2010; 15:1427–1448. doi:http://dx.doi.org/10.1007/s00500-010-0681-0.
3. Paul Marrow. Nature inspired computing technology and applications. BT Technology Journal. 2000; 18: 13–23. doi:http://dx.doi.org/10.1023/A.1026746406754.
4. Peter Rodemann H, Hans-Oliver Rennekampff. Functional diversity of fibroblasts. Springer Science+Business Media B.V. Chapter 2. 2011; 23–36. doi:http://dx.doi.org/10.1007/978-94-007-0659-0_2.
5. Howard Stebbings. Cell motility. Encyclopedia of Life Sciences. 2001; 1–6.
6. Steven McDougall, John Dallon, Jonathan Sherratt, Philip Maini. Fibroblast migration and collagen deposition during dermal wound healing: mathematical modeling and clinical implications. Philosophical Transactions of the Royal Society. 2006; 364: 1385–1405. doi: http://dx.doi.org/10.1098/rsta.2006.1773.
7. John Dallon C, Jonathan Sherratt A. A mathematical model for fibroblast and collagen orientation. Bulletin of Mathematical Biology. 1998; 60:101–129.
8. Stephanou A, Chaplain M A J, Tracqui P. A mathematical model for the dynamics of large membrane deformations of isolated fibroblasts. Bulletin of Mathematical Biology. 2004; 66:1119–1154. doi:http://dx.doi.org/10.1016/j.bulm.2003.11.004.
9. John Dallon, Jonathan Sherratt, Philip Maini, Mark Ferguson. Biological implications of a discrete mathematical model for collagen deposition and alignment in dermal wound repair. IMA Journal of Mathematics Applied in Medicine and Biology. 2000; 17: 379–393.
10. Momin Jamil, Xin-She Yang. A literature survey of benchmark functions for global optimization problems. International Journal of Mathematical Modelling and Numerical Optimisation. 2013; 4:150–194. doi:http://dx.doi.org/10.1504/IJMMNO.2013.055204.
11. James McCaffrey D. Simulated protozoa optimization. IEEE IRI. 2012; 179–184.
12. https://en.wikipedia.org/wiki/Mean_squared_error.
13. James Kennedy, Rui Mendes. Population structure and particle swarm performance. Proceedings of the Congress on Evolutionary Computation (CEC). 2002; 2:1671–1676. doi:http://dx.doi.org/10.1109/CEC.2002.1004493.

Evolvable Hardware Architecture Using Genetic Algorithm for Distributed Arithmetic FIR Filter

K. Krishnaveni, C. Ranjith and S.P. Joy Vasantha Rani

Abstract The aim of the paper is to design evolvable hardware (EHW) architecture for Finite Impulse Response Filter using Genetic Algorithm. Evolvable hardware refers to hardware that can change its behaviour (parameters such as coefficients) according to the changes in its environment. To update the filter coefficients adaptively, genetic algorithm was used. The proposed filter architecture was implemented with Xilinx Spartan 6 FPGA (XC6SLX45-CSG324) Trainer Kit. Hardware design was synthesized using the EDK (Embedded Development Kit) platform and the genetic algorithm was implemented in SDK (Software Development Kit) of Xilinx Platform Studio tool (XPS) 14.6.

Keywords Evolvable hardware · Genetic algorithm · FIR filter · Distributed arithmetic · FPGA · VLSI

1 Introduction

The evolvable hardware is a method of designing circuits based on the reconfigurable structures. Evolvable hardware uses the evolutionary algorithms (EA) to design the circuits without manual engineering [1]. Evolvable hardware can change its architecture and behaviour dynamically and autonomously according to its changes in environment [2]. The evolvable hardware design deals with the design of circuit based on the evolutionary techniques. The circuit design using this techniques use the mathematical model either of reconfigurable structure or of system with variable parameters. These techniques are used to find new and innovative solution to the circuits. Then the resulting circuit can be implemented as real hardware [3].

K. Krishnaveni (✉) · C. Ranjith · S.P. Joy Vasantha Rani
Electronics Engineering Department, MIT Campus, Anna University, Chennai, India
e-mail: krishnaveni.k.2012@gmail.com

S.S. Dash et al. (eds.), *Artificial Intelligence and Evolutionary Computations in Engineering Systems*, Advances in Intelligent Systems and Computing 517,
DOI 10.1007/978-981-10-3174-8_26

The search for the suitable configuration of the evolvable structures is entirely guaranteed, by using the EA [4]. In this paper, genetic algorithm (GA) is used as the evolutionary algorithm and the programmable devices are preferably field programmable gate array (FPGA) [5]. The configuration bits for FPGA are the chromosomes of the GA. The evolvable finite impulse response (FIR) filter acts as the backbone of adaptive noise cancellation.

Genetic Algorithm is the adaptive algorithm used to update the FIR filter coefficients. Thus implementing evolvable FIR filters using GA for noise cancellation and thereby extracting the original signal. The FIR filter was designed using distributed arithmetic (DA) method using VHDL [6]. For this filter design the optimal coefficients are generated using GA.

In this paper, the GA operations are performed on the soft core Micro Blaze processor of the Spartan 6 FPGA for finding the fittest chromosome by configuring the FPGA to design an optimized system. The Micro Blaze is a 32 bit soft core processor provided as an intellectual property (IP) core by the Xilinx vendor [7].

The rest of the paper is organized as follows. Section 2 discusses the proposed system architecture of the evolvable hardware architecture of FIR filter. Section 3 describes the implementation of the GA for EHW of FIR filter. Results are discussed in the Sect. 4. Then the conclusions are given in Sect. 5.

2 Proposed System Architecture

The block diagram of the evolvable system is shown in Fig. 1. The FIR filter implemented in VHDL is added as an IP core and is integrated with GA written in C which is operated on Micro Blaze processor of FPGA. The FIR filter is designed using DA method. The evolvable system is used for changing the behaviour of the filter. The GA finds the fittest configuration bits (coefficients) for the FIR filter [6].

This GA program is imported into Micro Blaze soft core processor and the generations with their fitness values are displayed on the console window of the software development kit(SDK) using the UART peripheral [5].

2.1 FIR Filter

Digital filter consists of interconnection of filter taps connected in certain topology which operates on discrete-time signals. Each tap holds a filter co-efficient. FIR filter output is computed as a weighted sum (finite) of the past, present and perhaps future values of the filter input. FIR filter is chosen for this work because it has more stability and reliability. It can be used for the study of the effects of evolution on adaptability.

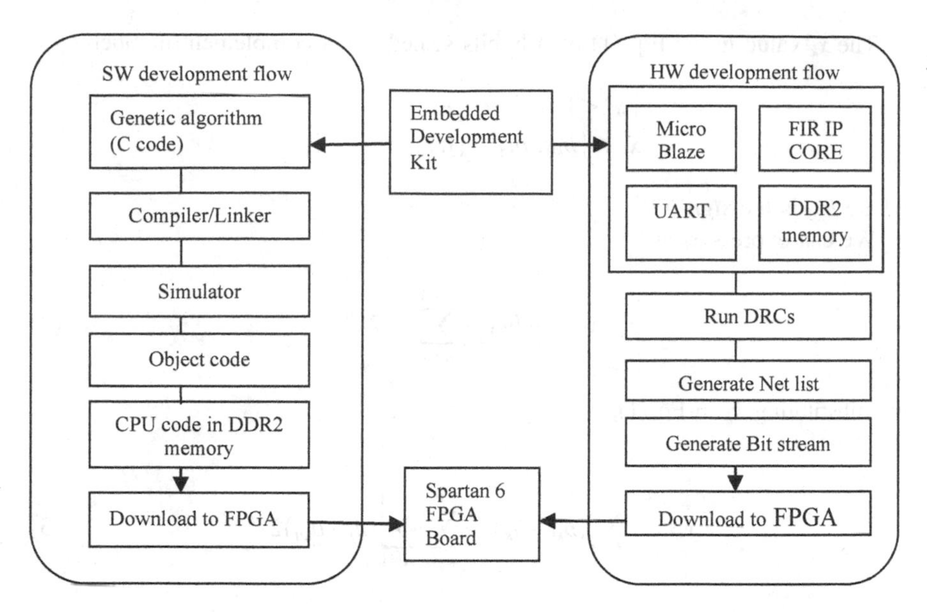

Fig. 1 Block diagram of evolvable hardware architecture. *DRC* Design Rule Check; *SW* Software; *HW* Hardware

FIR filter is described by equation,

$$y = \sum_{k=0}^{n-1} h_x x_{n-k}, \tag{1}$$

where x is the input signal, y is filter output, k is the number of co-efficients of the filter, h is filter coefficient, n is number of taps.

The transpose form of FIR filter topology is shown in Fig. 2. This filter structure is designed using DA method. Distributed arithmetic is the extension of multiply and accumulate unit (MAC). It is the efficient method for calculating the inner product or sum of products and accumulates to the filter output [8].

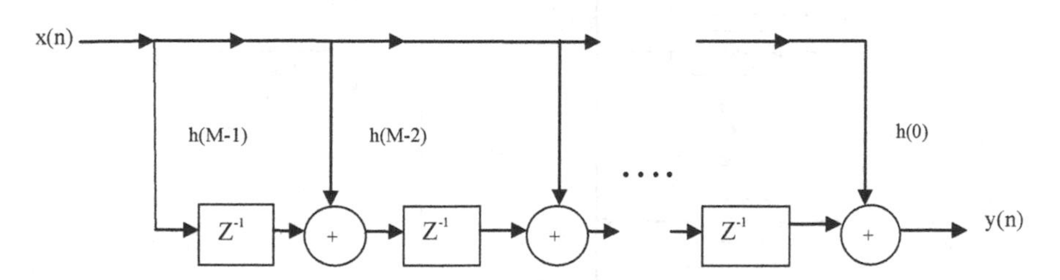

Fig. 2 Transpose form FIR Filter

The x_k value in the Eq. (1) be a N-bits scaled two's complement number

$$|x_k| < 1$$
$$x_k : \{b_{k0}, b_{k1}, b_{k2}..., b_{k(N-1)}\},$$

where b_{k0} is the *sign bit*

We can express x_k as

$$x_k = -b_{k0} + \sum_{n=1}^{N-1} b_{kn} 2^{-n} \tag{2}$$

by substituting x_k in Eq (1),

Therefore,

$$y = -\sum_{k=1}^{K} (b_{k0} \cdot h_k) + \sum_{k=1}^{K} \sum_{n=1}^{N-1} (h_k \cdot b_{kn}) 2^{-n} \tag{3}$$

$$y = -\sum_{k=1}^{K} ((b_{k0}) \cdot h_k) + \sum_{n=1}^{N-1} [h_k \cdot b_{1n} + h_k \cdot b_{2n} + \cdots + h_k \cdot b_{2n}] 2^{-n} \tag{4}$$

Equation (4) is the final equation of the DA.

$$\sum_{k=1}^{K} h_k b_{kn} \tag{5}$$

Equation (5), can be pre-calculated for all possible values of b_{1n}, b_{2n} ... b_{Kn}. These values are stored using a look-up table of 2^K words addressed by *K-bits*.

The Distributed arithmetic FIR filter structure is shown in Fig. 3. Carry save accumulation is used for shift accumulation process [9].

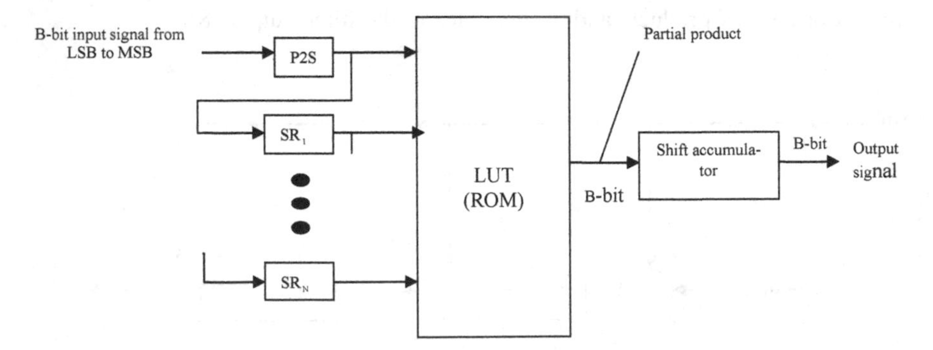

Fig. 3 Distributed arithmetic FIR filter structure. *P2S* parallel to serial converter, *SR* shift register, *LUT* look-up table

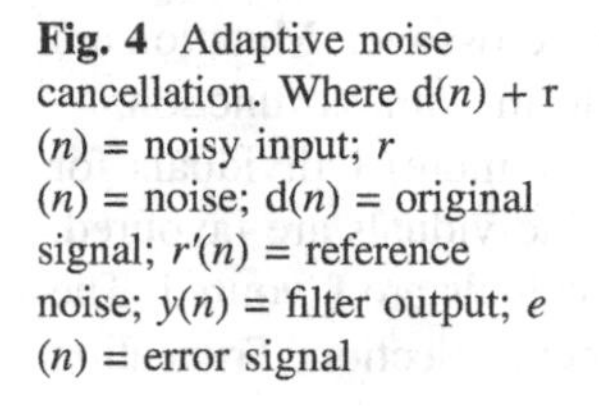

Fig. 4 Adaptive noise cancellation. Where $d(n) + r(n)$ = noisy input; $r(n)$ = noise; $d(n)$ = original signal; $r'(n)$ = reference noise; $y(n)$ = filter output; $e(n)$ = error signal

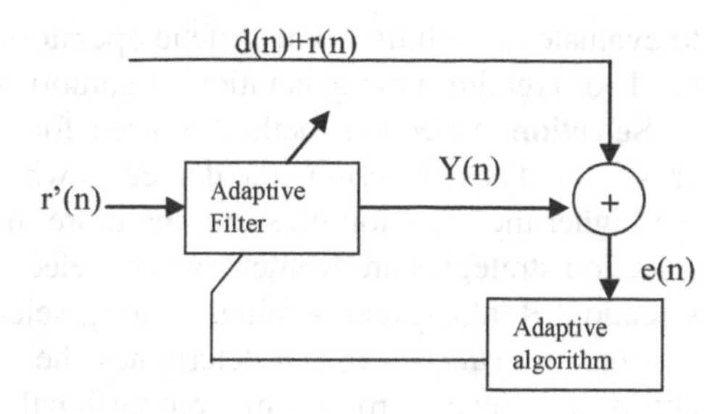

Adaptive Filter. Adaptive filters are self-designing using recursive algorithm and used where knowledge of environment is not available. Adaptive filters are used for noise cancellation application. Adaptive noise cancellation, is used to filter out an interference component by identifying a linear model between measurable noise source and the corresponding immeasurable. Figure 4 shows the adaptive noise cancellation system. The GA used as the adaptive algorithm [10].

2.2 Genetic Algorithm

Genetic algorithm (GA) is a particular class of EA, categorized as global search heuristics. GAs is heuristic search algorithms based on the mechanism of natural selection and genetics. The general flow of GA is shown in Fig. 5. Each individual in the population is called a chromosome (or individual), representing a solution to the problem. Chromosome is a string of symbols either it can be a binary or real-valued bit string [11].

The chromosomes evolve through successive iterations called generations. During each generation, the chromosomes are evaluated, using some measures of fitness. The next generation is created by, new chromosomes, called offspring, are formed by crossover, mutation operators [3]. Then new generation is formed by using selection operation, based on the fitness values. Fittest chromosomes have the highest probabilities of being selected. GAs converges to the best chromosome which represents the optimal solution to the problem after several generations. GA requires the following, a genetic representation of the solution and a fitness function

Fig. 5 Pseudo code for GA

Choose initial population
Evaluate each individual's fitness
Repeat
Select best-ranking individuals to reproduce
Mate the pairs at random
Apply crossover operator
Apply mutation operator
Determine the population's average fitness until termination condition (until one individual has the optimal desired fitness or number of generations has passed)

to evaluate the solution. The genetic operators are selection, crossover, Mutation are used for creating new generation of solutions according to the fitness function.

Selection. Selection method is used for finding two (or more) individuals for crossover. The selection is the degree to which the better individuals are favoured: the higher the selection pressure, the more the better individuals are favoured. The selection strategies are Roulette wheel selection, Tournament selection, Truncation selection, Ranking and Scaling, Sharing selection.

Roulette wheel selection determines the selection probability or survival probability for each chromosome proportional to the fitness value. The Selection probability P_{select} of each chromosome is calculated by:

$$p(x)_{select} = \frac{J(x)}{\sum J}, \quad (6)$$

where $J(X)$ is the fitness score of the chromosome $X1$, $X2$, $X3$ … Xn and J is the sum of all fitness scores in the population.

Crossover. Crossover is an artificial mating in which chromosomes from two individuals are combined to form chromosome for next generation. Crossover is performed by splicing two chromosomes from two different individuals (solutions) at a crossover point and swapping the corresponding spliced parts. This process will provide a better solution represented by the new chromosome.

Crossover types are single point crossover, two point crossover, and uniform crossover. Single point crossover is shown in Table 1. In single point crossover on both parents' organism strings is selected. All bits after that point in either organism string is swapped between the two parent organisms. Crossover probability decides how often crossover will be performed. The crossover probability Pc has to be always high which will guarantee rapid search for the solution.

Mutation. Mutation is a random adjustment in the genetic composition. The mutation operator changes the current value of a gene to a different one. For bit string chromosome this change amounts to flipping a 0 bit to a1 or vice versa. Table 2 shows an example for random mutation. Mutation probability refers how often parts of chromosome will be mutated. Mutation generally prevents the GA from falling into local extremes.

Table 1 Single-point crossover

Parent A	11001|001
Parent B	11011|101
Offspring A	11001|101
Offspring B	11011|001

Table 2 Random Mutation

Chromosome after crossover	11011001
Chromosome after mutation	11**1**110**1**1

3 Implementation

This section explains about the implementation of evolvable system design. A hardware platform was developed using Xilinx platform Studio tool with the available library components, such as Micro Blaze processor [12], UART (Universal Asynchronous Receiver Transmitter), DDR2 (Double data rate synchronous dynamic random access memory). The UART and DDR2 components are interface with Micro Blaze processor using a common clock module. This hardware design is done using embedded development kit (EDK) platform [9].

Now, the FIR filter structure using DA method implemented in VHDL is imported as an IP core in the above hardware platform and is interfaced with the Micro Blaze processor. The GA is implemented in the SDK platform. Then an embedded system is formed by integrating the SDK platform with the hardware platform implemented in EDK. The GA is coded using C language. The specifications of GA:

Selection method: Roulette wheel selection in which each individual gets a slice of the wheel, but more fit ones get larger slices than less fit ones.
Crossover: Single point crossover.
Probability of crossover: 70%
Mutation: Random mutation
Probability of Mutation: 0.5%
Population size: 4(number of coefficients)
No. of iteration: 1000(maximum)

The FIR filter specifications: The cut-off frequency of the filter was set to 0.5 kHz and the filter order 3 with 4 filter coefficients. The reason for the number of coefficients was chosen small is mainly reduces the computation of the code.

4 Results and Discussions

The performance results of the proposed evolvable hardware architecture were compared with Matlab results. The hardware platform was designed using XPS tool and the GA was implemented using the SDK platform of the Xilinx Spartan 6 FPGA. Then this design was fused to the FPGA Kit and the terminal of the SDK platform was connected to PC through serial communication port UART.

The Fig. 6 shows the implementation of GA configuration in SDK platform in FPGA Spartan 6 Trainer Kit. The final set of normalized coefficients was displayed on the console window of the SDK platform through the use of UART peripheral.

The Fig. 7 shows the original signal with frequency of 100 kHz and the sampling frequency is set to 300 kHz. 100 number of samples was taken for simulation (Iterations). The original signal is corrupted by white noise with variance $\sigma^2 = 0.5$. The sinusoidal signal with additive white noise is illustrated in Fig. 8.

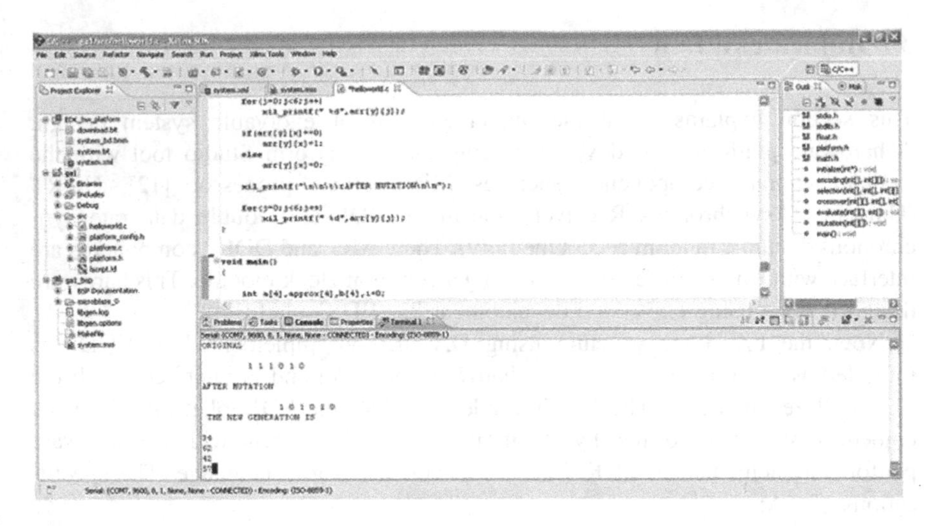

Fig. 6 SDK Platform for GA configuration

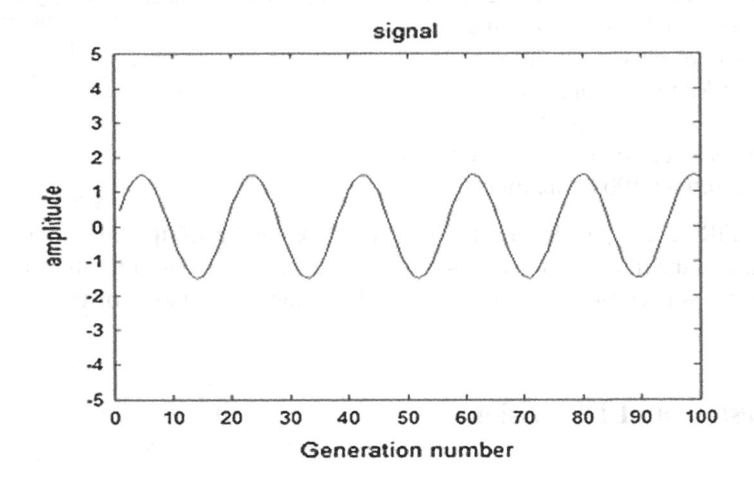

Fig. 7 Original signal

The aim of the simulation was the mean square error minimisation. Figure 9 shows the enhanced signal for the crossover probability of 0.7 and mutation probability of 0.005 of GA [13]. The better results are achieved by a crossover probability of range between 0.65 and 0.9. So crossover probability of 0.7 was considered. Similarly for mutation rate the range was 0.5–1%. So Mutation probability of 0.5% was considered.

The error was reduced significantly and reaches the global optimum value after 80th generation. The final set coefficients from GA were (0.0075, 0.0021, 0.0021

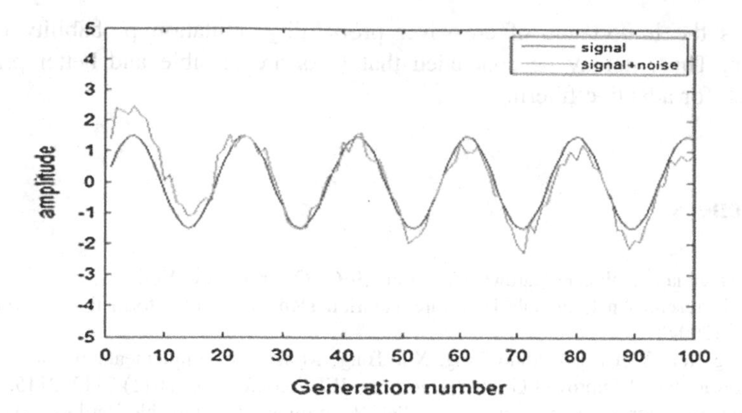

Fig. 8 Original signal and original signal + noise

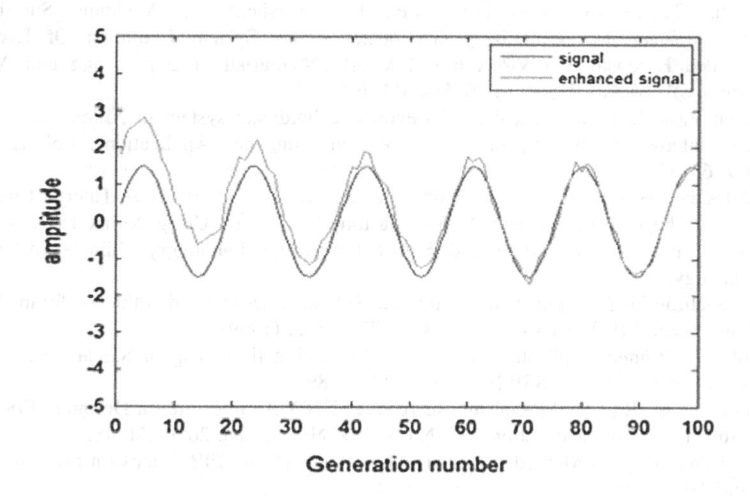

Fig. 9 Signal and enhanced signal by genetic algorithm

and 0.0071). It can be observed that the filter output gradually gets closer to the original signal after 80th generation.

5 Conclusion

The use of adaptive algorithm for noise cancellation has been presented in this paper by using the method of updating the coefficients of FIR Filter (Evolvable Filter) using GA. The design of the FIR IP Core in VHDL was done and the coefficients were updated using GA (implemented in C) in FPGA. This paper also describes the importance of crossover probability, mutation probability of GA program. Thus, it may be concluded that GAs are feasible and better practical approach for adaptive filtering.

References

1. L. Sekanina.: Evolvable Hardware Tutorial. GECCO 2007, New York.
2. Jim Torresen.: An Evolvable Hardware Tutorial. Department of Informatics. University of Oslo. (2004).
3. Aifeng Ren, Wei Zhao, Shuo Tang, Xin Tong, Ming Luo.: Implementation of Evolvable Hardware based Improved Genetic Algorithm. IEEE conference (2011) 2112–2115.
4. Ruben Salvador, Lukas Sekanina et. al.: Self-Reconfigurable Evolvable Hardware system for Adaptive Image Processing. IEEE transactions on computers vol. 62, No. 8, (Aug. 2013) 1481–1493.
5. Ranjith, C., Joy Vasantha Rani, S.P., Priyadharsheni, B., Medhuna Suresh and Madhusudhanan, M.: Optimizing GA operators for System Evolution Of Evolvable Embedded Hardware On Virtex 6 FPGA. ARPN Journal of Engineering and Applied Sciences vol. 10, No. 11, (June 2015) 4908–4914.
6. Zdenek Vasicek, Lukas Sekanina.: An evolvable hardware system in Xilinx Virtex II Pro FPGA. International Journal on Innovative Computing and Applications, Vol. 1, No. 1, (2009) 63–73.
7. Rod Jesman, Fernando Martinez Vallina and Jafar Saniie.: Micro Blaze Tutorial Creating a Simple Embedded System and Adding Custom Peripherals Using Xilinx EDK Software Tools. Embedded Computing and Signal Processing Laboratory, Illinois Institute of Technology.
8. Syed Shahzad Shah, Saquib Yaqub and Faisal Suleman.: Distributed arithmetic for the Design of High Speed FIR Filter using FPGAs. UET, Taxila, (1999).
9. Stanley A. White.: Applications of Distributed Arithmetic to Digital Signal Processing: A Tutorial Review. IEEE ASSP Magazine, (July 1989).
10. Uma Raja ram, Raja Paul Perinbam, Bharghava.: EHW Architecture for Design of FIR Filters for Adaptive Noise cancellation. IJCSNS, vol. 9, No. 1, (Jan. 2009) 41–48.
11. Ajoy Kumar Dey.: A Method of Genetic Algorithm (GA) for FIR Filter Construction. Vol. 1, (Dec. 2010) 87–90.
12. Yang Zhang, Stephen L. Smith, and Andy M. Tyrell.: Digital Circuit Design using Intrinsic Evolvable Hardware. Proceedings of NASA/DoD Conference on Evolution Hardware (EH'04), (2004).
13. Pradeep kaur, Simatpreet Kaur.: Optimization of FIR Filters using Genetic Algorithm. IJETICS vol. 1, No. 3 (2012) 228–232.

Dynamic Auditing and Deduplication with Secure Data Deletion in Cloud

N. Dinesh and I. Juvanna

Abstract As cloud computing technology becomes the major storage and sharing of data in Internet era, outsourcing cloud server for storing data becomes the latest trend. All small business and major social networking sites rely on cloud storage for dynamic data storage. Storing and maintaining cloud data are not easy, it require lots of effort and space to manage cloud data. Since the cloud is accessible by everyone connected in the network the security risk of cloud data is also high. To improve the cloud storage and make the storage more effective we are implementing the audit function in cloud data to reduce the duplication of data in cloud storage. This audit is done using file id and hash key generated for every file during the time of uploading. Proof of Ownership is used to track the owner of the files uploaded. Clustering of user data is included to monitor user file and improve service to the user for that particular type of file. To avoid data remanence attack in the cloud data, we are implementing a secure data deletion scheme which required key to delete the files which cannot be recovered during the attack.

Keywords Deduplication · Auditing · *K*-means clustering · Cloud storage · Encryption · Data remanence

1 Introduction

Cloud storage is an important data storage in Internet era where the data are stored in a virtual environment. Since the maintenance of cloud storage needs lot of money and work to manage the system, this is maintained by third parties. The storage is complex in all types of cloud. Using cloud storage instead of centralized system

N. Dinesh (✉) · I. Juvanna
Hindustan Institute of Technology and Science, Padur 603103, Chennai, India
e-mail: dineshnagarajan19@gmail.com

I. Juvanna
e-mail: ijuvanna@hindustanuniv.ac.in

S.S. Dash et al. (eds.), *Artificial Intelligence and Evolutionary Computations in Engineering Systems*, Advances in Intelligent Systems and Computing 517,
DOI 10.1007/978-981-10-3174-8_27

leads to easy access of data. Cloud data can be accessed from anywhere which allows maximum availability of data for the user.

The growth of social networks and smart devices was highly supported by the cloud system. Since social networks provide more features to share data among multiple users the duplicate data linked between every user will increase gradually. To avoid inefficient utilization of cloud storage deduplication of data is used to improve the storage. The data handled by today's cloud increasing day by day with large quantity. In coming years, this rate of cloud data will increase in large scale in an unexpected manner. A study on digital data by EMC shows that 75% of digital data are duplicate copies. Major cloud service providers provide features to remove the duplicate data in the cloud storage which improve the cloud efficiency and reduce storage cost for the user. This also makes the maintenance process easier for the service provider.

Storing data in the cloud is an easy task, but handling the data and storage in the cloud is much demanding task to attain. Some of the major challenge cloud storage face today is integrity auditing, deduplication and deleting the data in a secure way to avoid security flaws in the existing cloud storage deletion. Since the cloud is accessible by a large number of clients at a time, the number of file uploaded in cloud is also high comparing to previous storage system. This will create extra load for the cloud server. The domain used for storing data in the cloud is not under user control; it depends on the cloud service provider. After storing the data in the cloud the responsibility of the data owner become less and the owner need to depends on the service provider for the security and privacy of data. This raises the data integrity and security threat in both internal and external cloud users on cloud data. In some case cloud, service provider may clear the long time unaccessed data in cloud to improve the storage. In order to save money and space service provider follows such activities, even though this violates user data privacy. Keeping the large volume of data in cloud to outsource the integrity of the files cannot be verified by the client even though he needs it. Here, the client needs to fully depend on the cloud service provider which works only on trust-based understanding between client and service provider which is not reliable.

Increase in usage of cloud leads to new problem for service provider to maintain cloud data. The challenging problem is reducing duplicate data in cloud due to vast spread of cloud the duplicate data availability cannot be avoided without proper system. Deduplication is the effective approach used to avoid duplicate data in cloud server. Deduplication is done by keeping only one copy (or block) for every client and provides pointers for all clients who own the file. Unfortunately, deduplication leads to some security threat, for example a service provider telling a client about the availability of existing data reveals that some other client has the same file which leaks the data privacy and lead to loss of secure data. Since the client owned file is not involved based on a static short value.

Now cloud data can be removed by service provider after a period of time. Sometimes, this data is valuable for the user which is removed by the service provider without proper verification by the user. For example, the data in social network and cloud storage is removed by the service provider after a period of no access by the user this leads to loss of data to client. Data cleared from cloud also face attack from the hackers. This cleared data may contain sensitive information of the user. Data cleared from the cloud can be obtained by hackers using data remanence attack which recovers deleted files from the cloud system.

This paper aiming to achieve data integrity, dedupication, clustering, and secure data deletion in cloud.

Auditing using mapreduce cloud helps to generate data tags while uploading. Data tags can be used to ensure the integrity of data stored in cloud. This reduces the work of auditing in existing system. Security in this system prevent leakage of side channel information, proof of ownership [1] protocol is used between client and cloud service provider. This ensure client that they own the target data.

The rest of the page contains explained as follows: Sect. 2 explains the related work of secure auditing and deduplication with secure measures for preventing data remanence. Section 3 describes the preliminary work related to the project work. Section 4 introduces the proposed system on deduplication and secure deletion and Sect. 5 shows the performance evaluation of the system with comparing to existing system. In Sect. 6, detail of future enhancement and conclusion is explained.

2 Related Works

This section explains the auditing, deduplication, and secure file deletion in the cloud, respectively.

2.1 Secure Auditing

Provable data possession (PDP) [2] was first approached by Ateniese which ensures the cloud server that possess the file without holding the entire file. Meta data is analyzed to check the integrity of the data. To reduce the client side load proxy PDP is introduced by Zhu et al. [3] in public cloud, to ensure the retrievability of stored files in cloud proof of retrievability (POR). POR [4] possess the stored file in cloud and also assure file recovery from service provider. An improved work in new cloud architecture by Li et al. with two cloud server to perform integrity auditing prevent client load.

2.2 Deduplication

Deduplication is a technique used to improve the cloud storage in case of storage efficiency and cost-effective system. This allows the user to store only one copy of file in the system and avoid storing same file multiple time. Avoiding duplicated causes some data issues in database, for example, when removing duplicate data; the availability of data for the users will be reduced and this issue can be overcome by VMCloning introduced by Diego Perez-Botero. Another issue on deduplication is when a user tries to upload existing file in cloud, the server will not accept the file to be stored and with this it is understood that the same file is available and used by some other user and in some case it may be a sensitive data. This leads to leakage of side channel information which can be avoided by applying proof of ownership approached by Halevi et al. [6] holds the detail of the file client who holds the data. Cryptographic approach of convergent encryption is used to ensure data privacy in deduplication. Keelveedhi et al. introduced DupLess system in which file-based key to perform secure deduplication. The integrity of the system is maintained by the data id generated during the time of upload and the data encryption.

2.3 Clustering

Cloud storage comes up with huge volume and variety of data storage. Without proper storage structure the maintenance of such a huge volume data become hard. Clustering of data will make the process easy and effective to maintain the data in cloud storage. One of the most successful clustering algorithms used for effectiveness is *K*-Means algorithm which makes the clustering process more easy and effective for large volume of data.

2.4 Secure Data Deletion

Clearing data from cloud enables attackers to recover the cleared files with forensic analysis for deleted files. This attack in cloud data is common in recent days to retrieve the cleared files. Data remanence is the active attack which allows the attacker to recover the deleted files and access the confidential data from the retrieved file. To avoid attack in deleted cloud data, a secure file deletion approach is introduced in order to avoid data remanence attack. This can be avoided by cryptographical approach in which a secure key to remove the user data is given to the user and whenever the server needs to clear the data this, key is requested and once after the key is obtained, the data is deleted and the data location is over-ridded in order to avoid data recovery.

Data remanence is an active attack in cloud data, in which the attacker recover the cleared cloud data and analysis for sensitive information from the recovered cloud data.

3 Preliminary

An effective method for secure integrity, deduplication, and secure file deletion is explained in the following section. Each and every module with the technique used to solve the problem to attain the discussed problem is explained in detail.

The preliminary work explains the detail of method used to implement the system of our approach.

3.1 Deduplication

Deduplication can be achieved by many methods; some of the most commonly used method to achieve deduplications are

File-Based Compare. This technique is the basic one to achieve deduplication in file level. Comparison of files stored in the system with file-based algorithm. This algorithm is used to eliminate duplicate file in existing file storage. The effectiveness of this method is comparatively less than other methods used to remove duplicate files. This comparison is done based on file name, size of the file, type of the file, and modified information.

File Based Versioning. This versioning method looks inside file content and compares difference between the files and store the difference among the file as "delta" from the original file. This delta act as the difference between files; whenever the user requests the file, this particular version of the file merges with the original file and is displayed to the user. This method is effective until more user accesses the database at the same time. If the load become more, this method becomes less effective to find the duplicate files present in the storage.

File Based Hashing. File-based hashing creates unique hash value for the file this hash value is unique for each non-duplicate file. If any duplicate data is uploaded, then the generated hash value for the file will be same as file existing in the storage. By comparing the hash value of the file, the duplicate files can be identified effectively. Once same hash value is found among the database; those files can be removed to improve storage efficiency.

Block Versioning. Block versioning is similar to file-based versioning but the difference is the data stored on the disk does not need to know the details about files and the is being used. This can be used in both unstructured data and structured data. Block versioning is applied to the block level files (Fig. 1).

Block Hashing. Files stored in cloud are separated as blocks and then the hash generation is applied on the content, so that the duplicate data can be identified effectively. This applies hashing in more granular level.

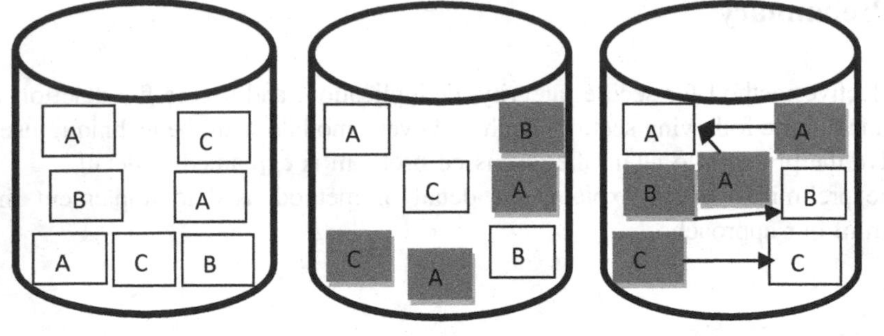

Data block with duplicate in cloud storage

Duplicate data are identified using hash value

Duplicate data replaced with pointer(improve storage)

Fig. 1 File-based hashing

3.2 Convergent Encryption

Convergent encryption is also called as content hash keying, used to provide confidentiality in the process of deduplicating data, without providing access to encryption key to the service provider. This allows the user to ensure data security. Identical cipher text is generated for plain text without allowing access to the encryption key. Because of convergent encryption, the storage capacity of cloud is increased and made it cost-effective.

3.3 K-Means

Cloud handles huge volume of data which is stored and accessed by large number of users at a time. In order to maintain the efficiency and reduce load in server during multiple access request at the same time clustering resolves the problem in which the data is grouped in small number of clusters. If the storage has n data value then X_i, $i = 1 \ldots n$ this n data is to be partitioned in k clusters [5]. The clustering is performed by assigning cluster to each data point. The steps involved in clustering data using k-means algorithm are

1. Define center for cluster.
2. Attribute closest cluster to each data.
3. Set position for each cluster to mean all data points belonging to that cluster.
4. Repeat steps 2 and 3 until convergence.

3.4 Data Remanence

Data remanence is a security issue in cloud system. In which the data cleared from the cloud storage will remain even after clearing the data and erase the data. Data remanence attack is performed mostly on uncontrolled environment and is an uncontrolled environment of storage with huge volume of data and so it is easy to collect and analysis data from cloud. Some of the effective methods used to prevent this attack is clearing, purging. Most effective methods are overwriting, encryption, and destruction.

4 Proposed System

4.1 System Overview

The major part of the paper is to provide data auditing, deduplication, and secure file deletion. First auditing is applied to the files uploaded by the user. Data integrity is ensured during the auditing process. Second, the audited files are verified and stored in cloud storage without duplicating exists in the file.

File-based hashing method is applied to ensure the deduplication before and after storage of file. This process improves the data storage and reduces cost for service provider. Since the cloud handles huge volume of data clustering of cloud data is important to organize and manage the data both by service provider and cloud user. So effective clustering of files is implemented using k-means algorithm (Fig. 2).

Clearing cloud data leads to security threats for sensitive data uploaded by user. Since the user did not get proper access after uploading the data, this valuable data is cleared without user knowledge and to avoid this a secure encryption key method is introduced to improve the user control on their data stored in cloud. After getting

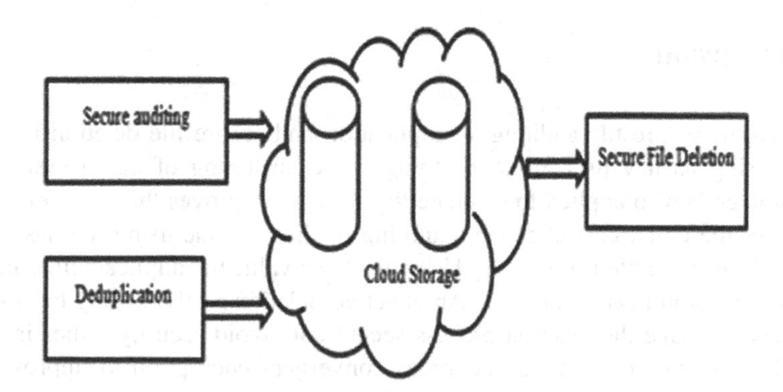

Fig. 2 System architecture

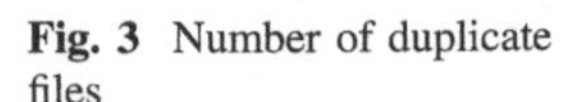

Fig. 3 Number of duplicate files

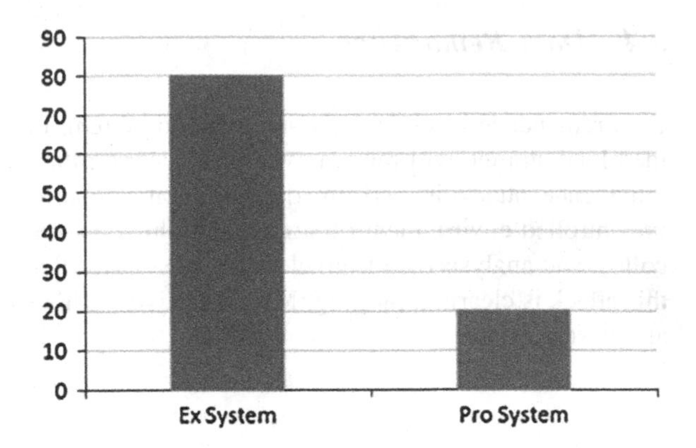

proper access from the user the data is cleared in a way that cannot be recovered during recovery. This avoids data remanence attack in cloud which is secure for the cloud users. This data is cleared and to avoid recovery overriding of data space with some other trash data is done in order to avoid recovery.

5 Evaluation

Proposed system evaluated based on the removal of duplicate file in the cloud. As the existing system efficiency of removing duplicate data is comparatively less this approach of auditing and deduplication can be applied to improve storage effectiveness. Since the data is encrypted throughout the process, the security of the system is ensured.

Figure 3 shows the avoidance of duplicate file in the system. The load of the cloud is also reduce in identifying duplicate files time taken to perform is reduced by applying PDP which improves the system efficiency.

6 Conclusion

In this paper, secure file auditing, deduplication and secure file deletion is implemented using hash values in cloud storage. The clustering of data stored in the cloud storage is also applied to evaluate the storage, improves the cloud efficiency and makes the cloud cost effective. Auditing of files is done using the hash value generated when the file is uploaded. Using the hash value the duplicate files are also identified in deduplication process. An effective deletion method using hash key is introduced to make the deletion process secure and avoid security issues in cloud data. The data stored in cloud undergoes convergent encryption to improve data confidentiality in deduplication. Comparing with previous work, the proposed system reduces the time of audit and increases the reliability of deduplication process and clustering of cloud data with secure data deletion makes the cloud storage more secure and avoids recovery of deleted unsecured cloud data.

References

1. S. Halevi, D. Harnik, B. Pinkas, and A. Shulman-Peleg, "Proofs of ownership in remote storage systems," in *Proceedings of the 18th ACM Conference on Computer and Communications Security*. ACM, 2011, pp. 491–500.
2. G. Ateniese, R. Burns, R. Curtmola, J. Herring, L. Kissner, Z. Peterson, and D. Song, "Provable data possession at untrusted stores," in Proceedings of the 14th ACM Conference on Computer and Communications Security, ser. CCS '07. New York, NY, USA: ACM, 2007, pp. 598–609.
3. Y. Zhu, H. Hu, G.-J. Ahn, and M. Yu, "Cooperative provable data possession for integrity verification in multicloud storage," IEEE Transactions on Parallel and Distributed Systems, vol. 23, no. 12, pp. 2231– 2244, 2012.
4. S. Keelveedhi, M. Bellare, and T. Ristenpart, "Dupless: Serveraided encryption for deduplicated storage,"in Proceedings of the 22Nd USENIX Conference on Security, ser. SEC'13. Washington, D.C.: USENIX Association, 2013, pp. 179–194.
5. A. Mahendiran, N. Saravanan, N. Venkata Subramanian and N. Sairam, "Implementation of *K*-Means Clustering in Cloud Computing Environment" in Research Journal of Applied Sciences, Engineering and Technology, 2012, pp 1391–1394.
6. J. Yuan and S. Yu, "Secure and constant cost public cloud storage auditing with deduplication," in *IEEE Conference on Communications and Network Security (CNS)*, 2013.

Hesitant Fuzzy Soft Set Theory and Its Application in Decision Making

T.R. Sooraj, R.K. Mohanty and B.K. Tripathy

Abstract There are several models of uncertainty found in the literature like fuzzy set, rough set, soft set and hesitant fuzzy set. Also, several hybrid models have come up as a combination of these models and have been found to be more useful than the individual models. In everyday life we make many decisions. Making efficient decisions under uncertainty needs better techniques. Many such techniques have been developed in the recent past. These techniques involve soft sets and fuzzy sets. In this paper we redefined the hesitant fuzzy soft sets (HFSS) with the help of membership function. We also provide a decision making algorithm.

Keywords Soft set · Fuzzy sets · Fuzzy soft sets · Intuitionistic fuzzy set · Hesitant sets · Intuitionistic fuzzy soft set · Decision making

1 Introduction

The notion of fuzzy sets introduced by Zadeh [1] in 1965 is one of the most fruitful models of uncertainty and has been extensively used in real life applications. In order to bring topological flavor into the models of uncertainty and associate family of subsets of a universe to parameters, Molodtsov [2] introduced the concept of soft sets in 1999. A soft set is a parameterized family of subsets. Many operations on soft sets were introduced by Maji et al. [3, 4]. Hybrid models are obtained by suitably combining individual models of uncertainty have been found to be more efficient than their components. Several such hybrid models exist in literature. Maji et al. [5] put forward the concept of fuzzy soft set (FSS) by combining the notions

T.R. Sooraj (✉) · R.K. Mohanty · B.K. Tripathy
School of Computing Science, VIT University, Vellore, Tamilnadu, India
e-mail: soorajtr19@gmail.com

R.K. Mohanty
e-mail: rknmohanty@gmail.com

B.K. Tripathy
e-mail: tripathybk@vit.ac.in

S.S. Dash et al. (eds.), *Artificial Intelligence and Evolutionary Computations in Engineering Systems*, Advances in Intelligent Systems and Computing 517,
DOI 10.1007/978-981-10-3174-8_28

of fuzzy set and soft set. Tripathy et al. [6] defined soft sets through their characteristic functions. This approach has been highly authentic and helpful in defining the basic operations like the union, intersection and complement of soft sets. Similarly, defining membership function for FSSs will systematize many operations defined upon them as done in [7]. Many of Soft set applications have been discussed by Molodtsov in [2]. An application of soft sets in decision making problems is discussed in [3]. Among several approaches, in [5], FSS and operations on it are defined. This study was further extended to the context of fuzzy soft sets by Tripathy et al. in [7], where they identified some drawbacks in [3] and took care of these drawbacks while introducing an algorithm for decision making.

The concept of hesitant fuzzy sets was introduced by Torra [8]. This is an extension of fuzzy sets. It is sometimes difficult to determine the membership of an element into a set and in some circumstances this difficulty is caused by a doubt between a few different values. The concept of hesitant fuzzy soft sets was introduced by Sunil et al. They also discussed an application of decision making. In this paper, we introduce hesitant fuzzy soft sets. Here, we follow the definition of soft set due to Tripathy et al. [6] in defining HFSS. Applications of various hybrid models are discussed in [7, 9–20]. The major contribution in this paper is introducing a decision making algorithm which uses HFSS for decision making and we illustrate the suitability of this algorithm in real life situations. Also, it generalizes the algorithm introduced in [5] while keeping the authenticity intact. Some operations on hesitant fuzzy sets are defined in [21]. In this paper, we introduce the concept of HFSS with the help of membership function. Also, we discuss an application of HFSS in decision making problems which uses the concept of negative parameters.

2 Definitions and Notions

In this section we present some of the definitions to be used in the paper. We assume that U is a universal set and E is a set of parameters defined over it.

Definition 2.1 A fuzzy set A is defined through a function μ_A called its membership function such that $\mu_A : U \to [0, 1]$.

Definition 2.2 An intuitionistic fuzzy set over U is associated with a pair of functions $\mu_A, \nu_A : U \to [0, 1]$ such that for any $x \in U$, $0 \leq \mu_A(x) + \nu_A(x) \leq 1$.

The hesitation function π_A is defined as $\pi_A(x) = 1 - \mu_A(x) - \nu_A(x)$, $\forall x \in U$.

Definition 2.3 A soft set over the soft universe (*U*, *E*) is denoted by (*F*, *E*), where

$$F : E \to P(U) \tag{2.1}$$

Here *P*(*U*) is the power set of *U*.

Let (*F*, *E*) be a soft set over (*U*, *E*). Then in [7] it was defined as a parametric family of characteristic functions $\chi_{(F,E)} = \{\chi^a_{(F,E)} | a \in E\}$ of (*F*, *E*) as defined below.

Definition 2.4 For any $a \in E$, we define the characteristic function $\chi^a_{(F,E)} : U \to \{0,1\}$ such that

$$\chi^a_{(F,E)}(x) = \begin{cases} 1, & \text{if } x \in F(a); \\ 0, & \text{otherwise.} \end{cases} \tag{2.2}$$

Definition 2.5 An interval valued fuzzy set *A* over *U* is given by a pair of functions $\mu_A : U \to D([0,1])$, where *D* ([0, 1]) is the set of all closed sub intervals on [0,1], such that $\mu_A(x) = [\mu^-_A(x), \mu^+_A(x)] \subset [0,1]$.

3 Hesitant Fuzzy Soft Sets

In this section, we introduce the notion of hesitant fuzzy soft sets and also introduce some of the operations of HFSSs.

Definition 3.1 A pair (*F*, *E*) is called a HFSS if $F : E \to \mathrm{HF}(U)$, where HF (*U*) denotes the set of all hesitant fuzzy subsets of *U*.

An HFSS H on *U* is defined in terms of its membership function $\mu_H : E \to P(\mathrm{HFS})$, $\nu_H : E \to P(\mathrm{HFS})$ such that $\forall a \in E$ and $\forall x \in U$, $\mu^a_H(x), \nu^a_H(x) \in P([0,1])$ such that $0 \leq \sup \mu^a_H(x) + \sup \nu^a_H(x) \leq 1$.

Given three HFEs in an HFSS is represented by *h*, h_1, and h_2. Then we can define union and intersection operations as follows.

Definition 3.1 For any two HFSSs (*F*, *E*) and (*G*, *E*) over a common universe (*U*, *E*), the union of (*F*, *E*) and (*G*, *E*) is the HFSS (*H*, *E*) and $\forall a \in E$ and $\forall x \in U$, we have

$$\begin{aligned} h^a_1 \cup h^a_2 &= \{(\alpha_1, \alpha_2) | \alpha_1 \in h^a_1, \alpha_2 \in h^a_2\} \\ &= \left\{\max\left\{\mu^a_{\alpha_1}(x), \mu^a_{\alpha_2}(x)\right\}\right\} \end{aligned} \tag{3.1}$$

where $\alpha_1 \in h^a_1, \alpha_2 \in h^a_2$. h^a_1 and h^a_2 denote the hesitant fuzzy set.

Definition 3.2 For any two HFSSs (*F*, *E*) and (*G*, *E*) over a common universe (*U*, *E*), the intersection of (*F*, *E*) and (*G*, *E*) is the HFSS (*H*, *E*) and $\forall a \in E$ and $\forall x \in U$, we have

$$\begin{aligned} h^e_1 \cap h^e_2 &= \{\min(\alpha_1, \alpha_2) | \alpha_1 \in h^a_1, \alpha_2 \in h^a_2\} \\ &= \left\{\min\left\{\mu^a_{\alpha_1}(x), \mu^a_{\alpha_2}(x)\right\}\right\} \end{aligned} \tag{3.2}$$

Definition 3.3 The complement of HFSS (*F*, *E*), represented as $(F, E)^c$, is defined as

$$h^c = \left\{\left(1 - \mu^a_{(F,E)}\right)\right\} \quad (3.3)$$

Definition 3.4 An HFSS (*F*, *E*) is said to be a null HFSS if and only if it satisfies

$$\mu^a_{(F,E)}(x) = 0 \quad (3.4)$$

Definition 3.5 An HFSS (*F*, *E*) is said to be an absolute HFSS if and only if it satisfies

$$\mu^e_{(F,E)}(x) = 1 \quad (3.5)$$

4 Application of HFSS in Decision Making

In [2] Molodtsov has given several applications of soft set. In [9] the decision making example given depends on the decision of a single person. Here we discuss an application of DM in HFSSs.

Many of researchers have tried to provide solutions for the decision making problems in lot many situations. Some of these approaches are preference ordering, utility values, preference values.

The parameters can be categorized as of two types [7].

To make decision from hesitant fuzzy soft set data, we need to consider three cases.

(i) Pessimistic: Lowest membership value of an object out of multiple hesitant values for that parameter.
(ii) Optimistic: Highest membership value of an object out of multiple hesitant values for that parameter.
(iii) Neutral

Neutral values are obtained by taking the mean of all hesitant values for that parameters.

$$\text{neutral value} = \frac{\sum \text{hesitant values}}{\text{No. of hesitant values}} \quad (8)$$

To get the normalized score from multiple scores, the following formula is used.

Let us consider a person want to buy a laptop. The person want to choose a laptop by considering parameters such as Processor, RAM, Display, Price, Size and Brand. Some parameters like price inversely affect the decisions. Those parameters are called negative parameters.

Algorithm

1. Input the parameter data table. The parameter data table contains all the information about the parameters.
2. Input the HFSS table.
3. Construct an FSS table from the values of HFSS table by taking the mean value of all hesitant values.
4. Multiply the priority values with the corresponding parameter values to get the priority table.
5. Construct the comparison table by finding the entries as differences of each row sum in priority table.
6. Construct the decision table by taking the row sums in comparison table.
   - 6.1. Assign rankings to each candidate based upon the score obtained.
   - 6.2. If there is more than one having same score than who has more score in a higher ranked parameter will get higher rank and the process will continue until each entry has a distinct rank.

Let U be a set of laptops given by $U = \{p1, p2, p3, p4, p5, p6\}$ and E be the parameter set given by $E = \{e1, e2, e3, e4, e5, e6\}$., where $e1$, $e2$, $e3$, $e4$, $e5$ and $e6$ represents Processor, RAM, Display, Price, Size and Brand respectively.

The parameter data table is given in the Table 1. It contains all the details about the parameters. The parameter rank is given by comparing the absolute priority value.

Let us consider a HFSS, (F, E) which shows the quality of laptops with maximum three opinions for the given set of parameters as shown in Table 2.

Any HFSS can be converted to FSS by computing the mean of all hesitant values. HFSS data in Table 2 is converted to FSS as shown in Table 3.

The priority table can be constructed by multiplying the values of FSS table to the corresponding priorities in parameter data table as shown in Table 4.

Comparison table can be computed by taking the entries as differences of each row sum of priority table as shown in Table 5.

Decision table can be constructed by total score of each object in comparison. The objects can be ranked according to the score obtained as shown in Table 6.

From the decision Table 6, it can be easily guess the quality of the product and the customer can buy accordingly.

Table 1 Parameter table

Parameter	$e1$	$e2$	$e3$	$e4$	$e5$	$e6$
Priority	0.2	0.4	0.2	−0.3	0.1	0
Parameter rank	3	1	3	2	5	6

Table 2 HFSS table

U	e1	e2	e3	e4	e5	e6
p1	0.1	0.3	0.6	0.3	0.1	0.9
	0.6	1	0.9	0.4	0.3	0.3
	0	0.7	1	0.3	0.3	0.3
p2	0.5	0.5	0	0.4	0.7	0.4
	0.3	0.3	1	0.1	0	0.6
	0.2	0	0.4	0	0.4	0.2
p3	0.4	0.3	0.2	0.8	0.6	0.4
	0.9	0.4	0.5	0.1	0.4	0.3
	0	0	0.7	0.1	0.2	0.9
p4	0.9	0	0.6	1	0.4	0.7
	0.6	0.3	0.8	0.7	0.2	1
	0.6	0.4	0	0.5	0.8	0.4
p5	0.7	1	0.5	0.9	0.3	0.2
	0.4	0.7	0.2	0.8	0.7	0.2
	0	1	1	0.7	0.3	0.4
p6	0.3	0.1	0.9	0.3	0.3	0.9
	0.2	0.3	0.1	0.5	0.2	0.1
	1	0	0.1	0.9	0	0.1

Table 3 FSS table

U	e1	e2	e3	e4	e5	e6
p1	0.233333	0.666667	0.833333	0.333333	0.233333	0.5
p2	0.333333	0.266667	0.466667	0.166667	0.366667	0.4
p3	0.433333	0.233333	0.466667	0.333333	0.4	0.533333
p4	0.7	0.233333	0.466667	0.733333	0.466667	0.7
p5	0.366667	0.9	0.566667	0.8	0.433333	0.266667
p6	0.5	0.133333	0.366667	0.566667	0.166667	0.366667

Table 4 Priority table

U	e1	e2	e3	e4	e5
p1	0.046667	0.266667	0.166667	−0.1	0.023333
p2	0.066667	0.106667	0.093333	−0.05	0.036667
p3	0.086667	0.093333	0.093333	−0.1	0.04
p4	0.14	0.093333	0.093333	−0.22	0.046667
p5	0.073333	0.36	0.113333	−0.24	0.043333
p6	0.1	0.053333	0.073333	−0.17	0.016667

Table 5 Comparison table

	*p*1	*p*2	*p*3	*p*4	*p*5	*p*6
*p*1	0	0.15	0.19	0.25	0.053333	0.33
*p*2	−0.15	0	0.04	0.1	−0.09667	0.18
*p*3	−0.19	−0.04	0	0.06	−0.13667	0.14
*p*4	−0.25	−0.1	−0.06	0	−0.19667	0.08
*p*5	−0.05333	0.096667	0.136667	0.196667	0	0.276667
*p*6	−0.33	−0.18	−0.14	−0.08	−0.27667	0

Table 6 Decision table

	Score	Rank
*p*1	0.973333	1
*p*2	0.073333	3
*p*3	−0.16667	4
*p*4	−0.52667	5
*p*5	0.653333	2
*p*6	−1.00667	6

5 Conclusions

In this article, we introduced a new definition of HFSS which uses the more authentic characteristic function approach for defining soft sets provided in [6]. This provides several authentic definitions of operations on HFSS and made the proofs of properties very elegant. We pointed out some of the flaws of FSS given by Maji et al. in [4] and provided solutions to rectify them in [7]. This made decision making more efficient and realistic. Here, we proposed an algorithm for decision making using HFSS which uses the concept of negative parameters. Also, an application of this algorithm in solving a real life problem is demonstrated.

References

1. L.A. Zadeh, "Fuzzy sets", Information and Control, vol. 8, 1965, pp. 338–353.
2. D. Molodtsov, "Soft Set Theory - First Results", Computers and Mathematics with Applications, vol. 37, 1999, pp. 19–31.
3. P.K. Maji, R. Biswas, and A.R. Roy, "An Application of Soft Sets in a Decision Making Problem", Computers and Mathematics with Applications, vol 44, 2002, pp. 1007–1083.
4. P.K. Maji, R. Biswas, and A.R. Roy, "Soft Set Theory", Computers and Mathematics with Applications, vol. 45, 2003, pp. 555–562.
5. P.K. Maji, R. Biswas, and A.R. Roy, "Fuzzy Soft Sets", Journal of Fuzzy Mathematics, vol. 9 (3), 2001, pp. 589–602.
6. B.K. Tripathy and K.R. Arun, "A New Approach to Soft Sets, Soft Multisets and Their Properties", International Journal of Reasoning-based Intelligent Systems, Vol. 7, no. 3/4, 2015, pp. 244–253.

7. B.K. Tripathy, Sooraj, T.R., RK Mohanty, "A new approach to fuzzy soft set and its application in decision making", Advances in intelligent systems and computing, vol. 411, 2016, pp. 305–313.
8. Torra V, Hesitant fuzzy sets and decision, International Journal of Intelligent Systems 25 (6),2010, pp. 395–407.
9. B.K. Tripathy, Sooraj T.R., RK Mohanty and K.R. Arun, "A New Approach to Intuitionistic Fuzzy Soft Set Theory and its Application in Decision Making", Advances in Intelligent Systems and Computing, 439, 2016, pp. 93–100.
10. Sooraj T.R., RK Mohanty and B.K. Tripathy, "Fuzzy Soft Set Theory and its Application in Group Decision Making", Advances in Intelligent Systems and Computing, 452, 2016, pp. 171–178.
11. Tripathy, B.K., Sooraj, T.R and Mohanty, RK, "Advances Decision Making usisng Hybrid Soft set Models", International Journal of Pharmacy and Technology, 8(3), pp. 17694–17721.
12. B.K. Tripathy, RK Mohanty, T.R. Sooraj, "On Intuitionistic Fuzzy Soft Sets and Their Application in Decision Making", Lecture Notes in Electrical Engineering, 396, 2016, pp. 67–73.
13. B.K. Tripathy, RK Mohanty, Sooraj, T.R., "On Intuitionistic Fuzzy Soft Set and its Application in Group Decision Making". In: Proceedings of the International Conference on Emerging Trends in Engineering, Technology and Science (ICETETS), Pudukkottai, 2016, pp. 1–5.
14. B.K. Tripathy, RK Mohanty, Sooraj T.R, A. Tripathy, "A modified representation of IFSS and its usage in GDM", Smart Innovation, Systems and Technologies, 50, pp. 365–375
15. RK Mohanty, Sooraj, T.R, B.K. Tripathy, "An application of IVIFSS in medical diagnosis decision making", International Journal of Applied Engineering Research, vol. 10, pp. 85–93.
16. B.K. Tripathy, Sooraj, T.R., RK Mohanty, "A New Approach to Interval-valued Fuzzy Soft Sets and its Application in Decision Making", Advances in Intelligent Systems and Computing, 509, 2017, pp. 3–10.
17. RK Mohanty, Sooraj, T.R, B.K Tripathy, "IVIFS and Decision making", Advances in Intelligent Systems and Computing, 468, 2017, pp. 319–330.
18. B.K.Tripathy, Sooraj, T.R, RK Mohanty. "A New Approach to Interval-valued Fuzzy Soft Sets and its Application in Group Decision Making", Proc. of Int. Conf. on Computer systems, Data Communication and Security, CDCS-2015, pp. 1–8.
19. B.K.Tripathy, RK Mohanty, Sooraj, T.R, "Application of Uncertainty Models in Bioinformatics", Handbook of Research on Computational Intelligence Applications in Bioinformatics, Chapter - 9, Publisher: IGI Global, Editors: Sujata Dash, Bidyadhar Subudhi, 2016, pp. 169–182.
20. B.K.Tripathy, Sooraj, T.R, RK Mohanty and Arun K.R, "Parameter Reduction in Soft Set Models and Application in Decision Making", Handbook of Research on Fuzzy and Rough Set Theory in Organizational Decision Making, Publisher: IGI Global, Editors: Arun Kumar Sangaiah., Xiao-Zhi Gao and Ajith Abraham, 2016, pp. 331–355.
21. Torra V, Narukawa Y, On hesitant fuzzy sets and decision, The 18th IEEE international conference on Fuzzy Systems, Jeju Island, Korea, 2009, pp. 1378–1382.

Privacy-Aware Set-Valued Data Publishing on Cloud for Personal Healthcare Records

Elizabeth Alexander and Sathyalakshmi

Abstract Nowadays, Cloud computing becomes a looming computing prototype. Users can get a variety of services such as high computation power, storage, etc. Thus, applications of users can be more cost-effectively put on cloud by utilizing various commodity computers together. But, cloud computing faces some security concerns even if they provides many services. Some of such important concerns are data security and privacy of data. Some personal data like personal healthcare records and financial records contain sensitive information which can be analyzed and mined for public researches although these records offer important human assets. Data should be privacy preserved because malicious cloud users or untrusted cloud providers can get the data with less effort. To deal with these problems, privacy-aware set-valued data publishing on cloud for personal healthcare records has been proposed. An efficient privacy-aware system, named PHKEM (Personal Healthcare *k*-anonymity Encryption Model), for eliminating privacy breaches in publishing of personal healthcare data on cloud as well as data querying is designed. A data anonymization technique, named *k*-Anonymity with extended quasi-identifier partitioning (EQI-partitioning), interactive differential privacy, and AES encryption is applied to preserving personal healthcare records to prevent unauthorized access. Therefore, the security is efficiently enhanced with a natural and expressive fashion.

Keywords Cloud computing · Set-valued data · Differential privacy · Personal healthcare records · Encryption

---

E. Alexander (✉) · Sathyalakshmi
Hindustan Institute of Technology and Science, Padur, Chennai 603103, India
e-mail: elizabetha@hindustanuniv.ac.in

Sathyalakshmi
e-mail: slakshmi@hindustanuniv.ac.in

S.S. Dash et al. (eds.), *Artificial Intelligence and Evolutionary Computations in Engineering Systems*, Advances in Intelligent Systems and Computing 517,
DOI 10.1007/978-981-10-3174-8_29

1 Introduction

Cloud computing is a distributed model of computing, which makes it possible to share hosted services over the internet. The resources of servers can be shared such as data, software, and on-demand services. Cloud computing gives several updating features in IT services, which are internet-based applications or tools. When the original data set is processed, many intermediate results will be produced in data comprehensive applications. Storing of such intermediate data sets on cloud reduces the cost of recomputing them. Privacy preserving of the intermediate data set is so important because they may contain sensitive data. Attackers may retrieve sensitive information by interpreting different intermediate data sets. So privacy protection of intermediate data sets is an asserting problem.

There are many concerns related to the cloud. Among them data protection is an important cloud concerns. To provide data confidentiality, the sensitive information stored in cloud need preservation. Since the cloud servers are out of users' supervision, threats would rise fiercely. The illegal usage of user's sensitive information leaks the most valuable data. This leads the personal information's are also at risk by revealing the authentication method for retrieving data from the cloud. Sensitive personal data can also be taken away by unauthorized users. The cloud computing system should be volatile in case of security threats.

In this research, the difficulties for publishing personal healthcare records on cloud are studied. Outsourcing of set-valued data on cloud must ensure the integrity of data. Many research works and studies have been investigating to protect the data integrity of the plaintext publishing. Privacy preserving of personal healthcare data is very challenging since healthcare data contains very sensitive information. Most of the privacy-aware techniques concentrate on anonymizing personal information.

There are many techniques available for preserving the data confidentiality of personal healthcare data. But most of the existing techniques incur large overheads on data publishing and querying. Some techniques depend on attaching fake words or records [1]. These methods would cause a high, irreversible data loss. So they are not relevant in many real applications. Such QI-partitioning-based works are very few. QI-partitioning is a method of partitioning, which splits quasi-identifiers into several parts. QI-partitioning [2]-based works illustrate the partitioning of QIs into several parts. Most of these works focus on *k-anonymity* method. Data querying is as important as data publishing. Data querying techniques must be powerful to prevent privacy attacks. There are many techniques for privacy protection while querying. Privacy preserving via differential privacy is comparatively good for data querying. Differential privacy can provide protection against background knowledge attacks. Differential Privacy model satisfies various needs of users having different roles. The agreements between data owner and authorized users are reflected by user roles. Data owners are those who provide distinct coarse-grained results for users with various roles. This is similar to fine-grained access control. Data owners give different results even if they face same data for the same data query.

A comprehensive, secure framework needs to be implemented to address the crucial problems occurring while publishing of personal healthcare data on the cloud. This framework explicitly considers the security of data while publishing and querying. This privacy preserving personal healthcare model includes two phases: data publishing and data querying. The data publishing phase focuses on a data partition technique called EQI-partitioning. In this modeling, EQI-partitioning strategy splits combinations among terms. It guarantees both data integrity and data confidentiality. Explicitly, k-anonymity is used as the base for privacy preserving. The information of each released records in the published table cannot be identified from at least $k - 1$ records if k records have the same information in the published table. Then such anonymized dataset is k-anonymous.

EQI-partitioning of personal healthcare data can be explicitly measured by data integrity parameters. This privacy metrics expresses the requirement that any quasi-identifier combinations, describing less than $k - 1$ record should be split into several chunks so that each one satisfies *k-anonymity*. Thus, it provides confidentiality and availability constraints of data. Another security metric of this framework is encryption. Here, only the sensitive information of personal healthcare data is encrypted, which provides additional privacy protection. Interactive differential privacy is employed on data querying phase to protect the data privacy. The other existing data anonymization techniques are liable to threats based on background information and it is different from noninteractive differential setting. In noninteractive differential method, noises are attached to the published real data. But in the interactive differential privacy, noise is added into the results of data query. It gives a powerful privacy mechanism that is independent of an attacker's background knowledge. In this research, the privacy assurance of the proposed method is proved. The privacy assurance of the proposed research work is justified rigorously. The validity and the practicality of the proposed model are demonstrated by pervasive experiments on cloud with real data set. The contributions of the research are summarized as follows:

(1) This is the research that characterizes the privacy-aware set-valued data publishing on cloud for personal healthcare records, and provides a complete system model. A generic data partition strategy that divides EQI into different chunks is proposed. This system assures that the sensitive data will not be disclosed to the public. In addition, an interactive differential privacy setting is employed into the proposed system that contributes powerful security protection.

(2) AES is employed as the essential primitive for encryption. Substitution and permutation techniques are used in the AES encryption method. This encryption is done not in a single round, but it is done in a described number of rounds. Thus, AES provides additional security in the proposed framework.

(3) The proposed scheme is implemented and gained limited data loss by testing the system over cloud on real datasets.

The rest of the paper is systematized as follows: Sect. 2 analyzes related works. Section 3 explains the problems identified the system model, and the security metric. Section 4 introduces the explanation of privacy-aware data publishing for personal healthcare records and gives a fair analysis of privacy assurance. Section 5 evaluates the quality of data and the performance of the model and Sect. 6 remarks the conclusion and future work.

2 Related Works

Privacy preserving of set-valued data for personal healthcare records has been studied and widely accepted as an important issue in data privacy protection studies. Latanya Sweeney [3] proposes an efficient algorithm for anonymizing set-valued data using *k-anonymity.* [4] introduces the EQI-partitioning, that is employed and continued in this study. The authors provide algorithms for achieving EQI-partitioning using k^m*-anonymity.* [5] studies the differential privacy strategy, which is used in this proposed work. The authors propose differential privacy strategy which is achieved by the help of taxonomy trees which is context-free. They show that high utility is maintained by using differential privacy. Ningui Li et al. [6] proposes a privacy technique for anonymizing set-valued data, named t-closeness. [1] explains about another data anonymization procedure named *l*-diversity. The authors describe the data is not generalized or suppressed in *l*-diversity technique. But the original data are disassociated by publishing the data separately to hide the combinations among them. Manolis Terrovitis et al. [7] proposes k^m-anonymity which focuses on generalization. The authors introduce algorithms such as optimal anonymization algorithm, direct anonymization heuristic for privacy preserving of set-valued data. Reference [2] introduces a framework for the secure access of personal healthcare records on cloud. The authors describe that privacy protection and secured access control is achieved using ABE method. Reference [8] uses RSA and AES algorithm to address the security in access control systems which are cryptographic role-based for preserving privacy of data storage on cloud.

3 Problem Statement

The proposed data anonymization model concentrates on privacy preserving personal healthcare data publishing on cloud. In this section, some background details on set-valued data, private cloud and personal healthcare records are briefly explained. In addition to these details, the security measures of this model are defined.

3.1 Background Details

Set-Valued Data Set-valued data have been generally in the form of collection of transactions such as shopping transactions, web query logs. A set of values in set-valued data are associated with private information. Since the set-valued data may contain sensitive information, the users can be easily identified by mining the set-valued data. In this research, the common case in which any term that might disclose private information and any term which can be used as a part of quasi-identifier for identifying a particular user are considered.

Private Cloud Private cloud is an important cloud storage model. It provides opportunity for secure data publishing. A single organization operates on this cloud infrastructure and a trusted third-party manages this internally and hosts it either internally or externally. In this work, the data owners partition the original records into different chunks. Then they send the partitioned chunks to private cloud after anonymization.

Personal Healthcare Records Personal healthcare records contain the health related data and information of patients and it is maintained by the patient. Personal healthcare records provide so many benefits for both patients and physicians. PHR provides a patient's medical history, such as personal details, diet plans, diagnosis list, medications, allergies, immunization, etc. PHR is not same as electronic medical record that is maintained by hospitals. Physicians can use PHR for giving better treatment to patients. In this work, the data owner partitions PHR into general identifiers, quasi-identifiers and sensitive information and they send the anonymized data to private cloud.

3.2 System Overview

A high-level model for the proposed system is described, as depicted in Fig. 1 that includes three elements: data owner, authorized users, and private cloud.

Data Owner The data owner can be an institution or a specific person. The data owner is those who save personal healthcare records, which are to be published to the private cloud.

Private Cloud Private cloud is the central part of the system that controls the system. The data owner set up the private cloud and the real personal healthcare dataset is preprocessed before publishing. The connection between the private cloud and the data owner is privacy protected. In this system, EQI-partitioning method is used to split the original dataset to break the combinations among the terms which can be used to directly identify users. Finally, these sanitized data are outsourced to private cloud. In addition, AES encryption is also used to protect the private information. The differential privacy technique employed on data querying gives

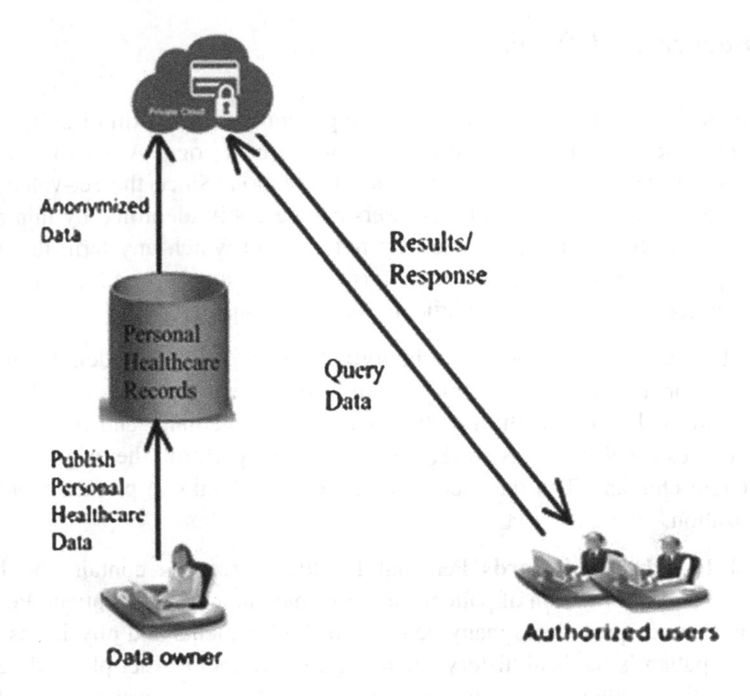

Fig. 1 The high-level system overview

more security by providing an interactive interface to users who are authorized. The privacy mechanism provided by interactive differential privacy has the capability to hide the correlations between the data owner and private cloud.

Authorized Users Authorized users, are those who are allowed to query on a published dataset. They are not responsible for the secure management of data. They can submit queries for data to private cloud and get their required results directly from private cloud. The access of data by unauthorized persons should be prevented. So Data access control methods have to be employed before data owner outsources data to private cloud.

The access of data by unauthorized persons should be prevented. So data access control methods have to be employed before data owner outsources data to private cloud. The authentication of users has been done properly in this work. When an authenticated user submits a query to the private cloud, this system links this query to the required results depend on the distribution of data in the cloud.

3.3 Security Methods

Many security methods such as *t*-closeness [6], *l*-diversity [1], and noninteractive differential privacy [5] have been proposed to protect the data from attackers. There are many techniques existing in case of set-valued data publishing. But the existing techniques would incur a high data loss which is irreversible. The techniques available for privacy preserving of personal healthcare records are not effective. So this system opts for *k-anonymity* and *AES* encryption. *k-anonymity* uses generalization and suppression.

***k*-anonymity** The information of each released records in the published table cannot be identified from at least $k - 1$ records if k records have the same information in the published table. Then such anonymized dataset is k-anonymous.

k-anonymity [3, 7] can be achieved by fragmenting any term combinations of quasi-identifiers, those can directly identify a record into different chunks. A single quasi-identifier cannot identify a record directly. But the combinations of quasi-identifiers can directly identify records. Therefore, the quasi-identifier combinations should be broken down. So in this work, the extended QI-partitioning using *k-anonymity* is adopted.

Generalization and Suppression *k-anonymity* uses generalization and suppression method to anonymize the quasi-identifiers whose combinations can directly identify records. These methods are more effective to preserve the truthfulness of the original information than other existing methods such as scrambling or swapping. The values of a field in the original dataset are substituted by more general values those can be assumed by that field. Suppression means not publishing a value at all.

4 Anonymization of Personal Healthcare Records

In this study, a privacy-aware set-valued data anonymization method based on EQI-partitioning for personal healthcare records is proposed. The proposed scheme splits the real data into several chunks. This partitioning is done vertically to hide

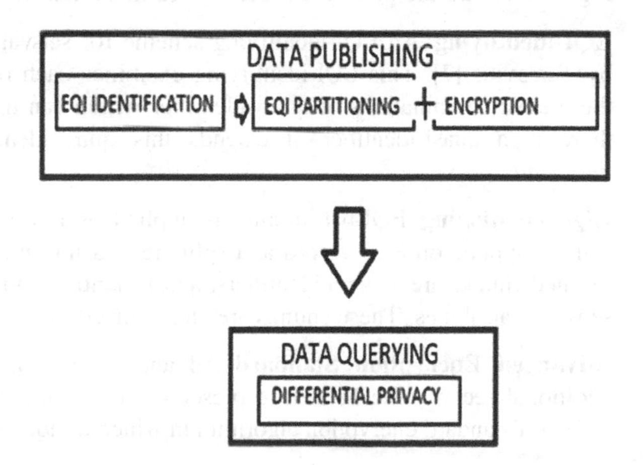

Fig. 2 Schema in PHKEM

Table 1 Original data

Personal healthcare records
{Jesse Wilson, 173, 51, Male, A+, Fever}
{Jose Diaz, 55, 180, Male, A+, Cold}
{Jeremy White, 45, 164, Male, A+, Heart attack}
{Harry Johnson, 57, 188, Male, A−, Diarrhea}
{Carlos Evans, 47, 187, Male, A−, Malaria}
{Daniel Hill, 41, 161, Male, A−, Asthma}
{Robert Carter, 56, 153, Male, O−, Food poison}
{David Nelson, 52, 123, Male, B+, Head ache}
{Claude Hutchinson, 43, 186, Female, B−, Hydrophobia}
{Kaitlin Teague, 43, 176, Female, AB+, Influenza}

the quasi-identifier combinations those can uniquely identify records. This method splits the real datasets into several chunks to shield the rare term combinations in the original records. The EQI-partitioning method used in this model can help to process big datasets parallelly. Table 1 shows the original dataset after preprocessing. Table 2 shows the partitioned data: general identifier, quasi-identifier and sensitive information. This example, contains $r1$–$r10$ where they are attacked by identifying the combinations of quasi-identifiers (age, gender, and blood group). Here, the quasi-identifiers are anonymized using *k-anonymity* where $k = 5$, which is illustrated in Table 2. This shows the sanitized dataset using generalization and suppression after partitioning.

The details of the technique are presented below. First, the basic techniques are described, which operates in two steps: EQI-identifying, and EQI-partitioning. In addition, the encryption technique AES and differential privacy are presented.

4.1 Primitive Schema

Figure 2 shows the primitive schema used in PHKEM.

EQI-identifying An EQI-identifying scheme for set-valued data is proposed first by Terrovitis [7]. This EQI-identifying examines each record to find out whether there exist combinations of quasi-identifier which can uniquely identify record. If there is a quasi-identifier, it extends this quasi-identifier and checks the *k-anonymity*.

EQI-partitioning EQI-partitioning is applied as a vertical partitioning strategy. This is applied on each record and splits record into different partitions. The partitioned chunks are general identifiers, quasi-identifiers that satisfy *k-anonymity* and sensitive attributes. These chunks are anonymized and outsourced in private cloud.

Advanced Encryption Standard Advanced Encryption Standard is used an additional security mechanism to preserve the privacy of the private information. AES is a standard encryption algorithm in which a block size of 128 bits is taken as

Table 2 Partitioned data with *k*-anonymity

General Identifier	Quasi-identifiers	Sensitive attributes
{Jesse Wilson, 173}	{>50, Person, Positive}	{Fever}
{Jose Diaz, 180}	{>50, Person, Positive}	{Cold}
{Robert Carter, 153}	{>50, Person, Negative}	{Food poison}
{David Nelson, 123}	{>50, Person, Positive}	{Head ache}
{Harry Johnson, 188}	{>50, Person, Negative}	{Diarrhea}
Carlos Evans, 187}	{>40, Person, Negative}	{Malaria}
{Daniel Hill, 161}	{>40, Person, Negative}	{Asthma}
{Jeremy White, 164}	{>40, Person, Positive}	{Heart attack}
{Claude Hutchinson, 186}	{>40, Person, Negative}	{Hydrophobia}
{Kaitlin Teague, 176}	{>40, Person, Positive}	{Influenza}

a symmetric block cipher. Since the data is encrypted, the persons who are authenticated only will be granted to get the data depends on their query. AES can use three different sizes of keys. The number of rounds of encryption is different depends on the key sizes. Since there is multiple numbers of rounds of encryption, the data on cloud is more secure.

Algorithm 1 EQI-partitioning using *k-anonymity*

Begin

Scan all the records.
Partition the records into chunks.
Split chunks into general identifiers, quasi-identifiers and sensitive attributes based on *k-anonymity*.
Begin

For each term
Test data chunk satisfies *k-anonymity*.
If exists, extends this chunk with the candidate term.
Else

Create new chunk.

End.

End

Interactive Differential Privacy Differential privacy is an anonymization method for preserving privacy of set-valued data with more guaranteed utility. It gives powerful privacy assurance independent of an attacker's background information. Two types of differential privacy settings are available. They are interactive and noninteractive. A sanitization mechanism is employed between users and the database in interactive differential privacy. Each query result is injected with a random noise. This provides privacy protection such that no adversary can infer the presence or absence of any record from the published noisy result even if they are aware of background details.

5 Evaluation

5.1 Experimental Setting

In the experiments, the performance of the system is examined. The experiment has done on cloud using personal healthcare records dataset. This dataset contains the personal details and the medical history of patients. The data are shown as general identifiers, quasi-identifiers, and sensitive attributes. The private information is selected and encrypted. The patients and doctors can view their details after decrypting using the secret key.

The anonymized form of information can only be seen by unauthorized users. The data are partitioned by a trusted authority. This trusted authority does the encryption. The personal healthcare dataset are uploaded by trusted persons/data owners.

The experiment is done by testing with users such as patients, doctors, etc. The patients can view their data only if they enter the correct secret key. The doctors can view their patient details only and they can edit only the medical details. The patients cannot edit their details.

5.2 Experimental Results

The results are analyzed using various personal healthcare records. The result graph shows the performance of the model with various real datasets compared to other existing systems using different techniques for preserving privacy of set-valued data. Figure 3 shows the overall performance in terms of efficiency.

The experiment uses different PHR datasets to test the results of partition applied. Figure 4 shows the performance in terms of execution time. The very important technique is the partition technique. The results shows that how the anonymous element *k* can affect the data partition quality.

In this research, a privacy-aware set-valued data publishing for personal healthcare records is presented. This work proposes an EQI-partitioning method in data publishing phase. EQI-partitioning is applied to each record and disassociates the records into several chunks. The AES encryption method is used to improve privacy in data publishing.

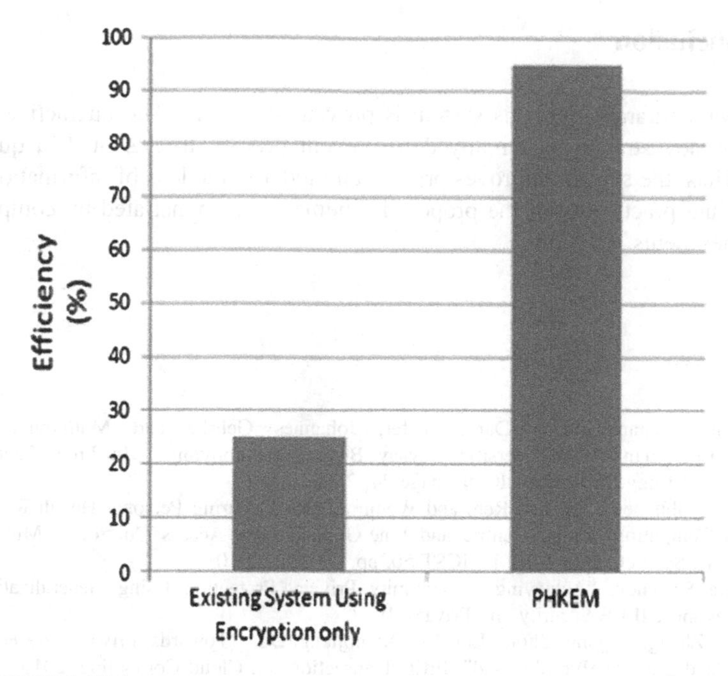

Fig. 3 PHKEM efficiency

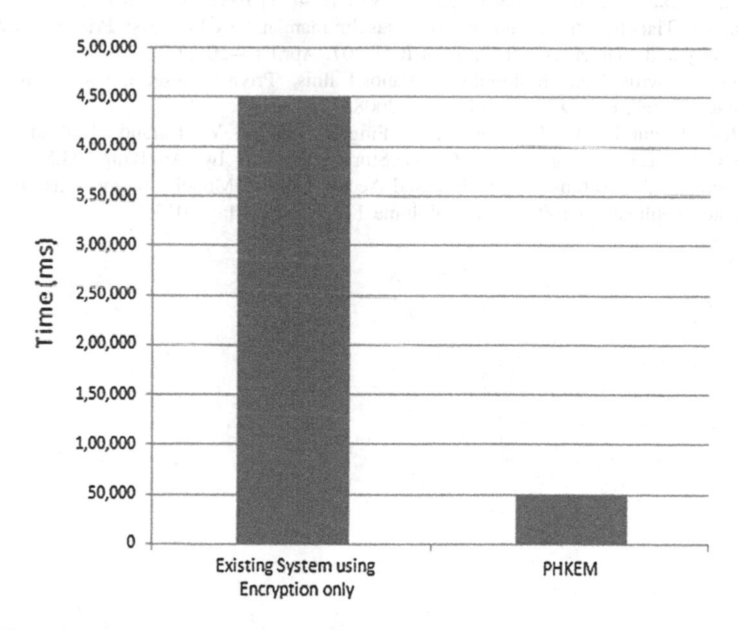

Fig. 4 Execution time performance of PHKEM

6 Conclusion

The privacy guarantee of this system is proven rigorously. The interactive differential privacy strategy is employed to prevent privacy threats in data querying phase. Thus, the system improves privacy and reduces the loss of information. The validity and practicality of the proposed scheme are demonstrated by comprehensive experiments.

References

1. Ashwin Machanavajjhala, Daniel Kifer, Johannes Gehrke and Muthuramakrishnan Venkitasubramaniam, "*l*-Diversity: Privacy Beyond *k*-Anonymity", In Proc. 22nd Intnl. Conf. Data Engg. In Proc. of ICDE, page 24, 2006.
2. Ming Li, Shucheng Yu, Kui Ren, and Wenjing Lou, "Securing Personal Health Records in Cloud Computing: Patient-Centric and Fine-Grained Data Access Control in Multi-owner Settings", SecureComm 2010, LNICST 50, pp. 89–106, 2010.
3. Latanya Sweeney, "Achieving k-anonymity Privacy Protection Using Generalization and Suppression", IEEE Security and Privacy 1998.
4. Hongli Zhang, Zigang Zhou, Lin Ye, Xianojiang Du, "Towards Privacy Preserving of Set-valued Data on Hybrid Cloud", IEEE Transactions on Cloud Computing, 2015.
5. Rui Chen, Noman Mohammed, Benjamin C. M. Fung, Bipin C. Desai, Li Xiong, "Publishing Set-Valued Data via Differential Privacy", PVLDB, 4(11):1087–1098, 2011.
6. Ningui Li, Tianchengh Li, Suresh Venkatasubramanian, "*t*-Closeness: Privacy Beyond *k*-Anonymity and *l*-Diversity", In Proc. of ICDE'07, April 15–20, 2007.
7. Manolis Terrovitis, Nikos Mamoulis and Panos Kalnis, "Privacy-preserving Anonymization of Set-valued Data", PVLDB, 1(1):115–125, 2008.
8. Bokefode Jayant D., Ubale Swapnaja A, Pingale Subhash V., Karande Kailash J., Apate Sulabha S, "Developing Secure Cloud Storage System by Applying AES and RSA Cryptography Algorithms with Role based Access Control Model", International Journal of Computer Applications (0975–8887) Volume 118-No. 12, May 2015.

A Decimal Coded Genetic Algorithm Recommender for P2P Systems

C.K. Shyamala, Niveda Ashok and Bhavya Narayanan

Abstract Research has witnessed a wide range of trust-based recommendation systems especially for P2P networks considering the open nature of networks. The recommendation systems have been devised with parameters, factors, and techniques to mitigate attacks on the recommendations. Though the ground functionalities are the same (trust management, reputation querying, recommendation filtering, and aggregation of fair recommendations) each of them have come up with different models which are unique in their own way. An important thread of research focuses on trust models for P2P systems based on the rule-based anomaly detection using genetic algorithm, evolving recommendations using genetic programming, etc. While, attempts have been made using binary coded GA, a decimal coded GA has not received attention so far. A decimal coded GA consumes less memory space and it gives accurate results. Our work focuses on evolving a self-organizing trust model for P2P systems using decimal coded genetic algorithm for mitigating service-based and recommendation-based attacks. A detailed discussion on responsiveness of P2P trust models to particularly malicious peer behavior, brings out the feasibility of the proposed model.

Keywords Trust · Recommendation · Reputation · Genetic algorithm · Decimal coded

1 Introduction

Peer-to-peer systems are gaining immense attention in recent days due to their potential of being deployed in versatile applications. Today due to the cheaper internet rates and high bandwidth availability, the commodity computers in most homes are scaled to P2P networks to facilitate increased resource sharing [17].

C.K. Shyamala · N. Ashok (✉) · B. Narayanan
Department of Computer Science and Engineering, Amrita School of Engineering, Amrita Vishwa Vidyapeetham, Amrita University, Coimbatore 641112 India
e-mail: nivedaashok@gmail.com

S.S. Dash et al. (eds.), *Artificial Intelligence and Evolutionary Computations in Engineering Systems*, Advances in Intelligent Systems and Computing 517,
DOI 10.1007/978-981-10-3174-8_30

These networks are not only confined to simple file sharing applications, but also find use in highly confidential military organizations. In peer-to-peer (P2P) systems, peers collaborate dynamically with other peers each time they wish to accomplish a task [11]. In pure P2P networks, each peer is autonomous and the network is highly decentralized in nature [5]. The absence of a central authority makes P2P networks as bait for maliciousness [23]. Peers can join and leave the network on their own will. This open nature of the of P2P systems poses a security threat as it creates attack opportunities for malicious peers in the network.

One obvious way to deal with these attacks is by creating a long term trust relationship among peers based on the past interactions and service histories. Trust is a social concept and often hard to measure numerically. Therefore, well-formed metrices are required in computational models to represent trust. Moreover, these metrices also have to be precise since peers are ranked based on their trustworthiness. Calculating trust in service context is based on the direct interaction with peers, whereas in recommendation context it differs. In the later context, there is a high chance that the information is deceptive if the recommender is malicious. This poses a challenge in assessing trustworthiness. Also the absence of a central authority in peer to peer systems makes them even more vulnerable. The key concern in these decentralized systems is to organize them in a fashion that enables them to store and manage trust information securely.

Having analyzed the problems in peer-to-peer systems and the extent to which solutions have been suggested by existing models, we propose a new model based on the genetic algorithm which is more promising in countering attacks on recommendations. Genetic algorithm is an optimization technique which is designed on the basis of the biological evolution process. Recent research trends show that GA has marked is venture into multidisciplinary spheres and has proven successful [2, 6, 7, 10, 12–14, 18]. A thorough study has been done on the feasibility of employing GA for trust modeling in P2P systems as part of our preliminary work. Having collected the evidences from our initial work we extent it by designing a self-organizing trust model for P2P systems using GA. The model is developed by modifying and enhancing the existing SORT model [1] for typical file download scenario. In the proposed work recommendations provided by peers are evolved using GA to achieve fairness in recommendations and efficient attack resiliency. Our benefactions are aligned as follows: Sect. 2 vividly discusses some of the related works. Section 3 explains the proposed systems methodology. The results are analyzed and depicted in Sect. 4. Section 5 discusses the conclusion and future work.

2 Related Work

Trust modeling for P2P systems has been in focus ever since the use of P2P applications for varied purposes. To achieve efficient attack mitigation during file exchange is an important issue and has been in concern in most research. Various trust models and the state of art in trust and reputation has been discussed in [16].

The paper analyses each models uniqueness and the commonalities in all of them. The authors of Eigen Trust propose a trust-reputation model which uses two trust ratings namely the local trust and global trust for making trusting decisions. The global trust rating is the systems trust based on the past interactions (such as upload history, data exchange, etc.), where as the local trust is each nodes trust on other nodes based on the previous interactions [9]. The authors of CubiodTrust have contributed a trust model based on the global reputation for P2P networks. It considers three trust factors namely—contribution of the peer to the system, peers trustworthiness, and quality of resource upon which it builds four relations [3]. Another trust-reputation model which uses feedback mechanism to determine nodes trustworthiness is discussed [22]. Here, the feedback analysis is based on the transactions carried out between nodes. The authors of Bayesian network-based trust model provides a multidimensional view of trust scores and employ reputation for the computation of receiver and sender nodes utility values [21]. A self-organized trust model where reputation and recommendation are computed in addition to trust is put forward in [1]. A three-tier trust relationship model to counter the difficulty of establishing direct trust computation is proposed by the authors of GroupRep [20]. A system which dynamically selects a few peers that are most reputable by using a distributed ranking mechanism has been suggested by the authors of PowerTrust [24]. An AntRep model where reputations of nodes are obtained using the swarm intelligence paradigm is discussed in [16]. Another class of models makes use of the evolutionary techniques for achieving trust modeling. GenTrust is a genetic programming-based trust model for peer to peer systems [19]. Based on peers features extracted from past interactions and recommendations with another peer its trustworthiness is computed. These features are evolved using genetic programming which provides mathematical functions to measure trust values of each peer. A novel trust model which detects anomalies based on peer profiling is suggested here [18]. Trust is established for each peer by comparing its own prior activities and current activities based with other peers on historical data. To detect anomalous behavior genetic algorithm is employed here. With every transaction peer profile gets updates dynamically using GA crossover and mutation operators. A rule-based network intrusion detection using genetic algorithm has been put forward here [14]. The goal of this model is to identify suspicious network connections.

A large number of trust models based on the recommendations and reputations have been proposed in recent days. Each of these models have been successful in mitigating attacks to a fair extend. But most of them tend to fail when the network becomes highly malicious because a highly malicious network often disseminates a high amount of misleading recommendations and there is also great possibility for collusion attacks. So far there is very scant work in literature which has proved capable of withstanding such tightly coupled attack environment. The aim of our work is to design a trust model which uses GA as an optimization technique to evolve recommendations, and in turn performs intelligently in case of attacks.

3 Methodology

The aim of our work is to investigate the extent to which decimal coded genetic algorithms impacts the efficiency of recommendations offered by trust models during attacks. Malicious peers have more attack opportunities in P2P systems due to lack of a central authority. Hence, efficiency and accuracy in finding malicious peers is still a challenge for trust models in peer-to-peer systems. Focus of our work is to design a model for content distribution in P2P network (file upload and file download) using Genetic Algorithm based on Peer loyalty which is efficient in mitigating attacks. The chromosomes which form the initial population are decimal coded which makes our model more precise and accurate, thereby reducing the number of misleading recommendations and service-based attacks.

System flow diagram depicting the GA-based recommendation evolution process is outlined in Fig. 1 illustrating the overall system working in case of a typical file download scenario.

3.1 Decimal Coded GA

GA is a well-known evolutionary algorithm was initially put forward by John Holland in the 1970s [4] to mimic the human evolution process. It is often used as a

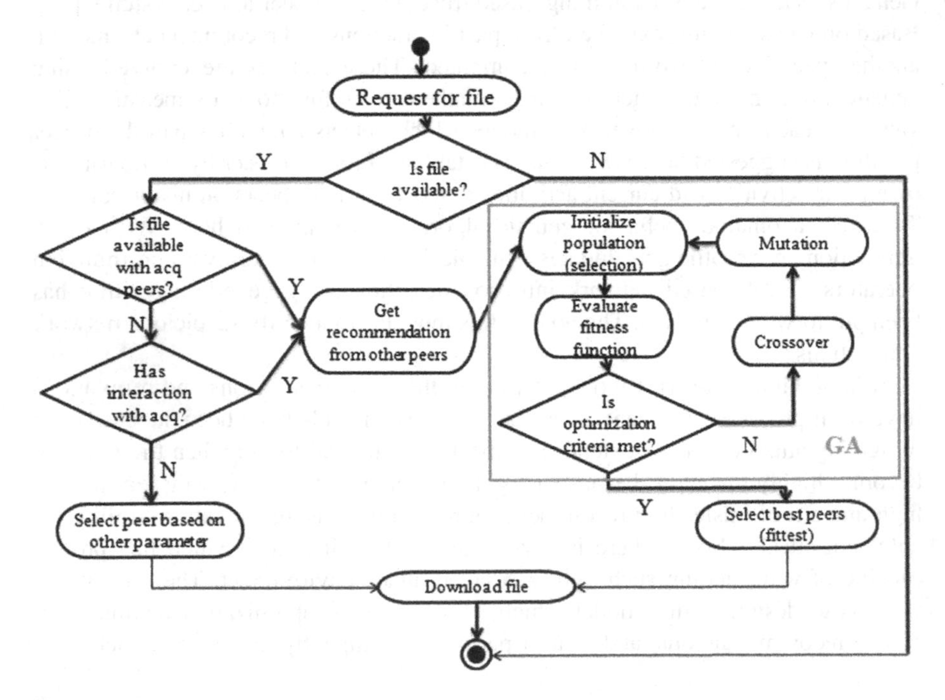

Fig. 1 System flow diagram

powerful optimization technique to provide a minimization or maximization solution for the problem in concern. The commonly used GA operators include Selection [8], Crossover [15] and Mutation [4] followed by evaluation of the fitness function. From recent research it is very evident that GA has found application in multidisciplinary spheres and has provided remarkable results [2, 6, 7, 10, 12–14, 18]. Majority of the work pertaining to GA are in its native binary form. But recently, decimal coded GA is gaining popularity due to the inherent advantages it offer such as lesser memory consumption, increased accuracy in results, etc.

From a thorough analysis carried out in our previous work it is justified that decimal coded GA optimization can be used to evolve recommendations for P2P trust modeling. The proposed system functionalities are explained in the subsequent sub sections.

3.1.1 Recommendation Manager

Recommendation Manager calculates and updates the trust values. All interactions to the peers are through recommendation manager. Recommendation Manager is responsible in handling all peer recommendations. The recommendation manager is designed based on the basic SORT model approach. The self-organizing trust model [1] proposes how the trustworthiness of a peer is evaluated on the basis of the past interactions and recommendations by another peer. Initially, SORT establishes and manages trust relationship among the peers without any prior knowledge. In SORT, the pre-existence of the trust is evaluated as strangers rather than trustworthy peers. To achieve trust relationship, peers are required to interact with each other. Peers are known as acquaintance once they provide services to other peer. The recommendations are provided by the acquaintance to other peers based on the trustworthiness of the peers. These recommendations can be the information collected from the acquaintance of the recommender peers, the experience of the recommender peers or the confidence of the recommenders on their recommendation.

Recommendation Aggregator

The key task of recommendation aggregator is in assisting the recommendation manager to manage recommendation related information based on peer interactions. It not only maintains recommendation information but also maintains service and reputation trust values. To facilitate the computation of trust three key metrices described by SORT, such as Service Trust Metric, Reputation Metric, and Recommendation Trust Metric have been used.

Service Trust is the peer's trustworthiness in service context. Service Trust is calculated by evaluating the competency belief and integrity belief on the basis of the service history information available. Competency belief is the efficiency with which the peer performs its past interactions. Integrity belief is the degree of confidence with which the future interactions are performed. The Service Trust metrics is evaluated as

$$STM_{mn} = \frac{SH_{mn}}{SH_{max}}\left(SCB_{mn} - \frac{SIB_{mn}}{2}\right) + \left(\frac{1 - SH_{mn}}{SH_{max}}\right) \times R_{mn} \quad (1)$$

where STM_{mn} denotes the P_m's service trust value about P_n, SH_{mn} denotes the Size of P_m's service history with P_n, SH_{max} represents the Upper bound of service history size, SCB_{mn} denotes the Competency belief of P_m about P_n in service context, SIB_{mn} denotes the Integrity belief of P_m about P_n in service context.

Reputation describes the trust of one peer on the other peers based on the experiences. Reputation Metrics is calculated as

$$RM_{mn} = \frac{\lfloor \mu_{mn} \rfloor}{SH_{max}}\left(PCB_{mn} - \frac{PIB_{mn}}{2}\right) + \frac{SH_{max} - \lfloor \mu_{mn} \rfloor}{SH_{max}} PR_{mn} \quad (2)$$

where RM_{mn} denotes the Reputation value of P_m about P_n, μ is referred to as the average level of first-hand experience of P_m's acquaintance, SH_{max} represents the Upper bound of service history size, PCB_{mn} is the estimated Competency belief of P_m about P_n, PIB_{mn} denotes the estimated Integrity belief of P_m about P_n, PR_{mn} represents the estimated Reputation.

Recommendation Trust Metrices is evaluated for choosing the acquaintance for the reputation queries and for recommendation evaluation. Recommendation trust is evaluated based on the past recommendation interactions and reputations. The Recommendation Trust value is evaluated as

$$RTM_{mo} = \frac{RH_{mo}}{RH_{max}}(RCB_{mo} - RIB_{mo}/2) + \frac{RH_{max} - RH_{mo}}{RH_{max}} R_{mo} \quad (3)$$

where RTM_{mo} denotes the Recommendation Trust value of P_m about P_o, RH_{mo} denotes the Recommendation history of P_m with P_o, RH_{max} refers to the Upperbound of recommendation history size, RCB_{mo} denotes the Competency belief of P_m about P_o in recommendation context, RIB_{mo} denotes the Integrity belief of P_m about P_o in recommendation context, RT_{mo} is the Reputation value of P_m about P_o.

3.2 Recommendation Filter

The recommendation filter is responsible for eliminating unfair recommendation with the help of decimal coded GA which in turn results in reducing the impact of misleading recommendations. The proposed model uses the SORT [1] parameters from Eqs. (1)–(3) in order to define the population. The chromosomes of the population are initialized with 10 real values in the range [0, 1] as depicted in the Fig. 2.

zSHmn	zSHMAX	SCBmn	SIBmn	Rmn	zRHmn	zRHMAX	RCBmo	RIBmo	Rmo

Fig. 2 Population parameters

The parameters such as SH_{mn}, SH_{max}, RH_{mn} and RH_{max} are normalized as zSH_{mn}, zSH_{max}, zRH_{mn} and zRH_{mn} to fall in the range [0, 1].

The fitness function is formulated as:

$$\text{Fitval} = \text{STM} + \text{RTM} \in [0, 2] \tag{4}$$

Initially, the fitness function is evaluated for each of the recommender peers which is part of the population. This is followed by a roulette wheel selection after which selection probability and actual count is derived for each of the selected parent recommender peer. The parents peer now undergoes a whole arithmetic recombination to produce new off springs [4]. This results in every pair of parents peers producing two new identical offspring's. In order to bring in diversity in the population a single bit mutation operation is performed on offspring peers randomly. The fitness function is evaluated at the end of each generation and those peers with Fitval lower than average initial population fitness fail to qualify for the next generation. The fitness value of the last generation is used to carry out a threshold mapping with the initial population. The closest match is chosen as the recommender peer.

3.3 *Ranker*

Having chosen a recommender peer, the ranker now carries out ranking of possible service providers on the basis on the service trust values. In case of equal service trust values, the service histories are compared to find the peer with more direct experience. If SHnm values are also found equal then SCB-SIB/2 values are compared again for greater value. If again the values are found equal the peer with greater competency belief is chosen. Still if the values are equal the peer with larger upload bandwidth is selected. If the tie still prevails one among the equal peers is randomly chosen.

Figure 3 gives the overall architecture of the proposed system. The downloader peer initially queries other peers for the file it wishes to download. Based on the response, recommendation about these peers is requested to the acquaintance peers. This is done by recommendation system on behalf of the downloader peer. The system aggregates the recommendations received, filters unfair recommendations and ranks service providers based on their service trust values. Finally, the file it downloaded from the service provider peer with highest service trust value.

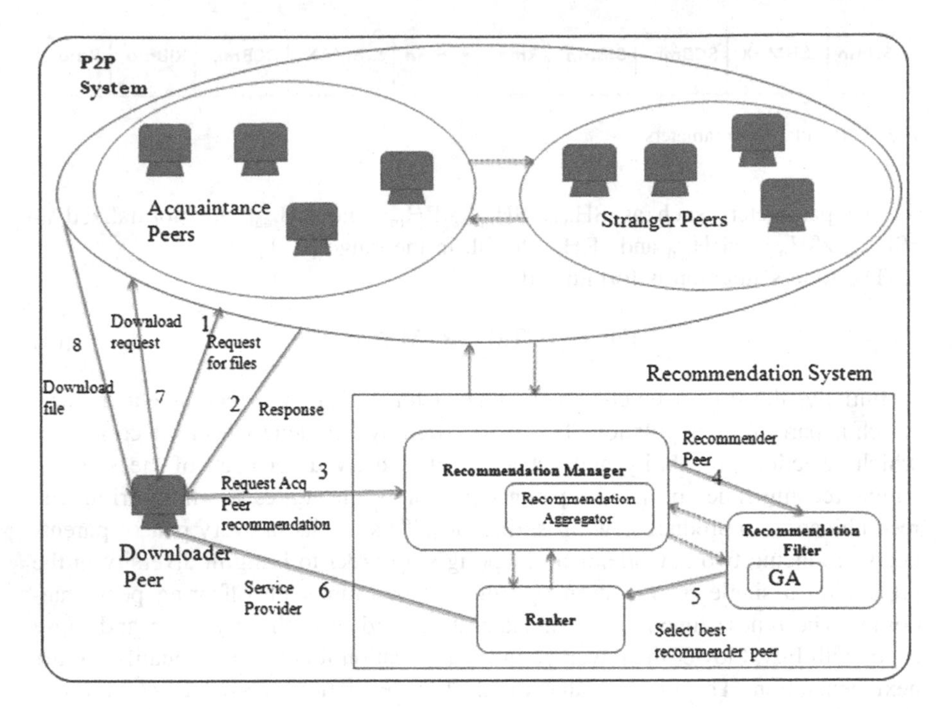

Fig. 3 Decimal coded GA model

4 Results and Analysis

4.1 Simulation

In order to show the efficiency of the designed decimal coded GA model two cases are considered. In the first case a number of recommendation values offered by recommenders based on their transactions at various time intervals are considered. In this scenario, the choice of a recommender peer is solely done based on recommendation trust values, no matter if values vary sharply. In the second case, a GA-based recommender selection is carried out which uses the fitness value to select the recommender. The fitness value takes account of both the service trust as well as recommendation trust values.

In Case 1 the choice of recommender is solely based on the recommendation trust values and hence often the peer with the highest recommendation trust value is chosen as the recommender. This may not always give promising results since malicious peers can induce misleading recommendations to the system. It can be observed from Fig. 4 that a sudden increase in recommendation trust of peer2 makes trusting decisions difficult. But in Case 2 the decimal coded GA evolutionary technique ensures that only the fittest individual make it to the final generation. Also a peer has to not only been a good recommender but also a decent service

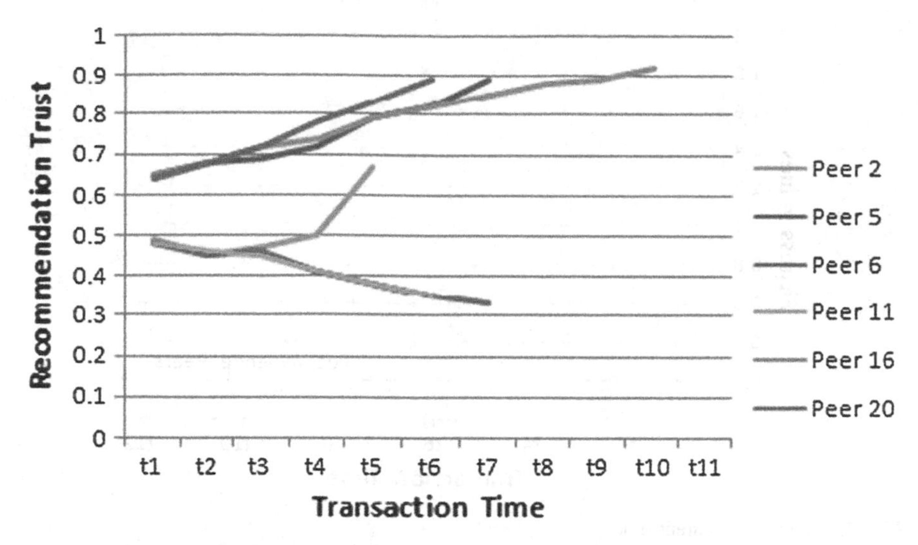

Fig. 4 Peer recommendation trust graph

provider to prove fit. The scatter plot in Fig. 5 shows the selection of fittest parent individuals in the first generation which are evolved to produce the best off springs in the final generation as shown in Fig. 6. The final generation witnesses 5 off springs which satisfy the fitness criteria from 10 generated off springs.

Case 1: Table 1.

Case 2: Table 2.

4.2 A GA Analogy

The GA-based evolution mechanism can be portrayed interestingly using a simple comparison process. Consider, the concept of dominant and recessive alleles where the former gene often suppresses the later during evolution. In order to explain the whole ideas let us take the reproduction case of a plant species, where in Tall (T) and Green (G) are dominant and short (t) and yellow (g) are recessive traits.

Generation 1

	TG	Tg	tG	tg
TG	TG	TG	TG	TG
Tg	TG	Tg	TG	Tg
tG	TG	TG	tG	tG
tg	TG	Tg	tG	tg

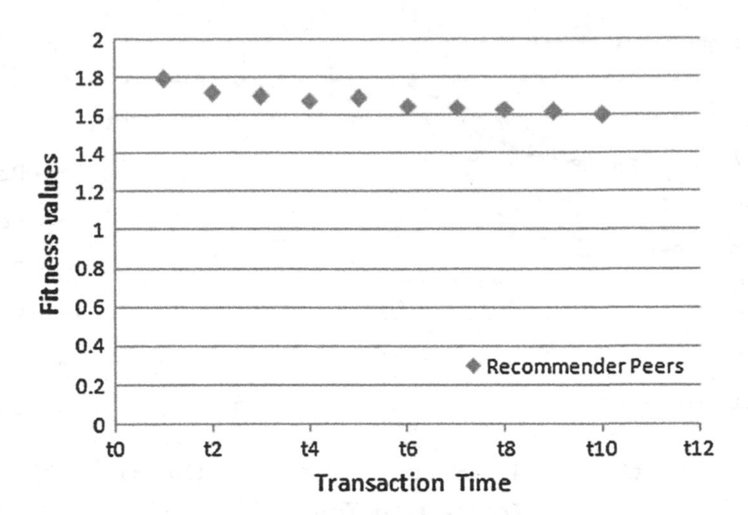

Fig. 5 Generation1 parent fitness

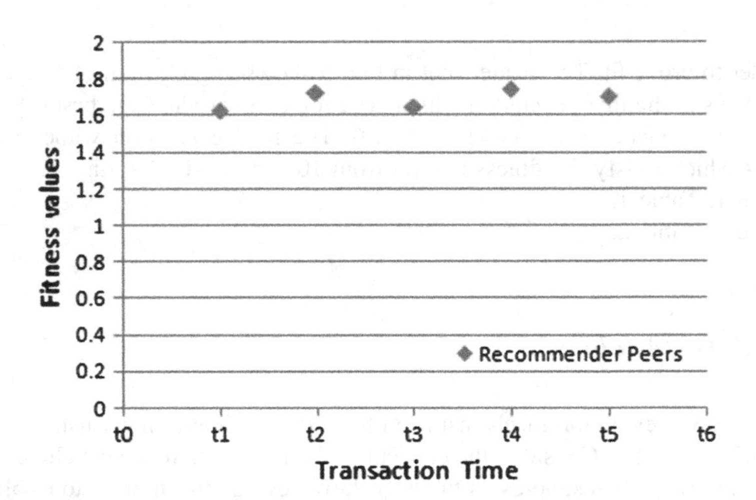

Fig. 6 Generation100 offspring fitness

Obtained probabilities:

$$\mathrm{TG} = 9/16 \quad \mathrm{Tg} = 3/16 \quad \mathrm{tG} = 3/16 \quad \mathrm{tg} = 1/16$$

It is understood from the evolution process that height and color are synonymous to the service trust and recommendation trust. In the proposed system, the dominant traits being High service trust and high recommendation trust where as the recessive

Table 1 Recommender peer selection based on recommendation values

Recommender peers	Interaction based recommendation values									
1	0.54	0.54	0.6	0.59	0.55	0.62	0.66	0.68		
2	0.49	0.45	0.47	0.5	0.67					
3	0.52	0.56	0.58	0.64	0.66					
4	0.33									
5	0.65	0.68	0.69	0.72	0.79	0.82	0.89			
6	0.48	0.45	0.46	0.41	0.38	0.35	0.33			
7	0.42	0.39	0.35	0.4	0.33	0.29				
8	0.52	0.58	0.59	0.65	0.67	0.66				
9	0.59	0.62								
10	0.57	0.59	0.65							
11	0.48	0.46	0.45	0.41	0.38	0.35				
12	0.6									
13	0.65	0.67	0.68	0.72	0.75	0.78	0.79	0.82		
14	0.47	0.7								
15	0.52	0.59	0.56	0.58	0.6					
16	0.65	0.68	0.72	0.74	0.79	0.82	0.85	0.88	0.89	0.92
17	0.45	0.42								
18	0.56									
19	0.54	0.59	0.62	0.63						
20	0.64	0.68	0.72	0.78	0.83	0.89				

trait being low service and recommendation trust. This clearly justifies the application of GA for peer selection during file download based on trust relationships giving higher chance of selection to the fittest peers. Moreover, those peers with a Fitval lesser than the average fitness of the initial population get eliminated after each generation.

Carrying forward the same example, the GA-based peer selection is improved when crossover and mutation is done for a large number of generations. Though the fittest offspring peers in the final generation may not be found in the initial population, it doesn't invalidate our results. A closeness computation factor is considered to make a choice of the best from the initial population which closely resembles any of the final generation offspring. Thus our system achieves fairness in peer selection by incorporating an imitation of natural evolution process using GA.

Table 2 Recommender peer selection based on fitness value

Recommender peers	Interaction based fitness values									
1	1	1.01	1.1	1.09	1.05	1.21	1.25	1.26		
2	0.62	0.73	0.75	0.77	0.5					
3	1.15	1.22	1.18	1.2	1.12					
4	0.7									
5	1.5	1.54	1.55	1.53	1.6	1.59	1.62			
6	0.8	0.82	0.83	0.84	0.86	0.88	0.9			
7	0.6	61	0.64	0.59	0.64	0.55				
8	1.22	1.34	1.35	1.4	1.39	1.43				
9	1.16	1.24								
10	1.25	1.23	1.3							
11	0.68	0.69	0.72	0.74	0.76	0.79				
12	0.6									
13	0.65	0.67	0.68	0.69	0.71	0.74	0.77	0.79		
14	0.6	0.5								
15	1.09	1.19	1.2	1.25	1.28					
16	1.5	1.56	1.57	1.59	1.62	1.64	1.68	1.69	1.7	1.79
17	0.7	0.72								
18	1.16									
19	1.14	1.19	1.21	1.29						
20	1.56	1.62	1.63	1.65	1.68	1.72				

5 Conclusion

The proposed system successfully models recommendation evolution having adapted GA as an optimization technique. Our system achieves adequacy over binary representation as a improved decimal coded representation is proposed. This also increases the accuracy in precision as opposed to binary coding where the only possible values are 0 or 1. Less memory is required as floating-point internal computer representations can be used directly, thereby improving memory usage greatly. Moreover, the system provides greater freedom to use different genetic operators. Overall the system succeeds in attaining better resilience to common attacks. As a future work we would like to investigate the extent to which our system withstands some toughest forms of attacks like Sybil attacks, collusion attacks, etc. and provide further improvisation as needed.

References

1. Can, Ahmet Burak, and Bharat Bhargava. "Sort: A self-organizing trust model for peer-to-peer systems." *Dependable and Secure Computing, IEEE Transactions on* 10.1 (2013): 14–27.
2. Chen, Buhua, et al. "A Fast Parallel Genetic Algorithm for Graph Coloring Problem Based on CUDA." *Cyber-Enabled Distributed Computing and Knowledge Discovery (CyberC), 2015 International Conference on*. IEEE, 2015.
3. Chen, Ruichuan, et al. "CuboidTrust: a global reputation-based trust model in peer-to-peer networks." *Autonomic and Trusted Computing*. Springer Berlin Heidelberg, 2007. 203–215.
4. Eiben, Agoston E., and James E. Smith. *Introduction to evolutionary computing*. Vol. 53. Heidelberg: springer, 2003.
5. Foster, I., &Kesselman, C. (Eds.). *The Grid 2: Blueprint for a new computing infrastructure*. Elsevier, 2003.
6. Gang, L., & Chun-ling, H. Research on Recommender System Based on Ontology and Genetic Algorithm. *Neurocomputing,* 2015.
7. Hao, Jin-Xing, et al. "A genetic algorithm-based learning approach to understand customer satisfaction with OTA websites." *Tourism Management*48 (2015): 231–241.
8. Jebari, K., &Madiafi, M. . Selection methods for genetic algorithms. *International Journal of Emerging Sciences*, *3*(4), 2013.
9. Kamvar, Sepandar D., Mario T. Schlosser, and Hector Garcia-Molina. "The eigentrust algorithm for reputation management in p 2p networks."*Proceedings of the 12th international conference on World Wide Web*. ACM, 2003.
10. Karayer, Erdem, and Muge Sayit. "A path selection approach with genetic algorithm for P2P video streaming systems." *Multimedia Tools and Applications* (2015): 1–19.
11. Kshemkalyani, Ajay D., and Mukesh Singhal. *Distributed computing: principles, algorithms, and systems*. Cambridge University Press, 2011.
12. Kumari, Sweta, and Shashank Pushkar. "A Genetic Algorithm Approach for Multi-criteria Project Selection for Analogy-Based Software Cost Estimation."*Computational Intelligence in Data Mining-Volume 3*. Springer India, 2015. 13–24.
13. Lee, Giyoung, et al. "A Genetic Algorithm-Based Moving Object Detection for Real-time Traffic Surveillance." *Signal Processing Letters, IEEE* 22.10 (2015): 1619–1622.
14. Li, Wei. "Using genetic algorithm for network intrusion detection." *Proceedings of the United States Department of Energy Cyber Security Group* (2004): 1–8.
15. Magalhães-Mendes, J. A comparative study of crossover operators for genetic algorithms to solve the job shop scheduling problem. *WSEAS transactions on computers*, 2014, *12*(4), 164–173.
16. Mármol, Félix Gómez, and Gregorio Martínez Pérez. "State of the art in trust and reputation models in P2P networks." *Handbook of Peer-to-Peer Networking*. Springer US, 2010. 761–784.
17. Marti, Sergio. "*Trust and reputation in peer-to-peer networks*. Diss. Stanford University, 2005.
18. Selvaraj, Chithra, and Sheila Anand. "Peer profile based trust model for P2P systems using genetic algorithm." *Peer-to-Peer Networking and Applications*5.1 (2012): 92–103.
19. Tahta, Ugur Eray, Sevil Sen, and Ahmet Burak Can. "GenTrust: A genetic trust management model for peer-to-peer systems". *Applied Soft Computing*34 (2015): 693–704.
20. Tian, Huirong, et al. "A group based reputation system for P2P networks." *Autonomic and trusted computing*. Springer Berlin Heidelberg, 2006. 342–351.
21. Wang, Yong, et al. "Bayesian network based trust management." *Autonomic and Trusted Computing*. Springer Berlin Heidelberg, 2006. 246–257.
22. Xiong, Li, and Ling Liu. "Peertrust: Supporting reputation-based trust for peer-to-peer electronic communities." *Knowledge and Data Engineering, IEEE Transactions on* 16.7 (2004): 843–857.

23. Yang, Yu, and Lan Yang. "A survey of peer-to-peer attacks and counter attacks." *Proceedings of the International Conference on Security and Management (SAM)*. The Steering Committee of The World Congress in Computer Science, Computer Engineering and Applied Computing (WorldComp), 2012.
24. Zhou, Runfang, and Kai Hwang. "Powertrust: A robust and scalable reputation system for trusted peer-to-peer computing." *Parallel and Distributed Systems, IEEE Transactions on* 18.4 (2007): 460–473.

Hybrid Paradigm to Establish Accurate Results for Null Value Problem Using Rough—Neural Network Model

Aishwarya Asesh and B.K. Tripathy

Abstract The systems in which missing values (NULL) occur are called incomplete information systems and computations on these may lead to biased conclusions. The structured difference of the datasets and importance of attributes compels us to depend on uncertainty-based approaches for finding the null values. This paper presents a hybrid approach for solving null value problems using the concepts of rough set theory and neural network. In this, complete tuple set is used for training the NN. The incomplete tuples are then tested using the model. Level of dependency is used to judge the importance of association rules [11]. Testing the dataset after reducing unwanted attributes, yields a reduced error percentage. The system produces result with better efficiency as observed by the values of accuracy, completeness, and coverage. Thus, the proposed algorithm can be suitably modified for different scenarios using the algorithm step-by-step to solve the null value problem.

Keywords Imputation · Neural network · Null value problem · Rough set theory · Training

1 Introduction

Major reasons for the incompleteness of real-world datasets may be due to lack of proper examination during entry of the data, improper manual data selection, corruption of the continuous data sectors, bad data recovery from historical sets or any such problems. As the problems are due to various reasons, the algorithm applied for solution should accommodate generalized solution strategies; otherwise

A. Asesh (✉) · B.K. Tripathy
School of Computer Science and Engineering, VIT University,
Vellore 632014, Tamilnadu, India
e-mail: a.asesh@gmail.com

B.K. Tripathy
e-mail: tripathybk@vit.ac.in

S.S. Dash et al. (eds.), *Artificial Intelligence and Evolutionary Computations in Engineering Systems*, Advances in Intelligent Systems and Computing 517,
DOI 10.1007/978-981-10-3174-8_31

erroneous results will be produced. Common methods applied to datasets for finding the missing values involves the concepts of rough sets, neural networks, interval-based and non-interval-based methods. For example, swarm intelligence method constitutes of evaluation function based on rough set theory [16]. Fuzzy neural network is also used in cases, where the dataset has a common characteristic in terms of attribute dependency. In general, rough set theory is the conceptual model for solving null value problem, but the variation in datasets adds to the restrictions of using the generalized method [22].

The proposed algorithm is divided into two stages. First, to train the model we only use the complete tuples from the dataset. The separated group of incomplete tuples is then tested to get the results. Thus the output is a set of decision rules that leads to the given decision of the rule, along with the relative and absolute error percentage. Now in the second stage, rough set theory is used to evaluate the rank factor for attributes, which gives the attribute importance. Once the decision for attribute importance is computed, we repeat the training and testing process after removing the unwanted attributes from the dataset. Thus proposed approach uses neural network for predicting the terms depending on evaluation function provided by rough set theory. The algorithm gives a better accuracy ratio compared to the generalized model using full attribute set which is reflected by the lower values of absolute and relative error rates in determining the decision class. Thus we get a better performance measure using above process, with respect to reliability and accuracy. The proposed algorithm steps can be applied for dealing with missing value problems in different domains using feature selection [12].

2 Related Work

Basic concepts of data mining and rough set theory govern the knowledge discovery in databases [20]. For example, the normal feature selection is based on the relative reduct concept of the Rough Set Theory [7]. Training of NN is done based on the present attributes in the dataset [9]. It can only be used in cases where the number of attributes is more than a certain limit. A method for estimating missing values in relational databases is based on auto clustering technique [3]. But automatic clustering technique requires a certain strong relation to be present amongst the attributes of a dataset otherwise it may give erroneous results [5]. A fuzzy concept learning system (FCLS) to build fuzzy decision trees using relational database structures and generate fuzzy rules from those fuzzy decision trees is an advanced approach for the solution of null value problem. Based on the generated fuzzy rules, estimation of null values in relational database is done [4]. Imputation of missing data is done using various classifiers such as Naive—Bayes, CLIP4, C4.5. Such techniques are well defined in cases where data is present as a group of Target Domain [18]. Learning Bees algorithm is based on random solution approach. After searching neighborhood sites and evaluating fitness, the fittest version of the solution is taken from each site [1]. Modern methods such as iterative imputation method (missForest) based on random forest are implemented on mixed data type databases [18]. Rough set ADR signaling methods are useful when spontaneous dataset changes occur and values are needed to be replaced in real time [13]. However, the accuracy is compromised in such cases. A method is also proposed for single imputation based on locally linear reconstruction (LLR) that improves the prediction of supervised learning in case of smaller datasets [10]. With this background, we propose an illustrative technique in this paper which is efficient in handling the null value problem as well as attribute reduction or feature selection.

3 Definitions and Notations

3.1 *Rough Set Theory*

Rough set theory was established by Z. Pawlak, in the year 1982. This theory governs the fundamental principle for artificial intelligence and reasoning sciences including all the decision-making aspects of expert systems. The concept is based on bound value tables. Lower bound values denote that the object's chances of belonging to a specific target class are certain. Upper bound values contain other points in the whole set which cannot be classified as belonging to the target set with a definite certainty. The classification of object as a part of the regular set or its complement or none is decided by boundary-line elements. Thus, we find that the uncertainty in relationships is not measured precisely by the membership values of properties but with the help of boundary value limits of a function. Crisp set is defined when the boundary value region is null, else any non-null values in boundary explains that the set is a rough set [15]. The concepts of rough set attribute reductions are basically used to deal with rule induction and feature selection problems. The accuracy in results and reduction of computational efforts using rough set theory is very practical, especially when dealing with data from large repositories. The ability to reduce unwanted attributes is one of the most important features of rough set theory [2].

3.1.1 Rough Set and Approximations

Here we take two finite, non-empty sets U and A, where U is the *universe*, and A is a set of *attributes*. With every attribute $a \in A$ we associate a set V_a, of its *values*, called the *domain* of a. Any subset B of A determines a binary relation $I(B)$ on U, which is called an *indiscernibility relation*, and is defined as follows [14]:

$x\ I(B)\ y$ if and only if $a(x) = a(y)$ for every $a \in A$,

where $a(x)$ denotes the value of attribute a for element x.

Here $I(B)$ is an equivalence relation. The family of all equivalence classes of $I(B)$, i.e., partition determined by B, is denoted by $U/I(B)$, or U/B; an equivalence class of $I(B)$, i.e., block of the partition U/B, containing x is denoted by $B(x)$.

If (x, y) belongs to $I(B)$ then x and y are B-indiscernible. Equivalence classes of the relation $I(B)$ are referred to as B-elementary sets.

The B-lower and B-upper approximations of a set X can be defined as follows:

$$B_*(X) = \{x \in U : B(x) \subseteq X\},\ B^*(X) = \{x \in U : B(x) \cap X \neq \phi\}, \text{respectively.}$$

The set $BN_B(X) = B^*(X) - B_*(X)$ is referred to as the *B-boundary region* of X.

If the boundary region of X is the empty set, i.e., $BN_B(X) = \phi$, then the set X is *crisp* (*exact*) with respect to B; in the opposite case, i.e., if $BN_B(X) \neq \phi$, the set X is to as *rough* (*inexact*) with respect to B.

Rough set can be also characterized numerically by the following coefficient [15]

$$\alpha_B(X) = \frac{|B_*(X)|}{|B^*(X)|}$$

called *accuracy of approximation*, where $|X|$ denotes the cardinality of X. Here $0 \leq \alpha_B(X) \leq 1$. If $\alpha_B(X) = 1$, X is *crisp* with respect to B, or else, if $\alpha_B(X) < 1$, X is *rough* with respect to B.

3.1.2 Dependency of Attributes

Dependency can be defined in the following way [21]. Let D & C be subsets of A.

D depends on C in a *degree k* ($0 \leq k \leq 1$), denoted $C \Rightarrow_k D$, if $k = \gamma(C, D)$.

If $k = 1$, *D depends totally* on C, and if $k < 1$, *D depends partially* (in *degree k*) on C.

If *D depends in degree k*, $0 \leq k \leq 1$, on C, then

$$\gamma(C, D) = \frac{|POS_C(D)|}{|U|}, \text{ where } POS_C(D) = \bigcup_{X \in U/I(D)} C_*(X)$$

The expression $POS_C(D)$, called a *positive region* of the partition U/D with respect to C, is the set of all elements of U that can be uniquely classified to blocks of the partition U/D, by means of C [15].

3.1.3 Rule Discovery

Decision-making in data computation is based on attributes present in the dataset. As all the attributes are not equally important for decision-making process, the core generation algorithm can be used to reduce the unwanted attributes. But the removal of attributes is a sensitive case, as any important decision classifiers must not be omitted. Rules that govern the dataset with low level of dependency are considered important.

Core Generating Algorithm [8]
Input: Decision table T (C, D). Output: Core attributes.
(1) Core $\leftarrow \phi$
(2) For each attribute A $\in$ C
(3) If Card ($\prod$ (C − A + D)) $\neq$ Card ($\prod$ (C − A))
(4) Then Core = Core U {A}
where C is the set of condition attributes, and D is the set of decision attributes.

Two Fundamental laws for rule importance can be defined as [12]:

Property 1. If a rule is generated more frequently across different rule sets, we say this rule is more important than other rules.
Property 2. Rule Importance Measure

$$\text{Rule Importance Measure} = \frac{\text{Number of times a rule appears in all rule sets}}{\text{Number of reduce sets}}$$

3.2 *Neural Network*

NN is often used to find out the missing values based on the currently present data. Predictions are based on various equations and thus whole system is evaluated based on solution of such equations. In general, prediction based on a predefined model is better as it accommodates changes for a particular case. For making the results of the NN more accurate, the system should compute the data based on the least possible number of attributes and maximum number of instances. Neural networks have an advantage over other leaning algorithms as the trained model can accommodate some noise in the data. But for best possible results, only complete tuples should be used while training a neural network. Error of prediction can be generally estimated while using the neural network.

3.2.1 Network Reduction

In case of missing values present in the dataset, one solution to the problem is to create a set of neural network classifiers, with each connection performing classification for random set of attributes. Thus complete set of data can be classified by presenting different case to the network with proper inputs [19].

3.2.2 Value Substitution

Another approach to the null value problem is to substitute the missing value with an estimated value. The estimated values are then substituted to give a complete set of inputs. Finally, the complete data set is diagnosed using classifier network [19].

3.3 Performance Measure

The performance measure can be made using the three evaluation measures namely,

$$\text{Accuracy} = \frac{\text{number of cases correctly classified}}{\text{Total number of cases}} \times 100\%$$

$$\text{Correctness} = \frac{\text{number of cases correctly classified}}{\text{Total number of cases classified}} \times 100\%$$

$$\text{Coverage} = \frac{\text{number of cases classified}}{\text{Total number of cases}} \times 100\%$$

The performance evaluation may be conducted on the training set or on a test set [17]. A high evaluation rate measured on training set does not mean that the performance of the classifier will be good in practice. The results on the test set give a better indication of how well the classifier is able to generalize on new examples.

4 Materials and Methods

4.1 Dataset

The Hepatitis Disease dataset has been taken from Machine Learning Repository. It contains 152 data instances, each having 20 attributes (Table 1).

4.2 Proposed Algorithm

PHASE 1

STEP 1: Changing the categorical values into numeric equivalents
It is the first step that applies to the system. In this stage the attribute class values of the dataset are converted from its categorical values into binary equivalents.

Table 1 Dataset description with the domain value of each attribute

Attribute information:	
1. Class: Die, Live	11. Spleen palpable: no, yes
2. Age: 10, 20, 30, 40, 50, 60, 70, 80	12. Spiders: no, yes
3. Sex: male, female	13. Ascites: no, yes
4. Steroid: no, yes	14. Varices: no, yes
5. Antivirals: no, yes	15. Bilirubin: 0.39, 0.80, 1.20, 2.00, 3.00, 4.00
6. Fatigue: no, yes	16. ALK phosphate: 33, 80, 120, 160, 200, 250
7. Malaise: no, yes	17. SGOT: 13, 100, 200, 300, 400, 500,
8. Anorexia: no, yes	18. Albumin: 2.1, 3.0, 3.8, 4.5, 5.0, 6.0
9. Liver big: no, yes	19. Protime: 10, 20, 30, 40, 50, 60, 70, 80, 90
10. Liver firm: no, yes	20. Histology: no, yes

The input matrix can be considered as a $n \times m$ matrix
Where n = number of instances, m = number of attributes

STEP 2: Data preprocessing

The whole dataset is divided into complete as well as incomplete tuple sets by automated value recognition.

Let us consider all number of tuples to be n, Then, $p + q = n$
where, p = no. of complete tuples, q = no. of incomplete tuples

STEP 3: Training the Model
The model is trained using the complete tuples dataset. This involves the concepts of neural network as described in the following sub-steps:
3.1 The data in the complete tuples are recorded as input to the training model.
3.2 Neural networks are constructed based on the requirements in the above step. As the set has complete tuple data, input–output pairing can be done to satisfy the requirements for each network. Here the input is complete tuple dataset = P. Considering a three-layer neural network, we can use the following pseudo code:

Activate layer (input, output)
for each r input neuron, calculate output r
for each s hidden neuron, calculate output s
for each t hidden neuron, calculate output t
output = {output t}

STEP 4: Imputation
Here the input is the incomplete data tuples = q. It is tested to find out the null values. The output results also include the absolute and relative error percentage with respect to the predicted decision class values.

PHASE 2

STEP 5: Analysis of results
We observe the confusion matrix for the predicted data to validate the rule generation for decision-making. The decision rule snippet is also generated as a part of each step.

STEP 6: Ranking the Attributes based on Level of Dependency
Level of dependency is used to rank the attributes as described in these steps:

6.1 The attribute with high level of dependency are considered unwanted attribute and ranked lower in the Table
6.2 The attributes with low level of dependency are considered the most important attributes for the sense of decision-making.

PHASE 3

STEP 7: Modification of the Datasets—Reduction
The lower ranked attributes are removed from the dataset for next iteration. Number of attributes that are removed depends on the total no. of attributes present in the dataset.

Newly formed matrix can be considered as a $n \times (m - d)$ matrix
Where n = no. of instances, m = no. of attributes, d = no. of deleted attributes

STEP 8: Training the new model
The new model is trained with the reduced number of attributes in the complete dataset tuple. This requires lower time complexity for training than the prior training step.

STEP 9: Imputation Step
Now, the testing is done on newly trained model, as described in STEP 4.

STEP 10: Analysis of results
A new decision rule snippet is generated. The report produced during the testing process reflects a better time complexity constrain than the previous model. We find the relative and absolute error getting reduced. Thus this adds a value in terms of better accuracy and reliability of the whole system.

5 Application

5.1 Experimental Setup

Due to space constraints, the following is a sample of the code snippet for decision rule generation based on 20-attribute dataset—(Fig. 1).

Fig. 1 Resultant output data for ranking of attributes with age, Bilirubin, SGOT, Alk phosphate, protine having lowest importance in terms of decision making

```
Ranked attributes:
 0.21832  14 Varices
 0.16255  13 Ascites
 0.10966   1 Class
 0.09778  12 Spiders
 0.07999  18 Albumin
 0.04551   5 Antivirals
 0.03709  11 Splean Palpable
 0.02925  10 Liver Firm
 0.02145   3 Sex
 0.02048   6 Fatigue
 0.01598   7 Malaise
 0.00664   9 Liver Big
 0.00417   8 Anorexia
 0.00303   4 Steroid
 0         2 Age
 0        15 Bilirubin
 0        17 SGOT
 0        16 Alk Phosphate
 0        19 Protine
```

```
Varices = Yes
| Ascites = Yes
| | Bilirubin < 2.85
| | | Spiders = Yes
| | | | Sex = Female : No (8/0)
| | | | Sex = Male
| | | | | Alk Phosphate < 112.5
| | | | | | Alk Phosphate < 88.5
| | | | | | | SGOT < 44.5
| | | | | | | | Bilirubin < 0.65 : Yes (2/0)
```

5.2 Results and Observations

A sample from the code snippet for the modified decision rule generation based on 15-attribute dataset—(Figs. 2, 3 and 4).

```
Time taken to build model: 0.06 seconds

=== Evaluation on training set ===
=== Summary ===

Correctly Classified Instances         140               100      %
Incorrectly Classified Instances         0                 0      %
Kappa statistic                          1
Mean absolute error                      0.1497
Root mean squared error                  0.1703
Relative absolute error                 30.4389 %
Root relative squared error             34.3395 %
Total Number of Instances              140

=== Detailed Accuracy By Class ===

               TP Rate   FP Rate   Precision   Recall  F-Measure   ROC Area  Class
                 1         0          1          1         1          1      No
                 1         0          1          1         1          1      Yes
Weighted Avg.    1         0          1          1         1          1

=== Confusion Matrix ===

  a  b   <-- classified as
 79  0 |  a = No
  0 61 |  b = Yes
```

Fig. 2 Building the training model based on complete data tuples containing 140 instances and 20 attributes

Class = Live
| Spiders = Yes
| | Sex = Female
| | | Antivirals = Yes
| | | | Malaise = Yes
| | | | | Liver Firm = Yes
| | | | | | Liver Big = No : No (1/0)
| | | | | | Liver Big = Yes : Yes (1/0)
| | | | | Liver Firm = No : No (1/0)
| | | | Malaise = No : No (2/0)
| | | Antivirals = No : No (4/0)

```
Time taken to build model: 0.03 seconds

=== Evaluation on test set ===
=== Summary ===

Correctly Classified Instances           7               58.3333 %
Incorrectly Classified Instances         5               41.6667 %
Kappa statistic                          0.1667
Mean absolute error                      0.4059
Root mean squared error                  0.4439
Relative absolute error                 81.1833 %
Root relative squared error             88.0841 %
Total Number of Instances               12

=== Detailed Accuracy By Class ===

               TP Rate   FP Rate   Precision   Recall  F-Measure   ROC Area  Class
                 0.5       0.333      0.6        0.5      0.545      0.722    No
                 0.667     0.5        0.571      0.667    0.615      0.722    Yes
Weighted Avg.    0.583     0.417      0.586      0.583    0.58       0.722

=== Confusion Matrix ===

 a b   <-- classified as
 3 3 | a = No
 2 4 | b = Yes
```

Fig. 3 Result—testing the incomplete data tuples-12 instances and 20 attributes

6 Comparative Analysis

Time Complexity

Support vector Machine uses a time complexity of $O(n^3)$ where iterations are assumed to be constant. Levenberg Marquardt based Neural Network require N*N system of equations in every iteration, which can consume time complexity of $O(n^3)$. Preprocessing step can improve the classification speeds with an extra cost of O (nd) (Table 2).

```
=== Evaluation on test set ===
=== Summary ===

Correctly Classified Instances          10               83.3333 %
Incorrectly Classified Instances         2               16.6667 %
Kappa statistic                          0.6667
Mean absolute error                      0.2563
Root mean squared error                  0.4219
Relative absolute error                 51.2636 %
Root relative squared error             83.7003 %
Total Number of Instances               12

=== Detailed Accuracy By Class ===

               TP Rate   FP Rate   Precision   Recall  F-Measure   ROC Area  Class
                 0.833     0.167      0.833     0.833     0.833      0.833    No
                 0.833     0.167      0.833     0.833     0.833      0.833    Yes
Weighted Avg.    0.833     0.167      0.833     0.833     0.833      0.833

=== Confusion Matrix ===

 a b   <-- classified as
 5 1 | a = No
 1 5 | b = Yes
```

Fig. 4 Result—testing the incomplete data tuples-12 instances and 15 attributes

Table 2 Comparison of results for phase 1 and phase 3 of the proposed algorithm

Labels	Phase 1	Phase 3
Correctly classified instances	58.3333	83.3333
Incorrectly classified instances	41.6667	16.6667
Kappa statistics	0.1667	0.6667
Mean absolute error	0.4059	0.2563
Root mean absolute error	0.4439	0.4219
Relative absolute error	81.1833	51.2636
Root relative squared error	88.0841	83.7003
Weighted average TP rate	0.583	0.833
Weighted average FP rate	0.417	0.167
Weighted average precision	0.586	0.833
Weighted average recall	0.583	0.833
Weighted average F-measure	0.58	0.833
Weighted average ROC area	0.722	0.833

7 Conclusion

Selection of the process for missing value imputation is directly concerned with the accuracy of the output result [6]. The proposed model demonstrates the null value solution process using two of the highly recognized computational concepts—

Rough set theory and neural network [15]. The solution using the above proposed model can be extended to large datasets as the time complexity versus accuracy rate is highly feasible. General testing and training concept can have high time complexity and can largely affect the resultant values due to presence of unwanted attributes. Removal of the lower importance attributes not only helps in achieving a better accuracy versus time ratio but also reduces the space complexity of the process.

References

1. Aghazadeh, F., & Meybodi, M. (2011). "Learning Bees algorithm for optimization", International Conference on Information and Intelligent Computing, IPCSIT, Vol. 18, IACSIT Press, Singapore.
2. Bazan, J., Son, N., Skowron, A. & Szczuka, M. (2003). "A view on rough set concept approximations", International Conference on Rough Sets, Fuzzy Sets, Data mining and Granular computing, Springer-Verlag Berlin Heidelberg, (pp. 181–188).
3. Chen, S. & Hsiao, H. (2005). "A new method to estimate null values in relational database systems based On automatic clustering techniques", International Journal of Information Sciences, 169(1), Elsevier, 47–69.
4. Chen, S. & Yeh, M. (1997). "Generating fuzzy rules from relational database systems for estimating null values", International Journal of Cybernetics and Systems, 28, Taylor & Francis, 695–723.
5. Chiu, H., Wei, T. & Lee, H. (2009). "A novel approach for missing data processing based on compounded PSO clustering", Journal of WSEAS Transactions on Information Science and Applications, 6(4), 589–600.
6. Farhangfar, A., Kurgan, L. & Pedrycz, W. (2004). "Experimental analysis of methods for imputation of missing values in databases", Proc. SPIE 5421, Intelligent Computing: Theory and Applications II, Orlando, FL.
7. Gómez, Y., Bello, R., Puris, A. & García, M. (2008). "Two step swarm intelligence to solve the feature selection problem", Journal of Universal Computer Science, 14(15), 2582–2596.
8. Hu, X., Lin, T. & Han, J. (2004). "A New Rough Sets Model Based on Database Systems", Fundamental Informatica, e 59, no. 2–3 (2004), pp. 135–152.
9. Kaiser, J. (2011). "Algorithm for Missing Values Imputation in Categorical Data with Use of Association Rules", International Journal on Recent Trends in Engineering and Technology, Vol. 6, No. 1.
10. Kang, P. (2013). "Locally linear reconstruction based missing value imputation for supervised learning", Neuro computing, Elsevier, 118(2013) 65–78.
11. Karegowda, A., Manjunath, A. & Jayaram, M. (2010). "Comparative Study of Attribute Selection using Gain Ratio and Correlation based Feature Selection", International Journal of Information Technology and Knowledge Management, July-December 2010, Volume 2, No. 2, pp. 271–277.
12. Li, J. & Cercone, N. (2005). "A rough set based model to rank the importance of association rules", Proceedings of the 7th International Workshop on New Directions in Rough Sets, Data Mining, and Granular-Soft Computing (RSFDGrC'99), Springer, Berlin, Germany, Lecture Notes in Computer Science 109–118.
13. Lin, W., Lan, L., Huang, F. & Wang, M. (2015). "Rough-set-based ADR signaling from spontaneous reporting data with missing values", Journal of Bio Medical In-formatics, 58, 235–246.

14. Pawlak, Z. (1982). "Rough sets", International Journal of Computer and Information Sciences, Vol. 11, No. 5.
15. Pawlak, Z. (2004). "Rough sets", Institute of Theoretical and Applied Informatics, Polish Academy of Sciences, ul. Bałtycka 5, 44 100 Gliwice, Poland.
16. Sadiq, A., Duaimi, M. & Shaker, S. (2013). "Data Missing solution using rough set theory and swarm intelligence", International Journal of Advanced Computer Science and Information Technology (IJACSIT) Vol. 2, No. 3, pp. 1–16, ISSN: 2296-1739.
17. Sharpe, P. & Solly, R. (1995). "Dealing with Missing Values in Neural Network-Based Diagnostic Systems", Neural Computing & Applications, June 1995, Volume 3, Issue 2, pp 73–77, Springer Verlag.
18. Stekhoven, D. & Bühlmann, P. (2011). "MissForest non-parametric missing value imputation for mixed-type data", Oxford University Press, Advance Access Publication.
19. Tibshirani, R. (1995). "A comparison of some error estimates for neural network models", Department of Preventive Medicine and Biostatistics and Department of Statistics, University of Toronto.
20. Tripathy, H., Tripathy, B.K., & Das, P. (2008). "An intelligent approach of rough set in knowledge discovery databases", International Journal of Electrical and Electronics Engineering, 2(5), 334–337.
21. Yao, Y. (1998). "Generalized rough set models", Rough Sets in Knowledge Discovery, Polkowski, L. & Skowron, A. (Eds.), Physica-Verlag, Heidelberg, pp. 286–318.
22. Zhao, J. & Zhang, Z. (2011). "Fuzzy Rough Neural Network and Its Application to Feature Selection", International Journal of Fuzzy Systems, Vol. 13, No. 4.

Network Protocol-Based QoS Routing Using Software Defined Networking

P. Shakthipriya and A. Ruhan Bevi

Abstract Software defined networking (SDN) is an incipient network paradigm in which the control plane is moved out of the individual network nodes and into a separate centralized controller which is capable of exploiting the complete knowledge of the network to optimize flow management. SDN is a promising way to support the dynamic nature of networks at present and in the future. OpenFlow is the most commonly used SDN protocol. OpenFlow protocol governs the communication between SDN controllers and the underlying network infrastructure. Routing not only implies mere forwarding of data packets, but also refers to the choosing of best path for the data traffic based on certain metrics. In this paper, a routing algorithm namely Network Protocol-based QoS Routing is proposed and simulated using the network emulation tool, Mininet. The working of the algorithm is verified to correctness using the network protocol analyser Wireshark. The performance analysis is done by considering the QoS parameters.

Keywords Mininet · OpenFlow · QoS · Routing · Software defined networking

1 Introduction

Software defined networking (SDN) is the networking trend that is viewed as a promising means to solve most of the problems that the existing, traditional networks suffer from. As opposed to the traditional networks, in which the control scheme and the topological scheme called the data plane are combined, the SDN have them separated. The control plane present with the individual network nodes in the topology, can limit the configurability of the network dynamically. Hence SDN, the control is moved out of the individual nodes in the network, to a centralized controller. This controller can be programmed to suit the dynamic changes that originate in the network [1].

P. Shakthipriya (✉) · A.R. Bevi
Department of ECE, SRM University, Kattankulathur, Chennai, India
e-mail: shakthipriya@outlook.com

S.S. Dash et al. (eds.), *Artificial Intelligence and Evolutionary Computations in Engineering Systems*, Advances in Intelligent Systems and Computing 517,
DOI 10.1007/978-981-10-3174-8_32

The work on programmable networks have spanned across three major stages of evolution namely active networks, separation of control and data plane, OpenFlow abstractions and Network operating systems.

Conventional networks once deployed by defining flows to the switches and routers were not changeable. Alterations to the network schema were limited by the need to define the flows at the firmware again, in all the switches and routers. This was a serious shortcoming due to the distributed control in all the nodes.

Active networking stage thus started with a vision to have programming functions in networking.

During mid-90s with the internet technology taking off, there were more protocols that were designed and tested. These had to be standardized by IETF (Internet Engineering Task Force) and this process was a very slow one. To accelerate this process, the conventional networks had to be improved to include programmability. Thus, Active networking concentrated mainly on programmable functions in the network to accelerate innovations.

Early 2000s, witnessed the increase in the users and the traffic volumes in the network called for the need to better management to network traffic and the field called traffic engineering came into the fore. This urge to better techniques to manage network traffic commenced two innovations—Open interface between the control and data plane and Logically centralized control [2].

Moving the control out of the network equipment makes more sense as the management is a network wide activity. SDN was subject to debate between its vision of fully programmable network and its viability for realistic deployment. The OpenFlow protocol arrived as a balance between the two in the stage 3. Initially OpenFlow protocol standardized a data plane model and control plane application programming interface (API) with the firmware of existing switches. By doing this, the vendors were spared the drill of upgrading the hardware in order to make it OpenFlow compatible. The main outcomes of the work in this period, is the generalizing of network devices to include more functions using OpenFlow.

The notion of node Operating System (Distributed approach) which was talked about in the Active networking era, was overpowered by the Network Operating System (Centralized approach) with OpenFlow. The emergence of NOS offered conceptual distinction of Network operation into three layers namely data plane which is the arrangement of the network equipment, state management which is the centralized controller and control plane where the processing of traffic in the network is outlined.

SDN focuses on key features like, Separation of the control plane from the data plane, presence of a centralized controller and view of the network, existence of open interfaces between the devices in the control plane (controllers) and those in the data plane and programmability of the network by external applications [3].

With SDN, the control plane is run on separate simple, off-the-rack equipment like PC called an Open Flow Controller and data plane is run on dumb (no L2 L3 intelligence) but powerful switches called the Open Flow Switch to forward data at line rate. The communication between the controller and switch, i.e., packets and commands happen over a secure channel using Open Flow Protocol [4–6].

The Open Networking Foundation (ONF) was formed to facilitate the standardization of the SDN principles. The OpenFlow protocol was an outcome of the efforts of the ONF.

The communication between the switch and the SDN controller happens using the OpenFlow protocol, where the messages are exchanged between the participating entities via a secure channel. The controller's default port for this connection is 6633. The authentication by exchange of certain key is performed between the controller and its designated switch on powering up.

OpenFlow switch is a basic forwarding element, which is accessible via OpenFlow protocol and interface. In addition to simplifying the switching hardware, flow-based SDN architectures, such as OpenFlow requires additional forwarding table entries, buffer space, and statistical counters that are not very easy to implement in traditional switches with application specific integrated circuits (ASICs). In an OpenFlow network, switches come in two flavors, hybrid (OpenFlow enabled) and pure (OpenFlow only). Hybrid switches support OpenFlow in addition to traditional operation and protocols (L2/L3 switching). Pure OpenFlow switches have no legacy features or onboard control, and completely rely on a controller for forwarding decisions. Most of the currently available and commercial switches are hybrids.

Open vSwitch (OVS) is a multilayer virtual switch licensed under the Apache license. It is designed to enable massive network automation through programmatic extension, while still supporting standard management interfaces and protocols.

The SDN controller is the entity that provides programmatic interface to OpenFow switches. OpenFlow switches consults an SDN controller each time a decision with respect to the flow table or actions has to be made. This centralized approach simplifies the network control as the reconfiguration policies required from time-to-time can be addressed only to the SDN controller. POX is one such open-source SDN controller that offers development platform for Python-based SDN control applications.

Section 2 of this paper discusses the routing algorithm for prioritizing the flow in a network. In Sect. 3, simulation of the algorithm using Mininet and the Wireshark verification is explained. Section 4 of the paper analyze performance of the algorithm.

2 Network Protocol-Based QoS Routing

Internet protocol (IP) specifies that the payload in the network must be transmitted in the form of datagrams. Figure 1 shows the frame level description of the IP datagram for IPv4. The network protocol field of the datagram determines what protocol should be used while passing the data to the upper layers. We discuss two such protocols called transmission control protocol (TCP) and user datagram protocol (UDP) that can be distinguished using network protocol numbers assigned to

Version	Internet Header Length	Type of Service	Total Length	
Identification			Flags	Fragment Offset
Time to Live	**Network Protocol**		Header Check sum	
Source Address				
Destination Address				
Padding				

Fig. 1 The frame level description of IP datagram

them. These serve as unique identifiers to recognize the IP payload at the datagram level. The network protocol number for TCP is 6 and UDP is 17.

TCP is known to be a very reliable connection oriented protocol due to the handshaking happening between the communicating agents in the form of synchronization (SYN) and acknowledgement (ACK). The sockets are established between the source and destination and the data transfer is possible in bidirectional way. The frames transmitted trigger an ACK from the destination on correct reception of data packets. This ensures the guaranteed delivery of data packets.

The pseudocode for the TCP header implementation below, shows the important fields in the TCP packet:

```
struct tcp_packet
{

uint32 src_addr;
uint32 dst_addr;
uint8 zero;
uint8 proto;
uint16 length;
uint8 seq_no
uint8 ack
uint8 chk_sum;

} tcp_header;
```

UDP is designed to be a best effort, connectionless protocol with no handshaking. It is mainly used in applications that require fast and efficient data transfer.

Large chunks of data are transmitted using UDP and there is no reliability of guaranteed delivery or the order of transmission.

Given below is the pseudocode for the UDP header:

```
struct udp_packet
{

uint32 src_addr;
uint32 dst_addr;
uint8 chk_sum;
uint16 length;

} udp_header;
```

Ability to prioritize critical messages to or from the network node over other traffic like P2P, FTP, Torrents, social networking applications, and other bandwidth consuming applications, is a significant feature a controller needs to be equipped with. This necessitates the network to route TCP and UDP when they occur on the same link of the network. The TCP data must be allowed to use a more reliable link for its transmission. The proposed algorithm uses the network protocol number as the metric for routing the data traffic and separates the traffic based on the frames. The flow chart of the algorithm is as presented below in Fig. 2.

The OVS documentation specifies that creation of flow tables by utilizing various parameters for packet forwarding. Unique feature of OVS is to distinguish the packets at frame level using the network protocol numbers. Further on exploring about similar kinds of algorithms, in [7], the authors suggest a scheme to improve the performance of TCP protocol transmission. The routing algorithm here, aims to find a suitable path with least expected transmissions. The frame based TCP-ETX parameter which calculates the expected frame transmissions for TCP is designed. ETX is the metric used to select a path with least expected transmission.

When the data traffic is passed at the ingress port of the OVS, the Virtual switch under the control of the SDN controller, distinguishes the packets based on the network protocol number and then segregates it into two links as illustrated by Fig. 2. If the network protocol number is 6, then it corresponds to TCP data and it is routed on link 2. Similarly, if the network protocol number is 17, then it corresponds to UDP data and is routed on link 1. Here the underlying assumption made is that link 2 is more reliable than link 1. The programmer can monitor some vital parameters and determine which link is more reliable and congestion free. This may involve reading the switch counters with respect to queuing, port data, and data flow.

The aim of the algorithm is to reduce the jitter in the way of transmission of TCP data by allocating a congestion free link for its transmission, to achieve improved QoS.

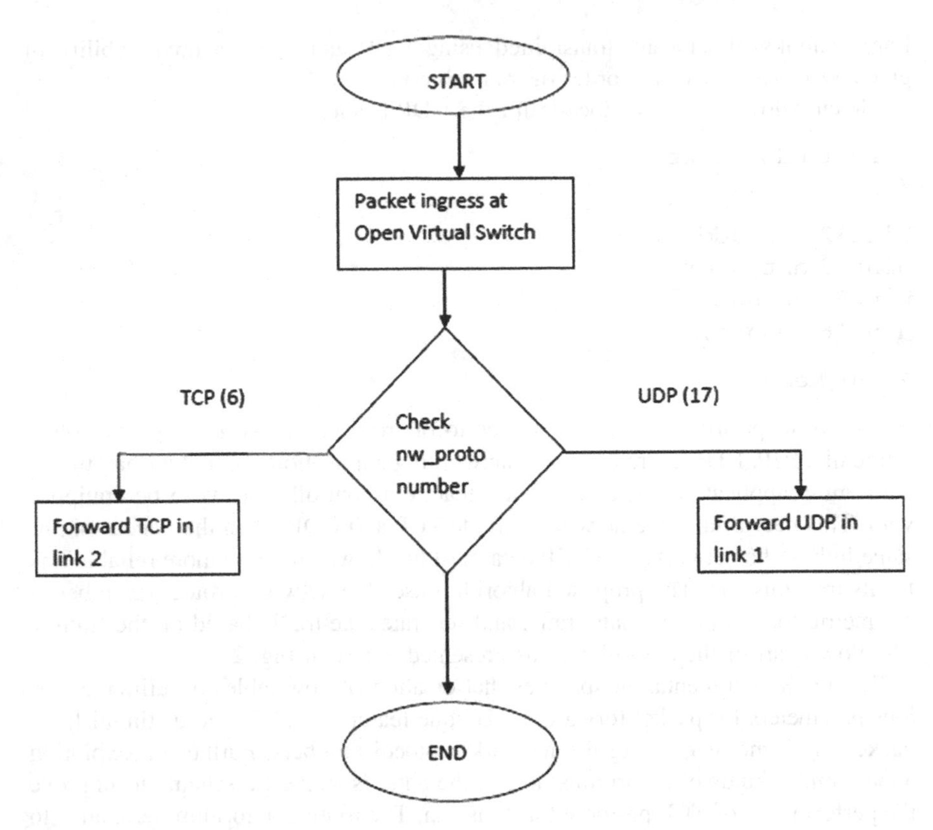

Fig. 2 Flow chart representation of the algorithm

3 Simulation of the Algorithm with Mininet

Networking simulators allows the experimentation of algorithms using virtual network elements. Many simulators were assessed to choose the most suitable simulator for SDN-based experiments. The key observation of the study is presented in Table 1.

3.1 Mininet

Mininet turns out to be the most viable open-source simulator for the simulation of the algorithm. Mininet is a network emulator which runs on Linux Operating Systems (OS). It runs multiple hosts and switches using single OS kernel. The OpenFlow switches and controllers are emulated in the user space. It is a great environment to develop, execute, test, and debug SDN applications. It comes with OVS version 1 by default and POX controller.

Table 1 Comparison of network simulators

	MININET	QUALNET	OPNET	NS-3	ESTINET
Vendor	ON.LABS	Freescale	OPNET TECHN.	NS3 projects	ESTINET TECHN.
OS	Linux	Linux/Windows	Linux/Windows	Linux	Linux
Open flow support	Yes	No	No (work for extension)	Yes	Yes
GUI	MINIEDIT	Animator	VND	VND	ESTINET GUI
Emulation mode	Yes	Yes	No	No	Yes
Simulation mode	No	Yes	Yes	Yes	Yes
Direct SDN controller	POX usable	No	No	No	Yes
Licence	Free	Commercial	(Single user) Commercial	Free	Commercial

3.2 Implementation Details

A simple tree topology is created in Mininet as shown in Fig. 3. It is the MiniEdit (the GUI counter part of Mininet) look of the topology. The tree topology is created in two levels with three OpenFlow switches (s1, s2, s3) and four end hosts (h1, h2, h3, h4) that are the end receivers of the data packets in the network. The switches s2 and s3 are connected to switch s1 and receive inputs from it. The hosts h1 and h2 are at the output of the switch s2 and the hosts h3 and h4 are at the output of the switch s3.

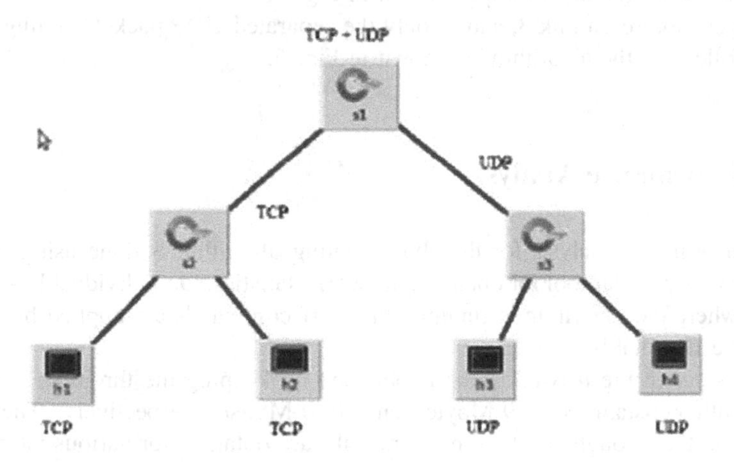

Fig. 3 Tree topology look from MiniEdit

A virtual Ethernet link is added to the switch s1 so as to pass the data traffic to the network. Artificial datagrams are crafted using the software, Ostinato and passed into the network. The stream is designed in such a way that it passes interleaved streams of TCP and UDP packets on the ingress port. The switches are OpenFlow compliant switches that are connected to POX controller which is invoked on a separate terminal and set to listen to the remote port. The switches run the shell script code for the Network Protocol-based QoS routing algorithm. Once the transmission starts, the OpenFlow switch populates the flow table by checking the match fields that has been given in the shell script code. Once the matches are found, they are forwarded from the open virtual switch.

When the xterms (offered by linux environment) of the end hosts are checked, there are only TCP packets on h1 and h2, and only UDP packets on h3 and h4, thereby confirming the segregation at switch s1. Further the working of the algorithm is verified using the network protocol analyzer Wireshark and the results are presented in Sect. 3.3.

3.3 Wireshark Verification

Wireshark is a network protocol analyzer that is used for capturing the packet transfers on desired links to view the type of packets that constitute the network traffic on that link. The algorithm is verified using Wireshark and the corresponding flow graphs are explained below.

Packet capture done on link 3 which is the ingress link at switch s1, shows the interleaved arrival of TCP and UDP packets and they are as shown in Fig. 4. The red bursts at periodic intervals are the UDP packets and the black bursts are the TCP data stream.

Packet capture on link 2, shows only the separated TCP packets getting routed on it, following the algorithm, as shown in Fig. 5.

Packet capture on link 3, shows only the separated UDP packets getting routed on it, following the algorithm, as shown in Fig. 6.

4 Performance Analysis

The performance analysis for the above routing algorithm is done using the tool iperf, a very popular tool for obtaining network statistics. On individual links of the OVS, where the algorithm is running, the iperf commands are applied by configuring the IP suitably.

The Jitter in the network links is obtained by keeping the throughput and the bandwidth constant at 1.19 Mbytes and 10.0 Mbits/s, respectively. The bandwidth and the throughput of the network links are obtained for various intervals of time.

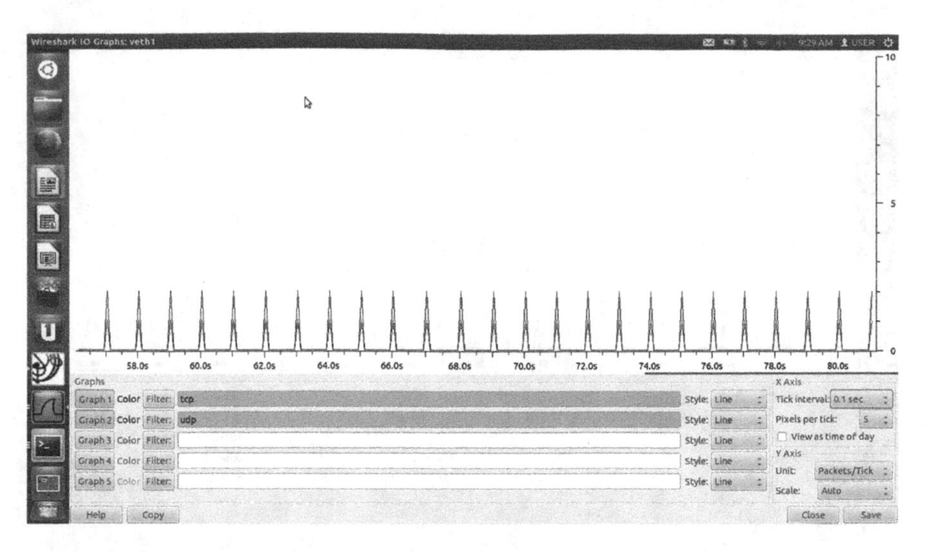

Fig. 4 The flow graph from Wireshark showing TCP and UDP packets

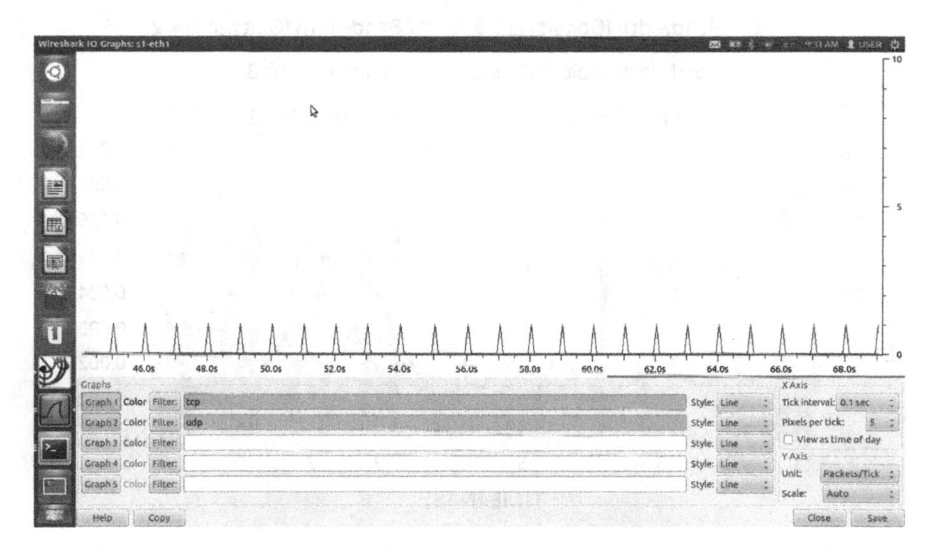

Fig. 5 The flow graph from Wireshark showing only TCP packets

The graph of bandwidth and jitter plotted against time is as shown in Fig. 7. It can be observed that the maximum jitter occurs in link 3. It is consequently reduced by 0.003 ms in both link 2 and link 3. The maximum bandwidth is achieved in case of link 2 and link 3 only as compared to link 1.

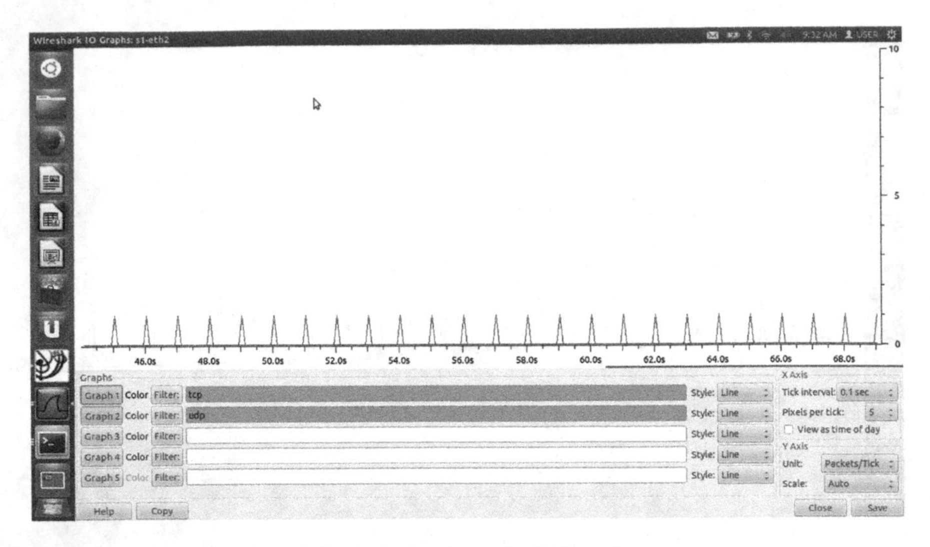

Fig. 6 The flow graph from Wireshark showing only UDP packets

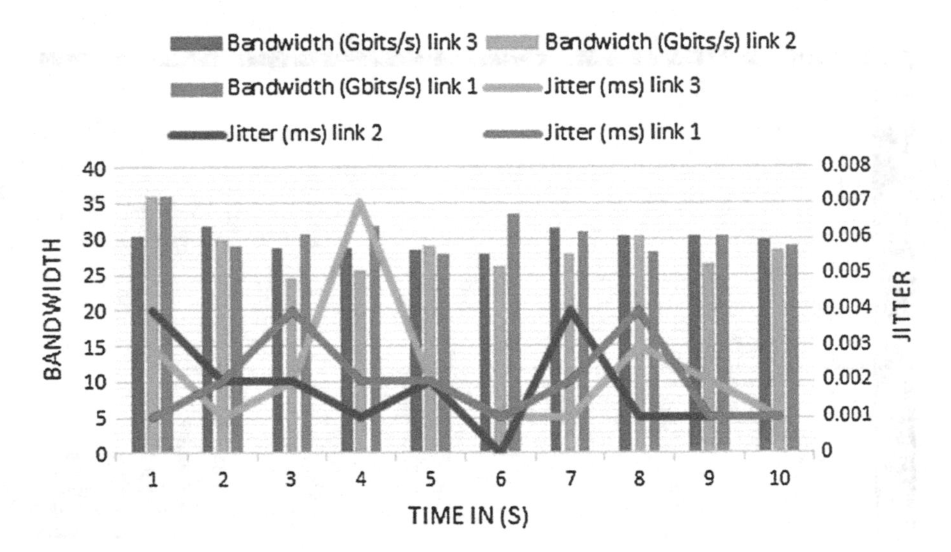

Fig. 7 The graph of bandwidth and jitter

Figure 8 shows the graph of throughput obtained for varying time interval. It is observed that the throughput of the egress link having TCP and UDP is improved as compared to the ingress link.

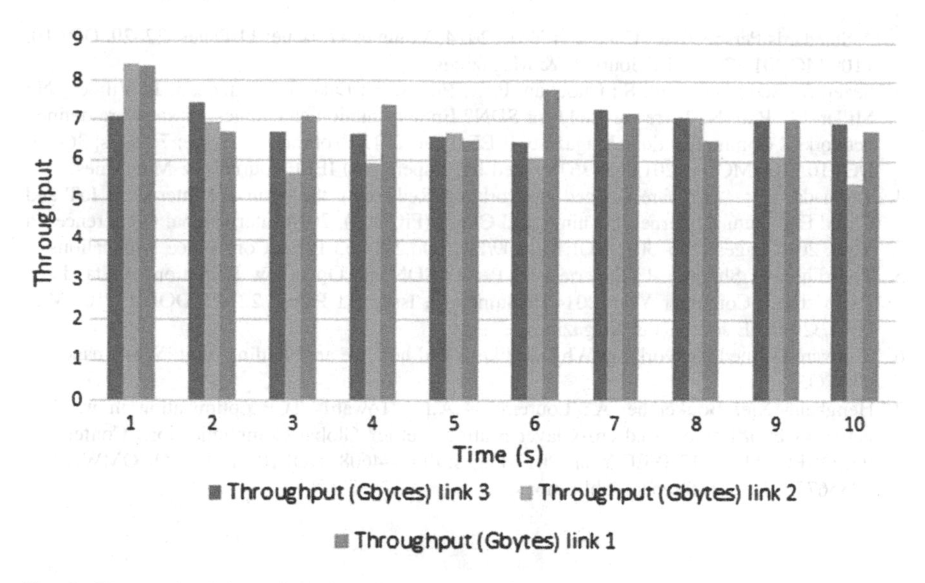

Fig. 8 The graph of throughput in different links

5 Conclusions

In this paper, we discussed the merits of SDN, the simulation and performance analysis of a routing algorithm designed using the same called Network Protocol-based QoS routing. The network statistics derived through the simulation of the network using Mininet, suggest that the traffic segregated following the algorithm yields improved bandwidth and reduced jitter for individual links as compared to everything routed through the same link. It is to be noted that the maximum Bandwidth attained in link 1 and link 2 have increased as compared to the non-SDN way in link 3. The maximum Jitter in the network has reduced by 0.003 ms in both link 1 and link 2.

The future work involves applying the above-mentioned algorithm in specific network scenarios like the Internet of Things and Information Centric Networks, to improve the ease of networkability in them.

References

1. Hyojoon Kim; Feamster, N. "Improving network management with software defined networking." Communications Magazine, IEEE Volume: 51, Issue: 2 Publication Year: 2013, Page(s): 114–119.
2. Nick Feamster, Jennifer Rexford, Ellen Zegura "The road to SDN an intellectual history" ACM SIGCOMM Computer Communication Review archive Volume 44 Issue 2, April 2014 Pages 87–98 8. Tourrilhes, J.; Sharma, P.; Banerjee, S.; Pettit "SDN and OpenFlow Evolution:

A Standards Perspective." Computer Year: 2014, Volume: 47, Issue: 11 Pages: 22–29, DOI:10.1109/MC.2014.326 IEEE Journals & Magazines.

3. Sezer, S.; Scott-Hayward, S.; Chouhan, P.K.; Fraser, B.; Lake, D.; Finnegan, J.; Viljoen, N.; Miller, M.; Rao, N. " Are we ready for SDN? Implementation challenges for software-defined networks." Communications Magazine, IEEE Year: 2013, Volume: 51, Issue: 7 Pages: 36–43, DOI:10.1109/MCOM.2013.6553676 Cited by: Papers (38) IEEE Journals & Magazines.
4. Tadinada, V.R. "Software Defined Networking: Redefining the Future of Internet in IoT and Cloud Era" Future Internet of Things and Cloud (FiCloud), 2014 International Conference on Year: 2014 Pages: 296–301, DOI:10.1109/FiCloud.2014.53 IEEE Conference Publications.
5. Tourrilhes, J.; Sharma, P.; Banerjee, S.; Pettit "SDN and OpenFlow Evolution: A Standards Perspective." Computer Year: 2014, Volume: 47, Issue: 11 Pages: 22–29, DOI:10.1109/MC.2014.326 IEEE Journals & Magazines.
6. Software Defined Networking: Why we like it and how we are building on it: White paper by CISCO.
7. Hengheng Xie; Boukerche, A.; Loureiro, A.A.F. "Towards TCP optimization in wireless networks by a frame based cross layer routing metric" Global Communications Conference (GLOBECOM), 2013 IEEE Year: 2013 Pages: 4603–4608, DOI:10.1109/GLOCOMW.2013.6855677 IEEE Conference Publications.

Comparison of GSA and PSO-Based Optimization Techniques for the Optimal Placement of Series and Shunt FACTS Devices in a Power System

Rajat Kumar Singh and Vikash Kumar Gupta

Abstract This paper presents the application of gravitational search algorithm (GSA) and particle swarm optimization (PSO)-based approach along with multiple FACTS (flexible AC transmission system) devices for the economic operation of an interconnected power system under different loading condition. Two different types of FACTS devices such as static Var compensator (SVC) and thyristor-controlled series capacitor (TCSC) are used in this paper. The location of the FACTS devices is obtained by the reactive power flow in the transmission lines. The reactive loading of the system have been increased from the base value to 110 and 120% of base reactive loading. Finally, results have been compared between both the techniques in terms of minimization of active power loss and operating cost.

Keywords FACTS devices · GSA · PSO · Reactive power flow · Operating cost

1 Introduction

In present scenario power system stability has become a challenging and complex area due to a reasonable increase in power supply demand as compared to power supply in recent years. To fulfill the needs of the population, it is required for the power system to work safely and efficiently. Voltage stability is one of the major concerns in power system stability. A heavily loaded power system network works close to voltage stability limit. To improve the voltage profile reactive power compensation on weak nodes is required, also by minimizing the reactive power flow the losses in the system can be reduced.

R.K. Singh (✉) · V.K. Gupta
Department of EEE, BIT, Mesra, Ranchi, Jharkhand, India
e-mail: rajat.6969@rediffmail.com

V.K. Gupta
e-mail: vikash1146@gmail.com

S.S. Dash et al. (eds.), *Artificial Intelligence and Evolutionary Computations in Engineering Systems*, Advances in Intelligent Systems and Computing 517,
DOI 10.1007/978-981-10-3174-8_33

The concept of FACTS was introduced by N.G. Hingorani. In [1], the concept of FACTS devices is discussed. FACTS devices are used to advance controllability and boost up the capability of the system to transfer the power. In [2], differential evolution technique is used to improve the power flow with the help of FACTS devices. The power flow in the system using FACTS devices considering network control variables is discussed in [3]. In [4], SVC models are advanced and studied for a load and power flow using N.R. method. Use of evolutionary algorithms for reduction of losses by proper management of reactive power flow is discussed in [5]. The concept of particle swarm optimization is applied on the FACTS devices along with the increase in loading of the system in addition to the consideration of cost is discussed by authors in [6]. The authors discussed the optimization of reactive power and voltage of a system using PSO in [7]. GSA is an optimization technique based on law of gravity is discussed in [8]. Optimization of the location of the FACTS device by the help of genetic algorithm is explained in [9]. A new algorithm combined of GSA and PSO is introduced in [10] by the authors. The economic advantage of genetic algorithm and PSO with FACTS devices for a transmission network at different loading is discussed in [11].

The major objective of this research work is to minify the cost of operation. To reduce the cost of operation, active power loss is minimized with the help of FACTS devices using evolutionary algorithm. The location of FACTS devices is determined on the basis of reactive power flow in the lines. PSO and GSA are the two techniques which are used here. The reactive power loading is increased up to 120%. Further, the comparison of the both techniques has been done on the basis of active power loss and cost of operation.

2 Mathematical Expression of FACTS Devices

FACTS devices are basically the controllers which are capable to enhance and increase the controllability of the system by changing one or many system variables such as voltage, current, impedance in synchronization with others. FACTS devices increase the power transmitting capability, thereby making the system more economic. From the number of FACTS devices TCSC and SVC are used in this paper, which are categorized in the first generation FACTS Devices.

2.1 *Thyristor-Controlled Series Capacitor (TCSC)*

Thyristor-controlled series capacitor (TCSC) provides an efficient way to control and increase power transfer level of a system by changing the apparent impedance of a desired transmission line.

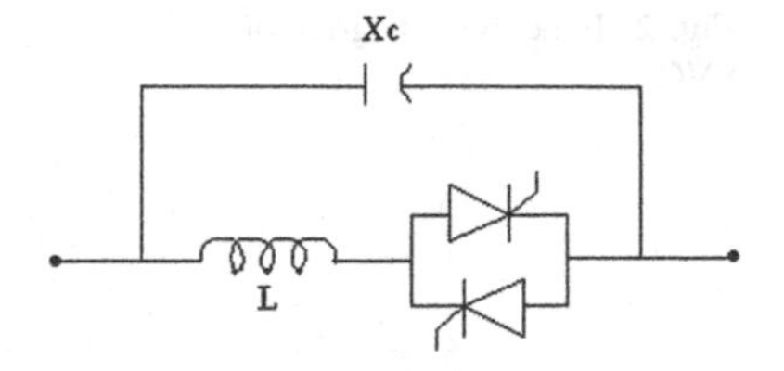

Fig. 1 Basic scheme of TCSC

The basic scheme of TCSC is shown in Fig. 1. It consists of the series controlled capacitor shunted by a thyristor-controlled reactor (TCR). The different value of impedance (X_{TCSC}) of TCSC depends upon the delay angle (α) of thyristor valves.

The TCSC operates in two ranges:

One is the $\alpha_{\text{Claim}} \leq \alpha \leq 180°$

where X_{TCSC} (α) is capacitive,

and the other is the $90° \geq \alpha \geq \text{L}_{\text{lim}}$

where X_{TCSC} (α) is inductive,

After providing the TCSC in the specifically determined line the line reactance will be calculated as

$$X_{ij\_\text{new}} = X_{ij} - X_{\text{TCSC}} \tag{1}$$

where

X_{ij_new} new line reactance
X_{ij} initial line reactance
X_{TCSC} reactance of TCSC

2.2 Static Var Compensator (SVC)

Static Var compensator (SVC) is a static device connected in shunt in the system. The main function of the SVC is to adjust the voltage at the bus where it is placed. SVC is basically consists of capacitor bank and thyristor-controlled reactor (TCR). TCR is connected in parallel with a fixed capacitor bank. A shunt transformer is used to connect the SVC with the transmission line as shown in Fig. 2.

By providing the SVC at the decided bus to improve the system stability, the reactance will be calculated as

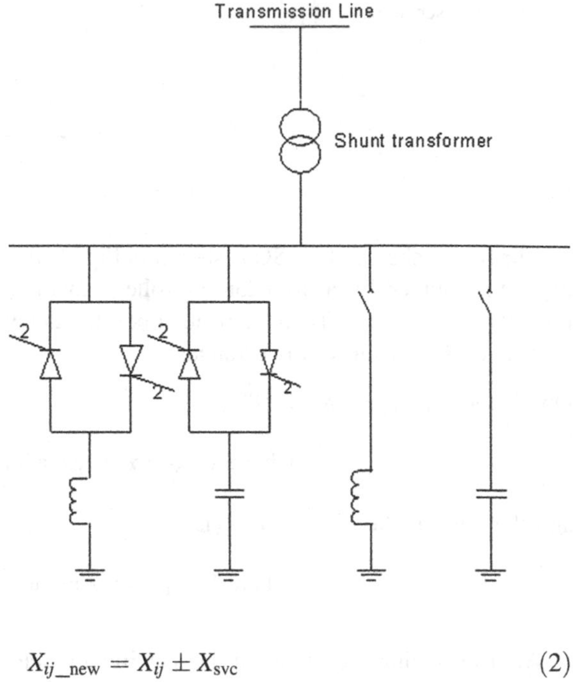

Fig. 2 Functional diagram of SVC

$$X_{ij\_new} = X_{ij} \pm X_{svc} \tag{2}$$

3 Weak Node Selection

The placement of the FACTS devices is one of the main concerns as the location is dependent on the system characteristics and its effect on the system variables. FACTS devices are generally placed at the lines which are loaded to a great extent to manage the flow of power in the line. The proper power flow control protects the line from getting overloaded and power gets distributed all over the system. The SVCs are mainly used for regulating voltage, improving power factor in addition to provide rapid change in reactive power in the transmission lines. Typically, the SVC's are placed near high or rapid changing loads or transmission lines having high reactive power flow. TCSC is a series connected FACTS device, which can be utilized to make the system operate above the power level it was initially intended without making it unstable. If the reactive power flow is increasing significantly in the confining transmission line, the TCSC or SVC placed in the network will provide the reactive power which in turn increases the active power flow in the lines. There are four SVCs placed at the end of the line and four TCSC located in the lines having high reactive power flow in the test system.

4 Problem Formulation

The objective function of the system is to minify the total operating cost. The mathematical expression of the active power losses can be formulated as in Eq. (3) subjected to constraints given below:

$$P_{\text{loss}} = \sum_{k=1}^{L} g_k (V_i^2 + V_j^2 - 2V_iV_j \cos\theta) \tag{3}$$

$Q_i^{\min} \leq Q_i \leq Q_i^{\max}$; (Reactive power constraints)
$V_i^{\min} \leq V_i \leq V_i^{\max}$; (Voltage magnitude constraints)
$Qg_i^{\min} \leq Qg_i \leq Qg_i^{\max}$; (Existing nodal reactive capacity Constraints)

The total operating cost can be given as

$$\text{Cost}_{\text{TOTAL}} = \text{Cost}_{\text{SVC}} + \text{Cost}_{\text{TCSC}} + \text{Cost}_{\text{OPERATING}} \tag{4}$$

where the cost of TCSC, SVC and operating cost can be calculated as

$$\text{Cost}_{\text{SVC}} = 0.0003(F)^2 - 0.305(F) + 127.38\ (\$/\text{kVar}) \tag{5}$$

$$\text{Cost}_{\text{TCSC}} = 0.0015(F)^2 - 0.7130(F) + 153.75\ (\$/\text{kVar}) \tag{6}$$

$$\text{Cost}_{\text{OPERATING}} = 0.06 \times 8760 \times 10^5 \times P_{\text{loss}}\ (\$/\text{kVar}) \tag{7}$$

Where F is operating values of FACTS devices

5 Proposed Approach

5.1 Particle Swarm Optimization

In power system optimization problems, there is a large search space which increases with problem size and takes a lot of time and unsuitable result in classical optimization approach. To overcome such problems, the evolutionary optimization technique can be used such as PSO. Particle swarm (PSO) is a biologically inspired optimization technique. This technique mimics the social behavior in group of animals having no leader. In PSO, a number of particles are randomly distributed in a search space. Every particle acquires its current location, velocity and the best position achieved by the particles in the given space. In particular, an individual particle in this technique is guided by the following equations:

$$v_i^{k+1} = w_i v_i^k + a_1 \text{rand} \times (\text{pbest}_i - s_i^k) + a_2 \text{rand} \times (\text{gbest} - s_i^k) \quad (8)$$

Equation (8) defines the velocity of each particle where, gbest is the globally best position in the given space and pbest is the personal best position obtained by the each particle till current iteration; a_1 and a_2 are taken as constants; rand is the random number in range of [0, 1]; w_i is the weight factor; and v_i and velocity of *i*th particle.

The weight factor w_i can be calculated as

$$w_i = w_{\text{max}} - ((w_{\text{max}} - w_{\text{min}})/\text{TI}) \times I \quad (9)$$

where w_{max} and w_{min} are the fixed values, TI is the total number of iteration to be carried out. I is the *i*th iteration.

The particle positions can be updated as

$$s_i^{k+1} = s_i^k + v_i^{k+1} \quad (10)$$

where

s_i^k Location of *i*th particle at *k*th iteration
s_i^{k+1} Location of *i*th particle at *k* + 1th iteration
v_i^{k+1} Velocity of *i*th particle at *k* + 1th iteration

By updating the position of the particles, the optimum value can be obtained as global best solution after the desired number of iterations.

5.2 *Gravitational Search Algorithm*

Gravitational search algorithm emulates the universal law of gravitation. In GSA, a number of variable strings are randomly distributed in the search space. Each particle experiences a force due to all other variable strings. Assuming there is '*N*' number of strings and each string has '*g*' dimensions. The force acting between two variable strings '*i*' and '*j*' of *k*th dimension at time '*t*' can be calculated as

$$F_{ij}^k(t) = G(t) \times \frac{m_{pi}(t) \times m_{aj}(t)}{D_{ij} + \varepsilon} \times (x_j^k(t) - x_i^k(t)) \quad (11)$$

where

$m_{aj}(t)$ Active mass of *j*th variable
$m_{pi}(t)$ Passive mass of *i*th variable
$x_j^k(t)$ Location of *j*th mass of *k*th dimension
$x_i^k(t)$ Location of *i*th mass of *k*th dimension

D_{ij} Euclidean distance between variable 'i' and 'j'
ε Low constant value.

The value of $G(t)$ can be obtained by the following expression:

$$G(t) = G_o e^{-\lambda \frac{t}{T}} \quad (12)$$

where G_o and λ are the predefined constant values and T is the total number of iterations. The total force acting on ith string variable of kth because of all other string variables of same dimension can be calculated as

$$F_i^k(t) = \sum_{j \in g_{\text{best}}, j \neq i}^{N} \text{rand}_j \times F_{ij}^k(t) \quad (13)$$

The overall mass of ith string can be formulated as

$$M_i(t) = \frac{m_i(t)}{\sum_{j=1}^{N} m_j(t)} \quad (14)$$

The value of the $m_i(t)$ is depends on the fitness value as given in equation below.

$$m_i(t) = \frac{\text{fit}_i(t) - \text{worst}(t)}{\text{best}(t) - \text{worst}(t)} \quad (15)$$

where

$\text{fit}_i(t)$ Fitness value at tth iteration
$\text{worst}(t)$ Worst fitness value in the tth iteration
$\text{best}(t)$ Best fitness value in the tth iteration

The velocity of the string can be obtained with the help of acceleration value. The acceleration of the ith string can be defined as

$$a_i^k = \frac{F_i^k(t)}{M_{ii}(t)} \quad (16)$$

From the obtained acceleration value the position of the string can be updated by updating the velocity of the string. The updated velocity of the ith string can be obtained as

$$\text{Vel}_i(t+1) = \text{rand}_i \times \text{Vel}_i(t) + a_i(t) \quad (17)$$

where the $vel_i(t)$ is the velocity of the variable of the ith string at tth iteration.
Similarly, the position of the ith string can be updated by the following equation.

$$X_i(t+1) = X_i(t) + \text{Vel}_i(t+1) \tag{18}$$

By updating the positions of the all the strings obtained from the GSA technique, we are able to minimize or maximize the objective function.

6 Result and Discussion

For IEEE 30 bus system, PSO and GSA are applied for minimization of losses and cost of operation. Table 1 shows the number of controlling variables. Table 2 shows the location of the SVCs and TCSCs. Similarly Table 3 shows the reactive power flow in the lines at 120% reactive loading with and without evolutionary techniques. Comparison of initial active power loss and minimized active power loss at different reactive loading by different techniques is tabulated in Table 4. The values of the operating cost with both the technique are given in Table 5. Figure 3 compares the reduction in active power loss by PSO and GSA at 120% reactive loading. The comparison of total cost with respect to generation by PSO and GSA is shown in Fig. 4.

Table 1 Number of controlled variables

TCSC	SVC	Transformer tap	Reactive generation of generator
4	4	4	5

Table 2 Location of SVC and TCSC in the network

Location of SVCs at the Buses				Location of TCSCs placed in lines			
21	7	17	15	25	41	28	5

Table 3 Reactive power flow at the SVC placed buses at 120% reactive loading using GSA and PSO

SVC lines	120% reactive loading		
	Without FACTS devices	Using GSA	Using PSO
27	0.1135	0.2645	0.0584
9	0.0793	0.0384	0.1104
26	0.0659	0.0429	0.0280
18	0.0675	−0.0250	−0.0448

Table 4 Active power loss at different reactive loading using GSA and PSO

Reactive loading (%)	Active power loss (p.u.)		
	Without FACTS devices	Using GSA	Using PSO
100	0.0711	0.0363	0.0436
110	0.0716	0.0369	0.0447
120	0.0721	0.0383	0.0450

Table 5 Operating cost at different reactive loading using GSA and PSO

Reactive loading (%)	Operating cost ($)		
	Without FACTS devices	Using GSA	Using PSO
100	3,737,016	1,936,961	2,321,663
110	3,763,296	1,968,533	2,379,500
120	3,789,576	2,041,065	2,395,244

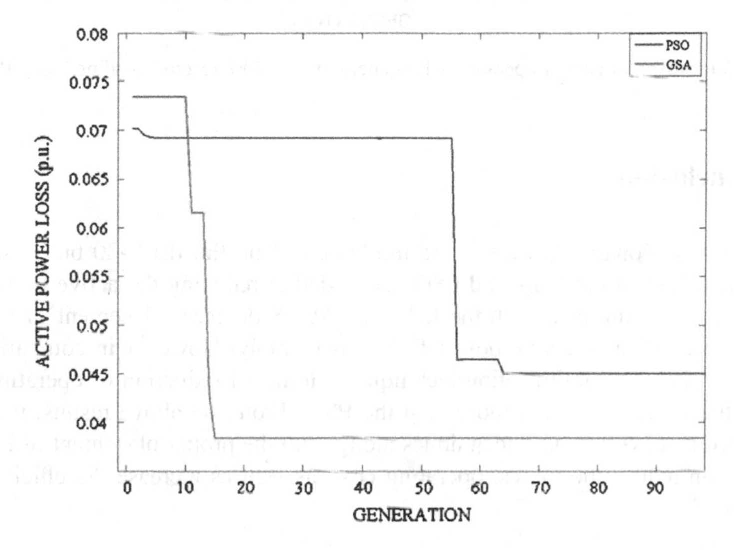

Fig. 3 Comparison of active power loss with generation at 120% reactive loading using PSO and GSA

The GSA and PSO techniques are applied to the test system. All the FACTS devices, transformer tap setting as well as reactive power generation of generator are treated as the variables. The reactive power flow is decreased in the lines where FACTS devices are placed as shown in Table 3, as the reactive power demand is supplied by the FACTS devices. The reactive power loading of the system is increased to 110 and 120%. The reduction in the active power losses can be seen in Table 4 with 100, 110% as well as 120% increase in reactive loading by applying both GSA and PSO. It can also be observed that operating cost is reduced considerably by both the techniques as observed in Table 5.

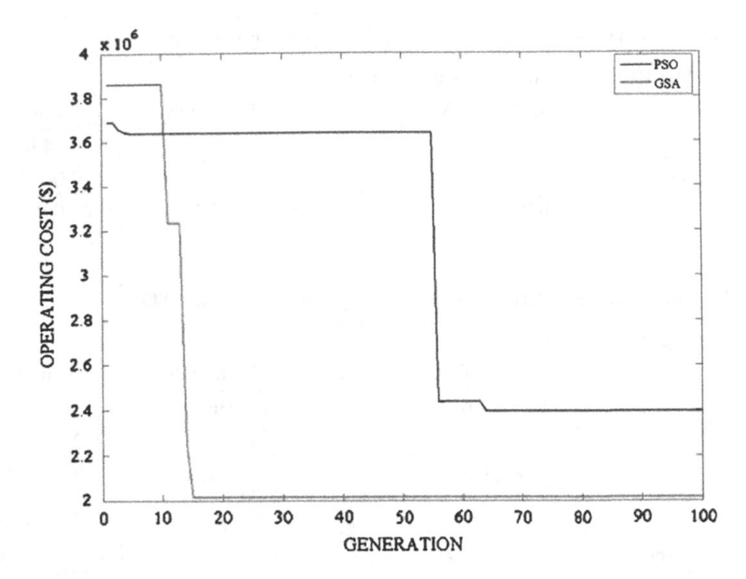

Fig. 4 Comparison of cost of operation with generation at 120% reactive loading using PSO and GSA

7 Conclusion

The two evolutionary algorithms are implemented on the IEEE-30 bus system in this paper. Both of the GSA and PSO succeeded in reducing the active power loss as well as operating cost with the help of FACTS devices. Placement of FACTS devices reduced the reactive power flow considerably. However in comparison to PSO, GSA is better optimization technique in terms of reduction of operating cost as it reduced the cost much more than the PSO. From the above results, it can be concluded that evolutionary techniques along with the proper placement of FACTS devices can reduce the losses, operating cost, as well as increase the efficiency of the network.

References

1. Hingorani, N.G., Gyugyi, L.: Understanding FACTS: Concepts and Technology of Flexible AC Transmission Systems. The institute of Electrical and Electronics Engineers, New York (2000).
2. Basu, M.: Optimal power flow with FACTS devices using differential evolution. Electrical Power and Energy Systems 30, 150–156 (2008).
3. Chung, T.S., Ge, S.: Optimal power flow incorporating FACTS devices and power flow control constraints. In: IEEE Conference, vol. 98, pp. 415–419 (1998).

4. Acha E., Ambriz-Perez H., Fuerte-Esquivel, C.R.: Advanced SVC Models for Newton-Raphson Load Flow and Newton Optimal Power Flow Studies. IEEE Transactions on Power Systems, 15(1), 129–136 (2000).
5. Bhattacharyya, B., Goswami, S.K., Bansal, R.C.: Loss Sensitivity Approach in Evolutionary Algorithms for Reactive Power Planning. Electric Power Components and Systems **37**(3), 287–299 (2009).
6. Bhattacharyya, B., Goswami, S.K., Gupta, V.K.: Particle swarm intelligence based allocation of FACTS controller for the increased load ability of power system. Int. J. Electr. Eng. Inform **4**(4) (2012).
7. Fukuyama, Y., Tahyama, S., Yoshida, H., Kawata, K., Nahnishi, Y.: A particle swarm optimization for reactive power and voltage control considering voltage security assessment. IEEE IRons. on Power Systems, 1232–1239 (2000).
8. Rashedi, E., Nezamabadi-pour, H., Saryazdi, S.: GSA: a gravitational search algorithm. Inf. Sc. **179**(13), 2232–2248 (2009).
9. Ippolito, L., Cortiglia, A.L., Petrocelli, M.: Optimal allocation of FACTS devices by using multi-objective optimal power flow and genetic algorithms. Int. J. Emerg. Elect. Power Syst. 7 (2) (2006).
10. Mirjalili, S., MohdHashim, S.Z.: A new Hybrid PSOGSA Algorithm for Function Optimization. In: IEEE International Conference on Computer and Information and Application (ICCIA), 374–377 (2010).
11. Bhattacharyya, B., Kumar, S., Gupta, V.K.: Enhancement of Power System Loadability with FACTS Devices. Journal of The Institution of Engineers (India): Series B 95(2), 113–120 (2014).

Neural Network-Controlled Wind Generator-Fed Γ-Z Source-Based PMSM Drive

A. Jaffar Sadiq Ali and G.P. Ramesh

Abstract This paper works out with the comparison of dynamic responses of closed-loop proportional integral derivation (PID) and artificial neural network (ANN) in wind generator-fed Γ-ZSI-controlled PMSM drive system. Voltage-type Γ-ZSI is proposed for PMSM drive for simulation. Γ-ZSI can boost the input voltage significantly and the speed of the drive is controlled using V/f control method. The drive system is developed using blocks of MATLAB/Simulink and the results are presented in terms of its rise time, settling time, and steady-state error.

Keywords Artificial Neural Network (ANN) · Wind generator · PMSM drive · Gamma ZSI

1 Introduction

In today's modern world, power electronics has seen tremendous growth with the systems that generate very low voltages from non-renewable sources. These voltages are fluctuating in nature and require voltage boosting when connected with the grid or utility. The AC–AC converters have been primarily developed to boost up the low voltage level using the VSI or CSI but suffer from some limitations like it cannot have an output AC voltage higher than the DC source voltage and also it is not allowed to have shoot-through states in a leg. The single-stage inverters further enhanced the boosting factor and found in [1, 2]. There was another inverter named

A. Jaffar Sadiq Ali (✉) · G.P. Ramesh
Department of Electronics and Communication,
St. Peter's University, Chennai, Tamil Nadu, India
e-mail: ajaffi1@gmail.com

G.P. Ramesh
e-mail: rameshgp@yahoo.com

S.S. Dash et al. (eds.), *Artificial Intelligence and Evolutionary Computations in Engineering Systems*, Advances in Intelligent Systems and Computing 517,
DOI 10.1007/978-981-10-3174-8_34

Z source inverter which has made rapid growth with its unique characteristics of both buck and boost conversion, which are studied. A survey was performed on various PWM switching techniques for Z source inverter [3]. Z source neutral point clamped (NPC) inverter uses the space vector modulation technique enables the entire performance to be optimized without any extra commutations [4]. Quasi-ZSI network along with DC link boost control method was introduced to reduce the impacts of disturbances on grids which helps reducing voltage fluctuations compared with the traditional wind generation power system [5]. By choosing optimum value for inductors and capacitors, the static states of operating cycles of Z source inverters are avoided [6]. Embedded EZ source inverters can produce the same gain as the Z source inverters but with smoother and smaller current/voltage maintained across the DC input source and within the impedance network [7]. These inverters have their DC sources inserted within their X-shaped impedance networks so as to achieve implicit source current or voltage filtering without requiring additional hardware, and therefore avoiding its accompanied control and resonant complications [8]. In comparison to the SL-ZSI, for the same input and output voltages, the proposed SL-qZSI provides continuous input current, a common ground with the DC source, reduced the passive component count, reduced voltage stress on capacitors, lower shoot-through current, and lower current stress on inductors and diodes [9]. The improved ZSI has inherent limitation to the inrush startup current and as soft start strategy was proposed [10]. A significant scheme using SVPWM was introduced for Quasi-ZSI to reduce voltage fluctuations when compared with the conventional wind energy power system [11]. The EZ source inverter produced the voltage gain same as that of basic ZSI but with smooth voltage and current maintained across DC input voltage [12]. Adapted SVPWM for T source inverter for renewable energy system was studied [13]. Transformer-based quasi-Z source inverter with high boost ability was studied in which replacing one of the two inductors in the quasi-Z source inverter with two-winding transformer, the proposed inverter produces a very high boost voltage gain [14]. The series Z source DC link combined with the ultra-sparse matrix converter topology provides a new converter with high voltage boosting with limited inrush current during startup [15]. By replacing the original two inductors found in the classical impedance network with two modified 2-terminal TL cells, output voltage range of the new inverter can be expanded with shorter shoot-through duration and get larger modulation index with better output [16]. The LCCT-ZSI has small number of components when compared to the normal ZSI which is operated with similar shoot-through ratio [17]. The new T source inverter has fewer reactive components and a common voltage source of the passive arrangement so as to couple power circuit with main circuit [18]. The transformer-based Z source network with enhanced voltage gain and less voltage stress in the voltage-fed trans-ZSIs and the expanded motoring operation ranges in the current-fed trans-ZSIs [19]. Gamma-shaped voltage-type Z source network for boosting their output voltage uses lesser components and a coupled transformer for producing both very high gain and good modulation ratio by tuning the transformer turn's ratio [20]. A survey on various types of Z-source inverters along with different types of wind generators is discussed [21]. Simulation of

Γ-shaped Z-source inverter with improved voltage gain controlling the speed of PMSM is discussed [22]. Simulation and experimental results of wind-based ZSI-type AC–AC converter-fed PMSM drive system are studied [23]. Comparison of PID and fuzzy logic-controlled wind generator-fed Γ-Z-source-based PMSM drive systems is performed in [24]. Simulation and experimental results of wind generator-fed Γ-ZSI-controlled PMSM drive are discussed [25]. Comparison of PI and PID-controlled wind generator-fed Γ-Z source-based PMSM drives is discussed in [26].

The above literature does not compare PID- and ANN-controlled closed-loop system implemented for PMSM. This work proposes ANN control to improve dynamic response of Γ-ZSI-controlled PMSM drive.

This paper is organized as follows: Sect. 2 deals with the working principle of Γ-ZSI; Sect. 3 deals with the proposed system of Γ-ZSI with ANN. Discussion about the simulation results of both Γ-ZSI-based PID and ANN closed-loop systems and its comparisons are presented in Sect. 4. Conclusion and the scope for future work are presented in Sect. 5.

2 GAMMA-Shaped (Γ) Z Source Inverter

Figure 1 depicts the Γ-shaped Z source network combined with three-phase PWM inverter. This network has got components same as that of Trans-Z source network. The placement of the winding of the transformer only is altered to form a gamma shape. This leads to increase the mutual inductance between the windings and drastically reduced the leakage flux. Hence, it has improved the voltage gain much more than Trans-ZSI by having low turn's ratio in the range that falls in between 1 and 2.

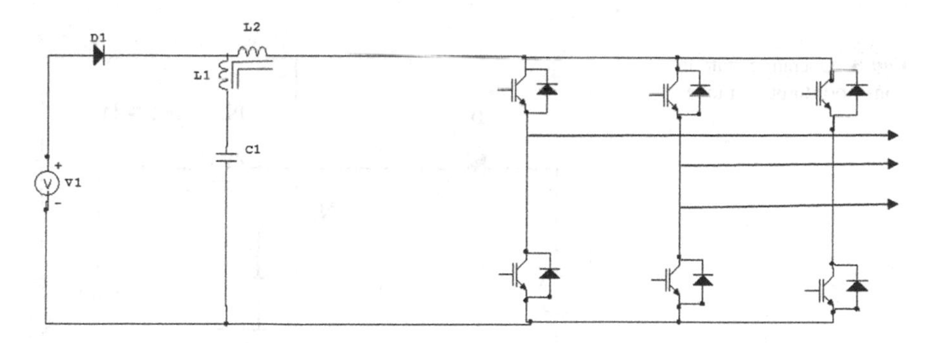

Fig. 1 Γ-Z source inverter

2.1 Working Principle of Γ-ZSI

Two modes of operations are explained as follows:

Shoot-through state:

Figure 2 depicts the equivalent circuit of shoot-through state of the gamma Z source inverter. This state can be attained by keeping any one of the phase legs shorted or all of them simultaneously. The diode D becomes reverse biased providing the path for the capacitor current to discharge into the closed path and the DC link voltage becomes zero.

Non-shoot-through state:

Figure 3 shows the circuit of non-shoot-through state mode of the gamma Z source inverter. The diode conducts and hence the source current passes through the capacitor leg and current source leg. The output from the Γ-Z source inverter is

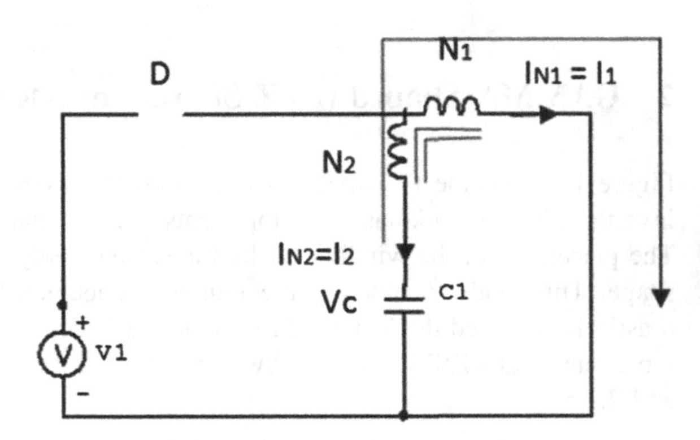

Fig. 2 Operating state in shoot-through mode

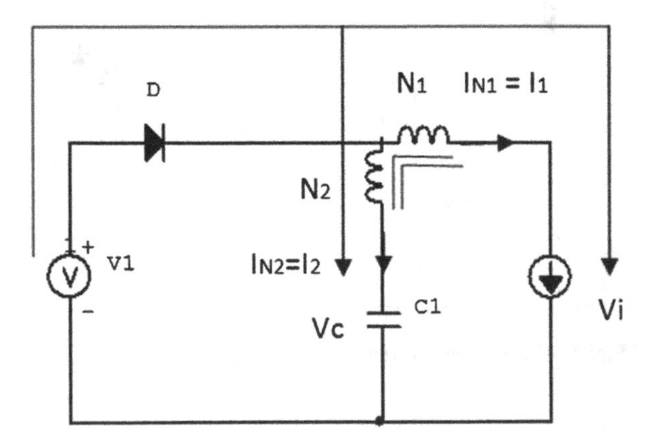

Fig. 3 Operating state in non-shoot-through mode

higher due to the coupled transformer network. It shows that the Γ-Z source currents have high peak to peak current during the shoot-through states identified by DC link zero voltage.

3 Proposed System Using Γ-ZSI with ANN

The proposed system employs Γ-ZSI as control element. The system consists of wind generator simulator which produces a very low AC output voltage and a three-phase rectifier which converts into very low DC output voltage. This voltage is boosted up with high voltage gain by Γ-ZSI network and three-phase PWM inverter converts DC into 3∅ AC output voltage with the help of control pulses generated by control units. The AC output voltage that runs the PMSM motor with high speed. The speed of the motor is taken as feedback and compared with reference speed. The difference of the speed is fed into ANN controller which uses error-back propagation algorithm to generate appropriate control action as input to the control unit which is given to both controlled rectifier and PWM inverter switches and thus the dynamic response of the motor is attained quickly (Fig. 4).

4 Simulation Results

Table 1 displays the list of simulation parameters as shown below.

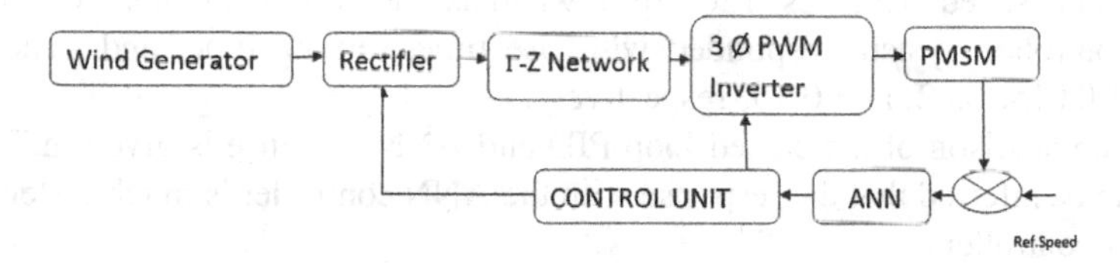

Fig. 4 Wind energy conversion system using Γ-ZSI with ANN control

Table 1 Simulation parameters with values

Parameter	Value
Wind turbine output voltage	230 V
Γ-ZSI self-resistance	5.8 Ω
Γ-ZSI self-inductance	$3e^{-3}$ Ω
Γ-ZSI mutual inductance	1 Ω
Γ-ZSI capacitance	$16{,}000e^{-6}$ μF
Resistance/phase	2.875 Ω
Inductance/phase	$8.5e^{-3}$ H
Switching frequency	50 kHz

Using MATLAB, the output voltage and current waveforms along with the measurement of PMSM motor speed and its developed torque were considered for comparison between PID and ANN systems.

4.1 Closed-Loop System with Proportional Integral Derivation (PID)

Simulink model of closed-loop system with PID controller is shown in Fig. 5a. The step speed and actual speeds are compared and the error is applied to a PID controller. The pulses are generated such that the speed is regulated. The output voltage and output current waveforms of inverter are shown in Fig. 5b, c, respectively. The speed response is shown in Fig. 5d. The torque response is shown in Fig. 5e. The speed achieves 1200 rpm whereas the torque attains 8 Nm. The PID-controlled system responded with rise time, settling time, and steady-state error of 0.05, 0.1, and 0.2, respectively.

4.2 Closed-Loop System with Artificial Neural Network (ANN)

Simulink model of closed-loop system with ANN controller is shown in Fig. 6a. The output voltage and output current waveforms of inverter are shown in Fig. 6b, c, respectively. The speed and torque responses are shown in Fig. 6d, e respectively. The speed achieves 1200 rpm whereas the torque attains 10 Nm. The ANN-controlled system responded with rise time, settling time, and steady-state error of 0.015, 0.02, and 0.15, respectively.

The comparison of the closed-loop PID and ANN response is given in Table 2. It can be inferred that the response with the ANN controller is much faster that of the PID controller.

5 Conclusions

The PID and the ANN-based wind generator-fed Γ-ZSI-controlled PMSM drives were simulated using Matlab/Simulink. The comparison of the results indicates that ANN-controlled drive provides smooth response with low steady-state error.

The proposed drive system has advantages like reduced rise time. The disadvantage of this system is that training of neural network takes much time. The present work deals with simulation of comparison of PID and ANN control system. The implementation of the comparison work can be done in future.

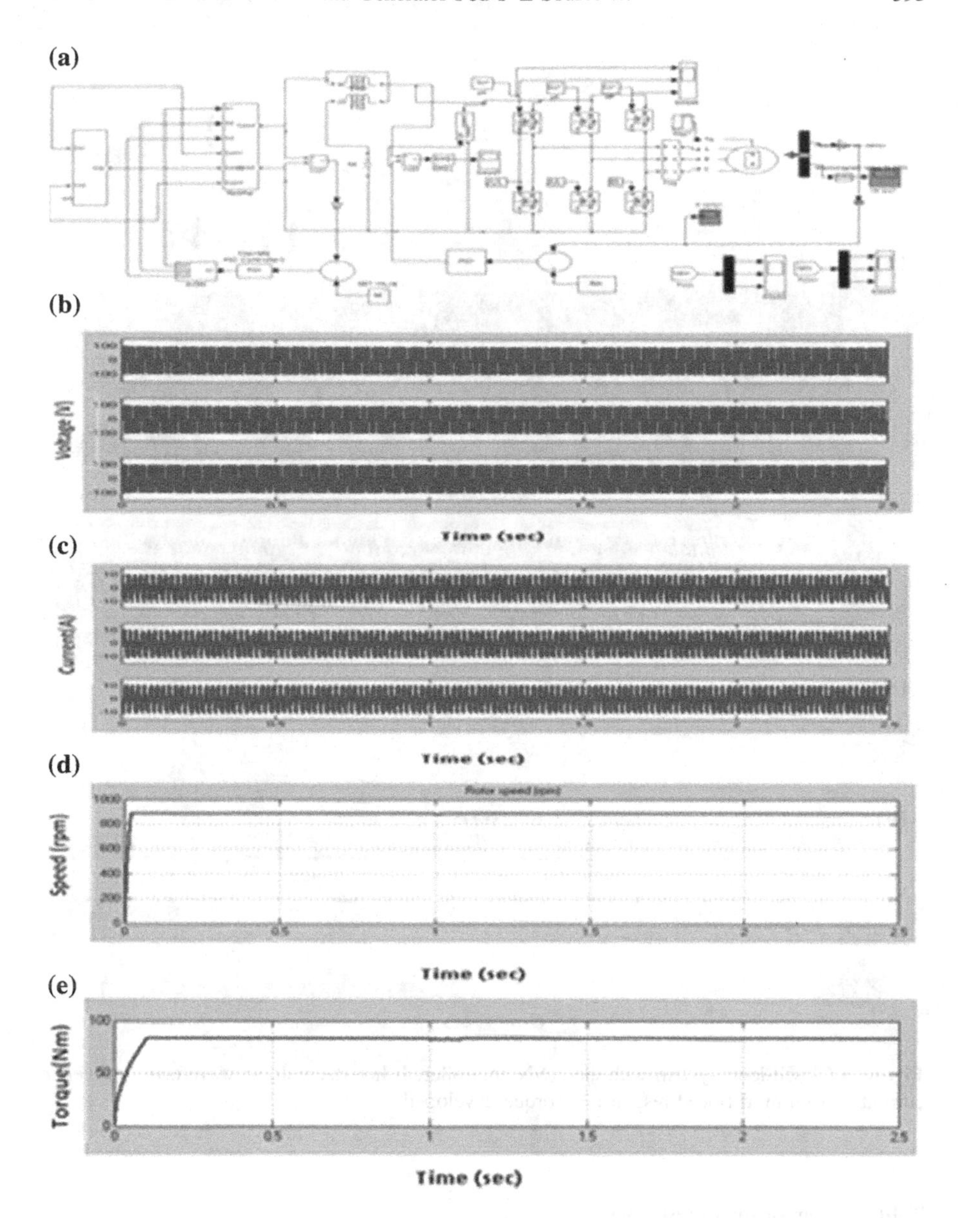

Fig. 5 **a** Closed-loop system with the PID controller. **b** Output voltage of the inverter. **c** Output current of the inverter. **d** Speed response. **e** Torque developed

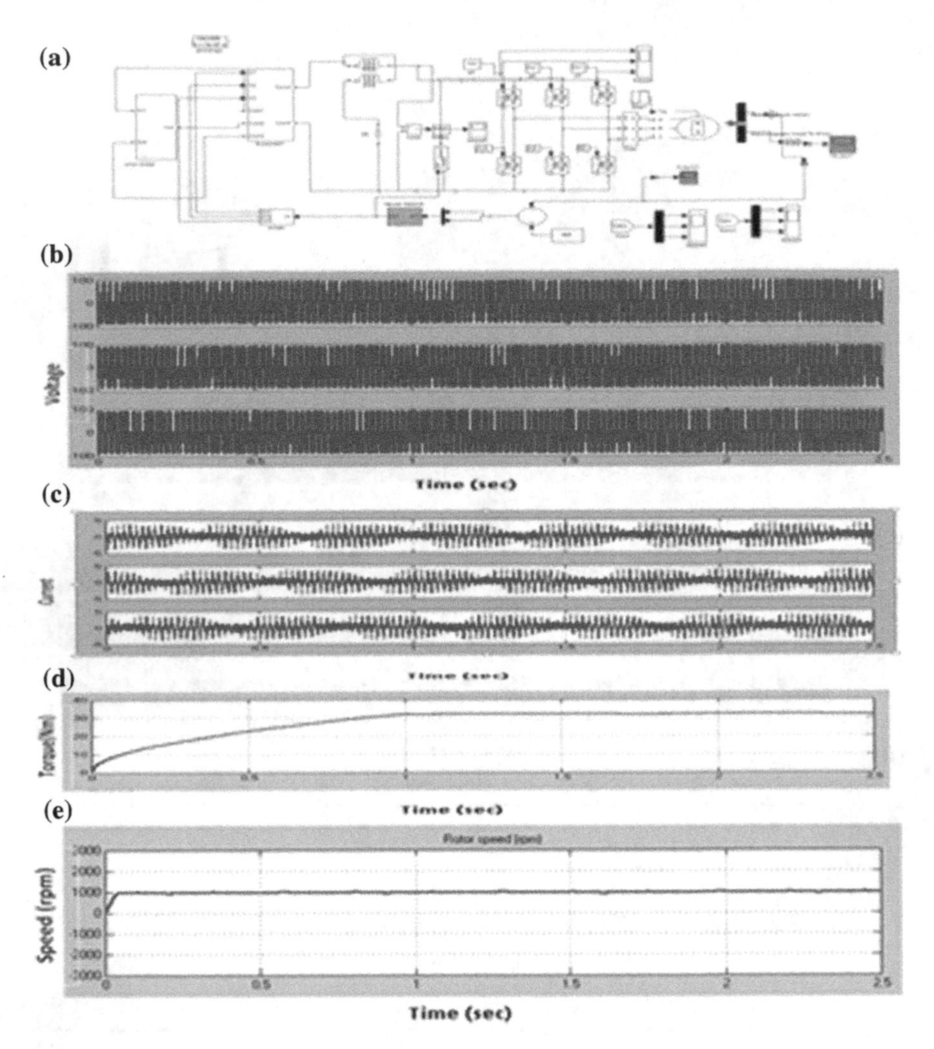

Fig. 6 **a** Closed-loop system with the ANN controller. **b** Inverter voltage waveform. **c** Inverter current waveform. **d** Speed response **e** Torque developed

Table 2 Comparison of responses

Controllers	Rise time (s)	Settling time (s)	Steady-state error (V)
PID	0.05	0.1	0.2
ANN	0.015	0.02	0.15

References

1. J. Kikuchi and T. A. Lipo, "Three phase PWM boost-buck rectifiers with power regenerating capability," *IEEE Trans. Ind. Appl.*, vol. 38, no. 5, pp. 1361–1369, Sep./Oct. 2002.
2. G. Moschopoulos and Y. Zheng, "Buck-boost type AC–DC single-stage converters," in *Proc. IEEE Int. Symp. Ind. Electron.*, Jul. 2006, pp. 1123–1128.
3. M.S. Bakar, N.A. Rahim, K.H. Ghazali, A.H.M. Hanafi, "Z-Source Inverter Pulse Width Modulation: A survey, International Conference on Electrical, Control and Computer Engineering", Pahang, Malaysia, June 21–22, 2011.
4. Francis Boafo Effah, Patrick Wheeler, Jon Clare, and Alan Watson, "Space-Vector-Modulated Three-Level Inverters With a Single Z-Source Network", IEEE Transactions On Power Electronics, Vol. 28, No. 6, June 2013, pp. 2806–2815.
5. G. Sen and M. E. Elbuluk, "Voltage and current-programmed modes in control of the Z-source converter," *IEEE Trans. Ind. Applicat.*, vol. 46, no. 2, pp. 680–686, Mar./Apr. 2010.
6. S. Rajakaruna and L. Jayawickrama, "Steady-state analysis and designing impedance network of Z-source inverters," *IEEE Trans. Ind. Electron.*, vol. 57, no. 7, pp. 2483–2491, Jul. 2010.
7. P. C. Loh, F. Gao, and F. Blaabjerg, "Embedded EZ-source inverters," *IEEE Trans. Ind. Appl.*, vol. 46, no. 1, pp. 256–267, Jan./Feb. 2010.
8. F. Gao, P. C. Loh, D. Li, and F. Blaabjerg, "Asymmetrical and symmetrical embedded Z-source inverters," *IET Power Electron.*, vol. 4, no. 2, pp. 181–193, Feb. 2011.
9. F. Z. Peng, M. Shen, and K. Holland, "Application of Z-source inverter for traction drive of fuel cell—Battery hybrid electric vehicles," *IEEE Trans.Power Electron.*, vol. 22, no. 3, pp. 1054–1061, May 2007.
10. Y. Tang, S. Xie, C. Zhang, and Z. Xu, "Improved Z-source inverter with reduced Z-source capacitor voltage stress and soft-start capability," *IEEE Trans. Power Electron.*, vol. 24, no. 2, pp. 409–415, Feb. 2009.
11. Liu Yushan, Ge Baoming, Abu Rub Haitham, Aníbal T. de Almeida, Fernando J. T. E. Ferreira, "Quasi-Z-Source Inverter based PMSG Wind Power Generation System", 978-1-4577-0541-0/11/$26.00 ©2011 IEEE, pp 291–297.
12. H. Itozakura, H. Koizumi, "Embedded Z-source inverter with switched inductor", IECON 2011—37th annual conference on IEEE industrial electronics society.
13. F. Gao, P. C. Loh, F. Blaabjerg, and C. J. Gajanayake, "Operational analysis and comparative evaluation of embedded Z-source inverters," in *Proc. IEEE Power Electron. Spec. Conf.*, Jun. 2008, pp. 2757–2763.
14. M. K. Nguyen, Q. D. Phan, Y. C. Lim, S. J. Park, "Transformer-based quasi-Z-source inverters with high boost ability", IEEE, Industrial Electronics (ISIE), 2013 IEEE International Symposium.
15. M. Zhu, K. Yu, and F. L. Luo, "Switched inductor Z-source inverter," *IEEE Trans. Power Electron.*, vol. 25, no. 8, pp. 2150–2158, Aug.2010.
16. M. Zhu, D. Li, P. C. Loh, and F. Blaabjerg, "Tapped-inductor Z-source inverters with enhanced voltage boost inversion abilities," in *Proc. IEEE Int. Conf. Sustainable Energy Technol.*, Dec. 2010, pp. 1–6.
17. M. K. Nguyen, Y. C. Lim, and G. B. Cho, "Switched-inductor quasi-Zsource inverter," *IEEE Trans. Power Electron.*, vol. 26, no. 11, pp. 3183–3191, Nov. 2011.
18. R. Strzelecki, M. Adamowicz, N. Strzelecka, and W. Bury, "New type T-source inverter," in *Proc. Compat. Power Electron.'09*, May 2009, pp. 191–195.
19. W. Qian, F. Z. Peng, and H. Cha, "Trans-Z-source inverters," IEEE Trans.
20. Poh Chiang Loh, Ding Li and Frede Blaabjerg "Γ Z-source inverters", IEEE Trans, Nov 2013, pp. 4880–4884.
21. Ali, A. Jaffar, and G. P. Ramesh. "Converters for wind energy conversion—a review" International Journal of Technology and Engineering Science (2013): 1019–1026.

22. A. Jaffar Sadiq Ali, and G. P. Ramesh "Simulation of Wind Generator Fed Γ-ZSI Controlled PMSM Drive"—International Conference on Green Peace Technologies Feb 2015, pp 12881–12885, Vol. 10 Issue. 17.
23. Jaffar Sadiq Ali. A, and Ramesh G.P. "Simulation and Experimental Results of Wind Based ZSI Type AC–AC Converter Fed PMSM Drive System", Australian Journal of Basic and Applied Sciences, 9(23) July 2015, Pages: 517–521.
24. Jaffar Sadiq Ali, A. and Ramesh, G. P. "Comparison of PID and Fuzzy Logic Controlled Wind Generator Fed Γ-Z Source Based PMSM Drive Systems" ARPN Journal of Engineering and Applied Sciences, ISSN 1819–6608, Vol. 10, No. 20, November, 2015.
25. Jaffar Sadiq Ali, A., and Ramesh, G.P. "Simulation and Experimental Results of Wind Generator Fed Γ-ZSI Controlled PMSM Drive are discussed", International Journal of Computer Theory and Applications (IJCTA), Vol. 8(4), Dec 2015, pp. 1495–1501.
26. A. Jaffar Sadiq Ali, and G. P. Ramesh "Comparison of PI and PID Controlled Wind Generator Fed Γ-Z Source Based PMSM Drives", Indian Journal of Science and Technology (INDJST), Vol. 9(1), January 2016.

Analysis of Power Management Techniques in Multicore Processors

K. Nagalakshmi and N. Gomathi

Abstract Power and performance have become significant metrics in the designing of multicore processors. Due to the ceasing of Moore's law and Dennard scaling, reducing power budget without compromising the overall performance is considered as a predominant limiting factor in multicore architecture. Of late technological advances in power management techniques of the multicore system substantially balance the conflicting goals of low power, low cost, small area, and high performance. This paper aims at ascertaining more competent power management techniques for managing power consumption of multicore processor through investigations. We highlight the necessity of the power management techniques and survey several new approaches to focus their pros and cons. This article is intended to serve the researchers and architects of multicore processors in accumulating ideas about the power management techniques and to incorporate it in near future for more effective fabrications.

Keywords Multicore processor · Power management · DVFS · Clock gating · Task scheduling · Task migration

1 Introduction

In today's technology, multicore architectures are becoming dominant design paradigm, which assimilates two or more processing elements (cores) in a single chip for higher performance computing. The proliferation of heavy computational

---

K. Nagalakshmi (✉)
Department of Computer Science and Engineering,
Hindusthan Institute of Technology, Coimbatore, Tamilnadu, India
e-mail: nagulaxmi@gmail.com

N. Gomathi
Department of Computer Science and Engineering,
Vel Tech Dr. RR & Dr. SR. University, Chennai, Tamilnadu, India
e-mail: gomathin@veltechuniv.edu.in

S.S. Dash et al. (eds.), *Artificial Intelligence and Evolutionary Computations in Engineering Systems*, Advances in Intelligent Systems and Computing 517,
DOI 10.1007/978-981-10-3174-8_35

requirements of real-time applications in the multicore system leads severe power and performance constraints to provide quality of service (QoS) to the users. For example, an ARM9 processor that supports 266 MHz clock frequency consumes power of 0.45 mW per MHz, and an ARM11 processor executing 550 MHz of speed, consumes power of 0.8 mW per MHz [1]. Thus, designing the multicore architectures to resolve demanding and competing power-performance tradeoff is a challenging endeavor. The research and design community have invested significant efforts in exploring several power management techniques to scale their performance and make sure reliability, prolonged existence, and acceptance in a wide range of applications.

As stated in Moore's law, the number of transistors fabricated in a single chip approximately doubles every 18–24 months [2], resulting in an exponential increase in transistor density. This indicates that the speed (clock frequency) of the processor will also double in every 18 months. However, we cannot enjoy this exponential growth continuously due to its increasing power density on-chip which prevents all the cores to be switched on simultaneously. This utilization barrier is called dark silicon problem, and is driving the emergence of heterogeneous multicore architectures [3].

Multicore processors can be categorized into three types: Homogeneous processors [4–6], heterogeneous processors [7, 8], and dynamic reconfigurable processors [9]. Conventionally, most of the general purpose multicore architectures are built with identical cores as shown in Fig. 1. All these cores consist of same micro-architectural innovations (i.e., cache memory, out-of-order execution, speculation, pipeline, branch prediction configuration, etc.) and are able to operate under same instruction set architecture (ISA). This type of architecture is called as homogeneous or symmetric multicore architecture (SMP). It is easy to design and implement as we just need to duplicate the core.

The multicore architectures: INTEL CORE i7 [4, 10], INTEL ATOM [11], AMD PHENOM [12, 13], and SUN NIAGARA [14, 15] are general purpose SMPs with large cache memories. These processors are designed for general purpose desktop and server applications where power is not a primary concern. In contrast, the homogeneous architectures XMOS-XS1 [5] and ARM-CORTEX [16] are specially designed for mobile devices, where power is an increasing concern. Some of the multicore systems are developed for high-performance computing. Therefore, they employ a larger number of cores. For example, AMD RADEON 700 GPU [6] contains 160 cores while NVIDIA G200 [17] contains 240 cores.

Though homogeneous cores are simple to design, easy to implement, and provide regular software environments; they cannot deliver required performance and energy efficiency for different real-time applications. Real-time applications with different QoS encourage the computer architects and software designers to exploit architecture innovations and design heterogeneous multicore processors.

Heterogeneous or asymmetric multicore processors (AMP) implement a mixture of non-identical processing elements that are asymmetric in their underlying principles and performance. The cores are varying in size and complexity, but they are designed to cooperate with each other to increase the performance of the system.

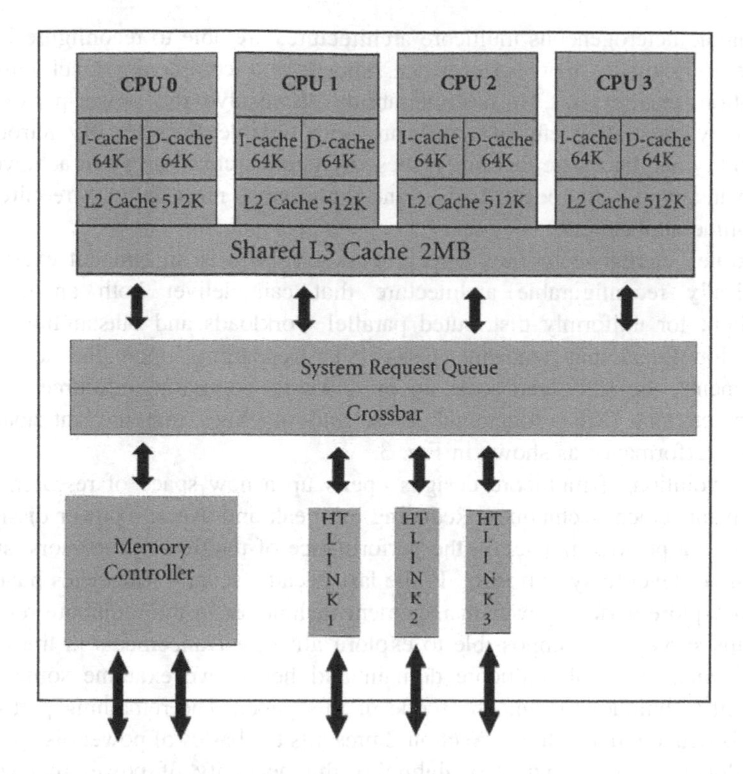

Fig. 1 Quad core homogeneous architecture

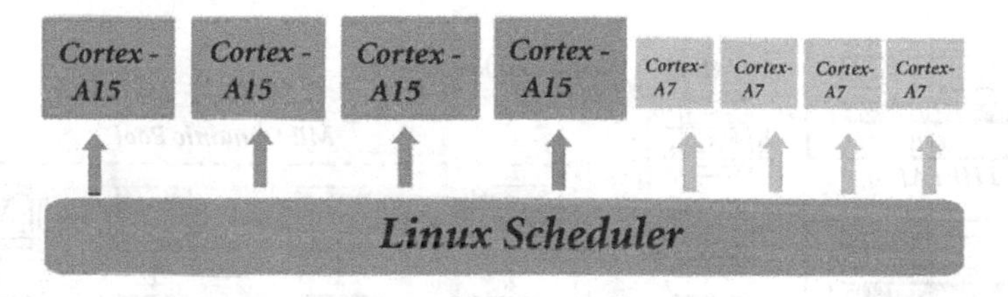

Fig. 2 ARM big.LITTLE asymmetric multicore processor

These designs provide an efficient solution for dark silicon era and also increase reliability, performance, and energy efficiency of the applications. A typical AMP will integrate small (slow) cores to process simple tasks in an energy-efficient way, and complex (fast) cores to provide higher performance.

Cell BE [7] and ARM big.LITTLE [8] are renowned examples for AMP architectures. Cell BE is extensively used in gaming devices and computing platforms aiming at high performance. ARM big.LITTLE is designed for mobile platforms where complex, performance driven, quad-core Cortex A15 assembly is combined with simple, power-optimized, quad-core Cortex A7 assembly to provide peak performance as shown in Fig. 2.

Dynamic heterogeneous multicore architectures are able to reconfigure itself at runtime to regulate their performance, speed, and complexity level based on application requirements. It has the ability to resolve the power-performance tradeoff by integrating efficient hardware with flexible software. By introducing adaptability and hardware flexibility, these dynamic architectures can achieve high performance within the power budget and therefore to meet the QoS requirements of real-time applications.

Flexible heterogeneous multicore processor (FMC) is an eminent example of dynamically reconfigurable architecture that can deliver both an increased throughput for uniformly distributed parallel workloads and outstanding performance for fluctuating real-time tasks [9]. Depending upon the application requirements, the FMC can scale up or down its computing resources such as memory engines (ME), functional units, and pipelines that are anticipated to improve performance as shown in Fig. 3.

The evolution of multicore designs opens up a new space of research called power management techniques. Reducing the peak and average power dissipation could have a positive impact on the performance of multicore processors, starting from circuit level to system level. In the last decade, several researches have been keen to explore various power management techniques in the multicore regime.

In this survey, it is impossible to explore all the advancements in the field of power management of multicore domain and hence, we examine some of the important techniques to limit the scope of this paper. The remaining part of our survey is structured as follows: Section 2 presents the basics of power dissipation in the multicore domain and also highlights the necessity of power management

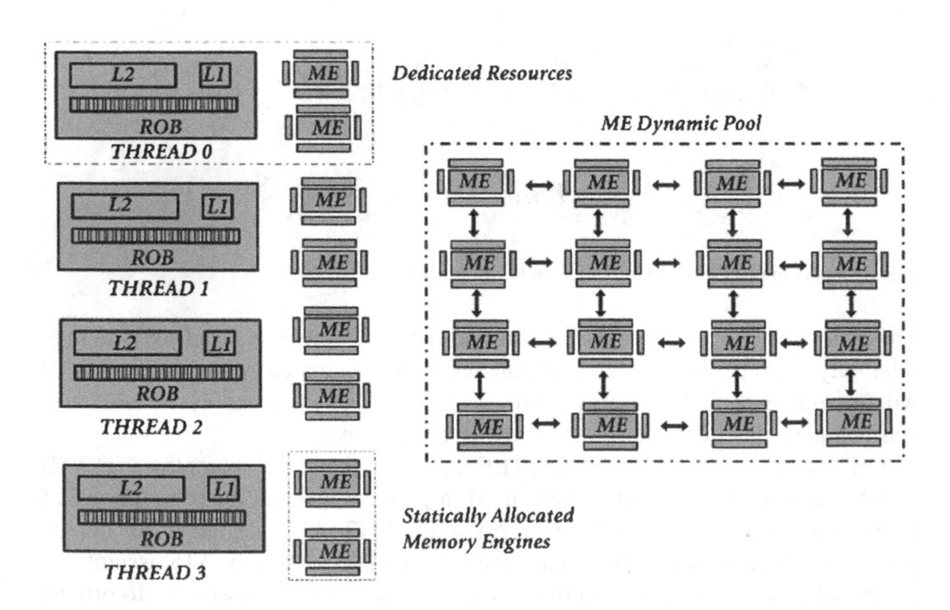

Fig. 3 Flexible heterogeneous multicore processor (FMC)

techniques. Section 3 provides an overview and taxonomy of classic power management techniques and explores some of these techniques in detail. Finally, Sect. 4 is arranged to provide the conclusions of our investigation.

2 Background

2.1 Basics of Power Dissipation

Nowadays, multicore is ubiquitous both in general purpose and application-specific computing systems starting from smartphones to commercial servers. Power dissipation is an important constraint because it increases the temperature and cooling costs, reduces reliability, and degrades performance. Power consumed by CMOS devices can be resolved into three parts [18] as shown in Eq. (1).

$$P_{tot} = P_{static} + P_{dynamic} + P_{short\ circuit} \tag{1}$$

where P_{tot} is total power dissipation, P_{static} is static or leakage power due to leakage of a transistor's bias currents. $P_{dynamic}$ is dynamic power due to switching of transistors. $P_{short\ circuit}$ is power dissipation related to concurrent conduction of *p* type and *n* type transistors.

$$P_{static} = V \times I_{Leak} \tag{2}$$

where V denotes source voltage, I_{Leak} is the transistor leakage current. Usually, leakage power contributes 20–40% of the total power dissipation [19]. Dynamic power is a prevalent factor as compared to other two components in Eq. (1). It can be denoted as follows V^2

$$P_{dynamic} = \beta \times C_{Load} \times V^2 \times f \tag{3}$$

where β is activity factor, C_{Load} indicates the effective load capacitance and f is the switching speed. To simplify Eq. (3), it is assumed that the clock speed is linearly proportional to the source voltage. If we apply this notion to the above Eq. (3), then the reduction in source voltage and switching speed lowers dynamic power cubically.

$$P_{dynamic} \propto V^3 \tag{4}$$

Short circuit power is calculated by the following equation

$$P_{short\ circuit} = V \times I_{SC} \tag{5}$$

Here I_{sc} represents the short circuit current flowing from supply to ground. Short circuit power is comparatively trivial for static CMOS circuits.

2.2 Failure of Dennard Scaling

For almost 30 years, the computing community has realized a stable performance evolution in the uniprocessor, motivated by Moore's Law [2] and Classical (Dennard) scaling [20]. But now this curve is slowed down and came to halt due to memory wall, power wall, and instruction level parallelism (ILP) wall [21]. The maximum power dissipation of a processor (power wall) is measured in terms of thermal power design (TDP) envelop. TDP is defined as the maximum power at which a processor chip can operate without overheating.

Under Dennard scaling, the power requirements per unit space can remain constant across semiconductor generations. According to this principle [20], with a linear dimension scaling ratio of 0.707, the transistor count could doubles (Moore's Law), frequency increases by 40%, but the power consumed per transistor is reduced by half keeping the total chip power constant in every two years [22]. From Eq. (3) the power density in a chip area A is measured as follows:

$$\text{Dynamic Power Density} = (\beta \times C_{\text{Load}} \times V^2 \times f)/A \tag{6}$$

As we move toward the next generation of IC manufacturing technology, the linear size of an IC gets scaled by 0.707. The same scaling ratio is applied for load capacitance and supply voltage while clock speed is scaled by 1/0.707. So the area of the chip is now 0.707^2 A. If we calculate a new power per unit space, we have $0.707C \times 0.707V^2 \times f/(0.707^3 \times A)$. Hence, the power per unit area becomes unchanged. But unfortunately, in 65 nm technology and below, this law ceased because of exponential growth in leakage current and reduction in supply voltage decreases the speed of the processor. Nonetheless, the high-performance demand is continued and this stimulates a shift from the single core to a multicore paradigm.

2.3 Need for Power Management

Today's innovations in semiconductor technology lead to not only an increase in the number of cores on a die but also an increase in power density and concomitant heat dissipation. Increasing power dissipation leads many negative impacts on power delivery, performance/watt (PPW) ratio, packaging and cooling costs, reliability, availability, and overall performance of the processors. So power consumption issues occasionally more important than speed of the processor.

Power optimization is essential in mobile electronics where devices are battery powered. For last few decades, processor performance has been accelerating at a rate faster than the evolutions in battery technologies. This has led to a considerable drop of the battery life in mobile devices. At the same time, modern computational intensive applications demand very high performance. These two conflicting requirements, the need to conserve power and the demand to deliver outstanding performance lead new approaches to resolve it. Existing researches to achieve optimal power budget have two significant guidelines. First is in what way to increase the processor's efficiency within a specified power limit. Second is in what way to decrease the power dissipation of computing devices without sacrificing performance.

3 Overview of Classic Power Management Techniques

In this section, we delve into the up-to-date techniques in power management of multicore systems. As of now, several hardware and software approaches have been adopted for alleviating power and energy costs. Through this investigation, we aim to demonstrate how the research community is trying to achieve a greater performance and energy efficiency. We can classify the power management techniques into three broad categories: Hardware approaches, hardware-enabled middleware approaches, and software approaches. Figure 4 shows the taxonomy of the power management techniques in multicore architectures.

3.1 Hardware Approaches

Several power saving techniques with dedicated controllers are embedded into the modern processor architectures to provide energy efficiency. Applications running in a multicore domain need a carefully tailored computing architecture to meet their QoS within the power budget. The architectural innovations in designing of core, memory and interconnection networks improve the energy efficiency significantly.

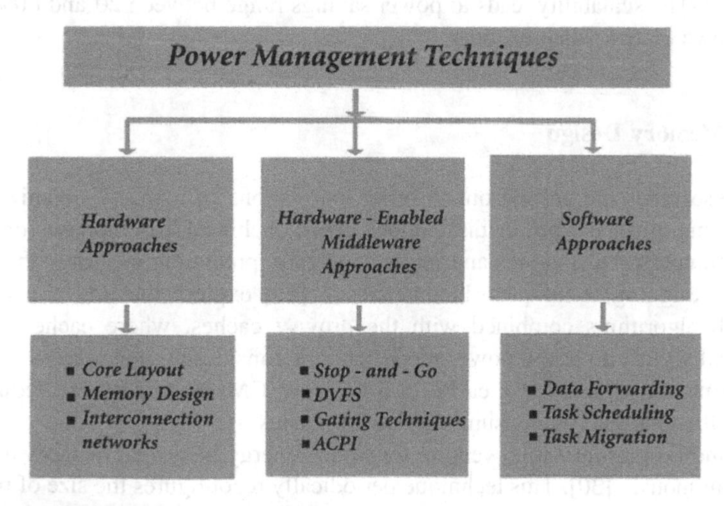

Fig. 4 Taxonomy of power management techniques

3.1.1 Core Layout

Puttaswamy et al. propose a 3D microarchitecture with thermal herding techniques, which provides outstanding PPW ratio [23]. Compared to a conventional planar processor, the proposed architecture can achieve 15–30% of active power reduction depending on the characteristics of the application. But 3D architecture incurs augmented power per unit area and associated temperature issues. This problem is resolved by Fazal Hameed et al. They present a thermal-aware 3D microarchitecture that effectively integrates the potential gains of dynamic architectural adaptation, fail-safe DVFS, and global migration [24]. Research shows that thermal-aware 3D architecture can achieve significant power reduction over 3D multicore processors [23], because it can reduce the active and the leakage power consumption simultaneously.

Kontorinis et al. introduce an adaptive processor with peak power guarantees, which can reduce peak power by table-driven reconfiguration [25]. Most of the functional units (e.g., ALU, L1 cache, register files, load-store units, and so forth) of the processor are dynamically organized for power conservation and maximum performance, whereas peak power constraints are assured. The adaptive processor can reduce peak power about 25% with a smaller amount of performance cost.

Rodrigues et al. propose dynamic core morphing (DCM) architecture for heterogeneous multicores [26]. The resources of the cores are morphed at runtime based on varying performance requirements. Depending on the computational need of the current workload, two cores may swap the execution units to maximize the PPW ratio.

Scalable stochastic processor developed by Narayanan et al. has been demonstrated as an auspicious way to tackle power dissipation problem for error-tolerant applications such as audio or video. Improved scalability is realized by substituting or augmenting conventional computational units by gracefully degrading functional units [27]. The scalability leads to power savings range between 20 and 60% in the well-known H.264 video encoder.

3.1.2 Memory Design

Many researches are carried out to bring innovations in memory organization in order to minimize the power dissipation. Smart caching [28] emphases on power saving computing techniques and implements way predicting caches with reduced leakage designing techniques. Flaunter et al. [29] explore the use of instruction pre-fetch algorithms combined with the drowsy caches, where cache lines are periodically put into a low power mode without considering their access histories. Implementation of a drowsy cache in a 0.07 μm CMOS process can reduce 50–75% of the total energy consumption in the caches.

Cai and Lu present a joint venture for saving energy in system memory and hard disk unanimously [30]. This technique periodically reconfigures the size of physical memory by adding or freeing up the allotted memory pages and uses a timeout policy for shutting down the hard disk. The suitable memory size and timeout are selected according to their proportionality with the average power consumption. This technique achieves energy savings higher than 50% over a fixed-timeout scheme.

3.1.3 Interconnection Network

Several researches show that the design choices for interconnection fabric have a significant impact on the power budget [31, 28]. The interconnection network itself is a power-consuming resource. The power consumption of the interconnection network for a 16 core processor is more than the combined power consumption of two cores. Rakesh Kumar et al. [31] demonstrate the need for a careful co-design of interconnection network and memory hierarchy. The power consumption of the core increases super-linearly with the number of connected units and the average length of wire. So a power-optimized architectural design needs a compact length of interconnection wire segments and appropriate routing algorithms [28].

3.2 Hardware-Enabled Middleware Approaches

The following techniques are employed as middleware and partially implemented in hardware. The hardware enables middleware to shutdown or slow down the functional units according to the operating temperature. Hardware-enabled middleware techniques including stop-and-go [32], dynamic voltage and frequency scaling [33], advanced configuration and power interface [34], and different gating techniques [35, 36] have attracted a great deal of attention.

3.2.1 Stop-and-Go

The stop-and-go is the simplest form of dynamic power management (DPM) technique [37]. The DPM techniques reduce the power consumption by shutting down or lowering the performance of idle cores. Stop-and-go can be realized on both global and local scale. In global approach, if one of the cores reaches its specified threshold temperature, this scheme shuts down the whole chip until its non-critical level has been recovered. If stop-and-go is realized locally, only the overheating core will be halted until it has cooled down. Global stop-and-go mechanism provides a smaller amount of control and less efficiency as a particular overheating core leads to the unwanted delaying of all other non-critical cores.

Donald et al. [32] implement 12 combinations of local and global stop-and-go policies with other power management techniques (i.e., DVFS and Task migration) for managing the temperature of multicore processors. They investigate the pros and cons of each combination by comparing their performance. Whenever peak temperature of the processor reaches 84.2 °C, the stop-and-go controller shutdowns the cores for 30 ms and allows the cores to cool down. Their implementation results show that local stop-and-go can outperform global schemes.

Chaparro et al. [38] propose a stop-and-go mechanism with clock gating mechanism. Whenever the core reaches its critical temperature, this combined technique halts the core, stores its current state information, and then shuts the core off completely. There is no dynamic as well as leakage power consumption in this technique. So it allows the overheating core to cool down quicker. As the current state of the core is saved before shutting down, this method would be the ideal choice of the options.

3.2.2 Dynamic Voltage and Frequency Scaling (DVFS) [33]

DVFS is the prevailing and powerful DPM technique, used to regulate the power consumption of the processor by dynamically scaling the level of supply voltage and clock frequency [33, 39–41]. The sub-threshold leakage current and gate-oxide tunneling leakage can be reduced by reducing supply voltage [42]. Reducing the clock frequency reduces the supply voltage linearly and decreases power consumption quadratically [43]. DVFS is widely used for memory-bound workloads and can be employed in two ways:

1. The Local (per-core) DVFS allow us to scale the voltage of individual cores so that the overheating core can cool down faster [38, 44, 45].
2. The Global DVFS allow us to adjust the voltages and frequencies of all cores uniformly and simultaneously. Similar to stop-and-go, a single hotspot on one of the cores could result in unnecessary performance penalty on all cores [38, 44, 46].

Local DVFS introduces more flexibility as each core can select its own voltage–frequency pair individually. However, it suffers from a large number of expensive inherent voltage regulators, the global DVFS can solve the thermal issues faster but the efficiency of DVFS is affected by limited flexibility to determining a single optimal voltage to all cores.

Weiser et al. present the first paper to suggest an interval-based DVFS for reducing the power dissipation in computing devices. Their work focuses on three scheduling algorithms: Unbounded-delay perfect-future (OPT), bounded-delay limited-future (FUTURE), and bounded-delay limited-past (PAST) [47]. The deployment of each algorithm controls the clock frequency and makes the scheduling decisions simultaneously. The PAST scheduling algorithm with a 50 ms adjustment interval can achieve power conservation of 50–70% based on circuit conditions.

Wonyoung et al. develop a fast, per-core DVFS mechanism with on-chip integrated voltage regulators [41]. This mechanism uses the potential benefit of both per-core voltage regulation and very fine-grained voltage switching. The inbuilt regulators can increase the energy efficiency opportunities of DVFS and result in 21% of energy savings over conventional global DVFS with off-chip regulators.

Many researchers have unified DVFS technique with thread migration policies to reduce energy consumption. Cai et al. develop a new thread shuffling algorithm, which integrates thread migration and DVFS techniques on a multicore system supporting simultaneous multithreading (SMT) [48]. Thread shuffling techniques dynamically migrate slower threads with same criticality degrees to a particular core and implements DVFS for other cores having fast threads. The proposed scheme realizes energy savings around 56% with no performance degradation.

Quan Chen et al. propose an energy-efficient workload-aware (EEWA) task scheduler, that consists of a work load-aware frequency adjuster and a preference-based task-stealing scheduler [49]. With the help of DVFS, the workload-aware frequency adjuster can accurately configure the frequencies of the cores in an efficient fashion based on the profiled workload statistics. The preference-based task-stealing scheduler can successfully distribute the tasks across various cores at runtime according to the preference list. The EEWA can save energy about 28.6% with only 0.9% of performance loss.

All the previous works cited above simply fail to consider the static power that has turn into a substantial portion of the total power consumption, unfortunately. LeSueur and Heiser [50] assess the factors influencing the efficiency of DVFS on AMD Opteron processors, using an extremely memory-bound benchmark. They illustrate that the ability of DVFS is retreating in modern digital systems due to escalating leakage power. Furthermore, their investigation reveals that switching off idle cores will facilitate greater energy savings. To reduce leakage power, Awan et al. [51] propose an enhanced race-to-halt (ERTH) approach. By integrating DVFS and slack management policies, ERTH can improve energy efficiency considerably.

When global DVFS is realized in multicore architecture, determination of optimal voltage that satisfies all cores is a challenging endeavor; some applications will suffer from performance penalty or overheads. This issue exacerbates as the running applications and number of cores in next-generation processors. From a hardware implementation point of view, local DVFS is more expensive than global DVFS, because of its costly inherent voltage regulators and phase-locked loops. However, the per-core DVFS provides a better tradeoff between performance and power.

3.2.3 Gating Techniques

Clock Gating [35] and Power Gating [36] are very useful methods for decreasing dynamic and static power correspondingly [52]. Gating techniques are realized by insertion of an additional logic between the clock source and clock input of the processor's circuitry. It diminishes power consumption by logically turning off (gating) the power to the portions of the core that are not useful to the current workload.

Clock Gating (CG)

The clock gating techniques employed in the Hexagon™ Digital Signal Processors (DSP) are analyzed by Bassett et al. [35]. The proposed four levels of clock gating and spine-based clock distribution allow switching off the power to the different regions, from single logic cell to entire chip. Further power reduction is achieved through a structured clock tree by distributing the clock signal across the chip with low skew and delay. This technique provides a reduction in power consumption by 8% for active mode and over 35% for sleep mode.

Hai et al. describe a deterministic clock gating (DCG) technique, which hinges on the advance knowledge about at what time the functional block will be idle in the upcoming cycles [53]. With this advance information, DCG can switch off the idle blocks that maximize the energy efficiency. By exploiting DCG to various functional units, the proposed technique achieves 19.9% of average diminution in power without any performance cost. However, for all these techniques, the effectiveness of gating is restricted by the granularity of components that can be gated, the failure to change the overall size and complexity of the processor. Also, these designs are still vulnerable to leakage inefficiencies.

Power Gating (PG)

Power gating (PG) is a circuit-level technique to reduce leakage power consumption by effectively turning off the source voltage to the idle elements. PG can be applied either at the core-level [36] or at the unit-level of the processor such as cache banks, ALUs, pipeline branches, etc. [54, 55]. Recently, Intel Core i7 processors use power gating transistors to turn off its idle cores [56].

Hu et al. develop a parameterized model based on analytical equations, which decides the breakeven point used for proper gating. They evaluate the dynamic power gating ability of the FPU (floating-point units) and FXU (fixed-point units) of POWER4 processor by three techniques namely ideal, time-based, and branch-misprediction-guided [54]. The implementation of these techniques in various execution units shows that a considerable decrease in static power consumption can be realized through power gating.

Lungu et al. propose a success monitor switch (SMS) and a token counting guard mechanism (TCGM) for applying predictive power gating technique in POWER6 processor [55]. By employing SMS, the control logic enables or disables the PG depends on the success of the policy. By implementing work for TCGM, this predictive power gating achieves a guarantee on the worst-case execution of the policy. Leverich et al. [36] propose a per-core power gating (PCPG) technique with DVFS for data center workloads. This combined technique can save up to 60% of energy consumption.

3.2.4 Advanced Configuration and Power Interface (ACPI) [34]

ACPI is an industry standard for the efficient handling of power management in computing devices. It is developed by the collaborative effort of Intel, Hewlett-Packard, Phoenix, Microsoft, and Toshiba [34]. ACPI provides platform-independent interfaces for power management and monitoring. These interfaces have the potential to work with existing DPM techniques [34]. ACPI relies on operating system-directed configuration and power management (OSPM), which defines four switchable C-states (CPU-idle states) C0, C1, C2, and C3 and n P-states (CPU-performance states) P0 to Pn for active power management. ACPI allows the processor to achieve fine tuning of the power consumption by moving idle devices into lower power states (sleeping state).

Bircher and John [57] point out the implicit and explicit performance impacts of various CPU-idle states and Performance states of AMD quad-core processors. They verify their results for both compute-bound and memory-bound applications with fixed and OS scheduling. They develop an enhanced hardware and operating system configurations that decrease average active power by 30% with 3% of performance loss.

3.3 *Software Approaches*

The performance per watt ratio of a multicore architecture depends on efficient built-in hardware and the ability of software to effectively control the hardware. Many up-to-date processors exploit software level power management techniques for energy efficiency. Recently, researchers have paid greater attention to the software power management policies because it can gain the power disparity statistics of processing threads on the fly with low cost. Software techniques can achieve predictable performance through transferring or scheduling tasks to minimize thermal gradients and hot spots. Software-based approaches include data forwarding [58, 59], task scheduling [60], and task migration [61].

3.3.1 Data Forwarding

Most modern processors use large size on-chip L1 caches with multiple ports. Such a cache consumes a substantial part of the overall power owing to its larger size and high-frequency access rate. Researches reveal that L1 data cache contributes 15% of the overall energy consumption of the processor [58]. Thus, it is essential to develop tactics for precluding large power consumption in the cache memories. Data forwarding is one of the appropriate solutions to reduce the energy consumption of L1 data cache.

Carazo et al. [59] propose a data cache filtering technique with forwarding predictor to reduce the power consumption of L1 data cache (DL1). This mechanism exploits an effective utilization of load-store queue (LSQ), which is responsible for providing the right data to load instructions by data forwarding method. Their experiments exhibit that the proposed cache filtering technique can achieve an average power saving up to 36% with a 0.1% of performance degradation.

To reduce the access rate of DL1, Nicolaesu et al. propose a cached LSQ (CLSQ) to maintain load and store instructions after their execution [58]. Hitting in the CLSQ is faster and wastes not as much of energy as a DL1 access. Thus, the significant savings in the frequency of accesses leads to 40% of energy reduction without any additional hardware complexity and performance penalty.

3.3.2 Task Scheduling

Task scheduling is another breakthrough technique in power management arises from software approach. Scheduling algorithms are designed to solve temperature issues by distributing tasks among different cores. There have been wide ranges of literature put out on scheduling algorithms to achieve more processor utilization, better power conservation, and more uniform power density without degrading the processor throughput. These algorithms schedule the tasks across cores based on predetermined temperature threshold.

Work proposed by Hsin-Hao Chu and Yu-Chon Kao is a perfect example of how an adaptive thermal-aware multicore task scheduling algorithm with multiple run-time controllers can mitigate the inter-core thermal costs and dynamic variations of task execution [60]. Implementations of runtime controllers increase the system complexity. To resolve this problem, the temperature-aware task scheduling algorithm, called low thermal early deadline first (LTEDF) is suggested by Wu et al. The LTEDF allocates tasks based on a novel history coolest neighborhood first allocation algorithm [62]. Simulation of the LTEDF algorithm demonstrates that it can satisfy the timing constraints for soft real-time tasks and minimize the thermal consequences simultaneously.

Power-aware task scheduling algorithms for multicore architectures can be classified into three types [63]: the global (dynamic binding) approaches [64], the partitioned (static binding) approaches [65], and the semi-partitioned approaches [66, 67]. In the global approach, any core in the multicore system may execute any task. Global scheduling saves tasks in the single priority-ordered queue, shared by all processors [64]. At every moment, the global scheduler chooses the highest-priority task for operation and the tasks are permitted to migrate between the processors. There are three types of priority assignment schemes for multicore architecture: fixed-priority, [65, 68], dynamic priority [69], and proportionate fair (PFair) priority [70].

Fisher et al. develop a thermal-aware global scheduling algorithm for sporadic real-time tasks based on two priority assignment schemes, namely the global earliest deadline first (EDF) and the global deadline-monotonic (DM) [71]. The suggested schemes can substantially lessen the peak temperature around 30–70 °C as compared to load-balancing strategies. Wang et al. develop a scheduling approach for hard real-time systems. They perform delay analysis for generic task arrivals using first-in-first-out (FIFO) scheduling and static-priority (SP) scheduling with reactive speed control techniques [72]. But Andersson and Baruah prove that the fixed-priority scheduling algorithms cannot achieve a utilization bound greater than 50% [65]. Some researchers handle this deficiency by fair priority assignment.

Baker addresses the aforementioned problem and demonstrates a schedulability test for preemptive deadline scheduling of periodic or sporadic real-time tasks [69]. Baruah and Shun-Shii Lin propose a new Pinfair algorithm that is very efficient in terms of runtime complexity and has a superior density threshold for a very large subclass of generalized pinwheel task systems [70]. Levin et al. propose a deadline partitioning algorithm, called DP-WRAP algorithm to handle sporadic task sets with arbitrary deadlines [73].

In the partitioned approaches, task set is partitioned and statically allocated to a designated processor. These task sets are executed by existing scheduling algorithms and migration across core is not permitted. Fan et al. present a partitioned scheduling algorithm with enhanced RBound (PSER) that exploits a flexible task set scaling technique and enhanced utilization bound for fixed-priority periodic real-time tasks [74]. This algorithm effectively improves the schedulability of the system. They combine PSER with harmonic aware partition scheduling (HAPS) in [75], which converts the complete task set into the harmonic set and takes the benefit of the harmonic relationship between tasks to achieve increased utilization bound up to 100% [76].

Andersson demonstrates global PFair and partitioned static-priority scheduling on multiprocessors [77]. Guan et al. develop two separate fixed-priority scheduling algorithms for light tasks and heavy tasks. The algorithm RM-TS/light (rate monotonic-task set for light loads) can execute light task sets with sustainable parametric utilization bound and the RM-TS algorithm can perform any task set, whereas the utilization bound is lower than a specified limit [78].

Recently, a significant portion of semi-partitioned approaches [66, 67, 79–83] have been proposed to minimize energy expenditure in multicores. Semi-partitioned algorithms allocate most of the tasks to one particular processor. But, limited tasks (i.e., less than the number of cores − 1) are partitioned into many subtasks and are allocated to various cores under some constraints.

Lakshmanan et al. introduce a highest-priority task-splitting (HPTS) algorithm to enhance the utilization bound of partitioned deadline-monotonic scheduling algorithms (PDMS) from 50 to 60% on implicit deadline task sets [82]. They can obtain 88% of average utilization with very low migration overhead for randomly generated implicit deadline task sets by extending this algorithm, which assigns the tasks in the decreasing order in terms of their size. Kato et al. [81] and Andersson et al. [67] present real-time scheduling algorithms with high schedulability. Similar to partitioned scheduling, the proposed algorithms assign each task to a specific processor but can divide a task into two processors if there is not sufficient capacity remaining on a processor.

The semi-partitioned approaches are more efficient as compared to the conventional global and partitioned approaches theoretically [82–84] and also suitable for practical implementations [80]. Zhang et al. show that the implementation complexity of semi-partitioned scheduling algorithm is relatively low. They investigate semi-partitioned approaches in the Linux OS and demonstrate their results on an Intel Core-i7 processor [85].

3.3.3 Thread Migration (TM)

Migrating threads across cores is a versatile approach to distribute power uniformly in the chip. Thread migration permits an arranged thread to migrate from an overheated core to a cold core based on the power profile of the thread. With the help of the thermal model, operating system, or thread scheduler is liable to switch threads from one core to another [32, 38]. The major challenge of task migration is to find suitable target core to decrease the migration frequency while decreasing the temperature. Several researches are proposed in thread migration for ensuring energy-efficient operation of the multicore system.

Heo et al. [61] introduce five different intriguing architectural alternatives for realizing activity migration (AM) in a superscalar processor. The AM technique replicates the functional units of the core (e.g., ALU and the register files). It makes the original units in use for a definite time period. After an interval, operation switches to the replicated units. Then the actual execution units are halted and employed in a sleep state. The authors report an average power reduction of 10% with 6% of area overhead.

To increase the utilization of computing resources, Gomaa et al. introduce an OS-based heat-and-run thread migration (HRTM) technique with heat-and-run thread assignment (HRTA) [86]. Using HRTA, the HRTM transfers threads from the hottest cores and schedules them across alternate cores. The HRTA enhances the processor utilization by co-scheduling threads across complementary resources.

Michaud et al. analyze the effectiveness of task migration approaches in multicore under temperature-constraint environment [87]. They demonstrate that the competence of task migration tempered by the increased thread count, their thermal features, and ambient temperature. They develop a novel thread migration technique that relies on swapping threads when a couple of hot and cold cores are sensed. By limiting the number of migrations, their proposed method can produce the similar results as HRTM.

Nevertheless, all the above-mentioned thread migration techniques do not care about real-time characteristics of tasks. Swaminathan et al. introduce a runtime method to minimize energy expenditure and meet the deadline of each task. They prove that mixed-integer linear programming (MILP) is an appropriate model to derive a precise solution for medium-sized tasks set [88]. They also present a low-energy earliest deadline first (LEDF) algorithm for larger problems and implement it to two real-life task sets.

4 Conclusions

High performance computing technology has transformed from single core to homogeneous multicore and into dynamic reconfigurable heterogeneous multicore at present. Current innovations in semiconductor technology lead to not only an increase in the number of cores on a die but also an increase in power density and concomitant heat dissipation. This adversely affects the system reliability and availability. Achieving high performance with low power consumption is imposing a new challenge on IC fabrication technology. This article presents various effective techniques for alleviating power dissipation of multicore architecture and its classification based on their attributes. We accept as true that our review will help the researchers and architects to acquire ideas into the next-generation multicore processors and encourage them to endorse new energy-efficient elucidations for fabricating competent architectures.

References

1. Silven, O., Jyrkka, K.: Observations on Power-Efficiency Trends in Mobile Communication Devices. EURASIP J. Embedded Systems, vol. 2007, no. 1, p. 17 (2007).
2. Gordon E. Moore. Cramming more components onto integrated circuits. vol. 86(1); pp. 82–85, IEEE (1998).
3. Christian Martin.: Post-Dennard Scaling and the final years of Moore's Law consequences for the evolution of multicore-architectures. Hochschule Augsburg University of Applied Sciences, Technical Report (2014).
4. Intel Corp.: Intel core i7-940 processor. Intel Product Information, 2009 [Online]. Available: http://ark.intel.com/cpu.aspx?groupId=37148.
5. May, D.: XMOS XS1 architecture and XS1 Chips. Micro IEEE; vol. 32(6); pp. 28–37, IEEE (2012).
6. Advanced Micro Devices Inc.: ATI Radeon HD 4850 & ATI Radeon HD 4870—GPU specifications. AMD Product Information, 2008. [Online] Available: http://ati.amdcom/products/radeonhd4800/specs3.html.
7. Gschwind, M., Hofstee, H.P., Flachs, B., Hopkin, M., Watanabe, Y., Yamazaki, T.: Synergistic Processing in Cell's Multicore Architecture, vol. 26(2); pp. 10–24, IEEE (2006).
8. ARM Ltd.: ARM Development Tools., 2011. [Online] Available: http://www.arm.com/products/tools/development-boards/versatileexpress/index.php.
9. Pericas, M., Cristal, A., Cazorla, F.J., Gonzalez, R., Jimenez, D.A., Valero, M.: A flexible heterogeneous multi-core architecture. In Proceedings of 16th International Conference on Parallel Architecture and Compilation Techniques, pp. 13–24, IEEE (2007).
10. Intel 64 and IA-32 Architectures.: Software Developer's Manual. Intel Developer Manuals, vol.3A, (2008).
11. Intel Corp.: Intel atom processor for nettop platforms. Intel Product Brief, 2008. [Online]. Available: http://download.intel.com/products/atom/319995.pdf.

12. Advanced Micro Devices Inc.: Key architectural features—AMD Phenom II processors. AMD Product Information, 2008 [Online]. Available: http://www.amd.com/usen/Processors/ProductInformation/0,30118153311591 7%5E15919,00.html.
13. Advanced Micro Devices Inc.: Software optimization guide for AMD family 10 h processors. AMD White Papers and Technical Documents, Nov. 2008 [Online]. Available: http://www.amd.com/us-en/assets/contentty_pe/whitepapers_and_tech_docs/40546.pdf.
14. Sun Microsystems Inc.: UltraSPARC T2 processor. Sun Microsystems Data Sheets, 2007, [Online] Available: http://www.sun.com/processors/UltraSPARCT2/datasheet.pdf.
15. Johnson, T., Nawathe, U.: An 8-core, 64-thread, 64-bit power efficient SPARC SoC (niagara2). In Proceedings of 2007 International Symposium on Physical Design (ISPD '07), pp. 2–2, ACM, (2007).
16. ARM Ltd.: The ARM Cortex-A9 Processors. ARM Ltd. White Paper, Sept. 2007 [Online] Available: http://www.arm.com/pdfs/ARMCortexA-9Processors.pdf.
17. NVIDIA Corp.: NVIDIA CUDA: Compute unified device architecture. NVidia CUDA Documentation, June2008. [Online] Available: http://developer.download.nvidia.com/compute/cuda/2_0/docs/NVIDIA_CUDA_Programming_Guide_2.0.pdf.
18. Zhuravlev, S., Saez, J.C., Blagodurov, S., Fedorova, A., Prieto, M.: Survey of Energy-Cognizant Scheduling Techniques, vol. 24(7); pp. 1447–1464, IEEE (2013).
19. David Chinnery, Kurt Keutzer.: Overview of the Factors Affecting the Power Consumption. In proceeding of Tools and Techniques for Low Power Design, pp. 11–53, Springer, (2007).
20. Dennard, R.H., Gaensslen, F. H., Yu, H., Rideout, V.L., Bassous,E., LeBlanc, A.R.: Design of ion-implanted MOSFET's with very small physical dimensions. In proceedings of IEEE Solid-State Circuits, vol. 12(1); pp. 38–50, IEEE (2007).
21. Hennessy, J.L., Pattersson, D.A.: Computer Architecture - A Quantitative Approach. Morgan Kaufmann, second edition (1996).
22. Hadi Esmaeilzadeh, Emily Blem, Renee St. Amant, Karthikeyan Sankaralingam, Doug Burger.: Dark silicon and the end of multicore scaling. In proceeding of 38th International Symposium on Computer Architecture (ISCA), pp. 365–376, IEEE (2011).
23. Puttaswamy, K., Loh.: Thermal Herding: Microarchitecture Techniques for controlling hotspots in high-performance 3D integrated processors. In proceeding of 13th International Symposium on High Performance Computer Architecture, pp. 193–204, IEEE (2007).
24. Fazal Hameed, Mohammad Abdullah Al Faruque, Jorg Henkel.: Dynamic thermal management in 3D multi-core architecture through run-time adaptation. In proceeding of Design, Automation & Test in Europe Conference & Exhibition, pp. 1–6, IEEE (2011).
25. Kontorinis, V., Shayan, A., Tullsen, D., Kumar, R.: Reducing Peak Power with a Table-Driven Adaptive Processor Core. In Proceedings of IEEE/ACM 42nd Annual International Symposium on Microarchitecture (MICRO-42), pp. 189–200, IEEE (2009).
26. Rodrigues, R., Annamalai, A., Koren, I., Kundu, S., Khan, O.: Performance per Watt Benefits of Dynamic Core Morphing in Asymmetric Multicores. In Proceedings of International Conference on Parallel Architectures and Compilation Techniques, pp. 121–130, ACM (2011).
27. Narayanan, S., Sartori, J., Kumar, R., Jones, D.: Scalable Stochastic Processors. In Proceedings of Design, Automation & Test in Europe Conference & Exhibition (DATE), pp. 335–338, IEEE (2010).
28. Kornaros G.: Multi-core Embedded Systems. Taylor and Francis Group, CRC Press, (2010).
29. Flautner, K., Kim, N., Martin, S., Blaauw, D., Mudge, T.: Drowsy Caches: Simple Techniques for Reducing Leakage Power. In Proceedings of 29th Annual International Symposium on Computer Architecture, pp. 148–157, IEEE (2002).
30. Le Cai, Yung-Hsiang Lu.: Joint Power Management of Memory and Disk. In Proceedings of the conference on Design, Automation and Test in Europe, pp. 86–91, IEEE Computer Society (2005).
31. Rakesh Kumar, Victor Zyuban, Dean M. Tullsen.: Interconnections in multi-core architectures: Understanding mechanisms, overheads and scaling. In Proceedings of 32nd International Symposium on Computer Architecture (ISCA), pp. 408–419, IEEE (2005).

32. Donald, D., Martonosi, M.: Techniques for Multicore Thermal Management: Classification and New Exploration. In Proceedings of 33rd International symposium on Computer Architecture, pp. 78–88, IEEE (2006).
33. Isci, C., Buyuktosunoglu, A., C.-Y. Chen, Bose, P., Martonosi, M.: An Analysis of Efficient Multi-Core Global Power Management Policies: Maximizing Performance for a Given Power Budget. In Proceedings of 39th Annual IEEE/ACM International Symposium on MICRO-39, pp. 347–358, IEEE (2006).
34. Hewlett-Packard, Intel, Microsoft, Phoenix Technologies, Toshiba: Advanced Configuration and Power Interface Specification, Revision 4.0a. April 2010, [Online]. Available: http://www.acpi.info/spec.html.
35. Bassett, P., Saint-Laurent M.: Energy efficient design techniques for a digital signal processor. In Proceedings of International Conference on IC Design & Technology, pp. 1–4, IEEE (2012).
36. Leverich, J., Monchiero, M., Talwar, V., Ranganathan, P.: Power management of data center workloads using per-core power gating. IEEE Computer Architecture Letters, vol. 8(2); pp. 48–51, IEEE (2009).
37. Young-Si Hwang, Ki-Seok Chung: Dynamic Power Management Technique for Multicore Based Embedded Mobile Devices. IEEE Transactions on Industrial Informatics, vol. 9(3); pp. 1601–1612 (2013).
38. Chaparro, P., Gonzalez, J., Magklis, G., Cai, Q. Gonzalez, A.: Understanding the Thermal Implications of Multicore Archtectures. IEEE Transactions on Parallel and Distributed Systems, vol.18 (8); pp. 1055–1065, IEEE (2007).
39. Hanumaiah, V., Vrudhula, S.: Energy-efficient Operation of Multicore Processors by DVFS, Task Migration and Active Cooling. Computers, vol. 63, no. 2, pp. 349–360, IEEE (2012).
40. B. de Abreu Silva, Bonato V.: Power/performance optimization in FPGA-based asymmetric multi-core systems. In Proceedings of 22nd International Conference on Field Programmable Logic and Applications (FPL), pp. 473–474, IEEE (2012).
41. Wonyoung Kim, Gupta M S, Gu-Yeon Wei, Brooks D.: System level analysis of fast, per-core DVFS using on-chip switching regulators. In Proceedings of 14th International Symposium on High Performance Computer Architecture, pp. 123–134, IEEE (2008).
42. Dongwoo Lee, Wesley Kwong, David Blaauw, Dennis Sylvester.: Simultaneous Subthreshold and Gate-Oxide Tunneling Leakage Current Analysis in Nanometer CMOS Design. In Proceedings of 4th International Symposium on Quality Electronic Design, pp. 287–292, IEEE (2003).
43. Burd, T., Pering, T., Stratakos, A., Brodersen, R.: A dynamic voltage scaled microprocessor system. In Proceedings of International Solid-State Circuits Conference, pp. 294–295, IEEE (2000).
44. Jayaseclan, R., Mitra, T.: A Hybrid Local-global Approach for Multi-core Thermal Management. In Proceedings of 2009 IEEE/ACM International Conference on Computer-Aided Design, pp. 314–320, IEEE (2009).
45. Pruhs, K., Van Stee, R., Uthaisombut, P.: Speed scaling of tasks with precedence constraints in approximation and online algorithms. In Proceedings of 3rd International conference on approximation and online algorithms, pp. 307–319, Springer (2006).
46. March, J.L., Sahuquillo, J., Hassan, H., Petit, S., Duato, J.: A new energy-aware dynamic task set partitioning algorithm for soft and hard embedded real-time systems. vol. 54, no. 8, pp. 1282–1294, The Computer Journal (2011).
47. Weiser, M., Welch, B., Demers, A., Shenker, S.: Scheduling for reduced CPU energy. In Proceedings of 1st USENIX conference on OSDI, pp. 13–23, ACM (1994).
48. Cai Qiong, Gonzalez, J., Magklis, G., Chaparro, P., Gonzalez, A.Q.:Thread shuffling: Combining DVFS and thread migration to reduce energy consumptions for multi-core systems. In Proceedings of 2011 International Symposium on Low Power Electronics and Design, pp. 379–384, IEEE (2011).

49. Quan Chen, Long Zheng, Minyi Guo, Zhiyi Huang.: EEWA: Energy-Efficient Workload-Aware Task Scheduling in Multi-core Architectures. In Proceedings of Parallel & Distributed Processing Symposium Workshops (IPDPSW), pp. 642–651, IEEE (2014).
50. Le Sueur E., Heiser G.: Dynamic voltage and frequency scaling: The laws of diminishing returns. In Proceedings of Hot Power'10: Workshop on Power aware computing and systems, pp. 1–8 (2010).
51. Awan, M.A, Petters, S.M.: Enhanced Race-To-Halt: A Leakage-Aware Energy Management Approach for Dynamic Priority Systems. In Proceedings of 23rd EUROMICRO Conference on Real-Time Systems (ECRTS), pp. 92–101, IEEE (2011).
52. Li Li, Ken Choi, Haiqing Nan.: Activity-driven fine-grained clock gating and run time power gating integration, vol. 21(8); pp. 1540–1544, IEEE (2013).
53. Hai Li, Swarup Bhunia, Yiran Chen, Vijaykumar T N, Roy K.: Deterministic Clock Gating for Microprocessor Power Reduction. In Proceedings of 9th International Symposium on High-Performance Computer Architecture, pp. 113–122, IEEE (2003).
54. Zhigang Hu, Buyuktosunoglu, A., Srinivasan, V., Zyuban, V., Jacobson Bose, P.: Micro architectural Techniques for Power Gating of Execution Units. In Proceedings of 2004 International Symposium on Low Power Electronics and Design, pp. 32–37, IEEE (2004).
55. Lungu, A., Bose, P., Buyuktosunoglu, A., Sorin, D.: Dynamic Power Gating with Quality Guarantees. In Proceedings of International Symposium on Low Power Electronics and Design (ISLPED), pp. 377–382, IEEE (2009).
56. Rajesh Kumar, Glenn Hinton.: A Family of 45 nm IA Processors. In Proceedings of IEEE International Solid-State Circuits Conference, pp. 58–59, IEEE (2009).
57. Lloyd Bircher, Lizy, W., John, K.: Analysis of Dynamic Power Management on Multi-Core Processors. In Proceedings of 22nd Annual International Conference on Supercomputing, pp. 327–338, ACM (2008).
58. Dan Nicolaescu, Alex Veidenbaum, Alex Nicolau: Reducing Data Cache Energy Consumption via Cached Load/Store Queue. In Proceedings of 2003 International Symposium on Low Power Electronics and Design, pp. 252–257, IEEE (2003).
59. Carazo, P., Apolloni, R., Castro, F., Chaver, D., Pinuel, L., Tirado F.: L1 Data cache power reduction using a forwarding predictor, vol. 6448; pp. 116–125, Springer (2011).
60. Hsin-Hao Chu, Yu-Chon Kao, Ya-Shu Chen.: Adaptive thermal-aware task scheduling for multi-core systems. vol. 99; pp. 155–174, ELSEVIER (2015).
61. Heo, S., Barr, K., Asanovi, K.: Reducing power density through activity migration. In Proceedings of the 2003 International Symposium on Low Power Electronics and Design, pp. 217–222, IEEE (2003).
62. Wu, G., Xu, Z., Xia, Q., Ren, J., Xia, F.: Task allocation and migration algorithm for temperature-constrained real-time multi-core systems. In Proceedings of 2010 IEEE/ACM International Conference on Green Computing and Communications, pp. 189–196, IEEE (2010).
63. Carpenter, J., Funk, S., Holman, P., Srinivasan, A., Anderson, J., Baruah, S.: A Categorization of Real-time Multiprocessor Scheduling Problems and Algorithms. In Handbook on Scheduling Algorithms, Methods, and Models, pp. 30.1–30.19 (2006).
64. Andersson, B.: Global Static-Priority Preemptive Multiprocessor Scheduling with Utilization Bound 38%. In Proceedings of ACM International Conference on Principles of Distributed Systems (OPODIS), vol. 5401; pp. 73–88, ACM (2008).
65. Andersson, B., Baruah, S., Jonsson, J.: Static-Priority Scheduling on Multiprocessors. In Proceedings of 2nd Real-Time Systems Symposium, pp. 193–202, IEEE (2001).
66. Kato, S., Yamasaki, N.: Semi-partitioned fixed-priority scheduling on multiprocessors. In Proceedings of 15th Real-Time and Embedded Technology and Applications Symposium, pp. 23–32, IEEE (2009).
67. Andersson, B., Bletsas, K, Baruah, S.: Scheduling Arbitrary Deadline Sporadic Task Systems on Multiprocessors. In Proceedings of Real-Time Systems Symposium, pp. 385–394, IEEE (2008).

68. Lundberg, L.: Analyzing fixed-priority global multiprocessor scheduling. In Proceedings of 8th Real-Time and Embedded Technology Symposium, pp. 145–153, IEEE (2002).
69. Baker, T.P.: An analysis of EDF schedulability on a multiprocessor. vol. 18(8); pp. 760–768, IEEE (2005).
70. Baruah, S.K, Shun-Shii Lin: Pfair scheduling of generalized pinwheel task systems. Transactions on Computers, vol.47 (7); pp. 812–816, IEEE (1998).
71. Fisher, N., J.-J. Chen, Wang, S. Thiele, L.: Thermal aware global real-time scheduling on multicore systems. In Proceedings of Real-Time and Embedded Technology and Applications Symposium, pp. 131–140, IEEE (2009).
72. Wang, S., Bettati, R.: Delay analysis in temperature constrained hard real-time systems with general task arrivals. In Proceedings of 27th IEEE International Real-Time Systems Symposium, pp. 323–334, IEEE (2006).
73. Levin, G., Funk, S., Sadowski, C., Pye, I., Brandt, S.: DP-fair: A simple model for understanding optimal multiprocessor scheduling. In Proceedings of 22nd EUROMICRO Conference, pp. 313, IEEE (2010).
74. Ming Fan, Qiushi Han, Gang Quan, Shangping Ren: Multi-core partitioned scheduling for fixed-priority periodic real-time tasks with enhanced RBound. In Proceedings of 15th International Symposium on Quality Electronic Design, pp. 284–291, IEEE (2014).
75. Ming Fan, Qiushi Han, Shuo Liu, Shaolei Ren, Gang Quan, Shangping Ren: Enhanced fixed-priority real-time scheduling on multi-core platforms by exploiting task period relationship. pp. 85–96, Elsevier (2014).
76. Liu, J.W.S.: Real-time systems. Prentice Hall (2000).
77. Andersson, B., Jonsson, J.: The utilization bounds of partitioned and pfair static priority scheduling on multiprocessors are 50%. In Proceedings of 15th EUROMICRO Conference on Real-time Systems, pp. 33–40, IEEE (2003).
78. Guan, N., Martin Stigge, Wang Yi, Ge Yu: Parametric Utilization Bounds for Fixed-Priority Multiprocessor Scheduling. In Proceedings of 26th International Parallel and Distributed Processing Symposium, pp. 261–272, IEEE (2012).
79. Anderson, J.H., Bud, V., Devi, U.C.: An EDF-Based Scheduling Algorithm for Multiprocessor Soft Real-Time Systems. In Proceedings of EUROMICRO Conference on Real-Time Systems (ECRTS), pp. 199–208, IEEE (2005).
80. Bastoni, A., Brandenburg, B.B, Anderson, J.H.: Is Semi-partitioned scheduling practical? In Proceedings of 23rd Conference on Real-Time Systems, pp. 125–135, IEEE, (2011).
81. Kato, S., Yamasaki, N.: Semi-partitioned fixed-priority scheduling on multiprocessors. In Proceedings of 15th Real-Time and Embedded Technology and Applications Symposium, pp. 23–32, IEEE (2009).
82. Lakshmanan, K., Rajkumar, R., Lehoczky, J.P.: Partitioned fixed priority preemptive scheduling for multi-core processors. In Proceedings of 21st EUROMICRO Conference on Real-Time Systems, pp. 239–248, IEEE (2009).
83. Guan, N., Stigge, M., Yi, W., Yu G.: Fixed-Priority Multiprocessor Scheduling with Liu and Layland's Utilization Bound. In Proceedings of IEEE Real Time and Embedded Technology and Applications Symposium (RTAS), pp. 165–174, IEEE (2010).
84. Guan, N., Martin Stigge, Wang Yi, Ge Yu: Fixed-Priority Multiprocessor Scheduling: Beyond Liu and Layland's Utilization Bound. In Proceedings of WiP Real-Time Systems Symposium (RTSS), pp. 1594–1601, IEEE, (2010).
85. Zhang, Y., Guan, N., Yi. W.: Towards the Implementation and Evaluation of Semi-Partitioned Multi-Core Scheduling. In Bringing Theory to Practice: Predictability and Performance in Embedded Systems. vol. 18; pp. 42–46, In Open Access Series in Informatics (2011).
86. Gomaa, M., Powell, M.D., Vijaykumar, T. N.: Heat-and-run: leveraging SMT and CMP to manage power density through the operating system. In Proceedings of ASPLOS, vol. 38(5); pp. 260–270, ACM (2004).

87. Pierre Michaud, Andre Seznec, Damien Fetis, Yiannakis Sazeides, Theofanis Constantinou: A study of thread migration in temperature constrained multicores. vol. 4(2); Article No. 9, ACM (2007).
88. Swaminathan, V., Chakrabarty, K.: Real-time task scheduling for energy-aware embedded systems. Journal of the Franklin Institute, vol. 338, pp. 729–750 (2001).

Comparative Study of Lookup Table Approach of Direct Power Control for Three-Phase DC/AC Inverter

Ami Vekariya, Tapankumar Trivedi, Rajendrasinh Jadeja and Praghnesh Bhatt

Abstract Direct Power Control (DPC) method for grid-connected Voltage Source Inverter is popular due to number of advantages such as elimination of inner current control loop, direct voltage vector selection, direct control of instantaneous active and reactive powers, and improved dynamic response. For voltage source inverter, effect of particular vector is to produce finite variations in instantaneous active and reactive power in a given sector. This paper investigates different switching patterns of DPC method for grid-connected Voltage Source Inverter. The performance of various switching tables such as Noguchi Table and Eloy-Garcia Table is studied. Based on the above tables, a modified table is proposed for 12-sector approach. The proposed method has less active as well as reactive power errors and can be applied to certain applications such as Active Power Filter and Unified Power Quality Conditioners. These methods are simulated and experimentally validated using RT-LAB + MATLAB Simulink®/Sim Power System tool.

Keywords DC/AC converter · Direct power control (DPC) · Instantaneous active/reactive power · Lookup table (LUT)

Nomenclature

U_g, V_g	Grid, converter voltage vectors
I_g	Converter current vector
v_a, v_b, v_c	Three-phase line–source voltages
i_a, i_b, i_c	Three-phase line current
L_g, R_g	Line inductance and resistance

A. Vekariya · T. Trivedi (✉) · R. Jadeja
Electrical Engineering Department, Marwadi Education Foundation's Group of Institutions, Rajkot, Gujarat, India
e-mail: tapankumar.trivedi@marwadieducation.edu.in

P. Bhatt
M & V Patel Department of Electrical Engineering, Charotar University of Science and Technology, Changa, India

S.S. Dash et al. (eds.), *Artificial Intelligence and Evolutionary Computations in Engineering Systems*, Advances in Intelligent Systems and Computing 517,
DOI 10.1007/978-981-10-3174-8_36

ω Frequency of grid
P_g, Q_g Active and reactive power
S_a, S_b, S_c Switching states

1 Introduction

Increase in power demand is a major concern for developing countries which require inclusion of alterative sources such as Solar Energy, Wind Energy Conversion systems, and Fuel cells. Voltage Source Inverter (VSI) is widely used for interfacing renewable energy sources with the grid. Such an interface ensures desired power transfer from source to grid [1]. Conventionally, control of VSI is done using current control technique [2] in order to control the amount of active and reactive power being exchanged between source and grid side. To ensure proper transfer of active and reactive power, different control techniques have been applied to three-phase grid-connected VSI for the control of current. Direct Power Control (DPC) and Indirect Power Controls (IPC) are two control techniques for grid-connected system. The indirect control also known as Voltage-Oriented Control (VOC) is similar to Field-Oriented Control (FOC) of induction machine where active and reactive power components are controlled by controlling the in phase and quadrature axis component of current. This approach uses either linear or nonlinear current controller. A linear controller is one which is represented by a modulator (pulse width modulation (PWM) or other) and synthesizes voltage in accordance with commanded current vectors by on/off times of a converter's switches along a switching period. This voltage reference is delivered by the controller, which makes the converter a dependent continuous voltage source. On the other hand, DPC technique establishes a direct relation between power to be controlled and state of the converter's switches [3].

In indirect control method, the grid currents are transferred into *d-q* axis components, which resemble active power and reactive power, respectively. Active and reactive powers are controlled by adjusting the decoupled *d-q* axis by PI controller. This method fully depends on tuning of PI parameters and grid voltage condition which is a major issue of concern or in other words main pitfall of the method. On the contrary, DPC method controls or realizes the instantaneous active and reactive powers using nonlinear control loop [4].

DPC is based on the principle of direct torque control (DTC) strategy for AC machines. In DTC, torque and flux are controlled to obtain the desired operation of the machine [5, 6], whereas in DPC active and reactive powers are controlled for appropriate transfer of power. Due to its dynamic capability and simple implementation, DPC has become an interesting control strategy. This nonlinear control strategy is defined as a direct control technique because it chooses the best suited converter's voltage vector without any modulation technique. In the DPC, instantaneous active- and reactive-power control loops are based on hysteresis regulators that select the appropriate voltage vector from a lookup table based on error of reference and active powers and angular position of the converter voltage vector [3]. The work henceforth mentioned in [3] will be termed as Noguchi's Table as the method works on the Lookup-Table (LUT) based approach. The main advantages of the method are that there is no need of internal control loop, coordinate transformation, avoiding coupling effect between transformed variable even in stationary reference frame [7]. It is mentioned in [8] that the purpose of control of grid-connected converter is to ensure power exchange between source and grid side. While most of the grid-connected applications requires active power transfer between DC side and AC grid side, many applications require reactive power control, e.g., D-STATCOM. For such a case, table is proposed in [9, 10]. In addition to this, various approaches of DPC are presented in [11, 12, 13, 14, 15, 16, 17] While Noguchi's Table primarily focuses on the minimization of active power error, Eloy-Garcia's Table suggests uniform reactive power variations so that it results in even distribution of reactive power error for a given period of time. In this paper, both the switching patterns for DPC are investigated and their response in terms of active and reactive power error is discussed. Based on the two tables, a modified switching table is formulated. The performance of various LUT methods are presented with their simulation results. The paper is organized as follows: Sect. 2 describes principle of DPC. Section 3 illustrates the dynamic behavior of a grid-connected DC-AC converter and the effect of converter voltage vector on active and reactive power variations. In Sect. 4, analysis of different switching table and their effect on power variation is done which is followed by simulation results and experimental validation.

2 Direct Power Control

Direct power control is one of the most popular control methods of grid-connected inverter in which various approaches are proposed in the literature. The block diagram of DPC method for grid-connected inverter is shown in Fig. 1, in which the DC power supply is connected to DC to AC converter, i.e., inverter. In an inverter, voltage vectors for given combination of switches occupy fixed position in voltage. Out of eight ($2^3 = 8$) vectors, six are active vectors and two are zero vectors as shown in Fig. 2. Since average value of the inverter voltage vector over given sample time T_s is either greater or less than grid voltage space phasor, this will result in instantaneous active and reactive power transfer from source to grid or vice versa. In this method, voltages v_a, v_b, v_c and/or current i_a, i_b, i_c are sensed/estimated and based on these the active and reactive power are calculated from following equations.

$$s = p + jq \tag{1}$$

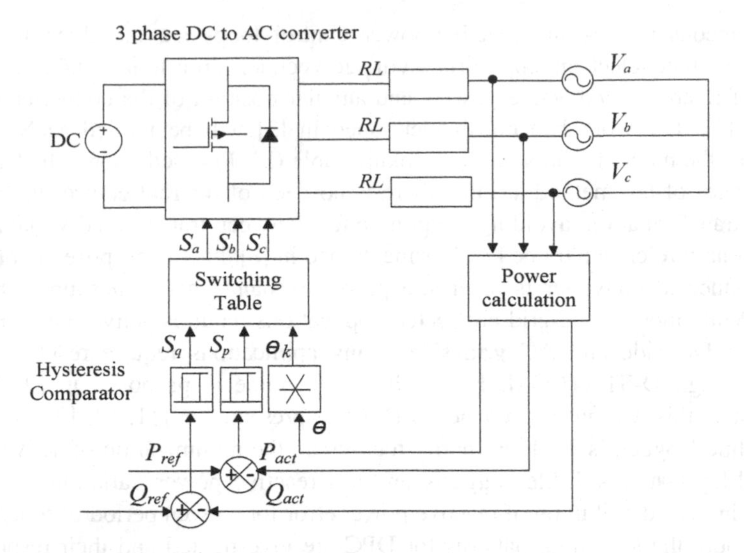

Fig. 1 Block diagram of direct power control

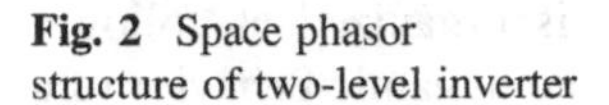
Fig. 2 Space phasor structure of two-level inverter

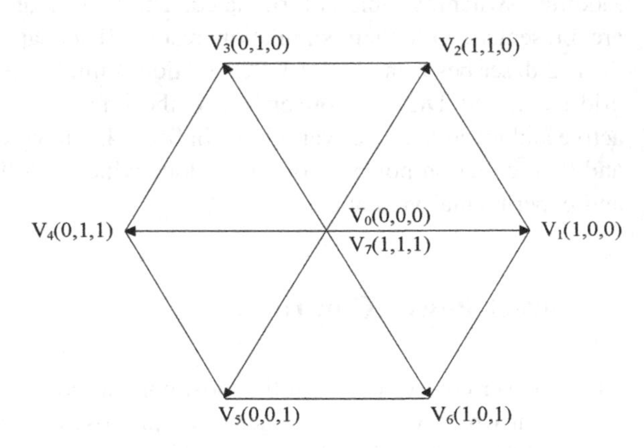

$$p_{\text{act}} = v_a i_a + v_b i_b + v_c i_c \tag{2}$$

$$q_{\text{act}} = \frac{1}{\sqrt{3}}\left[(v_b - v_c)i_a + (v_c - v_a)i_b + (v_a - v_b)i_c\right] \tag{3}$$

These p_{act} and q_{act} are first compared with predefined reference values of power based on application and then to hysteresis comparators. These hysteresis comparators generate quantized error signals S_p and S_q by comparing it with fixed band of power. With the help of this location of voltage vectors in 12-sector format and error signals S_p and S_q, the appropriate switching vector is selected from predefined switching table and the output of is given to the inverter as shown in Fig. 1.

3 Dynamic Behavior of Grid-Connected DC-AC Converter

In this section, relationship between grid voltage vectors, inverter voltage vector and power variation rates are presented in order to understand the formulation of table for the method. A DC-AC converter inverter connected to the grid through coupling inductor can be represented as shown in Fig. 3. To analyze the transfer of power, an equivalent circuit is represented in Fig. 4 in $\alpha\beta$-reference frame.

From Fig. 4, the relationship between the line current and voltage can be obtained in the $\alpha\beta$–reference frame as

$$U_g = V_g + L_g \frac{\mathrm{d}I_g}{\mathrm{d}t} + R_g I_g \quad (4)$$

For the grid voltage variation, the grid voltage in stationary frame of reference is given by

$$u_{g\alpha} = u_g \cos(wt) \quad (5)$$

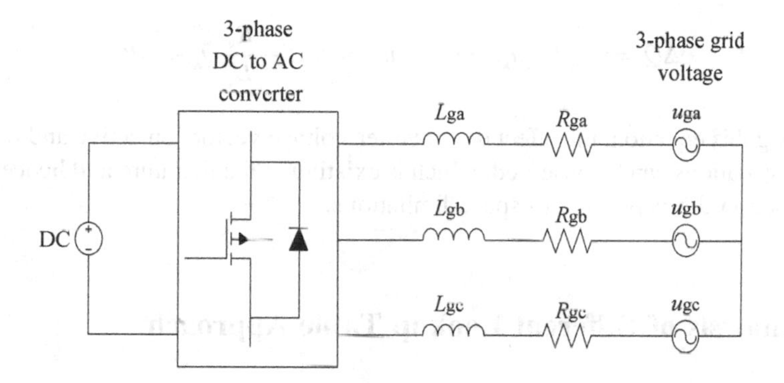

Fig. 3 Schematic diagram of three-phase inverter

Fig. 4 Equivalent circuit on AC side of a grid-connected inverter in stationary frame

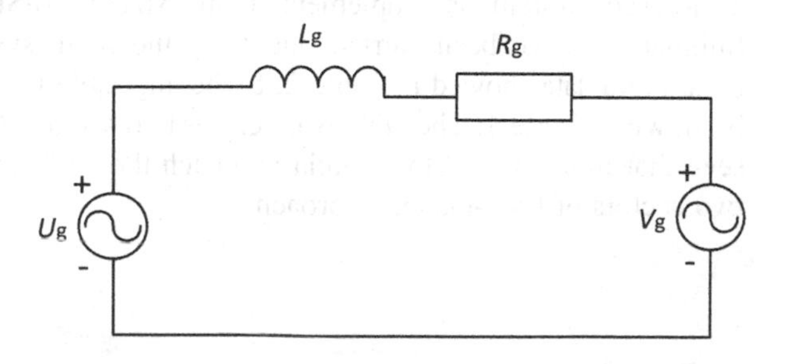

$$u_{g\beta} = u_g \sin(\omega t) \tag{6}$$

This grid voltage space phasor can also be represented by equivalent space phasor as

$$U_g = U_m \, e^{j\omega t} \tag{7}$$

Now inverter voltage in stationary frame references is given by

$$V_g = \sqrt{\frac{2}{3}} V_{dc} \left(S_a + S_b \, e^{j\frac{2\pi}{3}} S_c \, e^{j\frac{4\pi}{3}} \right) \tag{8}$$

$$V_{g\alpha} = \text{real}(V_g) \tag{9}$$

$$V_{g\beta} = \text{imag}(V_g) \tag{10}$$

Based on this equation the instantaneous rate of change of active power and reactive power is given by [3]:

$$\Delta p = -\frac{3}{2} L_g \left[\left(u_{g\alpha}^2 + u_{g\beta}^2 \right) - \left(u_{g\alpha} * v_{g\alpha} + u_{g\beta} * v_{g\beta} \right) \right] - \frac{R_g}{L_g} P_g + \omega Q_g \tag{11}$$

$$\Delta Q = -\frac{3}{2} L_g \left(u_{g\beta} * v_{g\alpha} - u_{g\alpha} * v_{g\beta} \right) - \frac{R_g}{L_g} Q_g + \omega P_g \tag{12}$$

Using this equation, the effect of converter voltage vectors on active and reactive power variations can be observed, which is existing in the literature and hence is not discussed in the paper due to space limitations.

4 Analysis of Different Lookup Table Approach

DPC method is a digital controller method and performance heavily relies on the sample time. To validate the method, it is essential that algorithm is run on the actual platform and performance is studied accordingly. A grid-connected VSI-based system is implemented in MATLAB/SIMULINK® environment. Simulations have been carried out using the main system parameters of power circuit and data showed in Table 2. Switching table for DPC for different schemes is shown in Table 1. The table is given for two sectors, (0° ≤ θ ≤ 60°). It can be seen that in modified Eloy-Garcia approach the switching states remains same for two sectors of Eloy-Garcia approach.

Table 1 Switching table of direct power control for different schemes ($0° \leq \theta \leq 60°$)

	Noguchi approach		Eloy-Garcia approach		Modified Eloy-Garcia approach	
	S-I	S-II	S-I	S-II	S-I	S-II
$S_p > 0, S_q < 0$	111	100	110	010	110	110
$S_p > 0, S_q > 0$	111	000	100	110	100	100
$S_p < 0, S_q < 0$	100	100	010	011	010	011
$S_p < 0, S_q > 0$	110	110	001	101	001	101

Table 2 System parameters of simulation

Supply voltage V_{dc}	650 V
Resistance of Reactor R	0.1 Ω
Inductance of reactor L	10 mH
Smoothing capacitor C	22 μF
Grid voltage V (Line)	400 V
Load (RL)	20 Ω, 35 mH
Average switching frequency	8 kHz

4.1 Noguchi Approach

This approach is presented in [3], which is considered as benchmark of DPC algorithms. The simulation results of Noguchi's approach are shown in Fig. 5a–d with reference values as P_{ref} = 5 kW and Q_{ref} = 2.5 kV AR. The system is simulated for the parameters given in Table 2. Where a constant DC voltage source is connected to the grid voltage through smoothening inductors. The method is a nonlinear digital controller, hence performance of the system largely depends upon sample time of the system. In the present work, sample time of system is taken as T_s = 10μs. The changes in active power variation and reactive power variations are detected and given to Hysteresis controllers having ±200 W and ±200 VAR limits. Hence, any change beyond this limit will cause alternative vector to switch resulting in reducing the effect of power variation. Output voltage and current are shown in Fig. 5a. Since aim of the method is to control the same, a step change in power is applied in terms of active and reactive powers respectively. The dynamic behavior of the active and reactive powers is shown in Fig. 5c. A close observation of Noguchi's Table suggests that the table is designed in order to minimize the commutation as well as active power variations. Figure 5d depicts the FFT analysis of the output current, which indicates presence of 4.12% THD in the system due to current distortions.

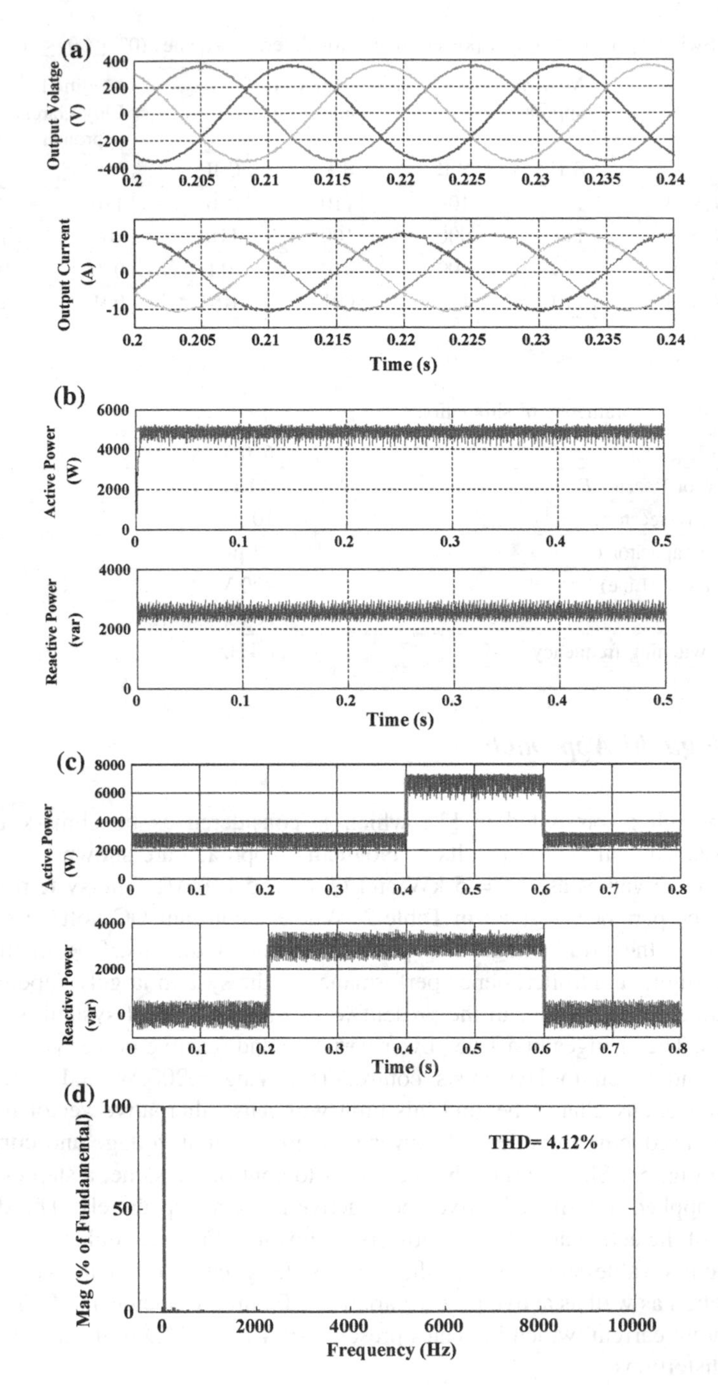

Fig. 5 Simulation results of Noguchi's table. **a** Output voltage and output current. **b** Active and reactive powers in steady state condition. **c** Active and reactive power in dynamic condition. **d** FFT analysis of output current

4.2 Eloy-Garcia Approach

This approach is presented in the form of Eloy-Garcia Table and is simpler than Noguchi's Table yet provides satisfactory results. Figure 6a–d shows the simulation results of this approach with same condition as the results present for Noguchi's approach. Results of both approaches are quite similar, however there exist some differences. It must be noted that Noguchi's Table aims at minimizing active component of power error whereas Eloy-Garcia Table aims at symmetrical variation in reactive powers. This is possible due to management of vector around the grid voltage vector. While voltage vector $\mathbf{v}_1$ and $\mathbf{v}_2$ are used for increase in active power error, vectors $\mathbf{v}_3$ and $\mathbf{v}_5$ are used for decrease in active power error. Since these vector are located at nearly symmetrical locations in geometry with respect to grid voltage vector position, uniform power variation can be obtained. However, selection of farthest vectors result in larger active power variation rates in either direction compared to Noguchi's Table which relies more on the nearest vectors. This can also be validated from Figs. 5d and 6d where THD of the Noguchi method is larger.

4.3 Modified Eloy-Garcia Approach

From the dynamic model of the grid-connected inverter, as well as Eqs. (8) and (9), it can be verified that instantaneous active and reactive power variation rates for the grid-connected inverter are different for particular location of vector in 12-sector approach, e.g., average instantaneous active power variation rates are maintained nearly constant for one inverter voltage vector in sectors $S_i = (2n; 2n + 1)$ whereas average instantaneous active power variation rates are maintained nearly constant for one vector in sectors $S_i = ((2n + 1); 2n)$ where $n = 0, 1, \ldots, 6$ and $S_i = , 1, \ldots, 11$. A modified switching table can be prepared using modified Eloy-Garcia Approach and the grid angle is divided into 12 sectors. This approach gives good results compared to above approach as shown in Fig. 7a–d. Figure 7b shows a small instantaneous power error in both active as well as reactive power due to proper selecting of voltage vector. The approach can be useful wherever control of both is required. It also shows that the THD of proposed table is much lesser than above approaches. Due to this modified approach the shortcoming of current problem of conventional DPC is overcome. It is worth noting that in this work, the performance of all the tables are compared at very low sample time. As the sample time is increased, the approach of Noguchi's Table can provide inferior results whereas results of Modified Eloy-Garcia Approach are comparable at higher sample time.

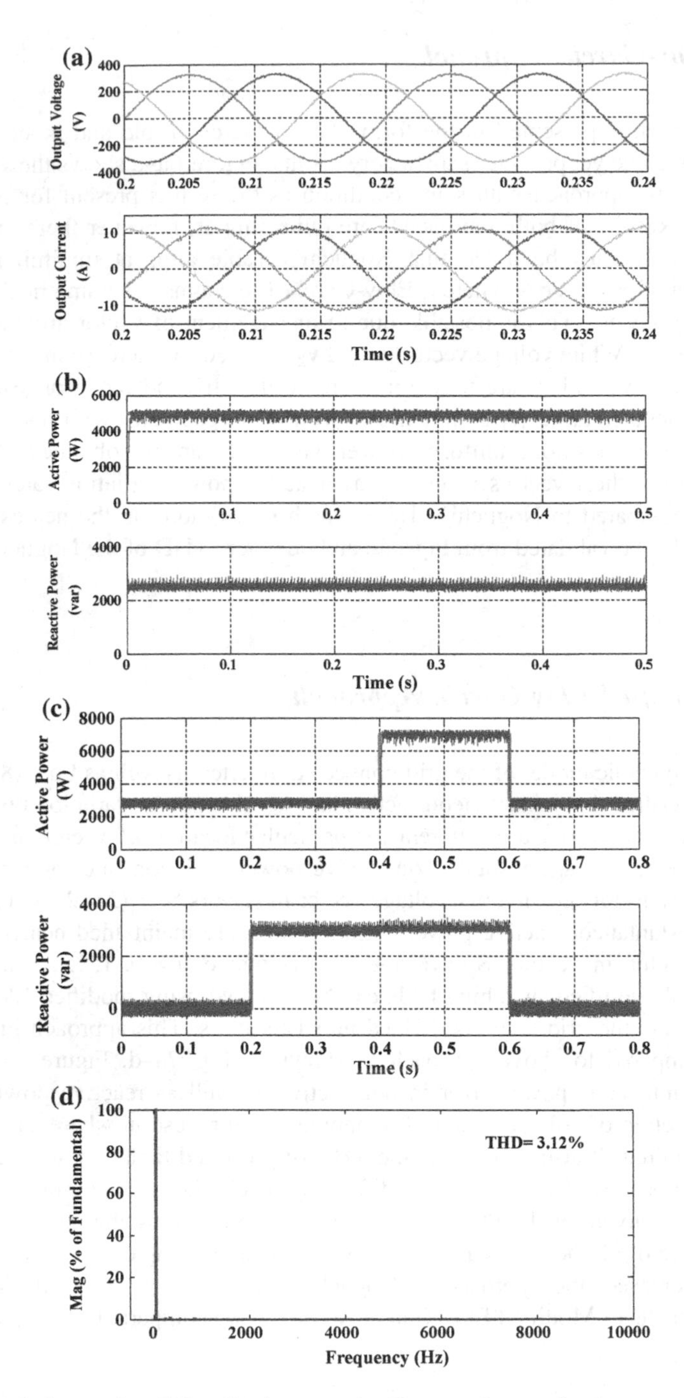

Fig. 6 Simulation results of Eloy-Garcia approach **a** Output voltage and output current. **b** Active and reactive powers in steady state condition. **c** Active and reactive power in dynamic condition. **d** FFT analysis of output current

5 Experimental Result of Different Lookup Table Approach

Several experimental tests were conducted to verify feasibility of the different approaches. The experimental results of different approaches using RT-lab is given in the below section. The result is taken at 100 V source phase voltage and the DC link voltage is 300 V. The references value of active and reactive power is 300 W and 100 VAR respectively. In the beginning of each control period, the control computes active and reactive power based on the instantaneous current and voltage measures. Then, the DPC algorithm selects the optimum pre-determined voltage vectors and the selected voltage vectors are applied during the computed application times, completing the control period. The control algorithm run under the SIMULINK®/MATLAB environment in OPAL RT-LAB and OP-4500 is used as a tool for real-time control prototyping. Both the power block and control block are divided into two subsystems, whereas parameters to be measured are put into another subsystem known as console as shown in Fig. 8. The control algorithm is run at sample time of 10 μs on a separate core whereas power circuit is run in different processor cores. All the variables to be observed are put in a console subsystem. The Target CPU interacts with the host CPU through TCP/IP protocol and variables are stored in the workspace of the host CPU. The result is measured in steady-state condition as well as in dynamic conditions. THD of the output current is recorded using Fluke 430 series Power Quality Analyzer.

5.1 Noguchi Approach

Figure 9a–e shows the experimental result of Noguchi approach under different conditions. An active and reactive power in steady state condition is shown in Fig. 9a. The current waveform has only lower order harmonic distortion. As shown in Fig. 9b, the reference of active power was step changed from 300 to 600 W with reactive power being 100 VAR, while in Fig. 9c the reactive power changed from 100 to 500 VAR with active power being 600 W. It is clearly seen that the DPC is successfully achieved in the transient state which shows that dynamic response of the system is good. Figure 9d depicts dynamic response of Noguchi approach for simultaneous step change in instantaneous active and reactive power which shows that active power and reactive power are independent from each other. A steady state performance is evaluated by total harmonic distortion (THD) measurements as shown in Fig. 9e. The output current is having a THD of 7.5%.

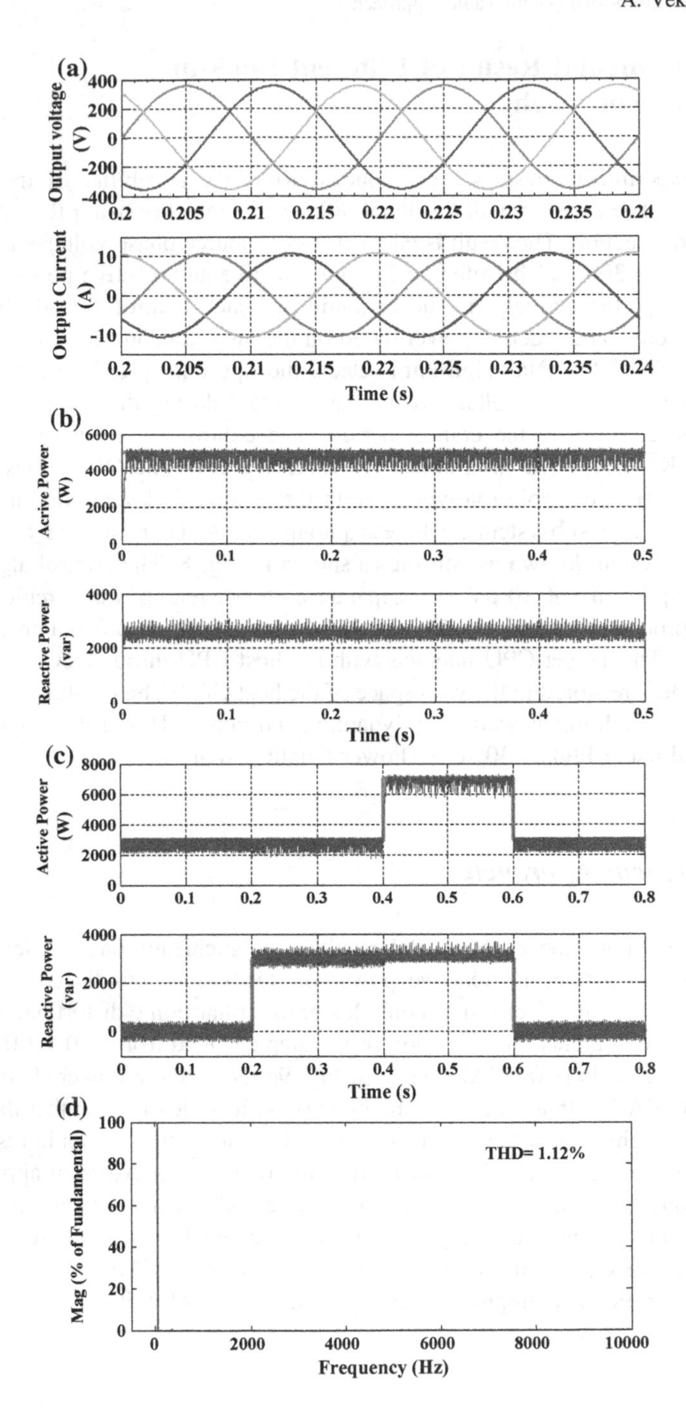

Fig. 7 Simulation results of modified Eloy-Garcia approach. **a** Output voltage and output current. **b** Active and reactive powers in steady state condition. **c** Active and reactive power in dynamic condition. **d** FFT analysis of output current

Fig. 8 Block diagram of simulation using OPAL RT-LAB simulator OP-4500

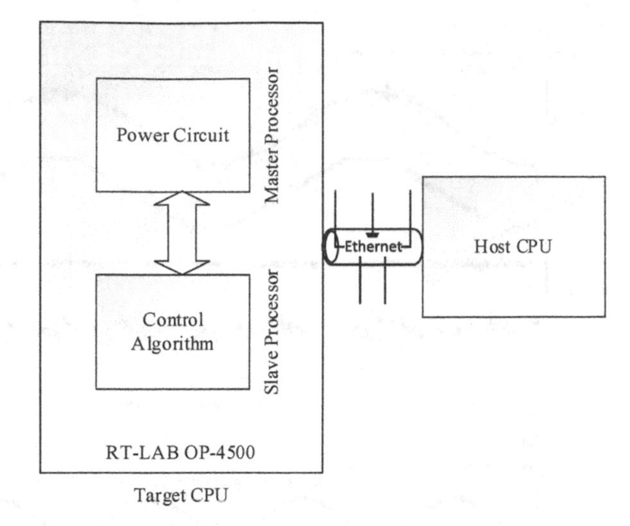

5.2 *Eloy-Garcia Approach*

Figure 10a–e shows the experimental result of Noguchi approach under different conditions. An active and reactive power in steady state condition is shown in Fig. 10a. As shown in Fig. 10b, the reference of active power was step changed from 300 to 600 W with reactive power being 100 VAR, while in Fig. 10c the reactive power changed from 100 to 500 VAR with active power being 600 W. It is clearly seen that, the dynamic response of the system is good. Figure 10d depicts dynamic response of Eloy-Garcia approach for simultaneous step change in instantaneous active and reactive power which shows that active power and reactive power are independent from each other. The THD of output current is 5.8% which is less compared to the Noguchi approach as shown in Fig. 10e.

5.3 *Modified Eloy-Garcia Approach*

Figure 11a–e shows the experimental results of Modified Eloy-Garcia approach. An active and reactive power in steady-state condition is shown in Fig. 11a. The current waveform has only lower order harmonic distortion. As shown in Fig. 11b, the reference of active power was step changed from 300 and 600 W with reactive power being 100 VAR, while in Fig. 11c the reactive power changed from 100 to 500 VAR with active power being 600 W. It is clearly seen that, the DPC is successfully achieved in the transient state which shows that dynamic response of the system is good. Figure 11d depicts dynamic response of modified approach for simultaneous step change in instantaneous active and reactive power which shows that active power and reactive power are independent from each other. A steady

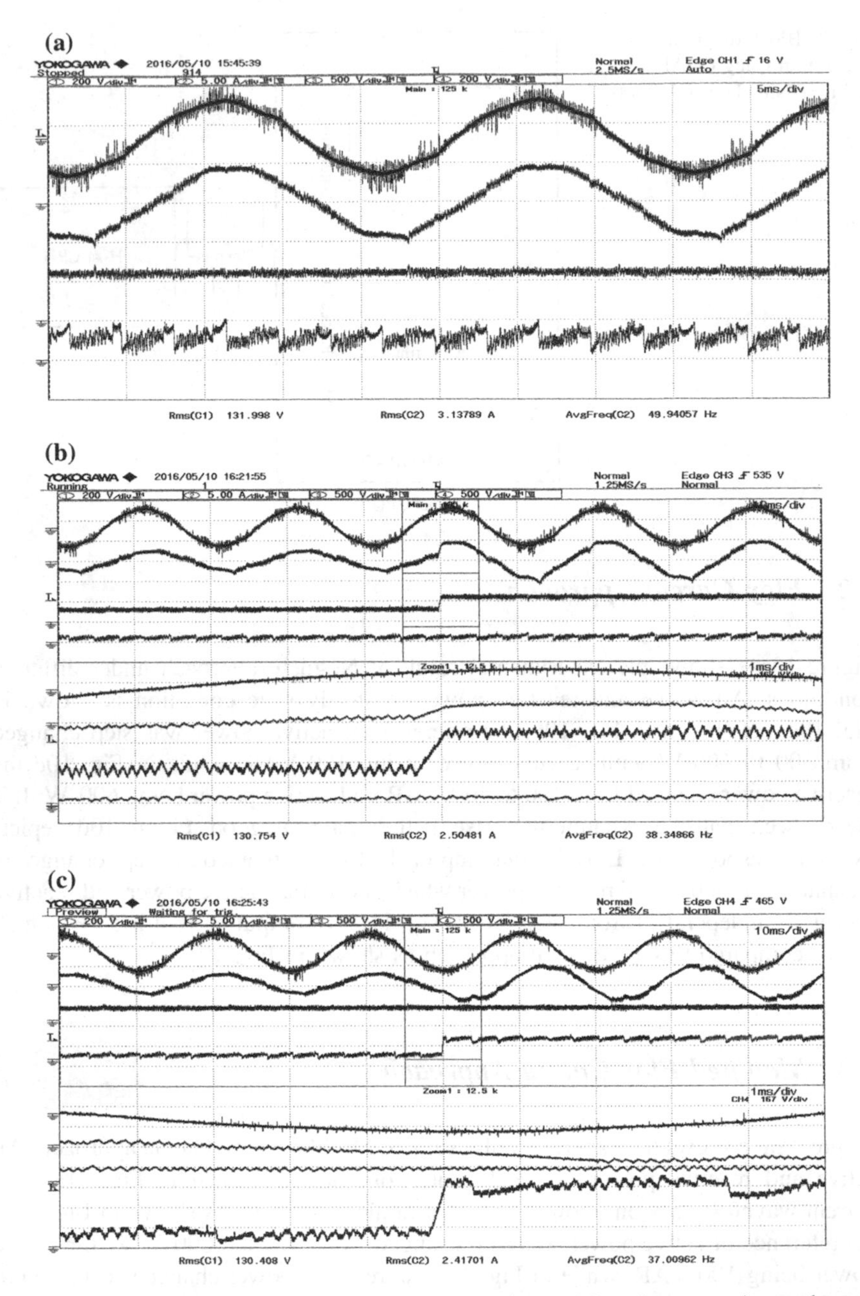

Fig. 9 Experimental results of Noguchi's table. **a** Active and Reactive power in steady state condition. **b** Dynamic Response for step change in active power from 100 to 600 W. **c** Dynamic response for step change in reactive power from 100 to 500 VAR. **d** Dynamic response for simultaneous step change in instantaneous active and reactive power. **e** THD of output current of Noguchi approach. CH 1: grid voltage (200 V/div), CH 2: Output current (5 A/div), CH 3: Active power (500 W/div), CH 4: Reactive power (500 VAR/div)

(d)

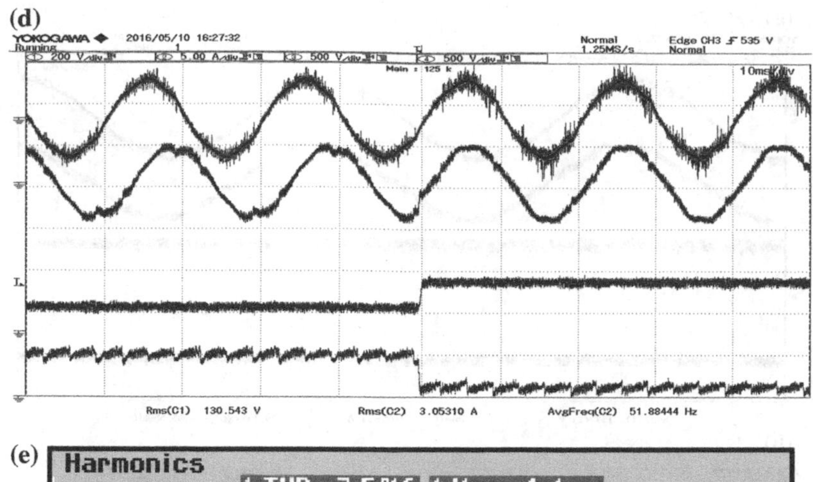

(e)

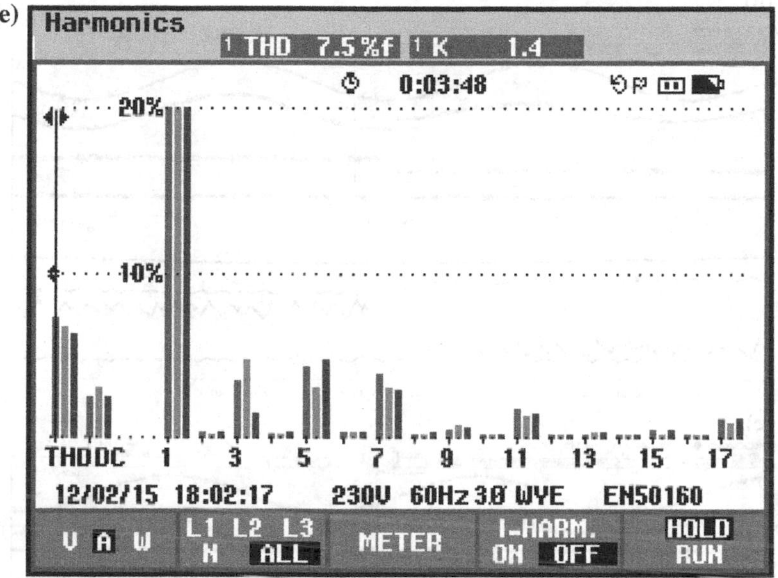

Fig. 9 (continued)

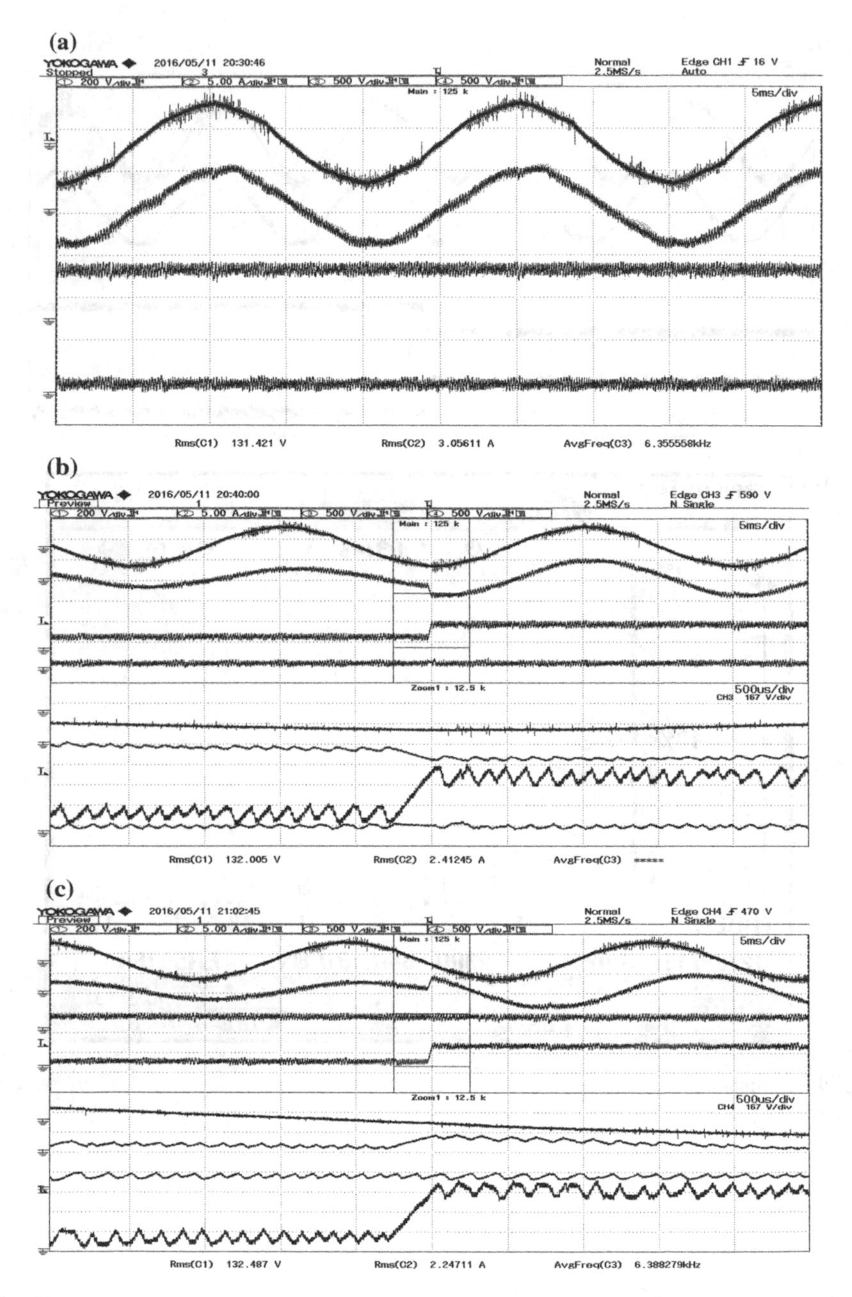

Fig. 10 Experimental results of Eloy-Garcia's table. **a** Active and reactive power in steady state condition. **b** Dynamic response for step change in active power from 100 to 600 W. **c** Dynamic response for step change in reactive power from 100 to 500 VAR. **d** Dynamic response for simultaneous step change in instantaneous active and reactive power. **e** THD of output current of Noguchi approach. CH 1: grid voltage (200 V/div), CH 2: Output current (5 A/div), CH 3: Active power (500 W/div), CH 4: Reactive power (500 VAR/div)

(d)

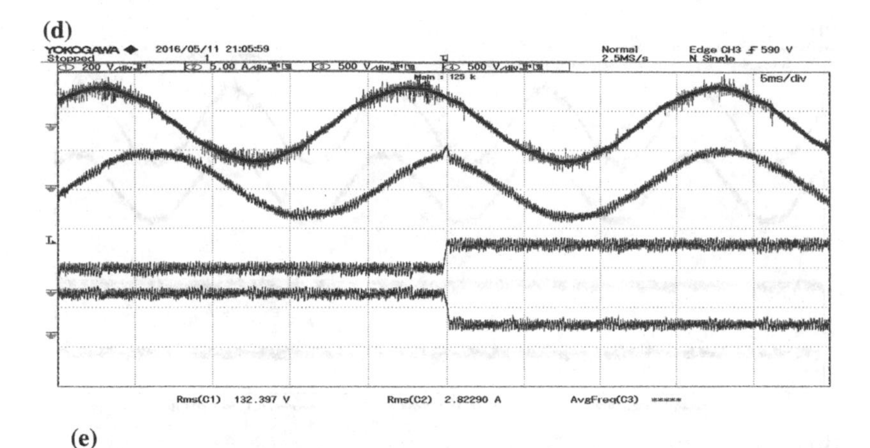

(e)

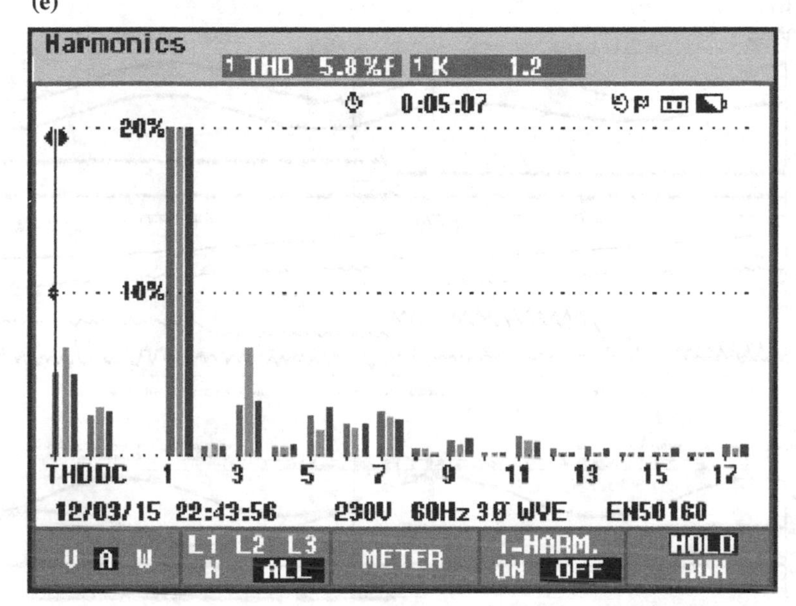

Fig. 10 (continued)

state performance is evaluated by THD measurements as shown in Fig. 11e. The output current is having a THD of 5.6%. The THD of modified Eloy-Garcia approach is less compared to above two approaches.

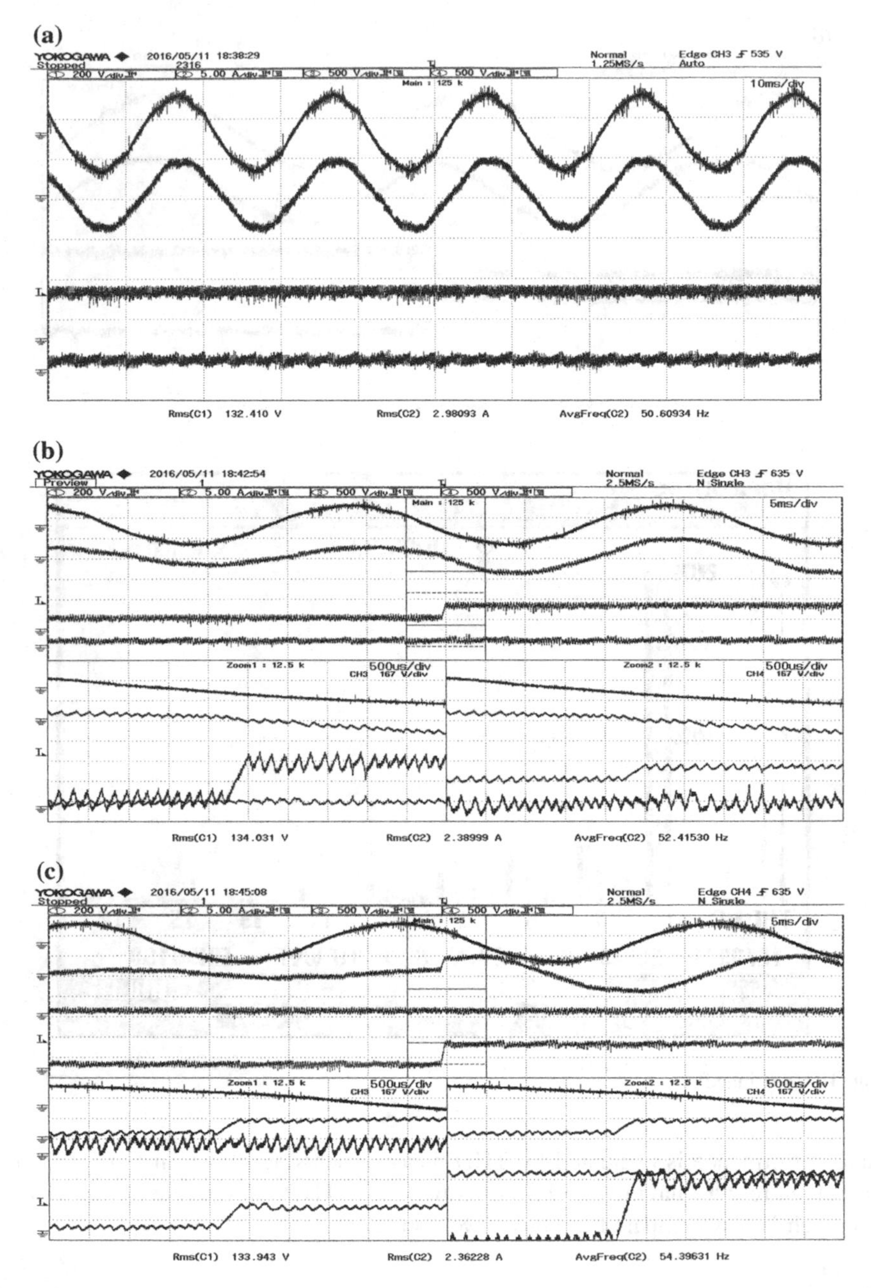

Fig. 11 Experimental results of modified Eloy-Garcia's table. **a** Active and reactive power in steady state condition. **b** Dynamic response for step change in active power from 100 to 600 W. **c** Dynamic response for step change in reactive power from 100 to 500 VAR. **d** Dynamic response for simultaneous step change in instantaneous active and reactive power. **e** THD of output current of Noguchi approach. CH 1: grid voltage (200 V/div), CH 2: Output current (5 A/div), CH 3: Active power (500 W/div), CH 4: Reactive power (500 VAR/div)

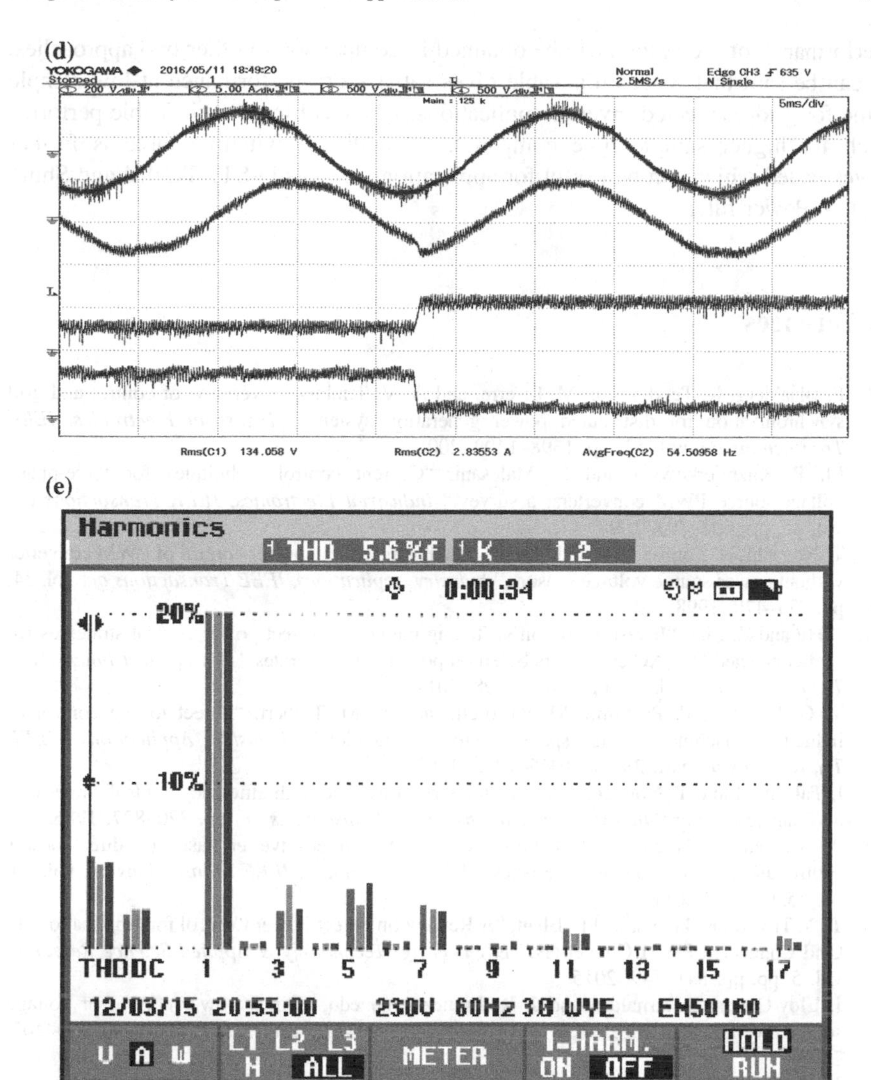

Fig. 11 (continued)

6 Conclusion

This paper investigates the different switching pattern of LUT approaches of DPC control strategies. These LUTs are based on variation rates of instantaneous active and reactive powers and are given by using each voltage vector in respective 12 different sectors. A modified switching table for DPC is presented which is based on Eloy-Garcia table. In this paper, the analysis of different lookup approaches is given and verified using simulation results. Due to modified table, improved performance of the system can be obtained in comparison to other two approaches. It can be said that Noguchi's Table gives satisfactory performance at low sample time for grid-connected inverter applications, whereas Eloy-Garcia Table performs well for higher sample time compared. A modified switching table is further investigated which can be useful for applications such as D-STATCOM and Shunt Active Power filter.

References

1. F. Blaabjerg, R. Teodorescu, M. Liserre, and A. V. Timbus, "Overview of control and grid synchronization for distributed power generation systems," *Industrial Electronics, IEEE Transactions on,* vol. 53, pp. 1398–1409, 2006.
2. M. P. Kazmierkowski and L. Malesani, "Current control techniques for three-phase voltage-source PWM converters: a survey," *Industrial Electronics, IEEE Transactions on,* vol. 45, pp. 691–703, 1998.
3. T. Noguchi, H. Tomiki, S. Kondo, and I. Takahashi, "Direct power control of PWM converter without power-source voltage sensors," *Industry Applications, IEEE Transactions on,* vol. 34, pp. 473–479, 1998.
4. J. Hu and Z. Zhu, "Investigation on switching patterns of direct power control strategies for grid-connected DC–AC converters based on power variation rates," *Power Electronics, IEEE Transactions on,* vol. 26, pp. 3582–3598, 2011.
5. T. G. Habetler, F. Profumo, M. Pastorelli, and L. M. Tolbert, "Direct torque control of induction machines using space vector modulation," *Industry Applications, IEEE Transactions on,* vol. 28, pp. 1045–1053, 1992.
6. I. Takahashi and T. Noguchi, "A new quick-response and high-efficiency control strategy of an induction motor," *Industry Applications, IEEE Transactions on,* pp. 820–827, 1986.
7. A. Sato and T. Noguchi, "Voltage-source PWM rectifier–inverter based on direct power control and its operation characteristics," *Power Electronics, IEEE Transactions on,* vol. 26, pp. 1559–1567, 2011.
8. T. A. Trivedi, R. Jadeja, and P. Bhatt, "A Review on Direct Power Control for Applications to Grid Connected PWM Converters," *Engineering, Technology & Applied Science Research,* vol. 5, pp. pp. 841–849, 2015.
9. J. Eloy-Garcia, S. Arnaltes, and J. Rodriguez-Amenedo, "Direct power control of voltage source inverters with unbalanced grid voltages," *Power Electronics, IET,* vol. 1, pp. 395–407, 2008.
10. J. Alonso-Martinez, J. E.-G. Carrasco, and S. Arnaltes, "Table-based direct power control: a critical review for microgrid applications," *Power Electronics, IEEE Transactions on,* vol. 25, pp. 2949–2961, 2010.
11. G. Escobar, A. M. Stanković, J. M. Carrasco, E. Galván, and R. Ortega, "Analysis and design of direct power control (DPC) for a three phase synchronous rectifier via output regulation subspaces," *Power Electronics, IEEE Transactions on,* vol. 18, pp. 823–830, 2003.
12. S. Vazquez, J. A. Sanchez, J. M. Carrasco, J. I. Leon, and E. Galvan, "A model-based direct power control for three-phase power converters," *Industrial Electronics, IEEE Transactions on,* vol. 55, pp. 1647–1657, 2008.
13. S. A. Larrinaga, M. A. R. Vidal, E. Oyarbide, and J. R. T. Apraiz, "Predictive control strategy for DC/AC converters based on direct power control," *Industrial Electronics, IEEE Transactions on,* vol. 54, pp. 1261–1271, 2007.

14. M. Malinowski, M. P. Kazmierkowski, S. Hansen, F. Blaabjerg, and G. Marques, "Virtual-flux-based direct power control of three-phase PWM rectifiers," *Industry Applications, IEEE Transactions on,* vol. 37, pp. 1019–1027, 2001.
15. M. Malinowski, M. Jasiński, and M. P. Kazmierkowski, "Simple direct power control of three-phase PWM rectifier using space-vector modulation (DPC-SVM)," *Industrial Electronics, IEEE Transactions on,* vol. 51, pp. 447–454, 2004.
16. M. Cichowlas, M. Malinowski, M. P. Kazmierkowski, D. L. Sobczuk, P. Rodríguez, and J. Pou, "Active filtering function of three-phase PWM boost rectifier under different line voltage conditions," *Industrial Electronics, IEEE Transactions on,* vol. 52, pp. 410–419, 2005.
17. A. Bouafia, J.-P. Gaubert, and F. Krim, "Predictive direct power control of three-phase pulsewidth modulation (PWM) rectifier using space-vector modulation (SVM)," *Power Electronics, IEEE Transactions on,* vol. 25, pp. 228–236, 2010.

Author Biographies

Ami Vekaria is PG Scholar at Marwadi Education Foundation's Group of Institutions. She completed B.E. from Indus Institute of Technology and Science, Ahmedabad. Her areas of interest include grid-connected converters, control of power electronic converter, and renewable energy interface.

Tapankumar Trivedi is Asst. Professor at Marwadi Education Foundation's Group of Institutions. He completed B.E. from Vishwakarma Government Engineering College, Chandkheda and M. Tech. from Indian Institute of Technology (IIT), Roorkee. His areas of interest include Power Quality Improvement devices, control of power electronic converters, and Multilevel Inverter-fed IM Drives.

Rajendrasinh Jadeja is Professor & Dean-Faculty of Engineering at Marwadi Education Foundation's Group of Institutions. His areas of interest include control of power electronic converters, AC drives, Pulse Width Modulated power electronic converters, and Power Quality Conditioning Equipments. He is an active member of Global Engineering's Dean Council (GEDC). He is also a member of Americal Society for Engineering Education and American Education Research Association.

Praghnesh Bhatt is associated with Charotar University of Science and Technology, Changa, Anand, India. His areas of interest include small signal stability analysis of Power System, Power System Dynamic, control of power electronic converter, and Artificial Intelligence Application to Power system.

Parallel Cache Management with Twofish Encryption Using GPU

S. Umamaheswari, R. Nithya, S. Aiswarya and B. Tharani

Abstract Only limited support is given to the resource management of graphics processing unit (GPU) by operating system in commodity software. Operating system manages the graphics processing unit as a peripheral device and restricts the use of graphics processing unit in various applications. GPUs are not only intended for graphics applications, but could also be used in applications that require high performance thereby giving more attention for its development in recent years. A desirable part of GPU memory is left unutilized in systems that are not used for GPGPU(general purpose GPU) computing. GPU memory of such systems can be viewed as disaggregated memory and by developing proper interfaces, it could be used as swap device or cache memory because of its low access latency to improve the operating systems performance. The proposed system manages GPU as buffer cache in operating systems and ensures security using Twofish encryption algorithm.

Keywords Graphics Processing Unit (GPU) · CUDA · Cache · Twofish

---

S. Umamaheswari · R. Nithya (✉) · S. Aiswarya · B. Tharani
Department of Information Technology, Madras Institute of Technology,
Anna University, Chennai, India
e-mail: nithyaragu1995@gmail.com

S. Umamaheswari
e-mail: uma_sai@mitindia.edu

S. Aiswarya
e-mail: aiswaryasekar8@gmail.com

B. Tharani
e-mail: tharaniboopathi1@gmail.com

S.S. Dash et al. (eds.), *Artificial Intelligence and Evolutionary Computations in Engineering Systems*, Advances in Intelligent Systems and Computing 517,
DOI 10.1007/978-981-10-3174-8_37

1 Introduction

Graphics processing units (GPU) was initially designed to process visual data, but it composes multiple cores that can process large blocks of data in parallel. GPU technology is used in various applications in recent years because of its highly parallel architecture resulting in more computational power. GPU is not only for graphics related applications, but could also be used to improve the performance of various applications. A large portion of GPU memory is idle in systems that are not intended for GPGPU computing. Transcendent Memory [1] provides an approach to collect the idle physical memory in a virtual environment and manage it with application program interfaces (API). These interfaces are used by the host systems to access the idle physical memory when required. We strongly support this approach in our system to utilize the idle GPU memory for improving the performance of OS.

We have implemented a system that manages GPU as cache memory for file system IO requests in operating systems.

Similar work is proposed in BAG(GPU as buffer cache) [2]. Our work is an extension to the BAG and we have added more security to that existing system. Data security is ensured in our system by encrypting and decrypting the data using Twofish algorithm [3], whereas GPU as buffer cache [2] used AES encryption algorithm. Twofish algorithm [3] encrypts 128 bits of data using key of varying size up to 256 bits. Symmetric keys are used by Twofish algorithm to encrypt the plain text and decrypt the cipher text. Twofish works faster than AES hence reducing the time for encryption and decryption.

2 Design and Implementation

2.1 System Architecture

Graphics processing unit is composed of thousands of computational cores that can process the data in parallel. Input and output requests sent by the user applications to the file system are processed by the operating system through system calls. We have developed a system that acts as an interface between the OS and User applications. Our system utilizes GPU memory as cache memory and stores the most recently accessed file content. Upon receiving the similar request, the content from the cache memory is used instead of fetching the data from file system.

Our system is composed of three main components: indirector, relay, and user space daemon. Indirector receives the read requests and write requests from the user application. A data block is identified by the logical block address and each request specifies its type (read or write) and the logical block address. Separation of read and write requests is the first step, which is followed by the lookup operation. Lookup operation is used to find the existence of the required block in the GPU

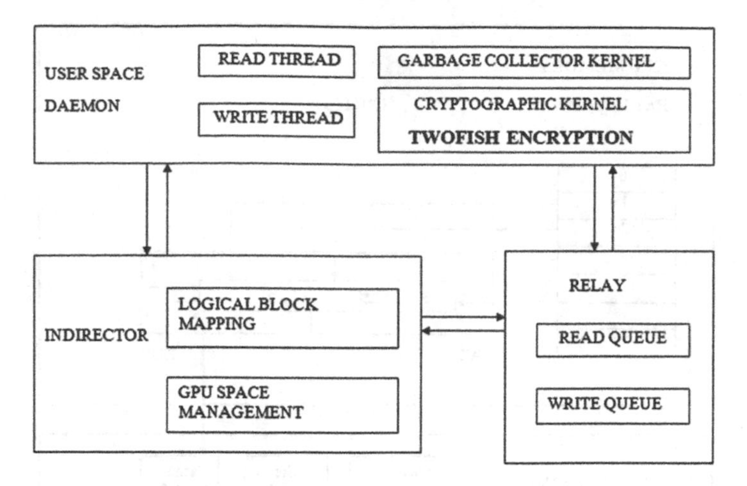

Fig. 1 Parallel cache management system architecture

memory, hence, divides the requests further into hit requests and miss requests. Read hit requests are sent to relay and the read miss and write requests are redirected to the OS. The architecture of our system is shown in Fig. 1.

Relay is responsible for maintaining read request queue and write request queue. Multiple threads are invoked to process the read hit requests in parallel and the data blocks from the GPU are transferred to the user space.

This system stores the file system data in GPU memory, hence, confidentiality of data is to be ensured. We implemented Twofish algorithm to encrypt the plain text before storing in GPU memory and decrypt the cipher text after fetching the data from GPU memory. Identical keys are used for encryption and decryption and this key is not stored in the GPU. Encryption and decryption are done by the CPU core to maintain the data confidentiality.

Garbage collector kernel revokes the memory that has older blocks makes it available for new threads. The system manages the GPU memory as circular buffer where the front end of the buffer stores the most recently accessed data block.

2.2 Hash Table and LRU List Data Structure

Existence of data block in the GPU is identified by the hash table. Hash table is implemented with 1D pointer array with direct addressing where the hash function returns the logical block address and hash value represents the GPU address in which the respective data block is cached. Garbage collector revokes the memory that stores the oldest block if required; hence, it is necessary to store the access order of the data blocks. Least recently used list is used for this purpose and a DLL (list that has nodes connected to both previous and next nodes) is used to keep track of the access order. Hash table and LRU list Data Structure is shown in the Fig. 2.

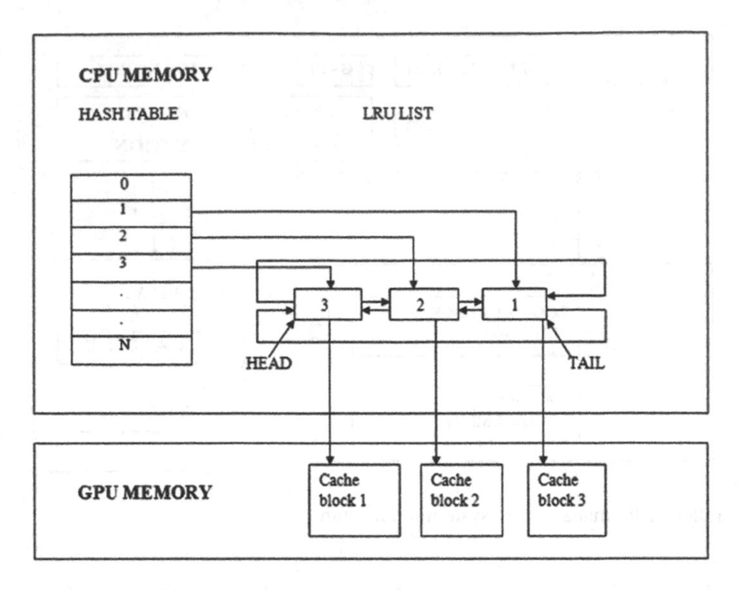

Fig. 2 Hash table and LRU list data structure

2.3 *Parallelizing the IO Request*

The IO requests received in our system are divided into read and write requests by the indirector. It separates the read hit requests from read miss and write requests. Splitting of IO requests is shown in Fig. 3.

IO requests are divided into batches. Number of IO requests in each batch is equal to the number of GPU blocks allocated for the implementation of the proposed system. Requests are processed batch wise. IO requests will have the logical block address and the lookup operation is performed on the hash table to find the existence of the corresponding block in the GPU. If the requested block is present in the GPU, then it is treated as read hit, otherwise read miss. Dividing the read requests is parallelized where multiple GPU threads will access the hash table simultaneously. First the read hit requests are processed and then the read miss requests. Retrieving the content from the GPU is done parallely by multiple threads simultaneously and each thread processes a single request in each batch.

Read miss requests are processed sequentially in CPU. The requested block is retrieved from the file system and then loaded directly in the GPU if it has enough space. If the GPU space is insufficient to load a new block, then cache eviction algorithm finds the victim block and replaces it with the new block. Least recently accessed block is treated as the victim. Tail node of the LRU list will always store the address of the least recently accessed block. Dirty bit is used to notify that the corresponding block is modified or not. If the dirty bit of the victim is zero, then it is directly replaced with the new block. Otherwise, the content of the victim block is updated in the file system and then replaced with the new block.

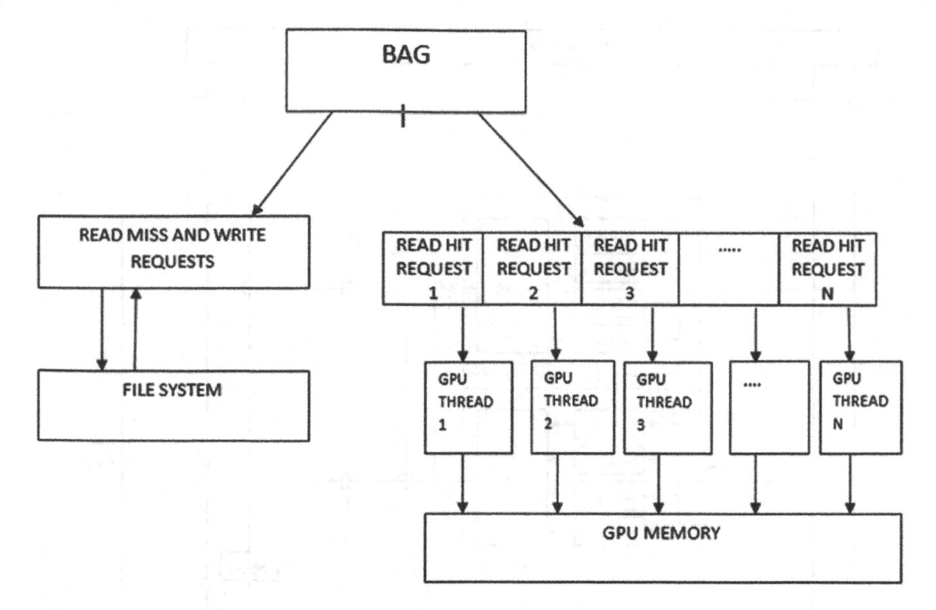

Fig. 3 Parallelizing the IO request

The write requests are processed sequentially in CPU. If the requested block already exists in the GPU then, it is over written and its dirty bit is set as one. Otherwise, the requested block is loaded in the GPU from the file system and updated with the new content. Dirty bit of the updated block is set as one. Updation is not reflected to the file system at the time of processing.

2.4 Cryptographic Kernel

Cryptographic kernel is used for encryption and decryption of data blocks. In the proposed system, Twofish algorithm [3] is used for encryption. Plain text is converted into cipher text and then it is stored in the cache. The Twofish architecture as given in "Twofish: A 128-Bit Block Cipher" is shown in the Fig. 4.

Twofish algorithm [3] encrypts 128 bits of data using key of varying size up to 256 bits. Same cryptographic keys are used by the Twofish algorithm to encrypt the plain text and decrypt the cipher text. It has a complicated algorithmic design where the plain text is preprocessed and function '*F*' is executed in a series of 16 rounds.

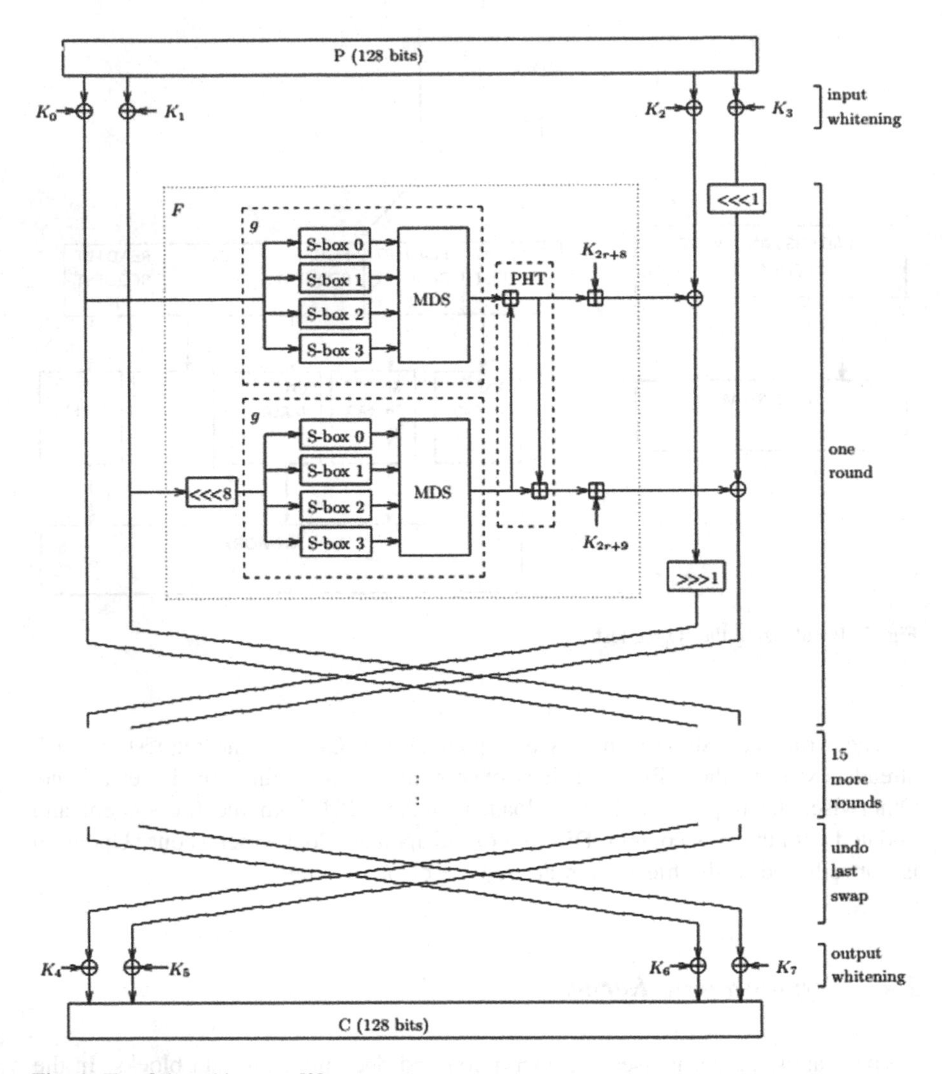

Fig. 4 Twofish architecture [3]

2.4.1 Feistel Network

Feistel network is a, onto function that converts the given function '*F*' into its permutation as defined in Eq. (1).

$$F : \{0,1\}^{n/2} * \{0,1\}^{N} \tilde{a} \{0,1\}^{n/2} \quad (1)$$

For every block of size *n* bits, it processes each *n*/2 bits of block with n bits of key and transforms it into a permutation string of size *n*/2 bits.

2.4.2 S-Boxes

Twofish is more secured since it uses four S-Boxes varies based on the cryptographic key and makes it more secured. Two fixed 8-by-8-bit permutations and cryptographic key are used to build these S-boxes and every data block is substituted based on the values stored in these S-Boxes nonlinearly.

2.4.3 MDS Matrices

It is the main diffusion mechanism that uses an MDS matrix of size 4 $\times$ 4. A matrix is said to be maximum distance separable if and only if the determinant of all possible square submatrices are nonsingular.

$$u = [\mathrm{MDS}][u] \quad (2)$$

It maps 'v' to 'u' as given in Eq. (2) by multiplying it with MDS matrix where 'u' and 'v' are vectors of four bytes each.

2.4.4 Pseudo-Hadamard Transform (PHT)

Pseudo-Hadamard Transform is fast and reversible diffusion operation. Given two inputs a and b, 32 bit PHT is defined as

$$a' = a + b \bmod 2^{32} \quad (3)$$

$$b' = a + 2b \bmod 2^{32} \quad (4)$$

two 32-bit Pseudo-Hadamard Transforms takes place parallely on the outputs from MDS.

2.4.5 Whitening

Whitening is a process of performing bit-wise XOR operation with key material before the first round and after the last round where the former is known as pre-whitening and the latter is known as post-whitening.

Twofish encryption algorithm splits the plain text (T) into four 32-bit words (T0, T1, T2, T3) and the input whitening step takes place individually. The first two words (T0, T1) on the left are used as inputs to g function. The g function contains 4 byte wide key dependent S-Boxes and a MDS matrix. The output of two *g* functions are combined using a PHT. The outputs of PHT are XORed with the other two words (T2, T3). The left (T0, T1) and the right (T2, T3) halves are swapped for the next round. These steps are repeated for the next sixteen rounds. The swap of last round is reversed and the four words (T0, T1, T2, T3) are XORed with four key words to produce cipher text.

2.5 Algorithm

Algorithm 1: Create a node in LRU list

```
1: if head=NULL
2:        head <- new node
3:        head->next <- head
4:        head-> prev <- head
5:        tail=head
6:else
7:        new_node->next <- head
8:        new_node->prev <- new_node
9:        head-> prev <- new_node
10:       tail->next <- new_node
11:       head<- new_node
12: end if
```

Algorithm 2: Rearrange LRU List

```
1: if tail=recent
2:        head <- recent;
3:        tail <- recent->prev;
4: else
5:        recent->prev->next <- recent->next;
6:        recent->next->prev <- recent->prev;
7:        recent->next <- head;
8:        recent->prev <- head->prev;
9:        head->prev->next <- recent;
10:       head->prev <- recent;
11:       head <- recent;
12:       tail <- head->prev;
13:endif
```

Algorithm 3: Read Hit

```
1:If hash[lba]!=NULL
2:        //Retrieving the content from GPU
3:        //allocate memory in host
4:        b=(cache_block*)malloc(sizeof(cache_block*))
5:        cudaMemcpy(b,hash[lba]->gpu_block,sizeof(cache_block*),
cudaMemcpyDeviceTo Host)
6:        Content : b->data.substr(1)
7:        Dirty Bit : b->data.substr(0,1)
8:        rearrange(hash[lba]);
9:endif
```

Algorithm 4: Read Miss without replacement

```
1: if filled_blocks<cache_blocks
2:        hash[lba] <- new node;
3:        hash[lba]->lba=lba;
4:        create(hash[lba]);
5:        //Allocating memory on GPU.
6:        cudaMalloc(&(hash[lba]->gpu_block),sizeof(struct cache_block*));
7:        //Copying the content of host block to GPU
8:        cudaMemcpy(hash[lba]-
>gpu_block,g,sizeof(g),cudaMemcpyHostToDevice);
9:endif
```

Algorithm 5: Read Miss with replacement

```
1: if filled_blocks = cache_blocks
2:        //GPU cache is full.
3:        LBA of Least Recently Used block : tail->lba
4:        ///Invalidating the victim and loading the new block.
5:        hash[lba] ] <- new node;
6:        hash[lba]->lba] <- lba;
7:        hash[lba]->gpu_block] <- hash[tail->lba]->gpu_block;
8:        hash[tail->lba] ] <- NULL;
9:        cudaMemcpy(hash[lba]-
>gpu_block,g,sizeof(g),cudaMemcpyHostToDevice);
10:       tail->prev->next] <- head;
11:       head->prev] <- tail->prev;
12:       tail] <- tail->prev;
13:       create(hash[lba]);
14:endif
```

Algorithm 6: Write block (overwrite an existing block)

```
1:if hash[lba] != NULL
2:        //updating the block in GPU
3:        Allocate memory for cache block b
4:        b<- hash[lba]->gpu_block
5:        b->dirty_bit="1";
6:        // modify the dirty bit
7:        b->data <-  new_data
8:        copy b back to GPU
9:        rearrange lru list;
10:endif
```

3 Simulation Environment

3.1 NVDIA GeForce 940M

It is a graphics processing unit that can support DirectX-11 technology. It has two Giga bytes of DD3_SDRAM and 384 shader units. It is built with Maxwell architecture which is more efficient than Kepler architecture. GeForce 940M is composed of smaller streaming multiprocessors with 128 ALUs and an optimized scheduler. It is better than GeForce 840M because of its higher clock rate.

3.2 CUDA Toolkit Version 6.5

CUDA toolkit provides an environment to develop software that can run on GPU cores and utilize GPU memory. CUDA is a programming language similar to C language with additional support for developing GPU codes that can access the GPU resources directly. CUDA program includes both device functions (called from GPU and executed on GPU) and global functions (called from CPU and executed on GPU). CUDA programs are executed only in computers with NVDIA Graphics Processing Units. Multiple GPU threads can execute a single global function parallely on GPU cores. The first API to access the GPU resources directly is CUDA. The C programs can invoke the CUDA program and the converse is also possible.

3.3 NVCC

The compiler driver for CUDA language is NVCC. The motive of NVCC is to hide the compilation details of CUDA from the developers. The non-CUDA code is compiled by the C++ compiler. The NVCC does these forwarding options and it takes care of other operations like splitting, merging and preprocessing steps of the CUDA code.

4 Evaluation

The proposed system is executed in a PC with 2GB NVDIA GeForce 940M graphics card. Operating System is Centos 6.5. Time taken to process 100 requests using Sequential and parallel cache is shown in Table 1. Time taken to process 500 requests using Sequential and parallel cache is shown in Table 2. Time taken to process 1000 requests using Sequential and parallel cache is shown in Table 3.

Tabulated results shows that parallel cache management with multithreading is better than sequential cache management for large number of IO requests. Parallel cache takes longer time to process fewer requests because of the overhead in dividing the read hit and read miss requests.

Table 1 Comparison of parallel and sequential cache with number of requests = 100

No. of cache blocks	Time (ms)—sequential cache	Time (ms)—parallel cache
3	130	140
10	120	140
15	150	140

Table 2 Comparison of parallel and sequential cache with number of requests = 500

No. of cache blocks	Time (ms)—sequential cache	Time (ms)—parallel cache
30	180	160
50	180	140
100	170	130

Table 3 Comparison of parallel and sequential cache with number of requests = 1000

No. of cache blocks	Time (ms)—sequential cache	Time (ms)—parallel cache
30	1060	160
50	890	140
100	590	130

5 Conclusion

A system that manages GPU as buffered cache in operating system is implemented. Data encryption provides security to the data present in the GPU. Twofish algorithm is used for encryption and decryption. With carefully designed data structures such as concurrent hash table, least recently used list, log structured data store this system achieves good performance under various workloads.

References

1. Curry,Gregerson N O, Buttenschon H N and Hedamad A, "Transcendent Memory", (2010) Available: https://oss.oracle.com/projects/tmem/.
2. Hao Chen, Jianhua Sun, Ligang He, Kenli Li and Huailiang Tan, "BAG: Managing GPU as Buffer Cache in Operating Systems" in IEEE TRANSACTIONS, (2014).
3. Schneier, John Kelsey, Doug Whiting, David Wagner, Chris Hall and Niels Ferguson, "Twofish: A 128-Bit Block Cipher", (1998) Available: http://www.counterpane.com/Twofish.html.

Enhanced Single Image Uniform and Heterogeneous Fog Removal Using Guided Filter

Pallawi and V. Natarajan

Abstract In this chapter, we propose an effective method to remove uniform and heterogeneous fog from the image using dark channel prior (DCP) and guided filter. Variation in thickness of the fog present in variety of image has helped to analyze and upgrade the dark channel prior image quality. Fog acts as veil which obscures the original scene radiance. A significant amount of fog could be observed around the edges of the objects in fog free image which is carried from the foggy image during intermediate processing steps. Efficient selection of dark channel prior kernel parameters helps us to carry minimum fog from the input foggy image to fog free image which has certainly reduced the halo effect around the edges of the objects in the fog free image. Also the use of guided filter to preserve edges of the objects in image is a fast and cost effective approach toward generation of fog less image.

Keywords Digital image processing · Fog removal · Image enhancement · Digital filter

1 Introduction

Today, the increasing demand of minute but significant real-time data and its analysis has given birth to too many fields of study like artificial intelligence, computer vision which is contributing toward mankind with its industrial application and feasibility.

ADAS (Advance driving assistance system), being one of its application has reached us at a very fast rate and vividly seen in most of the countries where there

Pallawi (✉) · V. Natarajan
Embedded System Technology, Electronics and Communication
Engineering, SRM University, Chennai 603203, India
e-mail: pallawi.ds@gmail.com

V. Natarajan
e-mail: vy.natarajan@gmail.com

S.S. Dash et al. (eds.), *Artificial Intelligence and Evolutionary Computations in Engineering Systems*, Advances in Intelligent Systems and Computing 517,
DOI 10.1007/978-981-10-3174-8_38

are well-maintained road infrastructures to deliver smooth running of computer vision and image processing algorithms.

A very important point to notice here is that performances of vision algorithms (e.g., feature detection, filtering, and sign recognition) are dependent on the inputs given to the sensors like camera or radar in form of images or signals are actively contributed by quality of road infrastructures like lane marking and their feature, milestones, speed limit posts, and many others. Unfortunately, due to poor visibility conditions the color fidelity of the input images are lost reason being presence of fog, haze, dust, or smoke in the atmosphere. This creates turmoil in the decision making abilities and sometimes a complete failure of system which might lead to catastrophic events too. Fog and haze deters the scene radiance and gives us insignificant and sometimes no information about the scene of interest. Thus fog removal is highly desired to increases the visibility of the scene and correct the color shifts caused by airlight. Fog removal can also produce depth information and benefit many vision algorithms.

Therefore, many algorithms have been proposed by using multiple images. In [1] fog removal and recovery of the depth map is done using two images of the same scene under different weather conditions. In [2, 3], the parameters required to recover the scene radiance is calculated at different degree of polarization. Color and contrast enhancement techniques like multiscale retinex (MSR) [4, 5] and Contrast limited adaptive histogram equalization (CLAHE) [6] are not suited for foggy images. Tripathi [7] uses anisotropic diffusion for refining airlight map which is an expensive method. Kaiming He [8] used soft matting for refining the transmission map which involves complex computation and consumes more time. Tan [9] observes that the fog free image must have higher contrast compared with the input fog image and he removes the fog by maximizing the local contrast of the restored images. The result is visibly compelling but may not be physically valid. In [10] vision enhancement in homogenous and heterogeneous fog is performed using set of synthetic images from FRIDA (Foggy Road Image Database) dataset which works on different types of fog introduced in the same scene. They use no-black-pixel constraint and the planar assumptions (NBPC + PA) in which the image is split into three regions: the sky, the objects out of road plane and the free space in the road plane. Each region is enhanced according to the requirement of application.

In this chapter, we propose a novel analysis-based method for using dark channel prior to remove homogenous and heterogeneous fog from the image efficiently. The use of guided filter to preserve the edges of objects in the fog free image is a simple and reliable method but it might produce sever halo around the edges of the object when the radius of objects are small.

The rest of this paper is organized as follows. Our Methodology is detailed in Sect. 2. Section 3 includes our results followed by conclusion in Sect. 4.

2 Methodology

2.1 Fog Model

Clouds are made up of tiny water droplets and ice crystals, due to condensation of the water vapors from the water bodies present around, these molecules are so small that they can float in the air. If these tiny water droplets become large, then they become low lying cloud or fog and when these water molecules become even larger, they fall as rain (or snow). Thus fog is widely been modeled in computer vision as a combination of two components direct attenuation and airlight.

Which degrades the image quality, gives very limited information about the scene, hinders smooth running of many computer vision and ADAS algorithms like sign recognition.

The model used in this paper to describe the formation of fog image is as follow:

$$\text{Fog} = \text{Direct Attenuation} + \text{Airlight}.$$

$$I(x) = [(J(x) \times t(x)) + A \times (1 - t(x))], \quad (1)$$

where I is the intensity of foggy image, J is the scene radiance, which is the intensity of a fogless image, t is medium transmission describing the portion of light that is not scattered and reaches the camera, and A is the airlight radiance at infinite distance.

Direct attenuation is the gradual loss in intensity of light when the light from the atmosphere decays in the medium, it is a multiplicative distortion of scene radiance. While airlight represents the shift of the scene color. The fog atmosphere consists of a mixture of molecules and particles of various sizes. These molecules add whiteness to the scene. The visible effect of airlight is due to the scattering of light by those fog molecules toward the viewer. Also airlight is an increasing function of scene point distance $d(x)$, where $d(x)$ is the depth of the xth pixel. Thus, when the atmosphere is homogenous, the transmission t can be expressed as:

$$t(x) = e^{-kd(x)}, \quad (2)$$

where k is the scattering coefficient [11], of the atmosphere. Equation (2) indicates the scene radiance is attenuated exponentially with the scene depth d. Figure 1 shows the uniform foggy image at the left and its depth map at the right.

2.2 Improved Dark Channel Prior

The dark channel prior (DCP) is based on the observation that in an RGB fog free outdoor image which mandatorily has infinity (sky), tells that in most of the

(a) Uniform Foggy Image (b) Our recovered depth map

Fig. 1 **a** Uniform foggy image **b** our recovered depth map

non-sky patches(regions), which could include infrastructures, vegetation, road there would be at least one color channel which will have low intensity at the pixels of non-sky patches in the fog free image. In other words, the minimum intensity in such a patch has a very low value. Dark channel prior is defined as,

$$J^{\text{dark}}(x) = min_{c\epsilon\{r,g,b\}}(\min_{y\in\Omega\delta(x)}((J^{C}(y)) \qquad (3)$$

where J^C the color channels of haze free image J and $\delta(\boldsymbol{x})$ is a local patch centered at x. DCP generated in many algorithms for fog removal have used a patch size of 15×15. In our algorithm, we have used a patch size of 5×5 to calculate the DCP.

In Fig. 2 we describes the reason behind selection of smaller window for DCP calculation by comparing effect of 15×15 and 5×5 window around the tree trunk which is surrounded by fog. When we use a 15×15 patch around the tree trunk it happens that the minimum intensity of the pixel among all the three channel

(a) (b)

Fig. 2 **a** DCP of a uniform foggy image with patch size 5×5 small window with less fog around the trunk of tree, **b** DCP of a uniform foggy image with patch size 15×15 which tends to accommodate fog around the tree trunk during DCP calculation

is contributed by one of the foggy pixel from the concerned patch around the tree trunk and when we use an overlapping window to calculate the minimum intensity among all the three channels around the edges of the objects whose radius is smaller than the used patch size then intensity around the tree remains same as that of fog and appears in the form of increased radius of object which appears as halo around its edges in haze free image. The effect of fog around the object would be more in case of thin branches of tree or the traffic poles.

The selection of a 5×5 window in fig. 2b takes care that less fog is accumulated at the edges of objects present in the scene during computation of DCP.

The low intensities in the DCP is mainly due to reasons like (a) dark object or dark road surface, e.g., dark color building, dark traffic light stands, dark color cars; (b) Shadow, e.g., Shadows of building, trees, cars, lamp post; (c) Colorful objects or surface lacking color in any one channel like a green tree, complete red, or blue bill board. As the natural outdoor image is full of shadows the DCP of these images are really dark.

So the DCP of the foggy image will have higher intensity at the region with denser fog and as the depth of the scene increases the presence of fog also increases.

2.3 *Airlight Radiance*

Airlight radiance is calculated from the most haze-opaque pixel at an infinite distance [12, 11, 13]. To find this we pick 0.1% of the brightest pixel present in the dark channel prior of the fog image and then pick the unique intensities from those 0.1% pixels of the dark channel prior followed by mapping the locations of those picked unique intensities onto the channels of RGB image individually and computing the average of intestines present at those mapped location in every channel and using them as the $A_{(\text{Red})}, A_{(\text{Green})}, A_{(\text{Blue})}$.

2.4 *Transmission Image*

After estimation of airlight radiance for the individual channels we find the transmission image $\tilde{t}(x)$ of the fog image [8] assuming that the transmission is constant in the patch $\delta(x)$.

Taking the min operation in the local patch of fog imaging Eq. (1) we get:

$$\min_{y \in \delta(x)}(I^c(y)) = \tilde{t}(x) * \min_{y \in \delta(x)}(J^c(y)) + A^C * (1 - \tilde{t}(x)). \tag{4}$$

This min operation is performed on three color channels independently. This equation is equivalent to:

$$\min_{y\in\delta(x)}\left(\left(\frac{I^c(y)}{A^c}\right)\right) = \tilde{t}(x) * \min_{y\in\delta(x)}(J^c(y)) + A^C * (1-\tilde{t}(x)). \quad (5)$$

Then we take minimum of three channels on the above equation:

$$\min_{y\in\delta(x)}\left(\frac{I^c(y)}{A^c}\right) = \tilde{t}(x) * \min_c\left(\min_{y\in\delta(x)}\left(\frac{J^c(y)}{A^c}\right)\right) + (1-\tilde{t}(x)) \quad (6)$$

According to the dark channel prior, the dark channel $J^{\mathrm{dark}}(x)$ of the fog free image radiance J should tend to be zero as the dark object will contribute to the reduced intensity in any one of the channel.

$$J^{\mathrm{dark}}(x) = \min_c\left(\min_{y\in\delta(x)}(J^c(y))\right) = 0. \quad (7)$$

And as A^c is always positive, this leads to:

$$\min_c\left(\min_{y\in\delta(x)}\left(\frac{J^c(y)}{A^c}\right)\right) = 0. \quad (8)$$

Substituting Eq. (8) into Eq. (6), we can estimate the transmission $\tilde{t}$

$$\tilde{t}(x) = 1 - \min_c\left(\min_{y\in\delta(x)}\left(\frac{J^c(y)}{A^c}\right)\right). \quad (9)$$

The color of the sky is very similar to the airlight radiance $A_{(\mathrm{Red})}, A_{(\mathrm{Green})}, A_{(\mathrm{Blue})}$ in a haze image and we have:

$$\min_c\left(\min_{y\in\delta(x)}\left(\frac{J^c(y)}{A^c}\right)\right) \rightarrow 1 \text{ and } \tilde{t}(x) \rightarrow 0, \text{ sky regions.}$$

On a clear day the atmosphere contains air molecules which are very small in size. So the fog is still present when we look at the objects present at a greater depth and while generating a fogless image, if we remove the fog completely from the image it would make the image look unnatural and the feeling of depth may be lost.

To make the fogless image look natural with the increasing depth we keep a very small amount of haze [8] into the image for distant object by introducing a parameter γ $(0<\gamma\leq 1)$ into Eq. (9).

This introduction of γ automatically introduces fog according to the depth of the scene. Value of γ is application based. We have used γ as 0.95.

$$\tilde{t}(x) = 1 - \gamma * \min_c\left(\min_{y\in\delta(x)}\left(\frac{J^c(y)}{A^c}\right)\right). \quad (10)$$

Due to the patch used in DCP calculation and using the same image to compute the transmission map carries the block effect and losses the edges of the objects. So we perform edge preserving operation onto the transmission image.

2.5 Guided Filter

We have used guided filter [14] with a window size of 40 × 40 for preserving the edges of the transmission image. Size of the window could also be varied according to the requirement of application to get a better edge preserved image.

The model of guided filter is

$$t(x) = [(a_k * I_i) + b_k], \forall_i \in W_k, \quad (11)$$

where $t(x)$ is the refined transmission image and I_i is the guiding foggy image. Where (a_k, b_k) are linear coefficient assumed to be constant in W_k.

To determine the linear coefficients, we seek a solution to (11) that minimizes the difference between $t(x)$ and the filter input which is the $\tilde{t}(x)$ specifically, we minimize the following cost function in the window:

$$a^k = \frac{\frac{1}{|W|}\sum_{i\in W_k}\left(I_i * \left(\widetilde{\overline{t(x)}}\right)\right) - \left(\mu_k * \widetilde{\overline{t(x)}}\right)}{\partial^2 + \rho}. \quad (12)$$

$$b^k = \left(\widetilde{\overline{t(x)}}\right) - a^k * \mu_k. \quad (13)$$

Here W_k is the window of window size $|W|$ to be considered for the image and (i) is the pixels present in the window W_k. μ_k and ∂^2 the mean and variance of the window W_k of the foggy image $I(x)$. ρ is the regularizing parameter we have used value for ϵ as 0.1000.

$$\left(\widetilde{\overline{t(x)}}\right) = \frac{1}{|W|}\sum_{i\in W_k}\widetilde{t(x)}\text{, is the mean of transmission image in } W_k.$$

Now we apply Eq. (11) for entire image. The $t(x)$ gives us the refined transmission image with smooth and refined edges. There are limitations of using this filter as it can make the halos generated during DCP calculation more visible when it undergoes smoothing and edge preservation [15, 16].

2.6 Scene Radiance

With the transmission Image we could recover the scene radiance according to Eq. (1). But the direct transmission could be very close to zero when the $t(x)$ is close to zero. The directly recovered scene radiance is prone to noise. Therefore we restrict the $t(x)$ to a lower bond t_0, which means a small certain amount of haze is preserved in very dense haze regions. Value take for t_0 is 0.1.

$$J(x) = \frac{I(x) - A^c}{\max(t(x), t_0)} + A^C. \quad (14)$$

Thus $J(x)$ is our fog free image.

3 Results

To evaluate the results of our algorithm, we need image with and without fog. However, obtaining such pairs of images is extremely difficult in practice since it requires to check that the illumination condition are same in the scene with and without fog. As a consequence, for the evaluation of the proposed fog removal algorithm and its comparison with existing algorithms, we use Synthetic fog from dataset [10] of FRIDA.

In our experiments, we perform the local min operator using Marcel van Herk's fast algorithm. The patch size is set to 5 × 5 for a 640 × 480. In Guidance filter we have used a patch size of 40 × 40. MATLAB R2014b takes 5–8 s to process a 640 × 480 pixel image on a PC with a 3.0 GHz Intel Pentium 7 Processor.

In Fig. 3 we present our result which gives a much clear visibility of the scene in uniform and heterogeneous fog. Algorithm like MSR creates an additional gray veil onto the scene [10]. DCP uses soft matting which takes larger amount of time and involves complex computation as compared to our edge preserving algorithm of Guided filters.

3.1 Quantitative Comparison

Table 1 is a quantitative comparison of the results of algorithms for MSR [4, 5], CLAHE [6], DCP [8], free-space segmentation [10], Non-black pixel constraint [10], Planer assumption [10], and our algorithm. This table represents absolute mean error over set of 18 images for each types of fog taken from dataset FRIDA [10]. From this table we can infer that in two of the cases those are when the fog is uniform and heterogeneous with the varying scattering coefficient of the atmosphere our result shows a huge improvement to generate the fog free image and provides

Fig. 3 Results of our proposed algorithm. From *left* to *right* the original synthetic image with various fog followed by our recovered fogless image

Table 1 Mean absolute error

No	Mean absolute error				
	Algorithm	Uniform fog	Variable *k* fog	Variable *A*	Variable *k* and *A* fog
1	MSR	47.5 ± 8.8	74.5 ± 21.7	47.6 ± 14.0	72.2 ± 20.4
2	CLAHE	53.4 ± 8.8	55.8 ± 9.4	47.1 ± 7.6	49.6 ± 7.8
3	DCP	32.8 ± 14.1	36.2 ± 10.2	34.9 ± 14.2	36.9 ± 11.5
4	FSS	38.2 ± 7.3	34.7 ± 8.1	32.4 ± 6.5	30.1 ± 5.9
5	NBPC	41.8 ± 6.7	43.0 ± 6.4	35.8 ± 5.3	36.5 ± 4.8
6	PA	29 ± 5.9	31.5 ± 6.8	27.3 ± 5.7	29 ± 6.7
7	Proposed algorithm	20.9486 (Maximum)	25.7642 (Maximum)	52.2971 (Maximum)	54.4628 (Maximum)

better visibility. In other case due to overestimation of depth of the foggy image while generating the transmission image leads to black pixels and thus deters the scene radiance.

4 Discussion and Conclusion

In this chapter, we have proposed a new aspect of using dark channel prior with guided filter to remove uniform and heterogeneous fog from the foggy image. Our approach to use a smaller window would work well for the objects of small radius and has certainly improved the quality of image with less halo or fog artifacts in the fog free image. The visibility distance has also improved as compared to other algorithms. Our work also has artifacts due to the fog present around edges between two completely different intensity of objects. DCP has its own limitations and could only work well for daytime fog. We believe that these limitations could be taken as another research problem in future.

Acknowledgments We would like to express our deep gratitude to our technical guide Ajit Kulkarni, Anoop Pathayapurakkal and Avinash Nagaraj from company Continental Automotive India Pvt. Ltd., for their patient guidance, encouragement and wide vision which gave us a chance to explore, reason, and express our ideas. Their insightful advice and constructive criticisms at different stages of our research were thought-provoking and they helped us focus on our ideas.

References

1. Srinivasa. G. Narsimhan, Shree K. Nayar, "Contrast restoration Of weather degraded Images," IEEE Transaction on pattern analysis and Machine Intelligence, vol. 25, No. 6, June 2003.

2. Yoav Y. Schechner, Srinivas G. Narsimhan, Shree K. Nayar, "Instant Dephasing of Image using polarization." Computer vision and pattern recognition, 2001. CVPR 2001. Proceedings of the 2001 IEEE Computer Society Conference.
3. Atul Gujral, Aditi, Shailendra Gupta, Bharat Bhushan, "A comparison of various defogging techniques," International Journal of signal processing image processing and character recognition, Vol. 7, No. 3 (2014), pp. 147–170.
4. D. Jobson, Z. Rahman, and G. Woodell, "A multiscale retinex for bridging the gap between color images and the human observation of Scenes," Image Processing, IEEE Transactions on Image Processing, vol. 6, no. 7, pp. 965– 976,1997.
5. Weixing Wang, Lian Xu, "Retinex algorithm on changing scale for haze removal with depth map," International Journal of Hybrid Information Technology, Vol. 7, No. 4 (2014), pp. 3.353–364.
6. K Zuiderveld, "Contrast limited adaptive histogram equalization," In Graphics gems IV. San Diego, USA: Academic Press Professional, Inc., 1994, pp. 474–485.
7. A.K. Tripathi, S. Mukopadhyay, "Single image fog removal using anisotropic diffusion," IET Image Process, 2012, Vol. 6, Iss. 7, pp. 966–975I.
8. K. He, J. Sun, and X. Tang, "Single image haze removal using dark Channel prior," IEEE Transactions on Pattern Analysis and Machine Intelligence, vol. 33, no. 12, pp. 2341–2353, December 2010.
9. R. Tan. "Visibility in bad weather from single image," CVPR.
10. Jean-Philippe Tarel, Nicolas Hautiere, Laurent Caraffa, Aurelien Cord, Houssam Halmaoui, Dominique Gruyer, "Vision enhancement in homogenous and heterogenous fog," HAL archives-ouvertes.
11. Srinivasa. G. Narsimhan, Shree K Nayar, "Vision and the atmosphere", Intenational Journal of computer vision, 48(3), pp. 233–254, 2002.
12. Shree K. Nayar, S G Narsimha "Vision in bad weather," ICCV, page 820 1999.
13. Terek El. Gaaly, "Measuring atmospheric scattering form digital images of urban scenery using temporal polarization based vision", Thesis, American University in Cairo school of science and engineering.
14. K. He, J. Sun, and X. Tang. "Guided Image Filtering," IEEE Trans. Pattern Anal. Mach. Intell., vol. 35, no. 6, pp. 1397–1409, Jun. 2013.
15. Jean-Philippe Tarel, Nicolas Hautiere "Visibility restoration from a single color or gray level image," Computer Vision 2009, IEEE 12 International Conference.
16. Qingsong Zhu, Jaiming Mai, Ling Shao, "A fast single image haze removal algorithm using color attenuation prior", IEEE Transaction on Image Processing, VOL. 24, NO. 11, November 2015.

Multi-objective Genetic Algorithm-Based Sliding Mode Control for Assured Crew Reentry Vehicle

Divya Vijay, U. Sabura Bhanu and K. Boopathy

Abstract A reentry control system is proposed for an assured crew reentry vehicle (ACRV), where the control law is tuned using multi-objective genetic algorithm-based sliding mode controller. The controller designed guarantees the robustness properties with respect to parametric uncertainties and other disturbances. The system state remains in the neighborhood of a reference attitude and the control signal is close to a well-defined equivalent control. The amplitude of the sliding mode controller is tuned using an evolutionary optimization technique, i.e., Genetic algorithm. Multi-objective optimizer is used for the controller as it is to minimize the error in the Bank Angle (degree), Angle of Attack (degree), and Sideslip Angle (degree). The reference attitude is obtained in terms of the outputs given by the trajectory controller and the navigational system. A pulse width pulse frequency (PWPF) modulator is designed to modulate the attitude controller through the thrust torque developed. The simulation results show the effectiveness of the proposed method.

Keywords ACRV · Coordinate transformations · Attitude control · Quaternion · PWPF thrusters

1 Introduction

The Human Space Mission is providing a great deal of opportunities in space research activities which are oriented toward the design and development of space stations. Space stations are the permanent manned presence in space. Space stations

Divya Vijay (✉) · U. Sabura Bhanu · K. Boopathy
Department of EEE, B.S. Abdur Rahman University, Chennai, Tamil Nadu, India
e-mail: divya.vijay@bsauniv.ac.in

U. Sabura Bhanu
e-mail: sabura.banu@bsauniv.ac.in

K. Boopathy
e-mail: boopathy@bsauniv.ac.in

S.S. Dash et al. (eds.), *Artificial Intelligence and Evolutionary Computations in Engineering Systems*, Advances in Intelligent Systems and Computing 517,
DOI 10.1007/978-981-10-3174-8_39

always met with the problem of low-cost transportation. This can be achieved by using an assured crew reentry vehicle [1–3]. In this paper, the attitude control of an assured crew reentry vehicle is illustrated. An attitude controller is planned to track the reference attitude obtained from the trajectory controller and navigational system of an assured crew reentry vehicle. This allows the vehicle to experience the aerodynamic force required to follow the reference trajectory path. The trajectory controller computes the attitude required and attitude controller computes the torque required to generate the attitude.

The controller tuning is complicated and in recent past, many modern heuristic optimization algorithms like genetic algorithm, simulated annealing, tabu search, swarm intelligence techniques, like particle swarm optimization, bacterial foraging, bee colony optimization, fire fly algorithm are investigated [4]. Heuristic algorithms are techniques used to obtain global optimal solution [5]. Genetic Algorithm has been used extensively to obtain the conventional PID controller due to its capability of not getting struck in the local minima. Genetic algorithm has been implemented for many processes like control of reverse osmosis [6], single link flexible manipulator in vertical motion [7], etc. Various performance measures like minimum rise time, settling time, overshoot, integral absolute error, integral square error, and integral time absolute error has been considered as the objective function by many authors [8].

The attitude controller is designed using genetic algorithm-based sliding mode controller. quaternions are used to describe the kinematics of the ACRV [9, 10]. Finally, thrusters are incorporated to provide the torque required by the attitude controller. These are modulated using a PWPF thruster modulator. Simulations are performed by using the trajectory controller output as reference.

Section 2 discusses about the coordinate systems highlighting the coordinate transformation followed by the Quartenion formulations. Section 4 presents the mathematical model of ACRV and Sects. 5 and 6 discusses in detail about the controller design and optimization of the parameters using genetic algorithm-based sliding mode controller followed by the Design of Thrusters. Finally the simulation results are presented.

2 Coordinate Systems

To describe the model of an assured crew reentry vehicle, a suitable coordinate system is to be adopted. The various coordinate systems are:

- Inertial Coordinate System: System is having the origin at the Earths center, with the X and Y axis in the equatorial plane and the Z axis aligned with the planet polar axis, positive toward the south.
- Local Geocentric System: System is having the Z axis along the line through the center of mass of the body, positive toward the Earth center; X axis is

orthogonal to Z axis and positive toward the north and Y axis completes the right-handed orthogonal frame.

- Wind Axis System: System is having the X axis along velocity vector corresponding to the local geocentric frame. The Z axis is aligned with the lift force, with opposite direction and Y axis completes the right-handed orthogonal frame.
- Body Axes System: System will be having X axis along the geometric longitudinal axis of the body, positive toward the capsule nose and Y axis is along symmetry plane of the body pointing downwards entry and perpendicular to the X axis. The Z axis completes the right-handed orthogonal frame.

For obtaining the suitable co-ordinate system for describing the assured crew reentry vehicle dynamics, various coordinate transformations are done.

2.1 Coordinate Transformations

α, β, σ are the trajectory controller outputs. The other angular values required for the generation of quaternions are provided by the navigational system. Various coordinate transformations are done to achieve the required input parameters. The coordinate transformations are:

The transformation matrix for transforming the inertial coordinates to local geocentric coordinates is:

$$G = \begin{bmatrix} -\sin(\phi) * \cos(B) & -\sin(\phi) * \sin(B) & -\cos(\phi) \\ \sin(B) & -\cos(B) & 0 \\ -\cos(\phi) * \cos(B) & -\cos(\phi) * \sin(B) & \sin(\phi) \end{bmatrix} \tag{1}$$

The transformation matrix for transforming local geocentric coordinates to wind axis coordinate system is:

$$W = \begin{bmatrix} \cos(\gamma)\cos(\phi) & \cos(\gamma)\sin(\phi) & -\sin(\gamma) \\ (-\sin(\phi)\cos(\sigma)) + (\sin(\gamma)\cos(\phi)\sin(\sigma)) & (\cos(\phi)\cos(\sigma)) + (\sin(\gamma)\sin(\phi)\sin(\sigma)) & \cos(\gamma)\sin(\sigma) \\ (-\sin(\phi)\sin(\sigma)) + (\sin(\gamma)\cos(\phi)\cos(\sigma)) & (-\cos(\phi)\sin(\sigma)) + (\sin(\gamma)\sin(\phi)\cos(\sigma)) & \cos(\gamma)\cos(\sigma) \end{bmatrix} \tag{2}$$

The transformation matrix for transforming wind axis coordinate system to body axis coordinate system is:

$$B = \begin{bmatrix} \cos(\alpha) * \cos(\beta) & \sin(\beta) & \sin(\alpha) * \cos(\beta) \\ -\sin(\beta) * \cos(\alpha) & \cos(\beta) & -\sin(\beta) * \sin(\alpha) \\ -\sin(\alpha) & 0 & \cos(\alpha) \end{bmatrix} \quad (3)$$

The direction cosine matrix 'E' is obtained as:

$$E = G * W * \mathrm{inv}(B) = \begin{bmatrix} e11 & e12 & e13 \\ e21 & e22 & e23 \\ e31 & e32 & e33 \end{bmatrix} \quad (4)$$

From the direction cosine matrix, the commanded quaternions for the ACRV are generated.

3 Quaternion Formulation

Quaternions are the most effective method for identifying the rotational motion. Quaternions are the vectors in 4D space algebra. Instead of the trigonometric functions of Euler angles, the quaternion uses algebraic relations. Hence quicker computation is achieved and use of quaternions also helps to eliminate the possibility of a singularity problem [11, 12].

From the direction cosine matrix as in Eq. (4), the commanded quaternions are calculated as follows:

$$q_{4c} = 0.5 * \sqrt{1 + e11 + e22 + e33} \quad (5)$$

$$q_{1c} = \left(\frac{1}{4 * q_{4c}}\right)(e23 - e32) \quad (6)$$

$$q_{2c} = \left(\frac{1}{4 * q_{4c}}\right)(e31 - e13) \quad (7)$$

$$q_{3c} = \left(\frac{1}{4 * q_{4c}}\right)(e12 - e21) \quad (8)$$

The quaternions have two parts, i.e., a quaternion magnitude given by Eq. (5) and quaternion orientation represented by Eqs. (6), (7), and (8).

The reference angular velocities can also be computed from the reference quaternions. It is obtained as:

$$\hat{\omega} = 2 * \Omega(\hat{q}_c)^{\mathrm{T}} * \dot{\hat{q}}_c \quad (9)$$

where

$$q_c = [q_{1c} \quad q_{2c} \quad q_{3c} \quad q_{4_c}]^{\mathrm{T}}.$$

4 Mathematical Model of ACRV

Mathematical model of ACRV is derived from the Rigid Body Dynamic Equation and then the Rigid Body Kinematic Equation [13, 14].

4.1 ACRV-dynamic Equation

$$J * \dot{\omega} + S(\omega) * J * \omega = M_t + M_a \tag{10}$$

where

$$J = J_0 + \Delta J \quad \text{and} \quad \omega = (\omega_x, \omega_y, \omega_z)^{\mathrm{T}} \tag{11}$$

The skew symmetric matrix is given by,

$$S(\omega) = \begin{pmatrix} 0 & -\omega_z & \omega_y \\ \omega_z & 0 & -\omega_x \\ -\omega_y & \omega_x & 0 \end{pmatrix} \tag{12}$$

4.2 ACRV-kinematic Equation

The kinematic description of the rigid body motion involves three parameters known as the Euler angles. But in case of large angle maneuvers, Euler angles can exhibit singularities. In such cases quaternions are also used. The quaternion parameters are defined as,

$$q = [q_1 \quad q_2 \quad q_3 \quad q_4]^{\mathrm{T}} \tag{13}$$

where q_1, q_2 and q_3 represents the angular part and q_4 represents the magnitude part.

The rigid body kinematic equation is given by,

$$\dot{q} = 0.5 * \Omega(q) * \omega, \tag{14}$$

where

$$\Omega(q) = \begin{bmatrix} q_4 & -q_3 & q_2 \\ q_3 & q_4 & -q_1 \\ -q_2 & q_1 & q_4 \\ -q_1 & -q_2 & -q_3 \end{bmatrix}$$

5 Controller Design

The attitude controller for the ACRV will track the capsule along the prescribed path.

The results of the trajectory controller are the commanded angles $\alpha_c, \beta_c, \sigma_c$ [13, 14]. The commanded angles, along with the data composed from the navigational systems are used to generate the reference angular rates using Eq. (9).

For obtaining a sliding mode controller [13], consider the system

$$\dot{x} = Ax + Bu, \quad x \in R_1, \quad u \in R^3$$

Let $s : R_1 * [0, \infty)$ be a continuously differentiable map and define a related sliding manifold as in Eq. (15). Let the function g be defined by Eq. (16).

$$S = \{(x, t) \in R_1 * [0, \infty) : s(x, t) = 0\}. \tag{15}$$

$$p(x, u, t) = \frac{\partial}{\partial x} s(x, t) + \frac{\partial}{\partial x} s(x, t)[Ax + Bu] \tag{16}$$

The system of ordinary differential equations is expressed as:

$$\begin{aligned} \dot{x} &= Ax + Bu \\ \zeta * \dot{u} &= g(x, u, t), \quad \zeta < 0 \end{aligned} \tag{17}$$

For obtaining the solution of Eq. (17), a theorem is stated that for any pair $(x, t) \in I$, the unique solution $u^*(x, t)$ of the algebraic equation $g(x, u, t) = 0$ is called equivalent control for the problem.

It is possible to show that uniquely the equivalent control is,

$$u^*(x, t) = -\left[\frac{\partial}{\partial x} s(x, t) B\right]^{-1} \times \left[\frac{\partial}{\partial t} s(x, t) + \frac{\partial}{\partial x} s(x, t) A(x) x\right] \tag{18}$$

Let, $\hat{x} = (\hat{q}^{\mathrm{T}}, \hat{\omega}^{\mathrm{T}})^{\mathrm{T}}$ be a reference state attitude for the reentry vehicle. The problem can be formulated as to design a feedback controller in such a way that, the error vector, $e = \hat{x} - x$ satisfies Eq. (19), for a given $\beta > 0, \delta > 0$.

$$\|e\| \leq \delta + A * e^{-\beta t} \tag{19}$$

where A is a constant depending on the data. To solve the attitude control problem, the sliding surface s is defined as in Eq. (20).

$$s(x, t) = H(q)[e - f(x_0, t)] \tag{20}$$

where, $x_0 = x(0)$ and $H(q)$ is a matrix of suitable dimensions to be chosen. Let,

$$H(q) = \begin{bmatrix} H_1' * \Omega^{\mathrm{T}}(q) & H_2 \end{bmatrix} \tag{21}$$

where $H_2 * J$ is nonsingular. The $f(x_0, t)$ in Eq. (20) is given by,

$$f(x_0, t) = e^{Ct} * e \tag{22}$$

where C is a stability matrix. Now a theorem can be stated as follows to define the equivalent control [10, 13].

Theorem *Let $x_0 \in R_1$ be given. Assume that, $Re\lambda_{\min}(H_2 J^{-1}) > 0$ and H_1' be a matrix such that $H_2^{-1} \times H_1'$ is symmetric and positive definite. Let, $Re\lambda_{\max}(C) < -\lambda_{\min}(H_2^{-1} \times H_1')$ then there exists an $\xi_0 > 0$ such that the solution of [x, u] satisfying the initial conditions and $\hat{x}(t) - x(t, \xi) < \delta + A * e^{-\beta t}$.*

Then the controller is obtained as in Eq. (23).

$$u(t, \xi) = (1/\xi)\begin{bmatrix} H_1'\Omega^{\mathrm{T}}(q) & H_2 \end{bmatrix} \times \left[\hat{x}(t) - x(t, \xi) - e^{Ct}(\hat{x}_0 - x_0)\right] + u_0 \tag{23}$$

where, $t \geq 0$ and A are β positive constants depending on H_1', H_2, C.

For the proposed model of ACRV, the constant matrices are selected as, $C = \mathrm{diag}(C_q, C_w)$, $H_1' = 10 I_3$, $H_2 = \mathrm{diag}(100, 50, 50)$ $C_q = -0.1 * I_4$, $C_\omega = -0.1 * I_3$, $\zeta = 0.001$. The initial conditions are selected as;

$$\begin{aligned} \hat{q}_0 &= (0.7272, 0.1207, -0.6426, -0.2091) \\ q_0 &= (0.7268, 0.14, -0.6274, -0.2419) \\ \hat{\omega}_0 &= (0.0008, 0.001, -0.002), \quad \omega_0 = (0, 0, 0) \end{aligned}$$

For the proposed model, thrusters are considered to provide the torque required by the attitude controller.

6 Multi-objective Genetic Algorithm-Based Sliding Mode Controller

The genetic algorithms are an evolutionary optimization technique which can find the global optimal solution in complex multidimensional search space. It is an iterative and population-based algorithm. It involves minimization or maximization of an objective function and determination of the parameters of interest.

The algorithm of genetic algorithm is as follows

Step-1 Initialize the parameters. In the proposed work, seven parameters are considered and an initial population is created.
Step-2 Compute the objective function.
Step-3 Check for stall limits or max Generation.

1. Reproduce the next generation
   - Crossover operation
   - Mutation operation
2. Compute objective function
3. Selection of the offspring for the next generation using Roulett wheel.

The multi-objective function used to obtain the optimal controller parameters so as to minimize is given by

$$J = \min\{0.33 * \text{ISE(Bank Angle)} + 0.33 * \text{ISE (Angle of Attack(degree))} + 0.33 * \text{ISE(Sideslip Angle(degree))}\}$$

Controller parameter tuned using genetic algorithm

$$C = \begin{bmatrix} 0.3566 & 0 & 0 & 0 & 0 & 0 & 0 \\ 0 & 0.100 & 0 & 0 & 0 & 0 & 0 \\ 0 & 0 & 0.3863 & 0 & 0 & 0 & 0 \\ 0 & 0 & 0 & 0.2528 & 0 & 0 & 0 \\ 0 & 0 & 0 & 0 & 0.597 & 0 & 0 \\ 0 & 0 & 0 & 0 & 0 & 0.100 & 0 \\ 0 & 0 & 0 & 0 & 0 & 0 & 0.2406 \end{bmatrix}$$

7 Design of Thrusters

The output of the attitude controller is provided to the thrusters, which will be modulated. The thrusters provided are based on the pulse width pulse frequency (PWPF) technique [15]. The command inputs to the controller and the output are

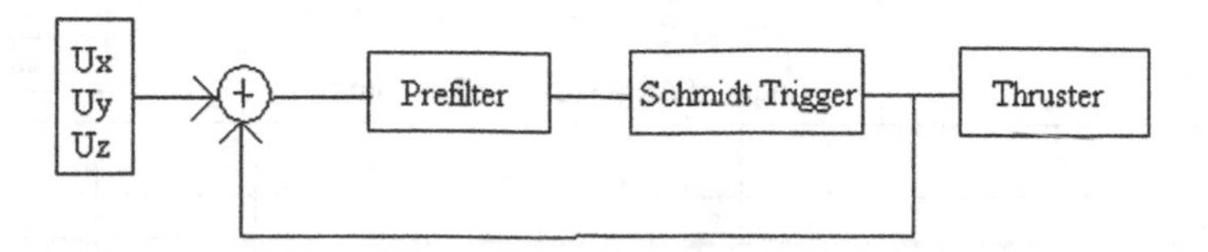

Fig. 1 Modulator scheme

continuous in nature. The actual control for operating the thrusters for steering the capsule along the prescribed path is in the pulse mode. Hence modulators are used. They will generate the pulsed mode output. The modulator scheme proposed has an integrator and a nonlinear element as in Fig. 1.

8 Simulation Results

To check for robustness, the simulations are carried out in the developed model in the presence of aerodynamic disturbance uncertainties. The simulation results are shown in Figs. 2, 3 and 4.

The side slip angle error in Fig. 2 depicts that, the error value of multi-objective genetic algorithm-based sliding mode control method is much close to zero and is having fewer disturbances.

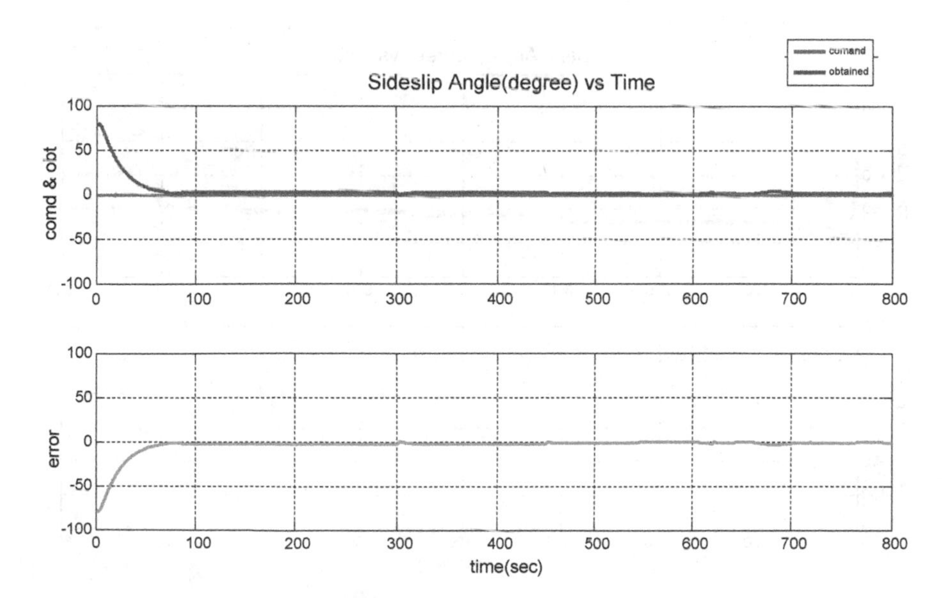

Fig. 2 Sideslip angle of multi-objective genetic algorithm-based sliding mode control

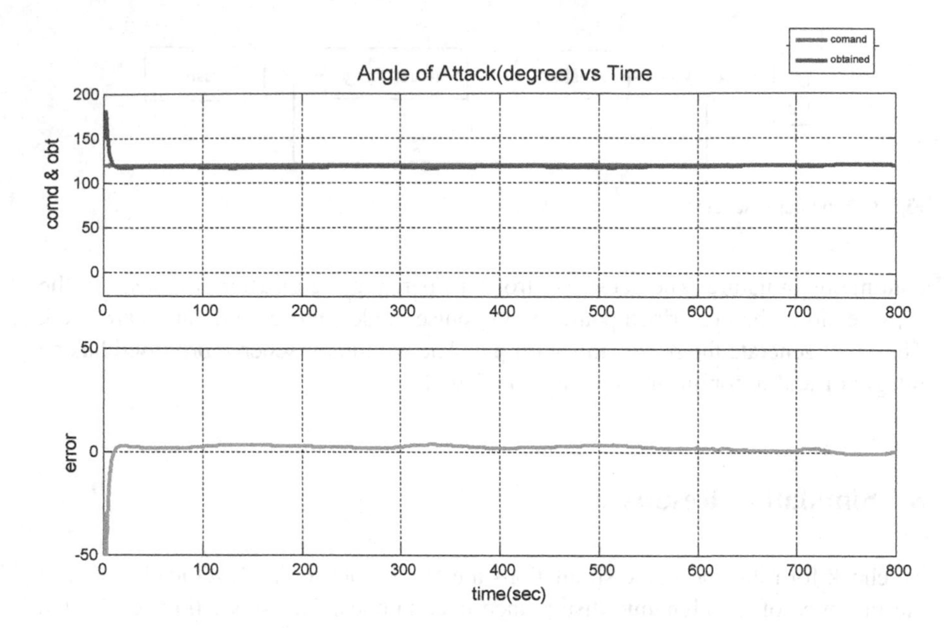

Fig. 3 Angle of attack of multi-objective genetic algorithm-based sliding mode control

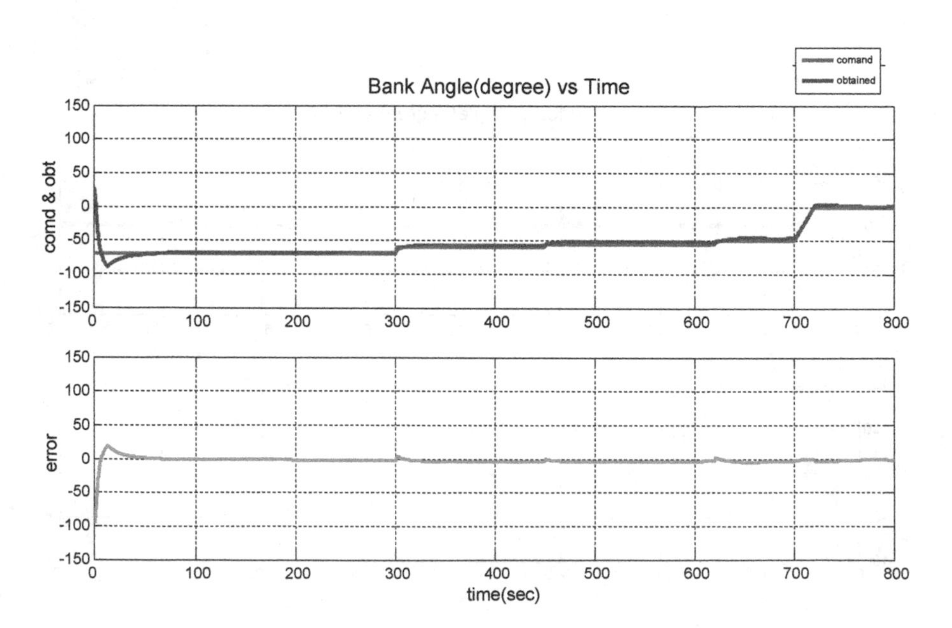

Fig. 4 Bank angle of multi-objective genetic algorithm-based sliding mode control

Tracking of angle of attack for a commanded value is done smoothly with very less amount of ripples and reduced disturbances, when the controller is a genetic algorithm-based sliding mode control. This shown in Fig. 3.

In Fig. 4, it is seen that, the control variable bank angle error and the amount of disturbance is also reduced when the controller is multi-objective genetic algorithm-based sliding mode control.

9 Conclusions

An attitude control system is formulated for a reentry vehicle. The attitude control law is based on multi-objective genetic algorithm-based sliding mode control. The objective function is formulated using the minimization function of Integral Square Error including side slip angle, angle of attack, and bank angle. The step-by-step procedure involved in multi-objective controller is reducing the error in side slip angle, angle of attack and bank angle considerably.

References

1. S.F. Wu, R.R. Costa, Q.P Chu, J.A. Mulder, "Nonlinear Dynamic Modeling and Simulation of the Re-entry Crew Return Vehicle", *European Space Agency international Conference on Spacecraft Guidance, Navigation and Control Systems, February 2000.*
2. H. P. Lee, S. E. Reiman and C. H. Dillon, "Robust Nonlinear Dynamic Inversion Control for a Hypersonic Cruise Vehicle," *AIAA Guidance, Navigation, and Control Conference and Exhibit, August 2007.*
3. Reding, J.P and Svendsen, "Lifting Entry Rescue Vehicle Configuration", *Journal of Spacecraft and Rockets, Volume 27, Number 5, pp. 606–612, 1990.*
4. P.B. de Moura Oliveira, "Modern heuristics review for PID control optimization: A teaching experiment", *Proceeding of International Conference on Control and Automation, Budapest, Hundary, pp. 828–833, 2005.*
5. C.R. Reeves, "Modern Heuristic Techniques for combinatorial problems", *McGraw Hill, London, 1995.*
6. J.S. Kim, J.H Kim, J.M. Park, S.M. Park, M.Y. Choe and H. Heo, "Auto tuning PID controller based on improved genetic algorithm for reverse osmosis plant", *International Journal of Intelligent systems and technologies, vol. 3, No. 4, pp. 232–237, 2008.*
7. B.A.M. Zain, M.O. Tokhi and S.F. Toha, "PID based control of a single link flexible manipulator in vertical motion with genetic optimization", *Proceeding of the 3rd UKSim European Symposium on Computer modeling and simulation, Athens, Greece, pp.355–360, 2009.*
8. F.Yin, J. Wang and C. Guo, "Design of PID controllers using Genetic Algorithms approach for Low Damping slow response plants", *Springer-Verilog, Berlin Heidelberg, 2004.*
9. Pablo Ghiglino, Jason L. Forshaw, Vaios J. Lappas, Oqtal, "Optimal Quaternion Tracking Using Attitude Error Linearization", *IEEE Transactions On Aerospace And Electronic Systems Vol. 51, No. 4 October 2015.*
10. Wertz. J, "Spacecraft Attitude Determination and Control, Reidel, Dordrecht, The Netherlands", *Springer, 1986.*

11. B. Wie, H. Weiss, A. Arapostathis, "Quaternion Feedback Regulator for Spacecraft Eigenaxis Rotations", *Journal of Guidance, Control and Dynamics, Volume 12, Number 3, pp. 375–380, 1989.*
12. Federico Thomas, "Approaching Dual Quaternions From Matrix Algebra", *IEEE Transactions On Robotics, Vol. 30, No. 5, October 2014.*
13. Alberto Cavallo, Giuseppe De Maria and Ferdinando Ferrara, "Attitude Control for Low Lift/Drag Re-Entry Vehicles", *Journal of Guidance, Control and Dynamics, Volume 19, Number 4, July 1999.*
14. Jie Jia, Fengqi Zhou, Jun Zhou, "Re-Entry Attitude Two-Loop SMC of the Spacecraft and its Logic Selection", *Systems and Control in Aerospace and Astronautics, 2006. ISSCAA 2006, 0-7803-9395-3 IEEE Publisher.*
15. Gangbing Song, Nick V Buck, Brij N Agrawal, "Spacecraft Vibration Reduction Using Pulse-Width Pulse Frequency modulated Input Shaper", *Journal of Guidance, Control and Dynamics, Volume 22, Number 3, 1999.*

Author Biographies

Mrs. Divya Vijay received her Bachelor's Degree from Cochin University of Science and Technology in "Electrical and Electronics Engineering" in 2007 and her Master's in "Control Systems" from Kerala University in 2011. Currently, she is pursuing her Ph.D. in the area of "NonLinear Control System for Atmospheric Re-entry Vehicle." She is currently an Assistant Professor with the Department of Electrical and Electronics Engineering at B.S. Abdur Rahman University, Chennai, India. Her research interests include nonlinear system control, sliding mode control, robust control, etc.

Dr. U. Sabura Banu completed B.E./ICE in Arulmigu Meenakshi Amman College of Engineering in May 1999. Completed MS (By Research) in Faculty of Electrical Engineering from Anna University in July 2004 and Ph.D. in the Faculty of Electrical Engineering from Anna University in August 2009. Worked as lecturer, Assistant Professor, Associate Professor, and Professor in B.S. Abdur Rahman University. Published more than 60 research papers in leading Journals and Conferences like Science Direct, ACTA press, Instrument Science and Technology, IEEE, IET, etc.

Dr. K. Boopathy received his B.E. from Madras University (Electrical & Electronics Engineering) in 1999, M.E. in Electronic Engineering from Anna University (MIT campus) in 2003, and Ph.D. in Power Electronics from Anna University, Chennai in 2012. Currently he is working as a Associate Professor in Dept. of EEE, B.S. Abdur Rahman University Chennai. He has published several National and International Journals and Conferences. His area of interest is Power Electronics and Drives, Soft Computing Technique Application to Power Converter and Electrical Machines.

A Study on Mutual Information-Based Feature Selection in Classifiers

B. Arundhathi, A. Athira and Ranjidha Rajan

Abstract Multilabel classification is a classification technique in which each sample may be related with more than one class labels. This paper deals with a comparative study of mutual information (MI) technique and other methods in different classifiers. MI is a technique in filter approach type feature selection. It is a fine indicator, which measures the information or data that common between two variables, it audits and evaluates how one of the variables reduces the uncertainty of the other. We consider other two classifiers for the study; they are Naïve Bayesian (NB) and ID3. The experiments were done using MI and compared with the two classifiers for different benchmark data set from UCI repository flag, music, and yeast. The results were verified using evaluation measures. An accurate precision and recall value can be obtained in MI technique rather than using the classifiers NB and ID3.

Keywords Feature selection · Filter approach · Mutual information (MI) · ID3 · Naïve Bayesian (NB)

1 Introduction

The training data set that is used for evaluation may contain hundreds of attributes or variables, in which most of them are insignificant and repeating. Although it may be likely for a user to select some of the useful attributes, it is a crucial and time consuming task especially when their behavior is not well known [1]. Many

---

B. Arundhathi (✉) · A. Athira · R. Rajan
Department of CS & IT, Amrita School of Arts and Sciences Kochi,
Amrita VishwaVidyapeetham, Amrita University, Kochi, India
e-mail: arundathib1992@gmail.com

A. Athira
e-mail: athiraajith4@gmail.com

R. Rajan
e-mail: ranjidha@yahoo.com

S.S. Dash et al. (eds.), *Artificial Intelligence and Evolutionary Computations in Engineering Systems*, Advances in Intelligent Systems and Computing 517,
DOI 10.1007/978-981-10-3174-8_40

unwanted aspects may be available in the data that is to be mined which need to be taken out. Moreover, many algorithms that are used for mining do not perform well with large amount of features. Therefore, some techniques need to be applied before any kind of mining algorithm is enforced. The feature selection is a step to select a set of data or attributes that overcomes all these problems. To start with, the accidental attributes needs to be clean before applying any technique. The main aim of feature selection is to improve the performance of a data, provide speed and accurate dataset [2].

The process of filtering is done using feature selection techniques that involve filter, wrapper, and embedded technique.

- **Filter methods** are independent ones that avert the dependencies of features and avoid a common way which affect the classifier with one another. It includes mutual information (MI) technique, which tells about the data that is shared between two variables [3]. The key measures of the information are entropy, which is the measure of defilement.
- **Wrapper methods** consider the selection of a set of features as a problem, where various combinations are prepared, figured, and compared with others. Wrapper method aims at wrapping up a classifier in algorithms.
- **Embedded method** has been proposed to reduce the classification of learning and it decides which attributes can be assigned to find the accuracy of the model while it is being developed.

In decision tree learning, ID3 is an algorithm, which is the predecessor to the C4.5 algorithm [4], and is used to produce a decision tree. In machine learning, Naive Bayesian (NB) is the family of simple probabilistic classifier, which analyzes that the existence/truancy of a specific feature of a class are unlike to the existence/truancy of any other feature.

2 Background Study

"An Introduction to Variable and Feature Selection", the variable or feature selection has become an inevitable criterium in areas of research for which datasets with numerous variables are available. The aim of variable selection is threefold: First is to revamp the performance of the prediction. Second is to provide fast and more cost-effective predictors. Third is to provide a better understanding of the underlying process that generated the data [2].

This paper deals with several feature selection methods viz. filter, wrapper, and hybrid methods (which use combinations of filter and wrapper techniques). Even though many filter methods are found to give no increase in accuracy for the classifier, MI technique is reliable since it relies up on entropy, which measures the defilement [3].

This paper introduces a new technique to takeout feature selection in multilabel classification problems. This allows acknowledging the common dependencies during the feature selection process between the class labels and the features [5].

"Feature selection with dynamic MI". The paper demonstrates feature selection that plays an important role in data mining and other reorganization techniques primarily for large training data set. Since MI is one of the mostly used measurement in feature selection, its general technique is recommended, which gives more accurate information measurement [6].

"A Tutorial on Multi-label Classification Technique". The paper defines about the divergent techniques to solve the problems with demos that illustrate the technique because most of the classification task has problems of associating a single class to each instance. However, there are many classification tasks in which each attributes can be concord with more than one class [7].

"Feature Selection based on Information Gain". The paper describes about an algorithm to reduce attributes based on discernibility matrix and Information gain. Most of the features are unimportant and repeating ones that may affect the classification but the efficiency of classification can be improved by discarding the unimportant and repeating data [1].

3 Materials and Methodologies

In classification techniques, the most commonly evaluated criterion is accuracy. In order to find the accuracy some metrics have been used. In this paper, the precision and recall is used as the evaluation metrics [8]. A multilabel dataset can be used for feature selection and it is done by the classifiers NB, ID3, and MI. After that the resultant features are analyzed to find the highest precision and recall value. The dataset that is used to find the result is given in Table 1.

Table 1 Data about nations and national flags

Landmass	Zone	Population	Language	Religion	Red	Green	Blue	Yellow	White	Black	Orange
5	1	16	10	2	1	1	0	1	1	1	0
3	1	3	6	6	1	0	0	1	0	1	0
4	1	20	8	2	1	1	0	0	1	0	0
6	3	0	1	1	1	0	1	1	1	0	1
3	1	0	6	0	1	0	1	1	0	0	0
4	2	7	10	5	1	0	0	1	0	1	0
1	4	0	1	1	0	0	1	0	1	0	1
1	4	0	1	1	1	0	1	1	1	1	0
2	3	28	2	0	0	0	1	0	1	0	0
2	3	28	2	0	0	0	1	1	1	0	0

3.1 Feature Selection

Feature selection plays a significant role in data mining, especially for large volume of datasets [9]. It also known as subset selection, a process commonly used in machine learning, where a subset of features is selected from the available dataset and apply in learning algorithms. It can be devoted to both supervised and unsupervised learning; The main goal of feature selection is to decide a minimum data with the features for the classification, it shows a high exactness to represent the original features. Some of the advantages of feature selection are [9].

- It reduces extensity.
- Limit Storage Space.
- Removes irrelevant, redundant, and noisy data.
- Increases the accuracy of resulting model.

The Feature selection uses three types of techniques, namely filter, wrapper, and embedded methods. Here we are using a technique from filter approach to do the comparison with the classifiers. Filter methods is the one that apply a mathematical measure for scoring or ranking of each feature, which helps in classification [10]. Once this ranking has been calculated, features set with best features get developed.

3.1.1 Naive Bayesian

The NB classifier is a straightforward and fastest learning algorithm commonly used method for supervised learning. It provides a comfortable way for dealing with any number of attributes or classes, and is based on probability theory. The NB classifier is used in multilabel dataset to predict the class that has the highest value. In this application it calculates the probability of classes and attributes. From the dataset the probability of labels are calculated, then applying certain conditions the probability of each attribute with class are found. Finally, the highest value of the class is determined.

3.1.2 ID3 Algorithm

Decision trees are familiar for gaining information for the purpose of decision making. The basic idea of ID3 algorithm is to develop a decision tree by employing a top down greedy search through the given sets to test each attribute at every tree node. Moreover, ID3 uses a statistical property called information gain to select the feature to test at each node in the tree. Information gain measures how well a given feature separates the training samples according to their target classification, and also it measures the expected reduction in entropy by partitioning the instances according to their features. Entropy is a measure in the information theory, which describe the different features and the impurity of an arbitrary collection of data files. The training file lists the attributes and their possible values. It doesn't deal with continuous, numeric data which means we have to discretize them. The ID3 classifier used to create a decision tree for the multilabel dataset. Entropy for each labels is found, and they are added together to get entropy of parent by using the formula.

$$H(L) = -\sum_{\mathrm{i}} p_i(\log_2 p_i) \tag{1}$$

After finding the entropy, the information gain for each attributes is calculated using the formula.

$$IG(A, L) = H(L) - \sum (p(t_1)H(t_1) + p(t_2)H(t_2) + \cdots + p(t_n)H(t_n)) \tag{2}$$

The highest value is determined, and a decision tree is constructed.

3.1.3 Mutual Information

The idea of MI is elaborately linked to the random variable. It is a basic idea on information theory that tells the amount of data or information that available in a random variable or feature. MI is a good indicator, which measures the information or data that shared between two variables, it checks and measures how one of the variable reduces the uncertainty of the other. Using MI, the classes and the attributes of the multilabel [11] dataset get compared from the formula.

$$I(R, S) = \frac{n_{tc}}{n}\log_2\frac{nn_{tc}}{n_t.n_c} + \frac{n_{0c}}{n}\log_2\frac{nn_{0c}}{n_0.n_c} + \frac{n_{t0}}{n}\log_2\frac{nn_{t0}}{n_t.n_0} + \frac{n_{00}}{n}\log_2\frac{nn_{00}}{n_0.n_0} \tag{3}$$

In this formula, n is the count of documents in each multilabel data set. The values e_t and e_c are indicated by two subscripts 0 and 1. It computes the utility measure *I* (*R, S*) to select the largest value for the nth term. *R* is a random variable that take the value $e_t = 1$ (the document contain the attribute *t*) and $e_t = 0$ (the document does not contain the attribute *t*). *S* is the random variable that take the value $e_c = 1$ (the document contain the class *c*) and $e_c = 0$(the document does not contain the class *c*). Here n_{0c} defines $t = 0$ and $c = 1$, n_{t0} defines $t = 1$ and $c = 0$ (Fig.1).

Here the MI measures how much information an attribute contain about the class. MI reaches its maximum value if the term is a perfect indicator for class membership. The term with high MI score gets selected and the selected item is of predictable utility. This help in decision making for their respective classes and others get removed (e.g., Landmass, zone, language for the class Green) (Table 2).

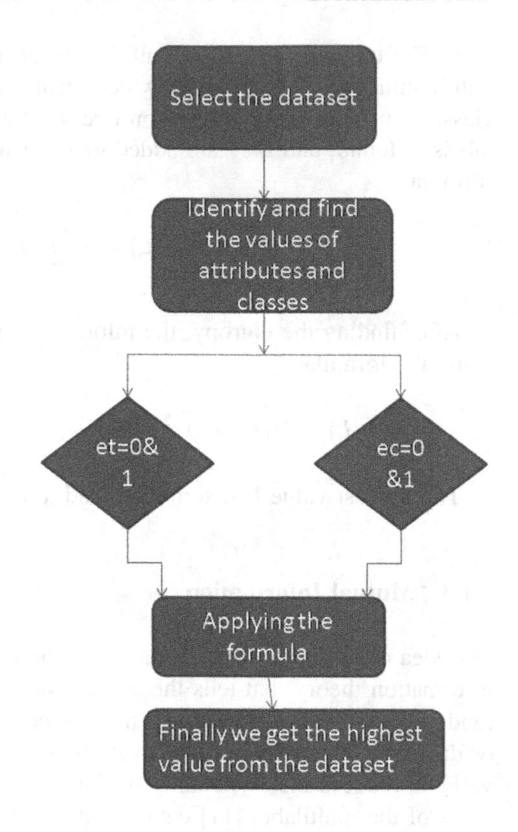

Fig. 1 Flowchart describes steps of mutual information

Table 2 Shows mutual information for classes red, green and blue

Table 3.2: Red		Table 3.3: Green		Table 3.4: Blue	
Landmass	0.300	Landmass	0.301	Landmass	0.301
Zone	0.300	Zone	0.301	Zone	0.301
Population	0.313	Population	0.275	Population	0.275
Language	0.300	Language	0.301	Language	0.301
Religion	0.3S1	Religion	0.254	Religion	0.254

4 Experimental Result

The resultant graph after applying the technique in filter approach and the classifiers are shown in Figs. 2, 3, and 4);

Fig. 2 Mutual information

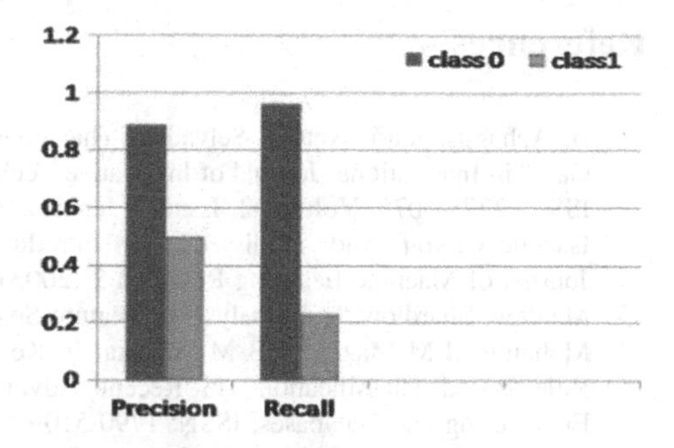

Fig. 3 Naive Bayesian

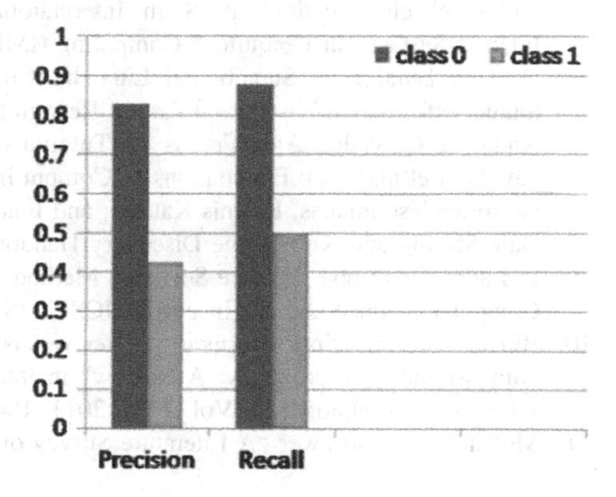

Fig. 4 ID3

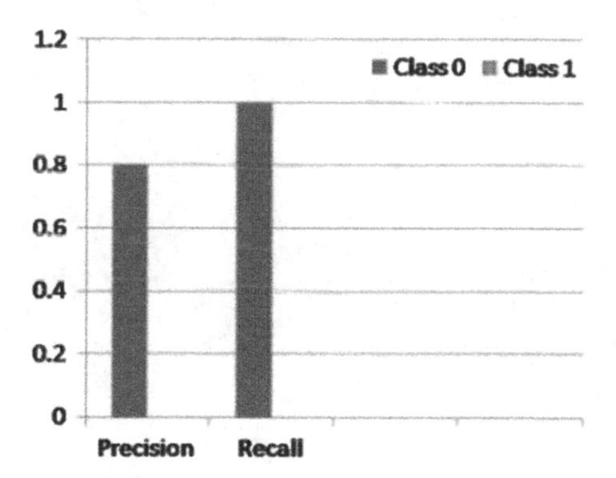

5 Conclusion and Future Work

The filter approach for ranking the features is highly reliable because it handles huge amount of data set than other techniques in feature selection. Comparing the experimental steps performed above clearly indicates that MI gives better results than the classifiers ID3 and NB. In future, we are looking onto implementing the MI method in ensembles of classifiers.

References

1. B. Azhagusundari, Antony Selvadoss Thanamani "Feature Selection based on Information Gain" in International Journal of Innovative Technology and Exploring Engineering (IJITEE) ISSN: 2278-3075, Volume-2, Issue-2, January 2013.
2. Isabelle Guyon, Andr´e Elisseeff "An Introduction to Variable and Feature Selection" in Journal of Machine Learning Research 3 (2003) 1157–1182.
3. Matthew Shardlow "An Analysis of Feature Selection Techniques".
4. Mohammed M Mazid, A B M Shawkatali, Kevin S Tickle "Improved C4.5 Algorithm for Rule Based Classification" in Recent Advances in Artificial Intelligence, Knowledge Engineering and Databases, ISSN: 1790-5109.
5. Gauthier Doquire, Michel Verleysen "Mutual information-based feature selection for multilabel classification" in Sixth International Conference on Innovative Mobile and Internet Services in Ubiquitous Computing (IMIS-2012), Volume 122.
6. Huawen Liua, Jigui Suna,b, Lei Liua, b, HuijieZhangc "Feature selection with dynamic mutual information" in journal Pattern Recognition archive, Volume 42 Issue 7, july 2009.
7. Andre de Carvalho, Alex Freitas "A Tutorial on Multi-label Classification Techniques" in Abraham et al.(Eds.): Foundations of Comput.Intel. Vol. 5, SCI 205, pp. 177–195.
8. Grigorios Tsoumakas, Ioannis Katakis, and Ioannis Vlahavas, "Mining Multi-label Data" in Data Mining and Knowledge Discovery Handbook, pp. 667-685, 2nd Edition.
9. L. Ladha, T. Deepa "Feature Selection Methods and Algorithms" in International Journal on Computer Science and Engineering (IJCSE), ISSN: 0975-3397 Vol. 3 No. 5 May 2011.
10. Binate Kumari, Tripti Swarnkar* "Filter versus Wrapper Feature Subset Selection in Large Dimensionality Micro array: A Review" in International Journal of Computer Science and Information Technologies, Vol. 2 (3), 2011, 1048–1053.
11. Mohammad S Sorower "A Literature Survey on Algorithms for Multi-label Learning".

Solar PV-Based VFD Fed Permanent Magnet Synchronous Motor for Pumping System

Vishnu Kalaiselvan Arun Shankar, Subramaniam Umashankar, Soni Abhishek, Bhimanapati Lakshmi Anisha and Shanmugam Paramasivam

Abstract This paper determines the incorporation of photovoltaic source and battery with a PMSM motor fed with a centrifugal pump based on conventional Direct Torque Control technique (DTC). For the boosting of solar output, interleaved DC-DC converter with fuzzy logic-based MPPT control is studied and in order to convert the DC output of converter for the required form of motor input, three-phase inverter has been taken. With the help of a bi-directional converter, regulation of battery is carried out which depends on load and source. Output results of motor as well as pump are obtained using MATLAB Simulink environment.

Keywords PMSM · Fuzzy logic · PV array · Battery · MPPT · DTC

1 Introduction

Due to the cutting-edge growths in power, the concern for small distributed power generation schemes like photovoltaic array (PV) systems, which are connected to motors for pumping applications is increasing rapidly [1]. In this paper, PV panel is taken as the source which supplies the system and it is integrated to a lead-acid battery [2]. When the demand of the output is less, batteries are used to store the surplus energy and stock when required [3]. As the output of solar array is DC type, power electronics converters are realized so that DC type power can be converted

V.K. Arun Shankar (✉) · S. Umashankar · S. Abhishek · B. Lakshmi Anisha
Department of Energy and Power Electronics, VIT University-Vellore, Vellore, India
e-mail: arunshankarvk@gmail.com

S. Paramasivam
Danfoss Industries Pvt. Ltd., Chennai, India
e-mail: param@danfoss.com

S.S. Dash et al. (eds.), *Artificial Intelligence and Evolutionary Computations in Engineering Systems*, Advances in Intelligent Systems and Computing 517,
DOI 10.1007/978-981-10-3174-8_41

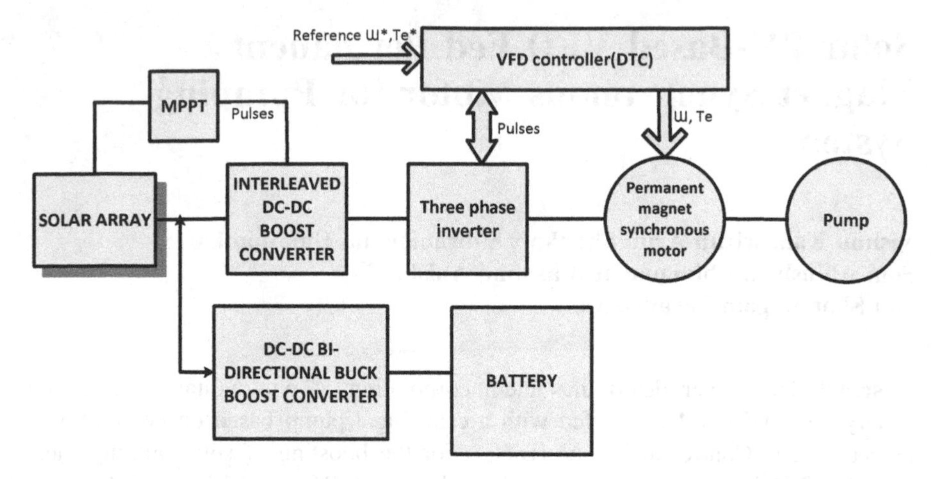

Fig. 1 Block diagram

into AC type in order to supply the required loads [4]. Bi-directional DC-DC converters [5] are coupled with battery in order to avoid the usage of number of batteries and to control the battery output.

Generally, solar photovoltaic systems employ two stage converters i.e., boost converter in order to boost the output of converter [6, 7] to the required output and a pulse width modulated inverter which converts DC output to required input of three-phase PMSM motor which is controlled by conventional DTC technique [8].

IGBT-type switches are employed due to the existence of low impedance and high switching frequency. This paper intends the incorporation of solar PV array with a VFD controlled PMSM motor coupled with centrifugal pump [9]. The block diagram of the proposed micro grid EMS system is shown in Fig. 1.

2 PV Array

PV cell is considered as the input source in order to deliver the power to load and charge the battery in normal conditions. Ambient temperature of PV cell is considered as 25 °C and irradiation level of 1000 Wm^{-2} is considered. At normal conditions, it yields an output of 43 V. PV array is further coupled with a DC-DC interleaved boost converter and a three-phase inverter. It is connected to a battery (lead-acid) by means of a bidirectional DC-DC buck–boost converter. Output of three-phase inverter is connected to a PMSM motor. PV module is shown in Fig. 2 (Table 1).

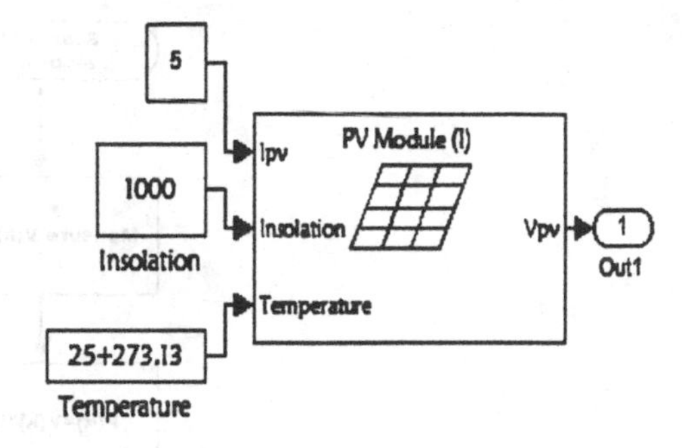

Fig. 2 PV module

Table 1 Factors considered for designing PV array

Factor	Value
Open-circuit voltage	21.24 V
Short circuit current	4.74 A
Parallel resistance	100 Ω
Series resistance	0.924 Ω
Diode saturation current	3.83e−10 A
Irradiation	1/100

3 Maximum Power Point Techniques Used

3.1 Perturb and Observe Method

In order to track the maximum power point of a solar array, P&O is a simple and efficient method which necessitates solar array's output current and voltage. Algorithm of P&O is shown in Fig. 3 where the perturbation at every step is used to find MPPT.

3.2 Fuzzy Logic-Based MPPT

In order to avoid the oscillatory behavior of the maximum power point, fuzzy logic controllers are castoff. Fuzzy logic controllers do not necessitate exact model knowledge but they require complete knowledge of PV system operation by the designer. FLC will have easy mathematical calculations.

FLC comes in three stages, i.e., fuzzification, inference, and de-fuzzification.

a. Fuzzification

Procedure of conversion numerical variable into a linguistic variable is fuzzification by means of seven fuzzy subsets called negative large (NL), negative average (NA), negative small (NS), zero(Z), positive small (PS), positive average (PA), positive large (PB) values of membership functions are allocated to the linguistic variables. Fuzzy rules for the proposed system are given in Table 2. Input variables

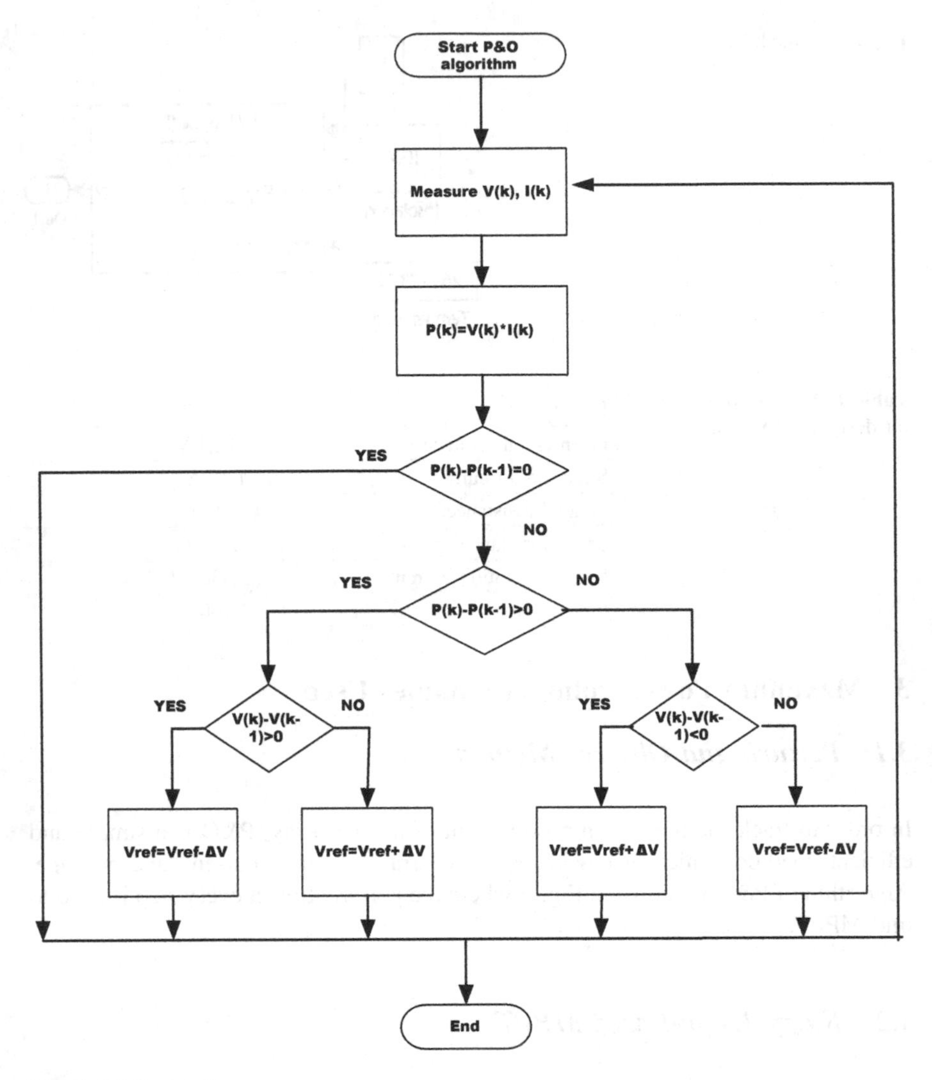

Fig. 3 P&O algorithm

that are considered are error (e); which is the difference between actual voltage (V_0) and reference voltage (V_r) and change in error (Δe); change in error of sample interval. Reference signal that is given to the PWM generator is the output membership function.

b. Inference

Assembly of if-then rules that includes data for the controlled parameters is the fuzzy rule base. Usually used method for fuzzy inference is Mamdani's. Here mamdani's method is used with Min-Max setups. On fuzzy rule base system, fuzzy inference is depended. Inside the inference engine block, fuzzy rules are framed

Table 2 Fuzzy rules

E/ΔE	NL	NA	NS	Z	PS	PA	PL
PL	Z	PVS	PS	PS	PL	PL	PL
PA	NVS	Z	PVS	PS	PA	PA	PL
PS	NS	NVS	Z	PVS	PA	PL	PS
Z	NA	NS	NVS	Z	PVS	PS	PA
NS	NL	NA	NS	NVS	Z	PVS	PS
NM	NL	NL	NA	NS	NVS	Z	PVS
NL	NL	NL	NL	NA	NS	NVS	Z

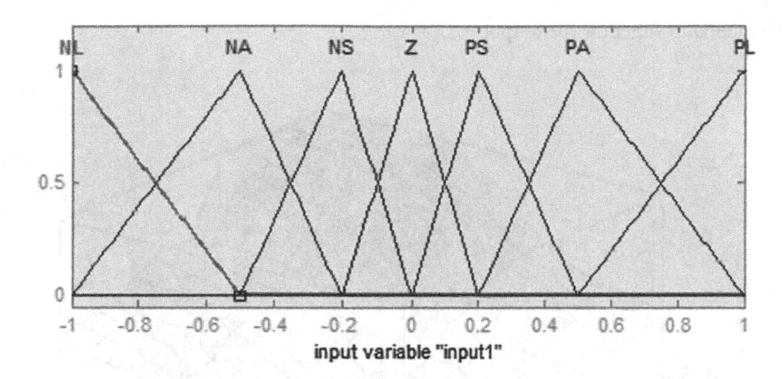

Fig. 4 Membership function for input1 (error)

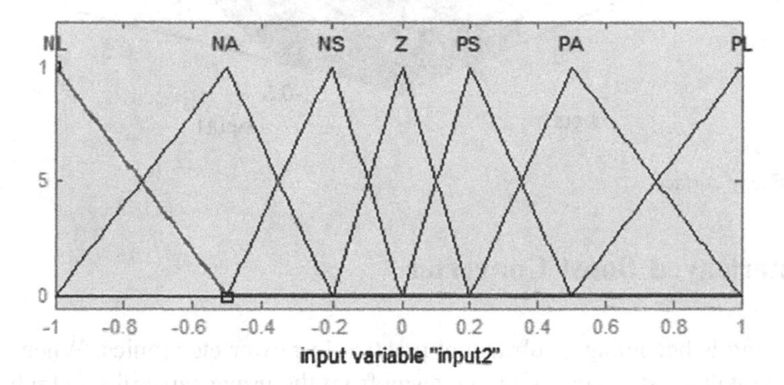

Fig. 5 Membership function for input2 (change in error)

[10, 11]. Control surface behavior which is governed by the set of rules, relates the input and output variables.

c. Defuzzification

Conversion of fuzzy value into crisp value is de-fuzzification, i.e., the inverse of fuzzification. Generally centroid method is used for de-fuzzification which is given by $\int \frac{x \times \mu(x) \mathrm{d}x}{\mu(x) \mathrm{d}x}$ (Figs. 4, 5, 6 and 7).

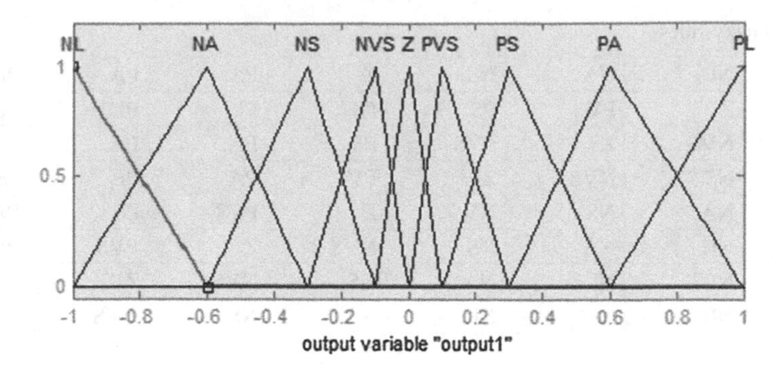

Fig. 6 Membership function for output

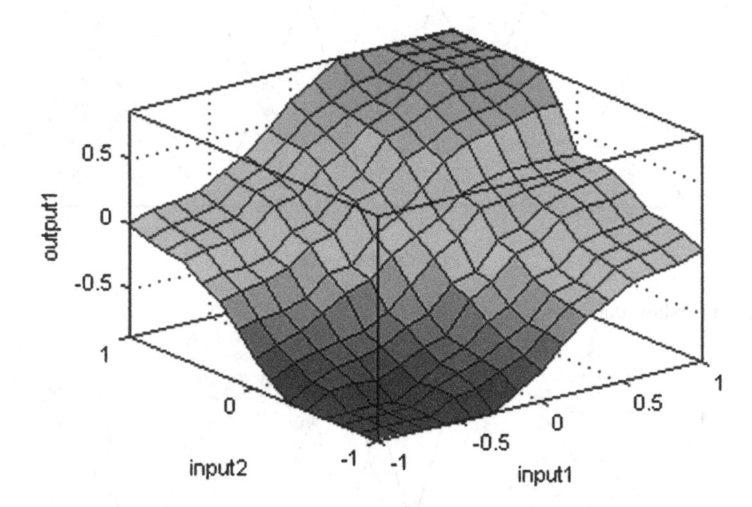

Fig. 7 Fuzzy surface

4 Interleaved Boost Converter

Interleaving is becoming a substantial practice in power electronics. When current stress and voltage stress capability go away from the usage capability in high power systems then interleaving of power converters can be used instead of using several power devices in series or parallel. Two-phase interleaved boost converter (IBC) is considered which encompasses of two parallel united boost converter units. With 360°/n phase shift, every single unit is controlled, where n signifies number of parallel units. 180° phase shift is taken for two units in parallel (Figs. 8 and 9).

Inductor values of IBC are calculated from the equation

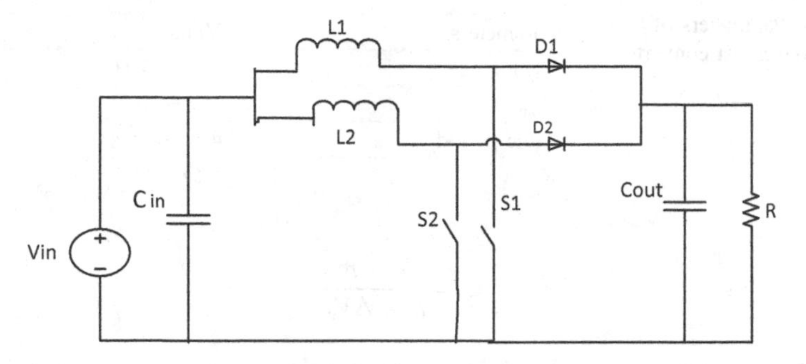

Fig. 8 Circuit of IBC

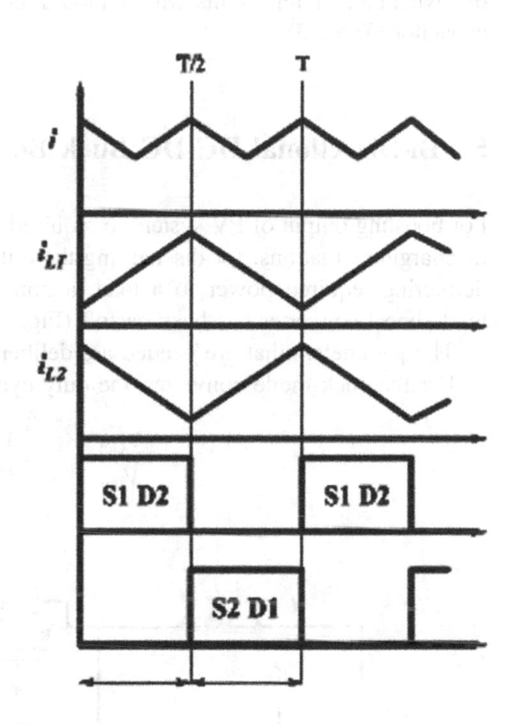

Fig. 9 Theoretical waveforms of IBC

$$L = \frac{V_{in} \times m}{f_s \times \Delta I_L} \quad (1)$$

where, L represents inductance, V_{in} represents Voltage at input, f_s represents switching frequency, ΔI_L represents Inductor current ripple, m represents Modulation Index.

Capacitor values of IBC are calculated from the equation

Table 3 Parameters of interleaved boost converter

Parameters	Value
L_1, L_2	7.63e−4 H
C_{out}	35e−4 F
Time period	4×10^{-5} s
R	1 Ω

$$C = \frac{I_0 \times m}{f_s \times \Delta V_c} \tag{2}$$

where C represents capacitance, I_0 represents output current, f_s represents frequency of switching, m represents Modulation Index, ΔV_c represents voltage ripple of capacitor (Table 3).

5 Bi-Directional DC-DC Buck Boost Converter

For boosting output of PV system to required voltage of battery in charging; and in discharging situations, for discharging the output of voltage to necessary value for delivering required power to a load, a non-isolated-based bi-directional DC-DC buck–boost converter has been castoff (Fig. 10).

The parameters that are needed are deliberated as follows:

For the buck mode converter, the duty cycle (D) is specified by

$$\frac{V_{\text{Out}}}{V_{\text{Inp}}} = \frac{1}{1 - D_{\text{boost}}}$$

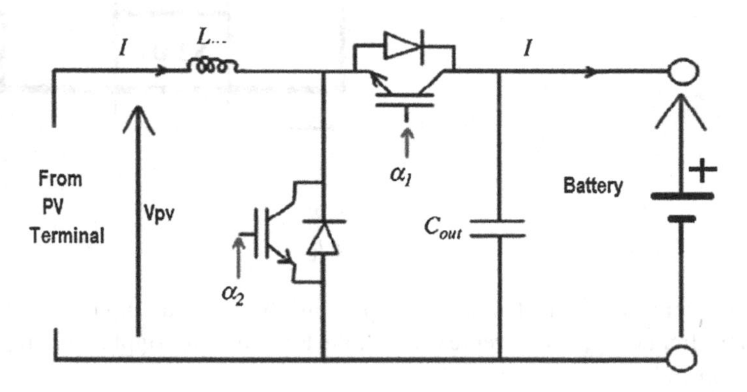

Fig. 10 Circuit of bi-directional buck boost converter

Table 4 Factors taken for bi-directional converter

Factor	Value
Inductance	6.46e−6 H
Pulse width of IGBT 1	21%
Pulse width IGBT 2	79%
Capacitance	7.8e−1 F

Table 5 Factors taken for lead-acid battery

Factor	Value
Rated capacity	6.5 Ah
Nominal voltage	100 V
Nominal discharge current	1.3 A
Voltage at full charge	108.8816

The current is specified by

$$I = \frac{DT_s}{2L}(V_{\text{out}} - V_{\text{inp}})$$

$$L = \frac{(2 \times 10^{-5}) \times 0.209 \times (110 - 23)}{2 \times 30} = 6.46 \times 10^{-6}$$

Factors that are taken for bi-directional converter and battery in this paper are presented in Table 4 and Table 5 respectively.

6 Modeling of PMSM

The D-Q axes coordinates PMSM is illustrated in Fig. 11.

The output power and torque equations of PMSM motor is written as $P_{out} = \frac{3}{2}\frac{p}{2}\omega_m(\lambda_{pm}I_q + (L_d - L_q)I_dI_q)$

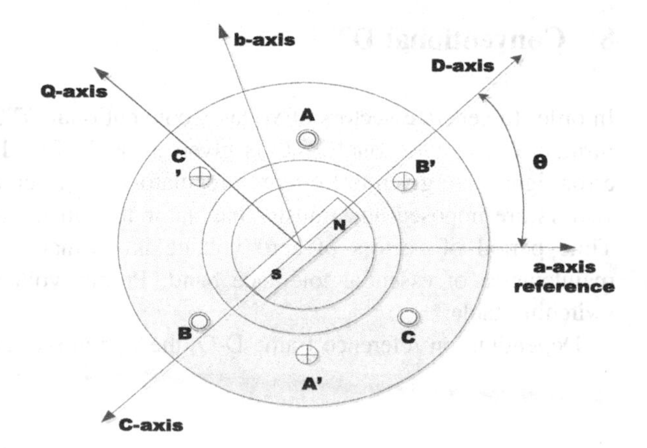

Fig. 11 PMSM D-Q axes frame

From the above equation the output torque equation of PMSM motor can be calculated as

$$T_e = \frac{3}{2}\frac{p}{2}(\varphi_d I_q - \varphi_q I_d);\ \varphi_d = L_d I_d + \lambda_{pm},\ \varphi_q = L_q I_q$$

7 Modeling of Pump

Most frequently used pump is centrifugal pump in many applications. The equations that have been used in modeling of centrifugal pump are written as

Affinity law equations are

$$\frac{Q_1}{Q_2} = \frac{D_1}{D_2}, \frac{Q_1}{Q_2} = \frac{D_1}{D_2}, \frac{H_1}{H_2} = \left(\frac{D_1}{D_2}\right)^2, \frac{H_1}{H_2} = \left(\frac{N_1}{N_2}\right)^2$$

Flow rate is given as $Q = \frac{T\omega}{gH*1000}$
Hydraulic power is given as $P_{\text{hyd}} = \rho Q g H$
Mechanical power at pump driving shaft is given as $P_{\text{mech}} = P_{\text{hyd0}} + P_{\text{fr}}$
Torque at driving shaft is given as $T = \frac{P_{\text{mech}}}{\omega}$
Pump total efficiency is given as $\eta = \frac{P_{\text{hyd}}}{P_{\text{mech}}}$
Where P_{hyd} = Reference hydraulic power
P_{mech} = Power at driving shaft
P_{fr} = Friction losses
p = Pressure
ρ = Density
g = gravitational force = 9.8 m/s^2

8 Conventional DTC

In order to generate vectors of voltage, conventional DTC usages switching tables. Structure of conventional DTC is given away in Fig. 12. Stator flux and torque error signals are given to hysteresis regulators, in which the time period of voltage vectors are imposed and transfer the stator flux in the direction of the reference. Time period of vectors of zero voltage is estimated by torque regulators, for maintenance of essential tolerance band. Proper voltage vector is chosen from switching table.

Depending on reference frame D-Q, the currents of D-Q axis are written as

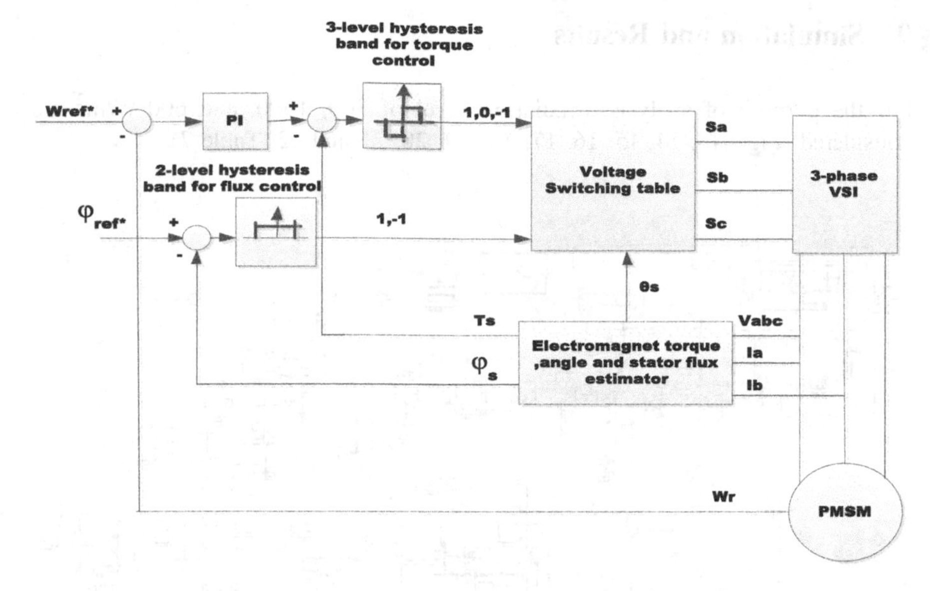

Fig. 12 DTC structure

$$\left.\begin{aligned} I_d &= i_{dc}\cos\delta - i_{qc}\sin\delta \\ I_q &= i_{dc}\sin\delta + i_{qc}\cos\delta \end{aligned}\right\}$$

$$T_e = \frac{3}{2}p\varphi_s\varphi_{pm}\frac{1}{L_s}\sin\delta$$

Above equation is the expression of PMSM torque. By this, by varying magnitude of flux linkage of stator, control of torque is carried out.

Stator flux linkage is as follows:

$$\varphi_s = \int (V_s - R_s i_s)\mathrm{d}t$$

By selecting appropriate voltage vectors, stator flux linkage can be controlled. The switching table for selecting voltage vectors is given in Table 6.

Table 6 Conventional DTC switching table

Flux $\Delta\varphi_s$	Torque ΔT_e	Voltage 1	2	3	4	5	6
+1	+1	v_2	v_3	v_4	v_5	v_6	v_1
	0	v_7	v_0	v_7	v_0	v_7	v_0
	−1	v_6	v_1	v_2	v_3	v_4	v_5
−1	+1	v_3	v_4	v_5	v_6	v_1	v_2
	0	v_0	v_7	v_0	v_7	v_0	v_7
	−1	v_5	v_6	v_1	v_2	v_3	v_4

9 Simulation and Results

For the purpose of analysis, irradiation level of 300, 1000, and 600 W/m^2 are considered (Figs. 13, 14, 15, 16, 17, 18, 19, 20, 21 and 22, Table 7).

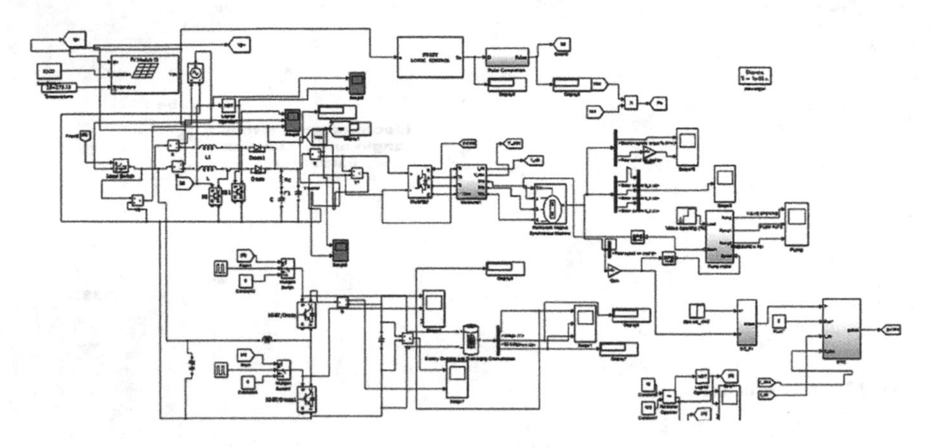

Fig. 13 Simulation diagram of proposed system

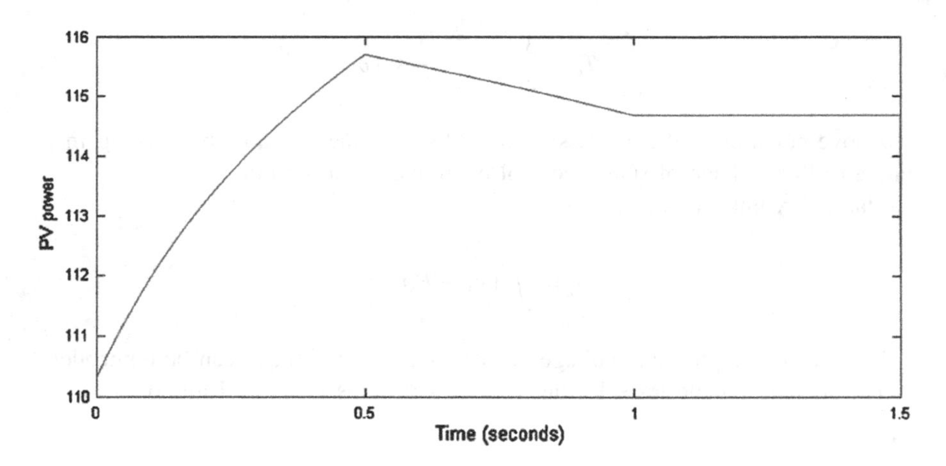

Fig. 14 Output of PV power

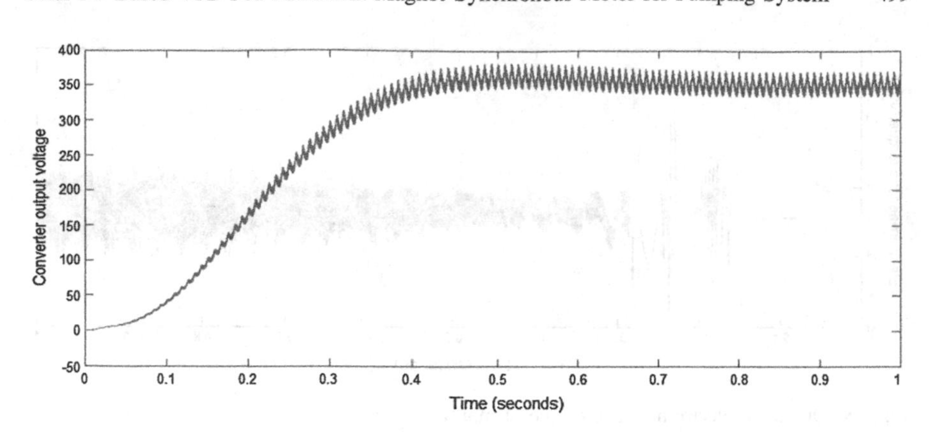

Fig. 15 Output of DC-DC interleaved boost converter

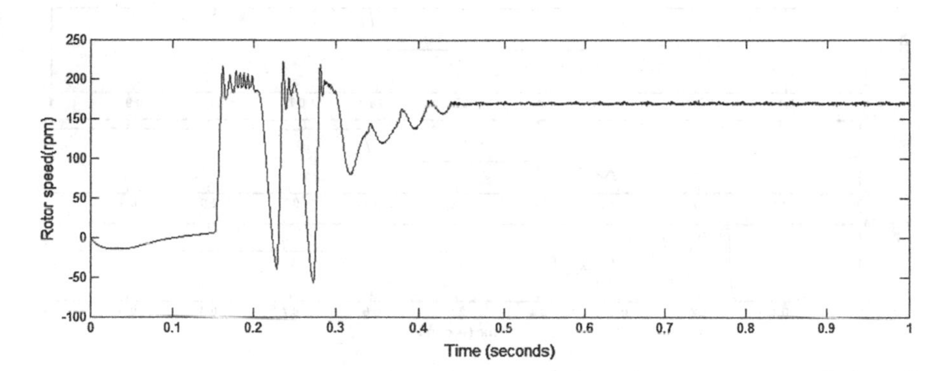

Fig. 16 Output of rotor speed of motor

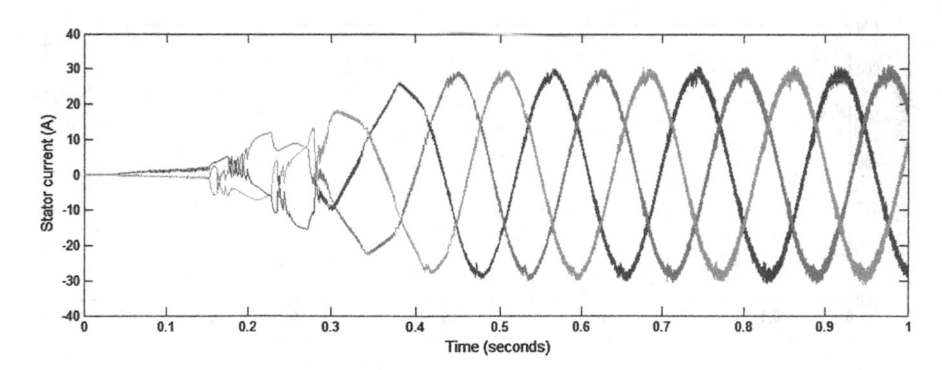

Fig. 17 Output of PMSM stator current

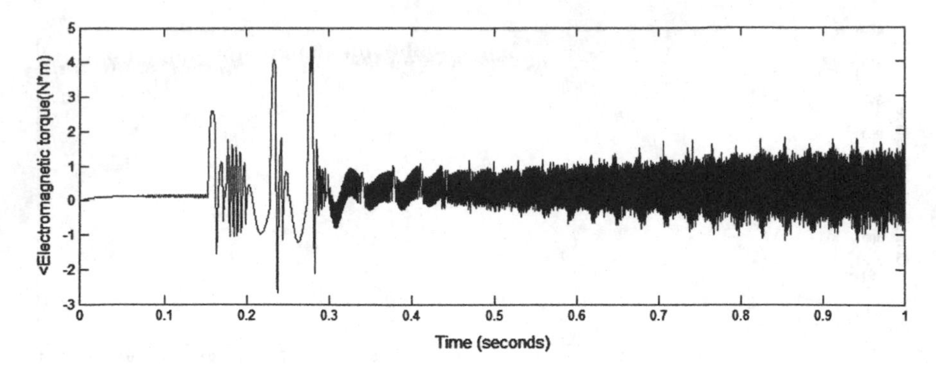

Fig. 18 Output of electromagnetic torque of motor

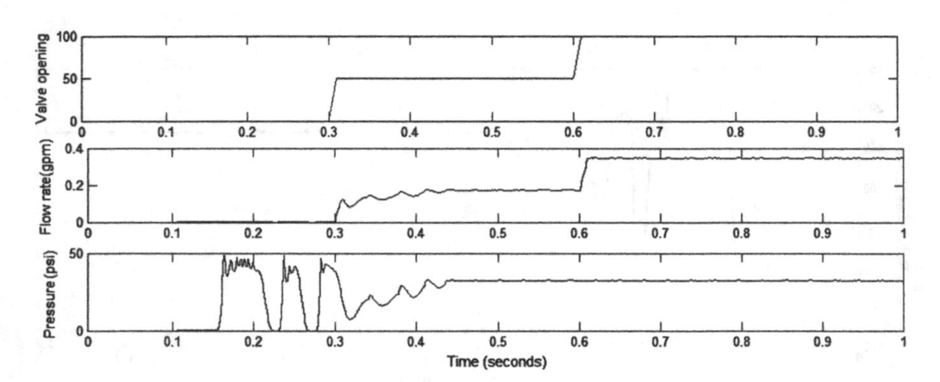

Fig. 19 Output of pump showing valve opening, flow rate, pressure

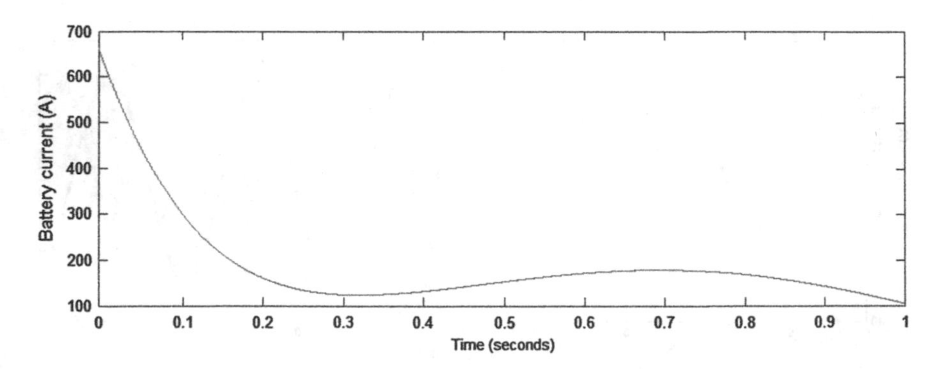

Fig. 20 Output of battery current

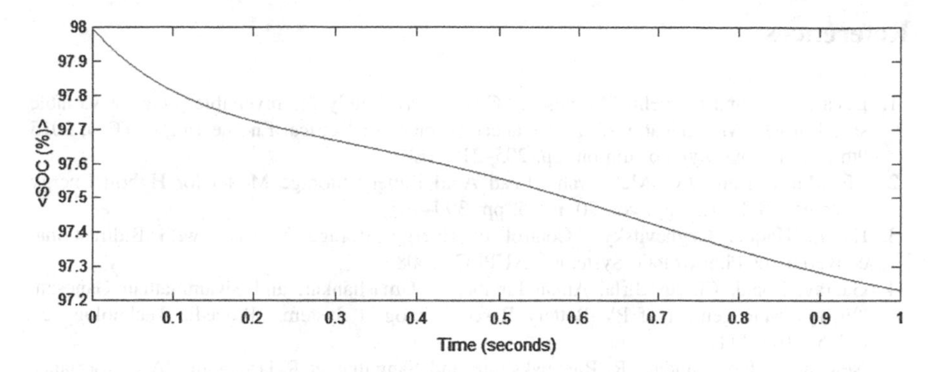

Fig. 21 Output of state of charge of battery

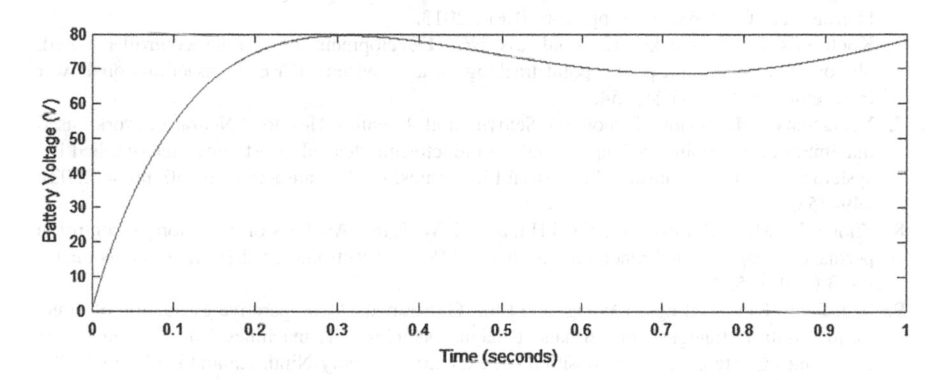

Fig. 22 Output of battery voltage

Table 7 Comparison between P&O and fuzzy based MPPT

Parameter	P&O technique	FLC-based MPPT
Maximum power point	Less	More
Response	Slow compared to FLC-based MPPT	Fast
Fluctuations	More	Less
Output smoothness	Not smooth when compared to FLC-based MPPT	Smooth

10 Conclusion

Fuzzy logic-based MPPT for interleaved boost converter which has less output voltage ripple is observed and performance of solar based PMSM motor which is coupled to a centrifugal pump has been evaluated. With the results, it can be concluded that the transient output has ripples but as it reaches steady state ripple diminishes. Pump outputs flow rate and pressure is smooth as motor reaches towards reference speed.

References

1. Ducar, Ioan, and Corneliu Marinescu. "Comparative study for reversible pump at variable speed in PMSM applications." In Advanced Topics in Electrical Engineering (ATEE), 2015 9th International Symposium on, pp. 205–210. IEEE.
2. J.F. Manwell and J.G. McGowan, "Lead Acid Battery Storage Model for Hybrid Energy Systems", Solar Energy, vol. 50, no. 5, pp. 399–405,
3. Hasan, Haque, Negnevitsky, "Control of Energy Storage Interface with Bidirectional converter for Photovoltaic Systems", AUPEC, 2008.
4. Gaurav, Sonal, Chirag Birla, Aman Lamba, S. Umashankar, and Swaminathan Ganesan. "Energy Management of PV–Battery Based Microgrid System." Procedia Technology 21 (2015): 103–111.
5. Newlin, D. Jeba Sundari, R. Ramalakshmi, and Sanguthevar Rajasekaran. "A performance comparison of interleaved boost converter and conventional boost converter for renewable energy application." In Green High Performance Computing (ICGHPC), 2013 IEEE International Conference on, pp. 1–6. IEEE, 2013.
6. Koutroulis E, Kalaitzakis K, Voulgaris NC. Development of a microcontroller based, photovoltaic maximum power point tracking control system. IEEE Transactions on Power Electronics 2001; 16(1):46–54.
7. Veerachary, Mummadi, Tomonobu Senjyu, and Katsumi Uezato. "Neural-network-based maximum-power-point tracking of coupled-inductor interleaved-boost-converter-supplied PV system using fuzzy controller." Industrial Electronics, IEEE Transactions on 50, no. 4 (2003): 749–758.
8. Zhong, Li, Mr F. Rahman, W. Yetal Hu, and K. W. Lim. "Analysis of direct torque control in permanent magnet synchronous motor drives." Power Electronics, IEEE Transactions on 12, no. 3 (1997): 528–536.
9. Niu, Feng, Kui Li, Bingsen Wang, and Elias G. Strangas. "Comparative evaluation of direct torque control strategies for permanent magnet synchronous machines." In Applied Power Electronics Conference and Exposition (APEC), 2014 Twenty-Ninth Annual IEEE, pp. 2438–2445. IEEE, 2014.
10. Bouzeria, H., C. Fetha, T. Bahi, I. Abadlia, Z. Layate, and S. Lekhchine. "Fuzzy Logic Space Vector Direct Torque Control of PMSM for Photovoltaic Water Pumping System." Energy Procedia 74 (2015): 760–771.
11. Reza, C. M. F. S., Md Didarul Islam, and Saad Mekhilef. "A review of reliable and energy efficient direct torque controlled induction motor drives." Renewable and Sustainable Energy Reviews 37 (2014): 919–932.

A Survey on Recent Approaches in Person Re-ID

M.K. Vidhyalakshmi and E. Poovammal

Abstract In the field of video surveillance, person re-ID (reidentification) is an assignment of identifying an individual caught by various cameras in the system at various place and time. With the developing security dangers, this undertaking has an incredible significance in observation out in the open spots like air terminal, railroad station, shopping buildings, and so forth. This assignment recognizes the individual of interest among the gathering of individuals caught by various cameras in the system put at various places and tracks the individual in various camera views. This undertaking faces numerous difficulties and has pulled in the scientists to this field to discover the answer for defeat the difficulties. In this paper, we have discussed about the latest research works that have been made to assault these difficulties.

Keywords Person reidentification · Feature extraction · Metric learning

1 Introduction

In the video observation framework, numerous cameras are utilized to screen a spot. The fundamental objective of the person re-ID is to recognize an individual among a gathering of individuals and track the individual when he moves from one field-of-perspective of camera to another. This errand stays testing in light of the fact that the images of the same individual caught at different place and distinctive time differ essentially. At the point when re-ID is done physically, it requires arduous exertion yet at the same time stays mistaken. With the expansion being used of video observation out in the open places the enthusiasm for computerized

M.K. Vidhyalakshmi (✉)
Department of Electronics and Communication Engineering, Tagore Engineering College, Chennai 600127, India
e-mail: mkvlakshmi@yahoo.co.in

E. Poovammal
Department of Computer Science and Engineering, SRM University, Chennai 603203, India

S.S. Dash et al. (eds.), *Artificial Intelligence and Evolutionary Computations in Engineering Systems*, Advances in Intelligent Systems and Computing 517,
DOI 10.1007/978-981-10-3174-8_42

re-recognizable proof is developing. The ordinary method for distinguishing an individual in group is finished by face recognition, yet this technique is impractical as it is exceptionally hard to get the points of interest required for extraction of face components. On the other hand, other visual elements like apparel, items related can be utilized for re-ID. Yet at the same time, the visual appearance features stay extremely feeble because of number of reasons. To begin with, the cameras utilized for surveillance are altered at a separation and the situations to be caught are highly uncontrollable. Second, when the disjoint cameras are utilized it is difficult to alter the travel time between the cameras as it fluctuates from individual to individual. Third, the features extracted from clothing are not distinct because there is a chance that many people wear same color clothes. Other important reasons include the variation in lighting condition, view angle, occlusion, and background clutter.

In recent years, many researchers work on re-ID problem to find the solution to these challenges. These efforts can be grouped into two main aspects. First aspect is based on emerging new feature representations, which are distinctive and reliable even with the variations in lighting condition, view angle. Second aspect is methods based on model learning.

2 Person Reidentification Steps

Robotized individual re-ID process has two noteworthy stages as appeared in Fig. 1. The primary stage is training stage and subsequent stage is testing stage. In the training stage reliable, robust features are extracted from probe images and gallery images of a person and are learned. A suitable representation is given to the features. In the testing stage, the features are extracted from given pair of images from camera A and camera B and then suitably represented. These features are then matched using some algorithm. When only one pair of image of person is available for matching, then the method is called single-shot method. If the multiple sets of images are available for matching, the method is called multi-shot method.

3 Feature Extraction Techniques

In person re-re-ID process former step to feature extraction is detecting the person from the video. The boundary box method [1] of segmentation aims to segment the person's image (foreground) from its back ground. A model-based cascade detector [2] is used for creating a bounding box over a holistic human body on the video. Holistic feature often undergo misalignment in pose. Li [3] used part based feature extraction from holistic human body. Here features are extracted from local patches of the human body part. Kostinger et al. [4] used regular grid to partition the image.

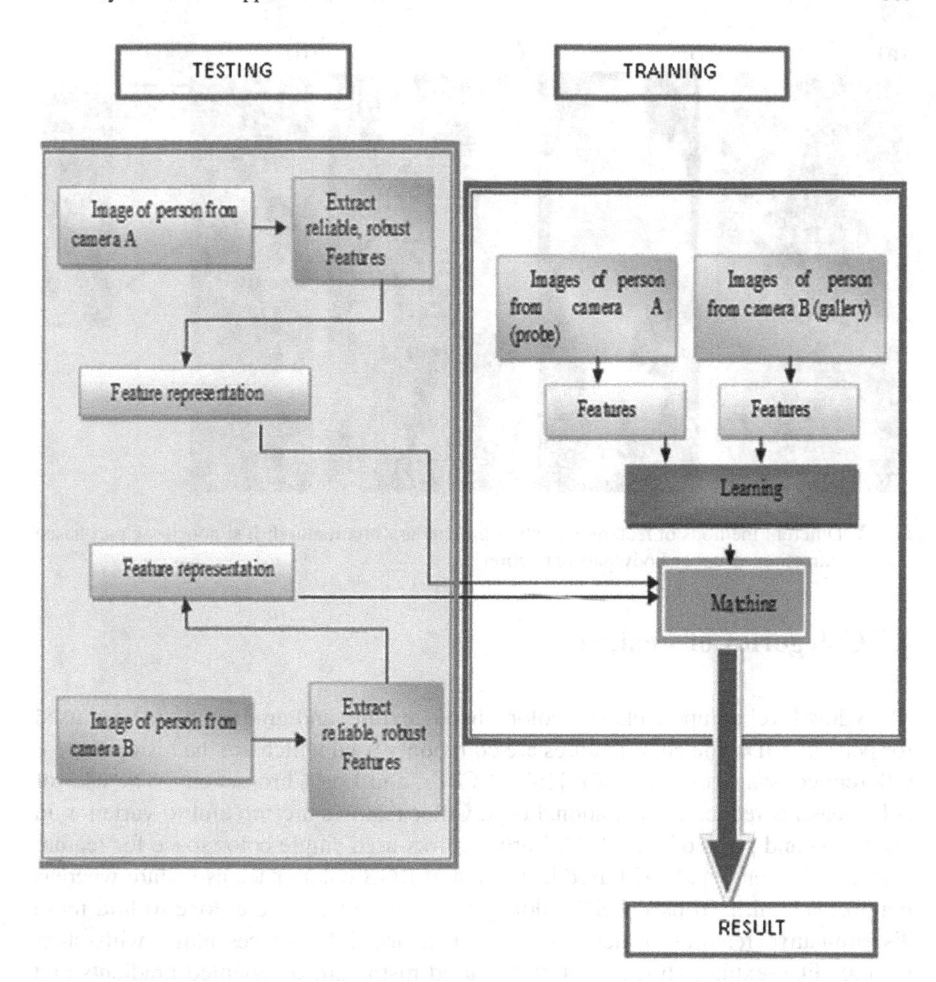

Fig. 1 Steps in person reidentification

The grid is divided into overlapping blocks and color and texture components are separated from these blocks. Tao et al. [5] used uniform grid with 8, 4 pixels spacing in horizontal and vertical direction, respectively, for feature separation. Kviatkovsky et al. [6] used bounding boxes and silhouettes to detect the person in the surveillance videos. After detection, two patches are extracted using mask. Some authors [7, 8] used stripes along the horizontal dimension and extract feature from each stripe and concatenated them. In [9], only upper body part alone is used to extract the features. This reduces the dimension of feature vector. Bak et al. [10] divided the person's body regions into head, chest, and limbs. Each of these parts is represented using color, texture, and spatial coordinates. Figure 2 shows the different methods of feature extraction.

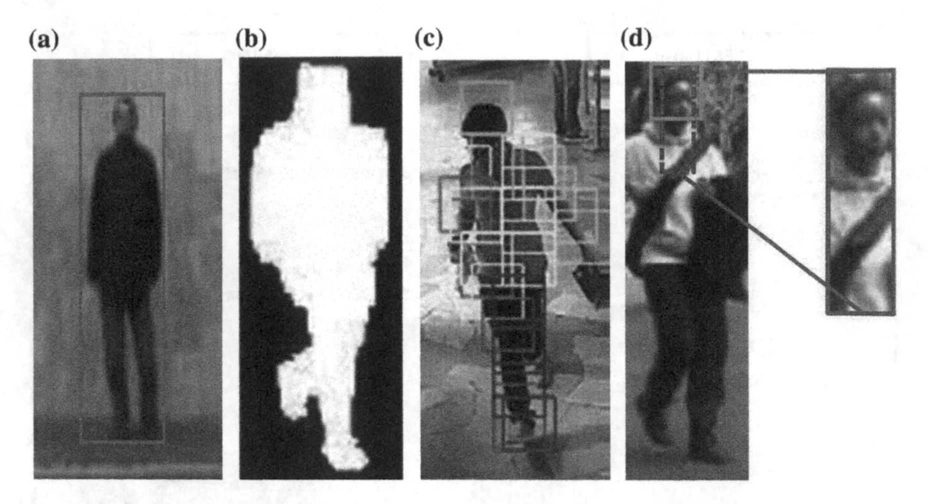

Fig. 2 Different methods of feature extraction **a** boundary box method; **b** silhouette; **c** part based feature extraction; **d** upper body part extraction

4 Categories of Feature

Many low level components like color, shape, texture, and gradient have been used for person re-ID. The color features are commonly used which are the histograms of different color spaces like RGB, HSV, YCbCr, and Log Chromaticity. The uses of color features reduce computational cost. Color features are forceful to variation in resolution and point of view [11]. Earlier works used single color space for feature extraction. Cheng et al. [12] used histogram of RGB color space as feature whereas Farenzena et al. [13] used HSV color space. Many works were done to find more discriminative features, which use more than one color spaces along with other features like texture, HOG. Li et al. [3] used histogram of oriented gradients and color histogram in HSV space as feature representation. Tahir [9] used features that are extracted only from the upper body using RGB, YCbCr, and HSV color spaces for dimension reduction. Liu [14] used Joint HSV histogram as the feature descriptor. Tao et al. [5] concatenated LBP descriptor, RGB histogram, and HSV histogram that were separated from overlapping blocks. LBP (local Binary Pattern) is a texture feature. Wang et al. [15] used RGB, HSV histograms concatenated with LBP descriptors. An [16] extracted color features quantized mean values of various channels in the HSV, Lab color space, the semantic color names, and texture feature LBPs. Gabor and Schmid filters are not sensitive to variation in lighting conditions and are used in many re-ID works. Zheng et al. [17] uses RGB, YCbCr, HSV color features along with texture features obtained from Schmid and Gabor filters to surge discriminative power.

As opposed to utilizing color values Kviatkovsky [6] utilizes parts SC signature an element which speaks to the relationship between the hues in the objective. Parts SC is gotten from log Chromaticity Color space. Tian [18] utilizes spatiogram

Table 1 Categories of features used for person Re-ID

Categories of features	Descriptor	References
Single color space	RGB	[12]
	HSV	[13]
Multiple color space	RGB, YCbCr, HSV	[9]
	Joint HSV	[14]
Texture feature	SIFT	[21]
	SURF	[22, 23]
Multiple color space, texture, attribute	HSV and HOG, clothing attributes	[3]
Multiple color spaces and texture feature	LBP, HSV, RGB	[5]
	RGB, HSV concatenated with LBP	[15]
	HSV, Lab, semantic color names, LBP	[16]
	RGB, YCbCr, HSV, Schmid Gabor filters	[17]
Spatial information	PartsSC signature	[6]
Histogram and texture	Spatiogram	[18]
	Weighted color histograms, SIFT, PHOG, Haralick	[19]
	Local Eigen dissimilarities	[20]

multi-channel spatial histogram (MCSH) as a descriptor which is a combination of shading histograms with first and second request spatial data. García [19] actualized the two arrangements of components, initial one comprises of weighted color histograms, maximally stable color regions and repetitive high-organized patches; second one comprises of SIFT elements, a pyramid of histograms of orientation gradients (PHOG) and Haralick texture elements. In a large portion of the calculations while extricating the elements some data is lost, to beat this Martinel [20] proposed LED (local eigen dissimilarities) between two images are utilized as an feature for re-ID. In [21], scale invariant feature transform (SIFT) and in [22, 23] speeded up robust features (SURF) are utilized as texture elements. Table 1 shows the outline of various sorts of components utilized as a part of re-ID.

5 Metric Learning Techniques

In case of distance learning, a set of training information is used to train the classifier. The training data will be a set of feature components extracted from each individual. The feature set of same person from different cameras form intra class and the feature set of different person form inter class. The distance learning problem hunt down for a distance that minimizes intraclass distance and maximizes interclass distances. Recently, keep it simple and straightforward (KISS) metric learning proposed by Kostinger et al. [4] is based on an hypothesis that pairwise differences are Gaussian disseminated. It performs well for person re-ID tasks when the training samples are large in number but with less training samples covariance matrices estimated by KISS become biased. To overcome this, Tao [5] proposed new distance metric learning algorithm called MCE-KISS (minimum classification error-based keep it simple and straight forward) and introduced the smoothing technique to increase the estimates of the lesser eigenvalues of a covariance matrix and give the better estimate of covariance matrix. Zheng [17] utilized relative distance comparison (RDC) figuring out how to learn similarity measure between a couple of individual image which expands the probability of a couple of genuine matches. RDC models are more vigorous against appearance variations in person's image and less liable to model overfitting compared to other distance learning models.

An et al. [16] utilized robust canonical correlation analysis (ROCCA) strategy alongside shrinkage estimation and smoothing method to address the issue of accessibility of small samples for training a high dimension feature. The relaxed margin component analysis (RMCA) proposed by Liu [14] considers the data from nearest point of intra- and interclasses to produce margins that are more adaptable. Keeping in mind the end goal to build the exactness they proposed kernalized relaxation margin component analysis (KRMCA). In another work, Datum-adaptive local metric learning strategy [24] is utilized to learn one local feature projection and map the samples in a typical segregate space. The authors additionally utilized local coordinate coding for enhancing computational productivity. Anchor points are produced from clusters in the feature space and projection bases are learned for these anchor points. A feature is represented as weighted linear combination of anchor points and the local feature projection represented by weighted linear combination of projection bases. Ma [25] proposed multitask maximally collapsing metric learning model taking into account the Mahalanobis distance metrics to conquer the challenges like lighting variation, background clutter, and perspective variation. In [26], an adaptive ranking support vector machine is used for person re-ID, which cope ups the issue of nonavailability of label information of target images by utilizing only the matched and unmatched image pairs from source cameras. This method discovers positive mean of target images using negative and unlabeled data at target domain and labeled data from source domain. Table 2 shows the types of metric learning techniques used in person re-ID.

Table 2 Types of metric learning techniques

Type of metric learning	Remarks	References
KISS (keep it simple and straightforward)	Simple and no iterative procedure Small sample size problem for calculating the inverse of covariance matrices for small training samples	[4]
MCE-KISS (minimum classification error based keep it simple and straight forward)	Increase the small Eigen values of the expected covariance matrix and are reliable Training speed is slower than KISS	[5]

(continued)

Table 2 (continued)

Type of metric learning	Remarks	References
RDC (relative distance comparison)	Computational cost is less than Adaboost and RankSVM It is a iterative process	[17]
ROCCA (robust canonical correlation analysis)	Robustly estimate the covariance matrices with limited training samples	[16]
RMCA (relaxed margin components analysis)	More flexible margin	[14]
KRMCA (Kernalised relaxation margin component analysis)	Increasing the accuracy	[14]
Datum-adaptive local metric learning method	Used when training set is small and the model learning becomes difficult	[24]
Multi task maximally collapsing metric learning model	Overcome the challenges like illumination variation, view angle variation and background clutter Distance metrics need to be relearned to tackle the variations in photographic and weather conditions	[25]
Adaptive ranking support vector machine based person re-ID method	Solves the problem of nonavailability of label information of target images by using only the matched and unmatched image pairs from source cameras It may not perform, when a large amount of training data is available	[26]
RMCA (relaxed margin components analysis)	More flexible margin	[14]

6 Conclusion

This paper is focused on the problem of people re-ID. We addressed the challenges faced by this problem and discussed the steps involved in the person re-re-ID. We surveyed the recent methods of feature extraction and metric learning techniques that are involved in this process. The problem with the metric learning methods is that it requires large number of training samples. For the problem like person re-id getting training samples is difficult in real time. Still methods has to emerge to solve this problem. We hope that still better feature representation and improvement in metric learning methods will enhance person re-ID.

References

1. S. Bak, E. Corvee, F. Bremond, and M. Thonnat, "Person Reidentification Using Haar-Based And DCD-Based Signature," in Proc. AVSS, Boston, MA, USA, 2010, pp. 1–8.
2. P. F. Felzenszwalb, R. B. Girshick, and D. McAllester, "Cascade Object Detection With Deformable Part Models," in Proc. IEEE Conf. Comput. Vis. Pattern Recognition, Jun. 2010, pp. 2241–2248.
3. Annan Li, Luoqi Liu, Kang Wang, Si Liu, and Huicheng Yan, "Clothing Attributes Assisted Person Re-identification", IEEE Transactions On Circuits And Systems For Video Technology, Vol. 25, No. 5, May 2015, pp. 869–878.
4. M. Kostinger, M. Hirzer, P. Wohlhart, P. M. Roth, and H. Bischof, "Large Scale Metric Learning From Equivalence Constraints," in Proc. IEEE CVPR, Providence, RI, USA, 2012, pp. 2288–2295.
5. Dapeng Tao, Lianwen Jin, Yongfei Wang, and Xuelong Li, "Person Re-ID by Minimum Classification Error-Based KISS Metric Learning", IEEE Transactions On Cybernetics, Vol. 45, No. 2, February 2015, pp. 242–252.
6. Igor Kviatkovsky, Amit Adam, and Ehud Rivlin, "Color Invariants for Person Reidentification", IEEE Transactions On Pattern Analysis And Machine Intelligence, Vol. 35, No. 7, July 2013, pp. No. 1622–1634.
7. D. Gray and H. Tao, "Viewpoint Invariant Pedestrian Recognition With An Ensemble Of Localized Features," in Proc. 10th ECCV, Marseille, France, 2008, pp. 262–275.
8. N. D. Bird, O. Masoud, N. P. Papanikolopoulos, and A. Isaacs, "Detection Of Loitering Individuals In Public Transportation Areas," IEEE Trans. Intell. Transp. Syst., vol. 6, no. 2, pp. 167–177, Jun. 2005.
9. Syed Fahad Tahir and Andrea Cavallaro, "Cost-Effective Features for Re-IDin Camera Networks", IEEE Transactions On Circuits And Systems For Video Technology, Vol. 24, No. 8, August, 2014, pp. 1362–1374.
10. S. Bak, E. Corvee, F. Bremond, and M. Thonnat, "Person Re-IDUsing Spatial Covariance Regions of Human Body Parts," Proc. IEEE Int'l Conf. Advanced Video and Signal Based surveillance, pp. 435–440, 2010.
11. D. Gray, S. Brennan, and H. Tao, "Evaluating Appearance Models For Recognition, Reacquisition, And Tracking," in Proc. 10th PETS, 2007.
12. Y. Cheng, W. Zhou, Y. Wang, C. Zhao, and S. Zhang, "Multi-Camera Based Object Handoff Using Decision-Level Fusion," in Proc. 2nd Int. Congr. Image Signal Process, Tianjin, China, Oct. 2009, pp. 1–5.
13. M. Farenzena, L. Bazzani, A. Perina, V. Murino, and M. Cristani, "Person Re-IDBy Symmetry-Driven Accumulation Of Local Features," in Proc. CVPR, San Francisco, CA, USA, Jun. 2010, pp. 2360–2367.
14. Hao Liu, Meibin Qi, and Jianguo Jiang, " Kernelized Relaxed Margin Components Analysis for Person Re-ID", IEEE Signal Processing Letters, Vol. 22, No. 7, July 2015, pp. 910–914.
15. Yimin Wang, Ruimin Hu, Chao Liang, Chunjie Zhang, and Qingming Leng, "Camera Compensation Using a Feature Projection Matrix for Person Reidentification", IEEE Transactions On Circuits And Systems For Video Technology, Vol. 24, No. 8, August 2014, pp. 1350–1361.
16. Le An, Songfan Yang, and Bir Bhanu, "Person Re-ID By Robust Canonical Correlation Analysis ", IEEE signal processing letters, Vol. 22, no. 8, August 2015.
17. Wei-Shi Zheng, Member, IEEE, Shaogang Gong, and Tao Xiang, " Reidentification by Relative Distance Comparison", IEEE Transactions On Pattern Analysis And Machine Intelligence, Vol. 35, No. 3, March 2013 pp. 653–668.
18. Chang Tian, Mingyong Zeng, and Zemin Wu, " Person Re-ID Based on Spatiogram Descriptor and Collaborative Representation", IEEE Signal Processing Letters, Vol. 22, No. 10, October 2015 pp. 1595–1599.

19. Jorge García, Alfredo Gardel, Ignacio Bravo, and José Luis Lázaro, "Multiple View Oriented Matching Algorithm for People Reidentification", IEEE Transactions On Industrial Informatics, Vol. 10, No. 3, August 2014, pp. 1841–1851.
20. Niki Martinel, and Christian Micheloni, "Classification Of Local Eigen-Dissimilarities For Person Re-Identification", IEEE signal processing letters, vol. 22, no. 4, April 2015.
21. D. G. Lowe, "Distinctive image features from scale-invariant key points", Int. J. Comput. Vis., vol. 60, no. 2, pp. 91–110, Nov. 2004.
22. O. Hamdoun, F. Moutarde, B. Stanciulescu, and B. Steux, "Person Re-ID In Multi-Camera System By Signature Based On Interest Point Descriptors Collected On Short Video Sequences," in Proc. ICDSC, Stanford, CA, USA, 2008, pp. 1–6.
23. X. Liu et al., "Attribute-Restricted Latent Topic Model For Person Reidentification," Pattern Recognition, Vol. 45, no. 12, pp. 4204–4213, 2012.
24. Kai Liu, Zhicheng Zhao, Member, IEEE, and Anni Cai"Datum-Adaptive Local Metric Learning For Person Re-Identification", IEEE signal processing letters, vol. 22, no. 9, September 2015.
25. Lianyang Ma, Xiaokang Yang, and Dacheng Tao, " Person Re-ID Over Camera Networks Using Multi-Task Distance Metric Learning", IEEE transactions on image processing, vol. 23, no. 8, August 2014.
26. Andy J. Ma, Jiawei Li, Pong C. Yuen, Senior Member, IEEE, and Ping Li, " Cross-Domain Person Re-ID Using Domain Adaptation Ranking SVMs", IEEE transactions on image processing, vol. 24, no. 5, May 2015.

An Analysis of Decision Theoretic Kernalized Rough C-Means

Ryan Serrao, B.K. Tripathy and A. Jayaram Reddy

Abstract There are several algorithms used for data clustering and as imprecision has become an inherent part of datasets now days, many such algorithms have been developed so far using fuzzy sets, rough sets, intuitionistic fuzzy sets, and their hybrid models. In order to increase the flexibility of conventional rough approximations, a probability based rough sets concept was introduced in the 90s namely decision theoretic rough sets (DTRS). Using this model Li et al. extended the conventional rough c-means. Euclidean distance has been used to measure the similarity among data. As has been observed the Euclidean distance has the property of separability. So, as a solution to that several Kernel distances are used in literature. In fact, we have selected three of the most popular kernels and developed an improved Kernelized rough c-means algorithm. We compare the results with the basic decision theoretic rough c-means. For the comparison we have used three datasets namely Iris, Wine and Glass. The three Kernel functions used are the Radial Basis, the Gaussian, and the hyperbolic tangent. The experimental analysis by using the measuring indices DB and D show improved results for the Kernelized means. We also present various graphs to showcase the clustered data.

Keywords Radial basis kernel · Hyper tangent kernel · Gaussian kernel · Decision theoretic rough set · Clustering

R. Serrao (✉) · B.K. Tripathy
School of Computer Science and Engineering, VIT University,
Vellore, Tamil Nadu 632014, India
e-mail: ryangodwin.serrao2013@vit.ac.in

B.K. Tripathy
e-mail: tripathybk@vit.ac.in

A. Jayaram Reddy
School of Information Technology and Engineering, VIT University,
Vellore 632014, Tamil Nadu, India
e-mail: ajayaramreddy@vit.ac.in

S.S. Dash et al. (eds.), *Artificial Intelligence and Evolutionary Computations in Engineering Systems*, Advances in Intelligent Systems and Computing 517,
DOI 10.1007/978-981-10-3174-8_43

1 Introduction

A cluster represents similarity in items of data collected with some consideration. Clustering is the process of forming clusters from a given data set and is crucial in data analysis. The early clustering algorithms are crisp by nature. It has been observed that uncertainty in data has become an inherent feature. This necessitated the development of uncertainty based algorithms and the hybrid algorithms. The DTRS model in a sense is an improvement of basic rough set model such that it includes probabilistic approaches where the restrictions on the approximations of conventional rough set model (RSM) has been relaxed. Using it with proper lost function, it is possible to obtain many significant rough set models. The decision theoretic rough c-means (DTRCM) was introduced recently and has been extended to the context of fuzzy setting in [1] and intuitionistic fuzzy setting in [2]. The predictions from clustering depend upon the formation of clusters. So, in an attempt in this direction, in this paper we introduce Kernel based DTRCM. For this purpose we take three kernels; namely the Gaussian, the hyper tangent and the radial basis for our study in order to compare their efficiencies. For the experimental purpose we use three data sets of different characteristics from the UCI repository. The experimental analysis and graphical representations provide a reflection of the efficiency of the algorithms.

2 Definitions and Notations

The following part gives an explanation of definitions and notations adopted by us. The RSM was made known by Pawlak [3] as another model to handle uncertainty. The definition depends upon the classification of the universe. For mathematical reasons Pawlak has taken equivalence relations which automatically induce classifications on the universe. In the next section we introduce this notion. Zadeh introduced fuzzy sets concept [4] earlier. These two were thought to be competing models. But Dubois and Prade [5] showed that these are complementary in nature.

2.1 Rough Sets

The notion of Rough set introduced by Pawlak depends upon the mathematical notion of equivalence relation (ER). We take W as the universe of discourse and T as an ER over it. The generated equivalence classes for every element z in V are denoted by $[z]_T$. With every subset M of U we associate two crisp sets $\underline{T}M$ and $\overline{T}M$, called the lower and upper approximations of M with respect to T and are defined as

$$\underline{T}M = \{y \in U | [y]_T \subseteq M\} \text{ and } \overline{T}M = \{ y \in U | [y]_T \cap M \neq \phi\}$$

Definition 2.1 The set M is said to be M-rough if and only if $\underline{T}M \neq \overline{T}M$ and is said to be T-definable otherwise.

The region of uncertainty of M with respect to T is called the T-boundary of M. It is denoted by $BUN_T(M)$ and is the difference of upper and lower approximations of M with respect to T. It may be noted that $\underline{T}M \subseteq \overline{T}M$ for all M and T.

We say that X is rough with respect to R if and only if $\underline{R}X \neq \overline{R}X$, equivalently $BN_R(X) \neq \phi$. X is said to be R-definable if and only if $\underline{R}X = \overline{R}$, or $BN_R(X) = \phi$.

The decision theoretic (DT) rough sets (RS) was introduced in [6]. We know that the Rough C-means clustering algorithm introduced in [7] where, the data set W is classified into clusters W_i, $i = 1, 2, \ldots n$ has the following properties:

$$\underline{T}(W_i) \subseteq \overline{T}(W_i) \subseteq W, \forall W_i \subseteq W \tag{1}$$

$$\underline{T}(W_i) \cap \underline{T}(W_j) = \phi, \forall W_i, W_j \subseteq W, i \neq j \tag{2}$$

$$\underline{T}(W_i) \cap \overline{T}(W_j) = \phi, \forall W_i, W_j \subseteq W, i \neq j \tag{3}$$

An element $w \notin \underline{T}W_i$ for any I may be a part the boundary regions of more than one $W_i s$. (4)

2.2 DTRS Model

In order to incorporate probabilistic approximation in the DTRS model the Bayesian decision procedure is used [6].

Let $F = \{s_1, s_2, \ldots s_n\}$ and $R = \{r_1, r_2, \ldots r_m\}$ be finite sets of states and actions respectively. We denote the cost (or loss) functions by $\mu(s_i|r_j)$ for taking action s_i under the state r_j. The probability (conditional) of an object z in state r_j be denoted by $P(r_j|z)$. We obtain the formula for expected loss denoted by

$$S(r_k|z) = \sum_{k=1}^{m} \mu(s_i|r_k) \,.\, P(r_k|z)$$

In DTRS model, $F = \{R, R^C\}$ represents a pair of sets, where R denotes the set of states where an object is in R and R^C denoting the set of states where an object is not in R. Suppose, we take three actions r_1, r_2, r_3 in R such that positiveset(R), negativeset(R) and the boundary region (R) are $\underline{S}(R), U \backslash \overline{S}(R)$ and $\overline{S}(R) \backslash \underline{S}(R)$, respectively. $P(R|[z])$ and $P(R^C|[z])$ be the probabilities of z being in.

The probabilities $P(R|[z])$ and $P(R^C|[z])$ are the probabilities that an object z is in the equivalence classes in R or R^C accordingly. We denote by $S(r_i|[z])$; i = 1, 2 … 3, the expected loss corresponding actions given by the formulae:

$$S(r_i|[z]) = \mu_{i1} P(R|[z]) + \mu_{i2}.P(R|[z]), i = 1, 2, 3$$

The DTRCM algorithm was proposed and studied in [6, 8]. However, in this paper we are proposing an improved version of this algorithm with the Euclidean distance being replaced by the kernels called DTKRCM. For the presentation we require the following additional concepts and notations. It may be noted that some hybrid algorithms like the rough fuzzy c-means have been introduced in literature [9, 10].

2.3 Calculating Risk

The lower approximation $L(x_\ell)$ of $x_\ell \in X$ is defined as:

$$L(x_i) = \{x \in X : D(x, x_i) \leq \delta \wedge x \neq x_i\}, \tag{5}$$

where

$$\delta = \frac{\min_{1 \leq k \leq c} D(x_\ell, v_k)}{p} \tag{6}$$

Let the conditional probability of x_ℓ in C_i be as

$$P(C_i|x_\ell) = \frac{1}{\sum_{j=1}^{c} \left(\frac{D(x_\ell, v_i)}{D(x_\ell, v_j)}\right)^{\frac{2}{m-1}}} \tag{7}$$

We denote the set of points similar to z by T_z.

$$T_x = \{C_i \in C : P(C_i|x) > d^{-1}\} \tag{8}$$

Continuing with the set of actions R, where r_j is the action that a data point is assigned to C_j. The amount of loss for assigning action r_j for x_1 when x_1 belongs to C_i is represented by $\mu_{z1}(r_j|C_i)$, and defined as:

$$\mu_{z_i}(r_j|c_i) = \mu_{C_i}^{r_j}(z_\ell) + \sum_{z \in L(z_i)} \gamma(z) \mu^{r_j}(z), \tag{9}$$

where

$$\mu_{C_i}^{r_k}(z_\ell) = \begin{cases} 0, & \text{if } i = k; \\ 1, & \text{if } i \neq k. \end{cases} \quad (10)$$

$$\mu^{r_j}(z) = \begin{cases} \dfrac{|r_j - T_z|}{r_j} = 0, & \text{if } r_j \in T_z; \\ \dfrac{|r_j - T_z|}{r_j} = 1, & \text{if } r_j \notin T_z. \end{cases} \quad (11)$$

$$\gamma(z) = \exp\left(-\frac{d^2(z, z_\ell)}{2\sigma^2}\right) \quad (12)$$

The risk related with taking action r_j for z_i is represented by $S(r_j/z_i)$ and is defined as

$$S(b_j/z_i) = \sum_{i=1}^{k} \mu_{z_i}(b_j/c_i)P(c_i/z_i) \quad (13)$$

For each data point z_i let $r_k = \arg\min_{r_i \in A}\{S(r_i/z_i)\}$. The index J_D is a measure of closeness of r_k through risk value and is given by

$$J_D = \{j|\{S(r_j/z_i)/S(r_k/z_i)\} \leq 1 + \varepsilon \wedge j \neq k\} \quad (14)$$

If $J_D = \phi$, z_i is assigned to C_k. Otherwise, $\forall j \in J_D, z_i \in b_n(C_j)$

2.3.1 DTRS

Let the classes of a classification π of W be denoted as $\pi = \{A_1, A_2, \ldots, A_m\}$. The two approximations are given by:

$$\underline{apr_{(\gamma,\delta)}}(A_i) = POS_{(\gamma,\delta)}(A_i) = \{z \in W|P(A|[z]) \geq \gamma\}$$
$$\overline{apr_{(\gamma,\delta)}}(A_i) = POS_{(\gamma,\delta)}(A_i) \cup BND_{(\gamma,\delta)}(A_i) = \{z \in U|P(A|[z]) \geq \delta\}$$

The approximations of a partition π in terms of the approximations of $A_i, i = 1, 2, \ldots, m$ are defined as follows:

$$POS_{(\gamma,\delta)}(\pi) = \cup_{1 \leq i \leq m} POS_{(\gamma,\delta)}(A_i), BND_{(\gamma,\delta)}(\pi) = \cup_{1 \leq i \leq m} BND_{(\gamma,\delta)}(A_i), NEG_{(\gamma,\delta)}(\pi) = U - POS_{(\gamma,\delta)}(\pi) \cup BND_{(\gamma,\delta)}(\pi)$$

The three regions defined above may not be mutually exclusive but together they form a covering for W.

2.4 Similarity Metrics

Similarity measures are used to cluster data into groups. There are various similarity measures used for this purpose. The easiest similarity measure is the Euclidean distance which takes into account the distance of every data point from the centroids. This also forms a limitation as the initial cluster centroids assigned play a crucial role in the clustering process. Another limitation is that this method cannot cluster non linear data. In such cases, kernel based similarity measures play a vital role. Kernal metrics convert the data represented on a ordinary plane to a feature plane of higher dimensional data, also called as the kernel plane. This transformation is done by using some nonlinear mapping function. Various similarity metrics are explained in this section.

Definition 2.4.1 (Euclidean distance)

Suppose $x = (x_1, x_2,\ldots, x_n)$ and $y = (y_1, y_2,\ldots, y_n)$ are two points in the n-dimensional Euclidean space. Then the Euclidean distance $d(x, y)$ between x and y is:

$$d(x,y) = \sqrt{(x_1 - y_1)^2 + (x_2 - y_2)^2 + \ldots + (x_n - y_n)^2}$$

Definition 2.4.2 (Kernel distance) [11]

Let 'x' represent a point. The transformation of 'x' to a higher dimensional feature plan is represented by $\Phi(x)$. The inner product space is defined as $K(x,y) \leq \Phi(x),\ \Phi(y)$. Let $x = (x_1, x_2, \ldots, x_n)$ and $y = (y_1, y_2, \ldots, y_n)$ represent points in the n-dimensional space.

Kernel functions applied for clustering of data in the study are defined as follows:

(a) Radial basis kernel:

$$R(x,y) = \exp\left(-\frac{\sum_{i=1}^{n} (x_i^p - y_i^p)^q}{2\sigma^2}\right)$$

The value of parameters set in the implementation of the above kernel is $p = 2$, $q = 2$

(b) Gaussian kernel: (RBF with $p = 1$ and $q = 2$)

$$G(x, y) = \exp\left(-\frac{\sum_{i=1}^{n}(x_i - y_i)^2}{2\sigma^2}\right)$$

(c) Hyper tangent kernel

$$H(x, y) = 1 - \tanh\left(-\frac{\sum_{i=1}^{n}(x_i - y_i)^2}{2\sigma^2}\right),$$

where $\sigma^2 = \frac{1}{N}\sum_{i=1}^{N}\|x_i - x'\|^2$ and $x' = \frac{1}{N}\sum_{i=1}^{N}x_i$.

N denotes the total number of existing data points and $\|a - b\|$ denotes the Euclidean distance between points a and b which pertain to Euclidean metric space. By [10, 12–15]

$D(x, y)$ represents the form of kernel distance function where $D(x, y) = K(x, x) + K(y, y) - 2K(x, y)$ and when similarity property (i.e., $K(a, a) = 1$) is applied, we get $D(x, y) = 2(1 - K(x, y))$.

3 DTKRCM

In this section we present our proposed algorithm Decision Theoretic Kernelized Rough C-Means.

3.1 *Basic Idea*

In this algorithm, we modify the Decision theoretic Rough C means algorithm by using various different distance metrics such as the use of the Gaussian kernel, Radial Basis kernel, and the hyper tangent kernel.

3.2 Algorithm Description

Input: The given data set Z = $\{z_1, z_2, ... z_n\}$, c, w_ℓ, ε, p, σ, m, S_{max}

Output: Clustering result: $(\underline{C_1}, \overline{C_1}), \ldots, (\underline{C_c}, \overline{C_c})$ and cluster centroids

1 randomly assign the initial centroid v_i for C_i, i = 1, 2 . . . d;
2 repeat
3 for i ← 1 to n do
4 for a data point $z_i \in$ Z, calculate P (C j | z_i), j = 1. . . d, by using Eq. (7);
5 determine z_i's neighbouring points set L (z_i) by using Eq. (5);
6 for every data point z ∈ L(z_i), determine T_z by using Eq. (6);
7 calculate $S(r_j | z_i)$, j = 1 , . . . , d, by using Eq. (7); to Eq. (10);
8 find the action with minimal risk. r_h = argmin $r_j \in$ A $\{S(r_j | z_i)\}$;
9 assign z_i to $\overline{C_h}$, i.e. $z_i \in \overline{C_h}$;
10 find the index set J_D with respect to ah by using Eq. (2.3.8);
11 if $J_D = \emptyset$ then
12 assign z_i to $\underline{C_h}$, i.e. $z_i \in \underline{C_h}$;
13 else
14 assign z_i to the upper approximations of the clusters determined by J_D, i.e. $z_i \in \overline{C_j}$ $\forall j \in J_D$;
15 end
16 end
17 calculate the new centroid for each cluster using eq. (9) in [5]
18 until the termination criterion are met;

4 Experimental Results

The following experimental study conducted by us was performed by implementing the algorithm in python programming language using the software Canopy. The detailed analysis of the results is discussed in this part. Here we have taken three datasets from [16]; namely Iris, Wine and Glass and obtained the results. These results are displayed in the graphs obtained from the Iris dataset only as other results cannot be put in the graphical format.

First we show the DB [17] and D [18] values for the three different data sets in Table 1 for Iris. Table 2 for Wine and Table 3 for Glass data sets. These values show that the Kernelized versions of the algorithm DTRCM taking any one of the three kernels show better result. That is the DB value is lower and D value is higher for them.

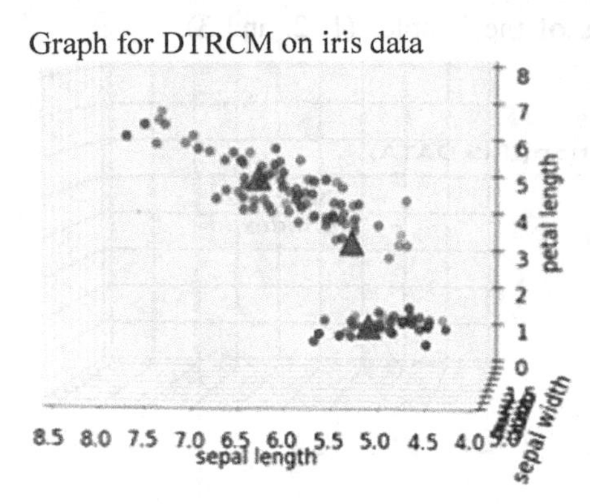

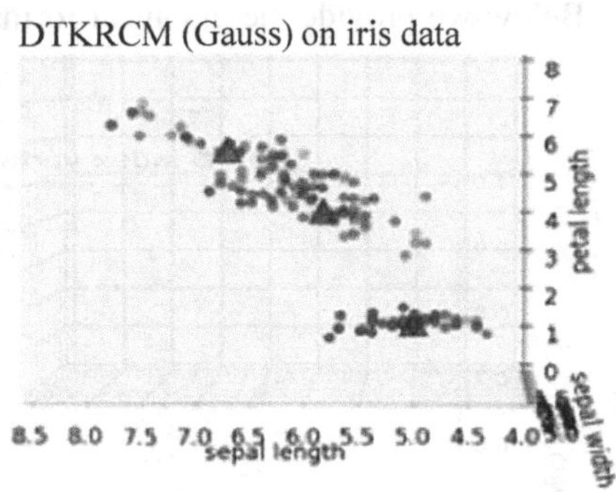

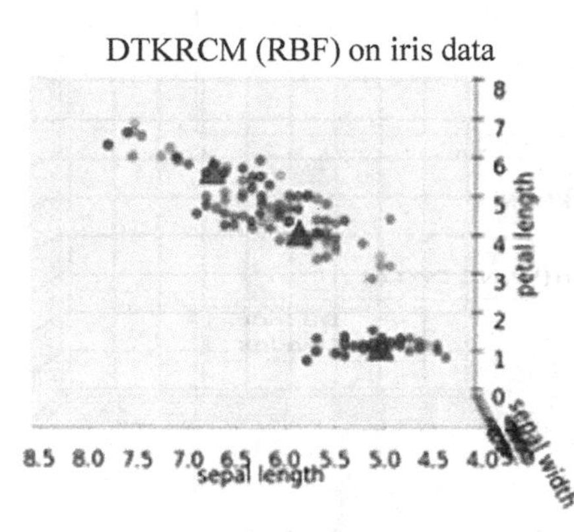

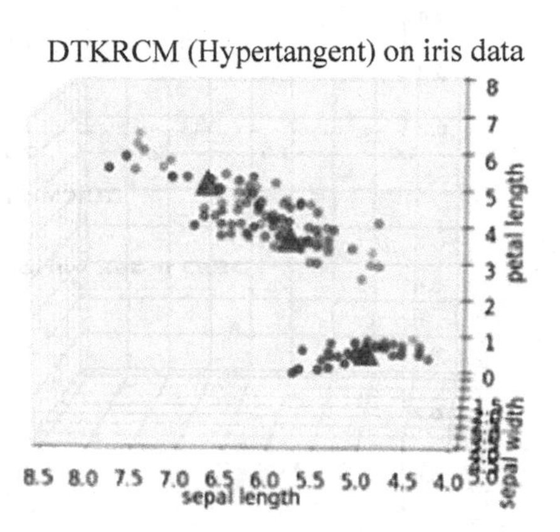

Table 1 Iris dataset

Algorithms	DB	D
DTRCM	0.689303	0.5245
DTKRCM(Gauss)	0.504824	0.68408
DTKRCM(RBF)	0.515994	0.66536
DTKRCM(Hypertangent)	0.5014841	0.68408

Table 2 Wine dataset

Algorithms	DB	D
DTRCM	0.53304	0.38037
DTKRCM(Gauss)	0.5279	0.55061
DTKRCM(RBF)	0.52912	0.55090
DTKRCM(Hypertangent)	0.52794	0.55091

Table 3 Glass dataset

Algorithms	DB	D
DTRCM	0.86649	0.29484
DTKRCM(Gauss)	0.874732	0.29116
DTKRCM(RBF)	0.85450	0.29787
DTKRCM(Hypertangent)	0.84330	0.313792

Below we provide the graphical forms of the 3 tables (1, 2, and 3).

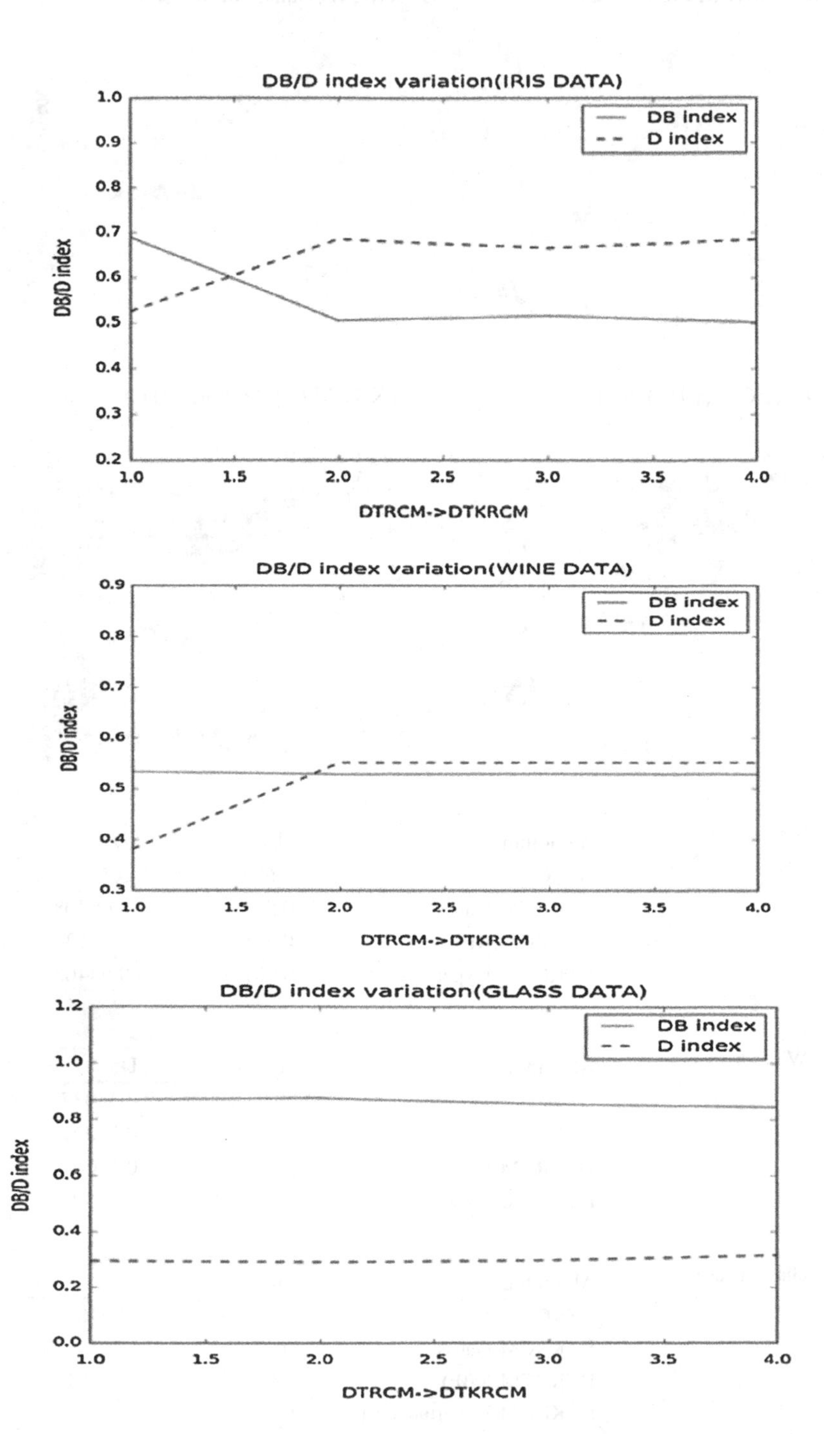

5 Conclusions

A novel algorithm called the DTKRCM, which is the Kernelized version of DTRCM was introduced in this study. Three kernels were taken for comparison and tested on three data sets from [16]. We have provided the graphical representation of the clustered datasets. For the comparison purpose we have taken two standard indices, the DB index and the D index. The results obtained are given in the tabular format for comparison. It has been observed that the Kernelized algorithms provide better index values in comparison to the normal algorithm.

References

1. Agrawal, S. and Tripathy, B.K.: Decision theoretic rough intuitionistic fuzzy C-means algorithm, Smart Innovation Systems and Technologies, 50, (2016), pp.71-82.
2. Agrawal, S. and Tripathy, B.K: Decision theoretic rough fuzzy C-means algorithm, In: Proceedings of 2015 IEEE International Conference on Research in Computational Intelligence and Communication Networks, ICRCICN 2015, Article number 7434234, pp.192-196.
3. Pawlak, Z.: Rough Sets: Theoretical aspects of Reasoning about Data, Kluwer academic publishers, (1991).
4. Zadeh, L.A.: Fuzzy sets, Information and control, vol. 8, (1965), pp. 338–353.
5. Dubois, D. and Prade, H.: Rough fuzzy set model, International journal of General systems, vol. 46, (1990), no. 1, pp. 191–208.
6. Yao, y.y., and Wong, S.K.M: A decision theoretic framework for approximating concepts, International journal of Man-machine studies, 37(6), (1992), pp. 793–809.
7. Lingras, P. and West, C.: Interval set clustering of web users with rough k-means, journal of intelligent information systems, 23 (1), (2004), pp. 5–16.
8. Yao, Y.Y.: decision-theoretic rough set models, in: Yao, J., Lingras, P. Wu, W.Z. et al (Eds.), proceedings of the second international conference on rough sets and knowledge discovery, LNCS, Vol. 4481, (2007), pp. 1–12.
9. Maji, P. and Pal. S.K.: A hybrid clustering algorithm using rough and fuzzy sets, Fundamenta Informaticae, 80(4), (2007), pp. 475–496.
10. Mitra, S., Banka, H. and Pedrycz, W.: Rough- fuzzy collaborative clustering, IEEE Transactions on Systems, Man and Cybernetics, 36(4), (2006), pp. 795–805.
11. Zeng, H., and Ming C.Y.: feature selection and kernel learning for local learning based clustering, IEEE Transactions on Pattern Analysis and Machine Intelligence, 33(8), (2011), pp. 1352–1547.
12. Tripathy, B.K., Ghosh, A. and Panda, G. K.: Adaptive K-means Clustering to Handle Heterogeneous Data using Basic Rough Set Theory, In the Proceedings of Second Intl. Conf. (CCIT-2012), Bangalore, Advances in Computer Science and Information Technology Network Communications, LNICST series, Springer, 84 (2012), pp. 193–201.
13. Tripathy, B.K., Ghosh, A. and Panda, G. K.: Kernel Based K-Means Clustering Using Rough Set, Proceedings of 2012 International Conference on Computer Communication and Informatics (***ICCCI* - 2012**), Jan. 10–12, (2012), Coimbatore, INDIA, pp. 1–5.
14. Tripathy, B.K., Tripathy, A., Govindarajulu, K. and Bhargav, R.: On kernel Based rough Intuitionistic Fuzzy C-means algorithm and a comparative analysis, Smart innovation systems and technologies, vol. 27, (2014), pp. 349–359.

15. Tripathy, B.K., Mittal, D.: Hadoop based Uncertain Possibilistic Kernelized C-Means Algorithms for Image Segmentation and a Comparative analysis, Applied Soft Computing Journal, vol. 46 (2016), pp. 886–923.
16. Blake, C.L. and Merz, C. J: UCI repository of machine learning databases, 1998. http://www.ics.uci.edu/mlearn/mlrepository.html.
17. Davis, D.L. and Bouldin, D.W.: Clusters separation measure, IEEE transactions on pattern analysis and machine intelligence, vol-Pami-1, no. 2, (1979), pp. 224–227.
18. Dunn, J.C.: Fuzzy relative of ISODATA process and its use in detecting compact well-separated clusters, Journal of Cybernetic, vol. 3, (1974), pp. 32–57.

Hybrid Techniques for Reducing Total Harmonic Distortion in a Inverter-Fed Permanent Magnet—SM Drive System

G. Kavitha and I. Mohammed Rafeequdin

Abstract AC drives fed by inverters find applications in a number of fields which are uncountable. Inverter performance can be enhanced by controlling the harmonics at various levels which can be at the input, output, and within the inverter. Hence, a hybrid method of controlling the harmonics on all the three levels can be put together in a inverter-fed permanent magnet system. A novel method of implementing a harmonic filter and multipulse converter on the input side and a space vector modulation technique used within the inverters will be a better solution in reducing the level of harmonics so as to improve the quality of output power fed to a permanent magnet synchronous motor drive system. This paper mainly focusses on improving the quality of power by various hybrid techniques which also involves the design of transformers using dimetrical connection for a multipulse system. The results are proved by reduced total harmonic distortion using PSIM and Matlab softwares.

Keywords SM—Synchronous motor · THD—Total harmonic distortion · MPC—Multipulse converter

1 Introduction

High power synchronous motor drives are more popular especially among the industries as they find a number of applications which includes mills, fans, pumps, and compressors by a wide spectrum of industries, for example, paper, mining, rolling mills, water treatment plants, and petrochemical plants [1]. These drives can be utilized at a maximum potential only when there is a possibility of speed control

G. Kavitha (✉) · I. Mohammed Rafeequdin
Department of Electrical and Electronics Engineering, B. S. Abdur Rahman University, Chennai, India
e-mail: kavitha@bsauniv.ac.in

S.S. Dash et al. (eds.), *Artificial Intelligence and Evolutionary Computations in Engineering Systems*, Advances in Intelligent Systems and Computing 517,
DOI 10.1007/978-981-10-3174-8_44

which needs a converter–inverter system which in other words can be called as a nonlinearity producing load toward the system [2]. These nonlinearities in turn distort the wave shape as the expected wave is a sinusoidal wave. Adding to this discussion these nonlinearities introduce harmonics to the system. This paper mainly concentrates on various novel techniques involved in reducing the harmonics at various levels in order to improve the quality of power so as to enhance the shape of the wave both at the source end and load end [3].

Enhancing the quality of power plays a major role especially among the industrial customers. Hence, it becomes the talk of the industry about the various methods of increasing the quality of power which includes (1) SVPWM techniques; (2) Multipulse converter and diode-chopper system; (3) Harmonic filter, multipulse converter, and diode-chopper system [3].

2 Topologies for PQ Improvement

2.1 Multipulse AC–DC Converter

Multipulse systems are greatly encouraged in order to achieve more number of pulses as this is observed to reduce harmonics at an appreciable rate. Generally these converters are put into two categories: an isolated and a non-isolated type of MPCs [1].

Isolated type of MPCs uses multi-winding transformers before the converter system. The ratings of such transformers are found to be high and hence it must be reduced. However, by the use of non-isolated MPCs, the size, cost, weight, and losses of magnetic components can be reduced drastically. These MPCs use different winding connections, namely star, delta, zigzag, polygon, hexagon, T-connection, tapped winding, plurality of winding of isolated multi-winding transformers, and autotransformers for the purpose of phase shifting in order to reduce harmonics at the AC mains [4].

Pulse multiplication topology, as reported in the literature, generates higher number of pulses in the multiple of 12 pulses. This topology achieves only pulse doubling for uncontrolled converters using one reactor and two diodes, whereas it generates 24, 36, and 48 pulses for controlled converters using a reactor and two, three, or four thyristors, respectively. For non-isolated topologies based on pulse multiplication technique, additional zero-sequence blocking transformer (ZSBT) and IPT are used [1].

Many numbers of works have been in progress in reducing the size of transformers used for multipulse generation so as to make the system economically feasible as MPCs are finding better solutions for harmonic control. Multipulse converters not only are superior in eliminating harmonics but also reduce electromagnetic interference, radio frequency interference and switching losses [4].

MPCs have been reported to reduce THD of AC mains current well below 2% or around and almost ripple-free DC output voltage to feed variety of loads with various configurations of AC–DC converters. But these configurations of MPCs have the drawback of higher DC-link voltage as compared to six-pulse converters, which makes them non-suitable for retrofit applications [2].

2.2 Transformer with Rectifier-Chopper System

Active filters with rectifiers have some drawbacks on adding complexities since designing a control block for such a system is difficult which also increases the cost thereby reducing the overall system efficiency. Hence, hybrid methods which combine active and passive with harmonic filters on the input side can be found as a best alternative for the existing system [5, 6].

This hybrid topology finds an efficient way of controlling harmonics both at the supply end and the load end. Also it reduces the overall VA consumption as compared to a normal phase-controlled converter system [5, 6].

2.3 SVPWM for Harmonic Control

Harmonic distortions are mainly due to the use of power electronic systems namely a converter–inverter which can be called as a nonlinear load as they do not draw a sine wave. These nonlinearities change wave shapes which are not appreciable among all the levels of customers as these variations may lead to an uncountable no of power quality issues. The solution for this would be using SVPWM techniques with transformer, rectifier-chopper system, which is superior to a PWM-controlled existing system as the THD level when measured is 1.55% which is as per the IEEE norms [7] (Figs. 1, 2, 3, 4 and 5).

2.4 Design of Transformers

In certain applications like thyristors and rectifiers six-phase supply is required. Therefore, it becomes necessary to convert three-phase a.c. supply into six-phase supply. Using three identical single-phase transformers suitably interconnected this can be achieved. The primary winding is connected in delta, whereas its secondary winding is split up into two halves. Thus conversion from three-phase to six-phase supply can be obtained by having two similar secondary windings for each of the primaries of the three-phase transformer. This is shown in Fig. 6 [8].

In some application do to the economical considerations three-phase to multiple-phase conversions are required. Using three single-phase transformers

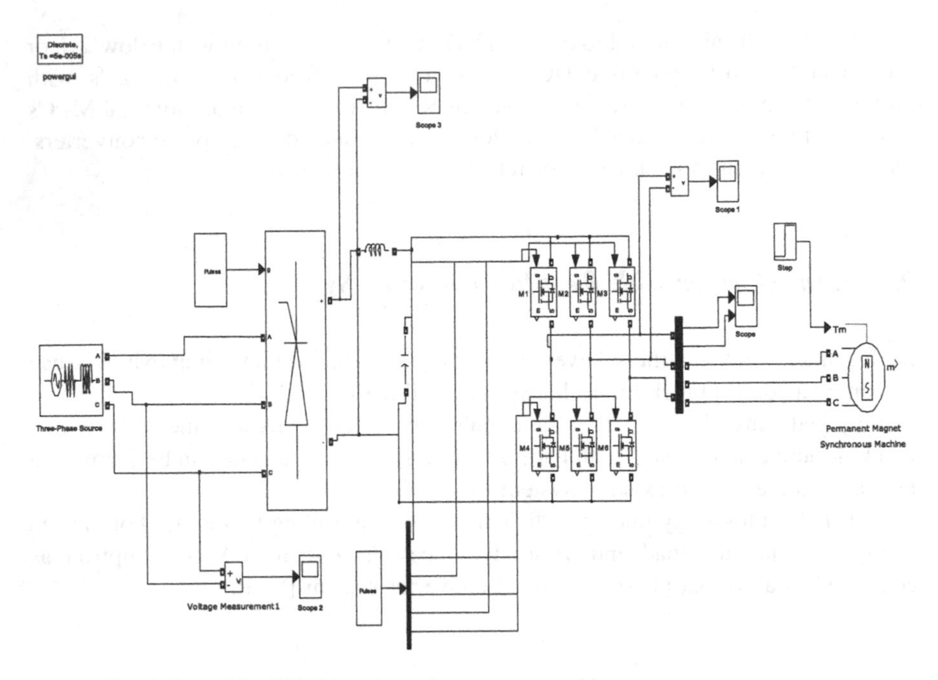

Fig. 1 Simulation of the PWM-based synchronous motor drive

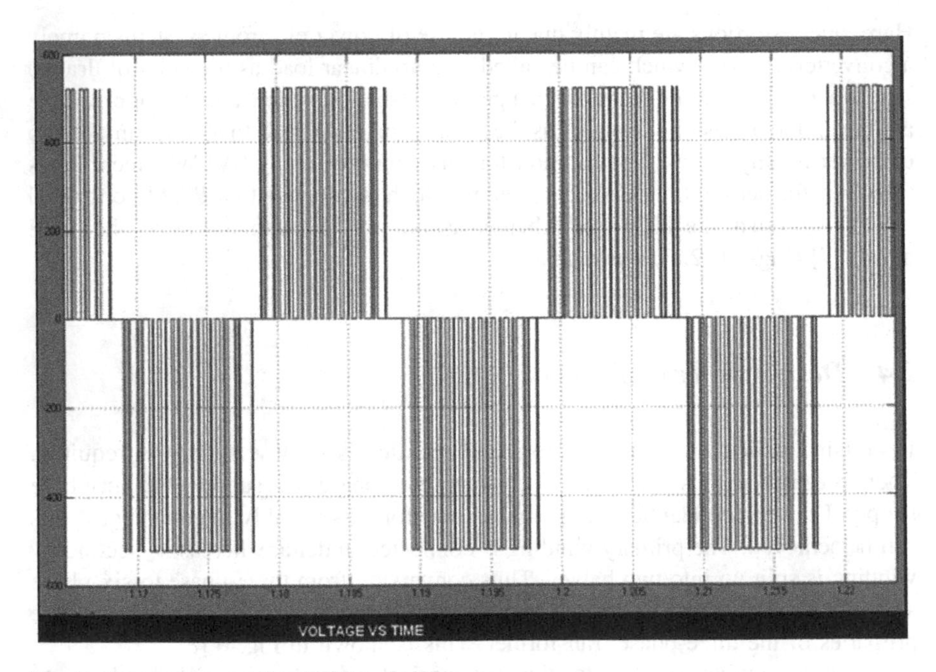

Fig. 2 Inverter output using PWM technique

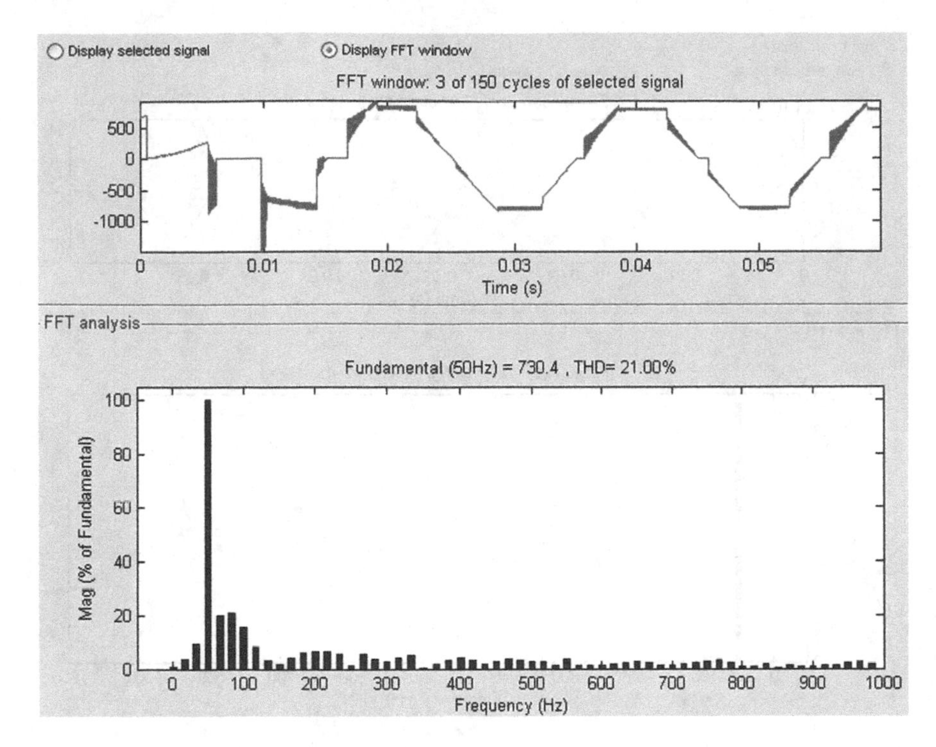

Fig. 3 Harmonic levels using PWM technique

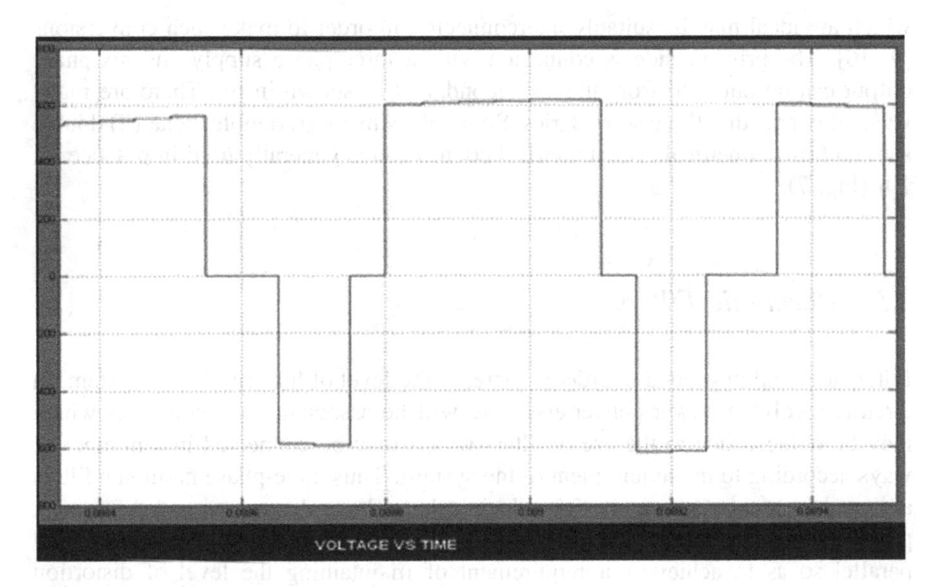

Fig. 4 Generation of SPWM waveform

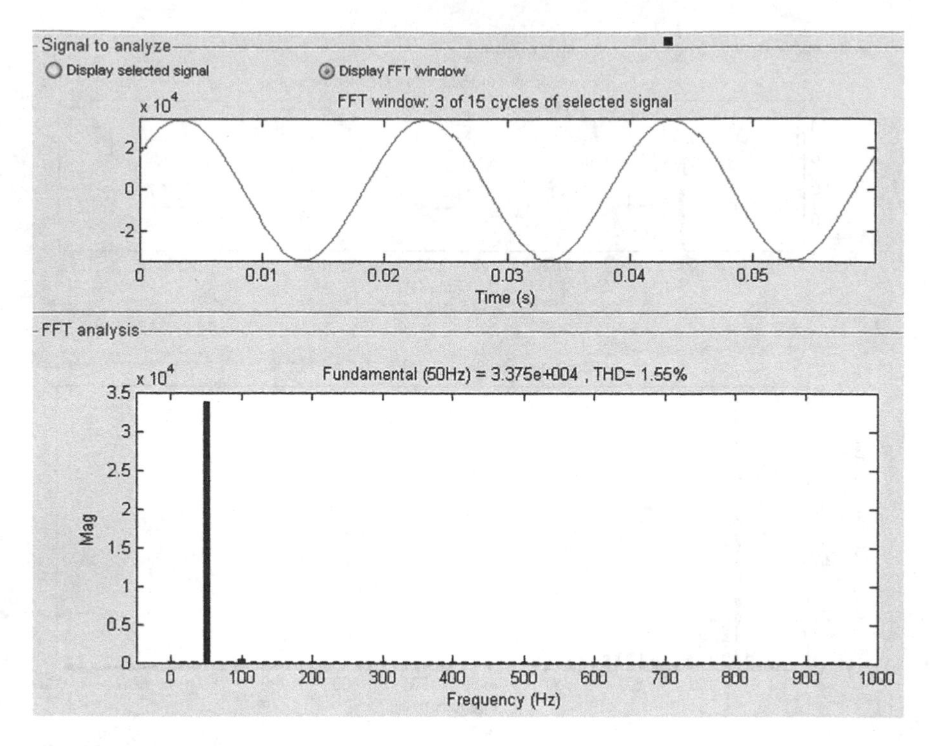

Fig. 5 Harmonic levels using SPWM technique

which are ideal may be suitably interconnected in order to make such conversions [9, 10]. The primary side is connected with a three-phase supply and six-phase output can be obtained from the six secondaries as shown in fig. There are many ways of connecting these secondaries. Some of them are (i) double delta, (ii) double star, and (iii) dimetrical. The dimetrical connection is generally used in practice [8, 10] (Fig. 7).

2.5 Harmonic Filters

Filters are used in general in order to increase the level of linearity in any system. In circuits involving power converters, there will be association of harmonics which may be reduced due to the use of filters which may be connected in a number of ways according to the requirement of the system. Thus three-phase harmonic filters are used for dual-purpose one for reducing the voltage distortions and the second purpose being the correction of power factor. Banks of filters are connected in parallel so as to achieve our requirement of maintaining the level of distortion within permissible limits. These filters when used before a multipulse converter

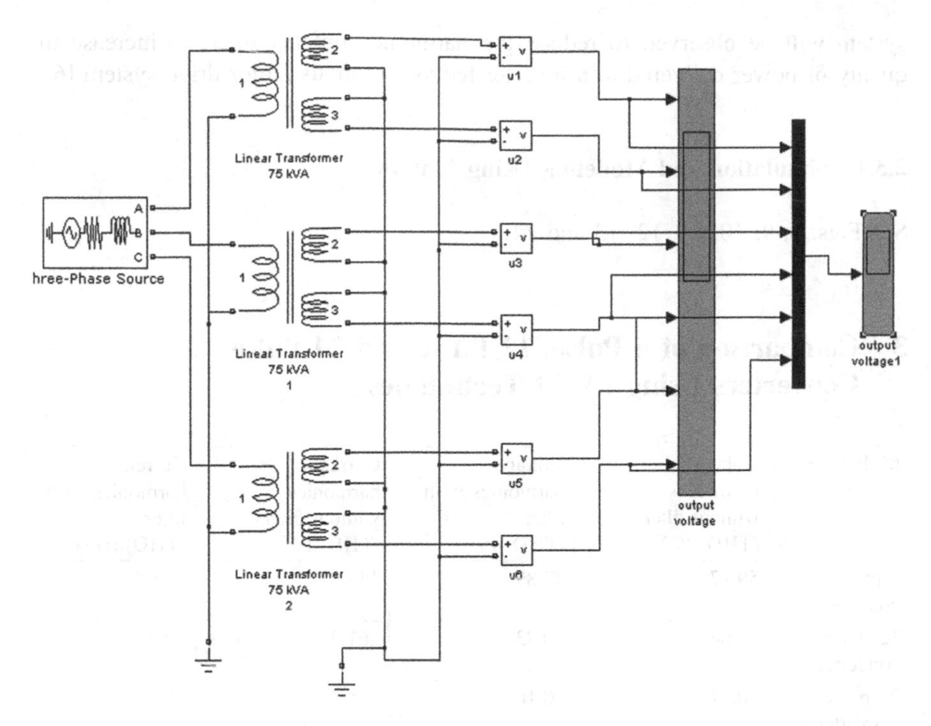

Fig. 6 Simulation for three-phase to six-phase conversion dimetrical connection transformers

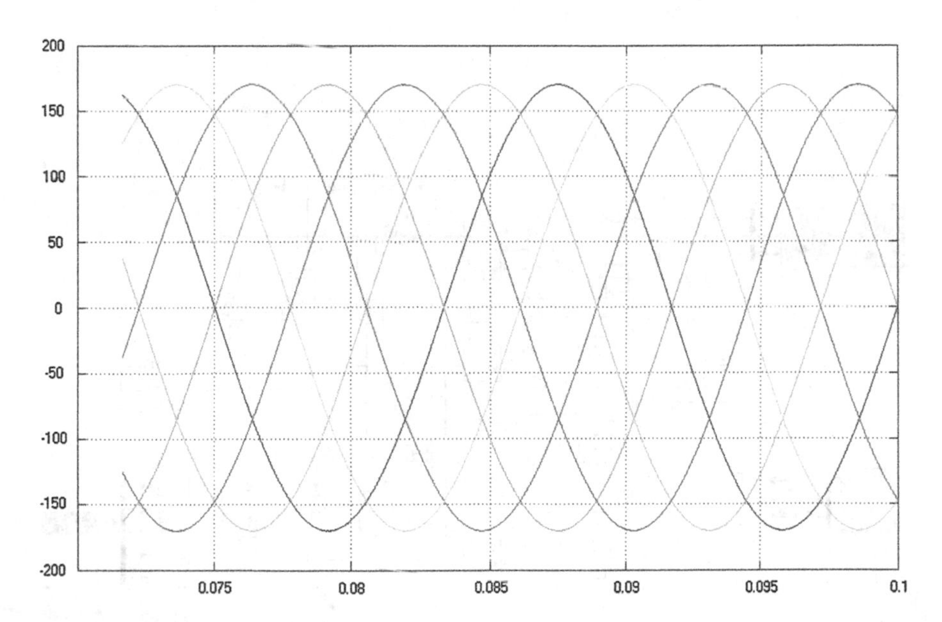

Fig. 7 Dimetrical connection for three-phase to six-phase transformers output

system will be observed to reduce the harmonic contents so as to increase the quality of power delivered to a inverter-fed synchronous motor drive system [6].

2.5.1 Simulation and Modeling Using Matlab

See Figs. 8, 9, 10, 11, 12, 13 and 14.

3 Comparison of 6 Pulse, 12 Pulse and 24 Pulse Converters Using PWM Techniques

Multipulse converter	Voltage harmonics without filter (THD) (%)	Voltage harmonics with filter (THD) (%)	Current harmonics without filter (THD) (%)	Current harmonics with filter (THD) (%)
6 pulse converter	59.47	37.89	143.21	65.99
12 pulse converter	27.03	20.83	61.38	40.47
24 pulse converter	0.31	0.10	38.13	0.30

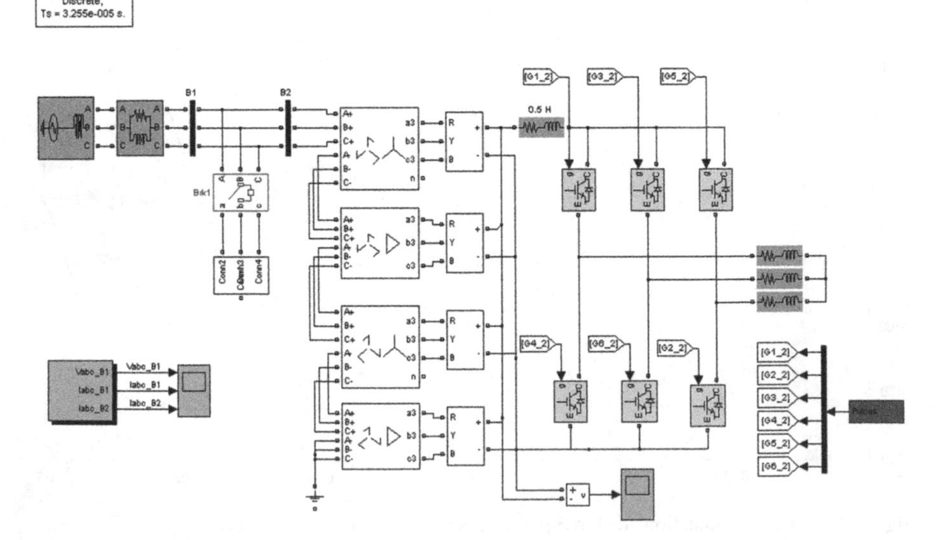

Fig. 8 Simulation circuit for 24-pulse converter

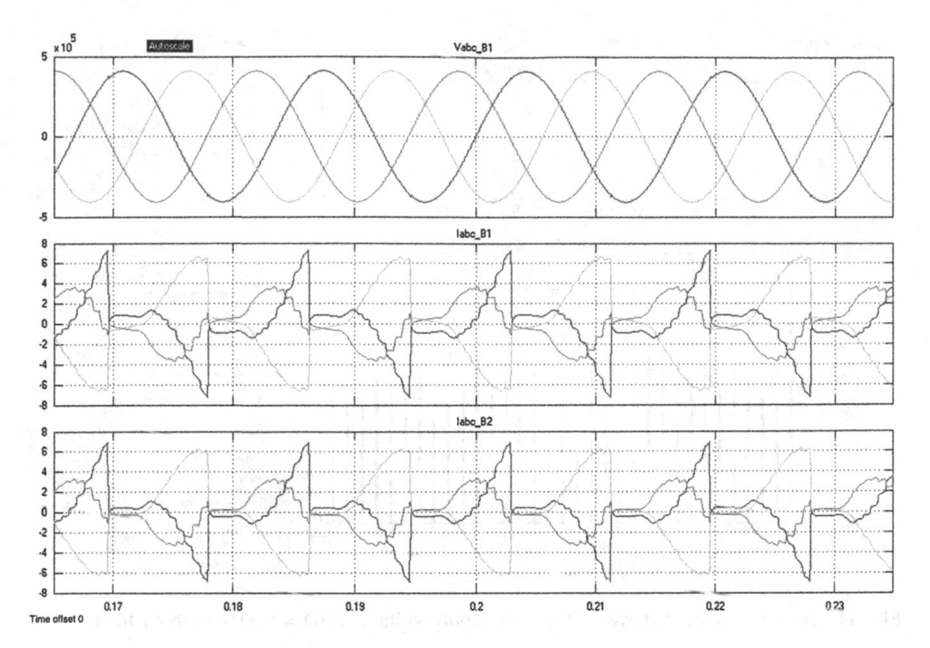

Fig. 9 Voltage and current waveform of three-phase AC supply without passive filter in 24-pulse converter

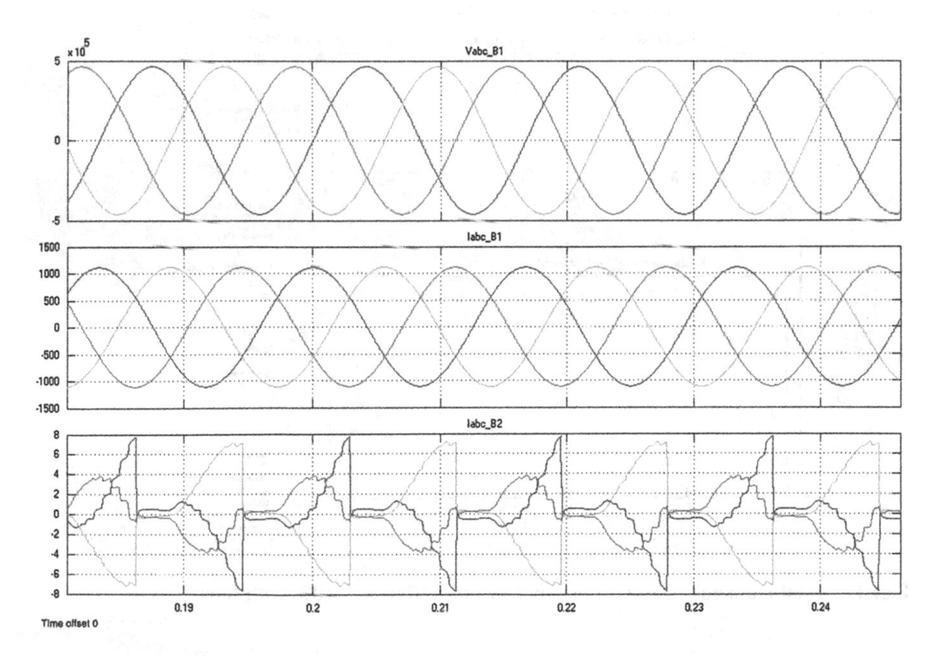

Fig. 10 Voltage and current waveform of three-phase AC supply with passive filter in 24-pulse converter

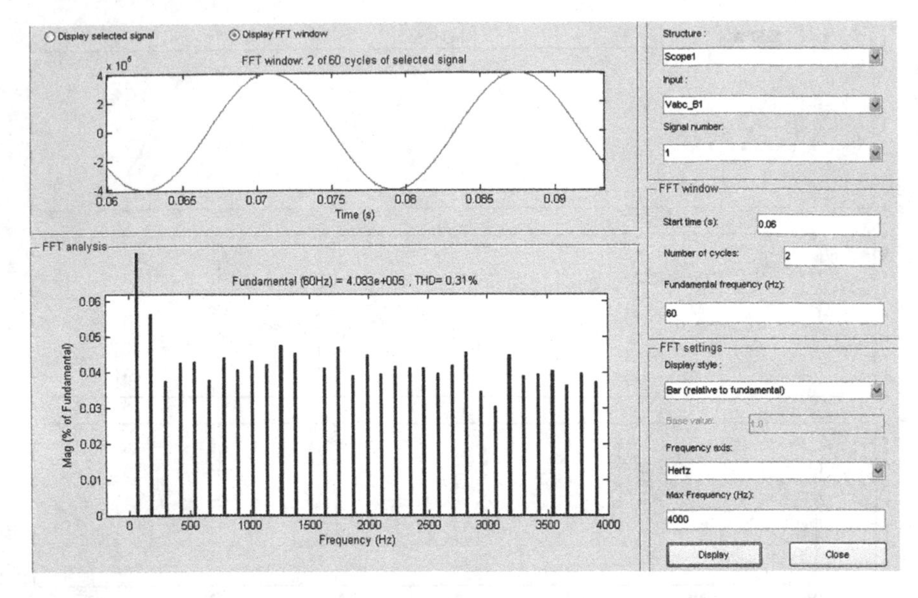

Fig. 11 Total harmonic distortion (THD) in supply voltage THD = 0.31% without filter

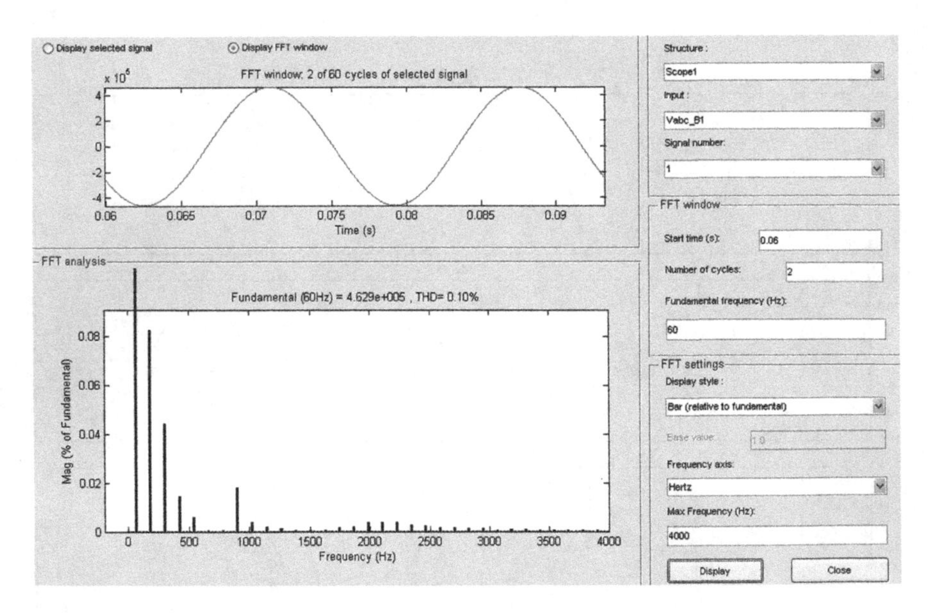

Fig. 12 Total harmonic distortion in supply voltage THD = 0.10% with filter

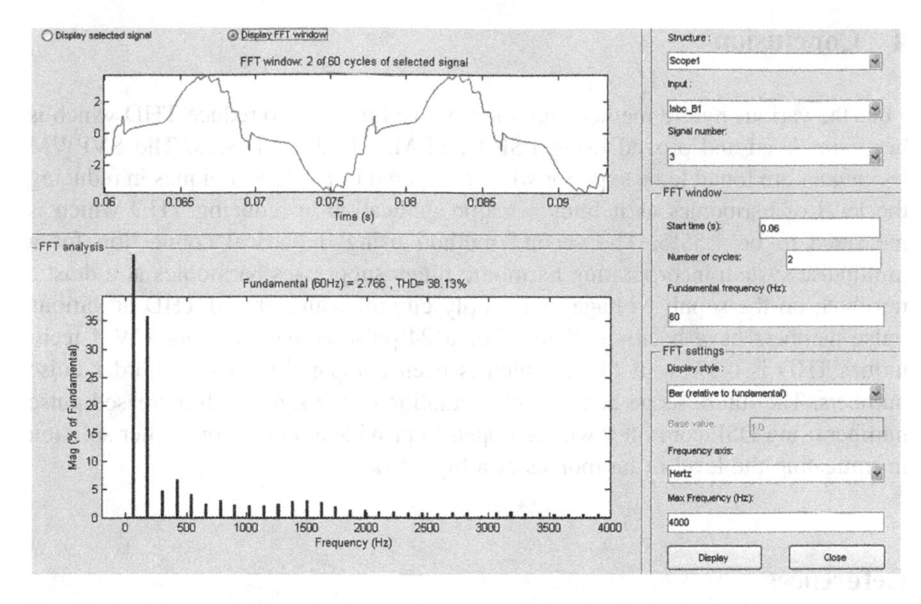

Fig. 13 Total harmonic distortion in supply current THD = 38.13% without filter

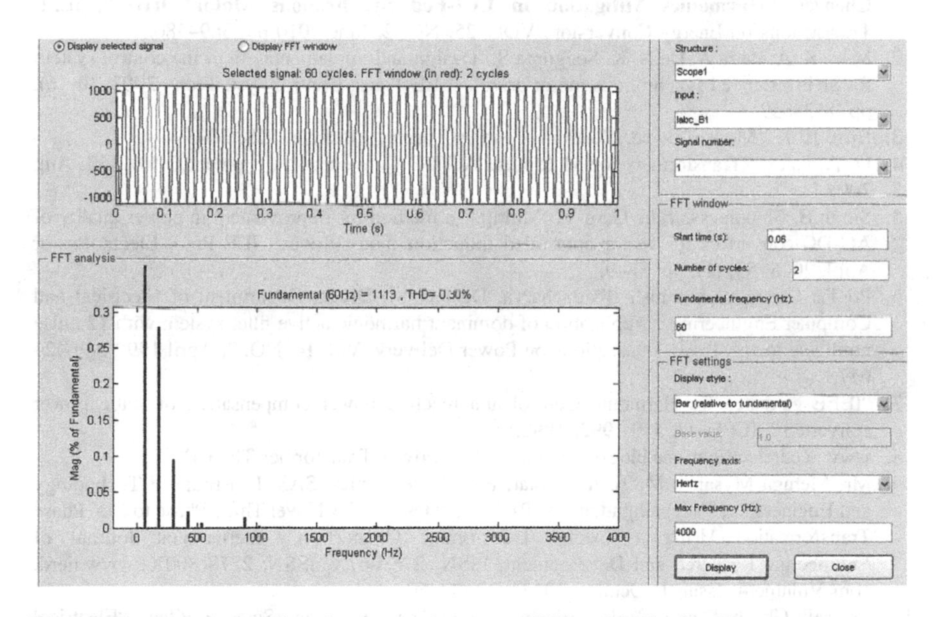

Fig. 14 Total harmonics distortion in supply current THD = 0.30% with filter

4 Conclusion

Thus the various hybrid methods are incorporated in order to reduce THD which is been simulated and proved using PSIM and MATLAB softwares. The SVPWM techniques are found to be superior when compared to PWM techniques in reducing the level of harmonics as it finds a major application in reducing THD which is measured to be 1.55%. The second method using dimetrical connection for a multipulse system incorporating harmonic filters suppresses harmonics at a drastic rate both on the supply voltage and supply current with reduced THD at various pulse numbers have been simulated. For a 24-pulse converter using PWM techniques THD is 0.1% with filters which is been compared with lower order pulse numbers. The future scope being implementation of the same with increased pulse number using DSP controller will be hoped to provide a much more better solution in mitigating the level of harmonics at a higher rate.

References

1. *Bhim Singh*, Fellow, IEEE, Sanjeev Singh, Student Member, IEEE, and S. P. Hemanth Chender, **"Harmonics Mitigation in LCI-Fed Synchronous Motor Drives",** IEEE Transactions on Energy Conversion, VOL. 25, NO. 2, June 2010 pp. 369–380.
2. Mary K. A, Patra A, De N. K, Sengupta S, 'Design and implementation of the control system for an inverter-fed synchronous motor drive', IEEE Trans. Control Syst. Tech., 2002, 10, (6), pp. 853–859.
3. Bose B. K, 'Modern power electronics and AC drives' (Pearson, US, 2003).
4. D. A. Paice, "Transformers for Multipulse AC/DC Converters", US patent 6 101 113, Aug 2008.
5. Singh B, Bhuvaneswari G, Garg V, 'Multipulse method for improvement in power quality of AC-DC convertors for vector controlled induction motor drives', IEE Proc. Electr. Power Appl, 2006, 53, (1), pp. 88–9.
6. Po-Tai Cheng, Subhashish Bhattacharya, Deepak M. Divan, Department of Electrical and Computer Engineering, "Application of dominant harmonic active filter system with 12 pulse nonlinear loads" IEEE Transactions on Power Delivery, Vol. 14, NO. 2, April 1999 pg 642–647.
7. "IEEE guidance for Harmonic Control and reactive power compensation of static power converters", IEEE std, 519-1992, 1992.
8. www.Yourelectrichrome.blogspot.com, "Label: Power Transformer Theory".
9. Mr. Merugu Mysaiah, M. Tech Scholar, Power Electronics, SAS, I Institute of Technology and Engineering, Tadepalligudem (A. P), India "Design of A Novel Three Phase to SIX Phase Transformation Using a Special Transformer Connection" International Journal of Engineering Research and Development, ISSN: 278 067X, ISSN: 2278 800X, www.ijerd.com Volume 4, Issue 1 (October 2012), PP. 39–50.
10. Smarajit Ghosh, Dept of EEE, Birla Institute of Technology and Science, Pilani, "Electrical Machines", Pearson Publication, Second edition 2008.
11. Villablanca M, Fichlmann W, Flores, Cuevar C, Armijo P, 'Harmonic reduction in adjustable speed synchronous motors', IEEE Trans. Energy Convers., 2001, 16, (3), pp. 239–245.

12. Neidhofer G. J, Troedson A. G, 'Large converter-fed synchronous motors for high speeds and adjustable speed operation: Design features and experience', IEEE Trans. Energy Convers, 1999, 14, (3), pp. 633–636.
13. Mohan N, Undeland T, Robbins W, 'Power electronics: Converters, Applications and Design' (Wiley, 1995, 2nd edn.).
14. Villablanca M, Abarca J, Cuevas C, Valencia A, Roias W, 'Adjustable speed synchronous motors, part I: system harmonic reduction', IEEE Trans. Ind. Appl., 1992, IA-28, (5), pp. 1072–1080.
15. Kim S, Enjeti P. N, Packbush P., Pitel I. J, 'A new approach to improve power factor and reduce harmonics in a three phase diode rectifier type utility interface', IEEE Trans. Ind. Appl, 1994, 30, (6), pp. 1557–1564.

Author Biographies

Mrs. Kavitha. G was born in the year 1979. She obtained B.E Electrical and Electronics Engineering from Madras University in 2001, and M.E (Applied Electronics) from Madras University in the year 2002. She is currently pursuing Ph.D in Power Electronics and drives in the Department of Electrical and Electronics Engineering at B.S. Abdur Rahman University. Her area of interests includes electrical machines and drives.

Dr I. Mohammed Rafeequdin B.E, M.Sc. (Engg.), Ph.D. is currently working as a visiting professor at B.S. Abdur Rahman University. He obtained his B.E Electrical and Electronics Engineering from College of Engineering, Guindy in the year 1967, and M.Sc. (Engg.) (Electrical Power systems) from College of Engineering, Guindy in the year 1969 and Ph. D from IIT Madras. He has worked in colleges such as Goa college of Engineering, College of Engineering Guindy, Delhi college of Engineering and University of Aden. He has got an overall teaching experience of 49 years. He has more than 50 international journal and international conference publications. His areas of research interests being in Electrical machines and power systems.

Palm Print and Palm Vein Biometric Authentication System

J. Ajay Siddharth, A.P. Hari Prabha, T.J. Srinivasan and N. Lalithamani

Abstract The modern computing technology has a huge dependence on biometrics to ensure strong personal authentication. The mode of this work is to increase accuracy with less data storage and providing high security authentication system using multimodal biometrics. The proposed biometric system uses two modalities, palm print and palm vein. The preprocessing steps begin with image acquisition of palm print and palm vein images using visible and infrared radiations, respectively. From the acquired image, region of interest (ROI) is extracted. The extracted information is encrypted using encryption algorithms. By this method of encryption, after ROI extraction, the storage of data consumes less memory and also provides faster access to the information. The encrypted data of both modalities are fused using advanced biohashing algorithm. At the verification stage, the image acquired is subjected to ROI extraction, encryption and biohashing procedures. The biohash code is matched with the information in database using matching algorithms, providing fast and accurate output. This approach will be feasible and very effective in biometric field.

Keywords Biometrics · Unimodal · Multimodal · Region of interest · Encryption

J. Ajay Siddharth (✉) · A.P. Hari Prabha · T.J. Srinivasan · N. Lalithamani
Department of Computer Science and Engineering, Amrita School of Engineering, Amrita Vishwa Vidyapeetham, Amrita University, Coimbatore, India
e-mail: ajaysiddharth23@gmail.com

A.P. Hari Prabha
e-mail: prabhaharini.16@gmail.com

T.J. Srinivasan
e-mail: srini.rs95@gmail.com

N. Lalithamani
e-mail: n_lalitha@cb.amrita.edu

S.S. Dash et al. (eds.), *Artificial Intelligence and Evolutionary Computations in Engineering Systems*, Advances in Intelligent Systems and Computing 517,
DOI 10.1007/978-981-10-3174-8_45

1 Introduction

Biometrics is the study of physical or behavioral characteristics of a person, in order to establish unique identity of an individual. There are many features such as face, iris, finger print, vein, hand geometry, voice, etc., that are used in biometrics. We have high uniqueness of physical characteristics which makes this field even more interesting. When a single feature alone is used to create an identity, then the system is called as unimodal biometric system. There are certain disadvantages in using unimodal system, like forging, creating fake identity, etc. In order to avoid this, a multimodal biometric system can be used, in which instead of using a single modality, the fusion of multiple features are used to create identity, which enhances the security. Biometrics has many advantages, it reduces the stress of memorizing passwords or carrying key cards around and also reduces administrative costs. Fusion at feature extraction level is most effective and hard to perform simultaneously (features collected from various identifiers must be independent and in same measurement scale, which would represent an identity in more discriminating feature space). Fusion at score level (combination of similarity scores by biometric matcher provides higher identification precision) has high precision. Decision-level fusion after each system performs its recognition at every stage.

In this work, we are using two features—palm vein and palm print. Palm is easy to be presented as a biometric system and we can take images of palm vein and palm print using single camera and a light controller, at a time. Then from the image, the region of interest is extracted and encrypted. Then the encrypted data of palm vein and palm print are used for further verification and identification process, which is stored in database.

2 Related Works

Biometrics is the technique of science which is used to identify a person using one's physiological or behavioral characteristics. The physiological traits include physical features, i.e., hand geometry, finger print or vein, palm print, face, iris, retina, etc., and behavioral traits include behavior features, i.e., gait, voice, signature, etc.

The basic purpose of biometrics is to provide security. Currently, unimodal analysis of palm print and palm vein has gained a lot of research efforts. Problems like noise in data, variations in class, degrees of freedom, and error rates make most of the unimodal methods to a restricted performance. Fusion of several modalities may be required for a robust identification system. Comparatively, multimodal system has got an upper hand over any of its individual components.

Requirement of palm print and palm vein simultaneously for authentication make the system more secure and gives higher protection against spoof attacks. Under visible light a palm print acquisition device acquires three different features.

- Principal lines: Three dominant lines on palm

- Wrinkles: More irregular and weaker lines
- Ridges: Patterns of raised skin similar to finger print patterns.

The wrinkles and principal lines can be acquired using a resolution about 100 dots per inch (100 dpi). Ridges features are hard to detect and hence they require higher resolution around 500 dots per inch (500 dpi). Taking cost effectiveness into consideration Charge-Coupled Device (CCD) captures low resolution images for systems based on palm prints. So far many feature extraction and matching algorithms has been proposed. Despite the great success achieved by the palm print recognition systems, there are some factors which still makes it not good enough as highly secured authentication system. Multimodal systems involving fusion of two different kinds of biometric data for the authentication process will be solution for the problem stated in unimodal palm print authentication systems.

Two kinds of traits are required by the system. So each trait is obtained by a unique sensor. More features such as palm vein can improve the anti-spoofing and distinctiveness capability of the system. The inner structure of the vessels present beneath the skin is referred to as the palm vein. The two ways to obtain palm vein images are

- Far Infrared (FIR)
- Near Infrared (NIR)

Intuitively, since both traits are obtained from a single organ, a convenient multimodal biometrics system can be established, which helps to use the two traits simultaneously and the system is made more accurate and robust to spoof attack with the help of the complementary information provided by the fusion of two traits. For personal recognition using the fusion of palm print and palm vein, number of studies have been conducted. Hence, in this work, we proceed with fusion of palm print and palm vein traits by providing a robust result from the information obtained, with a faster access and yields accurate output.

3 Proposed System for Authentication

For both palm print and palm vein, the two main process followed are Registration and Data Storage process and Matching and Verification process.

First, Image Acquisition process is carried out. In the Enrolment procedure, both palm print and palm vein images are given as inputs which is further segmented and stored in database. ROI extraction is performed. The ROI of an input image is obtained using Harris Corner point detection algorithm and the image is ready for further verification process. Then the image features are obtained using Gray scale and Gabor filter methods, where the precision between two input images are calculated. Palm print is verified by providing two input images and is processed, the Euclidean distance between two images are calculated and the status of the image is recognized. Further in the identification process of both palm print and palm vein,

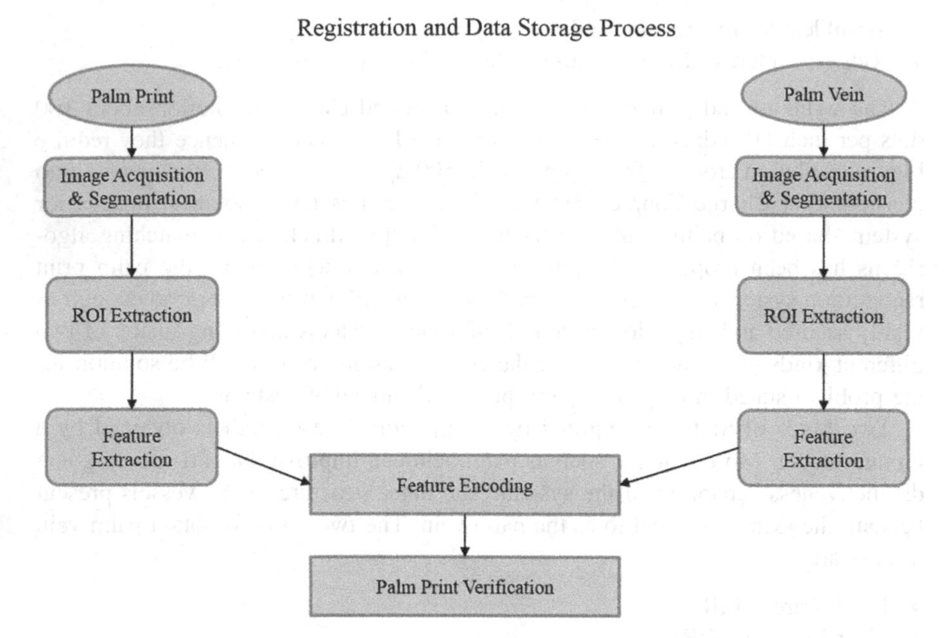

Fig. 1 Registration and data storage

the input image is verified with the already existing images of the database. If any image of the database have a calculated value matching, then the authentication is successful if not exists then its result is vice versa. The explanation about the proposed system is shown in Figs. 1 and 2.

4 Results

For the evaluation of methods described here, a database of palm print and palm veins are used with 28 images from 7 persons. Hence, image processing techniques are used. The images from palm print and palm vein are extracted where Image acquisition and Segmentation are done. The regions of interest are extracted, the region of interest of input image is subjected to feature encoding and Gabor filter process is performed to get feature extraction done. Once the features are extracted, the verification processes are done by comparing the image of the given sample with the dataset mutually. The verification procedure is considered as (1:1 match),

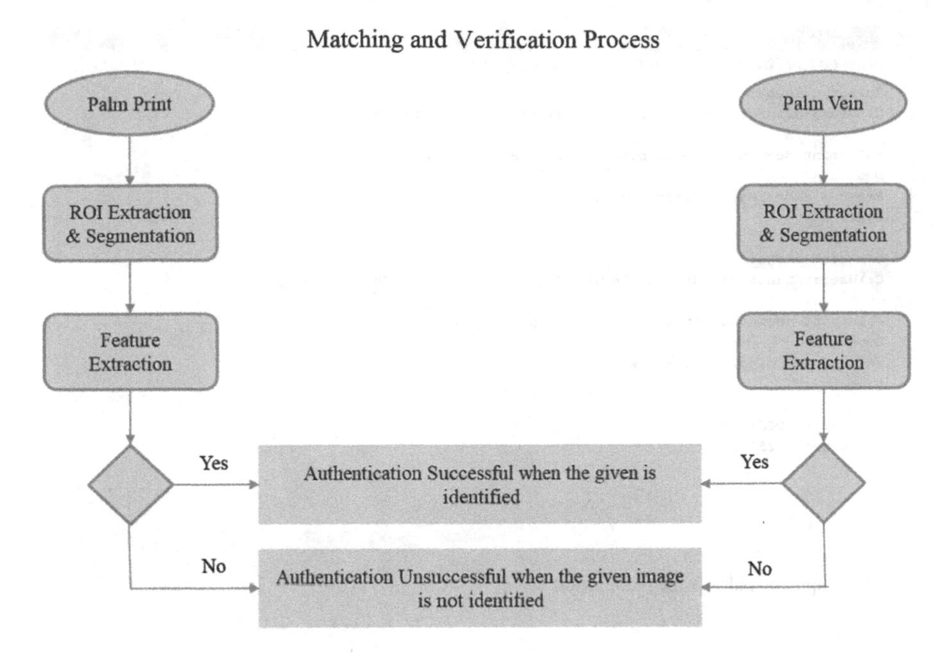

Fig. 2 Matching and verification

were input image is verified with another input image for a matching result. In the verification process, the Euclidean distance between both the images is calculated. The appropriate results of the distance lies between values 0 and 0.35 and were considered to be matching and hence access is authenticated. If the value exceeds the limit, then the status remains not recognized and the access gets denied. This scaling factor is applicable for both the test cases (i.e., for both the palm traits).

In the identification procedure, also called as (1:n match), where the input image is compared with the images that is already stored in the database for existence. If match is not found, it displays a value greater than the scaling factor. By these experimental results, the palm traits are identified and verified.

In Fig. 3, the output obtained after the verification process and the percentage error in the matching process is shown. Status = recognized indicates the successful verification of the palm print and Distance indicates the error percentage between the image in the database and the matching palm print.

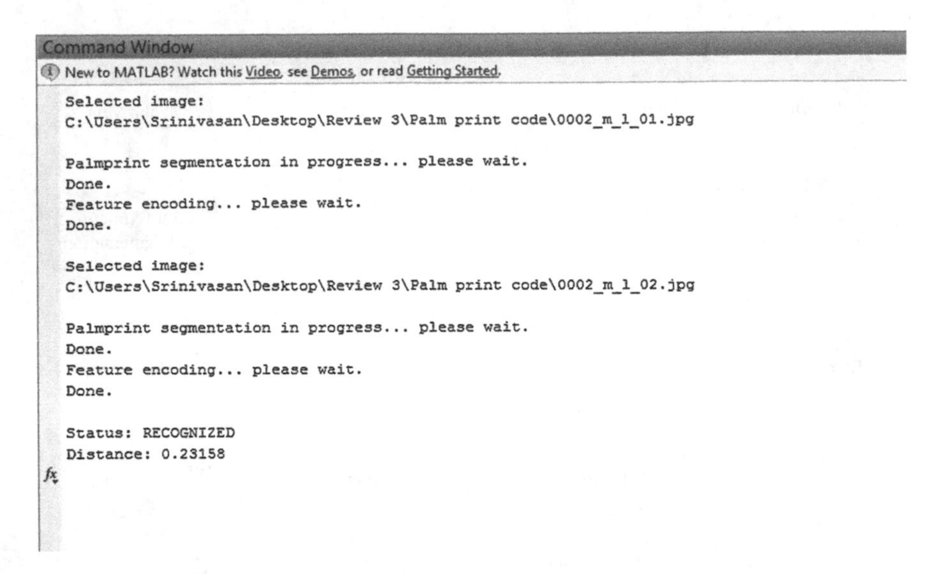

Fig. 3 Output of palm print recognition

5 Conclusions

In this work, we have discussed about creating biometric system where the main topic discussed here is about acquiring biometric data, extracting the information from images, storage and protection of biometric information, and their comparison. Major algorithms used are Harris corner, Gabor filter, and Binary masking. This biometric system helps us to resolve many security issues faster and also helps to identify the duplication of traits with higher precision level. By this process, illegal activities from hackers can be limited to an extent.

6 Future Works

Based on the experimental results of palm print and palm vein, we are planning to design a delay trap authentication system, which will be helpful for private sectors from unauthorized intruders. The delay trap authentication works on the principle where it traps a particular unauthorized person inside and also triggers a security alert to all authorized persons. The trap gets de-activated only when the authorized person performs a proper authentication process. These applications are very helpful for public and private sector works where the data should be highly secured.

References

1. Z. Sun, T. Tan, Y. Wang, S.Z. Li, Ordinal palm print representation for Personal identification, in: Proceedings of IEEE International Conference on Computer Vision and Pattern Recognition, (2005), pp. 279–284.
2. W.K. Kong, D. Zhang, Competitive coding scheme for palm print verification, in: Proceedings of the 17th International Conference on Pattern Recognition, (2004), pp. 520–523.
3. D. Zhang, W.K. Kong, J. You, M. Wong, Online palm print identification, IEEE Trans. Pattern Anal. Mach. Intell. 25 (2) (2003) 1041–1050.
4. W.K. Kong, D. Zhang, Feature-level fusion for effective palm print authentication, in: Proceedings of the First International Conference on Biometric Authentication, Lecture Notes in Computer Science, vol. 3072, (2004), pp. 761–767.
5. Fujitsu-Laboratories-Ltd, Fujitsu Laboratories Develops Technology for World's First Contactless Palm Vein Pattern Biometric Authentication System, (2003).
6. L. Wang, G. Leedham, A thermal hand vein pattern verification system, in: Proceedings of International Conference on Advanced Pattern Recognition, Lecture Notes in Computer Science, vol. 3687, (2005), pp. 58–65.
7. L. Hong, R. Bolle, On-line fingerprint verification, IEEE Transactions on Patten Analysis and Machine Intelligence 19 (4) (1997) 302–313.
8. L. Hong, A. Jain, Integrating faces and fingerprints for personal identification, IEEE Transactions on Pattern Analysis and Machine Intelligence 20 (12) (1998) 1295–1307.
9. A.K. Hrechak, J.A. McHugh, Automated fingerprint recognition using structural matching, Pattern Recognition 23 (8) (1990) 893–904.
10. A.K. Jain, S. Prabhakar, L. Hong, Filterbank-based fingerprint matching, IEEE Transactions Image Processing 9 (5) (2000) 846–859.
11. W. Jia, D.S. Huang, D. Zhang, Palmprint verification based on robust line orientation code, Pattern Recognition 41 (5) (2008) 1504–1513.
12. S. Ben-Yacoub, Y. Abdeljaoued, E. Mayoraz, Fusion of face and speech data for person identity verification, IEEE Transactions on Neural Networks 10 (5) (1995) 1065–1074.
13. M. Faundez-Zanuy, Data fusion in biometrics, Aerospace and Electronic Systems Magazine, IEEE 20 (1) (2005) 34–38.
14. R. Fuksis, A. Kadikis, M. Greitans, Biohashing and fusion of palmprint and palm vein biometric data, in: International Conference on Hand-Based Biometrics (ICHB), (2011), pp. 1–6.
15. A. Goshtasby, S. Nikolov, Image fusion: advances in the state of the art, Information Fusion 8 (2) (2007) 114–118.
16. Z.H. Guo, D. Zhang, L. Zhang, W.M. Zuo, Palmprint verification using binary orientation co-occurrence vector, Pattern Recognition Letters 30 (13) (2009) 1219–1227.
17. Y. Hao, Z. Sun, T. Tan, C. Ren, Multispectral palm image fusion for accurate contact-free palmprint recognition, in: IEEE International Conference on, Image Processing (ICIP), (2008), pp. 281–284.
18. K. Nandakumar, Y. Chen, S.C. Dass, A. Jain, Likelihood ratio-based bio-metric score fusion, IEEE Trans. Pattern Anal. Mach. Intell. 30 (2) (2008) 342–347.
19. L. Nanni, A. Lumini, S. Brahnam, Likelihood ratio based features for a trained biometric score fusion, Expert Sys. Appl. 38 (1) (2011) 58–63.
20. K.A. Toh, H.L. Eng, Y.S. Choo, Y.L. Cha, W.Y. Yau, & K.S. Low, "Identity verification through palm vein and crease texture," International Conference on Biometrics, (2005), pp. 546–553.

15. [illegible]

Study of Different Boundary Constraint Handling Schemes in Interior Search Algorithm

Indrajit N. Trivedi, Amir H. Gandomi, Pradeep Jangir and Narottam Jangir

Abstract Real-world problems usually have bounded search space and, therefore, the performance of optimization algorithms on them is also related to choose a proper boundary constraint handling methods. There are several classical approaches in the literature to handle the bounds. Usually, optimization algorithm use does not pay attentions to choose a proper boundary constraint handling scheme. In this study different boundary constraint handling schemes such as evolutionary scheme and classical schemes (including reflecting scheme, absorbing scheme, toroidal scheme, and random scheme) are evaluated on a recent evolutionary algorithm called interior search algorithm (ISA). In this paper, all the boundary constraint handling approaches have been adopted to ISA to solve a wide set of global numerical benchmark problems. The conclusions are made based on statistical results which clearly show the importance of different boundary constraint handlings in the searching process. The results obtained using the evolutionary boundary constraint handling scheme are better than the ones obtained by the other well-known approaches and it seems this scheme is suitable for a wider range of evolutionary optimization problems with very good convergence rate.

Keywords Classic scheme · Boundary constraint · Interior search algorithm · Benchmark · Evolutionary scheme · BCH

I.N. Trivedi (✉)
Department of Electrical Engineering, G.E. College, Gandhinagar, Gujarat, India
e-mail: forumtrivedi@gmail.com

A.H. Gandomi
BEACON Center for the Study of Evolution in Action, Michigan State University, East Lansing, MI 48824, USA
e-mail: a.h.gandomi@gmail.com

P. Jangir · N. Jangir
Department of Electrical Engineering, L. E. College, Morbi, Gujarat, India
e-mail: pkjmtech@gmail.com

N. Jangir
e-mail: nkjmtech@gmail.com

S.S. Dash et al. (eds.), *Artificial Intelligence and Evolutionary Computations in Engineering Systems*, Advances in Intelligent Systems and Computing 517,
DOI 10.1007/978-981-10-3174-8_46

1 Introduction

Recently, metaheuristic optimization algorithms have been developed in many areas such as hybridization, multi-objective version, binary version, training multi-layer perceptron, and improved several ways as levy flight, chaos theory, and genetic operator. Most of these improvements happened because they uses both deterministic and evolutionary components [1]. A good combination of global and local search has intensive local exploration and global exploration [2].

In optimization, a very good algorithm should also be able to deal with limits and therefore adopting the best boundary constraint handling (BCH) scheme from the current schemes (Reflect scheme, Random scheme, Toroidal scheme, absorbing scheme and evolutionary schemes) and also trying different penalty handling methods (Deb penalty, static penalty, dynamic penalty and MQM penalty) are very important for each optimization algorithm. In the optimization algorithm if boundary constraint handling and penalty handling are incorrect then optimization algorithm can give wrong answers. Therefore, dealing with the limits such as bound limits is very important [3] and most researcher have motivation to find out the best boundary constraint handling scheme and penalty method to have the most efficient algorithm.

Huang and Mohan [4] evolved a scheme for BCH in particle swarm optimization algorithm (PSO), called damping scheme. Damping method is same as well-known reflecting scheme except which amount of reflect is randomly determined. Xu and Rahmat-Samii [5] devolved some hybrid schemes for BCH in PSO algorithm. Chen and Chi [6] devolved boundary constraint handling (BCH) in PSO algorithm by step by step reducing the velocity. Chu et al. [7] and Helwig et al. [8] did experimental analysis of bound handling techniques in PSO algorithm. Gandomi and Yang [9] proposed evolutionary BCH (EBCH) and compared it with random, reflecting and absorbing schemes for benchmark optimization problems. EBCH have been successfully used in some studies. In a recent study performance of Cuckoo Search Algorithm [10] is improved by using EBCH. Among BCHs, EBCH performs better result and faster convergence in general.

In this paper, we have used different schemes for BCH in a recent and robust metaheuristic. For comparison of different boundary handling schemes and validations of result, we study of Reflect, Random, Toroidal, Absorbing and Evolutionary schemes for BCH in Interior Search Algorithm (ISA) [11]. The paper is divided into the following sections: First, we briefly review the interior search algorithm, and then we review of different BCH schemes. After then explain in detail the reflect, random, toroidal, absorbing, and evolutionary schemes. Finally, we compare reflect, random, toroidal, absorbing, and evolutionary approaches of BCH with test benchmark problems using interior search algorithm. Result and simulations suggest that the optimal solutions are obtained by evolutionary scheme as a BCH method. EBCH results are better than the best solutions obtained by the reflect, random, toroidal, and absorbing schemes and clearly outperform these schemes.

2 Interior Search Algorithm

ISA is a combined optimization analysis divine to the creative work or art relevance to interior or internal designing [11]. It consists of two stages: first stage is composition stage where a number of solutions are shifted toward to get optimum fitness, and second stage is reflector or mirror inspection method where mirror is placed in the middle of every solution and best solution to yield a fancy view [11] to design, satisfying all control variables to constrained design problem.

2.1 Algorithm Description [11]

1. However, the position of acquired solution should be in the limitation of maximum bound and minimum bounds, and later to estimate their fitness amount.
2. Evaluating the best value of solution, the fittest solution has maximum objective functions whenever aim of optimization problem is minimization and vice versa is always true. Solution has universally best in jth run (iteration).
3. Remaining solutions are collected into two categories: mirror and composition elements in respect to a control parameter α. Elements are categorized based on the value of random number (all used in this paper) ranging from 0 to 1. If $\text{rand}_1()$ is less than equal to α, it moves to mirror category else moves toward composition category. For avoiding problems α must be carefully tuned.
4. Being composition category elements, every element or solution is, however, transformed as described below in the limited uncertain search space:

$$X_i^j = \text{lb}^j + \left(\text{ub}^j - \text{lb}^j\right) * r_2, \tag{1}$$

   where x_i^j represents ith solution in jth run, ub^j, lb^j upper and lower range in jth run, whereas its maximum and minimum values for all elements exist $(j-1)$th run and $\text{rand}_2()$ ranging from 0 to 1.
5. For ith solution in jth run spot of mirror is described as

$$X_{m,i}^j = r_2 X_i^{j-1} + (1 - r_3) * X_{gb}^j, \tag{2}$$

   where $\text{rand}_3()$ ranging from 0 to 1. Imaginary position of solutions is dependent on the spot where mirror is situated defined as

$$X_i^j = 2X_{m,i}^j - X_i^{j-1}. \quad (3)$$

6. It is auspicious for universally best to little movement in its position using uncertain walk defined:

$$X_{gb}^j = X_{gb}^{j-1} + r_n * \lambda \quad (4)$$

where r_n vector of distributed random numbers having same dimension of x, λ = (0.01*(ub − lb)) scale vector, dependable on search space size.

7. Evaluate fitness amount of new position of elements and for its virtual images. If its fitness value is enhanced, then position should be updated for next design. For minimization optimization problem updatings are as follows:

$$X_i^j = \begin{Bmatrix} X_i^j & f(X_i^j) < f\left(X_i^{j-1}\right) \\ X_i^{j-1} & \text{Else} \end{Bmatrix}. \quad (5)$$

8. If termination condition is not fulfilled, again evaluate from second step.
   A. Parameter tuning
   A curious component in algorithm [11] is α. For unconstrained benchmark test function, it can be fixed as 0.25, but requirement is to increase its value ranging from 0 to 1 randomly as increment in maximum number of runs selected for particular problem. It requires to shift search emphases from exploration stage to exploitation optimum solution toward termination of maximum iteration.

2.2 *Pseudocode of Algorithm [11]*

Initialization
while any stop criteria are not satisfied

  find the x_{gb}^j
  for i = 1 to n

  if x_{gb}
  Apply Eq. $X_{gb}^j = X_{gb}^{j-1} + r_n * \lambda$

else if $r_1 > \alpha$

Apply Eq. $X_i^j = lb^j + (ub^j - lb^j) * r_2$

Else

Apply Eq. $X_{m,i}^j = r_2 X_i^{j-1} + (1 - r_3) * X_{gb}^j$

Apply Eq. $X_i^j = 2X_{m,i}^j - X_i^{j-1}$

end if

Check the boundaries except for decomposition elements.

end for

for i = 1 to n

Evaluate $f(x_i^j)$

Apply Eq. $X_i^j = \left\{ \begin{array}{ll} X_i^j & f(X_i^j) < f\left(X_i^{j-1}\right) \\ X_i^{j-1} & Else \end{array} \right\}$

end for

end while

3 Different Types of Boundary Constraint Handling Schemes

3.1 Absorbing Scheme

Absorbing scheme is one of the most popular used schemes for BCH to avoid violating upper and lower limits of variables. Absorbing scheme can be mathematically represented as

$$f(Z_i \to X_i) = \left\{ \begin{array}{l} \mathrm{lb}_i \to \text{if} \to Z_i < \mathrm{lb}_i \\ \mathrm{ub}_i \to \text{if} \to Z_i > \mathrm{ub}_i \end{array} \right\}, \tag{6}$$

where lb_i and ub_i are the ith upper bound and lower bound, respectively, and z_i and x_i are the out of space variable and corrected variable, respectively.

3.2 Random Scheme

Random scheme is another popular BCH scheme used in many researches to avoid violating upper or lower limits of variables. If upper and lower limits of variables are outside of domain, then random scheme replaces that violated solutions with a random number. Random scheme can be mathematically represented as

$$f(Z_i \rightarrow X_i) = \mathrm{lb}_i + \mathrm{rand} * (\mathrm{ub}_i - \mathrm{lb}_i) \tag{7}$$

$$\text{if} \rightarrow Z_i < \mathrm{lb}_i \quad \text{or} \quad Z_i > \mathrm{ub}_i, \tag{8}$$

where r_1 is a random number between 0 and 1.

3.3 Reflect Scheme

Reflect scheme is another scheme for BCH that avoids violating upper and lower limits of variables. When any variable goes outside of its upper or lower bounds, then in reflect scheme that variable can be replaced with a mirror image related to the boundary. Reflect scheme can be mathematically represented as

$$f(Z_i \rightarrow X_i) = \left\{ \begin{array}{c} \mathrm{lb}_i - (Z_i - \mathrm{lb}_i) \rightarrow \text{if} \rightarrow Z_i < \mathrm{lb}_i \\ \mathrm{ub}_i - (Z_i - \mathrm{ub}_i) \rightarrow \text{if} \rightarrow Z_i > \mathrm{ub}_i \end{array} \right\}. \tag{9}$$

3.4 Toroidal Scheme

Another BCH scheme is toroidal scheme which connected upper limit with lower limit; therefore, the variable extends over upper limit or lower limit, and it re-enters search space through another limit. Toroidal scheme can be mathematically represented as

$$f(Z_i \rightarrow X_i) = \left\{ \begin{array}{c} \mathrm{lb}_i + (Z_i - \mathrm{lb}_i) \rightarrow \text{if} \rightarrow Z_i < \mathrm{lb}_i \\ \mathrm{ub}_i + (Z_i - \mathrm{ub}_i) \rightarrow \text{if} \rightarrow Z_i > \mathrm{ub}_i \end{array} \right\} \tag{10}$$

3.5 *Evolutionary Scheme*

Evolutionary scheme is recently proposed by Gandomi and Yang [9]. It seems this is one of the most robust schemes for BCH. When any variables go outside of its upper and lower bounds, then it is replaced with a random variable between its upper and lower limit and similar component of the best solution so far. So this best value has important for generate current solution and update information about search domain and solutions. Evolutionary scheme very simple can be mathematically represented as

$$f(Z_i \rightarrow X_i) = \left\{ \begin{array}{c} \alpha * \text{lb}_i + (1-\alpha)X_i^b \rightarrow \text{if} \rightarrow Z_i < \text{lb}_i \\ \beta * \text{ub}_i + (1-\beta)X_i^b \rightarrow \text{if} \rightarrow Z_i > \text{ub}_i \end{array} \right\}, \quad (11)$$

where X_i^b is the related component of the best solution, and α and β are random numbers between 0 and 1. The use of mutation and the current, evolving best solution, makes the scheme having the "evolutionary" nature, thus the name of the scheme. Evolutionary scheme for constrained handling scheme does not require memory and easily used with any algorithm methods with current solution that from purely random solution that can be improved by convergence rate in benchmark function.

4 Numerical Examples

In this section, we have applied ISA with different BCH schemes for optimization of numerical benchmark problems that can be found in Appendix 1. Convergence plot and statistical results of ISA with different BCH schemes are presented in Fig. 1 and Tables 1, 2 and 3. For simulation of result, we set number of iterations equal to 500, population size equal to 25, and each BCH runs 10 times for each benchmark optimization problem. Results obtained for the case study F1–F16 using different BCH schemes show that evolutionary scheme can perform better compared to all other BCH schemes.

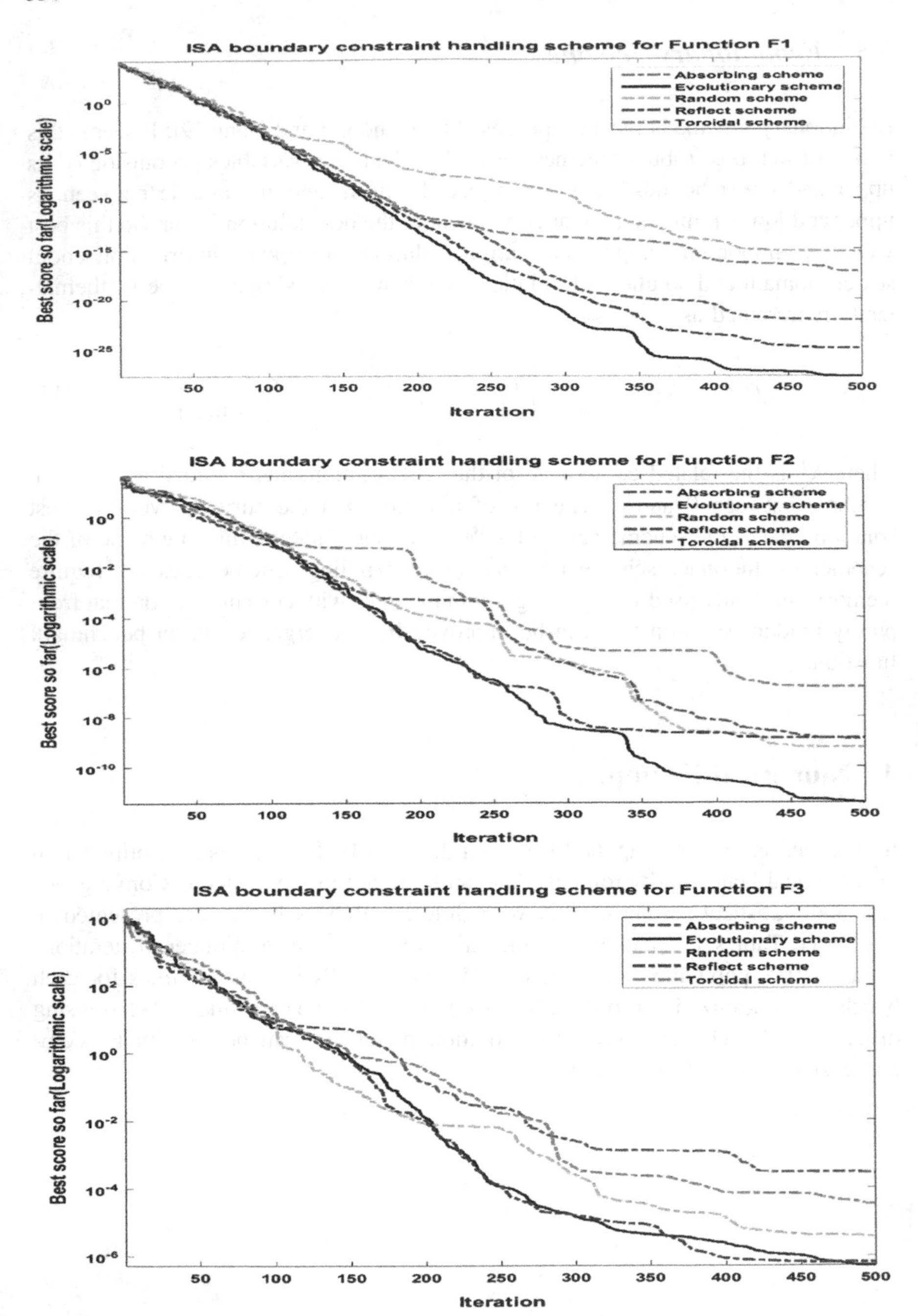

Fig. 1 Best score versus iterations for the different boundary constraint handling schemes

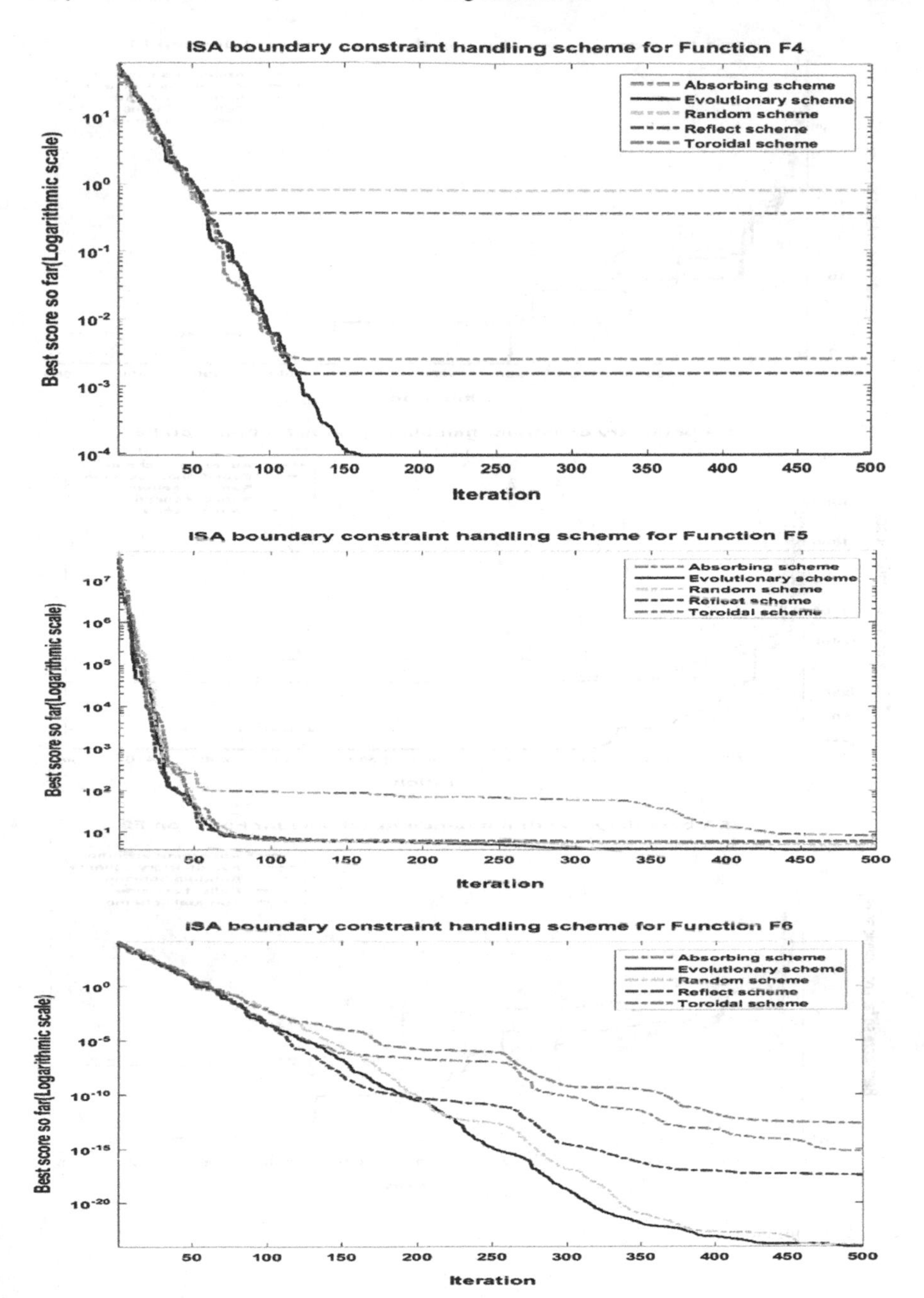

Fig. 1 (continued)

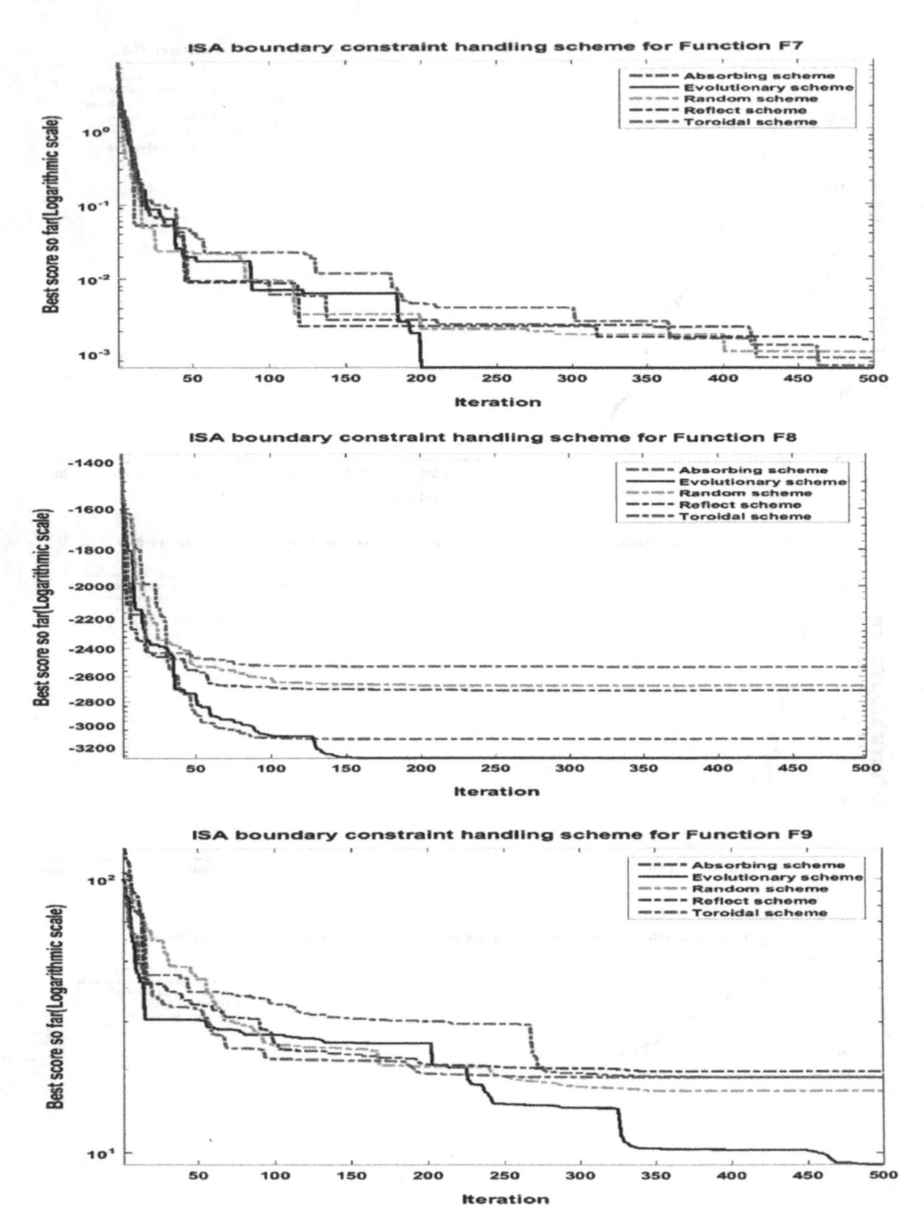

Fig. 1 (continued)

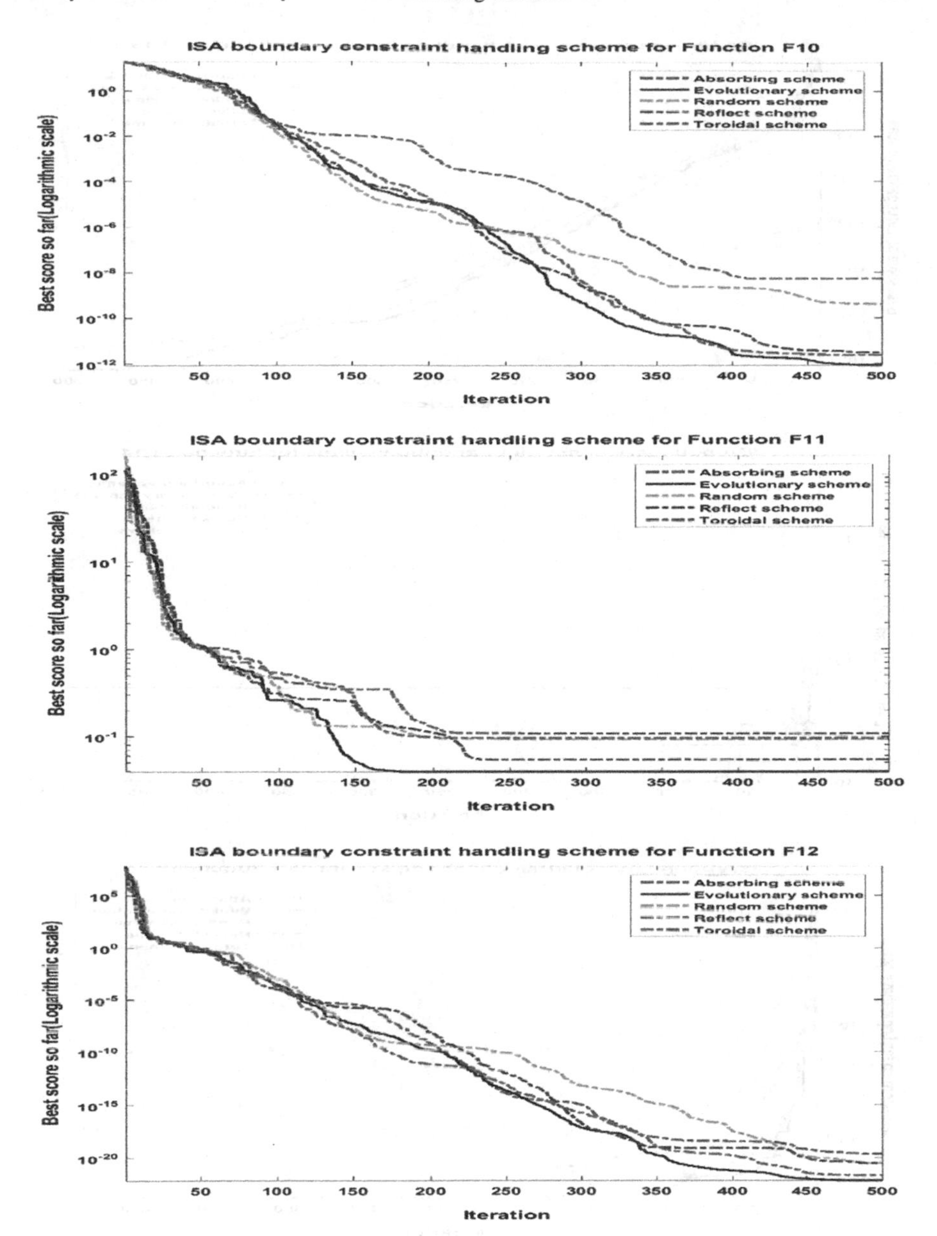

Fig. 1 (continued)

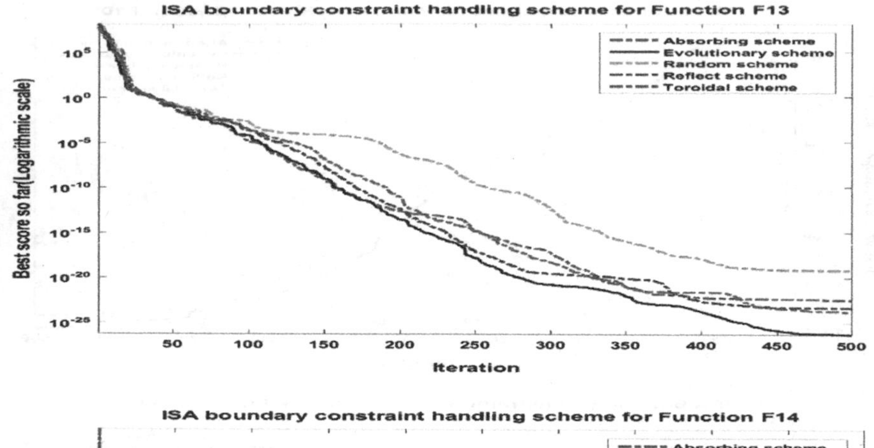

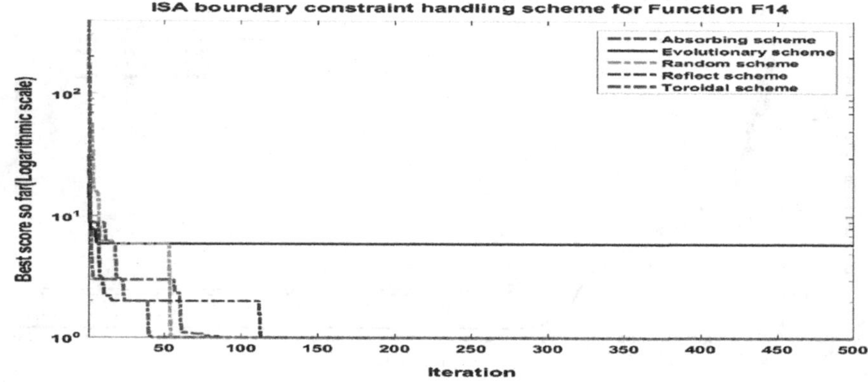

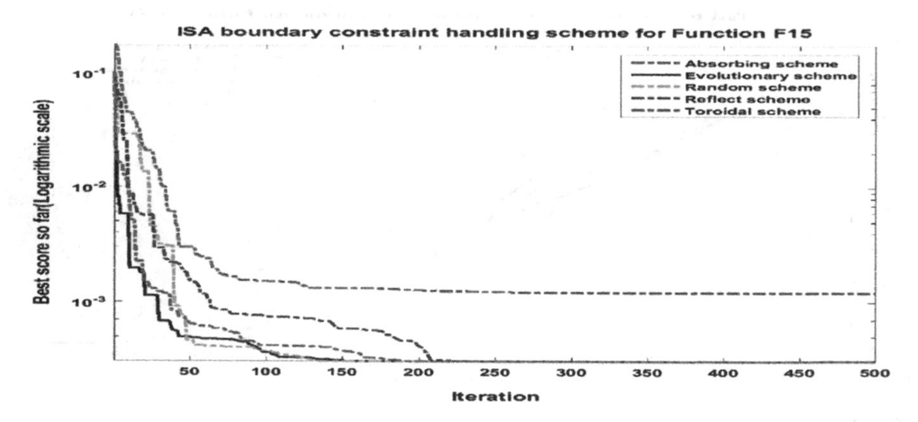

Fig. 1 (continued)

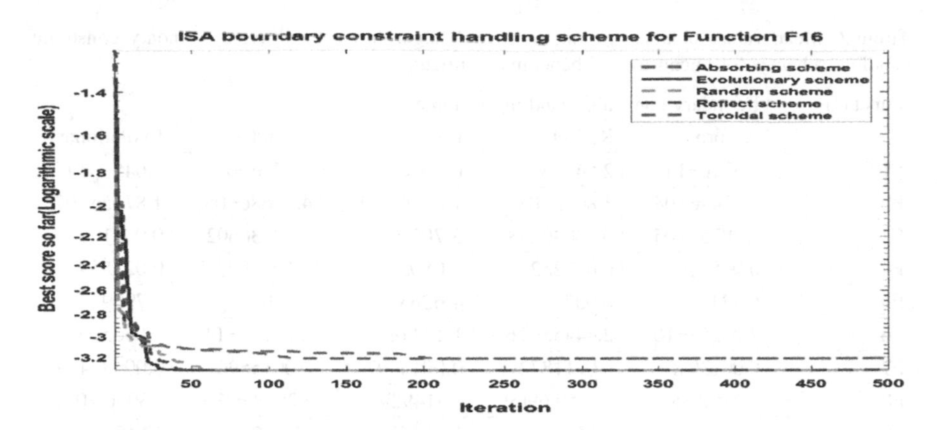

Fig. 1 (continued)

Table 1 Best values obtained by interior search algorithm with different boundary constraint handling schemes for unconstrained benchmark function

Function ID	Boundary constraint handling scheme				
	Absorb	Reflect	Random	Toroidal	Evolutionary
F1	7.152e−23	1.2734e−25	7.3669e−26	5.1687e−26	2.0734e−28
F2	1.5839e−11	5.409e−12	1.262e−12	1.1424e−12	3.5542e−12
F3	4.9222e−07	4.7183e−07	1.3286e−07	1.6685e−06	3.7813e−08
F4	2.1638e−07	3.5848e−05	7.4366e−05	1.1865e−05	3.1345e−05
F5	4.4925	4.6502	0.5528	4.6278	3.9452
F6	6.4282e−28	1.2876e−23	6.0325e−25	5.1436e−28	1.3188e−25
F7	0.00061927	0.00033212	0.00028329	0.00033774	0.00032617
F8	−3380.3839	−3597.6043	−4189.8289	−3535.0649	−3832.9968
F9	7.9597	6.0177	6.9647	7.9597	9.0625
F10	9.3259e−14	7.1498e−13	1.2169e−11	8.8551e−13	4.3521e−14
F11	0.017241	0.01969	0.0098573	2.5878e−09	0.017236
F12	3.9562e−27	3.8723e−28	2.8736e−26	4.4605e−26	1.4545e−24
F13	5.8979e−26	1.7979e−27	8.3884e−27	2.3364e−25	5.8664e−27
F14	0.998	0.998	0.998	0.998	0.998
F15	0.00030749	0.00030749	0.00030749	0.00030749	0.00030749
F16	−3.322	−3.322	−3.322	−3.322	−3.322

Table 2 Mean values obtained by interior search algorithm with different boundary constraint handling schemes for unconstrained benchmark function

Function ID	Boundary constraint handling scheme				
	Absorb	Reflect	Random	Toroidal	Evolutionary
F1	6.572e−14	2.6471e−14	6.093e−15	5.174e−08	7.045e−14
F2	2.4744e−08	3.807e−08	7.6726e−09	4.9868e−08	1.8735e−07
F3	7.7875e−05	0.00030298	3.707e−05	0.00030602	0.0001061
F4	0.045129	0.062252	0.080578	0.10459	0.023559
F5	6.4241	6.0074	4.6263	6.3603	5.7909
F6	2.2135e−16	2.4445e−16	1.1311e−14	2.2471e−14	1.8665e−12
F7	0.0013945	0.0011872	0.0013497	0.0015596	0.00087439
F8	−2953.5489	−2959.6969	−3149.3871	−2878.6436	−3056.4097
F9	16.2179	18.5113	16.5163	17.0325	17.6216
F10	2.6449e−09	2.7542e−08	2.8245e−09	5.618e−10	0.11551
F11	0.073077	0.067182	0.083003	0.081673	0.069612
F12	2.2427e−19	4.0744e−16	4.8947e−16	9.1448e−16	6.0209e−13
F13	8.8115e−12	3.3931e−19	1.2798e−15	2.6318e−20	0.0010987
F14	0.998	0.998	0.998	1.9842	1.4911
F15	0.00039905	0.00030749	0.00030749	0.00030749	0.00039905
F16	−3.2744	−3.3101	−3.3101	−3.2744	−3.3101

Table 3 Median values obtained by interior search algorithm with different boundary constraint handling schemes for unconstrained benchmark function

Function ID	Boundary constraint handling scheme				
	Absorb	Reflect	Random	Toroidal	Evolutionary
F1	3.4454e−19	5.2109e−21	1.9628e−21	4.2618e−20	1.7264e−20
F2	2.8689e−10	8.977e−10	5.0897e−10	6.9895e−10	1.8735e−07
F3	3.6805e−05	2.457e−05	5.5038e−06	2.866e−05	4.8289e−05
F4	0.0015757	0.0039122	0.0009562	0.0016597	0.00017871
F5	6.1648	6.2395	5.0655	5.8625	5.2337
F6	9.9532e−23	9.4761e−19	4.0801e−22	2.1171e−20	1.595e−19
F7	0.0012492	0.001183	0.0011857	0.0014751	0.000797
F8	−2975.8059	−2885.4659	−2995.4936	−2788.1827	−2995.4887
F9	14.9244	18.9042	17.4118	17.4119	16.4168
F10	3.759e−10	6.444e−10	4.5415e−10	2.3204e−11	2.9232e−10
F11	0.070076	0.061574	0.086078	0.084913	0.049188
F12	7.6774e−22	9.486e−20	1.1092e−19	4.2317e−20	3.0912e−17
F13	8.4607e−20	2.8582e−23	1.146e−21	4.3239e−23	5.454e−23
F14	0.998	0.998	0.998	0.998	0.998
F15	0.00030749	0.00030749	0.00030749	0.00030749	0.00030749
F16	−3.322	−3.322	−3.322	−3.322	−3.322

5 Conclusion

In this study, reflect, random, toroidal, absorbing, and evolutionary schemes have been compared for boundary constraint handling in interior search algorithm to solve benchmark optimization problems. The evolutionary scheme performs better when compared with other schemes. Evolutionary boundary constraint handles highly suitable for most optimization problems with improving convergence property. In addition, evolutionary scheme is very simple and can be easily implemented with any optimization algorithm.

For future scope, we are used to evolutionary scheme for boundary constraint handling with different recent devolved and efficient algorithms because this scheme does not require any memory, less parameter, and adoptable with all optimization algorithm.

Appendix 1

F1: Sphere Function
n (dimension): 10, lb: −50, ub: 100, fmin: 0

$$f(x) = \sum_{i=1}^{n} x_i^2 * R(x)$$

F2: Schwefel 2.22 Function
n (dimension): 10, lb: −5, ub: 10, fmin: 0

$$f(x) = \sum_{i=1}^{n} |x_i| + \prod_{i=1}^{n} |x_i| * R(x)$$

F3: Schwefel 1.2 Function
n (dimension): 10, lb: −50, ub: 100, fmin: 0

$$f(x) = \sum_{i=1}^{n} \left(\sum_{j-1}^{i} x_j \right)^2 * R(x)$$

F4: Schwefel 2.21 Function
n (dimension): 10, lb: −50, ub: 100, fmin: 0

$$f(x) = \max_i \{ |x_i|, 1 \le i \le n \}$$

F5: Rosenbrock's Function
n (dimension): 10, lb: −15, ub: 30, fmin: 0

$$f(x) = \sum_{i=1}^{n-1} \left[100\left(x_{i+1} - x_i^2\right)^2 + (x_i - 1)^2 \right] * R(x)$$

F6: Step Function
n (dimension): 10, lb: −50, ub: 100, fmin: 0

$$f(x) = \sum_{i=1}^{n} ([x_i + 0.5])^2 * R(x)$$

F7: Quartic Function
n (dimension): 10, lb: −0.64, ub: 1.28, fmin: 0

$$f(x) = \sum_{i=1}^{n} i x_i^4 + \text{random}[0, 1) * R(x)$$

F8: Schwefel 2.26 Function
n (dimension): 10, lb: −250, ub: 500, fmin: (−418.9829 * 5)

$$F(x) = \sum_{i=1}^{n} -x_i \sin\left(\sqrt{|x_i|}\right) * R(x)$$

F9: Rastrigin Function
n (dimension): 10, lb: −2.56, ub: 5.12, fmin: 0

$$F(x) = \sum_{i=1}^{n} \left[x_i^2 - 10\cos(2\pi x_i) + 10\right] * R(x)$$

F10: Ackley's Function
n (dimension): 10, lb: −16, ub: 32, fmin: 0

$$F(x) = -20\exp\left(-0.2\sqrt{\frac{1}{n}\sum_{i=1}^{n} x_i^2}\right) - \exp\left(\frac{1}{n}\sum_{i=1}^{n}\cos(2\pi x_i)\right) + 20 + e * R(x)$$

F11: Griewank Function
n (dimension): 10, lb: −300, ub: 600, fmin: 0

$$F(x) = \frac{1}{4000}\sum_{i=1}^{n} x_i^2 - \prod_{i=1}^{n} \cos\left(\frac{x_i}{\sqrt{i}}\right) + 1 * R(x)$$

F12: Penalty 1 Function
n (dimension): 10, lb: −25, ub: 50, fmin: 0

$$F(x) = \frac{\pi}{n}\left\{ \begin{array}{l} 10\sin(\pi y_1) + \sum_{i=1}^{n-1}(y_i - 1)^2 \\ \left[1 + 10\sin^2(\pi y_{i+1})\right] + (y_n - 1)^2 \end{array} \right\} y_i = 1 + \frac{x_i + 1}{4}$$

$$u(x_i, a, k, m) = \begin{cases} k(x_i - a)^m & x_i > a \\ 0 & -a < x_i < a \\ k(-x_i - a)^m & x_i < -a \end{cases}$$

F13: Penalty 2 Function
n (dimension): 10, lb: −25, ub: 50, fmin: 0

$$F(x) = 0.1\left\{ \begin{array}{l} \sin^2(3\pi x_1) + \sum_{i=1}^{n}(x_i - 1)^2 \\ \left[1 + \sin^2(3\pi x_i + 1)\right] \\ + (x_n - 1)^2\left[1 + \sin^2(2\pi x_n)\right] \end{array} \right\} + \sum_{i=1}^{n} u(x_i, 5, 100, 4) * R(x)$$

F14: De Joung (Shekel's Foxholes) Function
n (dimension): 2, lb: −32.768, ub: 65.536, fmin: 1

$$F(x) = \left(\frac{1}{500} + \sum_{j=1}^{25}\frac{1}{j + \sum_{i=1}^{2}\left(x_i - a_{ij}\right)^6}\right)^{-1}$$

F15: Kowalik's Function
n (dimension): 4, lb: −2.5, ub: 5, fmin: 0.00030

$$f(x) = \sum_{i=1}^{11} a_i - \left[\frac{x_i\left(b_i^2 + b_i x_2\right)}{b_i^2 + b_i x_3 + x_4}\right]^2$$

F16: Shekel 1 Function
n (dimension): 4, lb: 0, ub: 10, fmin: −10.1532

$$f(x) = -\sum_{i=1}^{5} \left[(X - a_i)(X - a_i)^T + c_i\right]^{-1}$$

References

1. Yang XS (2010) Nature-inspired metaheuristic algorithms, 2nd Ed., Luniver Press, Bristol.
2. Yang XS (2010) Engineering optimization: an introduction with metaheuristic applications. Wiley, London.
3. Michalewicz Z (1995) A survey of constraint handling techniques in evolutionary computation methods. In: proceedings of 4th annual conference on evolutionary programming, MIT Press, Cambridge MA, pp 135–155.
4. Huang T, Mohan AS (2005) A hybrid boundary condition for robust particle swarm optimization. IEEE Antennas Wirel Propag Lett 4:112–117.
5. Xu S, Rahmat-Samii Y (2007) Boundary conditions in particle Swarm optimization revisited. IEEE Trans Antennas Propag 55(3):112–117.
6. Chen TY, Chi TM (2010) On the improvements of the particle swarm optimization algorithm. Adv Eng Softw 41:229–239.
7. Chu W, Gao X, Sorooshian S (2011) Handling boundary constraints for particle swarm optimization in high-dimensional search space. Inf Sci 181(20):4569–4581.
8. Helwig, S., Branke, J., Mostaghim, S.: Experimental analysis of bound handling techniques in particle swarm optimization. IEEE Trans. Evol. Comput. 17(2), 259–271 (2013).
9. Gandomi A.H., Xin-She Yang. Evolutionary boundary constraint handling scheme. Neural Comp. & App., Springer, 21(6), 1449–1462, 2012.
10. Gandomi AH, Kashani AR, Mousavi M (2015) Boundary Constraint Handling Affection on Slope Stability Analysis. chaper 18 in Engineering and Applied Sciences Optimization, Springer, Switzerland, CMAS 38, 341–358.
11. Gandomi A.H. Interior Search Algorithm (ISA): A Novel Approach for Global Optimization. ISA Transactions, Elsevier, 53(4), 1168–1183, 2014.

Emotion Recognition from Videos Using Facial Expressions

P. Tamil Selvi, P. Vyshnavi, R. Jagadish, Shravan Srikumar and S. Veni

Abstract In recent days, automatic emotion detection is a field of interest and is used in fields such as e-learning, robotic applications, human–computer interaction (HCI), surveillance, ATM monitoring, mood-based playlists/YouTube videos, psychological studies, medical fields like supporting blind and dumb people, for treating autism in children, entertainment, animation, etc., The proposed work describes detection of human emotions from a real-time video or image with the help of classification technique. The major part of human communication constitutes of facial expression, which is around 55% of the total communicated information. The basic facial expressions that are considered by the psychologists are: happiness, sadness, anger, fear, surprise, disgust, and neutral. The proposed work aims to classify a given video into one of the above emotions using efficient facial features extraction techniques and SVM classifier. The author's contribution is to increase the efficiency in emotion recognition by implementing the above mentioned superior feature extraction and classification methods.

Keywords Appearance model · Emotion recognition · Feature extraction · Gabor filter · MATLAB · Occlusions · SVM classifier

1 Introduction

The proposed work utilizes human facial expressions as the features to detect different human emotions. The "Enterprise05" video database is chosen for the purpose which consists of 44 subjects who utter five sentences each for six different emotions, which are happiness, sadness, anger, fear, surprise, and disgust. The

P. Tamil Selvi (✉) · P. Vyshnavi · R. Jagadish · S. Srikumar · S. Veni
Department of Electronics and Communication Engineering, Amrita School of Engineering, Coimbatore, Amrita Vishwa Vidyapeetham, Amrita University, Coimbatore 641112, India
e-mail: tamiltheone67@gmail.com

S. Veni
e-mail: s_veni@cb.amrita.edu

S.S. Dash et al. (eds.), *Artificial Intelligence and Evolutionary Computations in Engineering Systems*, Advances in Intelligent Systems and Computing 517,
DOI 10.1007/978-981-10-3174-8_47

videos from the database are converted into frames (i.e., images). Preprocessing is an important process before feature extraction for better results. As part of preprocessing, the images are resized and normalized for further process. The preprocessed images' faces are separated from its background using Viola–Jones face detection algorithm. Two types of features are calculated in this work, which are geometric-based features and texture-based features. The Gabor filter technique is used for texture-based feature extraction from the images and the 63 fiducial point detection method is used for geometric-based feature extraction. The extracted features are given as the input to the classifier. The classification is done by SVM classifier algorithm for exact emotion recognition.

2 Related Work

The process of locating the region of interest and extracting the features play a major role in determining the efficiency of the classifier. The choice of the features to be extracted also affects the efficiency of the classifier. The features that can be chosen are geometric features and texture features. Priya Sisodia et al. [1] have used more than one features increases the efficiency of the classifier. However, the experiment was carried out with texture-based features using Gabor filters. The extracted texture features are given to a SVM classifier. They have concluded that Gabor filter not only gives a better performance under noise and intensity differences in the image but also gives efficiency when compared all other traditional methods.

Li Zhang and Ming Jiang [2] have used facial action coding system for emotion recognition in humanoid robots. A combination of two artificial neural networks is used to locate the action units (AUs) and SVM-based classifiers are used to detect the emotions shown in real time. The two artificial neural networks detect upper and lower facial AUs. Six AUs were extracted and given as inputs to the SVM classifier which detects the emotion of a person.

Happy and Aurobinda Routray [3] have stated other methods like PCA (principle component analysis) method followed by LDA (linear discriminant analysis) technique for feature extraction. And also have implemented Harris detection method for corner point detection, which is also one of the efficient methods for extracting features other than the method employed in this work, i.e., 63 fiducial facial point detection. Harris corner point detection and 63 fiducial facial point detection methods are explained in Sect. 6. This proposed work proves that the 63 fiducial point method is more efficient than the Harris corner detection method.

3 Methodology

The author implements the recognition process by first deducing the frames or images by sampling the videos that are needed to be analyzed. The number of frames is dependent on the duration of the video and it is also notable that high frequency of the sampling increased the efficiency of recognition but in turn increases the complexity of processing. Later, these images are to be individually taken for detecting the faces in each image. The face detection algorithm adopted in this proposed work is Viola–Jones algorithm. The emotions in the face are evidently concentrated in the mouth and the eye region. Thus, mouth and eye detections are also done using Viola–Jones algorithm itself. The detected mouth and eye images from the face are used to extract the texture features from them.

This paper imposes a double degree feature extraction process, i.e., two types of features are extracted from the images namely, texture feature extraction and geometric feature extraction. The Gabor filters are used for extracting texture based features from the image and 63 fiducial facial point detection methods for geometric-based feature extraction. The all extracted features are tabled together in a.csv file (excel file). Further, these features are combined and carried as a single excel sheet to the classifier to train the system so as to predict the test images properly. The classifier used in here in SVM classifier (Fig. 1).

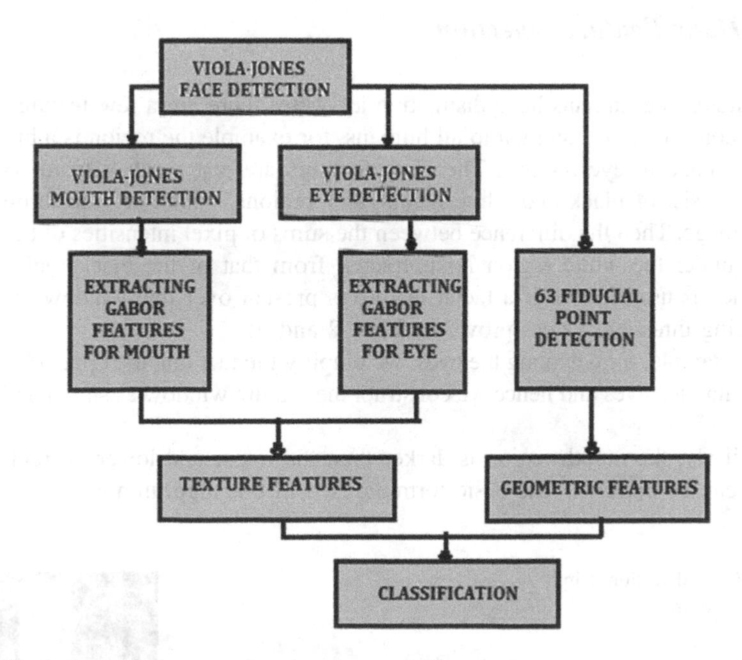

Fig. 1 Block diagram of proposed system for recognition

4 Face Detection

Viola–Jones algorithm was proposed by Paul Viola and Michael Jones in the year 2001. Although this algorithm is used for face detection purposes it can also be used as a classifier where other objects can also be detected once after training. Viola–Jones is a very robust framework designed to work in a real-time system. It has a very high detection rate and a very low false rate making it an ideal framework for face detection in image processing applications. Viola–Jones algorithm is used to detect human faces from an image for the proposed method. This step helps in eliminating the face from its background. But one setback in this system is that the face needs to be upfront with no tilt in the face and facing the camera directly.

The Viola–Jones algorithm has four main steps:

- Haar Feature Selection
- Integral Image Creation
- Adaboost Training
- Cascading Classifiers

The basic concept behind this framework is the addition of pixel intensities in predefined windows and comparison of these added intensities with the adjacent windows added intensities. Also the image needs to be converted to gray scale before proceeding to the Viola–Jones algorithm.

4.1 Haar Feature Selection

Even though all humans have distinctive identities there are a few features which can be considered as common to all humans, for example the region is a bit lighter than its adjacent eye regions. The Haar features are just a set of fixed windows which consist of black and white rectangular regions, which are run through the entire image. Then the difference between the sums of pixel intensities of the image region under the white region is subtracted from that of the black region. This difference is used to find if a facial feature is present over that window or not by comparing threshold values provided (Figs. 2 and 3).

For example, for detecting the eyes, we employ the fact that the upper cheeks are lighter than the eyes and hence we construct the feature window as shown in Figs. 4 and 5.

Similarly, the mouth region is darker than the upper and lower lip regions, so mouth can be detected. The basic formula used in this algorithm is

Fig. 2 Nose detection using the Haar feature

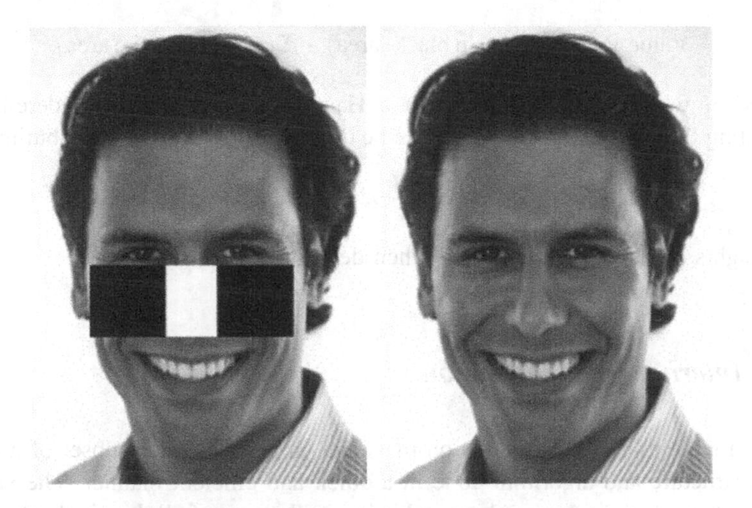

Fig. 3 Nose Haar feature window over the image

Fig. 4 Eyes detection using the Haar feature

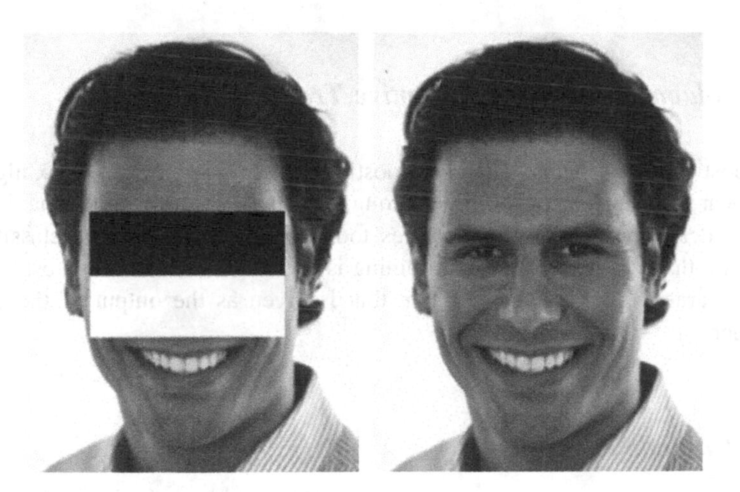

Fig. 5 Eyes Haar feature window over the image

$$\text{Value } V = \Sigma(\text{pixels in black area}) - \Sigma(\text{pixels in white area}) \tag{1}$$

Size of white or black regions in a Haar window is also considered while evaluating '*V*'. Weights with $\omega_i > 0$ can be used for adjusting the contributing sums

$$V = \omega 1 \cdot \text{SW1} + \omega 2 \cdot \text{SW2} - \omega 3 \cdot \text{SB} \tag{2}$$

Weights ω_i need to be specified when defining a Haar wavelet.

4.2 Internal Image Creation

An area table that is the summation of values in a rectangular subset of a grid is a data structure and algorithm done in a quick and efficient manner. The value at any point (x, y) in the summed area table is just the sum of all the pixels above and to the left of (x, y), inclusive as given below.

$$I_{\sum}(x,y) = \sum_{\substack{x' \le x \\ y' \le y}} i(x',y') \tag{3}$$

Once the Haar feature is selected, they are convoluted with the given face image and the results are stored as many subimages and given to Adaboost training for training the system.

4.3 Adaboost Training (Adaptive Training)

Adaboost basically means 'adaptive boosting' where a number of weak algorithm in a linear manner in order to form a strong algorithm. The best part is that it can be used to detect objects other than faces too. It just depends on the classification algorithm that you give. Adaboost training is done by combining the results of the "weak learners" into a weighted sum that is given as the output of the boosted classifier.

4.4 Cascade Classifiers

The value found after the application of the Haar feature windows is given to the cascade classifiers. The classifier then classifies the image as a face or non-face. In the case of eyes and mouth, it classifies the convoluted images as eyes or not eyes

Fig. 6 Input and output of face detection

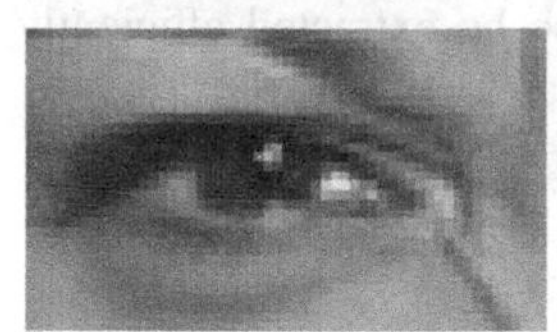
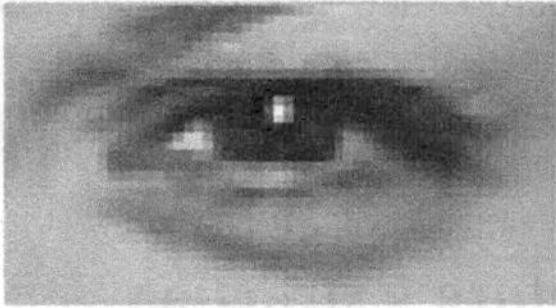
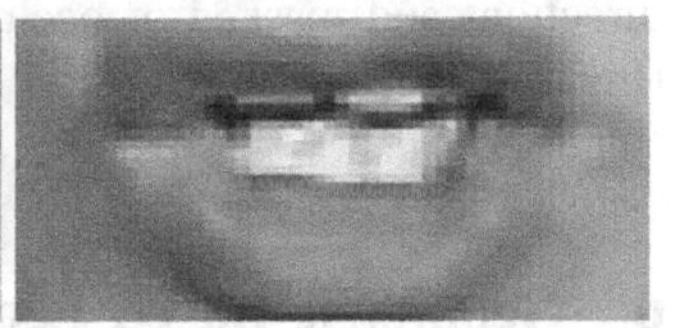

Fig. 7 Output of left eye, right eye, mouth detection

and nose or not nose depending on the "value" calculated in the above step. The classified region is then cropped out from the original image (Figs. 6 and 7).

5 Texture-Based Feature Extraction

The technique used for geometric-based feature extraction is Gabor filter method. The Gabor filter is usually the multiplication-convolution of a sinusoid wave with Gaussian function. Here, as Gabor filter helps in extracting useful data from the images.

$$G(x,y) = \frac{1}{2\pi\sigma^2} \mathrm{e}^{-\frac{(x^2+y^2)}{2}} \mathrm{e}^{-i(2\pi f(x\cos(\theta)+y\sin(\theta)))} \tag{4}$$

where f represents the frequency of the texture, that is, requires θ is the orientation that can be varied for accessing multiple directions, σ can be varies to change the size of the region of the image that is analyzed.

This advantage of differently available orientations and frequencies makes this method efficient for texture feature extraction. The images with faces along different

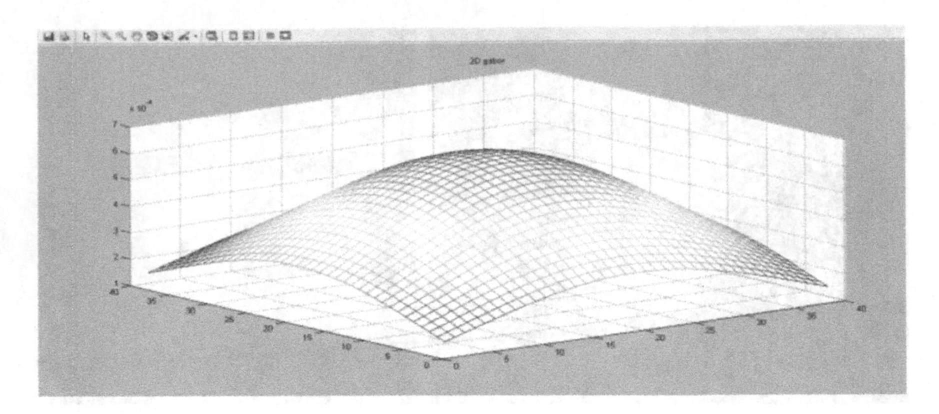

Fig. 8 2D-Gabor Filter Bank Magnitude plot

directions and masked or occlusion face features can also be extracted efficiently (Fig. 8).

6 Geometric-Based Feature Extraction

6.1 Harris Corner Detection

Harris corner detector is a mathematical operator used in computer vision systems and image processing in order to extract some very useful features. The basic working of this system can be described as the comparison of intensity values in a given small window of an image. First, a small window is to be selected with respect to the size of the image. Every window has a window function with the '*x*' axis and '*y*' axis parameters as its input. The window function can either be a Gaussian function or a constant function, but in either case the function value must be positive only over the window and must have a zero value over all other values of '*x*' and '*y*'. Now a comparison of intensities is done in the given window using the mathematical Eq. (5).

$$E(u,v) = \sum w(x,y)[I(x+u,y+v) - I(x,y)]^2 \tag{5}$$

We can understand from the above equation that the difference between the shifted intensity and the intensity of the current pixel will be very less or tending to zero for a window that is over a region with no corner point or an edge, but will be very high otherwise. So we need to only consider the regions over which the window gives very high values of *E*(*u*, *v*). Now the Eigen values of M are taken into consideration in order to differentiate a corner point from an edge.

$$\lambda_1 \approx 0 \quad \text{and} \quad \lambda_2 \approx 0 \Rightarrow \text{no features of interest.} \tag{6}$$

$$\lambda_1 \approx 0 \quad \text{and} \quad \lambda_2 = \text{large positive number} \Rightarrow \text{edge} \tag{7}$$

$$\lambda_1 \quad \text{and} \quad \lambda_2 = \text{large positive values} \Rightarrow \text{corner} \tag{8}$$

Equation (6) helps to find non-face images. Equation (7) draws all edges in the given image. Equation (8) marks all corner points in the processed image. However, not all non-faces, edges, and corner are detected efficiently by this method. Hence, this paper employs the high-efficient corner point detection named 63 fiducial facial point detection.

6.2 63 Fiducial Facial Point Detection

The technique is used for face recognition, pose estimation, and also to trace the corner points on the face. This method is said to be efficient because it helps in spotting maximum number of corner points on face which are countered on areas of the face that are salient for showing distinct features for extraction. Thus, the technique makes this system real time in nature. This method begins with the generation of initial graphs for all training images, one graph for each orientation, here 13 different orientations are used for every face in the database. The set of 13 oriented faces for a single image is called a facial bunch graph (FBG). Once the system has plotted graphs for different pose of face in training set, graphs for new test images need not be generated separately as they can be simply drawn automatically by elastic graph matching technique, which helps in adjusting the existing train image graphs elastically to match the test image characteristics.

The graphs are constructed like: At first fiducial points are marked on the given image depending on the image profile. The number of points marked for use as features in this project is 63 and these are numbered automatically for further use as shown in Fig. 10. Then the relation between the points on the face is realized to construct a general shape model or graph of the face which is shown as the red line in Fig. 9. In the graphs shown below, the red line drawn on the face is actually the geometric appearance when all the facial landmarks/points are connected. There are 63 such points that are detected and marked as shown in Fig. 11. Once these 63 points are marked on the given face image their coordinates are noted. But, only optimal number of points are itself sufficient for classification. For the purpose, only few important facial landmarks are taken into consideration like eye corner points, mouth corner points, eyebrow corner points, and the tip of the nose. The distance between these corner points is measured and these measurements are the extracted features from the images, Likewise six feature distance values are taken as features. The distance between any of these corner points are calculated using Euclidian distance formula.

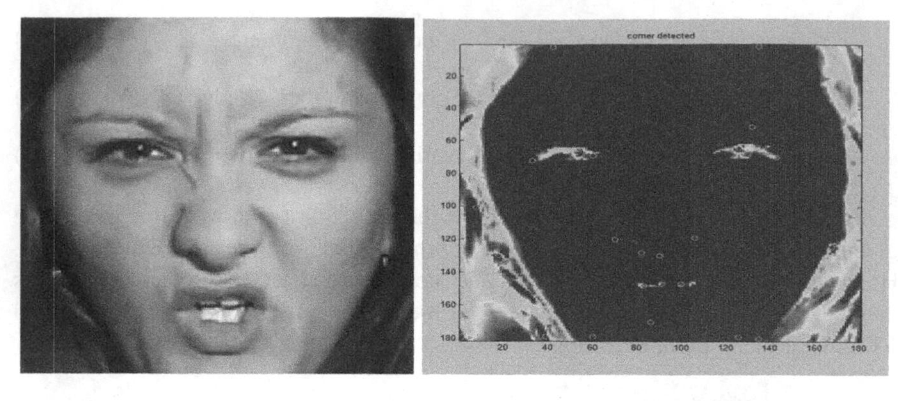

Fig. 9 Results for Harris corner detection

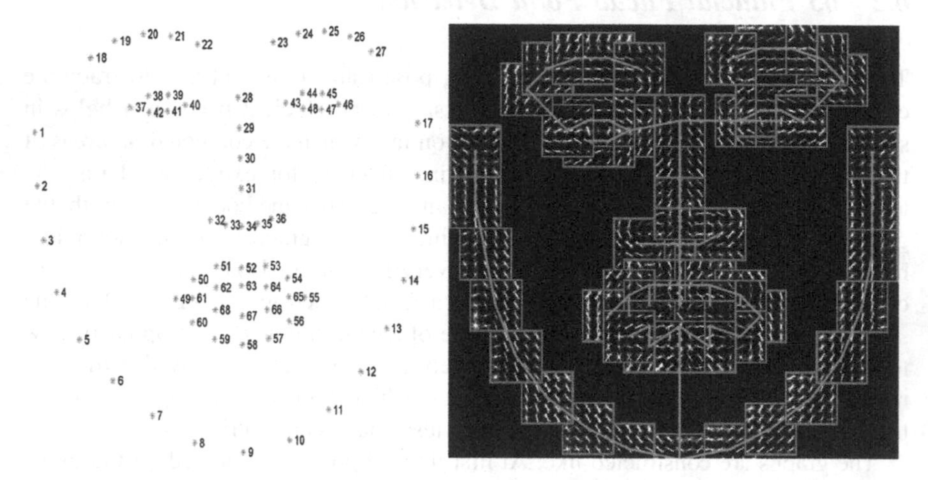

Fig. 10 Fiducial point graph model fitting

7 Emotion Classifications

The classifier used for the proposed work is SVM classifiers. SVM classifier is a nonprobabilistic linear classifier. Along with "kernel trick" it can even perform non-linear classification efficiently. Mapping a problem into a space with a high or large dimensions makes it more likely that the problem will become linearly separable. Classification involves the training of the SVM classifier based on the extracted features from the training samples and then testing the classifier using test samples. This task involves six emotions, i.e., it requires a multiclass SVM classifier. In SVM (choosing one-against-all approach), they are plotted as points in a graph and decision boundaries between different classes are obtained. From the whole database, 80% of the images are taken as training images and 20% of the

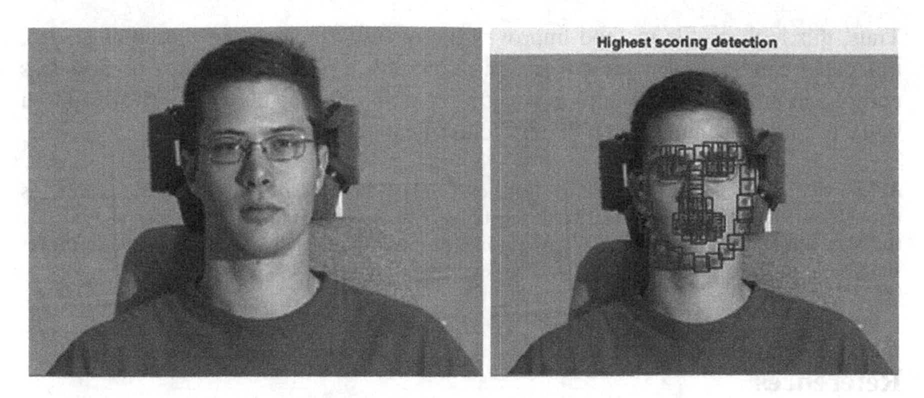

Fig. 11 Fiducial point method results

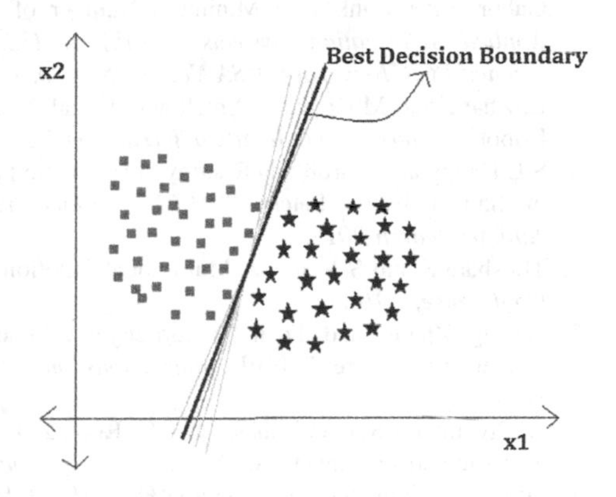

Fig. 12 Best decision boundary decision

images are taken as testing. These testing and training images are given to the classifier. The features are first marked in feature space hence each sample is represented by a point in the feature space. SVM classifier builds a decision boundary between two classes in the feature space. Later, the new test data plotted is classified depending upon its location among the boundaries of the graph. During the training phase, the equation of the best decision boundary hyperplane is found. The decision hyperplane is constructed by maximizing the margin between two class feature points (Fig. 12).

8 Conclusion

The Gabor and fiducial points combine feature extraction-classification and make the paper more efficient than its previous methods. Both Gabor and fiducial points are promising techniques in texture feature and geometric feature, respectively.

Thus, this high profile method improves the overall efficiency of the output. Also, this paper can be implemented easily in numerous applications as they possess less complexity in the processing phase making it simple for the classifier. The proposed work produces accuracy of 87% after classification.

Acknowledgment We would like to thank our class adviser Mr. Gandhiraj R for guiding us through-out the project. We also thank the university and lab staff for providing us with the software and all other necessities. We would like to thank our friends and college department for supporting us for completing the project successfully.

References

1. Priya Sisodia, Akhilesh Verma, Sachin Kansal, "Human Facial Expression Recognition using Gabor Filter Bank with Minimum Number of Feature Vectors," *International Journal of Applied Information Systems (IJAIS) – ISSN: 2249-0868 Foundation of Computer Science FCS, New York, USA Volume 5 – No. 9, July 2013.*
2. Li Zhang and Ming Jiang, "Intelligent Facial Action and Emotion Recognition for Humanoid Robots," *International Joint Conference on Neural Networks (IJCNN) July 6–11, 2014.*
3. S L Happy and Aurobinda Routray, "Automatic Facial Expression Recognition Using Features of Salient Facial Patches," *IEEE transactions on affective computing, vol. 6, no. 1, January-March 2015.*
4. Thushara S and S Veni, "A Multimodal Emotion Recognition System from Video," *ICCPCT Conference, 2016.*
5. Ligang Zhang and Dian Tjondronegoro, "Facial Expression Recognition Using Facial Movement Features," IEEE *transactions on affective computing, Vol. 2, no. 4, October–December 2011.*
6. K. Sreenivasa Rao and Shashidhar G. Koolagudi, "Recognition of emotions from video using acoustic and facialfeatures," *Springer-Verlag London 2013.*
7. https://en.wikipedia.org/wiki/Viola%E2%80%93Jones_object_detection_framework.
8. https://en.wikipedia.org/wiki/Gabor_filter.

Parse Tree Generation Using HMM Bigram Model

S. Arun, V.V. Vishnu and N. Arun Kumar

Abstract Parse tree generation from supervised sentence is crucial as well as essential task necessary for natural language processing. Syntax analyzers play a vital role in natural language processing. Prevailing system for parse tree generation is subject to compromise on ambiguous situation. Our approach is to generate a parse tree using Hidden Markov bigram model from tagged sentence results in improved parse tree generation. The user will be given the sentence to be parsed after part of speech tagging. If the sentence is syntactically correct, the parser will generate a correct parse trace.

Keywords Bigram translation probability · Hidden markov model · Phrase alignment · Statistical machine translation · Translation model · Word alignment

1 Introduction

Parsing [1, 2] is a technique of studying a string in computer languages or in natural language conforming to the guidelines of a formal grammar. It is one of the vital areas of natural language processing. Parsing of sentences takes into account to be a necessary intermediate stage for linguistics analysis in language process (NLP). In NLP; syntactic parsing is the method of studying and figuring out the textual content which is made from a sequence of tokens with respect to a given formal grammar. As human language is ambiguous, whose utilization is to deliver one-of-a-kind semantics, it's far a good deal tough to layout the functions for natural language processing constraints [3]. The main challenge is that the inherent quality of linguistic phenomena that build it onerous to symbolize the powerful

S. Arun (✉) · V.V. Vishnu · N. Arun Kumar
Department of CS&IT, Amrita School of Arts and Sciences,
Amrita Vishwa Vidyapeetham, Kochi, India
e-mail: arun_punalur@yahoo.com

S.S. Dash et al. (eds.), *Artificial Intelligence and Evolutionary Computations in Engineering Systems*, Advances in Intelligent Systems and Computing 517,
DOI 10.1007/978-981-10-3174-8_48

options for the goal progressing to understand models. Our objective is to create a parse tree and also create some rules that help to present a transfer methodology for each of the English sentences.

This paper deals with the HMM Bigram [4] model that's used for automated alignment of words and terms in parallel textual content. Parameters of a statistical phrase [5] alignment version are envisioned from parallel text and the model is used forward alignment in the same text utilized in estimation, but the various modern gadgets converting the document using the direct approach that converts each word of the phrase. In many cases direct translation is not possible because the structure of languages is different. Here we need new approach. HMM Bigram model is more suitable, which can find maximum probable word from corpus and can generate good structured parse tree.

2 How Parsers Work

Lexical analysis [6] or token generation is the first stage by which the input string is split into significant symbols outlined by a descriptive linguistics of standard expressions. The second step is parsing or syntactical analysis that is inspecting each token kind in the string and comparing with associate allowable tagged expression [7]. This is sometimes through with regard to a context-free descriptive linguistics that recursively defines parts which will form up associate degree expression and the order during which they need to seem. In the case of type validity and proper declaration of identifiers, the programming languages cannot make proper rules for context-free grammar. Attribute grammars [8] are used to express these rules formally. Semantic parsing or analysis is the final stage which takes suitable actions based on the validation of expressions. In parsing there are four stages which include reading the sentence, POS-tagging syntax tree generation and translation. POS-tagging is that the method of changing every word within the given sentence into tag word. For achieving this may use direct method or bigram approach.

3 Tree Generation Using HMM Bigram

The Hidden Markov model (HMM) [4] is a popular statistical tool or model, which model the system as a Markov process with hidden state. In terms of NLP, HMM has a wide range of application mainly in speech reorganization and signal process, however, have additionally been applied successfully to low level human language technology tasks like POS-tagging [3, 9], phrase configuration and extracting information from documents POS-tagging is that the method of changing every word within the given sentence into a tag word parse tree is associate degree order tree that shows the grammatical arrangement of a single line with a number of context-free descriptive linguistics. Parse tree as a part of computational linguistics. Parse trees vary from the abstract syntax tree, in that their structure and parts additional concretely mirror the sentence structure of the input language. They are additionally dissimilar from the sentence bigrams [8] generally used for descriptive linguistics teaching in colleges. Consistency grammars or dependency relation of consistence grammars are the two tree construction grammars, which is the base for the creation of parse tree. Take apart the tree is that the complete structure, starting from S and ending in each of the leaf nodes. The subsequent tags are commonly used in the creation of tree. Consider the example: "RAJU HIT THE BALL."

$$\mathbf{S(NP(NN(RAJU)), VP(V(HIT), NP(D(THE)), NN(BALL))))}$$

VP for verb part, V for verb (hit), D for determiner (the), NP for noun part, Penn Treebank [10] derived 36 different tags and 12 symbols and punctuations. Some of them are in Fig. 1.

The tree contains root node and leaf nodes. Each node may be a root node or may be leaf node which is depending on the rule associated with it. That is, each node is drawn based on the rules that reside in the corpus. Some of them are Fig. 2.

4 Algorithm for Generating Parse Tree

STK is a stack that contain tagged input sentence, ruled, un_ruled, and Tree are three data structures.

1. Tree<-NULL, ruled<-NULL, un-ruled<-NULL;
2. Read each tag from top of stack until 'S' part is reached
3. if (STK [top] have single rule.)

Fig. 1 Penn treebank sample tag set

CC	Coordinating conjunction
DT	Determiner
JJ	Adjective
NN	Noun, singular or mass
TO	to
VB	Verb, base form

Fig. 2 Sample rule list

S=NP+VP	SBAR=IN+S
S=NP_SBJ+VP	SBAR_ADV=IN+S
PP CLR=IN+NP	VP=VBD+NP+NP

Call single_rule_function (STK [top]);
if (VP is reached)
Call check_combination_rule ();

Else

Call check_combination_rule ();

4. If ‘S’ part is reached Combine NP and VP to form ‘S’ part and update tree if more than one rule is there use bigram approach and find maximum possible solution.

Single_rule_function (rule)
{Ruled<-rule}
Check_Combination_rule ()
{Backtrack and check all combination if there is more than one combination use bigram approach and select maximum probable tree combination rule, and update rule list and un_ruled list based on new rule.}

5 Execution of Algorithm

Stack STK contain tagged sentence, the algorithm takes each tagged word from top and find single rule from rule list if there is no single rule then take combinations. Consider this example (Fig. 3):

STK=

S	NNP	NN	VBZ	DT	JJ	NN	TO	NN	CC	NN

1. **nn**:check for single rule: np=nn;
    1. rule list=np (nn)
    2. **tree: null**
2. **cc:**check for single rule: “no single rule”
    - combination rules: null=cc+nn;
    - null=cc+np
    - no combination rules: so it move to un_ruled list
    - set **tree=null**
3. **nn:** check for single rule: np=nn
    - but no combination rule associated with un_ruledlist

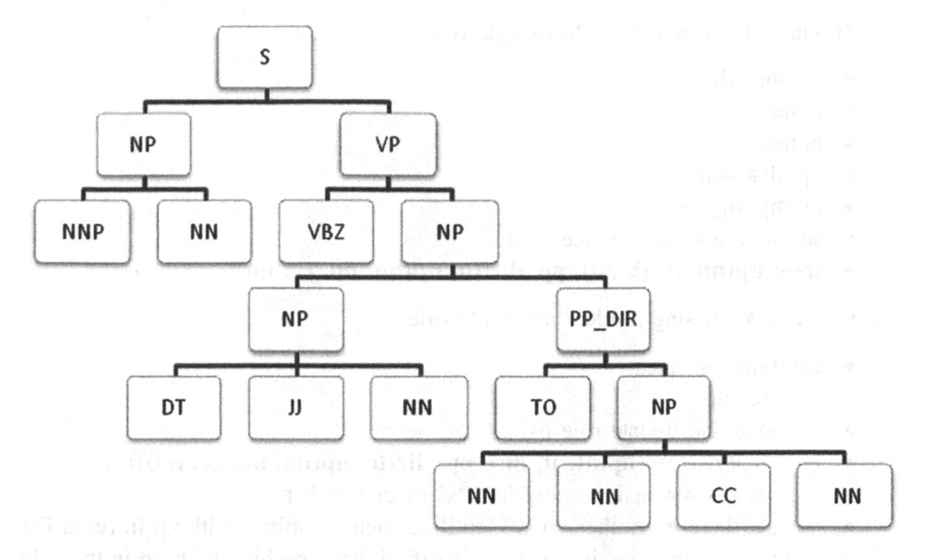

Fig. 3 Parse tree

4. **nn**: combination rule
    - np=nn+nn+cc+np
    - more than one rule list then use hmm bigram
    - so move ruled_list=unruled list
    - **tree: np (nn, nn, cc, np)**
5. **to**: check for single rule: "no single rule"
    - combination rules: pp_dir=to+np
    - so it move to ruled list
    - **tree: pp_dir (to, np (nn, nn, cc, nn))**
6. **nn:** check for single rule: np=nn;
    - move to un_ruled list
7. **jj:** check for single rule: "no single rule"
    - combination rules:
    - null=jj+np
    - np=jj+nn
    - np=np+pp_dir
    - so move and update rules list
    - **tree: np(np(jj, nn), pp_dir(to, np(nn, nn, cc, nn)))**

8. **dt**: check for single rule: "no single rule"
    - combination rules:
    - null=dt+np
    - null=dt+jj
    - np=dt+jj+nn
    - np=np+pp_dir
    - so move and update rule _list
    - **tree: np(np(dt, jj, nn), pp_dir(to, np(nn, nn, cc, nn)))**
9. **vbz:** check for single rule: "no single rule"
    - combination rules:
    - vp=vbz+np
    - so move and update rule list
    - **tree: vp(vbz, np(np(dt, jj, nn), pp_dir(to, np(nn, nn, cc, nn)))**
    - 'vp' part is reached, so next find 's' or end of list
    - note: if there is another vp is identified then combines with vp in ruled list and updates the tree. if s part is identified then combine with vp in the rule list and completing the tree.
10. **nn:** single rule: np=nn;
    - move to un-ruled list
11. **nnp**: np=nnp
    - move to un-ruled list
12. **s**: check for the combination rule.
    - np=np+np;
    - np=nnp+nn
    - np=np+np
    - np=nnp, np=nn
    - any one will be selected based on bigram, the probability np=nnp+np is higher than np=np+np
    - so np=nnp+np;
    - and then it combine with vp in the rule list and completing the tree
    - **TREE: S(NP(NNP, NN), VP(VBZ, NP(NP(DT, JJ, NN)PP_DIR(TO, NP(NN, NN, CC, NN))))**

6 Conclusion

This paper deals with English to syntax tree statistical machine translation and its alignment methodology supported computationally sensible Hidden Markov models. Statistical machine translation is predicated on chance and it produces additional correct result than alternative sorts and statistical alignment models improves the interpretation quality. HMM based mostly applied math alignment model is additional powerful.

References

1. Post, Matt & Daniel Gildea. "Parsers as language models for statistical machine translation." Proceedings of the Eighth Conference of the Association for Machine Translation in the Americas. 2008.
2. Ramshaw & M. Marcus. 1995 "Text Chunking using Transformation-Based Learning". Proceedings of the Third Workshop on Very Large Corpora. (1995).
3. K. M. Azharul Hasan, Al-Mahmud, Amit Mondal, & Amit Saha.: "Recognizing Bangla Grammar Using Predictive Parser" International Journal of Computer Science & Information Technology (IJCSIT) Vol 3, No 6, Dec 2011.
4. Vetrivel, S. & Diana Baby: "English to Tamil statistical machine translation and alignment using HMM." Proceedings of the 12th international Conference on Networking, VLSI and Signal Processing. 2010.
5. Och, Franz Josef, Christoph Tillmann, & Hermann Ney.: "Improved alignment models for statistical machine translation." Proc. of the Joint SIGDAT Conf. on Empirical Methods in Natural Language Processing and Very Large Corpora. 1999.
6. K. S. Narayana Pillai: "Adhunika Malayala Vyakaranam". The State Institute Of Languages, Kerala, Thiruvananthapuram-3, Second edition, 2003.
7. Antony P J. & Soman K P: "Kernel Based Part of Speech Tagger for Kannada", International Conference on Machine Learning and Cybernetics 2010, ICMLC 2010, Qingdao, Shandong, China.
8. Hongfei Jiang, Muyun Yang, Tiejunzhao, Sheng Li & Bowang.: "A statistical machine translation model based on a synthetic synchronous grammar." Proceedings of the ACL-IJCNLP 2009 Conference Short Papers. Association for Computational Linguistics, 2009.
9. Matt Post & Daniel Gildea: "Parsers as language models for statistical machine translation", Department of Computer Science University of Rochester, 2005.
10. Antony P J, Nandini. J. Warrier & Dr. Soman K P.: "Penn Treebank-Based Syntactic Parsers for South Dravidian Languages using a Machine Learning Approach" International Journal of Computer Applications (0975 – 8887) Volume 7– No. 8, October 2010.

[illegible] and statistical alignment models [illegible] applied math alignment [illegible]

References

1. [illegible] machine translation [illegible] Machine Translation [illegible]
2. [illegible] Based [illegible]
3. [illegible] & Information [illegible] Vol. [illegible]
4. [illegible] machine translation [illegible] using HMM [illegible] Conference on Networking, VLSI and Signal Processing [illegible]
5. [illegible] Improved alignment models for [illegible] Empirical Methods [illegible]
6. [illegible] Natural Language [illegible]
7. [illegible] International Conference [illegible]
8. [illegible] A statistical machine [illegible] Proceedings of the [illegible] Computational Linguistics [illegible]
9. [illegible] 2005.
10. [illegible] International [illegible]

Preserving Privacy in Vertically Partitioned Distributed Data Using Hierarchical and Ring Models

R. Srinivas, K.A. Sireesha and Shaik Vahida

Abstract The important aspect in research is to publish data by conserving one's privacy. Enormous techniques have been proposed to tackle this issue. The main concept is to concentrate on how cogently one protects individual privacy in publishing data without the exact attribute missing in the research. As per our knowledge is concerned, there raised a great scope for research on vertically partitioned distributed databases. This paper mainly concentrated on privacy which is addressed on vertically partitioned distributed data. It is the prime responsibility of the publisher to see that one's personal information is protected. In order to maintain such sustainability, organizations like hospitals, government agencies, etc., store information at multiple sites by vertically portioning attribute with a solution to integrate all these attributes without violation and distraction of any meaning. To implement this, good models are required to integrate information from multiple sites for publication or for research, and models proposed also require maintaining privacy. In this paper, modernistic models ring and hierarchical model are proposed to preserve privacy for vertically portioning distributed data.

Keywords Data · Distributed · Hierarchical · Personal information · Privacy · Ring · Vertically partitioned

---

R. Srinivas (✉) · S. Vahida
Department of CSE, ACET, Surampalem, India
e-mail: viceprincipal@acet.ac.in

S. Vahida
e-mail: vahida.shaik@acet.ac.in

K.A. Sireesha
Department of IT, ACET, Surampalem, India
e-mail: sireesha.kanduri@gmail.com

S.S. Dash et al. (eds.), *Artificial Intelligence and Evolutionary Computations in Engineering Systems*, Advances in Intelligent Systems and Computing 517,
DOI 10.1007/978-981-10-3174-8_49

1 Introduction

In this contemporary era, the quick expansion in networking, database, and computing technologies, a large amount of individual data can be incorporated and analyzed digitally, leading to a better use of data-mining tools to understand the current trends and patterns. This has raised universal concern about protecting the privacy of individuals. In recent years, publishing data has become very essential mostly for research. However, they have difficulties in releasing information which does not co-operate privacy between two parties. Many organizations (government agencies and medical authorities) are performing data mining on union of two or more private or public databases to fulfill their research needs. In this process organizations must publish (or) share person data with other parties. While publishing personal data with other parties' privacy should be provided to one's private data. The tradeoff between sharing information for analysis and keeping it secret, to preserve corporate trade secret and customer privacy is a growing challenge in the present age. So, while publishing data individual data should not be revealed. One should provide privacy to individual data before publication [1].

The word privacy is first introduced in 1988 by David Chaum to solve the "dinning cryptographers' problem" [2]. In latest years, number of methods has been proposed to publish data. Mainly these methods are related to single system. As per our knowledge is concerned no research was done on vertically portioning distributed database. So, there is a great scope for research on vertically portioning distributed data. The main reason is one should not store their personal data at one site. If so it is very easy for third party to obtain personal information. To overcome this, individual personal information should be stored at multiple sites by vertically portioning attributes with a provision to integrate all records if necessary. Such models are required to provide privacy for individuals' data. While partitioning attributes into multiple record sets a unique key is maintained as a common field for all the sets to link the sites together while publishing data. In this paper, two models are proposed namely ring and hierarchical models to publish vertically partitioned distributed data.

1.1 Basic Definitions

Micro data: The data released in tabular and statistical formats. And in other words the specific data which is capable of (re)identify the respondent. Example: social security number, name, race, sex, physical address, etc.

Key attributes: Key attributes are attributes which can uniquely identify an individual directly. Key attributes are always removed before releasing the micro-data. Examples: name, address, cell phone number.

Quasi identifiers: A set of attributes that can be potentially linked with external information to reidentify entities. Example: 5-digit ZIP code, Birth date, gender.

Sensitive attributes: These attributes which are always released and what the researchers need. It depends on the requirement. These can uniquely identify an individual directly. Example: Medical record, wage, etc.

2 Literature Survey

In early 1988, David Chaum proposed a method to provide anonymous communication between the persons paying bill in the hotel and is called "dinning cryptographers' problem." After this so many methods were proposed to provide privacy to individual data before publication.

In 2002, Sweeney proposed k-anonymity. In this model, the information for each person contained in the released table cannot be distinguished from at least $k - 1$ individuals whose information also appears in the release [3]. But have some drawbacks like Background knowledge Attack and Homogeneity Attack. To overcome the drawbacks of *K*-anonymity, *L*-diversity was proposed by Machanavajjhala et al. in 2006 [4]. In *L*-diversity, each equivalence class has at least l well-represented sensitive values. There are limitations in *l*-diversity like insufficient to prevent attribute disclosure and does not consider semantic meanings of sensitive values. To overcome the limitations of *L*-diversity a new model *T*-Closeness was proposed by Li et al. in 2007 [5]. In *t*-Closeness the overall distribution of sensitive values should be public information and the knowledge gain will be separated. Here also there are some risk factors with *t*-closeness and one cannot provide full privacy with *t*-closeness.

K^m-anonymization was defined by Maolis Terrovitis et al. in 2008 [6]. This model relies on generalization instead of suppression. K^m-Anonymity has been proposed for anonymous transactional database. K^m-Anonymity aim at protect the database against an adversary who has knowledge about almost m items in the transaction. After that, In 2009, Wong R.C proposed a (α, k) anonymity model to protect both identification and relationships to sensitive information in data and to limit the confidence of the implication from the quasi-identifier to a sensitive value to within α [7].

The existing *k*-anonymity property protects against identity disclosure, but it fails to protect against attribute disclosure. In 2010, Traian Marius Truta et al. proposed *p*-sensitive *k*-anonymity avoids this shortcoming [8]. Two necessary conditions to achieve *p*-sensitive *k*-anonymity property: first one is, minimum number of distinct values for confidential attributes must be greater than or equal top. Second one is, maximum allowed number of combinations of key attribute values in the masked micro data and the other method (*k*, *e*)-anonymity is presented by RCWong in 2006 [9].

In 2007, Xiaokui Xiao and yufei Tao developed a generalization-based m-invariance that effectively limits the risk of privacy disclosure in republication [10]. In 2005, Kristen LeFevre et al. proposed multidimensional *K*-Anonymity [11], which provides better privacy when compared with previous methods. In multi-dimensional k-anonymity data anonymization is done based on more than one record and attribute. Data growth rate is high nowadays in order to provide privacy for personal information author proposed technique for distributed data [12–15]. To provide better privacy for horizontally partitioned data in 2013 R. Srinivas et al. proposed Hierarchical Model [16, 17].

3 Methodology

3.1 Vertically Partitioned Distributed Data

To provide more privacy to individual private data an advanced approach is proposed for publishing vertically partitioned distributed data by preserving privacy using a secure join operation. In this paper, two models are used to publish vertically partitioned distributed data. The following major tasks are performed before publishing data.

(i) The data in the database is partitioned vertically by maintaining primary key as common in all partitions.
(ii) The partitions are distributed over all the sites
(iii) To publish data, data may be collected using ring or hierarchical model and published by applying Mondrian multidimensional *k*-anonymity method

For an example consider data given in Table 1. S. No. is considered as a primary key and maintains this primary key as a common field while vertically portioning data. The vertically partitioned data table shown in Table 2

Table 1 Sample hospital data

S. No.	First name	Last name	Address	Phone number	Age	Sex	Disease
1	John	Clinton	RJY	9967785423	85	M	Heart stroke
2	Abraham	Lincoln	KKD	9968885429	40	M	Bone break
3	Raheem	Sheik	HYD	9969985412	22	M	Kidney problem
4	Aditya	Chepuri	VJY	9966545420	60	M	Hepatitis
5	Harisha	Kondeti	KKD	9943255426	18	F	Malaria
6	Jyothi	Kuppa	VIZAG	9967756744	34	F	Asthma
7	Kavya	Upaala	GUNTUR	9967798765	78	F	Diabetes

Table 2 Vertically partitioned data with S. No. as a common field

(i)

S.No.	First Name
1	John
2	Abraham
3	Raheem
4	Aditya
5	Harisha
6	Jyothi
7	Kavya

(ii)

S.No.	Last Name
1	Clinton
2	Lincoln
3	Sheik
4	Chepuri
5	Kondeti
6	Kuppa
7	Upaala

(iii)

S.No.	Address
1	RJY
2	KKD
3	HYD
4	VJY
5	KKD
6	VIZAG
7	GUNTUR

(iv)

S.No.	Phone number
1	9967785423
2	9968885429
3	9969985412
4	9966545420
5	9943255426
6	9967756744
7	9967798765

(v)

S.No.	Age
1	85
2	40
3	22
4	60
5	18
6	34
7	78

(vi)

S. No.	Sex
1	M
2	M
3	M
4	M
5	F
6	F
7	F

(vii)

S. No.	Disease
1	Heart Stroke
2	Bone Break
3	Kidney Problem
4	Hepatitis
5	Malaria
6	Asthma
7	Diabetes

3.2 *Publishing Data Using Ring Model and Hierarchical Model*

In the proposed model, we consider that the sites can form ring or hierarchy models by considering some mechanism and site can also establish communication over the network to send and receive data securely. These models are shown Figs. 1 and 2.

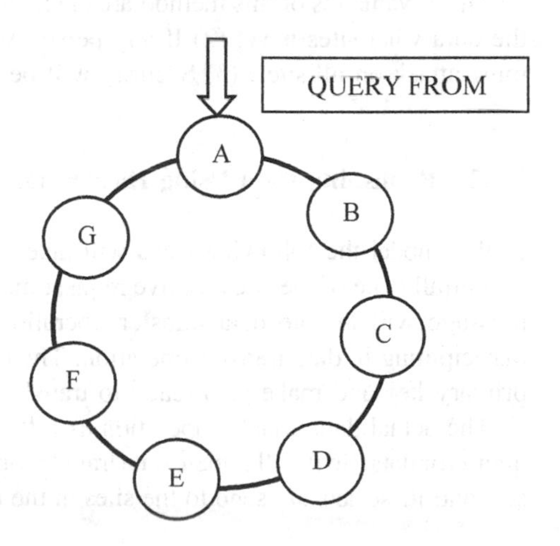

Fig. 1 Ring model

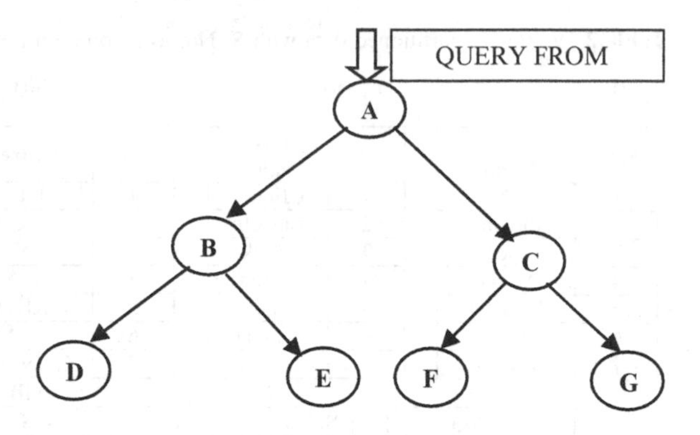

Fig. 2 Hierarchical model

3.2.1 Publishing Data Using Ring Model

In ring model the following steps will take place to publish data:

Initially one of the sites receive request message for data. The site which receives message will initiate data transfer operation by sending same query to all sites participating in data transfer operation. Then all these sites encrypt the data except primary key and make data ready to transfer.

The actual data transfer operation as follows. Here consider that site A received query for data. So it will initiate data transfer operation by sending its data to its next site say B as shown in Fig. 1 Site B receives data from A and performs secured union operation on data received and its content by using S. No. as key. Now B has data which contains fields S. No. and two encrypted attributes. After union operation B sends data to site C. C will also performs same operations. This process is repeated till all nodes involved in this process and the final result is send to A. At this stage A has data that contains all fields in encrypted form. It performs decrypt operation, perform Mondrian multidimensional *k*-anonymity on decrypted data and it will publishes data to third party.

The advantages of this method are (1) Sender and Receiver has no knowledge on the data what sites have. (2) If any person wants to know data content at sites he must attack on all sites. (3) Security will be high.

3.2.2 Publishing Data Using Hierarchal Network Model

In this model the following steps will take place to publish data:

Initially one of the sites receive request message for data. The site which receives message will initiate data transfer operation by sending same query to all sites participating in data transfer operation. Then all these sites encrypt the data except primary key and make data ready to transfer.

The actual data transfer operation as follows. Here consider that site A received query for data. So it will initiate data transfer operation by dividing its content into two sets and these sets are send to the sites in the next level., i.e., to the sites B and C as shown in Fig. 2 Site B receives one of the two data sets from A and performs secured union operation on data received and its content by using S. No. as key. Now B has data which contains fields S. No. and two encrypted attributes. After union operation B sends data to next level sites D and E by dividing its data into two sets. C will also performs same operations. This process is repeated for all levels. The data sets division is based on the number of child nodes. In this example we considered two child nodes so the data is divided into two sets. All the leave sites receive data from previous level and perform secured union operation on data received and its content finally results send to node A without dividing data into sets. At this stage A has data that contains all fields in encrypted form after performing union operation on all the sets received. It performs decrypt operation, performs Mondrian multidimensional *k*-anonymity on decrypted data and it will publishes data to third party.

The advantages of this method are (1) Data transmission load is reduced on the network (2) Sender and Receiver has no knowledge on the data what sites have. (3) If any person wants to know data content at sites he must attack all sites. (4) Security will be high. (5) Time required to transfer data in this network is less.

4 Results

4.1 Results for Ring Model

1. Consider vertically partitioned data shown in Table 2. These data sets are available at sites A, B, C, D, E, F, and G, respectively. Encrypted data sets are used for transmission. But for the sake of better understanding data in actual format is considered here.
2. First, Site A receives the query from third party.
3. At site A data shown in Table 3 is present
4. Site A sends this data to Site B. Now site B perform secured union on data set received and its own content. The data available after union operation at B is shown in Table 4
5. Now site B send encrypted data to site C. Site C has data set after union operation is shown in Table 5.

Table 3 Data at site A

S. No.	First name
1	John
2	Abraham
3	Raheem
4	Aditya
5	Harisha
6	Jyothi
7	Kavya

Table 4 Data at site B

S. No.	First name	Last name
1	John	Clinton
2	Abraham	Lincoln
3	Raheem	Sheik
4	Aditya	Chepuri
5	Harisha	Kondeti
6	Jyothi	Kuppa
7	Kavya	Upaala

Table 5 Data at site C

S. No.	First name	Last name	Address
1	John	Clinton	RJY
2	Abraham	Lincoln	KKD
3	Raheem	Sheik	HYD
4	Aditya	Chepuri	VJY
5	Harisha	Kondeti	KKD
6	Jyothi	Kuppa	VIZAG

6. This process is repeated for all the sites from D to G in circular fashion.
7. At last, Site G has data present in the Table 6 Site G send all encrypted data sets to A
8. At Site A, before publication of data remove sensitive attributes which reveals privacy of the person like phone number, names, etc., and Mondrian multidimensional k-anonymity method (WHERE $K >= 2$) applied on above data to provide better privacy.
9. After applying Mondrian multidimensional k-anonymity method, the final published data shown in Table 7.

Table 6 Data at site G

S. No.	First name	Last name	Address	Phone number	Age	Sex	Disease
1	John	Clinton	RJY	9967785423	85	M	Heart stroke
2	Abraham	Lincoln	KKD	9968885429	40	M	Bone break
3	Raheem	Sheik	HYD	9969985412	22	M	Kidney problem
4	Aditya	Chepuri	VJY	9966545420	60	M	Hepatitis
5	Harisha	Kondeti	KKD	9943255426	18	F	Malaria
6	Jyothi	Kuppa	VIZAG	9967756744	34	F	Asthma
7	Kavya	Upaala	GUNTUR	9967798765	78	F	Diabetes

Table 7 Final published data

S. No.	First name	Last name	Address	Age	Sex	Disease
1	John	Clinton	RJY–KKD	40–85	M	Heart stroke
2	Abraham	Lincoln	RJY–KKD	40–85	M	Bone break
3	Raheem	Sheik	HYD–RJY	22–60	M–F	Kidney problem
4	Aditya	Chepuri	HYD–RJY	22–60	M–F	Hepatitis
5	Harisha	Kondeti	KKD–HYD	18–34	F	Malaria
6	Jyothi	Kuppa	KKD–HYD	18–34	M–F	Asthma
7	kranti	Upaala	KKD–HYD	34–78	M–F	Diabetes

4.2 Results for Hierarchical Model

1. Consider vertically partitioned data shown in Table 2. These data sets are available at sites A, B, C, D, E, F, and G, respectively. Encrypted data sets are used for transmission. But for sake of better understanding data in actual format is considered here.
2. First, Site A receives the query from third party.
3. At site A data shown in Table 3 is present.
4. Site A sends the data to Site B and C by dividing its content into two sets as shown in 8. Now site B perform secured union on data set received and its own content. The data available at B is shown in Tables 8, 9 and 10
5. Now site B send encrypted data to sites D and E.

Table 8 Data at site A

S. No.	First name
1	John
2	Abraham
3	Raheem
4	Aditya
5	Harisha
6	Jyothi
7	Kavya

Table 9 Data at site B

S. No.	First name	Last name
1	John	Clinton
2	Abraham	Lincoln
3	Raheem	Sheik
4		Chepuri
5		Kondeti
6		Kuppa
7		Upaala

Table 10 Data at site A

S. No.	First name	Last name	Address	Age	Sex	Disease
1	John	Clinton	RJY–KKD	40–85	M	Heart Stroke
2	Abraham	Lincoln	RJY–KKD	40–85	M	Bone Break
3	Raheem	Sheik	HYD–RJY	22–60	M–F	Kidney Problem
4	Aditya	Chepuri	HYD–RJY	22–60	M–F	Hepatitis
5	Harisha	Kondeti	KKD–HYD	18–34	F	Malaria
6	Jyothi	Kuppa	KKD–HYD	18–34	M–F	Asthma
7	Kranti	Upaala	KKD–HYD	34–78	M–F	Diabetes

6. This process is repeated for all levels.
7. All leave sites receive data from previous level and perform secured union operation on data received and its content finally result send to node A without dividing data into sets.
8. At Site A, before publication of data, sensitive attributes which reveals privacy of the person like phone number names, etc., are removed and Mondrian multidimensional *k*-anonymity method (WHERE $K >= 2$) is applied on above data to provide better privacy.

The above process is illustrated in the below Fig. 3.

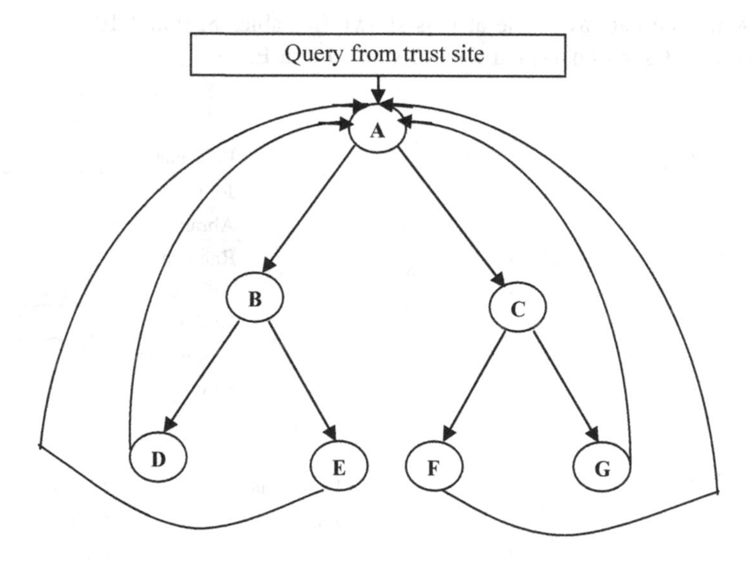

Fig. 3 Hierarchical topology model to publish data

5 Conclusions

In this paper, two models are proposed to safe-guard privacy for vertically portioning distributed data without revealing content at sites by using encryption methods. Using, these models the data transferred from various sites to one of the central site. Once data collected, Mondrian multidimensional k-anonymity is applied for providing better privacy to personal information before publishing data. In future, this work can be extended by providing anonymous communication between sites.

References

1. G kountouna, Olga, "A Survey on Privacy Preservation Methods", Technical Report, Knowledge and Database Systems Laboratory, NTUA, 2011, SECE, Knowledge and Data Base Management Laboratory (DBLAB), pp 1–30, 2011.
2. Chaum, David, "The dining cryptographers problem: Unconditional sender and recipient untraceability", Journal of cryptology 1.1, pp: 65–75, 1988.
3. Sweeney, Latanya. "k-anonymity: A model for protecting privacy", International Journal of Uncertainty, Fuzziness and Knowledge-Based Systems 10.05 pp: 557–570, 2002.
4. Machanavajjhala, Ashwin, et al., "l-diversity: Privacy beyond k-anonymity", ACM Transactions on Knowledge Discovery from Data (TKDD) 1.1,2007.
5. Li, Ninghui, Tiancheng Li, and Suresh Venkatasubramanian, "t-closeness: Privacy beyond k-anonymity and l-diversity", Data Engineering, 2007. ICDE 2007. IEEE 23rd International Conference on. IEEE, 2007.
6. Terrovitis, Manolis, Nikos Mamoulis, and Panos Kalnis, "Privacy-preserving anonymization of set-valued data", Proceedings of the VLDB Endowment 1.1 (2008): pp: 115–125.
7. R. C.W. Wong, J. Li, A. W.-C. Fu, and Ke Wang, "(α,k)-anonymity: an enhanced k-anonymity model for privacy-preserving data publishing", In Proceedings of the ACM KDD, pp: 754–759, New York, 2006.
8. Alina Campan, Traian Marius Truta, Nicholas Cooper, "P-Sensitive K-Anonymity with Generalization Constraints", Transactions On Data Privacy 3, pp: 65—89, 2010.
9. Wong, Raymond Chi-Wing, et al., "(α, k)-anonymity: an enhanced k-anonymity model for privacy preserving data publishing", Proceedings of the 12th ACM SIGKDD international conference on Knowledge discovery and data mining. ACM, 2006.
10. Xiao, Xiaokui, and Yufei Tao., "M-invariance: towards privacy preserving re-publication of dynamic datasets," Proceedings of the 2007 ACM SIGMOD international conference on Management of data. ACM, 2007.
11. LeFevre, Kristen, David J. DeWitt, and Raghu Ramakrishnan, "Mondrian multidimensional k-anonymity", Data Engineering, 2006. ICDE'06. Proceedings of the 22nd International Conference on. IEEE, 2006.
12. Xiong, Li, et al., "Privacy preserving information discovery on ehrs", Information Discovery on Electronic Health Records (2008): pp:197–225.
13. Jiang, Wei, and Chris Clifton, "Privacy-preserving distributed k-anonymity", Data and Applications Security XIX. Springer Berlin Heidelberg, 2005. Pp: 166–177.
14. LeFevre, Kristen, David J. DeWitt, and Raghu Ramakrishnan, "Incognito: Efficient full-domain k-anonymity", Proceedings of the 2005 ACM SIGMOD international conference on Management of data. ACM, 2005.

15. Bayardo, Roberto J., and Rakesh Agrawal, "Data privacy through optimal k-anonymization", Data Engineering, 2005. ICDE 2005. Proceedings. 21st International Conference on. IEEE, 2005.
16. Srinivas R., and A. Raga Deepthi. "Hierarchical Model for Preserving Privacy in Horizontally Partitioned Databases." Age 15: 35. IJETTCS, Volume 2, Feb 2013.
17. Srinivas R., K. PRASADA RAO, and V. SREE REKHA., "Preserving Privacy in Horizontally Partitioned Databases Using Hierarchical Model", IOSR Journal of Engineering May 2.5 (2012):pp: 1091–1094.

One-Class Text Document Classification with OCSVM and LSI

B. Shravan Kumar and Vadlamani Ravi

Abstract In this paper, we propose a novel one-class classification approach for text document classification using One-Class Support Vector Machine (OCSVM) and Latent Semantic Indexing (LSI) in tandem. We first apply t-statistic-based feature selection on the text corpus. Then, we apply OCSVM on the rows corresponding to the negative class of the document-term matrix of a collection of text documents and extract the Support Vectors (SV). Then, in the test phase, we employ LSI on the query documents from the positive class to compare them with the SVs extracted from the negative class and match score is computed using the cosine similarity measure. Then, based on a prespecified threshold for the match score, we classify the positive category of the text corpus. Use of SV for comparison reduces the computational load, which is the main contribution of the paper. We demonstrated the effectiveness of our approach on the datasets pertaining to Phishing, and sentiment analysis in a bank.

Keywords Text mining · Document classification · OCSVM · Latent semantic indexing · Sensitivity

1 Introduction

Text mining has several applications, including document classification [1], social network analysis [2], sentiment analysis, outlier detection, customer migration behavior [3], fraud detection in banking and insurance [4], churn prediction [3], customer feedback analysis [5], intelligence analysis, phishing emails and websites

B. Shravan Kumar · V. Ravi (✉)
Center of Excellence in Analytics, Institute for Development and Research in Banking Technology, Castle Hills, Road no. 1, Masab Tank, Hyderbad 500057, India
e-mail: rav_padma@yahoo.com

B. Shravan Kumar
School of Computer and Information Sciences, University of Hyderabad, Hyderabad 500046, India
e-mail: shravan.springer@yahoo.com

S.S. Dash et al. (eds.), *Artificial Intelligence and Evolutionary Computations in Engineering Systems*, Advances in Intelligent Systems and Computing 517,
DOI 10.1007/978-981-10-3174-8_50

detection [6, 7], spam detection [8], and Malware analysis [9]. Typically, text mining consists of text preprocessing resulting in an intermediate form called document-term matrix [10] upon which data mining tasks should be performed.

In this paper, our objective is to develop one-class classification models that detect phishing mails and websites even when they are occurring for the first time. Hence, we propose a novel method of document classification using t-statistic-based feature selection, OCSVM and LSI in tandem. The rest of the paper is organized in the following manner. Section 2 reviews the literature. We describe the methods employed in this work in Sect. 3 while Sect. 4 presents the proposed methodology. The experimental setup and description of the datasets are presented in Sect. 5. Discussion of results is presented in Sect. 6 followed by conclusions in Sect. 7.

2 Literature Review

We briefly review the work reported in applying text mining to phishing detection and other applications. Phishing detection and protection are the challenging application areas for text mining. A Multi-label Classifier-based Associative Classification (MCAC) was proposed for phishing detection by Abdelhamid et al. [6]. They experimented on 2100 websites with C4.5, JRip classifiers. Apart from these, other techniques viz., Blacklist-based, fuzzy rule-based, Machine learning techniques, CANTINA were also discussed in their work. Phishing detection using data mining techniques was carried out in the works of Abu-Nimeh et al. [11]. They experimented with the corpus size of 2889 emails (with the proportion of 59.5% legitimate, 40.5% phishing) and used 43 features. They employed different techniques including, Logistic Regression (LR), Naive Bayes (NB), Support Vector Machine (SVM), Random Forest (RF).

In another research work Garera et al. [12] explained the some of the key features (page-based, domain-based, type-based, and word-based) of phishing were identified from the URL's. They employed the LR for model construction. A method to identify features to find out the phishing web sites was proposed by Ludl et al. [13] who analyzed 1000 sites with two approaches (Blacklist and source analysis) and employed C4.5 classifier with 18 influential features. Phishing attacks concerning risk levels and the loss of market value was assessed by Chen et al. [14]. They analyzed 1030 phishing alerts of a public database using a hybrid method. They employed the Decision Tree (DT), SVM, and Neural Network (NN) models for classification and reported a prediction accuracy of 89%.

Classification of Web Pages into legitimate or phishing was carried out by He et al. [7]. They extracted 12 features from a web page and trained an SVM. They reported a true positive value of 97% and a false negative of 4%. Pandey and Ravi [15] employed various classifiers such as Genetic Programming (GP)/Multilayer Perceptron (MLP)/DT/Group Method Data Handling (GMDH)/SVM/LR/Probabilistic Neural Network (PNN) for phishing detection. Recently, Pandey and Ravi [16] performed the detection of phishing websites and spam using text mining. In this study, they built the various models with GP, LR, PNN, MLP, Classification And Regression Tree (CART), GP + CART and reported high sensitivity values. In both the studies (i.e., phishing email and websites) they reported high sensitivity value for GP. They also provided the statistical significance test for these experiments.

The Malware identification was performed by Lee and Stolofo [17] using Data Mining techniques. They extracted some of the features (system calls) for identifying the intrusion. Through the concept of Association Rule Mining, they found frequent system calls and generated the rule set. Intelligent Malware detection system (IMDS) using Association Rule Mining was done by Ye et al. [18], and further, they compared it with NB, SVM, and J4.8 classifiers. The accuracies of these models were 83.86%, 90.54%, and 91.49% respectively and 93.07% with proposed approach (IMDS).

Composite malware detection scheme using run-time API (Application Program Interface) calls of Windows were introduced in the works of Ahmed et al. [9]. They selected 237 API calls for both benign and Malware programs and grouped them into seven, based on their functionality. They reported a detection rate of 0.97. Malware detection using API calls through text mining was performed using Mutual Information as a feature selection method by Sundarkumar and Ravi [19]. They performed classification using DT/SVM/MLP/PNN/GMDH/OCSVM classifiers under 10-Fold Cross-Validation (10-FCV) and over-sampling the minority samples. Finally, they reported sensitivity values of 100% with OCSVM.

Latent Semantic Indexing (LSI) has been applied in various works from past years. MLP and Singular Valued Decomposition (SVD) method were used for document classification [20] on a subset Reuter-21578 dataset. Clustering of text documents by combining LSI and GA was reported in the works of Song and Park [21]. The proposed model was evaluated on Reuter-21578 and they concluded that the optimal number of clusters is formed with this approach. Recently, classification of research projects based on technologies of research funding organizations was done by Thorleuchter and Van den Poel [22]. Technologies that occur together are grouped into classes by LSI. Textual patterns are representatives for each class, and projects are assigned to these classes. This enables the assignment of each project to all technologies grouped by LSI.

3 Overview of Techniques Used

3.1 *One-Class Support Vector Machine (OCSVM)*

One-Class Support Vector Machine (OCSVM) is similar to SVM except that it is trained with data of one class. It builds the boundary space from other class examples. Some of the advantages using OCSVM include application to problems with high-dimensions where other methods fail, flexibility of the choice of the kernel. It also has the same disadvantages as that of SVM. OCSVM is being used in several applications including image retrieval [23], document classification [24], Face detection [25], Sound recognition system, [26] and Anomaly detection in videos [27].

3.2 *Latent Semantic Indexing (LSI)*

Singular Value Decomposition (SVD) was proposed by Berry et al. [28]. They identified the higher order structure, association of terms within documents by finding the SVD of Document-term matrix. They considered the matrix with largest singular vectors, in turn matched with query documents. This method is also called LSI proposed by Dumais et al. [29] in 1988. It is a well-known technique applied in the Information Retrieval (IR) field. It is a mathematical technique for extracting and inferring relations of expected contextual usage of words in passages of discourse. It is a method for discovering hidden concepts in the document data. The terms in a document are then expressed as vectors with elements corresponding to these concepts. Each element in a vector gives the degree of participation of the document or term in the corresponding concept. The goal is not to describe the concepts verbally, but to be able to represent the documents and terms in a unified way for exposing document-document, document-term, and term-term similarities. LSI was described in detail in the following works [30, 31].

3.3 *Feature Selection Method*

We used t-statistic feature selection method. It is one of the most popular feature selection techniques. Using this technique, we computed t- statistic values for all features using equation as in (1):

$$t = \frac{|\mu_1 - \mu_2|}{\sqrt{\frac{\sigma_1^2}{n_1} + \frac{\sigma_2^2}{n_2}}} \quad (1)$$

where μ_1 and μ_2 represent the mean values of features of two different classes respectively, σ_1 and σ_2 represent the corresponding standard deviations for each class and n_1 and n_2 represent the number of samples in each class. Accordingly, t-values are calculated for all features. The features with higher t-statistic value have more discriminating power compared to other features. Therefore, based on the t-value, we will select as many features as we want from the top.

4 Proposed Method

Latent Semantic Indexing has been applied in the past for text document classification on several occasions. However, when the number of documents (of the negative class) to compare against is large, which is normally the case in phishing and other applications, then LSI consumes a lot of computational time. In order to reduce the comparison time for determining phishing mail/website, we need to reduce the number of documents (of the negative class) considerably while ensuring that the reduced set of documents remain representative of the negative class. Thus, OCSVM trained on the negative class alone fits the bill exactly, wherein the support vectors (SV) extracted, while representing the negative class, are very few in number compared to that of the original negative class. Through this approach, we generated a model with less number of document comparisons for classifying the text documents. Then, the LSI is invoked to take care of the document comparison and eventually classifying the positive records, when they are presented in the test phase. The proposed method is depicted in Fig. 1.

Our proposed methodology mainly consists of three phases. First, we collected all the samples of the various types of data sets (i.e., phishing, customer feedbacks) and move forward for text preprocessing. After preprocessing (i.e., tokenization, stop words removal, stemming) phase, we formulate the Document-term matrix. Afterwards, we applied the t-statistic feature selection method to remove the irrelevant features.

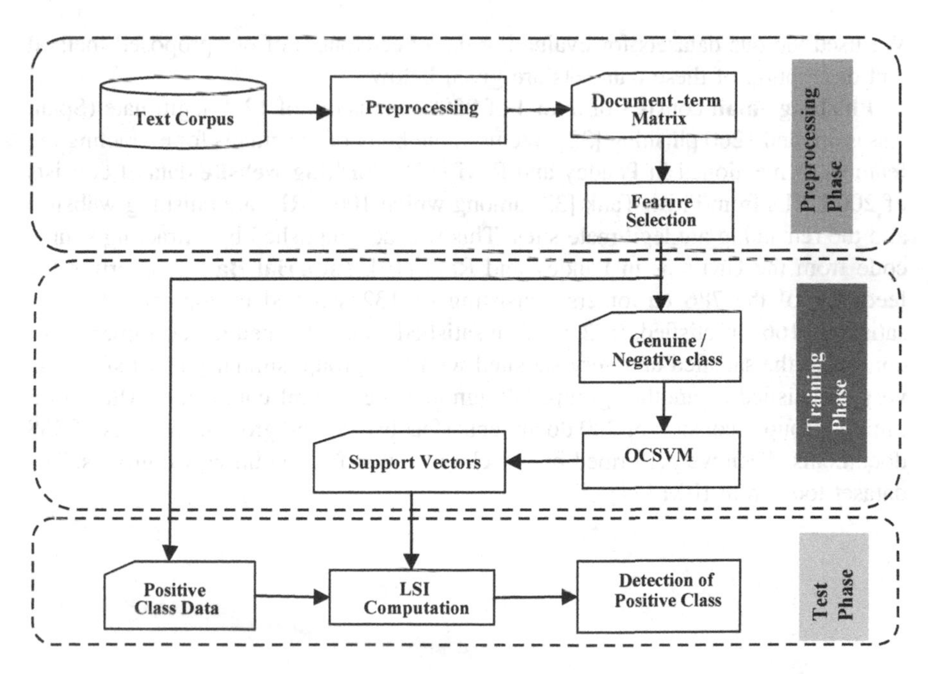

Fig. 1 Proposed methodology

In the second phase, we employed OCSVM and extracted SV for the negative class. Because of our assumption that we do not know the characteristics of any phishing email/Web site, we wanted to model the characteristics of the negative class. So, we extracted the most influential documents in the form of SV of negative class (i.e., legitimate/benign/genuine) from the Document-term matrix using OCSVM. After extraction of these SV, we apply the LSI technique in the final phase (test phase). In this test phase, first we decompose the support vector matrix into three matrices (U, S, V) using SVD. Later, we calculated the new document vector co-ordinates (which consist of Eigen values) in the new dimensional space. These co-ordinates are the individual support vectors/document vectors. After this step, we find the query vector for each test document in our positive class. Thereafter, we calculated the cosine similarity between query and document vectors.

After computation of similarity score, we kept threshold of 0.5 for classifying a document to the positive class. We compute the average value of each of the matching score with respect to number of documents and then, finally, we perform classification based on this value.

5 Experimental Methodology

5.1 Description of Datasets

We used various datasets for evaluating the effectiveness of our proposed method and description of these data sets are given below:

Phishing email consists of a total of 2500 emails out of 1240 legitimate (Spam Assassin) and 1260 phishing [32]. We used the body of the emails for extracting the features as mentioned in Pandey and Ravi [15]. **Phishing website** dataset consists of 200 URLs from Phish Tank [33] among which 100 URLs are phishing websites and the remaining are legitimate sites. This was accomplished by extracting source code from the URLs as in Pandey and Ravi [16]. **Imperial Bank** comprises the feedback of the 786 customers consisting of 132 satisfied customers, 148 very satisfied, 166 unsatisfied, 172 very unsatisfied, and 168 neutral customers. We combined the satisfied and very satisfied with one group, similarly unsatisfied and very unsatisfied as another group. We ignored the neutral comments. After combining, group-1 consists of 280 documents (feedbacks) and group-2 consists of 338 documents. Then we performed binary classification task on these two groups. The dataset took from IBM [34].

5.2 Experimental Procedure

In this process, first we collected all above-mentioned data sets from the various sources. Then we grouped into two categories, i.e., positive (phishing) and negative (Benign/Genuine). Later we performed text mining process. For text preprocessing to intermediate form, we used an open-source tool called Rapid Miner [35]. We followed the same process as that of [15, 16] with the same features generated by them for the sake of having proper comparison. We extracted the SV of legitimate (Negative class) samples from the document-term matrix. For this, we used the library called Library for Support Vector Machines (LIBSVM) [36]. After extraction of these SVs, we applied LSI. We used MATLAB [37] to implement LSI. We conducted the experiments on the machine with Intel i5 processor, 2.6 GHz, 8 GB RAM, 500 GB Hard Disk and 64-bit Windows 8 operating system.

6 Results and Discussion

Sensitivity (The true positive rate) is used as the performance measure in this paper, even though only the positive class is present in the hybrid model in the test phase. Because our goal is to detect phishing emails, phishing websites samples. We extracted the following features from the phishing website dataset which contained in the recent work [7, 16]. Similarly for Imperial Bank dataset, document-term matrix was generated with term frequency approach. We extracted 216 features from the Imperial Bank dataset. After construction of the document-term matrix and feature selection process, we obtained the SV for all data sets from negative class examples using OCSVM. The number of SV extracted from various data sets, viz., 621 from phishing emails, 54 from phishing websites (with feature selection) and 56 from phishing websites (without feature selection) and finally, 141 from Imperial Bank datasets respectively. Results obtained by through our proposed approach are listed in Table 1. We extracted SV with all kernels (Linear/Polynomial/Sigmoid/Radial Basis Function (RBF)). We experimented with all these kernels, and reported the best values compared to the previously reported results in the literature.

First, we will start our discussion from the Imperial Bank dataset results. We included the kernels and their respective values. In this, we reported the sensitivity in terms of percentage with respect to the kernels as follows: 66.44 (Linear), 81.3

Table 1 Results of the present study

Dataset	Sensitivity	
	Other approach	Proposed approach
Imperial bank	NA	99.11*
Phishing websites (with 9 features)	98 [16]	100*
Phishing email (with 23 features)	97.29 [15]	100*

(Polynomial), 99.22 (RBF), and 66.43 (Sigmoid). For phishing email these values are 100 (linear), 93.19 (polynomial), 99.83 (RBF), and 99.67 (sigmoid). Similarly for Phishing websites with 17 features, the sensitivity values with respect to kernel obtained are as follows: 92.73 (Linear), 98.21 (Polynomial), 96.11 (RBF) and 92.73 (Sigmoid). The sensitivity with respect to the kernel of the same dataset with nine features is 88.23 (linear), 89.28 (polynomial), 100 (RBF), and 88.23 (sigmoid). From the Table 1, it is observed that the results obtained by our method for Imperial Bank dataset sensitivity value is 99.11%. For Phishing email data we achieved 100% with 23 features, for phishing websites with nine features we reported the value of 100. Table 1 presents the results of [15] and [16] based on Binary classification. The best results obtained in [15] and [16] are with GP classifier on phishing emails and phishing websites respectively. In this proposed approach, both OCSVM and LSI are equally contributing toward the attainment of best results. These results are good with respect to the sensitivity.

7 Conclusion

In this paper, we proposed a methodology for text document classification using OCSVM and LSI. Primarily, we extracted the SV from negative class for each dataset and later we computed the latent score. The extraction of SV helped us to reduce the number of comparisons while computing the latent score. We tested the effectiveness of this method on various datasets like phishing email and websites and Imperial Bank dataset and achieved more than 99% of sensitivity on all of the datasets, outperforming other approaches in the literature.

References

1. Apte, C., Damerau, F., and Weissl, S. M.: Automated learning of decision rules for text categorization, ACM Transactions on Information Systems (TOIS) 12 (3), 233–251 (1994).
2. Bonchi, F., Castilo, C., and Gions, A.: Social Network Analysis and Mining for Business Applications. ACM Transactions on Intelligent Systems and Technology 2 (3), 1–37 (2011).
3. Dasgupta, K., Sigh, R., Viswanathan, B., Chakraborty, D., Mukherjea, S., Nanavati, A. A., and Joshi, A.: Social ties and their relevance to churn in mobile and telecom networks. In: 11th International Conference on Extending Database Technology (EDBT), March 25–30, Nantes, France, pp. 668-677 (2008).
4. Verbeke, W., Martens, D., and Baesens, B.: Social Network analysis for customer churn prediction. Applied Soft Computing 14 (C), 431–446 (2014).
5. Chakraborthy, G., Murali, P., and Satish, G.: Text mining and analysis: Practical methods, examples, and case studies. SAS Institute publisher (2014).
6. Abdelhamid, N., Ayesh, A., Thabtah, F.: Phishing detection based Associative Classification Data mining. Expert Systems with Applications 41(13), 5948–5959 (2014).
7. He, M., Horng, S-J., Fan, P., Khan, M. K., Run, R., Lai, J-L., Chen, R-J., and Sutanto, A.: An efficient phishing webpage detector. Expert Systems with Applications 38 (10), 12018–12027 (2011).

8. Metsis, V., Androutsopoulos, I., and Paliouras, G.: Spam Filtering with Naive Bayes - Which Naive Bayes?. In: 3rd Conference on Email and Anti-Spam (CEAS), July 27–28, Mountain View, California, USA (2006).
9. Ahmed, F., Hameed, H., Shafiq, Z., and Farooq, M.: Using Spatio temporal Information in API calls with Machine learning Algorithms for Malware detection. In: 2nd ACM workshop on Security and Artificial Intelligence (AISec), November 9th, Chicago Illinois, USA, pp. 55–62 (2009).
10. Salton, G., and McGill, M. J.: Introduction to Modern Information Retrieval. McGraw-Hill, Inc., New York, NY, USA, (1986).
11. Abu-Nimeh, S., Nappa, D., Wang, X., Nair, S.: A comparison of Machine Learning techniques for phishing detection. In: APWG eCrime Researchers Summit, October 4–5, Pittsburgh, PA, USA, pp. 60-69 (2007).
12. Garera, S., Provos, N., Chew, M., and Rubin, A. D.: A Framework for Detection and Measurement of Phishing Attacks. In: Special Interest Group on Security, Audit and Control (SIGSAC) Workshop On Recurring Malcode (WORM), November 2, Alexandria, Virginia, USA, pp. 1–8 (2007).
13. Ludl, C., Mcallister, S., Kirda, E., Kruegel, C.: On the effectiveness of techniques to detect phishing sites. In: Detection of Intrusions and Malware, and Vulnerability Assessment (DIMVA), July 12-13, Switzerland, pp. 20–39 (2007).
14. Chen, X., Bose, I., Leung, A. C. M., and Guo, C.: Assessing the severity of phishing attacks: A hybrid data mining approach. Decision Support Systems 50 (4), 662–672 (2011).
15. Pandey, M., and Ravi, V.: Detecting phishing e-mails using text and data mining, in: International Conference on Computational Intelligence & Computing Research (ICCIC). December 18–20, Coimbatore, India, pp. 249-255 (2012).
16. Pandey, M., and Ravi, V.: Text and Data mining to detect phishing websites and spam emails, in: Swarm, Evolutionary, and Memetic Computing (SEMCCO), December 19-21, Chennai, India, LNCS 8298 Part-II, pp. 559–573 (2013).
17. Lee, W., and Stolofo, J. S.: Data mining approaches for intrusion detection. In: USENIX Security symposium, January 26–29, San Antonio, Texas, pp. 1-6 (1988).
18. Ye, Y., Wang, D., Li, T., and Ye, D.: IMDS: Intelligent Malware Detection System. In: 13th KDD, August 12–15, San Jose, California, USA, pp. 1043-1047 (2007).
19. Sundarkumar, G. G., and Ravi, V.: Malware detection by text and data mining. In: International Conference on Computational Intelligence & Computing Research (ICCIC), December 26–28, Enathi, India, pp. 1-6 (2013).
20. Li, C. H., and Park, S. C.: An efficient document classification model using an improved back propagation neural network and singular value decomposition. Expert Systems with Applications 36 (2), 3208–3215 (2009).
21. Song, W., and Park, S. C.: Genetic algorithm for text clustering based on latent semantic indexing. Computers and mathematics with applications 57 (11), 1901–1907 (2009).
22. Thorleuchter, D., and Van den Poel, D.: Application based Technology Classification with Latent Semantic Indexing. Expert Systems with Applications 40 (5), 1786–1795 (2013).
23. Chen, Y., Zhou, X., and Huang, T. S.: One-class SVM for learning in image retrieval. In: International Conference on Image Processing, October 7-11, Thessaloniki, Greece, pp. 34–37 (2001).
24. Manevitz, L. M., and Yosef, M.: One-Class SVMs for document classification. Journal of Machine Learning Research 2, 139–154 (2001).
25. Jin, H., Liu, Q., Lu, H.: Face detection using one-class-based support vectors, In: 6th International Conference on Automatic Face and Gesture Recognition (FGR), 19th May, Seoul, South Korea, pp. 457–462 (2004).
26. Hempstalk, K., Frank, E., and Witten, I. H.: One-class classification by combining density and class probability estimation. In: ECML PKDD, September 15-19, Antwerp, Belgium, Part I, LNAI 5211, pp. 505–519 (2008).

27. Liu, C., Wang, G., Ning, W., Lin, X., Li, L., and Liu, Z.: Anomaly detection in surveillance video using motion direction statistics. In: 17th International Conference on Image Processing, September 26-29, Hong Kong, pp. 717–720 (2010).
28. Berry, M. W., Dumais, S. T., and Obrien, G. W.: Using Linear Algebra for Intelligent Information Retrieval. In: Society for Industrial and Applied Mathematics (SIAM) Review, 37 (4), pp. 573–595 (1995).
29. Dumais, S. T., Furnas, G. W., Landauer, T. K., Deerwester, S., and Harshman, R.: Using latent semantic analysis to improve access to textual information. In: CHI, April 18-23, Los Angeles, California, USA, pp. 281–285 (1988).
30. Deerwester, S. C., Dumais, S. T., Landauer, T. K., and Furnas, G. W.: Indexing by latent semantic analysis. Journal of the American Society for Information Science (JASIS), 391–407 (1990).
31. Furnas, G. W., Deerwester, S., Dumais, S. T., Landauer, T. K., Harshman, R. A., Streeter, L. A., and Lochbaum, K. E.: Information Retrieval using a singular value decomposition model of latent semantic structure. In: SIGIR, August 24-28, Grenoble, France, pp. 465–480 (1998).
32. Phishing corpus, http:// http://monkey.org/~jose/wiki/doku.php.
33. Phishtank, http://www.phishtank.com.
34. IBM SPSS, http://www-01.ibm.com/software/in/analytics/spss/products/data-collection/.
35. Rapid Miner (2012), https://rapidminer.com.
36. LIBSVM, http://www.csie.ntu.edu.tw/~cjlin/libsvm/#download. Visited 2014.
37. MATLAB (2012), www.mathworks.com.

System Identification: Survey on Modeling Methods and Models

A. Garg, K. Tai and B.N. Panda

Abstract System identification (SI) is referred to as the procedure of building mathematical models for the dynamic systems using the measured data. Several modeling methods and types of models were studied by classifying SI in different ways, such as (1) black box, gray box, and white box; (2) parametric and non-parametric; and (3) linear SI, nonlinear SI, and evolutionary SI. A study of the literature also reveals that extensive focus has been paid to computational intelligence methods for modeling the output variables of the systems because of their ability to formulate the models based only on data obtained from the system. It was also learned that by embedding the features of several methods from different fields of SI into a given method, it is possible to improve its generalization ability. Popular variants of genetic programming such as multi-gene genetic programming is suggested as an alternative approach with its four shortcomings discussed as future aspects in paving way for evolutionary system identification.

Keywords System identification · Modeling methods · Genetic programming · Computational intelligence

A. Garg
Department of Mechatronics Engineering, Shantou University,
Shantou 515063, China

K. Tai
School of Mechanical and Aerospace Engineering,
Nanyang Technological University, 50 Nanyang Avenue,
Singapore 639798, Singapore

B.N. Panda (✉)
Department of Industrial Design, National Institute of Technology,
Rourkela 769008, India
e-mail: biranchi.panda3@gmail.com

S.S. Dash et al. (eds.), *Artificial Intelligence and Evolutionary Computations in Engineering Systems*, Advances in Intelligent Systems and Computing 517,
DOI 10.1007/978-981-10-3174-8_51

1 Introduction on System Identification

System identification (SI) is referred to as the phenomenon of building mathematical models for the systems from the given input–output data. The notion of modeling is applied to model the systems so as to understand their hidden behavior and predict its characteristics. Modeling includes the systems, models, and modeling methods, which can also be studied under the field of SI. The systems modeled can be manufacturing processes such as turning, vibratory finishing, and additive manufacturing, etc., or chemical processes such as fuel cell, reactors, or such as those involving the study of mechanical and thermal properties of graphene and carbon nanotubes or the stock market and weather phenomenon, etc. Among these processes, additive manufacturing processes (processes involves the fabrication of products from CAD data automatically), machining processes (material removal processes), and vibratory finishing (material removal processes) are the potential ones (Fig. 1). The working mechanisms behind these systems are governed by multiple input and output variables, which make these operating mechanisms complex. The cost involved in the execution of such systems is reasonably high, and therefore it can be costly to measure the data. The motivation behind analyzing the data is to find out the information from the system that can be vital for optimizing the performance of the systems. Also, in an era of widespread development of capital intensive systems with their complex operating mechanisms, the need of modeling and optimization has been strengthened.

The work described in this manuscript is divided into four sections. Section 2 discusses the models and modeling methods classified under various fields of SI. Section 3 discusses the alternative methodology suggested in the latest trends in era of SI. Finally, Sect. 4 concludes with critical issues and future aspects in evolutionary SI.

2 Survey on Models and Modeling Methods

Models can be built based on analysis of variance (ANOVA), hypothesis tests or using statistical, finite element or computational intelligence methods, etc. These models are used in diversified fields from science to engineering to unveil the hidden information for the practical understanding and realization of the system. To formulate these models, a gamut of modeling methods such as regression analysis, response surface methodology (RSM), partial least square regression, genetic programming (GP), artificial neural network (ANN), fuzzy logic (FL), M5-prime (M5′), support vector regression (SVR), adaptive neuro-fuzzy inference systems (ANFIS), etc., can be applied [1–5]. The models should be accurate in prediction and also should satisfy the system characteristics, i.e., not violate the system constraints. The generalization of data obtained from manufacturing systems is a capability highly demanded by the industry. Higher generalization ability of the model indicates that it has rightly captured the physics behind the system.

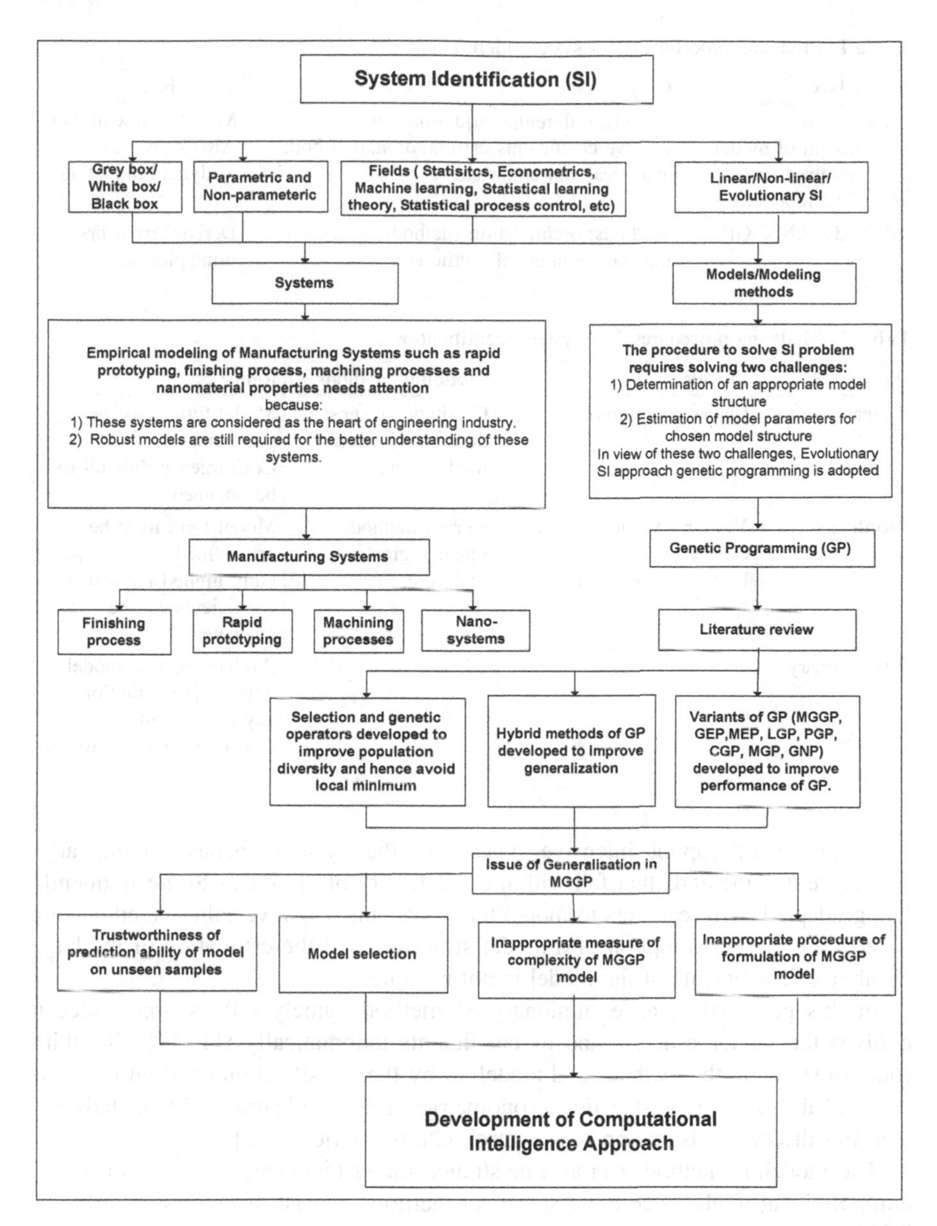

Fig. 1 Fields categorized under SI

Systems, models as well as modeling methods can be studied under various classifications of SI [6, 7]. For example, modeling methods can be studied under the three categories of modeling: gray box, white box, and black box (Table 1) [8].

Models and modeling methods can also be studied by the classification of SI into linear, nonlinear, and evolutionary (Table 2) [6, 9–13]. Due to advent in

Table 1 Modeling procedures 1 of system identification

Black box	Gray box	White box
Models: no assumption of model Also referred as empirical modeling	Models: differential equations, etc. Involve coefficients estimation and is both analytical and empirical	Models: Newton law Also known as analytical modeling
Methods: ANN, GP and FL	Methods: optimization methods, statistical methods, numerical methods	Derived from first principles

Table 2 Modeling procedures 2 of system identification

Type of SI	Models	Modeling methods	Remarks
Linear	Linear polynomial	Gradient, steepest descent, Newton method, etc.	Model form must be pre-defined Convergence difficult to be obtained
Nonlinear	Volterra, Weiner, block oriented systems, bilinear systems, etc.	Kernel method, linear regression, FFT, etc.	Model form must be pre-defined Only limited number of coefficients can be evaluated
Evolutionary	GP tree	GP	Produce explicit models without pre-definition of any of its form Generate variety of forms of models

development of capital intensive machines, the systems behave nonlinearly. Therefore, the methods that fall within the category of nonlinear SI are frequently being adapted by researchers to model these systems. However, these methods are based on a prior assumption of a model structure, and the estimation of the large number of coefficients of the model is not reliable.

In this perspective, an evolutionary SI method namely GP, is used, since it evolves the model structure and its coefficients automatically [11, 14]. The third route of studying the methods and models is by the classification of SI into various fields (Table 3) such as statistics, econometrics, machine learning (ML), statistical learning theory, statistical process control, chemometrics, etc. [7].

The modeling methods can also be studied under two categories: statistical and computational intelligence (CI). Statistical methods consist of regression analysis, RSM, partial least square regression, etc. [2, 15–17]. Statistical methods are based on various statistical assumptions (pre-definition of form of a model, residuals properties, etc.) which may not be a practical procedure to model the highly non-linear dynamic systems. CI methods comprise advanced heuristic and optimization methods such as GP, ANN, M5′, SVR, ANFIS, etc.

The study of modeling methods categorized under various classifications of SI reveals that the CI methods are extensively being applied by researchers for

Table 3 Modeling procedures 3 of system identification

Fields	Models	Modeling methods	Remarks
Statistics	*Z*, *t*, *F* and chisquare distributions	Regression, correlation, and factor analysis	Pre-assumption of the model structure, not suitable for modeling nonlinear systems
Econometrics	ARIMA, tobit, etc.	Mainly statistical methods	Need expertise for making decisions from the statistical models
Time series	ARIMA, GP, SVR and ANN	FFT, ANN, GP and SVR	Modern heuristic methods are mainly considered
Statistical learning theory	SVR model	Regularization networks and SVM	Includes new measure of performance of model such as ERM and SRM. Well known for providing generalization ability
Machine learning (ML)	Decision trees, ANN, SVR, GP and kNN	SVM, GP, M5, RIPPER, CN2, ANN and kNN	No pre-assumption of model structure Adapt to the nonlinearity of the systems. Implementation of methods require expert knowledge
Chemometrics	Polynomials, ANN, SVR, GP and kNN	DOE, signal processing PLS, PCA, MDS, ANN, SVR, GP and kNN	Emphasis on pre-processing of the data and validation of the model
Data mining (DM)	CRISP, SEMMA and six-sigma	Statistical charts, variable reduction methods and visualization methods	Finding hidden patterns in the data Highly crucial in banks and industries

modeling nonlinear systems because these methods have the ability to formulate the models from the given data without the need for incorporation of any other prior knowledge about the systems. More efficient CI methods have been developed by clustering the features of two or more methods. For example, the hybrid methods: GA-FL, GA-ANN, FL-ANN, particle swarm optimization (PSO)-ANN, etc., are able to predict the systems output accurately [16, 18–25]. The M5′ method used to build regression trees has the prediction accuracy on par with that of ANN but the researchers have not studied this method comprehensively [26, 27].

3 Genetic Programming for System Identification

Among CI methods, GP, also popularly known as evolutionary SI method, possesses a unique feature of generating the explicit models with just a prior adjustment of fewer settings such as the functional and terminal set. In addition, the mechanism of model formulation is not based on any statistical assumptions. The procedure is similar to GA and being extensively used as a structural optimization methodology. Extensive literature of GP in modeling of nonlinear systems is found. Researchers have developed hybrid approach of GP such as GA-GP [28], Clustering-GP [29], FEM-GP [30], GP-OLS [31], GP-SA [32], etc., for improving its generalization ability. New selection schemes and genetic operators for mutation, crossover, and reproduction have been developed [33]. In the past years, many variants of GP such as linear genetic programming, probabilistic genetic programming, multi-expression genetic programming, Cartesian based genetic programming (C-GP), gene expression programming (GEP), multi-gene genetic programming (MGGP), etc., have been developed [33]. The development of these variants involves the combination or extraction of the features from the other CI methods. Therefore, it can be hypothesized that by learning the features of several other modeling methods under the various modeling procedures of SI, it can thus provide a scope to develop a robust GP methodology. Among those variants developed, the MGGP method, which uses multiple sets of genes for the formulation of a model, is primarily focused.

4 Critical Issues and Future Aspects in Genetic Programming

Based on the preliminary applications of MGGP [34–38], it was found that the generalization is the main problem, due to which its applications have not gained much prominence. High generalization refers to the satisfactory performance of the model on the testing (unseen) data samples. The past applications of the MGGP include the manufacturing processes, soil mechanics, etc. [39]. However, it would be interesting to work on development of a robust MGGP methodology to model the highly dynamic and nonlinear systems such as the fuel cell systems, where there are high uncertainties in the system due to the external/uncontrollable factors such as the temperature.

In addition, while the authors were working closely with a major manufacturing industry (i.e., Rolls Royce), it is learned that the industry is keen to develop the robust functional expressions, which can take into account the uncertainties in the systems, easily optimized analytically and can be integrated into the system [40] for its improved monitoring. Moreover, there is also a demand for the development of user-friendly graphical user interface (GUI) software for the implementation of GP [33, 41].

Hence, the future work for authors is to work on different CI methods, specifically MGGP and develop robust (Fig. 1) GUI for it [42].

Acknowledgments This study was supported by Shantou University Scientific Research Funded Project (Grant No. NTF 16002)

Appendix

See Fig. 1.

References

1. M. Willis, H. Hiden, M. Hinchliffe, B. McKay, and G. W. Barton: Systems modelling using genetic programming, Computers & chemical engineering, 21, S1161–S1166 (1997).
2. M. Chandrasekaran, M. Muralidhar, C. M. Krishna, and U. Dixit, Application of soft computing techniques in machining performance prediction and optimization: a literature review, The International Journal of Advanced Manufacturing Technology, 46, 445–464 (2010).
3. E. Vladislavleva and G. Smits, Symbolic regression via genetic programming, Final Thesis for Dow Benelux BV.
4. U. Çaydaş and S. Ekici, Support vector machines models for surface roughness prediction in CNC turning of AISI 304 austenitic stainless steel, Journal of Intelligent Manufacturing, 23, 639–650 (2012).
5. S. N. Patra, R. J. T. Lin, and D. Bhattacharyya, Regression analysis of manufacturing electrospun nonwoven nanotextiles, Journal of Materials Science, 45, 3938–3946 (2010).
6. K. J. Astrom and P. Eykhoff, System identification–A survey, Automatica, 7, 123–162 (1971).
7. L. Ljung: Perspectives on system identification, Annual Reviews in Control, 34, 1–12 (2010).
8. S. Sette and L. Boullart: Genetic programming: principles and applications, Engineering applications of artificial intelligence, 14, 727–736 (2001).
9. X. Hong, R. Mitchell, S. Chen, C. J. Harris, K. Li, and G. Irwin: Model selection approaches for nonlinear system identification: a review, International Journal of Systems Science, 39, 925–946 (2008).
10. S. Billings, Identification of nonlinear systems a survey, 272–285(1980).
11. M. Affenzeller and S. Winkler, Genetic algorithms and genetic programming: modern concepts and practical applications, Chapman & Hall/CRC, 6 (2009).
12. Y. Ku and A. A. Wolf: Volterra-Wiener functionals for the analysis of nonlinear systems, Journal of The Franklin Institute, 281, 9–26 (1966).
13. L. A. Zadeh: From circuit theory to system theory, Proceedings of the IRE, 50, 856–865 (1962).
14. J. R. Koza: Genetic programming as a means for programming computers by natural selection, Statistics and Computing, 4, 87–112 (1994).
15. A. Garg, Y. Bhalerao, and K. Tai: Review of empirical modelling techniques for modelling of turning process, International Journal of Modelling, Identification and Control, 20, 121–129 (2013).
16. B. N. Panda, M. R. Babhubalendruni, B. B. Biswal and D. S. Rajput: Application of artificial intelligence methods to spot welding of commercial aluminum sheets (BS 1050). In Proceedings of Fourth International Conference on Soft Computing for Problem Solving, Springer India. 21–32 (2015).
17. A. Garg and K. Tai: Comparison of statistical and machine learning methods in modelling of data with multi-collinearity, International Journal of Modelling, Identification and Control, 18, 295–312 (2013).
18. Z. Yang, X. S. Gu, X. Y. Liang, and L. C. Ling: Genetic algorithm-least squares support vector regression based predicting and optimizing model on carbon fiber composite integrated conductivity, Materials & Design, 31, 1042–1049, (2010).

19. A. Garg, K. Tai, C. Lee, and M. Savalani: A hybrid\text {M} 5^\ prime-genetic programming approach for ensuring greater trustworthiness of prediction ability in modelling of FDM process, Journal of Intelligent Manufacturing, 1–17, (2013).
20. D. Umbrello, G. Ambrogio, L. Filice, and R. Shivpuri: A hybrid finite element method–artificial neural network approach for predicting residual stresses and the optimal cutting conditions during hard turning of AISI 52100 bearing steel, Materials & Design, 29, 873–883, (2008).
21. B. Wang, X. Wang, and Z. Chen: A hybrid framework for reservoir characterization using fuzzy ranking and an artificial neural network, Computers and Geosciences, 57, 1–10 (2013).
22. Y. G. Liu, J. Luo, and M. Q. Li: The fuzzy neural network model of flow stress in the isothermal compression of 300M steel, Materials and Design, 41, 83–88 (2012).
23. W. Li, Y. Yang, Z. Yang, and C. Zhang: Fuzzy system identification based on support vector regression and genetic algorithm, International Journal of Modelling, Identification and Control, 12, 50–55 (2011).
24. Garg, A., Lam L. S. Jasmine: Measurement of Environmental Aspect of 3-D Printing Process using Soft Computing Methods. Measurement, 75, 171–179 (2015).
25. Mukherjee, I. and Ray, P. K.: A review of optimization techniques in metal cutting processes, Computers & Industrial Engineering, 50, 15–34 (2006).
26. Quinlan, J. R: Learning with continuous classes, 343–348(1992).
27. Wang Y. and Witten, I. H: Induction of model trees for predicting continuous classes (1996).
28. Wei-Po, L., Hallam, J. and Lund, H. H: A hybrid GP/GA approach for co-evolving controllers and robot bodies to achieve fitness-specified tasks, in Evolutionary Computation, Proceedings of IEEE International Conference, 384–389 (1996).
29. Xie, H., Zhang, M. and Andreae, P: Population Clustering in Genetic Programming, Eds., ed: Springer Berlin Heidelberg, 3905, 190–201(2006).
30. Kumarci, K., Dehkordi, P. and Mahmodi, I: Calculation of Plate Natural Frequency by Genetic Programming, Journal of Applied Sciences, 10, 451–461, (2010).
31. Madár, J., Abonyi, J. and Szeifert, F: Genetic programming for the identification of nonlinear input-output models, Industrial & engineering chemistry research, 44, 3178–3186 (2005).
32. Folino, G., Pizzuti, C. and Spezzano, G: Genetic Programming and Simulated Annealing: A Hybrid Method to Evolve Decision Trees, in Genetic Programming. vol. 1802, R. Poli, W. Banzhaf, W. Langdon, J. Miller, P. Nordin, and T. Fogarty, Eds., ed: Springer Berlin Heidelberg, 294–303 (2000).
33. Garg, A. and Tai, K: Review of genetic programming in modeling of machining processes, Proceedings of International Conference in Modelling, Identification & Control (ICMIC), 653–658 (2012).
34. Garg A., Lam, J.S.L., Gao L: Energy conservation in manufacturing operations: modelling the milling process by a new complexity-based evolutionary approach, Journal of Cleaner Production, 108, 34–45(2015).
35. A. Garg and K. Tai: Selection of a robust experimental design for the effective modeling of the nonlinear systems using genetic programming, in Proceedings of 2013 IEEE Symposium on Computational Intelligence and Data mining (CIDM), Singapore, 293–298 (2013).
36. Garg A., Lam, J.S.L: Improving Environmental Sustainability by Formulation of Generalized Power Consumption Models using an Ensemble Evolutionary Approach, Journal of Cleaner Production, 102, 246–263 (2015).
37. Panda, Biranchi Narayan, MVA Raju Bahubalendruni, and Biswal, B.B: Comparative evaluation of optimization algorithms at training of genetic programming for tensile strength prediction of FDM processed part. Procedia Materials Science 5, 2250–2257 (2014).
38. Panda, B.N., Garg, A. and Shankhwar, K: Empirical investigation of environmental characteristic of 3-D additive manufacturing process based on slice thickness and part orientation. Measurement, 86, 293–300 (2016).
39. Panda, B., Garg, A., Jian, Z., Heidarzadeh, A. and Gao, L: Characterization of the tensile properties of friction stir welded aluminum alloy joints based on axial force, traverse speed,

and rotational speed. Frontiers of Mechanical Engineering, 1–10. (2016). doi:10.1007/s11465-016-0393-y.
40. U. M. O'Reilly, Genetic programming theory and practice II vol. 2: Springer-Verlag New York Inc, 2005.
41. A. Kordon, F. Castillo, G. Smits, and M. Kotanchek, Application issues of genetic programming in industry, Genetic Programming Theory and Practice III, pp. 241–258, 2006.
42. M. Deistler: System identification and time series analysis: Past, present, and future, Stochastic Theory and Control, 97–109 (2002).

Grid Integration of Small Hydro Power Plants Based on PWM Converter and D-STATCOM

Swayam Samparna Mishra, Abhimanyu Mohapatra and Prasanta Kumar Satpathy

Abstract In this paper, the problems posed during the integration of small hydro power generating units with the existing power grid are smoothly handled. The proposed scheme considers a pulse width modulated (PWM) converter and an improved D-STATCOM. The results obtained from MATLAB-Simulink-based simulation successfully validate the smooth interfacing of small hydro power units with the grid.

Keywords Small hydro integration · PWM converter · D-Statcom

1 Introduction

India is considered as a bulk producer of hydroelectricity in the global scenario. There are plenty of scopes in India for installation of small, mini, and micro-hydro power plants. However, the bottleneck lies with smooth exchange of the generated power by the renewable energy sectors into the existing system. Microgrids are often suitable for this purpose as they can operate either in grid-connected mode or in islanded mode of operation or both of them. Microgrids are also very flexible and reliable in their operation. However, the microgrid integration is also not free from disadvantages due to the presence of long distance transmission, variable losses in lines, frequent voltage fluctuation at the buses, and poor power quality. The major thrust behind the motivation for setting up of microgrid projects is to provide cheap and secured electricity in remote rural areas and at the same time to reduce the amount of environment pollution due to green house gas emission.

S.S. Mishra · A. Mohapatra · P.K. Satpathy (✉)
College of Engineering and Technology, Bhubaneswar, Odisha, India
e-mail: satpathy.pks@gmail.com

S.S. Mishra
e-mail: swayamsamparna.mishra@gmail.com

A. Mohapatra
e-mail: abhimanyumohapatra@yahoo.com

S.S. Dash et al. (eds.), *Artificial Intelligence and Evolutionary Computations in Engineering Systems*, Advances in Intelligent Systems and Computing 517,
DOI 10.1007/978-981-10-3174-8_52

Among various issues related to power quality problems, voltage sag or voltage swell is the most common occurrences in any power system. The magnitude of voltage sag and/or voltage swell usually lies in the range of 10–90% of nominal voltage with time period of one cycle to one minute.

Such events may be caused due to initiation of faults in the system or due to the event of starting of large industrial motor loads. Harmonic currents in distribution system can also affect power quality by causing harmonic distortion, low power factor, and additional losses in form of heating in the electrical equipment. Among the variety of methods available for improvement of power quality of the system, the use of distribution static compensator (D-STATCOM) is most effective.

The D-STATCOM can also sustain reactive current at low voltage, hence it can also be used as a voltage and frequency support for the system. The D-STATCOM is also capable of providing high speed control of reactive power to provide voltage stabilization in power systems. In this context, this paper mainly covers the various issues that will strengthen the economics of microgrids. This paper also identifies enabling technologies for efficient interfacing and smooth operation of microgrids in the smart grid environment. A new PWM-based control scheme has been implemented to control the electronic valves in the D-STATCOM.

2 Modeling of SHPs for Microgrid Interfacing

Rayes Ahmad et al. [1] presented the simulation model of a typical canal-type small hydroelectric power plant. The various components of small hydroelectric plant like open channel, governor and semi-Kaplan turbine, synchronous generator, exciter are being considered for modeling. Rama Rao et al. [2] presented the modeling of solar and hydro hybrid energy sources. Meshram et al. [3] carried out the simulation modeling of grid-connected DC-linked PV/hydro hybrid system. Almoataz et al. [4] provided control algorithm for a three-phase grid-connected photovoltaic system in which an inverter designed for grid-connected photovoltaic arrays can synchronize a sinusoidal current output with a voltage grid. Nagapadma et al. [5] presented the current control methods to produce a sinusoidal AC output. The main task of the control systems in current regulated inverters is to force the current vector in the three-phase load according to a reference trajectory. Satpathy et al. [6] explained the use of static VAR compensators (SVCs) to overcome these practical limitations encountered by SHPs so as to ensure smooth evacuation of every unit of real power generated by such units to the neighboring grid in a grid-connected power system scenario. Singh et al. [7] presented the brief description of the operation and control of microgrid, the voltage and power regulation and energy management in microgrid and highlighted the simulation works done for the different load models of microgrids. Kozhi et al. [8] highlighted the architecture of microgrids, which is driven by renewable sources to tackle the power outages. Ravichandrudu et al. [9] explained the small hydro generation unit and a wind farm that contains nine variable speed, doubly fed induction generator based wind turbines. Mithilesh et al.

[10] explained the role of D-STATCOM and located it at load side in the electrical distribution system so as to maintain the power quality. Kiran Kumar et al. [11] presented the enhancement of voltage sags/swell, harmonic distortion and low power factor using D-STATCOM. The model is based on the voltage source converter (VSC) principle. The D-STATCOM injects a current into the system to mitigate the voltage sags/swell to improve harmonic distortion and low power factor. Molavi et al. [12] proposed a method to extract the active and reactive parts of the positive and negative sequence component for generating reference values of current that needs to be injected into the D-STATCOM in order to compensate the voltage errors. Robert H. Lasseter et al. [13] explained that sources of microgrid could operate in parallel to the grid or in island, providing UPS services and could be intentionally disconnected when the quality of power from the grid falls below certain standards. Lopes et al. [14] described the evaluation of the feasibility of control strategies to be adopted for the operation of a microgrid when it becomes isolated.

Although the components of microgrid units are fairly well understood, the system as a whole involves lot of complications. When several components are assembled to form a microgrid, the system behavior becomes unpredictable. This being the case, modeling the system and simulating it in order to develop an appropriate control system, is the heart of microgrid research. Nowadays, several research groups around the world are investigating the feasibility and benefits that the microgrids may provide. Some problems are also encountered, which includes the effects due to unbalanced loads and harmonics associated with the system.

In this work, the authors intend to address such problems by detailing the modeling of the microgrid with a small hydro power plant interfacing for the investigation of the voltage, current active, and reactive power responses. The equivalent circuit and the phasor diagram are important tools to understand and study the power system stability phenomena. The equivalent per phase circuit of a generator connected to a strong grid is shown in Fig. 1 and the corresponding phasor diagram is shown in Fig. 2. The small hydro power plant considered in this paper also contains various subsystems for obtaining smooth control during grid integration of the plant. These subsystems are modeled as indicated in Figs. 3, 4, 5, and 6.

Given an infinite grid with voltage 1 pu, the current can be expressed as shown in Eq. (1). Knowing the current, the infinite grid voltage, and the q-axis synchronous reactance, the induced q-axis synchronous voltage can be found. The angles φ and δ are the angles of I and E_q, respectively, relative to V_s.

$$I = P - jQ \tag{1}$$

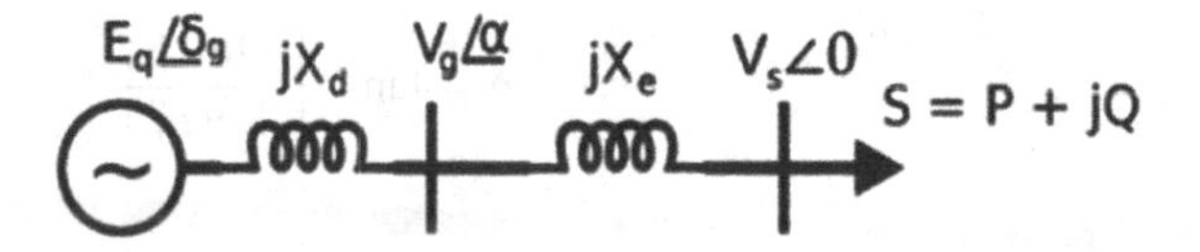

Fig. 1 Equivalent circuit of a generator connected to a strong grid

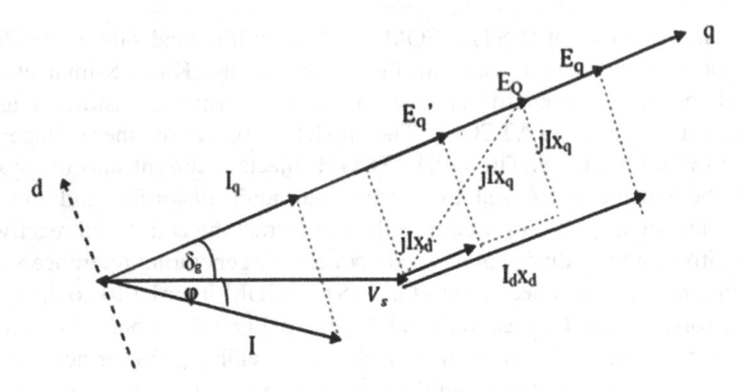

Fig. 2 Phasor diagram of a generator connected to a strong grid

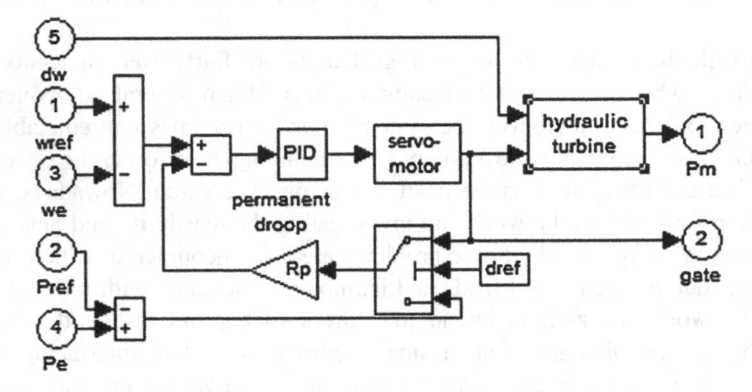

Fig. 3 Hydraulic turbine and governor subsystem

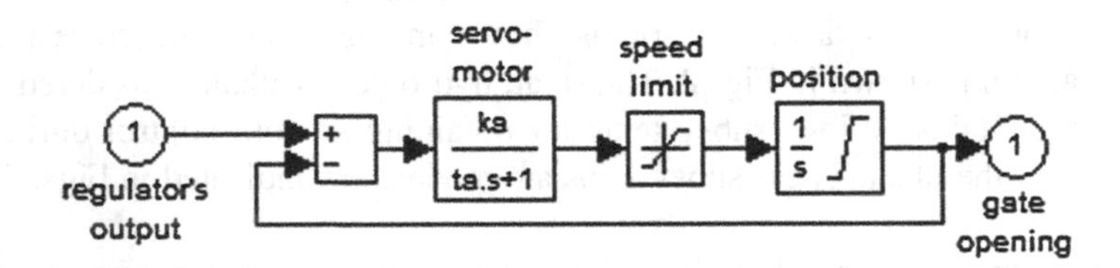

Fig. 4 Servo motor subsystem

$$E_q = V_s + jX_qI \tag{2}$$

$$\varphi = \tan^{-1}\left[\frac{-\text{Im}(I)}{\text{Re}(I)}\right] \tag{3}$$

$$\delta = \tan^{-1}\left[\frac{-\text{Im}(E_q)}{\text{Re}(E_q)}\right] \tag{4}$$

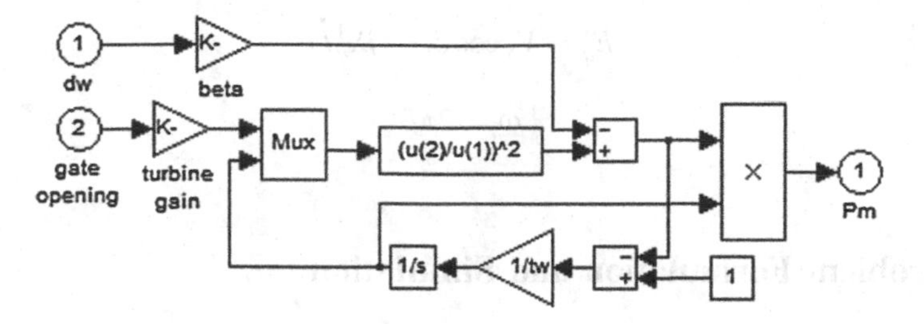

Fig. 5 Hydraulic turbine subsystem

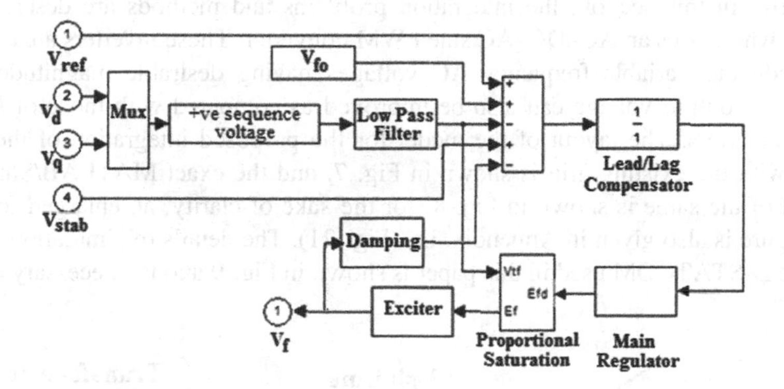

Fig. 6 Excitation subsystem

$$V_q = |V_g| \cos \delta_g \tag{5}$$

$$V_d = -|V_g| \sin \delta_g \tag{6}$$

$$I_d = -|I| \sin(\delta_g + \varphi) \tag{7}$$

$$I_q = |I| \cos(\delta_g + \varphi) \tag{8}$$

The d-axis and q-axis components of the terminal voltage and current can now be found, from the generator voltage V_g, the angles δ and φ and the current I. Knowing the d-axis component of the current and the d-axis reactance, the induced voltages E_q and E_q^1 in steady state and transient, respectively, can be found as follows:

$$E_q = V_s \cos \delta_g - jX_d I_d \tag{9}$$

$$E_q^1 = V_S \cos \delta_g - jX_d^1 I_d \quad (10)$$

$$\omega = 2\pi f \quad (11)$$

3 Problem Formulation and Simulation

The entire simulation system for the transient analysis of the proposed small hydroelectric power plant has been developed in a MATLAB/Simulink-based platform. In this section, the integration problems and methods are described in detail, which uses an AC–DC–AC sine PWM converter. These inverters are capable of producing variable frequency AC voltages having desirable magnitude. The quality of output voltage can also be improved as compared with those of square wave inverters. The layout of the model for the proposed integration of the SHP plant with the existing grid is shown in Fig. 7, and the exact MATLAB/Simulink model of the same is shown in Fig. 8 (for the sake of clarity, an enlarged form of this figure is also given in Appendix, i.e., Fig. 21). The details of simulation model for the D-STATCOM used in this paper is shown in Fig. 9 and the necessary details

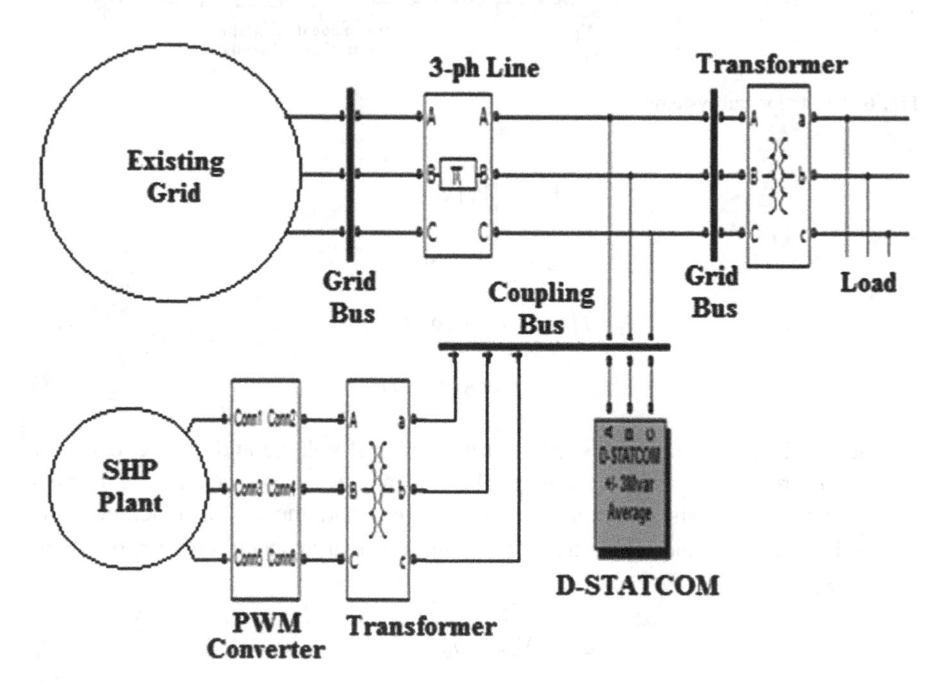

Fig. 7 Layout of the proposed model for Grid integration of SHP plant

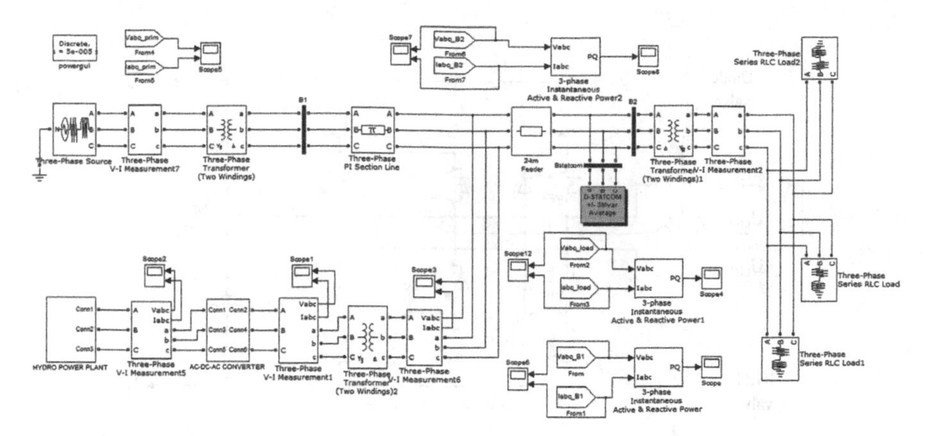

Fig. 8 MATLAB/Simulink model for the proposed model

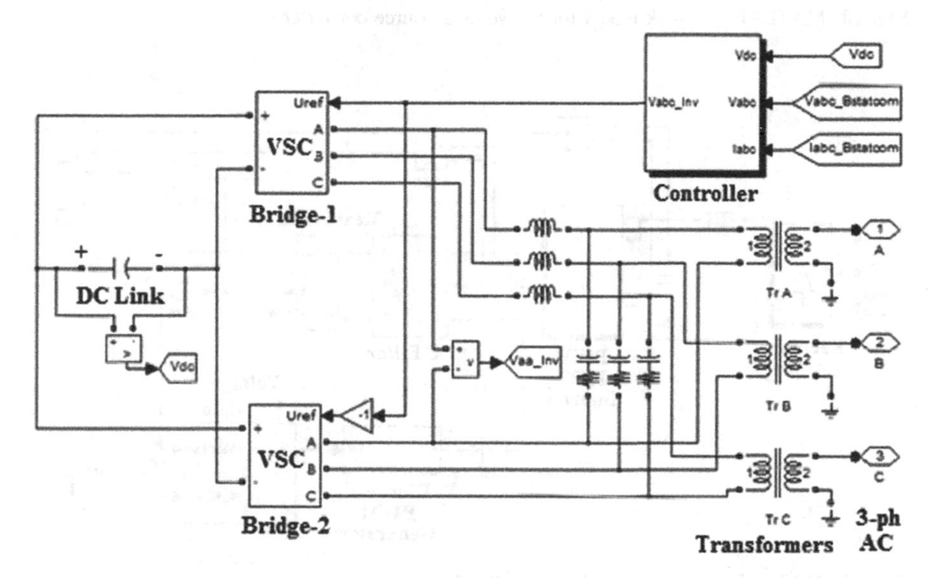

Fig. 9 MATLAB/Simulink model for the D-STATCOM

of simulation model for the three-level voltage source converter (VSC) used in this paper is shown in Fig. 10. The details of simulation model for the PWM converter used in this paper are shown in Fig. 11. The STATCOM consists of three-phase bus, shunt transformer, VSC, and DC capacitor. Figure 12 shows the equivalent circuit of the STATCOM.

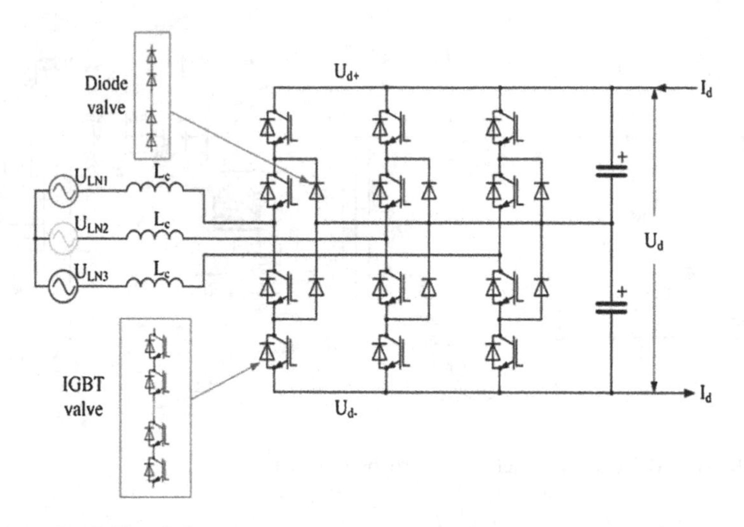

Fig. 10 MATLAB/Simulink model for the voltage source converter

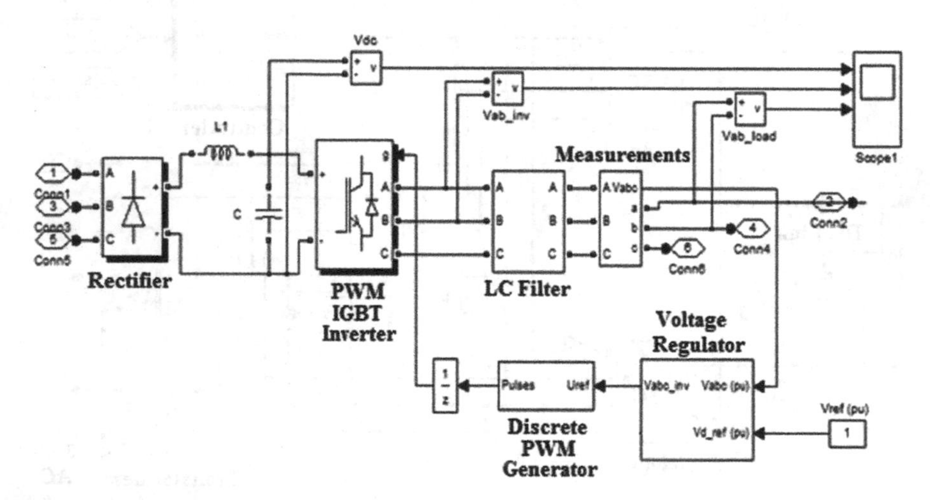

Fig. 11 MATLAB/Simulink model for the PWM converter

The loop equation for the STATCOM circuit is written in vector form as Eq. (12) and the output of the STATCOM is given by Eq. (13).

$$\frac{di_{abc}}{dt} = -\frac{\omega_s R_s}{L_s} i_{abc} + \frac{\omega_s}{L_s}(E_{abc} - V_{abc}) \tag{12}$$

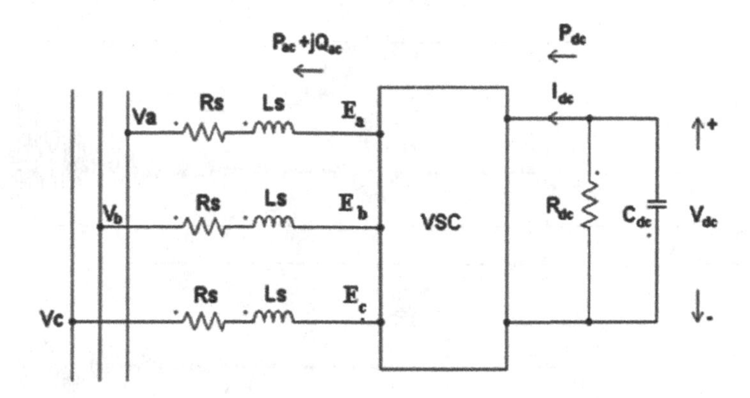

Fig. 12 Equivalent circuit model of STATCOM

$$E_a = V_{dc}\cos(\omega t + \alpha) \tag{13}$$

In the circuit shown in Fig. 12, R_s and L_s represent the STATCOM transformer resistance and inductance, respectively. Similarly, E_{abc} represents the converter AC side phase voltages, V_{abc} is the system-side phase voltages, i_{abc} is the phase currents, and V_{dc}, I_{dc}, P_{dc} are, respectively, the DC voltage, current, and DC power. Further, it may be noted that 'α' is the phase angle between bus voltage and converter output voltage. The STATCOM used in this model injects real and reactive power, which is represented by P_{ac} and Q_{ac}, respectively.

When a voltage source converter is connected to a grid as a STATCOM without energy storage, the power flow between the converter and the grid should be controlled such that the voltage at the connection point is maintained at a certain level and at the same time the converter DC side voltage is kept at a reasonable and relatively constant value to ensure successful converter operation. The power control of a three-phase PWM converter can be achieved either with or without an inner current control loop. Among all the current control strategies, deadbeat control is probably the most effective one and is widely used. Therefore, an inner current deadbeat control loop is employed in the converter control in this paper along with a novel flux modulation scheme for the switching control of the converter bridge. The simulation results in form of various responses are presented in the Appendix in form of Figs. 13, 14, 15, 16, 17, 18, 19, and 20. It may be observed from Figs. 13 and 14 that the SHP output voltage and current in the absence of the STATCOM are continuously increasing with respect to time, which is a highly unstable phenomenon. However, in presence of the STATCOM, the voltage and current waveform are more stable. From the simulation results of Figs. 15 and 16, we may conclude that the PWM converter injects almost harmonic less voltage to

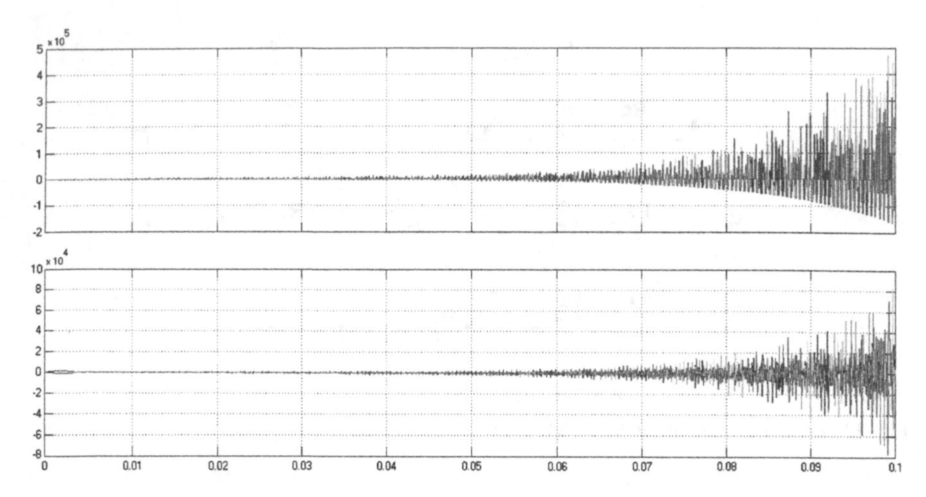

Fig. 13 Output *V* and *I* response of SHP without D-STATCOM

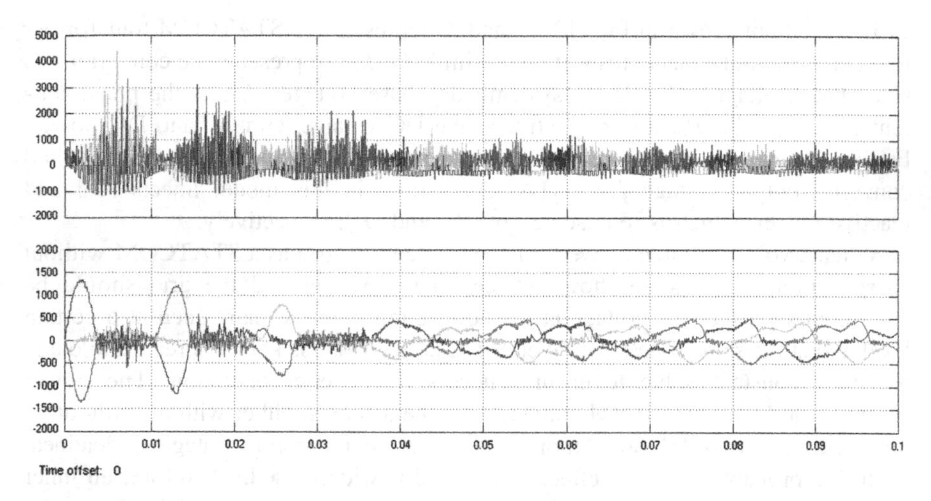

Fig. 14 Output *V* and *I* response of SHP with D-STATCOM

the main grid while the STATCOM is in operation. Hence the stability of voltage is almost achieved at the point of common coupling (PCC). From the active and reactive power responses shown in Figs. 17 and 18, it can be clearly verified that the STATCOM not only compensates the reactive power but also reduces the fluctuations in active power plot. Hence, the STATCOM improves the power

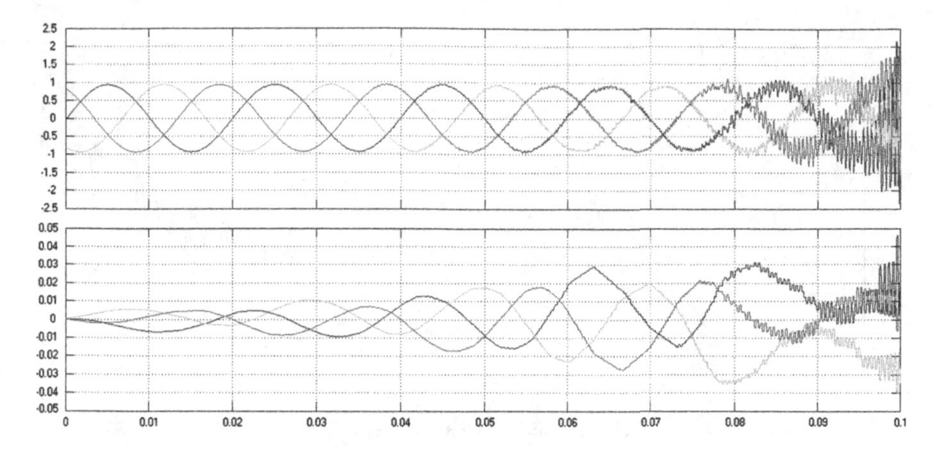

Fig. 15 Output *V* and *I* response of PWM converter without D-STATCOM

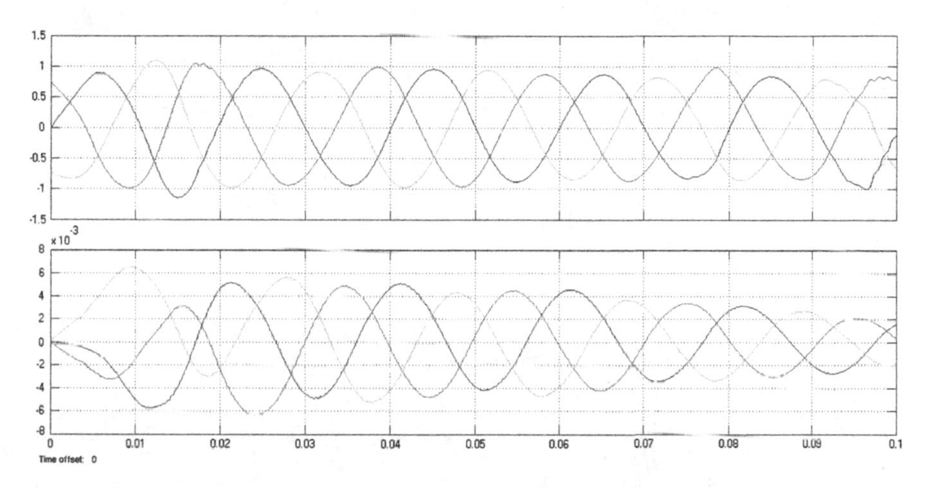

Fig. 16 Output *V* and *I* response of PWM converter with D-STATCOM

handling stability. Also, the presence of STATCOM improves the load side voltage and current profile, as indicated in Figs. 19 and 20. Therefore, the STATCOM is immensely helpful for improving voltage profile by reducing the harmonics.

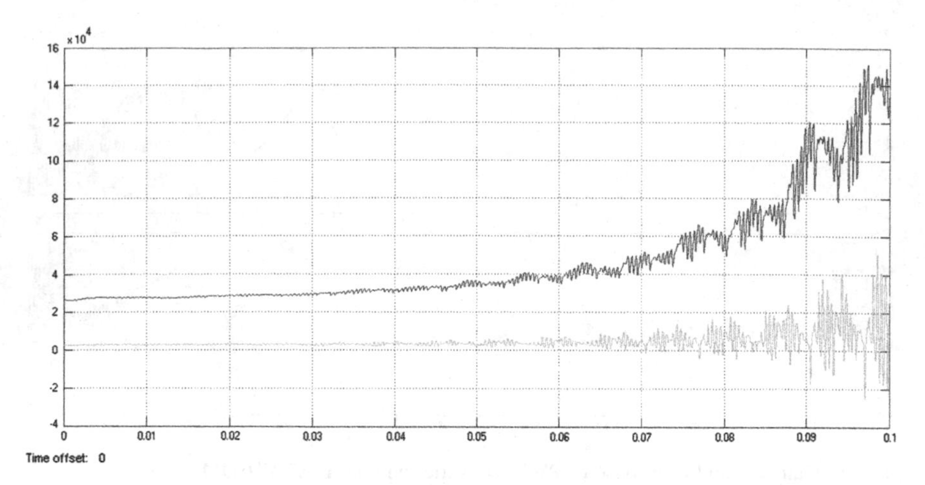

Fig. 17 Power at load side (*P* and *Q*) without D-STATCOM

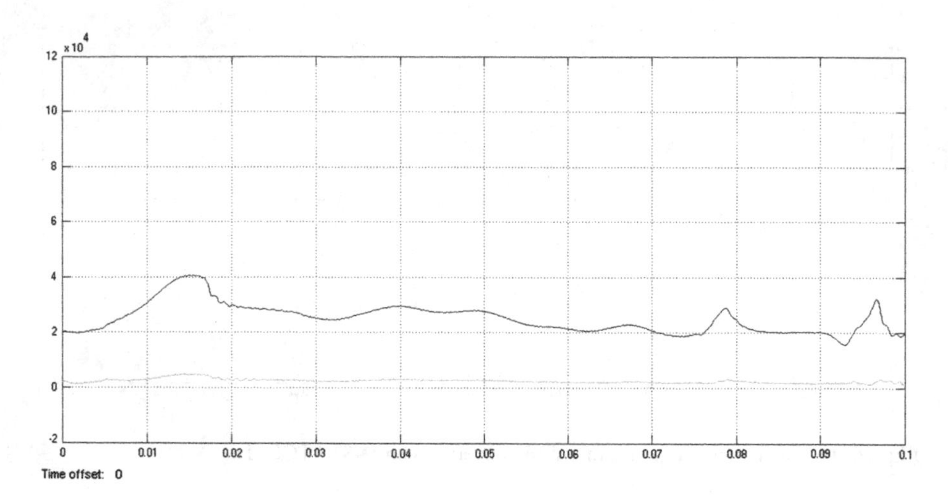

Fig. 18 Power at load side (*P* and *Q*) with D-STATCOM

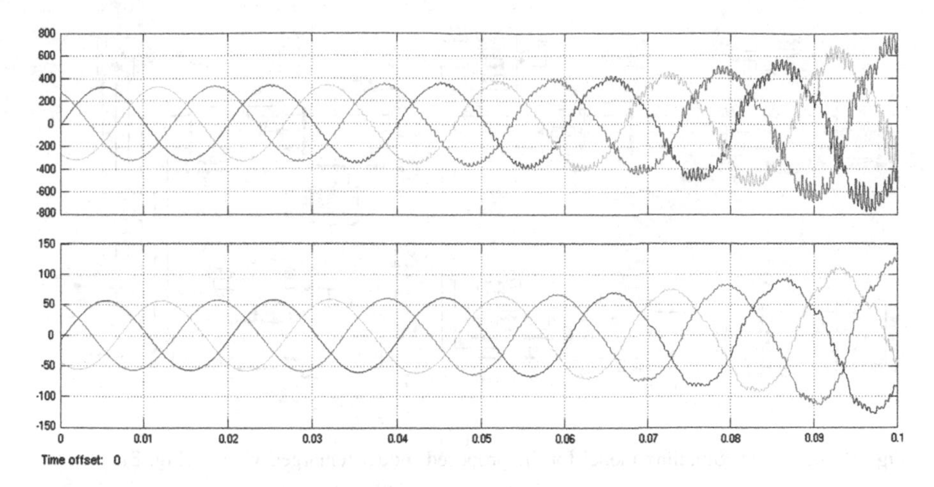

Fig. 19 *V* and *I* response at load side without D-STATCOM

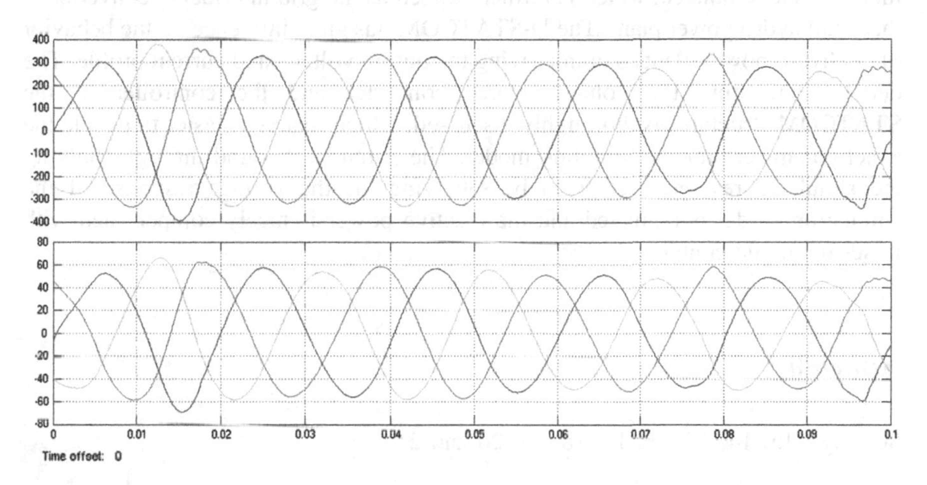

Fig. 20 *V* and *I* response at load side with D-STATCOM

4 Conclusion

The basic aim of this work is to model a small hydro power plant in order to integrate the same with the grid through a D-STATCOM. This scheme is very flexible in the sense that transfer of power between the plant and the grid becomes convenient with easy compensation of reactive power and the power quality also gets improved due to reduction of harmonics in the voltage and current waveforms. This scheme validates the above findings through the analysis of the true behavior of voltage, current, and the active/reactive power profiles in the simulated MATLAB platform. The PWM converter with a voltage regulator is perfectly

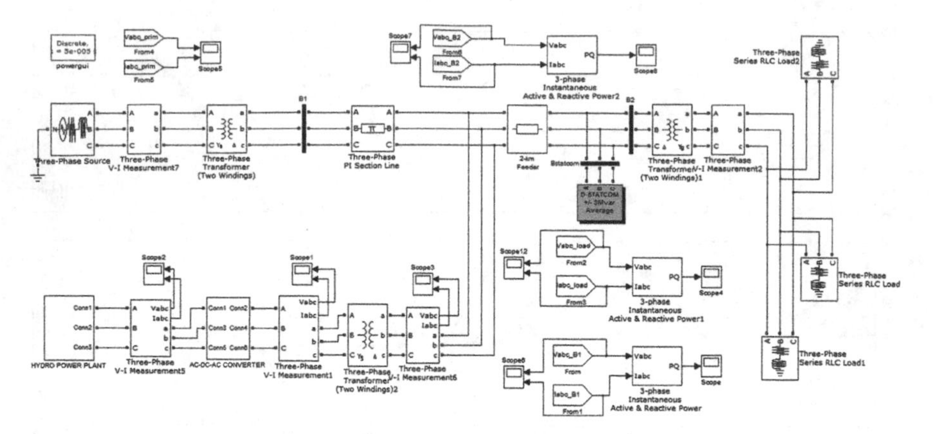

Fig. 21 MATLAB/Simulink model for the proposed model (enlarged view of Fig. 8)

modeled and simulated, which is further utilized as the grid interfacing converter of the small hydro power plant. The D-STATCOM has great influence on the behavior of the hydro power plant as in improving the output voltage and current profile. The details concerning the voltage source converter and the controller of the STATCOM structures is thoroughly discussed, which makes it easier to decide the exact parameters for the Simulink model. The parameters found through studying the simulated response seems to be satisfying, as the voltage response of the simulation model is regulated and the reactive power is nicely compensated with lesser harmonic contents.

Appendix

See Figs. 13, 14, 15, 16, 17, 18, 19, 20 and 21

References

1. R. Ahmad, 'Modeling and Analysis of Canal Type Small Hydro Power Plant and Performance Enhancement Using PID Controller', IOSR Journal of Electrical and Electronics Engineering, Vol. 6(2), pp. 06–14, 2013.
2. P. V. V. Rama Rao, B. Kali Prasanna, Y. T. R. Palleswari, 'Modeling and Simulation of Utility Interfaced PV/Hydro Hybrid Electric Power System', World Academy of Science, Engineering and Technology IJ Electrical, Computer, Energetic, Electronic and Comm. Engineering, Vol. 8(8), pp. 1309–1314, 2014.

3. S. Meshram, G. Agnihotri, S. Gupta, 'Modeling of grid connected dc linked pv/h ydro hybrid system', IJ Electrical and Electronics Engineering, Vol. 2(3), pp. 13–27, 2013.
4. Y. Almoataz, M. Ahmed, R.S. Jumaah. 'Simulation and Modeling of a Three-Phase Two-Stage Grid Connected Photovoltaic System', International Journal of Power Engineering and Energy, Vol. 6(2), pp. 512–520, 2015.
5. Rajnikanth, R. Nagapadma, 'Simulation of ac/dc/ac converter for closed loop operation of three phase induction motor', IJ Advanced Research in Computer and Communication Engineering, Vol. 3(6), pp. 7074–7079, 2014.
6. R.P. Panda, P.K. Sahoo, P.K. Satpathy, 'Smooth Evacuation of Power in Grid Connected Small Hydro Power Stations by Application of SVCs', International Journal of Renewable Energy Research, Vol. 5(2), pp. 622–628, 2015.
7. A. Singh, B. Singh, 'Microgrid: A review', IJ Research in Engineering and Technology, Vol. 3(2), pp. 185–198, 2014.
8. R. Kozhi, Meenikumari, L. Ramesh, 'Simulation Study of Renewable Energy driven Microgrid', Journal of Electrical Engineering, Vol. 15(1), pp. 78–88, 2015.
9. K. Ravichandrudu, M. Manasa, P. Yohan Babu, G.V.P. Anjaneyulu, 'Design of Micro-grid System Based on Renewable Power Generation Units', IJ Scientific and Research Publications, Vol. 3(8), pp. 1–5, 2013.
10. Mithilesh Kumar Kanaujia, S.K. Srivastava, 'Power Quality Enhancement With D-Statcom Under Different Fault Conditions', International Journal of Engineering Research and Applications, Vol. 3(2), pp. 828–833, 2013.
11. Sai Kiran Kumar. Sivakoti, Y. Naveen Kumar, D. Archana, 'Power Quality Improvement In Distribution System Using D-Statcom in Transmission Lines', IJ Engineering Research and Applications, Vol. 1(3), pp. 748–752, 2011.
12. H. Molavi, M. M. Ardehali, 'Application of Distribution Static Compensator (D-STATCOM) to Voltage Sag Mitigation', Universal Journal of Electrical and Electronic Engineering, Vol. 1 (2), pp. 11–15, 2013.
13. Robert H. Lasseter, 'Microgrids and Distributed Generation', Journal of Energy Engineering, American Society of Civil Engineers, Sept. 2007.
14. J. A. Peças Lopes, C. L. Moreira, A. G. Madureira, 'Defining Control Strategies for MicroGrids Islanded Operation', IEEE Transactions On Power Systems, Vol 21(2), 2006.

Pragmatic Investigation on Performance of Instance-Based Learning Algorithms

Bharathan Venkatesh, Danasingh Asir Antony Gnana Singh and Epiphany Jebamalar Leavline

Abstract In the recent past, there has been increasing usage of machine learning algorithms for classifying the data for various real-world applications such as weather prediction, medical diagnosis, network intrusion detection, software fault detection, etc. The instance-based learning algorithms play a major role in the classification process since they do not learn the data until the need of the developing the classification model. Therefore, these learning algorithms are called as lazy learning algorithms and implemented in various applications. However, there is a pressing need among the researchers to analyze the performance of various types of the instance-based classifier. Therefore, this chapter presents a pragmatic investigation on performance of the instance-based learning algorithms. In order to conduct this analysis, different instance-based classifier namely instance-based one (IB1), instance-based K (IBK), Kstar, lazy learning of Bayesian rules (LBR), locally weighted learning (LWL) are adopted. The performance of these algorithms is evaluated in terms of sensitivity, specificity, and accuracy on various real-world datasets.

Keywords Instance-based learning · IB1 · IBK · Kstar · LBR · LWL

B. Venkatesh (✉) · D.A.A.G. Singh
Department of Computer Science and Engineering, Anna University, BIT Campus, Tiruchirappalli, India
e-mail: venkateshaamec@gmail.com

D.A.A.G. Singh
e-mail: asirantony@gmail.com

E.J. Leavline
Department of Electronics and Communication Engineering, Anna University, BIT Campus, Tiruchirappalli, India
e-mail: jebilee@gmail.com

S.S. Dash et al. (eds.), *Artificial Intelligence and Evolutionary Computations in Engineering Systems*, Advances in Intelligent Systems and Computing 517,
DOI 10.1007/978-981-10-3174-8_53

1 Introduction

Nowadays, the data are becoming precious ones to analyze and understand the needs and wants of the people to render high quality services and to produce quality goods based on their needs and wants. In the last decade, the organizations generate data massively through the growth of information and communication technologies. The generated massive data are analyzed for making decision and prediction. Processing these massive data is a challenging task among the researches, and also the accuracy of the prediction and decision making is very important. Since, the inaccurate predictions or classification or decision making can induce physical or material losses in the business and organization. The prediction or classification can be carried out using supervised learning algorithms. These supervised classification algorithms can be categorized based on the nature of learning methodology used to learn the data for developing the model such as probabilistic-based, decision tree-based classification, rule-based, meta-based classification, etc.

The probabilistic-based learning algorithm predicts the class label from the sample input over a set of classes, it provides the classification with higher certainty. The best example for the probabilistic-based supervised learning algorithm is Bayesian learning algorithm. In Bayesian classification algorithm, the Bayes theorem is used to predict the class label. The main advantage of this method is that it takes lesser time and effort for classification. However, the major disadvantage is the computational overhead involved in specifying prior probability. In decision tree-based classification, the tree structure represents the classification of data. In the tree structure, the leaves represent class labels, branches represent conjunctions. This decision tree is used for prediction or classification or making decision on the given data. The main advantages of this method are easy to understand, it can adopt both numerical and categorical data. The major drawback of this method is that it is a NP complete problem and there are chances to create complex trees. In rule-based learning algorithm, the given data is classified based on IF-THEN rules. IF represents the precondition, THEN is the rule consequent. The instance-based learning algorithms uses the training data directly, the class label are predicted without using any rules. The major advantages are they do not use predefined rules for classification or prediction and only the training data is used to predict class label. The meta-based classification is a general strategy that combines a number of separate learning processes.

In the machine learning research area, different types of supervised machine learning algorithms have been developed by various researches especially in the instance-based learning algorithm. Identifying the suitable machine learning algorithm in the instance-based machine learning algorithm is a pressing need to the researches community. Hence, this chapter conducts the pragmatic investigation on various instance-based machine learning algorithms to identify the better learning algorithm for the classification task. The rest of the chapter is organized as follows: Section 2 presents the literature review of various machine learning algorithms.

Section 3 explains the experimental setup and experimental procedure. Section 4 discusses the results and Sect. 5 concludes this chapter.

2 Literature Review

This section presents various research works concerned to the development classification models that are developed by various researches for different applications. In general, some of the researchers deal with the probabilistic-based classification approach. Aggarwal et al. [1] provides the literature review on uncertain data mining and uncertain data management applications by considering the problem of classification and clustering. Wai Ho Au et al. [2] proposed a new data mining algorithm called data mining by evolutionary learning (DMEL) to handle classification problems. In this approach, the predictions can be estimated accurately. Akhlagh et al. [3] suggested that the temporal sequence along with a probabilistic approach based on Bayes theorem is applied to obtain prediction accuracy. Some of the researchers worked with the decision tree-based classification approaches. Ouyang et al. [4] focuses on continuous attributes handling for mining data stream with the concept of drift. Li et al. [5] proposed a decision tree classifier system that uses binary search trees to handle numerical attributes. Kumar et al. [6] compares the performance of decision tree (CHAID, QUEST & C5.0) and artificial intelligence neural network-based classifiers (ANN) in diversity of datasets.

Some of the researchers deal with instance-based classifiers. Hai Zhai et al. [7] discussed an instance selection method with supervised clustering with the intention to select an instance belonging to the boundary of clusters. Sane et al. [8] deals with a new wrapper method using the concept of instance to detect outliers with the use of neural network. Some of the researches compared the classification algorithm with other classification algorithms based on their performance such as rule-based approach. Lakshmi et al. [9] proposed two efficient techniques called PSTMiner and PSToswmine for prediction of data streams. Mohammead et al. [10] suggested the rule-based data mining for classification, which is good in predicting phishing websites from legitimate one. AIwen et al. [11] presented the rule-based gene expression data classifier called BSTC, handles datasets with any number of class types easily.

Some of the researchers deal with the meta-based approaches. Sun et al. [12] investigated the principle concepts and methodologies of mining the heterogeneous information networks. Mohapatra et al. [13] proposed design techniques that enable the hardware implementation of meta-functions to scale more gracefully under voltage over-scaling. Abe et al. [14] introduces a constructive meta-learning system to choose proper learning algorithms for the data classification. Some of the researchers deal with the Bayes-based approaches. Chen et al. [15] investigated the effects between multimedia data mining and text data mining. Gupta et al. [16] have applied Bayes classification method to classify large amount of data and the efficiency is also measured by their accuracy, F-measure, and number of iterations. Muhammed et al. [17] presented a predictive model as an artificial diagnosis for heart diseases based on the measure of individual using naïve Bayes classification. The authors discussed different approaches to improve the accuracy of the classifiers [18–22].

3 Experimental Setup and Experimental Procedure

The experiments are conducted using Weka data mining tool [23]. In order to validate the performance of the instance-based machine learning algorithms namely instance-based one (IB1), instance-based K (IBK), Kstar, lazy learning of Bayesian rules (LBR), locally weighted learning (LWL) on the various datasets collected from the UCI dataset repository [24]. The details of the datasets are listed in Table 1. The performance of the learning algorithms is evaluated in terms of sensitivity, specificity, and accuracy. The experiment is conducted as shown in Fig. 1. Initially, the dataset is given to the classification algorithms and the machine learning (ML) models are generated. The model is then evaluated in terms of sensitivity, specificity, and accuracy as expressed in Eqs. (1), (2), and (3). For the evaluation, the 10-fold cross validation method is used.

$$\text{Sensitivity (SN)} = \text{TP}/P = \text{TP}/(\text{TP}+\text{FN}) \tag{1}$$

Table 1 Details of the datasets

S. no	Dataset	Instances	Features	Classes
1	Breast-cancer	286	10	2
2	Diabetes	768	9	2
3	Ionosphere	351	35	2
4	Super market	4627	217	2
5	Weather-symbolic	14	5	2
6	Vote	435	17	2
7	Tic-tac-toe	958	9	2
8	Cylinder-bands	512	39	3
9	Lung-cancer	32	56	3
10	Sick	3772	29	2
11	Heart-statlog	270	13	2
12	Liver-disorders	345	7	2

Fig. 1 Experimental setup and procedure

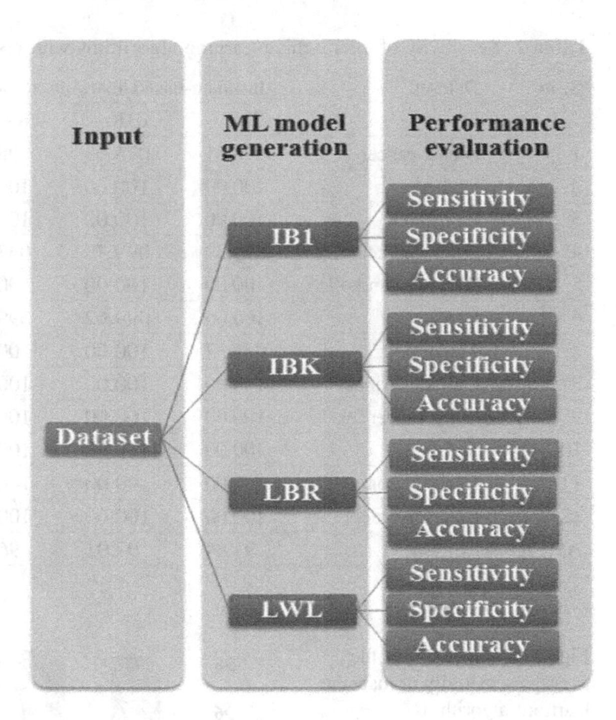

$$\text{Specificity (SP)} = \text{TN}/N = \text{TN}/(\text{TN}+\text{FP}) \quad (2)$$

$$\text{Accuracy(AC)} = (\text{TP}+\text{TN})/(\text{TP}+\text{FP}+\text{FN}+\text{TN}) \quad (3)$$

where

P Positive instances,
N Negative instances,
TP True positive,
TN True negative,
FP False positive, and
FN False negative.

4 Results and Discussion

This section shows the results obtained from the experiment and discusses the results. The results are obtained in terms of sensitivity (SN), specificity (SP), and accuracy (AC). Table 2 shows the sensitivity of the instance-based learning

Table 2 Sensitivity of the machine learning algorithms with respective dataset

S. no	Dataset	Instance-based learning algorithms				
		IB1	IBK	Kstar	LBR	LWL
1	Breast-cancer	098.01	075.63	098.02	079.81	080.60
2	Diabetes	100.00	100.00	100.00	100.00	081.98
3	Ionosphere	100.00	100.00	100.00	100.00	100.00
4	Super market	099.39	099.75	063.71	099.75	099.75
5	Weather-symbolic	100.00	100.00	100.00	090.00	100.00
6	Vote	100.00	099.62	099.62	096.26	098.06
7	Tic-tac-toe	074.37	100.00	100.00	099.65	056.47
8	Cylinder-bands	100.00	100.00	100.00	100.00	100.00
9	Lung-cancer	100.00	100.00	100.00	077.77	100.00
10	Sick	100.00	100.00	100.00	100.00	099.27
11	Heart-statlog	100.00	100.00	100.00	082.75	082.75
12	Liver-disorders	100.00	100.00	100.00	057.63	057.63
Average sensitivity		97.64	97.91	96.77	90.30	88.04

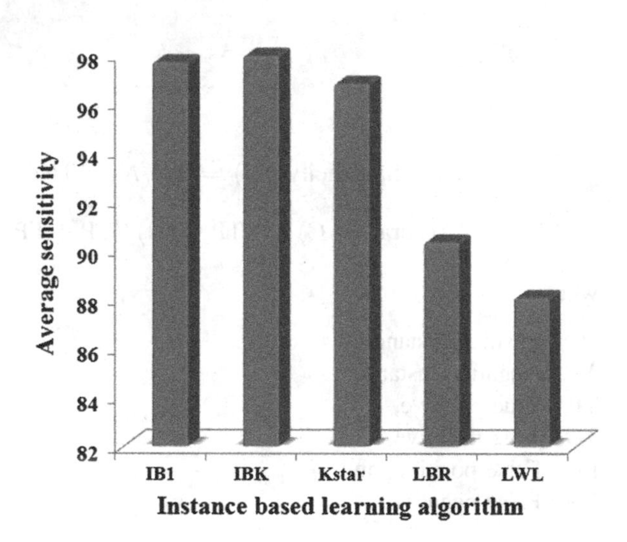

Fig. 2 Comparison of the average sensitivity of machine learning algorithms

algorithms IB1, IBK, Kstar, LBR, and LWL. Figure 2 shows the average sensitivity of the IB1, IBK, Kstar, LBR, and LWL. Table 3 shows the specificity of the instance-based learning algorithms IB1, IBK, Kstar, LBR, and LWL. Figure 3 shows the average specificity of the IB1, IBK, Kstar, LBR, and LWL. Table 4 shows the accuracy of the instance-based learning algorithms IB1, IBK, Kstar, LBR, and LWL. Figure 3 shows the average accuracy of the IB1, IBK, Kstar, LBR, and LWL.

Table 3 Specificity of the machine learning algorithms with respective dataset

S. no	Dataset	Instance-based learning algorithms				
		IB1	IBK	Kstar	LBR	LWL
1	Breast-cancer	096.42	056.25	097.59	060.29	074.07
2	Diabetes	100.00	100.00	100.00	100.00	065.32
3	Ionosphere	100.00	100.00	100.00	100.00	079.78
4	Super market	077.25	077.34	100.00	100.00	100.00
5	Weather-symbolic	100.00	100.00	100.00	100.00	100.00
6	Vote	099.40	100.00	098.23	094.61	092.09
7	Tic-tac-toe	099.81	099.84	100.00	099.84	074.80
8	Cylinder-bands	099.68	099.68	100.00	099.68	065.96
9	Lung-cancer	100.00	100.00	100.00	091.30	100.00
10	Sick	100.00	100.00	100.00	100.00	066.23
11	Heart-statlog	100.00	100.00	100.00	076.00	076.00
12	Liver-disorders	100.00	100.00	100.00	069.15	069.15
Average specificity		97.71	94.42	99.65	90.90	80.28

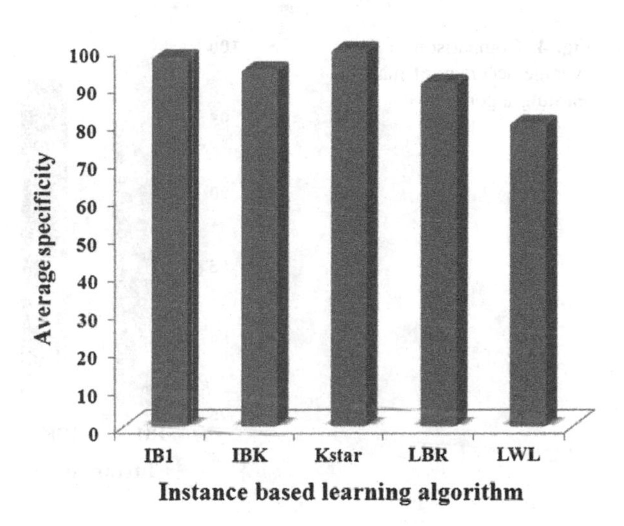

Fig. 3 Comparison of the average specificity of machine learning algorithms

Form Table 2 and Fig. 2, it is observed that the IBK performs better than other machine learning algorithms when compared in terms of average sensitivity. The LWL produces lesser average sensitivity than other machine learning algorithms. From Table 3 and Fig. 3, it is observed that Kstar performs better than the other machine learning algorithms in terms of average specificity. The LWL produces lesser average specificity than other machine learning algorithms. From Table 4 and Fig. 4, it is observed that IB1 performs better in terms of average specificity. LWL produces lesser specificity than other machine learning algorithms compared in terms of average specificity.

Table 4 Accuracy of the machine learning algorithms with respective dataset

S. no	Dataset	Classification algorithm				
		IB1	IBK	Kstar	LBR	LWL
1	Breast-cancer	097.55	072.37	097.90	075.17	079.37
2	Diabetes	100.00	100.00	100.00	100.00	076.04
3	Ionosphere	100.00	100.00	100.00	100.00	083.76
4	super market	089.08	089.34	073.34	099.85	099.85
5	Weather-symbolic	100.00	100.00	100.00	092.85	100.00
6	Vote	099.77	099.77	099.08	095.63	095.63
7	Tic-tac-toe	090.40	099.89	100.00	099.78	068.43
8	Cylinder-bands	099.81	099.81	100.00	099.81	070.18
9	Lung-cancer	100.00	100.00	100.00	087.50	100.00
10	Sick	100.00	100.00	100.00	100.00	096.55
11	Heart-statlog	100.00	100.00	100.00	079.62	079.62
12	Liver-disorders	100.00	100.00	100.00	064.34	064.34
Average accuracy		98.05	96.76	97.52	91.21	84.48

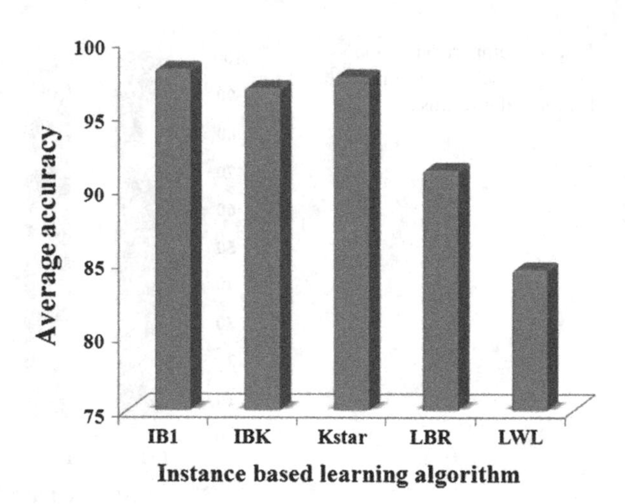

Fig. 4 Comparison of the average accuracy of machine learning algorithms

5 Conclusion

This chapter presented a pragmatic investigation on performance on instance-based machine learning algorithms for data classification. In order to observe the performance of various instances-based learning algorithms, various datasets are collected from the UCI repository. Then each algorithm is tested on the dataset in terms of sensitivity, specificity, and accuracy. From this empirical investigation, it is observed that the IBK, Kstar, and the IB1 produces better performance compared to other algorithms in terms of average sensitivity, specificity, and accuracy, respectively. The performance of LWL is poor compared to other algorithms in terms of sensitivity, specificity, and accuracy.

References

1. Aggarwal, Charu C. and Philip S. Yu. "A survey of uncertain data algorithms and applications." Knowledge and Data Engineering, IEEE Transactions on 5 (2009) 609–623.
2. Au, Wai-Ho, Keith CC Chan, and Xin Yao. "A novel evolutionary data mining algorithm with applications to churn prediction." Evolutionary Computation, IEEE Transactions on 7, no. 6 (2003) 532–545.
3. Akhlagh, Mojtaba Malek, Shing Chiang Tan, and Faiiaz Khak. "Temporal data classification and rule extraction using a probabilistic decision tree." In Computer & Information Science (ICCIS), 2012 International Conference on, vol. 1, pp. 346–351. IEEE, 2012.
4. Ouyang, Zhenzheng, Quanyuan Wu, and Tao Wang. "An efficient decision tree classification method based on extended hash table for data streams mining." In Fuzzy Systems and Knowledge Discovery, 2008. FSKD'08. Fifth International Conference on, vol. 5, pp. 313–317. IEEE, 2008.
5. Wang, Tao, Zhoujun Li, Yuejin Yan, Huowang Chen, and Jinshan Yu. "An efficient classification system based on binary search trees for data streams mining." In Systems, 2007. ICONS'07. Second International Conference on, pp. 15–15. IEEE, 2007.
6. Kumar, Pardeep, Vivek Sehgal, and Durg Singh Chauhan. "Performance evaluation of decision tree versus artificial neural network based classifiers in diversity of datasets." In Information and Communication Technologies (WICT), 2011 World Congress on, pp. 798–803. IEEE, 2011.
7. Zhai, Jun-Hai, Hong-Yu Xui, Su-Fang Zhang, Na Li, and Ta Li. "Instance selection based on supervised clustering." In Machine Learning and Cybernetics (ICMLC), 2012 International Conference on, vol. 1, pp. 112–117. IEEE, 2012.
8. Sane, Shirish S. and Ashok A. Ghatol. "Use of instance typicality for efficient detection of outliers with neural network classifiers." In Information Technology, 2006. ICIT'06. 9th International Conference on, pp. 225–228. IEEE, 2006.
9. K. P. Lakshmi; C. R. K. Reddy "Fast Rule-Based Prediction of Data Streams Using Associative Classification Mining" IT Convergence and Security (ICITCS), 2015 5th International Conference 24–27 Aug. (2015): 1–5
10. Mohammad, Rami M. Fadi Thabtah, and Lee McCluskey. "Intelligent rule-based phishing websites classification." Information Security, IET 8, no. 3 (2014): 153–160.
11. M. A. Iwen; W. Lang; J. M. Patel "Scalable Rule-Based Gene Expression Data Classification" Data Engineering, 2008. ICDE 2008. IEEE 24th International Conference 7–12 April (2008): 1062–1071
12. Sun, Yizhou, and Jiawei Han. "Mining heterogeneous information networks: principles and methodologies." Synthesis Lectures on Data Mining and Knowledge Discovery 3, no. 2 (2012): 1–159.
13. Mohapatra, Debabrata, Vinay K. Chippa, Anand Raghunathan, and Kaushik Roy. "Design of voltage-scalable meta-functions for approximate computing." In Design, Automation & Test in Europe Conference & Exhibition (DATE), 2011, pp. 1–6. IEEE, 2011.
14. Abe, Hidenao, Shusaku Tsumoto, Miho Ohsaki, and Takahira Yamaguchi. "Evaluating learning algorithms composed by a constructive meta-learning scheme for a rule evaluation support method." In Mining Complex Data, pp. 95–111. Springer Berlin Heidelberg, 2009.
15. Lin, Yuan, Yuqiang Chen, Guirong Xue, and Yong Yu. "Text-Aided Image Classification: Using Labeled Text from Web to Help Image Classification." In 2010 12th International Asia-Pacific Web Conference, pp. 267–273. IEEE, 2010.

16. Gupta, Amit, Naganna Chetty, and Shraddha Shukla. "A classification method to classify high dimensional data." In Computing, Communication and Security (ICCCS), 2015 International Conference on, pp. 1–6. IEEE, 2015.
17. Muhammed, LA-N. "Using data mining technique to diagnosis heart disease." In Statistics in Science, Business, and Engineering (ICSSBE), 2012 International Conference on, pp. 1–3. IEEE, 2012.
18. D. Asir Antony Gnana Singh1, E. Jebamalar Leavline, A Pragmatic Approach on Knowledge Discovery in Databases with WEKA, International Journal of Engineering Technology and Computer Research (IJETCR), Volume 2 Issue 7; December; 2014; Page No. 81–87
19. Danasingh Asir Antony Gnana Singh, Subramanian Appavu Alias Balamurugan and Epiphany Jebamalar Leavline, Improving the Accuracy of the Supervised Learners using Unsupervised based Variable Selection, Asian Journal of Information Technology Year: 2014| Volume: 13| Issue: 9| Page No.: 530–537
20. Danasingh Asir Antony Gnana Singh and E. Jebamalar Leavline. "An empirical study on dimensionality reduction and improvement of classification accuracy using feature subset selection and ranking." In Emerging Trends in Science, Engineering and Technology (INCOSET), 2012 International Conference on, pp. 102–108. IEEE, 2012.
21. Danasingh Asir Antony Gnana Singh, S. Balamurugan, and E. Jebamalar Leavline. "Towards higher accuracy in supervised learning and dimensionality reduction by attribute subset selection-A pragmatic analysis." In Advanced Communication Control and Computing Technologies (ICACCCT), 2012 IEEE International Conference on, pp. 125–130. IEEE, 2012.
22. D. Asir Antony Gnana Singh and E. Jebamalar Leavline, 'Data Mining In Network Security—Techniques & Tools: A Research Perspective, Journal of Theoretical and Applied Information Technology', 20th November 2013. Vol. 57 No. 2, 269–278
23. Mark Hall, Eibe Frank, Geoffrey Holmes, Bernhard Pfahringer, Peter Reutemann, Ian H. Witten (2009); The WEKA Data Mining Software: An Update; SIGKDD Explorations, Volume 11, Issue 1.
24. Lichman, M. (2013). UCI Machine Learning Repository [http://archive.ics.uci.edu/ml]. Irvine, CA: University of California, School of Information and Computer Science.

Identifying Sensitive Attributes for Preserving Privacy

Balakrishnan Tamizhpoonguil, Danasingh Asir Antony Gnana Singh and Epiphany Jebamalar Leavline

Abstract In recent past, the data are generated massively from medical sector due to the advancements and growth of technology leading to high-dimensional and massive data. Handling the medical data is crucial since it contains some sensitive data of the individuals. If the sensitive data is revealed to the adversaries or others then that may be vulnerable to attack. Hiding the huge volume of data is practically difficult task among the researchers. Therefore, in real life only the sensitive data are hidden from the huge volume of data for providing security since hiding the entire data is costlier. Therefore, the sensitive attributes must be identified from the huge volume of data in order to hide them for preserving the privacy. In order to identify the sensitive data to hide them for providing security, reducing the computational and transmission cost in secured data transmission, this chapter presents a pragmatic approach to identify the sensitive attributes for preserving privacy. This proposed method is tested on the various real-world datasets with different classifiers and also the results are presented.

Keywords Attribute selection · Sensitive attributes · Naïve Bayes classifier · K-star classifier

B. Tamizhpoonguil (✉) · D.A.A.G. Singh
Department of Computer Science and Engineering, Anna University, BIT Campus, Tiruchirappalli, India
e-mail: thamizhpoonguil@gmail.com

D.A.A.G. Singh
e-mail: asirantony@gmail.com

E.J. Leavline
Department of Electronics and Communication Engineering, Anna University, BIT Campus, Tiruchirappalli, India
e-mail: jebilee@gmail.com

S.S. Dash et al. (eds.), *Artificial Intelligence and Evolutionary Computations in Engineering Systems*, Advances in Intelligent Systems and Computing 517,
DOI 10.1007/978-981-10-3174-8_54

1 Introduction

In the recent past, the growth rate of medical sector is increased in terms of size and technology. Transmitting the medical related documents in a secured manner is a challenging task among the researchers since the data are of high volume. Providing security for the entire data is impractical for secure data transmission and it is a challenging task since hiding the huge volume of data is highly expensive and difficult. Hence, the necessary or sensitive data are identified from the huge volume of data. Only the sensitive data are hidden rather than the entire volume of data so that cost for providing security is reduced for secured data transmission. Thus, the privacy of the data is preserved. The sensitive attributes present in the huge volume of data can be identified using variable or feature selection.

The feature selection is a process of identifying the sensitive or relevant attributes from a high-dimensional space. The feature selection can be classified as filter, wrapper, and embedded methods. The filter method will not use any machine learning algorithm for identifying sensitive attributes. The wrapper method uses a machine learning algorithm for evaluating the selected sensitive attributes. The embedded method uses a part of the machine learning algorithm for identifying sensitive attributes from a high-dimensional data. This chapter presents a filter-based feature selection method to identify the sensitive attributes for preserving the privacy in data for secure data transmission. The rest of the chapter is organized as follows: Section 2 reviews the literature related to the proposed method. Section 3 details the proposed method and its implementation. Section 4 explores the experimental setup. Section 5 illustrates and discusses the results. Section 6 concludes this chapter.

2 Literature Review

Xinhua et al. presented a systematic framework for sharing a sensitive data in a secured manner on big data platform. In order to achieve the security in data transmission, they have used heterogeneous proxy re-encryption algorithm namely virtual machine monitor (VMM) and also illustrated the systematic framework for secured data transmission and storage on cloud [1]. Shucheng Yu et al. presented an improved cipher text policy attribute-based encryption (CP-ABE) scheme for secured data sharing. It avoids the chosen cipher text attacks. This scheme also is applicable for key policy attribute-based encryption for providing secured data transmission [2]. Sophia et al. presented a literature review on various cryptographic approaches for securing big data in cloud computing environment and also they presented a model for big data analytics with cloud computing. They presented various cryptographic techniques that are used for securing big data on cloud and the cryptographic techniques and compared in terms of adversary type, confidentiality, and integrity [3]. Asmaa et al. presented a scheme to share the data and knowledge in the health information technology. Adding and deleting the medical information often leads to leakage in the sensitive attribute. The privacy mechanism is provided to protect those data using generalization technique using K-Anonymization model [4]. Lei Eu et al. stated that the fast development of data mining technology has a serious threat on security on sensitive information. PPDM technique is to perform the data mining algorithms for giving the security on data. In data mining applications, there are four different types of users involved for exploring the privacy-preserving approach. Game theoretical approaches are proposed for analyzing the communications between the different user in a data mining and every user has their own valuation on the sensitive data [5].

Fang et al. presented that data may lead to leakage while storing and transmitting the data and gives a serious threat to organizational and personal security. Day by day, the data and the information are widely increasing, to protect the sensitive data by using the screening algorithm. This enables the privacy-preserving method to minimize the exposure of sensitive information when detecting and uses new mapreduce algorithm also for finding data leakage while computing collecting intersection [6]. Xuyun et al. presented a paper processes huge amount of data by mapreduce concept. A new techniquc is needed in the mapreduce to protect the sensitive information on cloud. More number of privacy sensitive information are scattered in the cloud. In the existing approaches, the privacy for processing data is provided by mapreduce using the access control mechanism or encryption technique. These techniques fail to protect the privacy cost consumption while sharing the data. They proposed a scalable and cost effective privacy-preserving layer (PPL) framework is applying on the mapreduce in cloud [7].

Ji-Jiang Yang et al. stored the individual patient details on cloud which raised a serious concern on privacy preserving. There are three categories followed in their work: privacy by policy, privacy by statistics, and privacy by cryptography. The requirement of data utilization and privacy concerns on various parts of the medical dataset is quite different. Also this paper proposed a vertical data partition, data merging, and integrity checking, and hybrid search across plaintext and cipher text that provides thc privacy for multiple paradigms for large scale medical data access and sharing [8]. Tamas et al. presented a scheme to protect multiple sensitive attributes by k-anonymity and l-diversity. The existing systems concentrated on single attribute only. So they proposed a new model of extended the k-anonymity and l-diversity method for protecting the multiple sensitive attributes [9].

Xinjing et al. presented the PPDM and set a new trend in privacy, security, and data mining research. This paper insisted on altering the original data in PPDM by developing the algorithms so the privacy data and the knowledge are kept private even after mining. They used the collision behavior in PPDM and it gives the collision resistant protocol in terms of the penalty function mechanism [10]. Prabhakaran et al. defined a cloud storage system which provides storing and retrieving a large amount of data. This paper is presented for managing the medical datasets on cloud. Personal health record (PHR) is useful for the patients throughout their life which is convenient and secured for selected caregivers. PHR machine gives a new architectural solution for health records. Patients can upload their data for accessing and sharing the same through virtual machine (VM) [11]. Cong Wang et al. presented a privacy-preserving public auditing system for data security in cloud computing. Third-party auditor (TPA) is used to check the integrity of data. TPA performs the auditing for many users efficiently. TPA would not have the knowledge of data content while auditing process [12].

Niyati et al. defined about a healthcare industry which is always large and sensitive. The health care data need to be handled carefully; so many data mining techniques are used in the industry. This paper proposed various classification techniques for providing better accuracy, sensitivity, and specificity percentage [13]. Nagendra kumar et al. presented a technique to preserve privacy of the sensitive data of individuals. This paper proposed a k-anonymity algorithm which considers all the sensitive values are the same; this may leads to some issues like information loss, data utility and privacy. So they need to provide a method with minimum data loss and maximum data utility. A sensitive attribute-based anonymity method is used to avoid the suppression to increase the data utility [14]. Yali Liu et al. presented a multilevel system known as sensitive information dissemination detection (SIDD) that is used to detect the dissemination of sensitive data by an insider. In the existing work the insider used the exfiltrate data but in the multilevel structure that provides a unified framework for detecting the leakage of information. A statistical and signal processing technique is applied for outbound sources to generate a signature. Then the signature is used to detect and prevent the leakage by analysis. They proposed the scheme for both analogy and misusage of detecting the information [15].

Ali Gholami et al. presented a survey of security and privacy in cloud computing. It evaluates various cloud computing technologies like virtualization and containers and also discussed various security challenges. Thus, the result of this paper is in the area of privacy that is based on the cloud provider activities that is orchestration, cloud provider service management layers, and physical resource [16]. Cong Wang et al. presented an investigation on cloud storage system. To ensure the data security in cloud, they proposed an effective and flexible distributed system along with altering, concatenating and deleting the data. For giving the redundancy parity and data dependent guarantee they proposed an erasure-correction code in the file distribution. The homomorphic token scheme is used to integrating the storage correctness values and data error localization, which means the corruption of data or misbehaving the servers are detected when storing the correctness verification has been done across the servers [17]. From the literature, it is observed that identifying the sensitive attributes is essential and preserving the privacy also be essential for secure data transmission. In this paper, the feature selection method is used to identify the sensitive attribute from a massive data for preserving privacy and secured data transmission with less computation cost. Danasingh Asir et al. presented several feature selection approaches for improving accuracy of the classifier [18–22].

3 Proposed Method and Implementation

This section explains the proposed method and its implementation. The flowchart representation of the proposed algorithm is illustrated in Fig. 1. The proposed method initially receives the dataset and calculates the chi-square statistical value for each attribute for determining the sensitiveness of the attribute with respective to the target class and calculates the threshold for separate the sensitive and the nonsensitive attribute. Then the sensitive and nonsensitive attributes are given to the classification algorithm to identify the level of significance in sensitive and non-sensitive attributes.

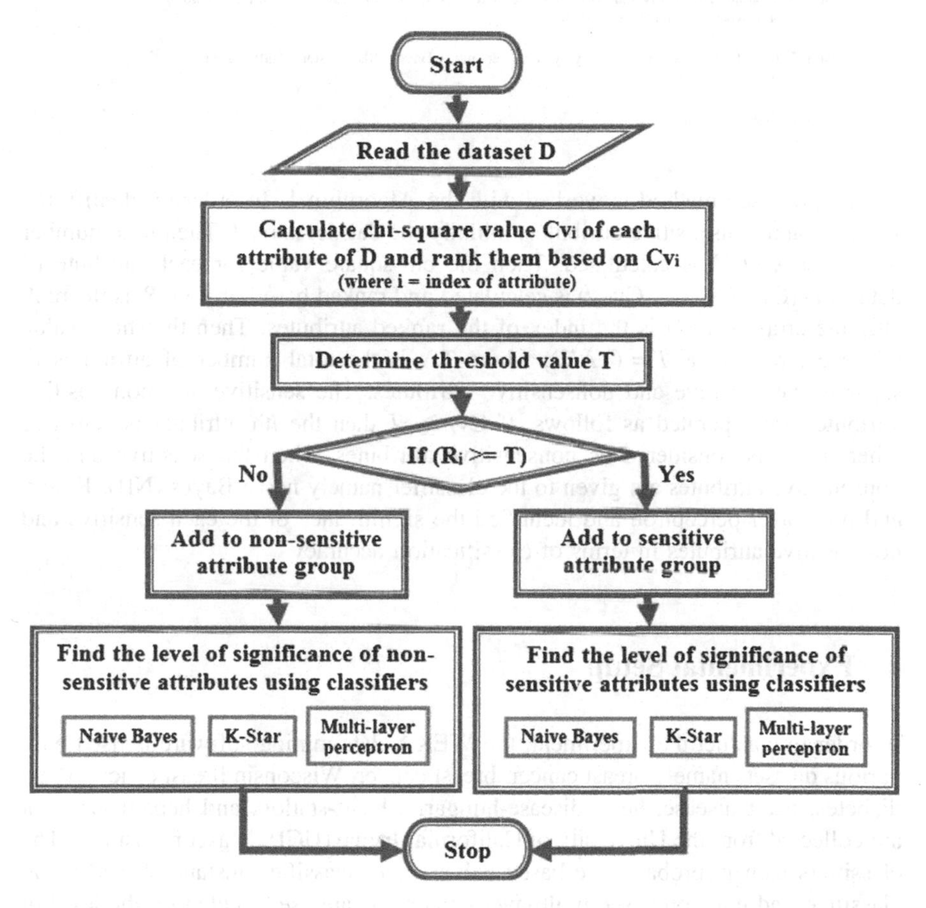

Fig. 1 Flowchart representation of the proposed algorithm

Algorithm 1 : Identifying sensitive and non-sensitive attributes

Input: Dataset (D)

Output: Selected sensitive and Non-sensitive attributes

Step 1: Begin

Step 2: Read the dataset (D)

Step 3: Find the total number of attributes (TA)

Step 4: Calculate the chi-square value for each attribute of dataset D (Cv_1, Cv_2,..., Cv_{TA}) and rank them based on Cv_i then the ranked indexed attributes are Ri where i=index of the ranked attribute

Step 5: Determine the threshold value (T) to separate the sensitive and non-sensitive attributes. T= (TA/2), where TA = Total number of attributes

Step 6: Separate the sensitive and non-sensitive by if the Cv_i => T then the ith attribute is sensitive otherwise non- sensitive attributes.

Step 7: Identify the level of sensitivity by classifiers naive Bayes (NB), K-star, Multi – layer perceptron. \

Step 8: End

The proposed method is worked with the Algorithm 1. In order to identify the sensitive and nonsensitive attributes, initially the dataset is read. Then total number of attributes (TA) is calculated. Then the chi-square value for each attribute of dataset D (Cv_1, Cv_2, …, Cv_{TA}) is calculated and ranked based on Cv_i. R_i is the rank attribute array where i is the index of the ranked attributes. Then threshold value (T) is determined as T = (TA/2), where TA is the total number of attributes to separate the sensitive and nonsensitive attributes. The sensitive and nonsensitive attributes are separated as follows. If $Cv_i \geq T$ then the ith attribute is sensitive otherwise it is considered as nonsensitive attributes. Then the sensitive and the nonsensitive attributes are given to the classifier namely naive Bayes (NB), K-star, and multilayer perceptron and identified the significance of the each sensitive and nonsensitive attributes in terms of classification accuracy.

4 Experimental Setup

In order to conduct the experiment, the WEKA data mining software is used with various datasets namely breast cancer, breast cancer, Wisconsin breast cancer, Pima diabetes, heart disease, heart disease-hungaria, heart-statlog, and hepatitisare that are collected from the University of California, Irvine (UCI) dataset repository. The classifiers namely probabilistic based naive Bayes classifier, instance based k-star classifier, and function based multilayer perceptron are used to identify the level of significance in sensitive and nonsensitive attributes. Initially, the datasets are read and the number of attributes present in the datasets is found. Then the threshold value is calculated to separate the sensitive and nonsensitive attributes. The sensitive and nonsensitive attributes are given to the classification algorithms to identify their significance in terms of classification accuracy.

5 Results and Discussion

This section discusses the results obtained from the experiments. Table 1 shows the experimental results in terms of classification accuracy with respect to sensitive and nonsensitive attributes of the datasets. Figure 2 illustrates the average classification

Table 1 Classification accuracy of classifiers with respect to sensitive and nonsensitive attributes of datasets

Dataset	NSA	SA	Naive Bayes		K-star		Multilayer perceptron	
			sensitive	Nonsensitive	sensitive	Nonsensitive	sensitive	Nonsensitive
Breast cancer	5	4	72.7273	68.1818	74.8252	70.6294	70.6294	65.0350
Wisconsin breast cancer	5	4	95.5651	93.4192	98.8498	93.9914	96.1373	94.2775
Pima diabetes	4	4	75.3906	66.5365	71.6146	63.8021	76.6927	67.1875
Heart disease	7	6	84.1584	69.967	79.538	62.0462	79.2079	64.6865
Heart disease-hungaria	7	6	82.9932	64.6259	82.9932	58.1633	78.9116	61.2245
Heart-statlog	7	6	85.1852	71.1111	80.7407	66.6667	78.8889	65.9259
Hepatitis	10	9	85.8065	76.7742	84.5161	73.5484	80.6452	77.4194

(*SA* Sensitive attribute, *NSA* Nonsensitive attribute)

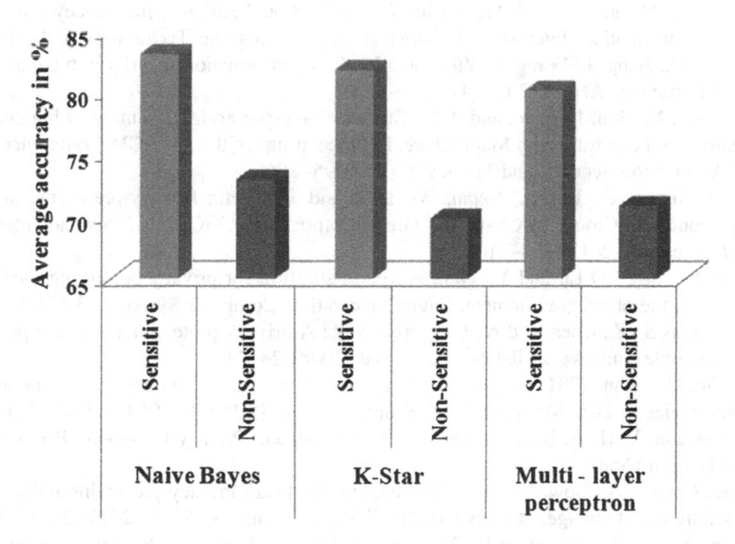

Fig. 2 Average classification accuracy of classifiers with respect to sensitive and nonsensitive attributes of datasets

accuracy of classifiers with respect to sensitive and nonsensitive attributes of the datasets. From Table 1 and Fig. 2, it is observed that the proposed method produces higher classification accuracy than the nonsensitive attributes.

6 Conclusion

This chapter presented a method for identifying the sensitive attributes for preserving the privacy in massive data for secured data transmission so that the cost spent for preserving the security is reduced. The experimental results show the difference between the sensitive and nonsensitive attributes in terms of the classification accuracy with the classifiers namely naive Bayes, K-star and multilayer perceptron.

References

1. Dong, Xinhua, R., Li, H., He, W., Zhou, Z., Xue, and H., Wu.: Secure sensitive data sharing on a big data platform. Tsinghua Science and Technology 20, 1 (2015), 72–80.
2. Yu, Shucheng, C., Wang, K., Ren, and W., Lou.: Attribute based data sharing with attribute revocation. In Proceedings of the 5th ACM Symposium on Information, Computer and Communications Security, ACM (2010) 261–270.
3. Yakoubov, Sophia, V., Gadepally, N., Schear, E., Shen, and A., Yerukhimovich, A.: Survey of cryptographic approaches to securing big-data analytics in the cloud. In High Performance Extreme Computing Conference (HPEC), (2014) 1–6.
4. Rashid, A., Hatem, and N.B M., Yasin.: Generalization Technique for Privacy Preserving of Medical information. International Journal of Engineering and Technology 6, 4 (2014) 262.
5. Xu, Lei, C., Jiang, J., Wang, J., Yuan, and Y., Ren.: Information security in big data: privacy and data mining. Access, 2 (2014) 1149–1176.
6. Liu, Fang, X., Shu, D., Yao, and A.R., Butt.: Privacy-preserving scanning of big content for sensitive data exposure with MapReduce. In Proceedings of the 5th ACM Conference on Data and Application Security and Privacy, (2015) 195–206.
7. Zhang, Xuyun, C., Liu, S., Nepal, W., Dou, and J., Chen.: Privacy-preserving layer over MapReduce on Cloud. In Cloud and Green Computing (CGC), 2012 Second International Conference on, (2012) 304–310.
8. Yang, Ji-Jiang, J.Q Li, and Y., Niu.: A hybrid solution for privacy preserving medical data sharing in the cloud environment. Future Generation Computer Systems 43 (2015) 74–86.
9. Gal, Tamas S., Z., Chen, and A., Gangopadhyay.: A privacy protection model for patient data with multiple sensitive attributes. IGI Global (2008) 28–44.
10. X., Ge, J., Zhu (2011). (Ed.).: Privacy-preserving data mining, "New Fundamental Technologies in Data Mining"-, Prof. Kimito Funatsu ISBN: 978-953-307-547-1, inTech.
11. Prabhakaran, V.M., S. Balamurugan, and S. Charanyaa.: Privacy Preserving Personal Health Care Data in Cloud.
12. Wang, Cong, S.S Chow, Q. Wang, K. Ren, and W. Lou.: Privacy-preserving public auditing for secure cloud storage. "Computers, IEEE Transactions" on 62, 2 (2013) 362–375.
13. Gupta, Niyati, A., Rawal, V.L., Narasimhan, and S., Shiwani.: Accuracy, sensitivity and specificity measurement of various classification techniques on healthcare data. IOSR J. Comput. Eng. (IOSR-JCE) 11, 5 (2013) 70–73.

14. Vasudevan, K., Prince.: Sensitive Attributes based Privacy Preserving in Data Mining using k-anonymity.
15. Liu, Yali, C., Corbett, K., Chiang, R., Archibald, B., Mukherjee, and D., Ghosal.: SIDD: A framework for detecting sensitive data exfiltration by an insider attack. In System Sciences, 2009. HICSS'09. 42nd Hawaii International Conference on, (2009) 1–10.
16. Gholami, Ali, and E., Laure.: Security and Privacy of Sensitive Data in Cloud Computing: A Survey of Recent Developments. (2016) 1601-01498.
17. C., Wang, Q., Wang, K., Ren, and W., Lou.: Ensuring Data Storage Security in Cloud Computing. In INFOCOM, (2010) 1–9.
18. Asir Antony Gnana Singh, D., Escalin Fernando, A., Jebamalar Leavline, E.: Performance Analysis on Clustering Approaches for Gene Expression Data. "International Journal of Advanced Research in Computer and Communication Engineering" 5, 2(2016) 196–200.
19. Danasingh Asir Antony Gnana Singh, S. Appavu Alias Balamurugan, and E. Jebamalar Leavline.: Literature Review on Feature Selection Methods for High-Dimensional Data. International Journal of Computer Applications, 136, 1 (2016) 9–17.
20. D. Asir Antony Gnana Singh, E. Jebamalar Leavline, E. Priyanka, C. Sumathi.: Feature Selection Using Rough Set for Improving the Performance of the Supervised Learner. International Journal of Advanced Science and Technology 87 (2016): 1–8.
21. D. Asir Antony Gnana Singh, E. Jebamalar Leavline, R. Priyanka, and P. Padma Priya.: Dimensionality Reduction using Genetic Algorithm for Improving Accuracy in Medical Diagnosis. I. J. Intelligent Systems and Applications (2016).
22. D. Asir Antony Gnana Singh, S. Appavu Alias Balamurugan, and E. Jebamalar Leavline.: A novel feature selection method for image classification. Optoelectronics And Advanced Materials-Rapid Communications. 9, 11–12 (2015) 1362–1368.

Enhanced Vehicle Detection and Tracking System for Nighttime

Ram Y. Wankhede and Diwakar R. Marur

Abstract The objective of this paper is to assist the driver during nighttime. Many papers were published which concerned with detection of vehicle during daytime. We propose a method which detects the vehicle during nighttime since it is problematic for human to analyze the shape of vehicles in nighttime due the limits of human vision. This project endeavors to implementation of vehicle detection based on vehicle taillight and headlight using technique of blobs detection by Center Surround Extremas (CenSurE). Blobs are the bright areas of the pixels of headlight and taillight. First stage is to extract blobs from region of interest by applying multiple Laplacian of Gaussian (LoG) operator which derive the response by manipulating variance between surrounding of blob and luminance of blob which are grabbed on road scene images. Compared to the automatic thresholding technique, Laplacian of Gaussian operator provides more robustness and adaptability to work under different illuminated conditions. Then vehicle lights which are extracted from the first stage are clustered based on the connected—component analysis procedure, to confirm that it is vehicle or not. If vehicle is detected, then tracking of vehicle is done on the basis of the connected components using bounding box and different tracking parameters.

Keywords Image processing · Computer vision · Vehicle detection · Advanced driving and assistance system · Intelligent transport system

1 Introduction

Every minute, on average, one person dies in a vehicle crash. Automobile Accident data includes at least 10 million people each year, two to three million are seriously injured. It is predicted that damaged property, hospital bill, and other costs will add

R.Y. Wankhede (✉) · D.R. Marur
Embedded System Technology, Electronics and Communication Engineering
SRM University, Chennai 603203, India
e-mail: ram1103w@gmail.com

S.S. Dash et al. (eds.), *Artificial Intelligence and Evolutionary Computations in Engineering Systems*, Advances in Intelligent Systems and Computing 517,
DOI 10.1007/978-981-10-3174-8_55

up 1–3 present of the world's Gross Domestic Product [1, 2]. Despite the fact, traffic at nighttime is much lower than during day time, 42% of all traffic accidents occur after dark and 58% are fatal [3].

In past vehicle detection is implemented using fusion sensor installed in the cars utilizing radar or sonar technology to detect the vehicle in front of the driver. However, with the advancement in embedded processors, and more sophisticated cameras, vision-based detection of vehicle became more prevalent. This is a part of areas of research under autonomous Vehicle driving and Advanced Driver Assistance Systems (ADAS).

It is easy to recognize vehicles during the daytime due to the human vision limits and other vision based devices. During nighttime it is not straightforward, ambient light sources contribute noise in the detection process, while the actual vehicle hull is not visible. For detection of vehicle in an image sequence various pattern recognition and image processing techniques are applied on grabbed image sequence. Searching specific pattern in image depends on shape, symmetrisation, or surrounding bounding box [4–6]. Until recently, most of this detection techniques were focused for daytime environment.

However, due to bad-illuminated conditions in nighttime the techniques which are available for detection in daytime are not effective, and become invalid in darker environment. At night or under the dark illuminated condition the only parameter that is available for detecting vehicle is headlight and taillight. However, in nighttime environment there are many illuminate sources are available with vehicle light, such as traffic light, road reflector plate placed on ground and street lamps. This illuminate source of light cause difficulties in detecting actual vehicle during nighttime.

In previous studies, the proposed method is binarization using simple threshold [7] and automatic multilevel thresholding [8]. In these methods, however it is difficult to segment the blob correctly due to various ambient light. Depending on the distance of vehicle, luminance varies which make vehicle detection difficult. Moreover, headlight and taillight have different luminances which make it further difficult to analysis the vehicle.

Laplacian of Gaussian (LOG) operator is responsible for extraction of brighter region by comparing with the surrounding rather than simply extracting the high luminance region. It is more vigorous than multilevel and simple thresholding to determine threshold using luminance of the whole image. Corresponding to various size of the vehicle light, multiple scale of LoG Operator is necessary to derive, but time for calculation is large for real-world implementation. Therefore, we are using CenSurE [9, 10], which is robust to operate in high-speed blob detection by deriving multiple scale LoG operator from integral images. In the previous studies for detection of vehicle during nighttime, they use methodology of Center Surround Extremas which is directly implemented on the image sequence [10]. But to reduce the computation cost we are using CenSurE technique on region of interest that is lower part of the vertical y-axis, we are not using the upper half because of illuminant object such as traffic lights and street lamps are present in that region.

Pairing of vehicle headlights of oncoming vehicle and taillight of preceding vehicle is obtained by set of identification rules applied on the brighter objects obtained by CenSurE, blob pairing is processed on the obtained blob to determine it as actual vehicle. The vehicle identification and tracking process is applied on blob pairing from the consecutive frames. While tracking we are improving the accuracy of the vehicle detection, by eliminating noise and correction of errors from the previous frame. The distance of each of the detected vehicle is estimated and reported. Goal of reducing processing time to process 30 frames/s, which is the frame rate of camera used for conducting tests.

2 Methodology

In the adopted system, an enhanced nighttime vehicle detection and tracking method is proposed which is capable of detecting analyzing vehicle Headlight and Taillight by using CenSurE, blob clustering is performed using response of CenSurE. Followed with the tracking process which has five parameters which analyzes the detection of vehicle, with minimizing errors. Figure 1 outlines the methodology of proposed algorithm.

2.1 *Selection of Region*

To screen out the non-vehicle illuminant source of disturbance in vehicle detection such as traffic lights and street lamps located on upper half of the vertical *y*-axis, that is "horizon." The region of less than 5 m from the current car and the region above 50% of the image height is ignored (virtual horizon) reducing the cost of computation, the blobs are only extracted from the remaining image.

As shown in Fig. 3 the image is cropped from original image (Fig. 2) to retrieve the result which will be obtained for reducing the cost and time complexity of computation.

2.2 *Blob Detection*

For detecting the brighter headlight of the vehicles, we are converting color image to gray-scaled image with NTSC conversion. For detection of taillight we are converting image to red emphasized image [11] with

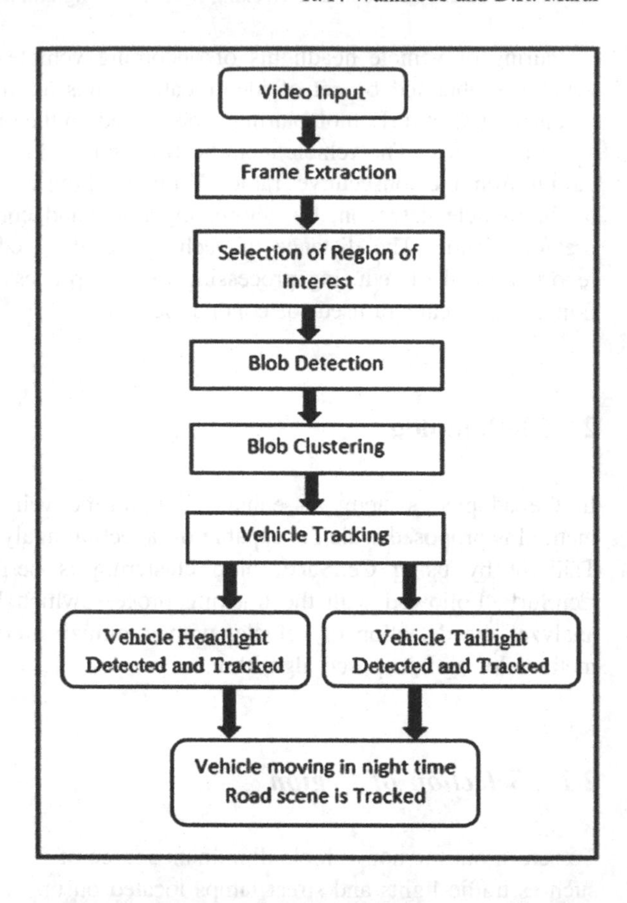

Fig. 1 Flowchart of proposed algorithm to detect and track vehicles at night

Fig. 2 A sample of nighttime image road scene

$$\text{RED} = (\text{R} + \text{G}) \times \alpha + \beta \tag{1}$$

where R is red value, G is green value of pixel of interest, and α, β are constant numbers ($\alpha = 1, \beta = 0$) since, detection of vehicle is tough from size and shape,

Fig. 3 Region selection for blob detection

vehicle are spotted on the basis of difference of blobs and its surrounding. Laplacian of Gaussian operator is combination of Gaussian operator and Laplacian operator which is expressed as

$$l(x, y, \sigma) = -\frac{x^2 + y^2 - 2\sigma^2}{2\pi\sigma^6} \exp\left(-\frac{x^2 + y^2}{2\pi\sigma^6}\right). \tag{2}$$

where σ is scale of the operator and (*x*, *y*) is distance from pixel of interest.

We cannot use LoG operator scale as constant, as the size of light increase in an image then output response of the LoG operator decreases so it is compulsory to use multiple operator for processing with different scales, computing cost to derive the operators output is high. So, CenSurE is used in detecting blobs using approximate LoG operator, enable the use of integral image in deriving high response in high speed.

The sum of any rectangular plan from an image can be calculated as follows

$$S = i(x_1, y_1) - i(x_2, y_2) - i(x_3, y_3) + i(x_4, y_4) \tag{3}$$

With the help of integral image, the response of the LoG operator is calculated with the constant time irrespective of its scale. More likely vehicle lights are in circular shape, the use of star operator is grouping of 45° rotated box operator and box operator. At this time to get 45° rotated operator we will rotate the original integral image by 45°. The sum of rotated box and original integral image is derived from above equation. The internal dimension of the box operator is determined as

$$I_k = 2 \times k + 1 \quad (1 \leq k \leq n) \tag{4}$$

And external size of box is determined as

$$O_k = 4 \times k + 1 \quad (1 \leq k \leq n) \tag{5}$$

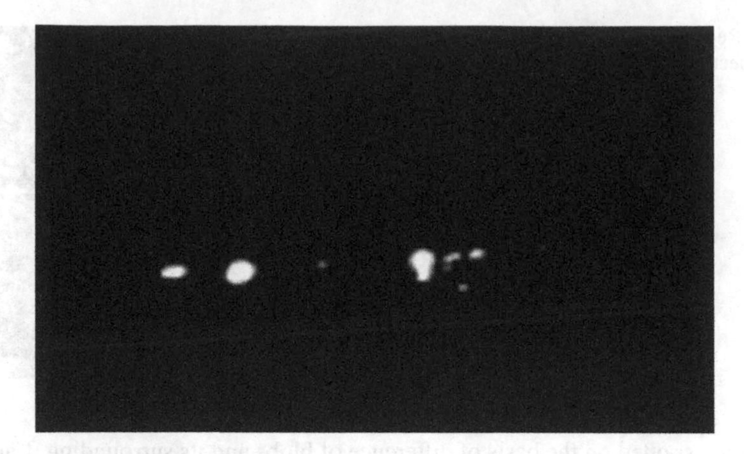

Fig. 4 Results of blob detection using CenSurE

After deriving the multiple scale operator of the star operator

- It finds size of the blobs and exact location of center, by finding the extreme point of the response.
- As in Fig. 4. The extreme point is detected using 3 continues scale as 1 set.
- Then, it searches maximum response among its neighboring pixel from smallest set of scale.
- The extreme scale '*n*' for detection was 8, but it is better to set the scale as 10 for the larger blobs.
- Set threshold to reduce the noise, as the response of noise turn into an extreme value.

Figure 4 shows the result for blob detection under high luminance during nighttime road scene.

2.3 Blob Clustering

Automobile has two front lights or rare lights, apart from Motorcycles. In this technique, after recognition of blobs they are paired (clustered) in group of two blob horizontally. Depending on the similarity of the blobs they are paired, but it would be tough if three lights are found on a horizontal axis. The lights of the same vehicle will nearly have a same shape, luminance, and size since distance between vehicle headlight or taillight from camera is almost same. Therefore, blob clustering is based on the response of derived scale of blob and CenSurE. The response of Center Surround Extremas is based on luminance and magnitude of blob. This technique is operative for pairing the blobs.

Similar Response of blobs using Center Surround Extremas is determine by

$$\mathrm{Sim}_{ij} = |\mathrm{Res}_i - \mathrm{Res}_J| \quad (6)$$

where Res_i is response of the blob i. When the similarity is minimum between blobs, which is derived from Eq. (6). When the change of their scale are in range of ± 1, then blobs are recognized as the lights of single vehicle. After recognition of pair of light we cannot discard single light, because we want detection of motorcycle as well.

2.4 *Vehicle Tracking*

The cluster of lights of automobile are obtained using aforementioned methods from each frame. However, complete features of potential vehicles are not possible to detect in a single frame, a tacking process is applied on the vehicle in the consecutive image frames to extract the information of the potential vehicle. This information is used to refine the detection result and can be used to correct the caused during CenSurE and clustering process. Tracking information can use to determine useful fact corresponding to the position, relative velocities, and the direction of movement of the potential vehicle entering under region of interest.

3 Potential Vehicle Tracking

If we consider only the potential vehicle in a single frame, the problems caused by the CenSurE and blob clustering might be:

- A former vehicle might be moving too close to another vehicle or street lamp parallely so during blob clustering they might obstruct.
- The host car can pass so close to the oncoming car that the reflection of light on the road is considered as a one huge connected region.
- Blob clustering cannot cluster all the pair of light in a single group, which are paired using taillight set and headlight set.

By associating motion of vehicle with the sequential frame we can refine the result of detection and efficiently handle above-mentioned complications.

The features needed for tracking are defined and described as follows:

(a) p_i^t Signifies *i*th possible vehicle which are seen through camera assister car in the frame t; and bonding box which encloses all blobs confined in p_i^t is represented as $B(p_i^t)$.

(b) p_i^t location is employed as a central position while tracking process, expressed by:

$$p_i^t = \left(\frac{l(B(p_i^t)) + r(B(p_i^t))}{2}, \frac{t(B(p_i^t)) + b(B(p_i^t))}{2}\right) \tag{7}$$

(c) Overlapping score of two possible vehicles p_i^t and p_j^t spotted at two different time t and t', is calculated using area of intersection,

$$s_O\left(p_i^t, p_j^t\right) = \frac{A\left(p_i^t \cap p_j^t\right)}{\mathrm{Max}(A(p_i^t) + A(p_i^t))} \tag{8}$$

(d) The size-ratio of enclosing bounding box of p_i^t, is defined as,

$$R_S(p_i^t) = \frac{W(B(p_i^t))}{H(B(p_i^t))} \tag{9}$$

(e) Symmetry score of the vehicle is found by,

$$S_s\left(p_i^t, p_j^t\right) = R_S(P_s)/R_S(P_l) \tag{10}$$

Here, P_s is smallest size among two vehicles $p_i^t p_j^t$ where P_l is the larger one.

In continuous tracking of vehicle in a successive frame the vehicle which is seen in arriving frame will be represented by $\boldsymbol{P}^t = \{p_i^t | i = 1, \ldots, k'\}$ and analyzed with respect to the vehicle which is already been traced in earlier frames represented by $\boldsymbol{P}^{t-1} = \{p_i^{t-1} | i = 1, \ldots, k'\}$ then group of traced vehicles $\boldsymbol{TP}^t$ will be updated with the following process.

Thru tracking of vehicle, a tracked vehicle might be in one of the five tracking stages which are applied on each vehicle in each frame. The operations applied on the tracked vehicle are as follows:

Update

Tracker is updates when potential vehicle $P_i^t \in \boldsymbol{P}^t$ in current frame is matched with the tracked potential vehicle $TP_i^{t-1} \in \boldsymbol{TP}^{t-1}$, A set of tracked vehicle TP^t by relating p_i^t with the tracker TP_j^t. The matching condition is;

$$m\left(P_i^t, TP_j^{t-1}\right) > 0.6 \tag{11}$$

where matching score is computed by

$$m\left(P_i^t, TP_j^{t-1}\right) = w_O \,.\, S_O\left(P_i^t, TP_j^{t-1}\right) + w_s \,.\, S_s\left(P_i^t, TP_j^{t-1}\right), \quad (12)$$

where w_O and w_s are weights of size-ratio feature and overlapping, set as 0.5 and 0.5 respectively.

Appear

Wherever new vehicle is detected condition $P_i^t \in \boldsymbol{P^t}$ cannot match with any existing $TP_i^{t-1} \in \boldsymbol{TP^{t-1}}$ then the new tracker is formed and append the updated set $\boldsymbol{TP^t}$.

Merge

One possible vehicle at present frame P_i^t matches with the multi-traced regions of the potential vehicles $TP_j^{t-1}, TP_{j+1}^{t-1} \cdots, TP_{j+n}^{t-1}$. This may be a situation that single vehicle is having multiple lights, so they can be grouped as a single vehicle light and correctly merged in the tracking process.

Split

One tracked vehicle can split into multiple regions in the existing frame, this may be a condition when the vehicle is moving parallel to another vehicle, reflected beam of road or street lamps, in this case matching condition is evaluated between each $TP_i^{t-1} \in \boldsymbol{TP^{t-1}}$ with $P_i^t, P_{i+1}^t \cdots, P_{i+m}^t$ if any one of the case match with TP_j^{t-1}, then potential vehicle is associated with the TP_j^t. Therefore potential vehicle otherwise non-potential vehicle, then create a new Tracker.

Disappear

If tracker of the potential vehicle TP_j^{t-1} cannot match the $P_i^t \in \boldsymbol{P^t}$ for three successive frames, at that point potential vehicle is judged as it disappears and tracker will be removed from the tracker set $\boldsymbol{TP^t}$.

Results of tracking process which we get in experimentation with camera which has frame rate of 30 frames/s are shown in Figs. 5, 6, 7 and 8.

Fig. 5 Results of headlight tracking of frame 51

Fig. 6 Result of headlight tracking of frame 94

Fig. 7 Result of taillight tracking of frame 156

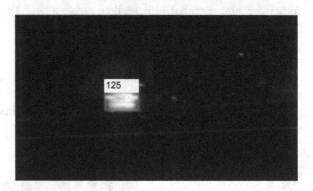

Fig. 8 Result of taillight tracking of frame 291

4 Discussion and Conclusion

In this paper we have proposed an Enhance Nighttime Vehicle Detection methodology which will reduce the cost of computation by using CenSurE in the region of interest, and we are tracking the vehicle under high luminance of vehicle light. The error cause while tracking in the sequential frame is minimized using five stages for tracking. This improve the tracking process which can be implemented to calculate the real-time parameters of vehicle like speed calculation and distance estimation more accurately in high speed scenario.

In future this proposed methodology can be implemented for driver assistance in nighttime by automatic control of headlights and can be used in driverless car where driving parameter will set according to image processing.

References

1. W. Jones, "Keeping Cars from Crashing," IEEE Spectrum, vol. 38, no 9, pp. 40–45, 2001.
2. W. Jones, "Building Safer Cars," IEEE Spectrum, vol. 39, no. 1, pp. 82–85, 2002.
3. New Headlight Sensors Make Night Driving Safer, Road and Travel Magazine, 2007.
4. A. Broggi, M. Bertozzi, A. Fascioli, C.G.L. Bianco, A. Piazzi, "Visual perception of obstacles and vehicles for platooning", IEEE Trans. Intel! Transport. Syst., vol. I, pp. 164–176, 2000.
5. S. Nedevschi, R. Danescu, D. Frentiu, T. Marita, F. Oniga, C. Pocol, R. Schmidt, T. Graf, "High accuracy stereo vision for far distance obstacle detection", in Proc. IEEE Intel! Vehicle Symp., pp. 292–297, 2004.
6. M. Betke, E. Haritaoglu, and L. S. Davis, "Real-time multiple vehicle detection and tracking from a moving vehicle", Mach. Vision Appl., vol. 12, pp. 69–83, 2000.
7. Y. L. Chen and Chuan-Yen Chiang, "Embedded on-road nighttime vehicle detection and tracking system for driver assistance," *Systems Man and Cybernetics (SMC), 2010 IEEE International Conference on*, Istanbul, 2010, pp. 1555–1562.
8. M. Agrawal, K. Konolige, and M. R. Blas, "CenSurE: Center surround extremas for real time feature detection and matching," Lecture Notes Comput. Sci., vol. 5305, pp. 102–115, 2008.
9. A. Lpez et al., "Temporal coherence analysis for intelligent headlight control," in Proc. 2nd Workshop Planning, Perception Navigat. Intell. Veh., Nice, France, 2008, pp. 59–64, 1996.
10. Sci., vol. 5305, pp. 102–115, 2008. N. Kosaka and G. Ohashi, "Vision-Based Nighttime Vehicle Detection Using CenSurE and SVM," in *IEEE Transactions on Intelligent Transportation Systems*, vol. 16, no. 5, pp. 2599–2608, Oct. 2015. doi: 10.1109/TITS. 2015.2413971.
11. N. Korogi, K. Ueno, K. Uchimura, and J. Hu, "Traffic flow analysis under day and night illumination," Tech. Rep. Instit. Electr on. Inf. Commun. Eng., vol. 104, no. 506, pp. 7–12, 2004.

A Testing + Verification of Simulation in Embedded Processor Using Processor-in-the-Loop: A Model-Based Development Approach

Kuman Siddhapura and Rajendrasinh Jadeja

Abstract This paper gives an idea of how to utilize the processor-in-the-loop (PIL) and reduce the development time. The rapid advancement of the computer system and user interactive software empowers the lot of simulation software which are available for technical applications. In this paper, the simple mathematical formulations using simulation block sets are formed. The simulated mathematical formulations are transformed into embedded C code. The generated code is launched into the embedded processor. The communication link between host PC and target processor is formed to view real-time variation where part of the simulation runs in PC and part in processor. The simulation results are compared with launched program executed by the processor. This approach unified the designer with a program and reduced the core technical manpower effort, simultaneously increases more functionality to test the program in processor. The simulation platform is used as MATLAB and Simulink, and STM32F4 Discovery board is used as a target to launch the program in processor.

Keywords MATLAB · Processor-in-the-loop (PIL) · Simulink · STM32F4

K. Siddhapura (✉)
School of Engineering, RK University, Rajkot, India
e-mail: kuman.siddhapura@darshan.ac.in

K. Siddhapura
Electrical Engineering Department, Darshan Institute of Engineering and Technology, Rajkot, India

R. Jadeja
Faculty of Engineering, Marwadi Education Foundation, Rajkot, India
e-mail: rajendrasinh.jadeja@marwadieducation.edu.in

S.S. Dash et al. (eds.), *Artificial Intelligence and Evolutionary Computations in Engineering Systems*, Advances in Intelligent Systems and Computing 517,
DOI 10.1007/978-981-10-3174-8_56

1 Introduction

Today's world, almost every application is running on embedded processor. Almost all analog application is transformed into digital and processor does all smart things with proper programming. Since the beginning of processor, first it modified to use as a microcontroller for a specific application. These microcontrollers gain more popularity in industry and chip manufacturer started developing more and advance microcontroller. With the application of telecommunication and control, the role of this processor is more on mathematical manipulation rather than simple execution of a program or logical operation. This requirement breeds another type of processor called as digital signal processor (DSP) and digital signal controller (DSC) [1]. There is much user-friendly software available to program this processor and software environment is created to program it easily and fast.

At the same time, there is rapid development in technical software such as MATLAB, ANSYS, Scilab, Rlab, LabVIEW, Mathematica, GNU Octave, PSCAD, PSIM, PSpice, AUTOCAD, etc. Over a span of decades, these platforms become strong enough that a lot of experience added in this to work towards the real technical phenomenon. The basic limitation of any system is nonlinearity in nature and to model such system high-level mathematics needed to involve. So, naturally a designer and engineer of a specific field or area feel more comfortable to learn this software and play with it, rather than writing a C code. With graphically interactive tools makes learning these tool to be faster and more autonomy is applied to change the variable and system parameters. This enables the designer to make a system with wide range of constraint to check the tangible condition while in the operation of actual application [2–13].

With the development of two parallel sides of advancement by their own, a gap is formed between these two kinds of developments. There are so many applications, in electrical, mechanical, electronics and communication, instrumentation, and civil engineering that they have to come across the both requirement to develop a product, design a system, control a close loop system or test an algorithm. Normally, a core engineer is acquainted with the technical environment software which is taken as an example as MATLAB in this study.

How to implement the developed system, model or algorithm for a researcher or engineer, in embedded processor is presented in a sequential pattern in this paper. It can be represented that this methodology goes by the first step to developing a simulation, model a system and testing of this model. This testing of the model in the simulation is termed as model-in-the-loop (MIL). The second step is a generation of code from this model, or complete simulation depends on system and algorithm. In this step one can verify the code by running the code in a simulation which is also called as software-in-the-loop (SIL). In the third step, the code generated will be implemented in a processor, and now processor and execution of processor is explored in this test. This test is called the processor-in-the-loop (PIL). Next step would be hardware-in-the-loop (HIL) is not covered in this study. The conceptual flow considering all the methodology is represented in Fig. 1. To check the program

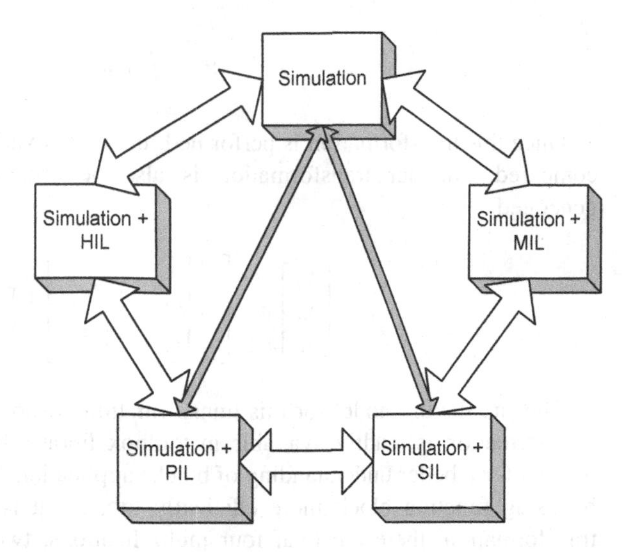

Fig. 1 Conceptual diagram of different co-simulation

and response of the processor, a PIL is presented in this study. The simulation results are visualized graphically with the tools available. With the given application programming interface (API), the same simulation system/model is implemented in embedded processor and launched in the target from host PC. The part of the system runs on PC and part in processor. This approach is also termed as co-simulation. A communication link established between processor and simulation software by API enables more visualization of response from the processor in real time.

2 System Development in Simulation and Model-in-the-Loop

The simulation study considered in this case is a mathematical transformation from one reference to other. The reason for choosing this transformation is it covers basic mathematical functions like multiplication, sin, cosine, integration, addition, gain, etc. So, with this example, it would be better in understanding to generate the code of this and implement in embedded processor.

The basic inputs are given as $x_d = 1$ and $x_q = 0$. One constant term x is also one of the input to generate the angle of sin and cosine functions.

Based on the given input the x_α and x_β can be generated as given by following formulae

$$\begin{bmatrix} x_\alpha \\ x_\beta \end{bmatrix} = \begin{bmatrix} \cos\theta & -\sin\theta \\ \sin\theta & \cos\theta \end{bmatrix} \begin{bmatrix} x_d \\ x_q \end{bmatrix} \tag{1}$$

$$\theta = \int x \, \mathrm{d}t \tag{2}$$

Once this transformation is performed, the output variables can be observed and compared. Another transformation is also performed as $x_a,\ x_b$ and x_c can be generated.

$$\begin{bmatrix} x_a \\ x_b \\ x_c \end{bmatrix} = \begin{bmatrix} 1 & 0 \\ -1/2 & \sqrt{3}/2 \\ -1/2 & -\sqrt{3}/2 \end{bmatrix} \begin{bmatrix} x_\alpha \\ x_\beta \end{bmatrix} \tag{3}$$

The simulation model for this transformation is shown in Fig. 2. Although this transformation is readily available in toolbox library, here it is done in a simple manner for a better understanding of blocks application. This can also be performed by using function block more efficiently too. As it is observed that in the first transformation, there are total four multiplications, two trigonometric, two summations, one gain, and one integration are performed using block sets. In the second transformation, another four multiplications, three gain, and two summations are involved.

The block diagram representation of simulation is shown in Fig. 3a. The simulation with model-in-the-loop is shown in Fig. 3b. The actual simulation system is shown in Fig. 3c. To observe the inputs and outputs, the scopes are used, and the entire variables are stored in the workspace as a structure with time. The simulation is also configured to visualize the data by simulation data inspector tool especially when it is required to compare the difference between two signals of same simulation run or two different output of the same signal with different simulation run.

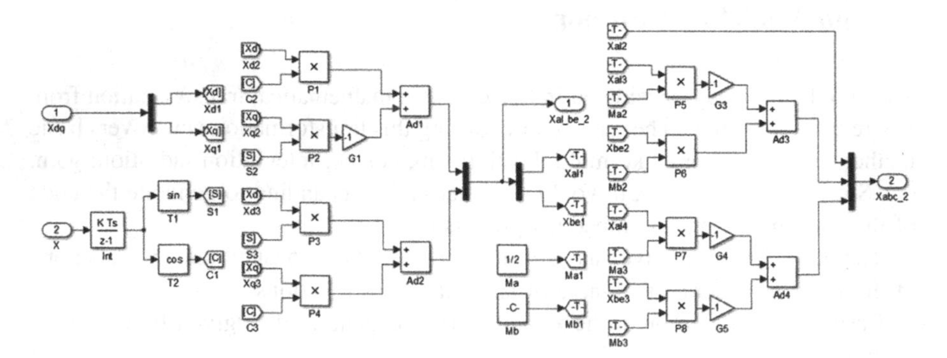

Fig. 2 Mathematical transformation model

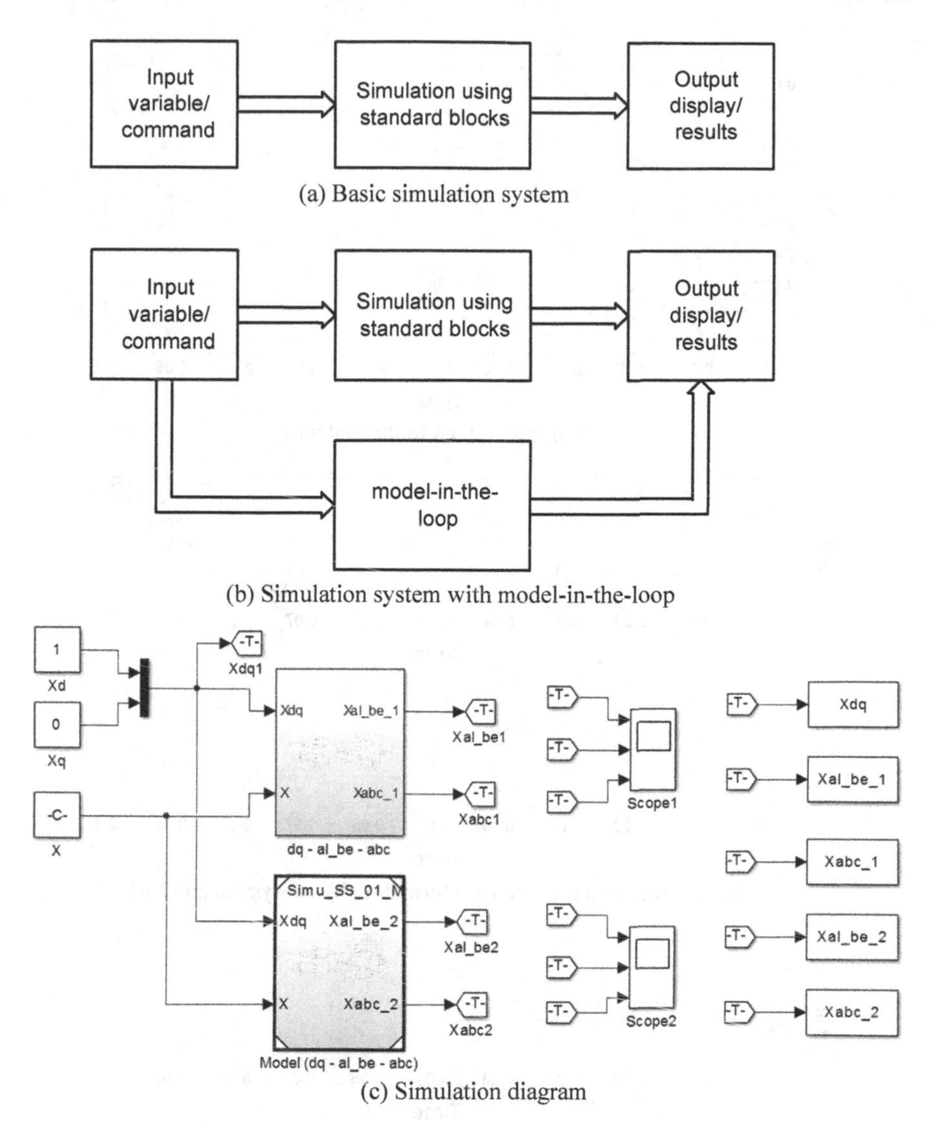

Fig. 3 **a** Basic simulation system. **b** Simulation system with model-in-the-loop. **c** Simulation diagram

As shown in Fig. 3c, the input variable is defined as X dq which is constant 1 and 0, respectively. The input variable in the plot can be seen in Fig. 4a. The constant x is taken value of 314.1593. One can take any value, based on this value; the frequency of output alternating quantities varies. The two outputs from each

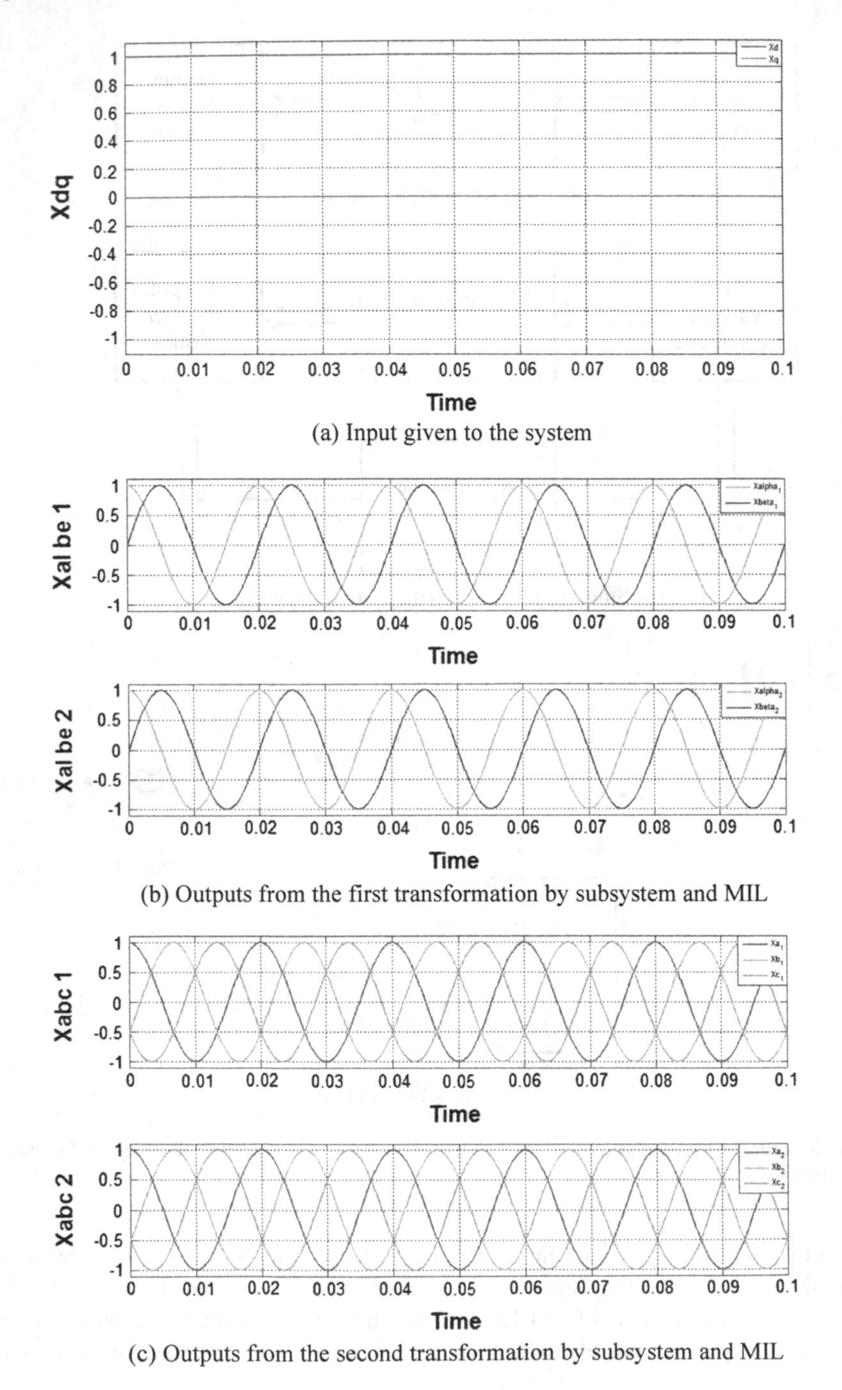

(a) Input given to the system

(b) Outputs from the first transformation by subsystem and MIL

(c) Outputs from the second transformation by subsystem and MIL

Fig. 4 **a** Input given to the system. **b** Outputs from the first transformation by subsystem and MIL. **c** Outputs from the second transformation by subsystem and MIL

system are represented as Xal_be_1, Xablc1 and Xal_be_2, Xabc2 respectively. Xal_be_1, Xablc1 are the quantities produced from subsystem and Xal_be_2, Xabc2 are the outputs from system considered as a model. Outputs from first transformation (Xal_be_1, Xal_be_2) by subsystem and model-in-the-loop are shown in Fig. 4b. Outputs from second transformation (Xabc1, Xabc2) by subsystem and model-in-the-loop are shown in Fig. 4c. The subsystem and model gave same results from both the transformation as the blocks and mathematical preliminaries used are same on the same platform.

3 Code Generation for the Processor

Once the specific algorithm, system or control design is implemented in model-based design and testing on the simulation gives the satisfactory result, the next step in model-based design and implementation is the code generation. There is specific API available base on the target device or board. As in this study, STMicroelectronics STM32F4 Discovery is as target and API are used from the manufacturer. The chip in this board has ARM cortex-M4 architecture, to launch the project; the embedded C code is generated from MATLAB embedded coder from this application. This code can be launched in the target, i.e., embedded processor. There are different development tool can be utilized during the code generation. One of the basic thing need to set is system target file, the target makes file, target tool chain, and a boot mode. The main code replacement libraries available are ANSI, ISO, GNU, and ARM CMSIS which are widely used. There is optimized library developed by the specific manufacturer to increase the efficiency of the code and execution time. The execution time can be tested, and effectiveness of the code can be checked in real time. The conceptual diagram for code generation and flow of code is shown in Fig. 5.

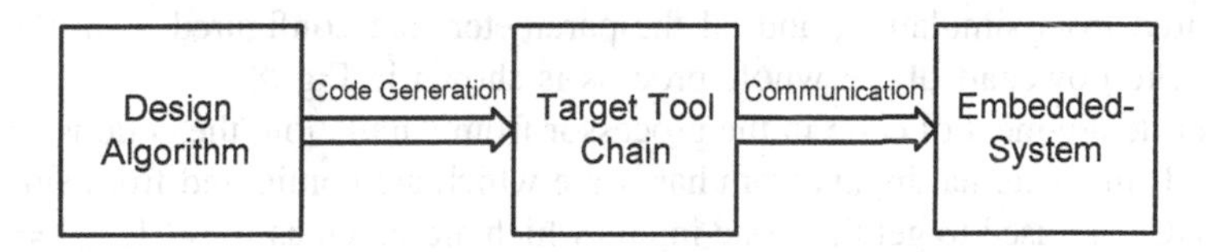

Fig. 5 Flow of code from model to processor

4 Co-simulation Based Simulation

Once the model is developed with any high degree system or algorithmic complexity, the code is generated to launch in the target device. To test the code, before putting in actual operation in real-life of operation of the product, PIL simulation run test is performed. This approach is also termed as co-simulation or on-target simulation. In this test, the code generated is launched in the processor and simulation run with the processor. To do so, a communication link has to establish between host and target. In the given board of study, UART communication is used with standard communication protocol as defined by the manufacturer. The port parameter settings are fixed from PC as specified by manufacturer and API. As STM board has several serial communication options, a Specific USART/UART communication port can be set from popup window before PIL executes and the same pin of Rx and Tx has to connect from the board. In the given study, USB to RS232 TTL UART Prolific PL2303 converter is used to communicate between PC and board. The conceptual diagram of PIL simulation is shown in Fig. 6a. The simulation from model-in-the-loop now configured as PIL is shown in Fig. 6b. The host, target device/board, and the communication link is shown in Fig. 6c.

The result of simulation subsystem and PIL are shown in Fig. 7. The inputs to both the system are same as shown in Fig. 4a. Outputs from the first transformation from subsystem and processor are given as Xal_be_1 and Xal_be_2 as shown in Fig. 7a. Outputs from the second transformation from subsystem and processor are given as Xabc1 and Xabc2 as shown in Fig. 7b. The result from simulation and processor are exactly same as expected and seen from model-in-the-loop. The execution time in seconds and execution tick in counts are shown in Fig. 7c. This plot is drawn as corresponding to each sample time. To reduce the execution time, the optimized code and alternative way of simulation can be performed with appropriate sampling time.

Once the result of MIL testing and PIL testing are satisfactory with the simulation are satisfactory, the completed code of algorithm is build, compiled, and downloaded to the processor with peripherals added as Simulink block. The chip is configured from simulation, and all the parameters are configured from the simulation. The flowchart of the whole process is shown in Fig. 8.

After deployment of code in the processor from simulation, the code is executed with real-time external inputs from hardware which are configured from simulation file. ADCs are used to get the basic inputs which are given as $x = 1$, $x_d = 1$ and $x_q = 0$ from hardware. To observe the output from transformation two DAC are used to observe the x_α and x_β. The outputs from DACs are observed in oscilloscope as shown in Fig. 9.

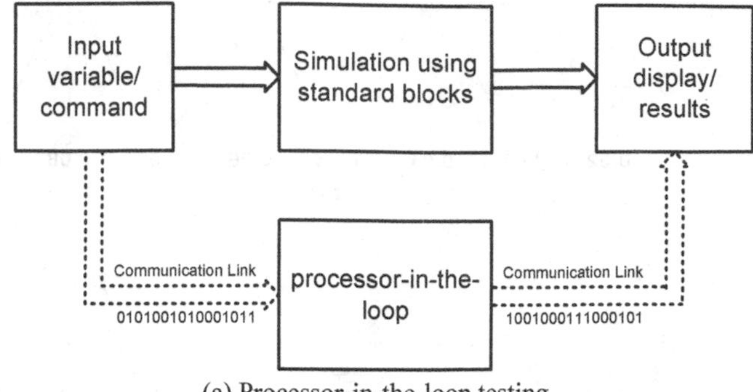

(a) Processor-in-the-loop testing

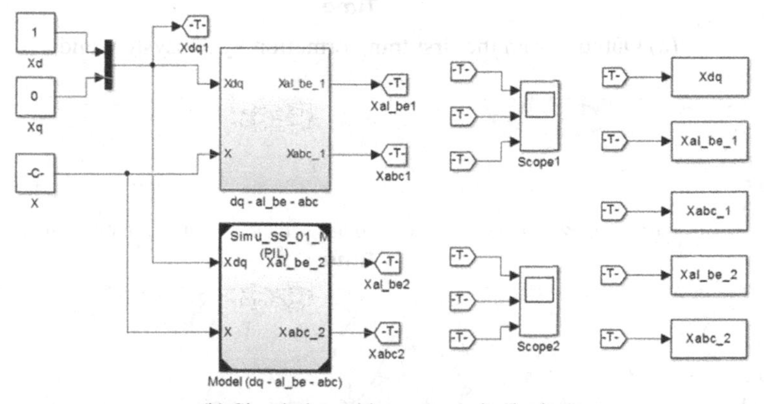

(b) Simulation with processor-in-the-loop

(c) Host, Target device/board and communication link

Fig. 6 **a** Processor-in-the-loop testing. **b** Simulation with processor-in-the-loop. **c** Host, target device/board and communication link

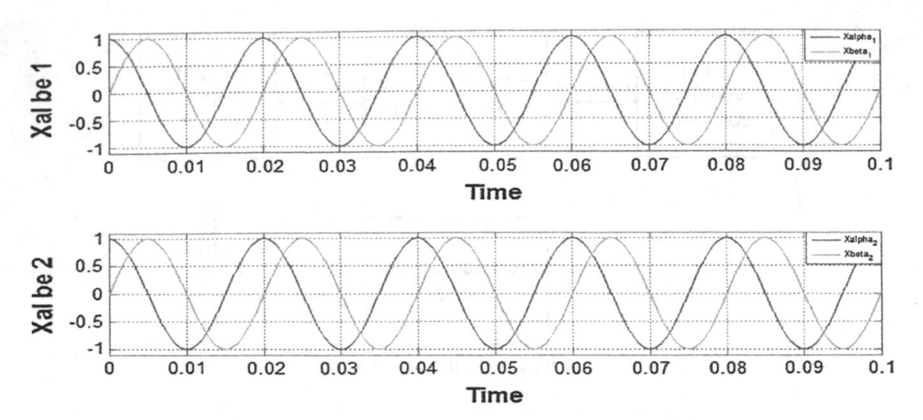

(a) Outputs from the first transformation by subsystem and PIL

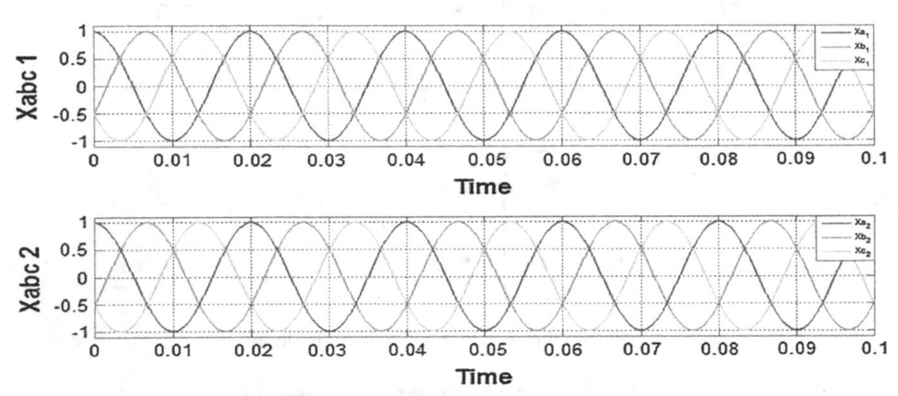

(b) Outputs from the second transformation by subsystem and MIL

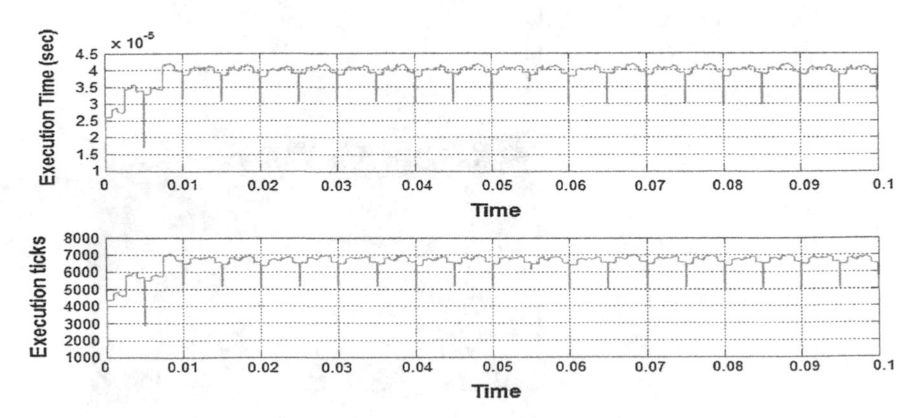

(c) Execution time in seconds and Execution tick in counts

Fig. 7 **a** Outputs from the first transformation by subsystem and PIL. **b** Outputs from the second transformation by subsystem and MIL. **c** Execution time in seconds and Execution tick in counts

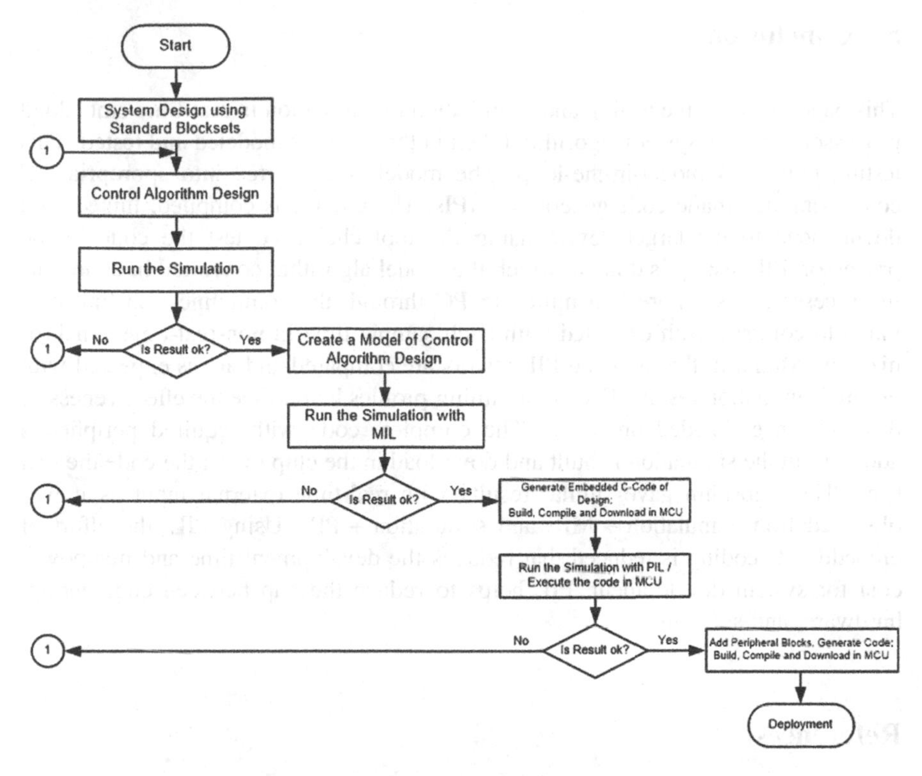

Fig. 8 Flow chart of process of simulation, MIL, and PIL

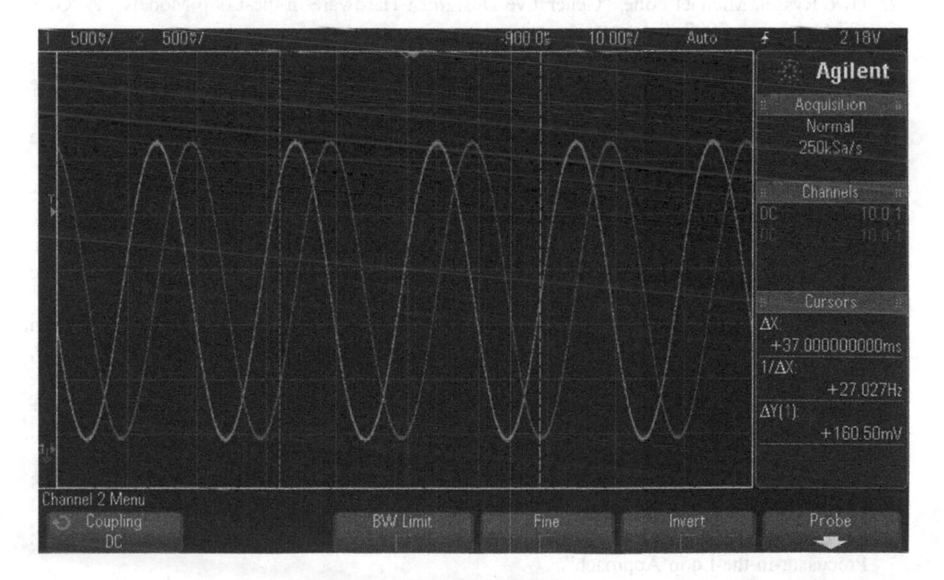

Fig. 9 Outputs from the first transformation observed from DACs in real-time execution

5 Conclusion

This paper presents the testing and verification of simulation in real-time embedded processor. First the system/algorithm taken in this study is modeled and tested. This testing termed as model-in-the-loop. The model is converted into appropriate C code using automatic code generation APIs. The C code is compiled, linked, and downloaded to the target device using the tool chain. To test the code in the processor, PIL testing is done in which the model algorithm converted in code runs in processor. Results are transmitted to PC through the communication link provided to compare with expected with each sample time in non-real-time synchronization. Model-in-the-loop and PIL results are compared and are as expected with original simulation results. Execution timing profiles help to see the effectiveness of the code in embedded processor. The complete code with required peripherals added from the simulation is built and download in the chip to test the code the real time. The algorithm gave similar results with real-time external input as it was observed from simulation + MIL and simulation + PIL. Using PIL, the effort of embedded C coding is reduced that reduces the development time and manpower cost for system development. PIL helps to reduce the gap between engineering, hardware, and software.

References

1. http://www.embedded-computing.com/pdfs/Motorola.Win03.pdf.
2. Uwe Ryssel, Michael Folie, "Generative Design of Hardware-in-the-Loop Models", APGES '07 October 4, 2007, Salzburg, Austria.
3. Zhenhua Jiang, Rodrigo Leonard, Roger Dougal, Hernan Figueroa, Antonello Monti, "Processor-in-the-Loop Simulation, Real-Time Hardware-in-the-Loop Testing, and Hardware Validation of a Digitally-Controlled, Fuel-Cell Powered Battery-Charging Station".
4. Girish Palan, Pavan K S, Rajani S R, "Simulator–in–the-Loop Environment for Autocode Verification", dSPACE User Conference 2012 – India, Sept 14th, 2012.
5. Mirko Tischer, Dietmar Widmann, "Model-based testing and Hardware-in-the-Loop simulation of embedded CANopen control devices", iCC 2012, CAN in Automation.
6. Dakai Hu, Feng Guo, Longya Xu, "Hardware-in-the-Loop Simulation Verification of a Smooth Transition Algorithm Between Maximum Torque per Ampere Control and Single Current Regulator", 978-1-4799-0336-8/13/$31.00 ©2013 IEEE.
7. P. Brandstetter, M. Dobrovsky, P. Krna, "Simulation of Sensorless Control of Induction Motor Using HIL Method", PIERS Proceedings, Kuala Lumpur, MALAYSIA, March 27–30, 2012.
8. Christian Alexandra, Ali M. Almaktoof, AK Raji, "Development of a Proportional + Resonant (PR) Controller for a Three-Phase AC Micro-Grid System", ISBN: 978-0-9891305-3-0 ©2013 SDIWC.
9. Jonathan Fielder, David Maclay, "Processor-in-the-Loop Co-Simulation of Model", US Patent 8 209 158 B1, Jun. 26, 2012.
10. S. Lentijo, A. Monti, R. Dougal, "Design of DC Motor Speed Control through Processor-in-the-Loop Approach".

11. Janica Zika, John Beale, Vimal Purushotham, "Real Time Processor-in-the-Loop Simulation using PCs".
12. Cristina Marconcini, "Functional verification on PIL mode with IAR Embedded Workbench".
13. Bogdan Alecsa, Marcian N. Cirstea, Alexandru Onea, "Simulink Modeling and Design of an Efficient Hardware-Constrained FPGA-Based PMSM Speed Controller", IEEE Transactions on Industrial Informatics, Vol. 8, No. 3, August 2012.

Interval Type-2 Mamdani Fuzzy Inference System for Morningness Assessment of Individuals

Debasish Majumder, Joy Debnath and Animesh Biswas

Abstract From the view point of revealing different preferences among the individuals, the assessment of individual typology, i.e., chronotype, is increasing largely in the recent past. Chronotype, recognized as a human attribute, is usually studied by self-reported instruments designed to find individual time of preference for daily activities in an easy manner. One of the criticisms of using these self-reported instruments includes the fact that total scores may not always reflect the actual chronotype of an individual. On the other hand, linguistic terms are used to address some of the items of these self-reported questionnaires. In this paper, an interval valued type-2 Mamdani fuzzy inference system has been proposed to assess chronotype of an individual. An illustrative case example is discussed for validation of the proposed model in the field chronobiology.

Keywords Chronotype · Morningness · Reduced version of morningness–eveningness questionnaire · Interval type 2 Mamdani fuzzy inference system

1 Introduction

In the present era, the study of chronotypes, commonly known as morningness–eveningness orientation, has been appeared as an emerging area of research due to the potential application in diverse fields such as the design of work-schedule [1, 2], athletic performance [3, 4], academic performance [5], and health and psychological well-being [6, 7]. Notable differences in several biological, psychological, and

D. Majumder
Department of Mathematics, JIS College of Engineering, Kalyani 741235, India
e-mail: debamath@rediffmail.com

J. Debnath · A. Biswas (✉)
Department of Mathematics, University of Kalyani, Kalyani 741235, India
e-mail: abiswaskln@rediffmail.com

J. Debnath
e-mail: joy6debnath@gmail.com

S.S. Dash et al. (eds.), *Artificial Intelligence and Evolutionary Computations in Engineering Systems*, Advances in Intelligent Systems and Computing 517,
DOI 10.1007/978-981-10-3174-8_57

behavioral variables, viz., body temperature, cortisol and melatonin secretion, mood, performance, alertness, appetite, sleep-wake, and usual meal times has been observed among the different chronotypes, viz., morning-type, intermediate type and evening type [8–10]. Phase differences, assessed over 24 h period of time for the variables listed above, signify different preferences among the individuals.

With the aim of achieving a reliable and valid measurement of chronotypes, several self-reported questionnaires, like morningness–eveningness questionnaire (MEQ) [11], circadian typology questionnaire (CTQ) [12], composite morningness questionnaire (CMQ) [13], etc., had been designed. Among them MEQ is the commonly used questionnaire to determine chronotypes. However, reduced version of MEQ (rMEQ) [14] has been identified as a quick and reliable measure of morningness with sufficient interitem correlation and validity [15].

In the existing crisp approaches for assessing morningness of an individual, morningness score of a person is evaluated by adding the quantified score of each item of the questionnaire assuming the fact that there exists linear relationship between the physiological factors, assessed through different questionnaires using preference scale, and morningness of the individuals. But, in real-life complex problems, where human knowledge and perception play important role, crisp approaches might not be able to capture the vagueness and uncertainty associated with the model, specially, in describing the different linguistic terms raised during the design phase.

Since its inception, fuzzy set theory proposed by Zadeh [16] in 1965, laid the foundation for computing with words and provides the platform for universal approximate reasoning [17–19] to quantitatively and effectively handle problems of this nature [20–22]. Extensive theoretical work [23, 24] may give the reader more insight into the foundation of fuzzy theory.

However, the study of morningness using fuzzy logic is very few in the literature. In the recent past, some research works have been done to generate different types of hybrid fuzzy inference systems (FISs) like adaptive neuro fuzzy inference systems (ANFIS) [25, 26], genetic algorithm-based Mamdani FIS (GAMFIS) [27], for assessing morningness of people.

It is worthy to mention here that the questions in rMEQ are predominantly subjective and involved different linguistic terms, viz., feeling best, very tired, fairly refreshed, rather morning than evening, whose interpretations are not straightforward. Again, there are some hypothetical situations, e.g., "Considering only your own feeling best rhythm, at what time would you get up if you were entirely free to plan your day?" etc., in the questionnaire that completely rely on individual perception and cognition. But human perceptions are vague and may vary over time. Again, concept of an individual is vastly depends on the context, depth of knowledge and individual sense [28]. Hence, the same concept may be interpreted differently by different individuals [29, 30].

Moreover, in FIS, the construction of rules suffers from linguistic uncertainty due to different interpretation of the same term used in describing antecedents and consequents of rules by different people [31]. Again, while a survey is carried out among a group of experts for assigning rule consequents, experts may differ in their opinion and hence the same rule may come up with different consequences. Hence, uncertainties related to antecedents or consequents may lead to rule uncertainties [32].

In Type-1 FISs (T1 FISs), type-1 (ordinary) fuzzy sets (T1 FSs) are used to describe antecedent and consequent of the rules. As a consequence, most of the time T1 FISs are incapable of handling directly rule uncertainties. But, since, the antecedent or consequent of type-2 FISs (T2 FISs) are described by using type-2 fuzzy sets (T2 FSs), it can handle rule uncertainties efficiently.

The concept of T2 FSs, introduced by Zadeh [33] as a generalization of T1 FSs, offers additional degrees of freedom in Mamdani FISs (MFISs) and Takagi-Sugeno-Kang FISs (TSKFISs). Thus T2 FISs have the potentiality to perform better than its counterpart in the context where large amount of uncertainties are observed [34–40].

However, the inclusion of footprint of uncertainty (FOU), as a third dimension in the definition of T2 FS, adds some computational complexity; and laid the path for interval valued T2 FISs (IT2 FISs), which are by far the most popular type of T2 FIS. IT2 FISs are characterized by interval valued type-2 fuzzy sets (IT2 FSs), which can be considered as a special case of general T2 FSs in which secondary membership grades are taken into account [41]. IT2 FISs are now widely used than T1 FISs due to their potentiality to capture and mitigate the uncertainty in effective manners [42–45].

In this article, an IT2 MFIS (IT2MFIS) has been designed to determine the chronotypes of an individual. During the design phase, intrapersonal and interpersonal uncertainties associated with different linguistic terms those raised in the context of morningness assessment are captured by trapezoidal type IT2 FSs [46].

The remaining part of this paper is organized in the following style. Section 2 provides a description of trapezoidal type IT2 FS, IT2 FIS. In Sect. 3, the construction of the proposed methodology is presented. Section 4, presents a case study on assessing morningness of a group of people to demonstrate the proposed morningness assessment methodology. Finally, Sect. 5 includes the concluding remarks of this work and a summary of the main benefits of using the proposed methodology in the process of morningness assessment.

2 A Brief View of Trapezoidal Type IT2 FS, and IT2 FIS

This section provides some preliminaries of trapezoidal type IT2 FS and IT2 FIS.

2.1 *Trapezoidal Type IT2 FS*

A trapezoidal type IT2 FS, denoted by $\tilde{A}$, is characterized as

$$\tilde{A} = \{((x, u), 1) : \forall x \in X, \forall u \in J_x \subseteq [0, 1]\} \tag{1}$$

where FOU of $\tilde{A}$, denoted by $\mathrm{FOU}(\tilde{A})$, is defined as

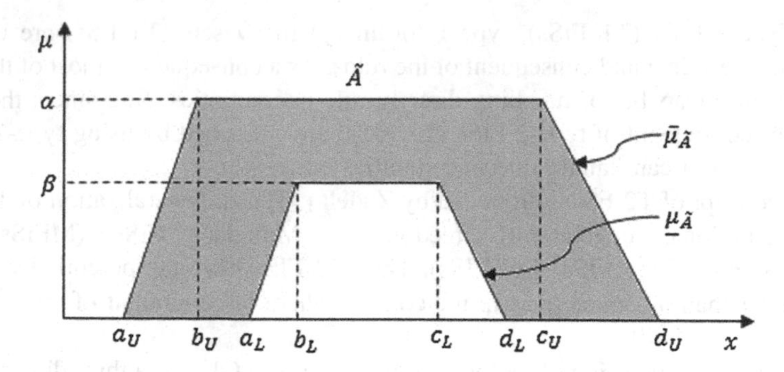

Fig. 1 Pictorial description of trapezoidal type IT2 FS

$$\mathrm{FOU}(\tilde{A}) = \{(x, u) : u \in J_x \subseteq [0, 1]\} \tag{2}$$

The upper membership function (UMF), denoted by $\bar{\mu}_{\tilde{A}}$, and lower membership function (LMF), denoted by $\underline{\mu}_{\tilde{A}}$, associated with $\mathrm{FOU}(\tilde{A})$ are defined as follows:

$$\bar{\mu}_{\tilde{A}}(x; a_U, b_U, c_U, d_U, \alpha) = \begin{cases} 0, x < a_U \\ \alpha(x - a_U)/(b_U - a_U), \quad a_U \le x < b_U \\ \alpha, b_U \le x < c_U \\ \alpha(d_U - x)/(d_U - c_U), \quad c_U \le x < d_U \\ 0, x > d_U \end{cases} \tag{3}$$

$$\underline{\mu}_{\tilde{A}}(x; a_L, b_L, c_L, d_L, \beta) = \begin{cases} 0, x < a_L \\ \beta(x - a_L)/(b_L - a_L), \quad a_L \le x < b_L \\ \beta, b_L \le x < c_L \\ \beta(d_L - x)/(d_L - c_L), \quad c_L \le x < d_L \\ 0, x > d_L \end{cases} \tag{4}$$

A trapezoidal type IT2 FS $\tilde{A}$ has been shown in Fig. 1.

2.2 IT2 FIS

IT2 FS is now becoming most acceptable T2 FS due to its computational simplicity in comparison with general type T2 FS. FIS in which such IT2 FSs are used is referred to IT2 FIS. In such FIS, rules may be generated by the domain experts, extracted from data or assumptions taken by the designers and are free from the type of FSs that are used to model their antecedent or consequent. The rule outputs are represented by IT2 FSs. The IT2 FSs are then converted into crisp numbers by performing two successive operations, viz., type-reduction and defuzzification. Type-reduction process converts the output IT2 FS into a T1 FS, and

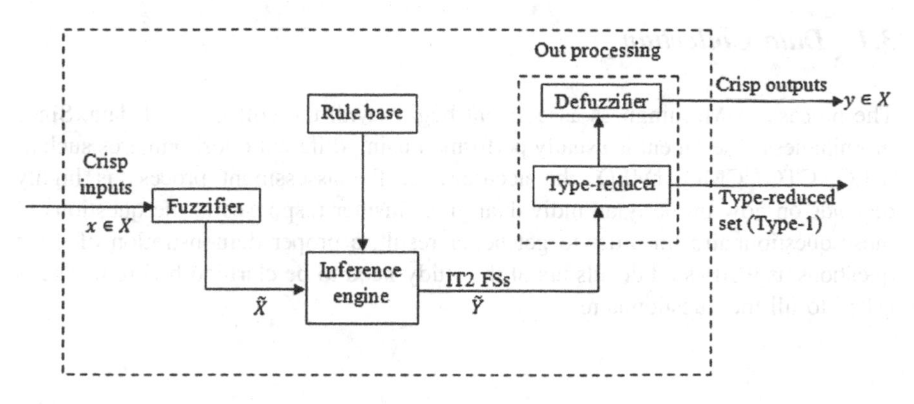

Fig. 2 Diagram of an IT2 FIS [41]

defuzzification transforms that T1 FS into a crisp number. Among several type-reduction processes, center-of-sets type-reduction method, developed by Karnik and Mendel [47] is popular one. When the type-reduction and defuzzification processes are applied to a general T2 FS the computational complexity increases enormously. In compare to T2 FS, computational complexity of IT2 FS reduces extensively, which is one of the major reasons for selecting IT2 FISs than T2 FISs [48]. An IT2 FIS has been depicted in Fig. 2.

In MFIS [49], the rule consequent is represented by a FS, whereas in TSKFIS rule consequent is expressed as a linear combination of linear or nonlinear functions of inputs. An IT2 FIS in which rule base consists of Mamdani type fuzzy if-then rules is known as IT2MFIS. In an IT2MFIS with m inputs, one output, and the rule base consists N rules, the nth rule, $\tilde{R}^n$, is given by

$$\tilde{R}^n : \text{IF } x_1 \text{ is } \tilde{A}_1^n \text{ and} \ldots \text{and } x_m \text{ is } \tilde{A}_m^n, \text{ THEN } y \text{ is } \tilde{B}^n . n = 1, 2, \ldots, N \quad (5)$$

where $\tilde{A}_i^n$ $(i = 1, 2, \ldots, m)$ and $\tilde{B}^n$ are IT2 FSs.

3 Proposed Methodology

In this section, an IT2 MFIS has been designed for assessing morningness of individuals on the basis of responses given to the MEQ as described in the following subsections.

3.1 Data Collection

The process of Morningness assessment begins with the collection of data. Since morningness assessment is usually performed using different questionnaires such as MEQ, CTQ, CMQ, rMEQ, the accuracy of the assessment process is highly depends on how properly an individual gives his/her responses to the questions of these questionnaire. In order to get better result, a proper demonstration of these questions as well as all details about the study need to be clarified before he/she is asked to fill the questionnaire.

3.2 Input Parameters Selection

Factor analysis, as a data reduction process, is often used to identify the most influential independent factors that explain the pattern of correlations with a set of observed variables. Moreover, a regression analysis may also be executed to estimate the unique effect of each factor. However, in this study, experts' knowledge has been used to identify the factors influencing the chronotype of individuals.

3.3 Determination of Fuzzifier

The qualitative descriptors of input and output parameters are characterized by trapezoidal type IT2 FSs that are derived by UMF and LMF as in Eqs. (3) and (4), respectively. A survey has been carried out to gather the experts' knowledge in order to determine the FOU of the IT2 FSs so that the model can capture more uncertainties by covering input/output domains with fewer FSs.

3.4 Development of Fuzzy Rule Base

A fuzzy rule base contains a collection of fuzzy if-then rules [50] which are used to capture the complex or nonlinear relationship between variables of different real-life problems. These fuzzy rules can be generated in various ways like data driven method, expert's judgment, and knowledge acquisition. In this study, experts' judgment and knowledge acquisition are used to derive the fuzzy rules which are generally represented using qualitative descriptors of input and output variables as described in Eq. (5). In an IT2 MFIS with m inputs and one output, if p qualitative descriptors are used to cover input/output domains, then the rule base consists of $N = p^m$ rules.

3.5 Morningness Assessment by IT2 MFIS

Morningness of an individual is generally assessed on the basis of his/her responses corresponding to the questions of different types of questionnaires. For an input vector $x' = (x_1, x_2, \ldots, x_m)$, the computational process in an IT2 MFIS involves the following steps:

Step 1 Determine the membership interval of x_i on each $\tilde{A}_i^n$, $\left[\underline{\mu}_{\tilde{A}_i^n}(x_i), \bar{\mu}_{\tilde{A}_i^n}(x_i)\right]$, $i = 1, 2, \ldots, m; n = 1, 2, \ldots, N$.

Step 2 Evaluate the firing interval of the nth rule (as in Eq. (1)), $F^n(x')$, as follows

$$\begin{aligned} F^n(x') &= \left[\underline{\mu}_{\tilde{A}_1^n}(x_1) \times \cdots \times \underline{\mu}_{\tilde{A}_m^n}(x_m), \bar{\mu}_{\tilde{A}_1^n}(x_1) \times \cdots \times \bar{\mu}_{\tilde{A}_m^n}(x_m)\right] \\ &= \left[\underline{f}^n, \bar{f}^n\right] \end{aligned} \tag{6}$$

In this approach, product t-norm has been used.

Step 3 The primary variable of IT2 FS, $\tilde{B}^n$, present in the consequent of the nth rule, are to be discretized as $y_t (t = 1, 2, \ldots, M)$ and determine the centroid of that IT2 FS, i.e., $C_{\tilde{B}^n} = \left[c_l^n, c_r^n\right]$ as described [47, 51] below:

$$c_l^n = \min_{\theta_t \in \left[\underline{\mu}_{\tilde{B}^n}(y_t), \bar{\mu}_{\tilde{B}^n}(y_t)\right]} \frac{\sum_{t=1}^{M} y_t \theta_t}{\sum_{t=1}^{M} \theta_t} = \frac{\sum_{t=1}^{k_l^n} \bar{\mu}_{\tilde{B}^n}(y_t) y_t + \sum_{t=k_l^n+1}^{M} \underline{\mu}_{\tilde{B}^n}(y_t) y_t}{\sum_{t=1}^{k_l^n} \bar{\mu}_{\tilde{B}^n}(y_t) + \sum_{t=k_l^n+1}^{M} \underline{\mu}_{\tilde{B}^n}(y_t)} \tag{7}$$

$$c_r^n = \max_{\theta_t \in \left[\underline{\mu}_{\tilde{B}^n}(y_t), \bar{\mu}_{\tilde{B}^n}(y_t)\right]} \frac{\sum_{t=1}^{M} y_t \theta_t}{\sum_{t=1}^{M} \theta_t} = \frac{\sum_{t=1}^{k_r^n} \underline{\mu}_{\tilde{B}^n}(y_t) y_t + \sum_{t=k_r^n+1}^{M} \bar{\mu}_{\tilde{B}^n}(y_t) y_t}{\sum_{t=1}^{k_r^n} \underline{\mu}_{\tilde{B}^n}(y_t) + \sum_{t=k_r^n+1}^{M} \bar{\mu}_{\tilde{B}^n}(y_t)} \tag{8}$$

where the switch points k_l^n and k_r^n are identified as

$$y_{k_l^n} \le c_l^n \le y_{k_l^n+1} \tag{9}$$

$$y_{k_r^n} \le c_r^n \le y_{k_r^n+1} \tag{10}$$

Step 4 Execute-type reduction to combine firing intervals of the rules $F^n(x')$ and the corresponding rule consequent centroids, $C_{\tilde{B}^n} = \left[c_l^n, c_r^n\right]$. Here center-of-sets type reducer [34, 52] has been used, which was derived from the extension principle [16]

$$y_l = \min_{k \in [1, N-1]} \frac{\sum_{n=1}^{k} \bar{f}^n c_l^n + \sum_{n=k+1}^{N} \bar{f}^n c_l^n}{\sum_{n=1}^{k} \bar{f}^n + \sum_{n=k+1}^{N} \underline{f}^n} = \frac{\sum_{n=1}^{L} \bar{f}^n c_l^n + \sum_{n=L+1}^{N} \bar{f}^n c_l^n}{\sum_{n=1}^{L} \bar{f}^n + \sum_{n=L+1}^{N} \underline{f}^n} \tag{11}$$

$$y_r = \max_{k \in [1, N-1]} \frac{\sum_{n=1}^{k} \underline{f}^n c_r^n + \sum_{n=k+1}^{N} \bar{f}^n c_r^n}{\sum_{n=1}^{k} \underline{f}^n + \sum_{n=k+1}^{N} \bar{f}^n} = \frac{\sum_{n=1}^{R} \underline{f}^n c_r^n + \sum_{n=R+1}^{N} \bar{f}^n c_r^n}{\sum_{n=1}^{R} \underline{f}^n + \sum_{n=R+1}^{N} \bar{f}^n} \tag{12}$$

where the switch points L and R are identified as

$$c_l^L \leq y_l \leq c_l^{L+1} \tag{13}$$

$$c_r^R \leq y_r \leq c_r^{R+1} \tag{14}$$

and $\{c_l^n\}_{n=1,2,\ldots,N}$ and $\{c_r^n\}_{n=1,2,\ldots,N}$ have been stored in ascending order, respectively.

Step 5 Calculate the defuzzified output as:

$$y = \frac{y_l + y_r}{2} \tag{15}$$

4 An Illustrative Case Example

An illustrative case example on morningness assessment of a group of students of University of Kalyani, Kalyani, West Bengal, India has been considered to demonstrate the proposed morningness assessment process. The students participated anonymously in voluntary and unpaid manners and provided their willingness prior to participation in the study. The University of Kalyani is situated at a semi urban area. Most of the students belong to the nearby three districts (Nadia, Murshidabad, and North 24 Parganas). The students either stay in hostels or day boarders. The study hours of University are in between 10.30 a.m. and 05.00 p.m. But in some departments the schedule time extends up to 07.00 p.m.

Among the different types of chronotypological questionnaires, the rMEQ is short in length, easy to understand. It is evidenced that rMEQ is the best morningness–eveningness predictor for Indian people [53] and hence it is used for this case study.

In the present case study, input parameters are selected on the basis of experts' knowledge. According to the expert, *Average Rising Time for Feeling Best* (X_1), *Feelings within First Half an Hour after waken up in the morning* (X_2), *Sleeping Time* (X_3), *Feeling Best Time in a Day* (X_4), and *Self-Assessment about Morningness* (X_5) are the best indicator of morning–evening orientation of an individual and hence are considered as input parameters to measure the output parameter *Morningness* (Y). It is worthy to mention here that in rMEQ the above defined variables are measured in terms of preference scales which fall within a specified range lying between 0 to 4 or 0 to 5 or 0 to 6. In defining IT2 FSs, the universe of discourse consists of those preference scale values. As a consequence, the universe of discourse of the IT2 FSs is considered as those above mentioned intervals.

On the basis of experts' knowledge, the qualitative descriptors of five input and one output parameters have been developed for the group of students of the said University. A survey has been carried out to gather the experts' knowledge in order to determine the FOU of the different IT2 FSs so that the model can capture more uncertainties by covering input/output domains with fewer FSs. Parameters of IT2 FSs corresponding to the qualitative descriptors of input and one output parameters are shown in the Table 1. Figure 3 shows the IT2 FSs corresponding to the qualitative descriptors of input and one output parameters.

Table 1 Parameters of IT2 FSs (trapezoidal) corresponding to qualitative descriptors of input and output fuzzy variables

Fuzzy variables		Preference scale	Qualitative descriptors	Parameters of IT2 FSs (trapezoidal)	
				$[a_U, b_U, c_U, d_U; \alpha]$	$[a_L, b_L, c_L, d_L; \beta]$
Input parameters	Average rising time for feeling best (X_1)	(0–5)	Early	$[3, 4, 5, 5; 1]$	$[3.2, 4.2, 5, 5; 0.8]$
			Moderate	$[1, 2, 3, 4; 1]$	$[1.2, 2.2, 2.8, 3.8; 0.8]$
			Late	$[0, 0, 1, 2; 1]$	$[0, 0, 0.8, 1.8; 0.8]$
	Feelings within first half an hour after waken up (X_2)	(0–4)	Good	$[2.5, 3, 4, 4; 1]$	$[2.6, 3.1, 4, 4; 0.9]$
			Fair	$[1, 1.5, 2.5, 3; 1]$	$[1.1, 1.6, 2.4, 2.9; 0.9]$
			Bad	$[0, 0, 1, 1.5; 1]$	$[0, 0, 0.9, 1.4; 0.9]$
	Sleeping time (X_3)	(0–5)	Early	$[3, 4, 5, 5; 1]$	$[3.3, 4.3, 5, 5; 0.7]$
			Moderate	$[1, 2, 3, 4; 1]$	$[1.3, 2.3, 2.7, 3.7; 0.7]$
			Late	$[0, 0, 1, 2; 1]$	$[0, 0, 0.7, 1.7; 0.7]$
	Feeling best time in a day (X_4)	(0–5)	Early	$[3, 4, 5, 5; 1]$	$[3.2, 4.2, 5, 5; 0.8]$
			Moderate	$[1, 2, 3, 4; 1]$	$[1.2, 2.2, 2.8, 3.8; 0.8]$
			Late	$[0, 0, 1, 2; 1]$	$[0, 0, 0.8, 1.8; 0.8]$
	Self-assessment about morningness (X_5)	(0–6)	Morning type	$[4, 5, 6, 6; 1]$	$[4.4, 5.4, 6, 6; 0.6]$
			Neither morning nor evening	$[1, 2, 4, 6; 1]$	$[1.4, 2.4, 3.6, 4.6; 0.6]$
			Evening type	$[0, 0, 1, 2; 1]$	$[0, 0, 0.6, 1.6; 0.6]$
Output parameter	Morningness (Y)	(0–6)	Morning	$[4, 5, 6, 6; 1]$	$[4.4, 5.4, 6, 6; 0.6]$
			Intermediate	$[1, 2, 4, 6; 1]$	$[1.4, 2.4, 3.6, 4.6; 0.6]$
			Evening	$[0, 0, 1, 2; 1]$	$[0, 0, 0.6, 1.6; 0.6]$

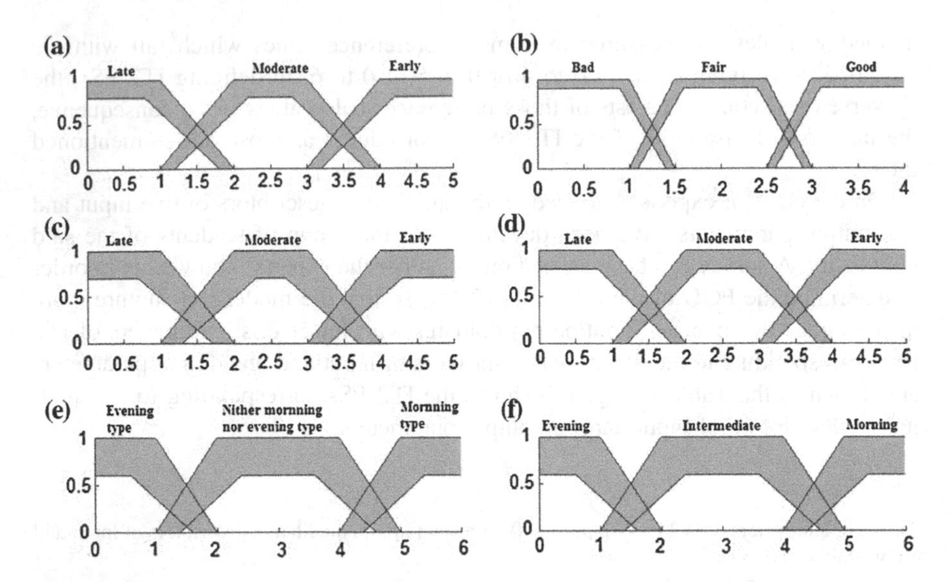

Fig. 3 Trapezoidal types IT2 FSs corresponding to linguistic terms associated with input and output variables **a** average rising time for feeling best (X_1), **b** feelings within first half an hour after waken up in the morning (X_2), **c** sleeping time (X_3), **d** feeling best time in a day (X_4), **e** self-assessment about morningness (X_5), and **f** morningness (Y)

4.1 Results and Discussions

The evaluation of Morningness scores of 10 individuals has been done as described in Sect. 3.5 and then individuals are categorized as "Morning type (M)," Intermediate type (I)'," and "Evening type (E)," along with their degree of belongingness. It is worthy to mention here that the rule base of the model has been developed on the basis of experts' knowledge. So, the performance of the model does not depend on the size of the sample. Hence, no difficulties can arise for considering larger sample size. The whole simulation process has been carried out using the *software* MATLAB (Ver. 2008b). The input values, corresponding morningness scores (crisp value) and their membership grades of associated chronotypes are shown in Table 2.

For the validation of the proposed IT2 MFIS, mean square error (MSE), as well as root mean square error (RMSE) between the predicted output $y(k)$ obtained by the proposed model and the desired output y_k recommended by the experts has been calculated, respectively, by Eqs. (16) and (17) as follows:

$$\text{MSE} = \frac{1}{n}\sum_{k=1}^{n}(y(k) - y_k)^2, \text{ where } n \text{ is the number of recorded data} \quad (16)$$

Table 2 Response in preference scale, defuzzified morningness scores and classification with membership grades

Individual	Input (responses in preference scale)					Morningness score (crisp)	Classification with membership grade					
							M		I		E	
	X_1	X_2	X_3	X_4	X_5		Lower	Upper	Lower	Upper	Lower	Upper
1	4	2	3	3	2	2.6266	0	0	0.6	1	0	0
2	3	2	3	2	0	2.6266	0	0	0.6	1	0	0
3	4	3	4	3	4	5.1190	0.4314	1	0	0	0	0
4	3	2	3	2	2	2.6266	0	0	0.6	1	0	0
5	5	4	4	5	6	5.1190	0.4314	1	0	0	0	0
6	3	1	3	2	0	0.5063	0	0	0	0	0.6	1
7	4	3	3	4	4	5.1190	0.4314	1	0	0	0	0
8	4	2	3	3	2	2.6266	0	0	0.6	1	0	0
9	4	3	4	3	4	5.1190	0.4314	1	0	0	0	0
10	5	3	3	3	4	2.6266	0	0	0.6	1	0	0

Note M: morning-type; I: intermediate type; E: evening type

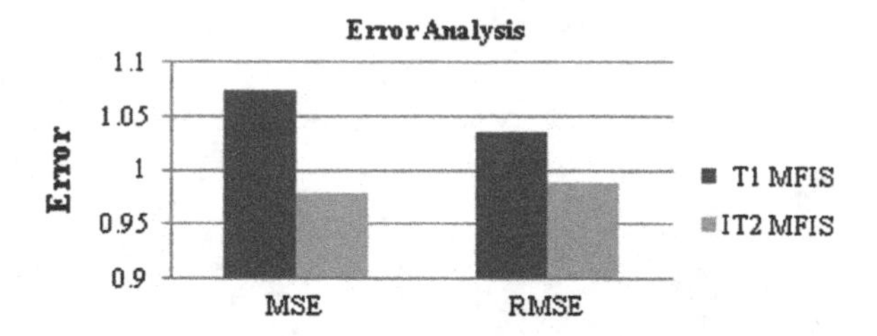

Fig. 4 Error analysis of T1 MFIS and IT2 MFIS

$$\text{RMSE} = \frac{1}{n}\sqrt{\sum_{k=1}^{n}\left(y(k) - y_k\right)^2},\ \text{where } n \text{ is the number of recorded data} \tag{17}$$

Both the error measures, MSE as well as RMSE, associated with the proposed IT2 MFIS are the compared with those of T1 MFIS. Figure 4 shows that the MSE as well as RMSE bars corresponding to IT2 MFIS are smaller than that of T1 MFIS and hence indicate that the proposed IT2 MFIS outperform its counterpart, T1 MFIS.

5 Conclusions

In this article, an IT2 MFIS has been proposed for morningness assessment of individuals. Based on domain experts, five input variables are selected on the basis of the items of rMEQ to evaluate morningness of individual. It is evident from the achieved results of the proposed model that IT2 MFIS can deal the inherent uncertainties present within the assessment process of morningness efficiently.

In the context of real-life complex decision making problems, like morningness assessment, where human knowledge and perception play important role to capture the vagueness and uncertainties associated with the model, especially, in describing different linguistic terms raised during the design phase, IT2 FSs provide extra flexibilities for managing those uncertainties due to the consideration of third dimension present in its membership functions. In such situation, IT2 FSs appeared as best alternative for the experts since it becomes easier for them to assign membership functions corresponding to the linguistic terms of input–output variables of FISs.

Moreover, in IT2 FIS, each bound of the type-reduced intervals is determined by using UMF and LMF of the same IT2 FS, simultaneously. As a consequence, an IT2 FIS can capture associated uncertainties more efficiently than its T1 counterparts; and is now being used widely for assessment purpose in different complex imprecisely defined circumstances.

Acknowledgments The authors remain grateful to Dr. S. Sahu, Associate Professor, Department of Physiology, University of Kalyani, Kalyani, India for his kind support and expert suggestions in developing the model. The authors are thankful to the reviewers for their valued comments and suggestions to improve the quality of the article.

References

1. Saksvik I, Bjorvatn B, Hetland H, Sandal GM, Pallesen S. (2010). Individual differences in tolerance to shift work – a systematic review. *Sleep Medicine Reviews.* 15:221–235.
2. Selvi FF, Karakaş SA, Boysan M, Selvi Y. (2015). Effects of shift work on attention deficit, hyperactivity, and impulsivity, and their relationship with chronotype, *Biological Rhythm Research.* 46(1):53–61.
3. Facer-Childs E, Brandstaetter R. (2015). The impact of circadian phenotype and time since awakening on diurnal performance in athletes. Current Biology. 25(4):518–22.
4. Thun F, Bjorvatn B, Flo E, Harris A, Pallesen S. (2015). Sleep, circadian rhythms, and athletic performance. *Sleep Medicine Reviews.* 23:1–9.
5. Taylor DJ, Clay KC, Bramoweth AD, Sethi K, Roane BM. (2011). Circadian phase preference in college students: relationship with psychological functioning and academics. *Chronobiology International.* 28:541–547.
6. Bohle P. (1989). The impact of night work on psychological wellbeing. *Ergonomics.* 32:1089–1099.
7. Mecacci, L, & Rocchetti, G. (1998). Morning and evening types: Stress related personality aspects. *Personality and Individual Difference.* 25(3), 537–542.
8. Emens JS, Yuhas K, Rough J, Kochar N, Peters D, Lewy AJ. (2009). Phase angle of entrainment in morning and evening types under naturalistic conditions. *Chronobiology International.* 26:474–493.
9. Adan A, Lachica J, Caci H, Natale V. (2010). Circadian typology and temperament and character personality dimensions. *Chronobiology International.* 27:181–193.
10. Selvi Y, Aydin A, Boysan M, Atli A, Agargun MY, Besiroglu L. (2010). Associations between chronotype, sleep quality, suicidality, and depressive symptoms in patients with major depression and healthy controls. *Chronobiology International.* 27:1813–1828.
11. Horne JA, Ostberg O. (1976). A self assessment questionnaire to determine morningness, eveningness. *International Journal of Chronobiology.* 4:97–110.
12. Folkard S, Monk TH, Lobban MC. (1979). Toward a predictive test of adjustment to shift work. *Ergonomics.* 22:79–91.
13. Smith CS, Reilly C, Midkiff K. (1989). Evaluation of three circadian rhythm questionnaires with suggestions for an improved measure of morningness. *Journal of Applied Psychology.* 74:728–738.
14. Adan A, Almirall H. (1991). Horne & Ostberg Morningness-Eveningness Questionnaire: a reduced scale. *Personality and Individual Differences.* 12:241–53.
15. Caci H, Deschaux O, Adan A, Natale V. (2009). Comparing three morningness scales: age and gender effects, structure and cutoff criteria. *Sleep Medicine.* 10:240–245.
16. Zadeh LA. (1965). Fuzzy Sets. *Information and Control.* 8(3):338–353.
17. Zadeh LA. (1975a). Fuzzy logic and approximate reasoning. *Synthese.* 30:407–428.
18. Zadeh LA. (1976). A fuzzy-algorithmic approach to the definition of complex or imprecise concepts. *International Journal of Man-Machine Studies.* 8:249–291.
19. Zadeh LA. (2008). Is there a need for fuzzy logic? *Information Sciences.* 178:2751–2779.
20. Bezdek JC, Pal SK. (1992). Fuzzy Models for Pattern Recognition. IEEE Press, USA.
21. Übeyli ED, Güler İ. (2005). Automatic detection of erythemato-squamous diseases using adaptive neuro-fuzzy inference systems. *Computers in Biology and Medicine.* 35(5):421–33.
22. Palma J, Juarez JM, Campos M, Marin R. (2006). Fuzzy theory approach for temporal model-based diagnosis: An application to medical domains. *Artificial Intelligence in Medicine.* 38(2): 197–218.
23. Dubois D, Prade H. (1980). *Fuzzy Sets and Systems.* Academic Press, London.
24. Zimmermann HJ. (1985). *Fuzzy Set Theory and Its Applications.* Kluwer Academic Publishers, Boston.

25. Biswas, A., Majumder, D., & Sahu, S. (2011). Assessing Morningness of a Group of People by Using Fuzzy Expert System and Neuro Fuzzy Inference Model. *Communications in Computer and Information Science*. 140:47–56.
26. Biswas A, Adan A, Haldar P, Majumder D, Natale V, Randler C, Tonetti L, Sahu S. (2014). Exploration of transcultural properties of the reduced version of the Morningness-Eveningness Questionnaire (rMEQ) using adaptive neuro fuzzy inference system. *Biological Rhythm Research*. 45(6):955–968.
27. Biswas A, Majumder D. (2014). Genetic algorithm based hybrid fuzzy system for assessing morningness. *Advances in Fuzzy Systems*, 2014, 1–9.
28. Moharrer, M., Tahayori, H., Sadeghian, A. (2013). Modeling complex concepts with type-2 fuzzy sets: the case of user satisfaction of online services. Sadeghian, A., Mendel, J.M., Tahayori, H. (Eds.), Advances in Type-2 Fuzzy Sets and Systems: Theory and Applications, *Studies in Fuzziness and Soft Computing*. 301:133–146.
29. Mendel, JM. (2007). Computing with words and its relationships with fuzzistics. *Information Sciences*. 177:988–1006.
30. Pedrycz, W. (2010). Human certainty in computing with fuzzy sets: an interpretability quest for higher order granular constructs. *Journal of Ambient Intelligence and Humanized Computing*. 1:65–74.
31. Mendel JM. (1999). Computing with words when words can mean different things to different people. In: Int'l. ICSC Congress on Computational Intelligence: Methods & Applications, Third Annual Symposium on Fuzzy Logic and Applications, Rochester, NY.
32. Liang Q and Mendel JM. (2000). Interval type-2 fuzzy logic systems: theory and design. *IEEE Transactions on Fuzzy Systems*. 8(5):535–550.
33. Zadeh LA. (1975b). The concept of a linguistic variable and its application to approximate reasoning-I. *Information Sciences*. 8:199–249.
34. Mendel JM. (2001). Uncertain Rule-Based Fuzzy Logic Systems: Introduction and New Directions, Prentice-Hall, Upper Saddle River, NJ.
35. Liang Q, Karnik NN, Mendel JM. (2000). Connection admission control in ATM networks using survey-based type-2 fuzzy logic systems. *IEEE Transactions on Systems, Man, Cybernetics, C, (Applications and Review)*. 30(3):329–339.
36. Melgarejo MC, Reyes AP, Garcia A. (2004). Computational model and architectural proposal for a hardware type-2 fuzzy system. In: Proceedings of the IEEE FUZZ Conference, Budapest, Hungary.
37. Melin P, Castillo O. (2004). A new method for adaptive control of nonlinear plants using type-2 fuzzy logic and neural networks. *International Journal of General Systems*. 33:289–304.
38. Wu D, Tan W. (2004). A type-2 fuzzy logic controller for the liquid-level process. In: Proceedings of the IEEE International Conference on Fuzzy Systems, Budapest, Hungary.
39. Wagner C, Hagras H. (2007). A genetic algorithm based architecture for evolving type-2 fuzzy logic controllers for real world autonomous mobile robots, In: Proceedings of the IEEE FUZZ Conference, pp. 193–198.
40. Zeng J, Liu ZQ. (2015). Type-2 Fuzzy Graphical Models for Pattern Recognition, Studies in Computational Intelligence. Springer-Verlag Berlin Heidelberg.
41. Mendel JM, John RI, Liu F. (2006). Interval Type-2 Fuzzy Logic Systems Made Simple. *IEEE Transactions on Fuzzy Systems*. 14(6):808–821.
42. Baklouti N, Alimi A. (2007). Motion planning in dynamic and unknown environment using an interval type-2 TSK fuzzy logic controller. In: Proceedings of the IEEE FUZZ Conference. 1848–1853.
43. Lin T, Liu H, Kuo M. (2009). Direct adaptive interval type-2 fuzzy control of multivariable nonlinear systems. *Engineering Applications of Artificial Intelligence*. 22:420–430.
44. Biglarbegian M, Melek W, Mendel J. (2011). On the robustness of type-1 and interval type-2 fuzzy logic systems in modeling. *Information Sciences*. 181:1325–1347.
45. Linda O, Manic M. (2011). Interval Type-2 fuzzy voter design for fault tolerant systems. *Information Sciences*. 181:2933–2950.

46. Lee LW, Chen SM. (2008). A new method for fuzzy multiple attributes group decision-making based on the arithmetic operations of interval type-2 fuzzy sets. In: Proceedings of the 2008 international conference on machine learning and cybernetics, China: Kunming. 3084–3089.
47. Karnik NN, Mendel JM. (2001). Centroid of a type-2 fuzzy set. *Information Sciences*. 132(1–4):195–220.
48. Mendel JM, Wu H. (2007). New results about the centroid of an interval type-2 fuzzy set, including the centroid of a fuzzy granule. *Information Sciences*. 177: 360–377.
49. Mamdani EH. (1974). Applications of fuzzy algorithms for simple dynamic plant. In: Proceedings of the IEEE. 121:1585–1588.
50. Huang C, Moraga C. (2005). Extracting fuzzy if–then rules by using the information matrix technique. *Journal of Computer and System Sciences*. 70(1):26–52.
51. Mendel JM, Liu F. (2007). Super-exponential convergence of the Karnik-Mendel algorithms for computing the centroid of an interval type-2 fuzzy set. *IEEE Transactions on Fuzzy Systems*. 15(2):309–320.
52. Mendel JM, Wu D. (2010). Perceptual Computing: Aiding People in Making Subjective Judgments. Wiley-IEEE Press, Hoboken.
53. Sahu S. (2009). An Ergonomic Study on Suitability of Choronotypology Questionnaires on Bengalee (Indian) Population. *Indian Journal of Biological Sciences*. 15:1–11.

A Novel 2D Face, Ear Recognition System Using Max–Min Comparison Technique for Human Identification

A. Parivazhagan and A. BrinthaTherese

Abstract Biometrics plays a major role in image identification field in the security environment, mainly in image recognition and detection. The pose variation in image leads to mismatch with its original image and it is one of the major frequently occurring problems. This proposed work explains new efficient techniques to recognize people with the help of their biometric features like Face and Ear. The main objective of this work is to recognize a person's face, ear image correctly, with respect to his/her pose varied images. Novel techniques named as, Max–Min Comparison, Red matrix Eigen vector recognition, are proposed for recognition processes. These techniques are tested on standard database images and its results are analyzed using mismatching and runtime parameters, and its accuracy yields high recognition rate of 99%.

Keywords Max–Min comparison · Red matrix Eigen vector recognition · Biometrics · Pose variation · Face recognition

1 Introduction

Biometric recognition, detection applications are present in our daily life; biometrics is the basic, reasonable identification authority to identify a person. Perfect systems are needed to process the biometric data and to overcome the difficulties present in it. In the biometric recognition process pose variation plays a major role, different poses of a same image produces wrong image matching results; and it produce wrong data. So it is needed to overcome this issue, in this work techniques are proposed to resolve the pose variation problem present in the recognition system.

A. Parivazhagan (✉) · A. BrinthaTherese
School of Electronics Engineering, VIT University, Chennai Campus, Chennai, India
e-mail: parivazhagan.a2013@vit.ac.in

A. BrinthaTherese
e-mail: abrinthatherese@vit.ac.in

S.S. Dash et al. (eds.), *Artificial Intelligence and Evolutionary Computations in Engineering Systems*, Advances in Intelligent Systems and Computing 517,
DOI 10.1007/978-981-10-3174-8_58

In this work, Face [1, 2] and Ear [3, 4] biometrics are used as biological parameters to identify a person. Face recognition [5, 6] is the foremost effective field in human recognition, but still there is a lag in accuracy perspective to overcome it and this research work is concentrated on accuracy very much. The Ear recognition [4] is one of the growing fields in human authentication, several research works are blooming for this technology, and here two standard databases of ears are used for recognition.

This paper proposed two recognition techniques and the rest of the paper is organized as, 2. Max–Min Comparison 3. Red matrix Eigen vector recognition. 4. Results and Discussions 5. Conclusions and References.

2 Max–Min Comparison

In this recognition technique, the maximum and minimum intensity values are extracted from the database images and the test image for a comparison process. (Fig. 1).

Step 1: The 2D images present in the database are transformed into 1D vectors, and a 2D matrix (Fig. 3) is formed with these 1D vectors. In this 2D matrix (database's Matrix) 1st column represents the first image, 2nd column represents the second image (and so on) of the database.
Step 2: Transform the 2D test image in to 1D vector. (Fig. 2). This1D vector is compared with all database images, i.e., Comparison takes place between the test image's 1D vector and database's Matrix, it is Row by Row comparison. Here the maximum intensity values are extracted, while comparing the rows between the test image's 1D vector and the database's Matrix. The extracted maximum intensity values are collected in a matrix called Maximum Matrix (MaxM) (Fig. 4).
Step 3: In this Maximum matrix the 1st column denotes the maximum intensity value extracted between the test image's 1D vector and the first image of the database's images (1st column of the database matrix), and so on up to last column.
Step 4: Similarly the minimum intensity values are collected between the database images and the test image. The steps are same as like as maximum intensity value detection. The above process is proceeded with an important change, (i.e.) here it is

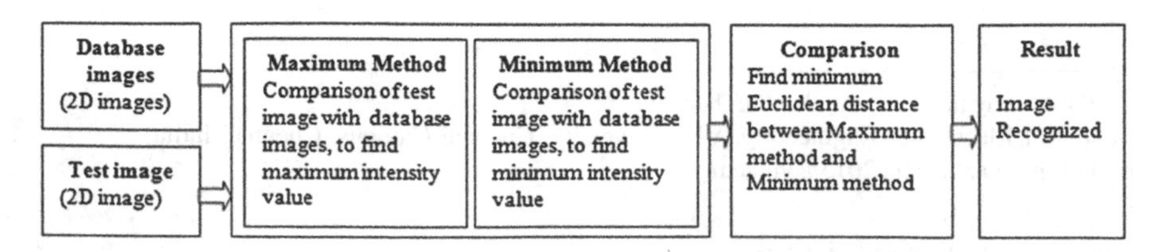

Fig. 1 Max–Min comparison block diagram

Fig. 2 Test image 1D vector

10
12
13
14
15

Fig. 3 Database images 1D vector (database's matrix) (model—not actual images matrix)

11	12	13	11	10
14	16	17	11	11
17	12	14	10	13
16	19	19	17	14
14	18	17	15	16

Fig. 4 Maximum matrix (MaxM)

11	12	13	11	10
14	16	17	12	12
17	13	14	13	13
16	19	19	17	14
15	18	17	15	16

Fig. 5 Minimum matrix (MinM)

10	10	10	10	10
12	12	12	11	11
13	12	13	10	13
14	14	14	14	14
14	15	15	15	15

to extract the minimum intensity values, instead of maximum intensity values. The extracted matrix from this step is called as Minimum matrix (MinM) (Fig. 5).
Step 5: At the end, this technique is concluded with the Euclidean distance phenomenon, (Fig. 6) to find the shortest Euclidean distance between the Maximum Matrix (MaxM) and Minimum Matrix (MinM). This shortest distance represents the

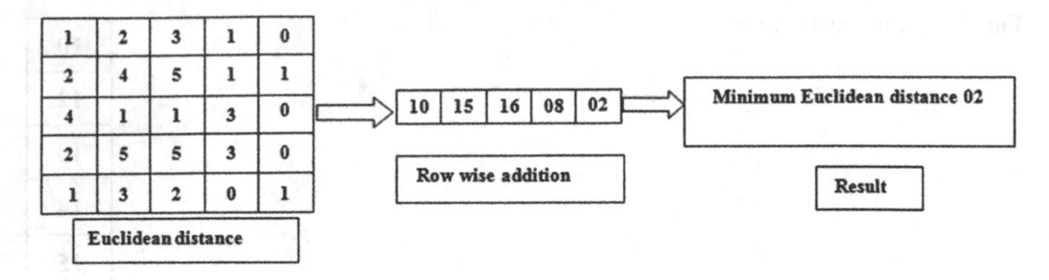

Fig. 6 Calculation of minimum euclidean distance value, between the maximum and minimum matrices

minimum variation between the test image and database images and it represents the correct database image that is matched with the given test image (Fig. 7). Euclidean distance for Max Min Comparison = $\sqrt{\left[(\mathrm{MaxM} - \mathrm{MinM})^2\right]}$.

3 Red Matrix Eigen Vector Recognition

The Principal component analysis [7], Eigen values and Eigen vectors [8] are the widely used algorithms for image recognition processes. In this proposed work it is combined with Maximum comparison technique for biometric object recognition. The overall process is shown in Fig. 8. And it is explained below.

Step 1: All the 2D database images Red component's are extracted and are converted into 1D vectors the converted 1D vectors of all the database images are collected in a matrix, called Red matrix. Why Red alone?: According to Electromagnetic spectrum, red is more visible to human eye, having higher wave length than green and blue.
Step 2: Find the mean for all database image's Red Component, and is called as Red component mean (Ψ).
Database images—$I_1, I_2, I_3, \ldots I_M$; Red Components—$R_1, R_2, R_3, \ldots R_M$.

$$\Psi = (1/M)\sum_{i=1}^{M} R_i$$

Step 3: Compare the Red matrix (R_M) with Red component mean (Ψ). While comparing the rows between the Red matrix and the Red component mean, the maximum intensity values, are detected.
The detected maximum intensity values are collected in a matrix. While comparing the first row between the Red matrix and the Red component mean, the maximum intensity values obtained are collected in a row, and so on, up to the last row. All the detected maximum intensity value rows are collected in a matrix called Maximum red matrix (M_{Red}) (Fig. 9).

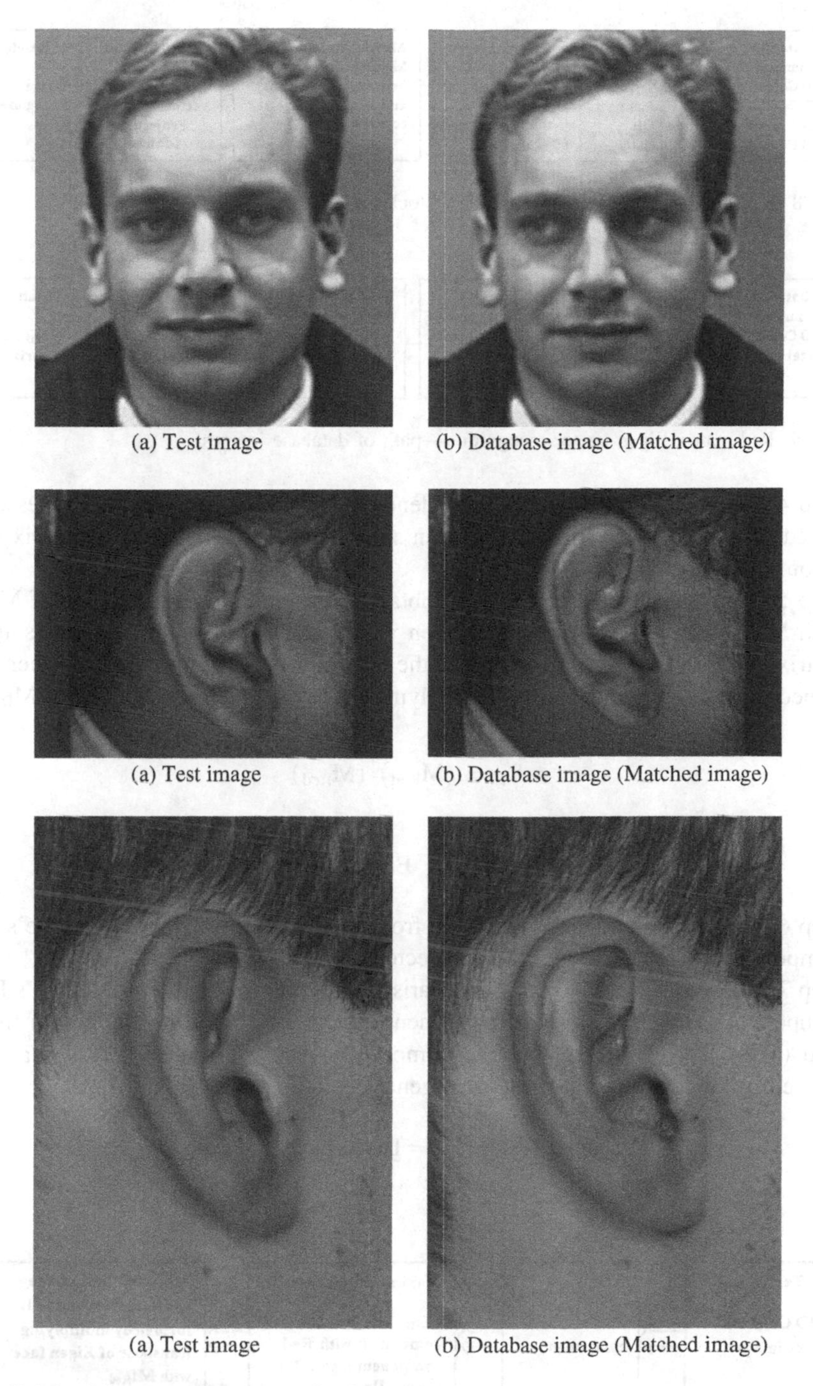

(a) Test image (b) Database image (Matched image)

(a) Test image (b) Database image (Matched image)

(a) Test image (b) Database image (Matched image)

Fig. 7 Max–Min comparison output for face and ear images

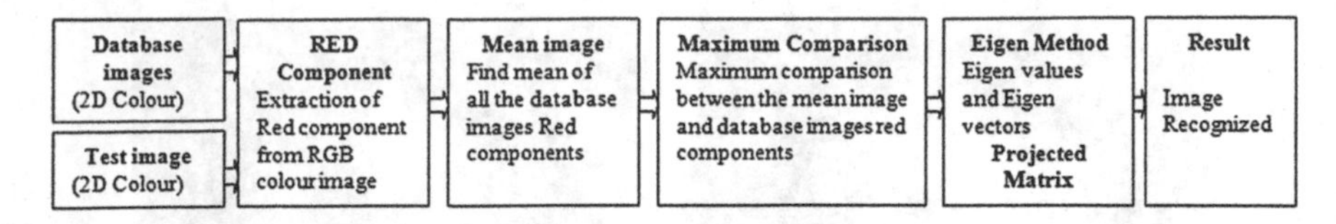

Fig. 8 Red matrix Eigen vector recognition block diagram

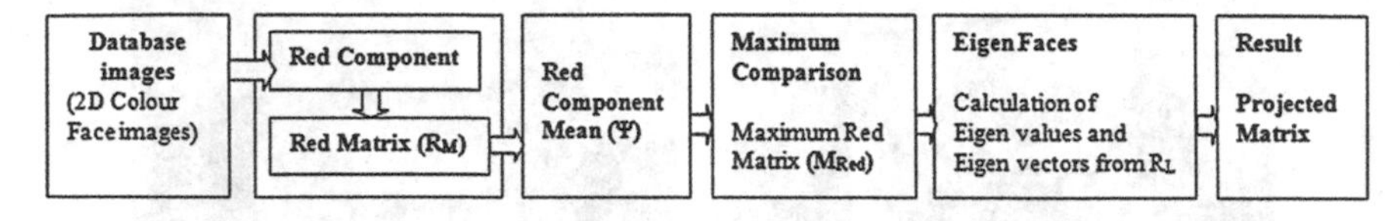

Fig. 9 Red matrix Eigen vector recognition—path of database images

Step 4: In this M_{Red} the first column denotes the maximum intensity values collected between the Red component mean and the first image of the Red matrix and so on up to last column.
Step 5: Square matrix R_L, (Fig. 9) is obtained by multiplying transpose of M_{Red} with M_{Red}, it is needed to find the Eigen values and Eigen vectors for this square matrix, the Eigen faces are formed with the help of these Eigen vectors, and then the projected matrix is produced by multiplying transpose of Eigen faces with M_{Red}.

$$R_L = (M_{Red})^T(M_{Red})$$

$$\text{Projected Matrix} = \text{Eigen faces}^T \times M_{Red}$$

Step 6: The red component is extracted from a 2D test image; this test image's red component is also converted into 1D vector.
Step 7: Maximum intensity value comparison is done between the test image's Red component 1D vector and Red component mean (Ψ), to get the Maximum Input Red (MI_{Red}). With the help of the comparison result, a projected test image is formed by multiplying transpose of Eigen faces with MI_{Red} (Fig. 10).

$$\text{Projected Test Image} = \text{Eigen faces}^T \times MI_{Red}$$

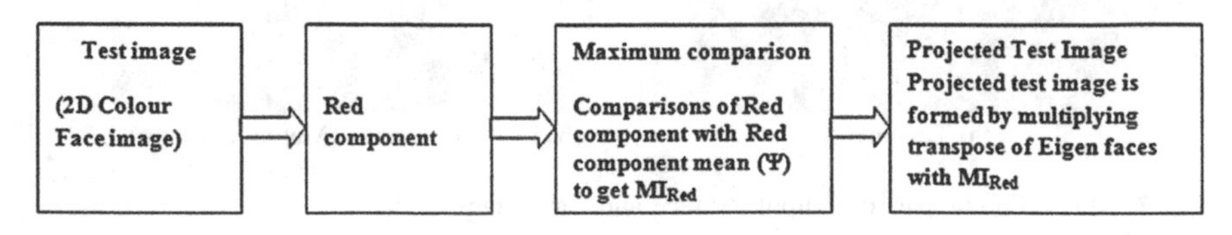

Fig. 10 Red matrix Eigen vector recognition—path of a test image

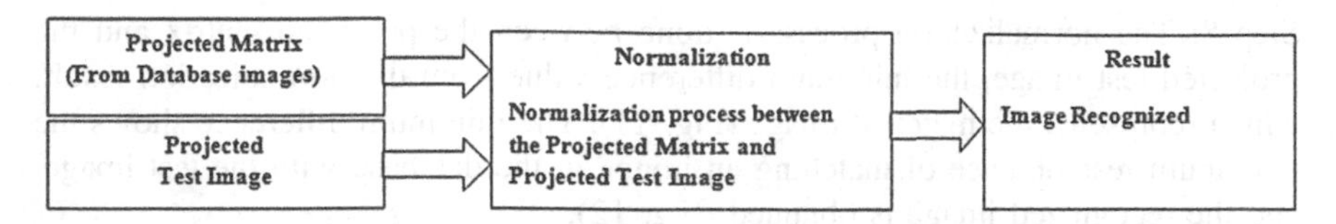

Fig. 11 Red matrix Eigen vector recognition—recognition part

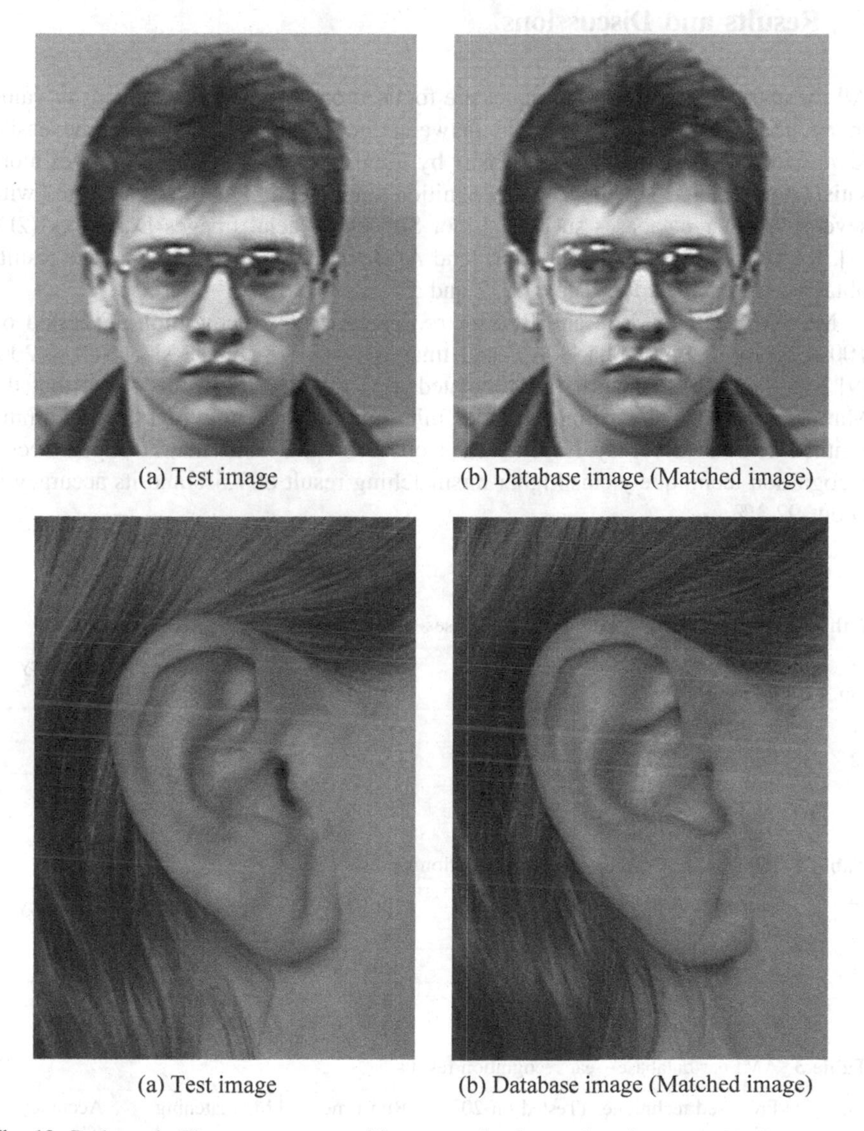

Fig. 12 Red matrix Eigen vector recognition output for face and ear images

Step 8: The normalization process is done between the projected matrix and the projected test image, the minimum difference value from the normalization result, which represents the matched image (Fig. 11). The minimum difference shows the maximum resemblance of matching an image in the database with the test image, and the recognized image is obtained (Fig. 12).

4 Results and Discussions

All the above recognition techniques are focused on each and every intensity value in an image, so small changes in between the group of images can be easily identified. Focusing on images intensity by intensity values comparison gives more satisfactory results. The proposed recognition and detection methods are tested with several face/ear databases such as, Libor Spacek's Facial Images Databases (2D) [9], IIT Delhi Ear database (2D) [10] and AMI Ear database (2D) [11]. The results obtained are tabulated as Tables 1, 2 and 3.

Max–Min Comparison and Red matrix Eigen vector recognition are tested on 100 subjects of Libor Spacek's Facial Images Database (Image size: 180 × 200) [9] and the results obtained are tabulated as Table 1. The results show that the Max–Min Comparison technique yields minimum mismatching of 1% in minimum runtime and its accuracy is 99%. It is observed that Red matrix Eigen vector recognition technique yields higher mismatching result of 7.5% and its accuracy is about 92.3%.

Table 1 Libor Spacek's facial images databases—face recognition results

S. no.	Proposed techniques (tested on 100 subjects)	Run time (s)	Mismatching (%)	Accuracy (%)
1	Max–Min comparison	2.1–2.3	1.0	99.0
2	Red matrix Eigen vector recognition	1.9–2.2	7.5	92.3

Table 2 IIT Delhi ear database—ear recognition results

S. no.	Proposed techniques (Tested on 100 subjects)	Run time (s)	Mismatching (%)	Accuracy (%)
1	Max–Min comparison	1.6–1.9	9.5	90.0

Table 3 AMI ear database—ear recognition results

S. no.	Proposed techniques (Tested on 20 subjects)	Run time (s)	Mismatching (%)	Accuracy (%)
1	Max–Min comparison	2.1–2.4	3.5	85.3
2	Red matrix Eigen vector recognition	2.2–2.4	5.8	74.2

Max–Min Comparison technique is tested on 100 subjects of IIT Delhi ear database (Image size: 272 × 204) [10] and the results are tabulated as Table 2, the Max–Min Comparison yields very minimum runtime and its mismatching level is 9.5%. It produces 90% of accuracy rate. It is a gray scale database so other proposed recognition technique is not tested.

The proposed two recognition techniques are tested on 20 subjects of AMI Ear database (Image size: 492 × 702) [11] and the results are tabulated as Table 3. The Max–Min Comparison technique yields minimum mismatching of 3.5% in minimum runtime and its accuracy rate is 85.3%. Red matrix Eigen vector recognition yields accuracy of about 74.2%.

5 Conclusions

This work is focused on a single issue called pose variation in the biometrics; to overcome this error two recognition methods are introduced. The proposed two recognition methods are tested with standard databases and it produces productive results. The Max–Min comparison process gives as much as maximum accuracy results of 99%. The Red Matrix Eigen vector also produces accuracy of about 92.3%. Thus, the proposed recognition system provides a path way with new recognition algorithms for the human face and ear recognition system; and these novel recognition techniques produces results in shortest run time with lesser mismatching.

References

1. Hadis Mohseni Takallou, Shohreh Kasaei: Multiview face recognition based on multilinear decomposition and pose manifold, IET Image Processing, 2014, Vol. 8, Iss. 5, pp. 300–309.
2. Dong Li, Huiling Zhou, Kin-Man Lam: High-Resolution Face Verification Using Pore-Scale Facial Features, IEEE Transactions on Image Processing, Vol. 24, No. 8, August 2015, pp. 2317–2327.
3. Ping Yan and Kevin W. Bowyer: Ear Biometrics Using 2D and 3D Images, Proceedings of the 2005 IEEE Computer Society Conference on Computer Vision and Pattern Recognition (CVPR'05), 1063-6919/05, © 2005 IEEE.
4. Mark Burge, Wilhelm Burger: Ear Biometrics in Computer Vision, 0-7695-0750-6/00 © 2000 IEEE.
5. Rajkiran Gottumukkal, Vijayan K. Asari: An improved face recognition technique based on modular PCA approach, Pattern Recognition Letters 25 (2004) 429–436.
6. Parivazhagan. A, BrithaTherese. A: A Novel, Location averaging, Linear equation and Exponential function techniques for face recognition in Human identification system, International Journal of Applied Engineering Research (IJAER), Special Issue, Volume 10, Number 87, 2015, pp. 21–26.
7. Martinez, A.M., Kak, A.C, PCA versus LDA. IEEE Trans. Patter Anal. Machine Intell. 23 (2), 2001, pp. 228–233.

8. Matthew Turk, Alex Pentland: Eigen faces for Recognition, Journal of Cognitive Neuroscience, vol. 3, no. 1, 1991.
9. Libor Spacek's Facial Images Databases: http://cmp.felk.cvut.cz/~spacelib/faces/.
10. IIT Delhi Ear Image Database version 1.0: http://www4.comp.polyu.edu.hk/~csajaykr/IITD/Database_Ear.htm.
11. AMI Ear database: http://www.ctim.es/research_works/ami_ear_database.

Secure Cloud Storage Based on Partitioning and Cryptography

Rakhi S. Krishnan, Ajeesha Radhakrishnan and S. Sumesh

Abstract Cloud-based storage is becoming a very good method for sharing any type of data in upcoming days. This method can reduce the burden of storing and retrieving data in local machines. But there exist many issues. The vendor itself can attack the user's data and it can even delete or hide the data; if the user did not access the data for a long period of time. It also faces problems for synchronizing and uploading files and the vast range of internal and external threats for data integrity. Another problem is based on data security in cloud storage, where it does not have a proper encryption technique. Storing data in the cloud is financially good for long-term large scale data storage, although it does not fully offer guarantee on data integrity and availability. In this paper we are focusing the integrity of the data using a technique that divides the data into number of pieces and encrypting it by newly introduced method. Through this technique cloud ensures more security and integrity.

Keywords Cloud · Cloud computing · Multiple location · Service provider · Data storage · Data security · Data correctness · Data availability · Data integrity

1 Introduction

Cloud computing is a model for enabling both user-friendly and on-demand network access to a shared pool of configurable computing resources (for e.g., networks, servers, storage, application, and services) that can be rapidly provisioned and released with minimal management struggle or service provider cooperation.

R.S. Krishnan (✉) · A. Radhakrishnan · S. Sumesh
Department of CS & IT, Amrita School of Arts and Sciences Kochi, Amrita Vishwa Vidyapeetham, Kochi, India
e-mail: rakhiskrishnan111@gmail.com

A. Radhakrishnan
e-mail: ajeesha694@gmail.com

S. Sumesh
e-mail: ssumesh@hotmail.com

S.S. Dash et al. (eds.), *Artificial Intelligence and Evolutionary Computations in Engineering Systems*, Advances in Intelligent Systems and Computing 517,
DOI 10.1007/978-981-10-3174-8_59

In this paper we are focusing on cloud storage. Cloud storage is a specific sub offering within Infrastructure as a Service (IaaS). Private data of users is gathered on multiple third-party providers with cloud storage technology rather than on the dedicated providers used in conventional networked data storage. The basic requirements for cloud storage system include mass storage and low expense [1]. However, consumers are ready to store their important and sensitive data in cloud only if it provides security and integrity.

Cloud storage system is facing many challenges nowadays. The major issue is concerned with the security of data which we store at servers and also it has threats from vendor who has cloud space. Another problem is in maintaining backup for the data. Apart from that, cloud also faces problems like Missing data (if the client did not access the data for a long period of time, the CSP automatically removes the data from its storage space), problem with synchronizing files, unknown errors (that is the attacks caused by the server itself is hidden from the users), and upload issues (when a user upload a file to the cloud space and he/she does not aware of where it is actually stored). These data may be stored in multiple locations like in India or in America. So, the users have to obey the legal rules of that particular country.

In this paper, we are introducing a new method to make the cloud more secure. Here we are dividing the entire data into small pieces and encrypting it using a new protocol. Cryptographic techniques play an important role in this technique. Then we are compressing each piece separately and storing it into different locations. At the time of fetching we are decompressing the pieces and decrypting it into the original text. Then it is arranged it into the order as in the initial state. After all these process it is given to the consumer. Through this technique, we can provide security and integrity to the cloud data. Even the vendor cannot access the consumer's data without user's permission.

2 Literature Review

Mainly cloud storage is subdivided into two types [2]. That is class A and class B. Class A is the cloud storage technique that is designed using cryptographic techniques but it is not in the cryptographic theory framework. Class B is the cloud storage technique designed using cryptographic techniques and also in the cryptographic theory framework. Here we survey existing application result of cloud storage mechanism and their main features. The following are some of the mechanisms used for data storage in cloud.

In this paper the method is based on not only the attribute-based encryption, but also cipher text—policy and which help to perceive the control of data for the requester's access and identity-based encryption that deliver a secure data communication. This allows requester's to access their own way and set different attributes to them. This scheme is called efficient and secure patient centric access control (ESPAC Scheme), which achieve desired security and good communication delay. This method with cryptographic techniques ensures highest security and privacy requirements [2].

Cryptography-based cloud storage is the first used symmetric encryption mechanism. It is designed to secure cloud storage architecture for both consumers and enterprise scenarios when the service provider of public cloud infrastructure is not completely trusted by consumers. It tells the basis of cryptographic cloud storage and it explains how to implement it using cryptographic techniques to satisfy the security and privacy requirements. The steps that take place are first indexes data and encrypt it with a symmetric encryption mechanism. Then it encrypts the index and the unique key under an appropriate policy. Finally, it encodes the encrypted data and index. So that later can verify their integrity [3].

The scheme ensures secure data repository service design on the top of a public cloud infrastructure. It uses digital signature assigned by the audit and file manager for the integrity and authentication of the data which is distributed as a file encryption and file signature key. Searchable encryption that is in the kryptonite client library without revealing the content of the file or without decrypting file is allowed with the help of the secure index manager [4].

The paper discusses about the problem of simultaneous public audit of data and data dynamics on cloud storage for the integrity of data [4]. They found a solution which provides the data dynamics efficiency. This is done by manipulating the classic Merkle Hash Tree construction and thereby improving the storage models. For achieving simultaneous multiple auditing task, their main goal, they have adopted bilinear aggregate signature [5].

Another paper tells the service providers who hide the data corruption occurred by the provider side itself [6]. This paper discusses the qualities of a protocol. They are public auditability, storage correctness, privacy preserving, batch-auditing, and lightweight. There are essentially two operation models. One is storage as a service and other is computing as a service. In the first model the file is created and opened at the side of cloud user. All the operations are taking place locally. Only the cloud is used for the storage purpose. The main functions are uploading and downloading. The paper in cloud storage security provides an algorithm for file uploading and downloading [7].

This strategy provides a new vision about flexible file distribution method. They focus on token pre computation to prove the precision of data [8].

3 Proposed System

In the proposed system, we are trying to solve the attacks from vendor side. Here we are proposing a new system to ensure the security in cloud storage from vendor side attacks. The system works in between the user and cloud storage provider. That is just like a middleware. When the user upload a file to the cloud, the original file is partitioned into several pieces based on certain algorithm. Then, these pieces are randomized using a key. This key is generated by a function and user part is the base of key generation. After that, each piece is encrypted by a key (which is generated by a predefined function) and then it is compressed. Finally it is stored into the cloud. There was maintained an array which have an identity of each piece and its location address. At the time of file retrieving the file name is given as the input and service provider maps this name to the servers and takes back the corresponding pieces from the storage space using the array (address stored array). Then it is decompressed and decrypted using corresponding key. After that each pieces is ordering in the correct format using the corresponding key. Finally it is combined to form the original file.

Here the encryption method is choosen dynamically. If the customer allows sharing then it is public key encryption. Otherwise it will be private key encryption. Also we are planning to allow the users to choose which encryption method they need as per there security level. A unique method is introduced for performing randomization.

The equation is,

$$SN = (\min * \text{seed} + \max)/\text{random}, \quad (1)$$

where SN is the new sequence number, min is lower sequence number of that particular file, max is higher sequence number of that particular file, seed and random are the dynamic variables which make the order unique. In seed variable, we are taking time and in random variable we are taking date.

The diagrammatical representation of proposed system is shown in Fig. 1.

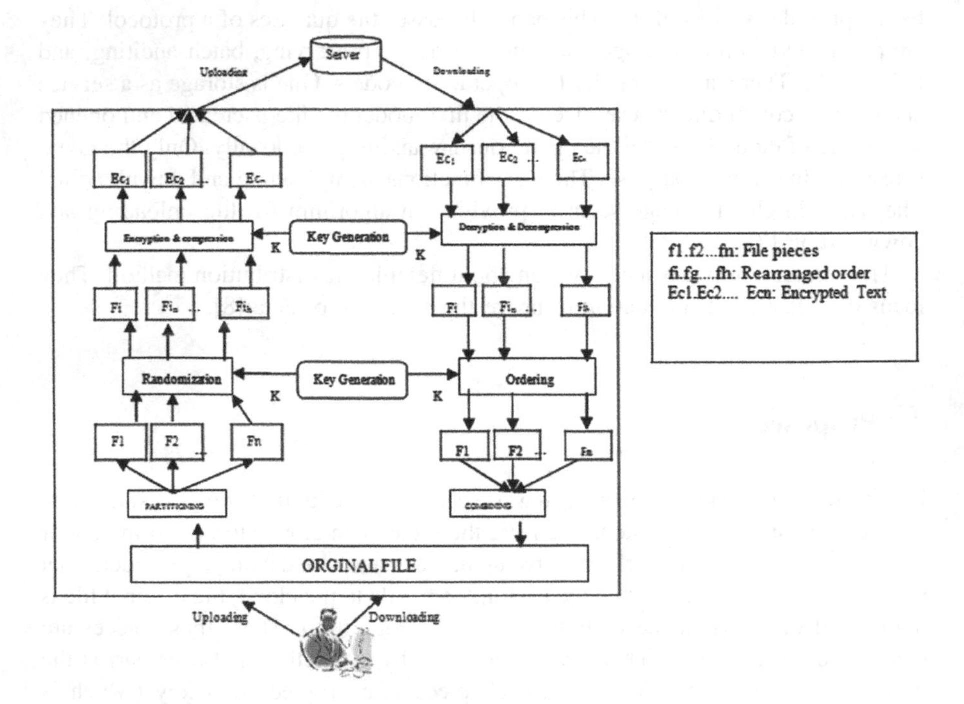

Fig. 1 Proposed system

4 Implementation

The proposed system is implemented in '.net'. There are two major modules in this system, i.e., uploading a file and downloading from the storage server. Once the user has registered the system using his/her email ID and password then using this mail ID, the system can generate a unique ID for each user.

4.1 Uploading a File into Storage Server

The next step is login using the unique ID and password. After that the user needs to specify the requirement. If the user is given with an upload option, then a browse page is displayed where we can browse the needed file. Thereafter, when we click on the OK button the file splits into different sub files and stored into a dynamically created folder. These are not visible to the user. Then the system asks for an algorithm to encrypt. The user can choose it according to their security level. When the user chooses one algorithm and clicks on the OK button then each file is encrypted using the specified algorithm and zipped together. The corresponding encryption method is used to encrypt the data which is stored. Finally, a confirmation message is given to the user.

4.2 Downloading a File from Storage Server

When the user has given download option, all the files uploaded by that user is displayed in a list box. The user can choose which one he/she needed. When he clicks on the download button the address fetches from the database and the specified folder is taken from that location. Then unzip the files and decrypt the file b using appropriate decryption method. This method is taken from the database where we store that at the time of file uploading. Then all the decrypted files are combined together into the original file. Then it is stored in the local space. Confirmation message is given to the user (Fig. 2).

Here we are analyzing how the time factor affects the efficiency of new cloud storage method with the old one. Table 1 shows the file uploading and downloading time in cloud using various encryption techniques.

A graph has been plotted from the analysis derived from the table. Though, the time constrain for both the system remains considerably the same, it is concluded that the security associated with the proposed system is more reliable than the existing one. Figure 3 shows a graph illustrating the file uploading and downloading time for different algorithms

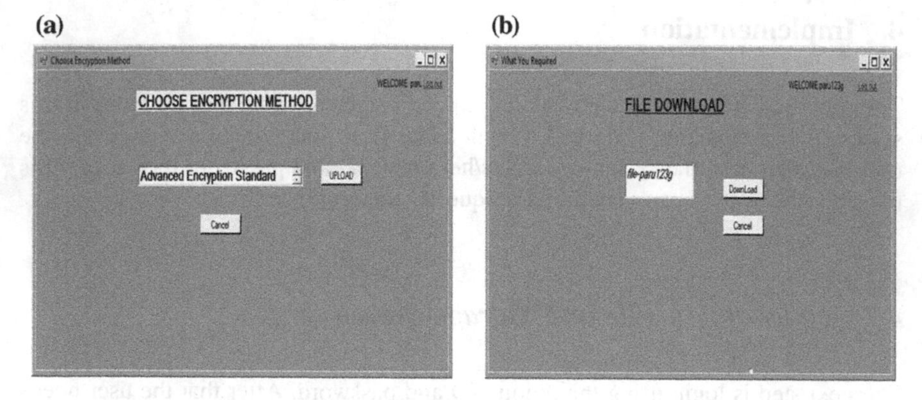

Fig. 2 **a** Shows uploading a file into the Storage Server, and **b** downloading a file from the Storage Server

Table 1 Time efficiency comparison of different encryption algorithms

Algorithm	Uploading time (s)	Downloading time (s)
Ceaser Cipher	2 s 96 ms	01 s 03 ms
AES	2 s 73 ms	71 ms
DES	2 s 51 ms	59 ms
RSA	2 s 65 ms	63 ms

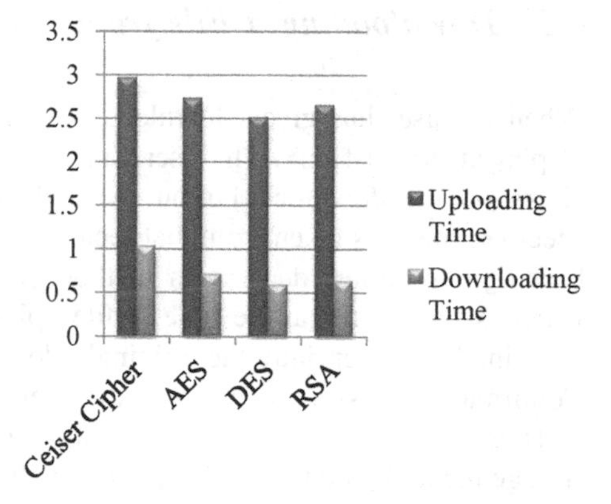

Fig. 3 A graph showing efficiency of the algorithm against the time factor

5 Future Work

Currently the system uses only few encryption techniques. The results obtained are promising and effective. In future, there exists more scope for improving the system by adding other advanced encryption methods. Also the system can improve data sharing feature based on owner permission.

6 Conclusion

The paper discusses about the integrity, efficiency, and security of the data in cloud storage. This storage system encounters several limitations such as vendor side attack, data hiding, file synchronization problem, etc. Being the most critical issue, vendor side attack has been concentrated here. In order to overcome this problem a new method is introduced which provides more reliability, efficiency, and security to the data.

References

1. Suruchee V. Nandgaonkar, Prof. A. B. Raut.: A Comprehensive Study on Cloud Computing. (2014).
2. Peng Yong, Zhao Wei, XIE Feng, Dai Zhong-hua, Gao Yang, Chen Dong-qing.: Secure cloud storage based on Cryptographic Techniques. (2012).
3. Kamara S, Lauter k.: Cryptographic cloud Storage. 14th international conference on financial cryptography and data security.
4. Kumbharc A, Simmhan Y, Prasanna V.: Designing a secure storage repository for sharing scientific databases using public clouds. 2nd international workshop on data intensive computing in the clouds.
5. Maulik Dave.: Data Storage Security in Cloud Computing. A survey- international Journal of Advanced Research in Computer Science and Software Engineering, Vol. 3, Issue 10. (2013).
6. Q. Wang, C. Wang, J. Li, K. Ren, W. Lou.: Enabling public verifiability and data dynamic for storage security in cloud computing. Springer-verlag (2009) pp. 355–370.
7. Ambika vishal pawar, Dr. Ajay r. Dani.: Design of privacy model for storing files on Cloud storage. Journal of Theoretical and Applied Information Technology. Vol. 67 No. 2. (2004).
8. Kalpana Batr1, Ch. Sunitha, Sushil Kumar.: An Effective Data Storage Security Scheme for Cloud Computing. International Journal of Innovative Research in Computer and Communication Engineering. Vol. 1. Issue 4 (2013).

Analysis of Power Quality in Hostel and Academic Buildings of Educational Institute

Mallakuntla Rajesh, K. Palanisamy, S. Umashankar, S. Meikandasivam and S. Paramasivam

Abstract In last decades, electronics and telecommunications have known unexpected development; the number of nonlinear loads has increased, along with this the disturbances in power quality also raised. Poor quality in power causes more losses and will cause malfunction, leading to replacement of particular connected load. This paper presents the preliminary results of Power quality survey done in academic and hostel buildings of VIT University. The power quality survey is necessary to record all events over an enough long time interval. Here we collected various parameter readings from all electrical feeders of for different buildings. This paper analyzes the complexities as well as consequences of power quality disruptions. In order to reduce the huge amount of data by recording and analyzing several electrical parameters over certain duration of time, some recording limits as per IEC standards are to be set. For particular feeders where limits exceed the set values, various strategies to enhance the quality of power and minimize the losses are implemented.

Keywords IEC standards · Network layout · Parameters · Power quality survey · PQ box · Voltage harmonics

1 Introduction

Nowadays, the quality of electric power has achieved more relevance with respect to operation and maintenance in various applications. In order to enhance the quality of power, steps like monitoring, recording, fetching the causes for losses, and solutions for the disruptions must be followed. The outcome of the survey might be associated with tremendous assistance in resolving the existing issues

M. Rajesh (✉) · K. Palanisamy · S. Umashankar · S. Meikandasivam
School of Electrical Engineering, VIT University, Vellore, India
e-mail: umashankar.s@vit.ac.in

S. Paramasivam
Danfoss Industries Pvt Ltd, Chennai, India

S.S. Dash et al. (eds.), *Artificial Intelligence and Evolutionary Computations in Engineering Systems*, Advances in Intelligent Systems and Computing 517,
DOI 10.1007/978-981-10-3174-8_60

Table 1 Loads in SJT power house with measured dates

Load type	S.no.	Corresponding feeder	Connected date
Academic loads	1	SJT lighting load (solar)	07/11/15 to 09/11/15
	2	SJT power & UPS	25/01/16 to 26/01/16
	3	SJT pump	21/01/16 to 22/01/16
Hostel loads	4	E-annex block lighting	20/11/15 to 21/11/15
	5	F-lighting	20/01/16 to 21/01/16
	6	F-power and AC feeder 1	22/01/16 to 23/01/16
	7	F-power and AC feeder 2	23/01/16 to 24/01/16
	8	E&F pump	23/11/15 to 24/11/15
Common load	9	Sewage treatment plant (STP)	26/11/15 to 27/11/15

associated with electric power, and possibly provide a foundation to take suitable step to avoid any kind of breakdowns resulting thereof [1]. The information extracted from the survey will allow simultaneous improvement of power quality of power through required methods and most significantly confirmation to particular standards of power quality enforced even in the presence of load [2].

The main objectives of the survey are (i) to analyze the quality of power at each individual load feeder, (ii) to make use of the outcomes of the survey regarding the development of the particular specifications of mitigating devices to be mounted in that specific feeder to boost up the quality of electric power. This survey is done in Silver Jubilee Tower (SJT) power house which shown in Fig. 2, using Win PQ Mobil −200 analyzer connected to all load feeders with duration of 24–48 h. With this analyzer we recorded the various power quality readings like current and voltage harmonics, sag/swell, flicker, etc., with interval of 1 s. The disruptions in power quality like voltage harmonics and inter-harmonics also recorded simultaneously as when they appeared. The data was collected from nine feeders which shown in Table 1.

2 Details and Layout Map of Load Distribution System

2.1 *University Map*

In this survey, we recorded various power quality issues like voltage harmonics, inter-harmonics, frequency fluctuations, flicker with the PQ analyzer at each feeder of particular load panel. For every event, the analyzer takes the readings of all parameters and magnitude along with duration of that event. These are shown in coming tables with corresponding load feeder readings. In all readings, one thing is to be noticed that all loads are affected with voltage harmonics. The power network of university is shown in Fig. 1.

The distribution of SJT power is connected to five loads as shown in Fig. 2, one academic and two hostel buildings, which again sub-divided as lighting feeder and

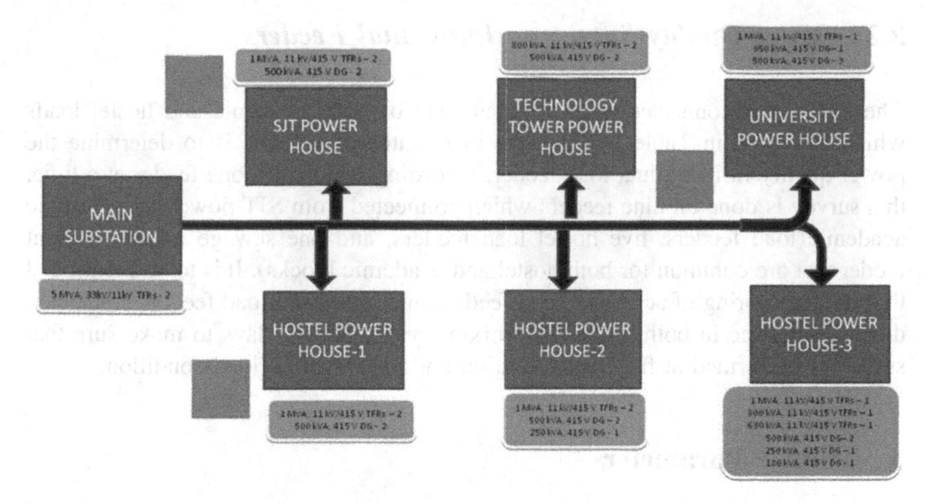

Fig. 1 Distribution system of University

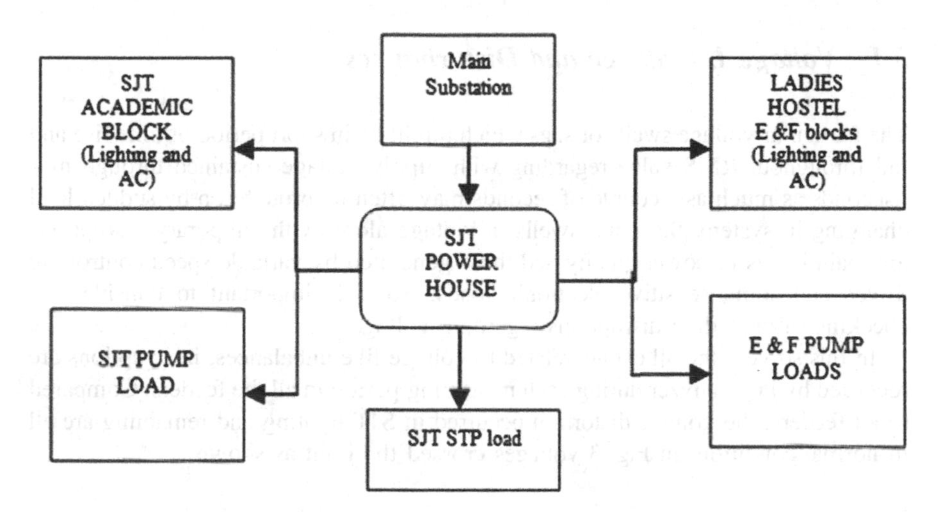

Fig. 2 Distribution system of SJT power house to various loads

power (AC and UPS) feeders, one pump load for academic and one for hostel buildings and a common sewage treatment plant load for both academic and hostel buildings.

2.2 *Power Quality Survey on Individual Feeders*

The PQ box is connected to each panel load of both academic and hostel loads which is shown in Table 1. Since the main intention of this is to determine the power quality in individual load feeder, recording is done on one feeder at a time, this survey is done on nine feeders which connected from SJT power house. (three academic load feeders, five hostel load feeders, and one sewage treatment plant feeder that are common for both hostel and academic blocks). It is to be mentioned that the monitoring of academic load feeders and the hostel load feeders connected during and done in both university working and weekend days to make sure that survey is performed at full load condition and also with no load condition.

3 Voltage Parameters

Here, the effects of voltage parameters on power quality are briefly explained.

3.1 *Voltage Unbalance and Disturbances*

The particular voltage swells or sags which might be in short period of time rise and fall throughout RMS value regarding with supply voltage sustained through milliseconds as much as a couple of seconds may often be brought on by sudden load changing in system. Sags and swells in voltage along with temporary disruptions are main issues in power quality and these generated by variable speed controlling drives and some sensitive electronic loads. So it is important to consider the checking of preceding disruptions regarding voltage.

In this survey, the all events related to voltage like unbalances, interruptions are recorded by PQ analyzer during each measuring period in all the feeders. Compared to all feeders, the voltage distortion occurred in SJT lighting and remaining are all in normal condition. In Fig. 3 voltages crossed the limit as shown.

3.2 *Voltage Harmonics*

The voltage harmonics in all feeders are recorded and some particular harmonics which crossing the limit are shown in Table 2. The main cause of appearance of voltage harmonics are nonlinear loads [3], in educational institute; the typical loads are

I. Projectors for teaching purpose
II. UPS and electrical ballasts for conventional purpose
III. Laboratories
IV. Air conditioning units

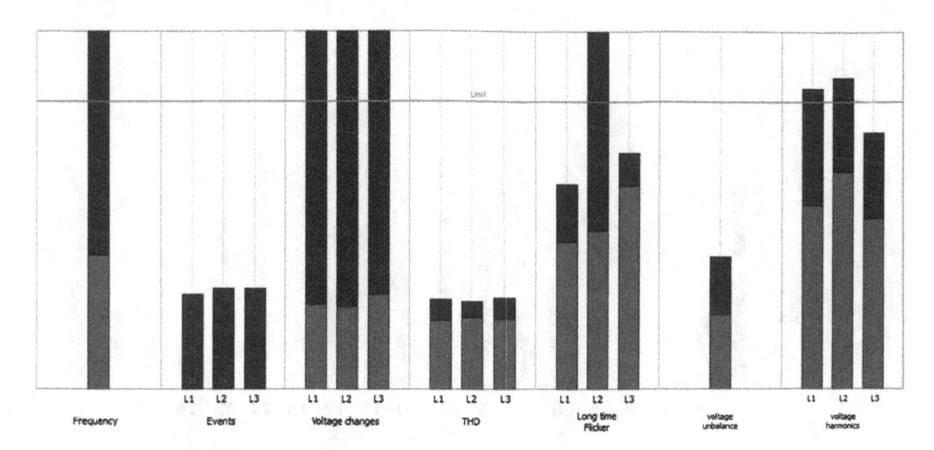

Fig. 3 Overview of SJT lighting feeder

Table 2 Voltage harmonics of all feeders

S. no.	Load type	Order of harmonics	Limit	Measured value	Particular line
1	SJT lighting	9	1.5	1.52	L1
2		11	0.5	0.52	L1
3		15	0.5	0.54	L2
4	SJT pump	6	0.5	0.53	L2
5		6	0.5	0.54	L3
6	SJT UPS	6	0.5	0.6	L2
7		9	1.5	2	L1
8		9	1.5	2.34	L2
9		9	1.5	1.78	L3
10	EF pump	9	1.5	1.54	L1
11		9	1.5	1.58	L2
12	F-lighting	6	0.5	0.72	L3
13		6	0.5	0.52	L1
14		9	1.5	1.74	L2
15	F-power feeder 1	9	1.5	1.52	L2
16		10	0.5	0.52	L1
17		15	0.5	0.54	L2
18	F-power feeder 2	6	0.5	0.56	L3
19		9	1.5	1.55	L3
20	No harmonics appeared in other feeders				

V. Power electronic converters, switch mode conversion
VI. Variable speed drives, microprocessor controlled drives.
VII. Lighting and lift loads.

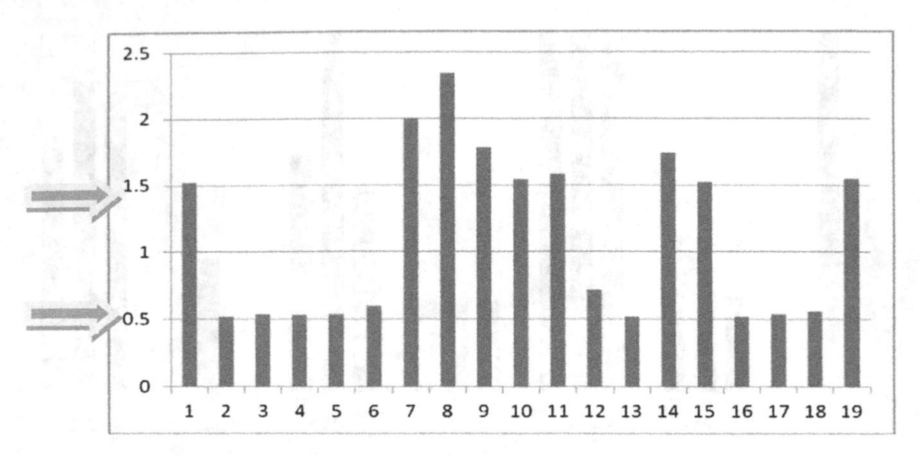

Fig. 4 Voltage harmonics of all feeders

These harmonics causes more effect on equipments like overheating of transformers, asynchronous machines, etc. Fig. 4 shows the THD of all load feeders which is also shown in Table 2.

The points marked in above chart are limits which given as per standards. The more voltage events observed in SJT lighting load fed with solar and those readings are plotted in power acceptability curve which is given by Information Technology Industrial Curve (ITIC) [4] is shown in Fig. 5; this also useful for information technology equipments and personal computer labs. From that plotted curve, the most voltage events like residual voltage, swell voltage is shown in Table 3.

4 Diesel Generator Effect

It is observed that due to power shutdown at measuring of SJT lighting load on November 7 and 9 from 2:00 to 4:00 PM and 10:02 AM to 11:43 AM, respectively, the power supplied by the diesel generators, of rating 500 KVA, 415 V, after connecting the load to diesel generator, the THD of voltage is raised to its limit and the frequency also deviated from its limit. The readings of all parameters are shown below.

4.1 Overview

The overview is shown in Fig. 6; here the frequency crossed the limit since the diesel generator interference that causes lot of disturbances in system. The main

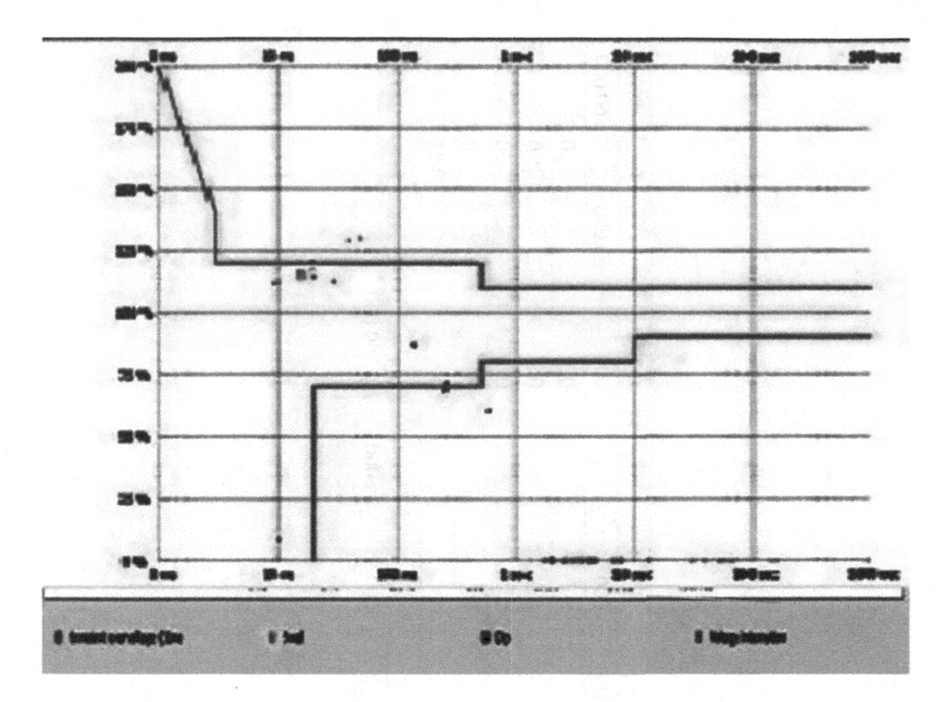

Fig. 5 Events in voltage compared with ITIC curve

effect in system are which causes the cross triggering of voltage and current, leads to power factor changes and negative power. The events that appeared while connecting, operating, and removing of the diesel generator are captured and some events are shown. The harmonic distortion in voltage is raised at the time of connecting and disconnecting of the diesel generator, this is shown in Fig. 7, where the arrow mark is the time that DG is operated, and in that the voltage harmonic distortion crossed the limits while connecting and removing from the load feeder.

4.2 *Frequency and Power Variations*

The frequency crossed the limit of standards due to sudden connection of main supply while diesel generator is operated due to past shutdown in supply which observed in overview. Figure 8 shows the frequency deviation of supply when the time sudden interference of main supply with diesel generator supply. The real power and power factor also came negative due to sudden operation of diesel generator, here the current crossed the 900 shift these shown in Fig. 11. In Figs. 9 and 10, the power factor and real power are shown where the value coming to zero

Table 3 Voltage events occurred in above curve

Residual voltage U %	Duration t (ms)					
	$10 \leq t \leq 200$	$200 \leq t \leq 500$	$500 \leq t \leq 1000$	$1000 \leq t \leq 5000$	$5000 \leq t \leq 60{,}000$	$t \geq 60{,}000$
$90 > U \geq 80$	1	0	0	0	0	0
$80 > U \geq 70$	0	3	0	0	0	0
$70 > U \geq 40$	0	2	1	0	0	0
$40 > U \geq 5$	0	0	0	0	0	0
$5 > U$	0	0	0	48	54	24
Swell voltage U %	Duration t (ms)					
	$10 \leq t \leq 500$	$500 \leq t \leq 5000$		$5000 \leq t \leq 6000$	$t \geq 60{,}000$	
$U \geq 120$	3	0		0	0	
$120 > U \geq 110$	15	0		0	0	

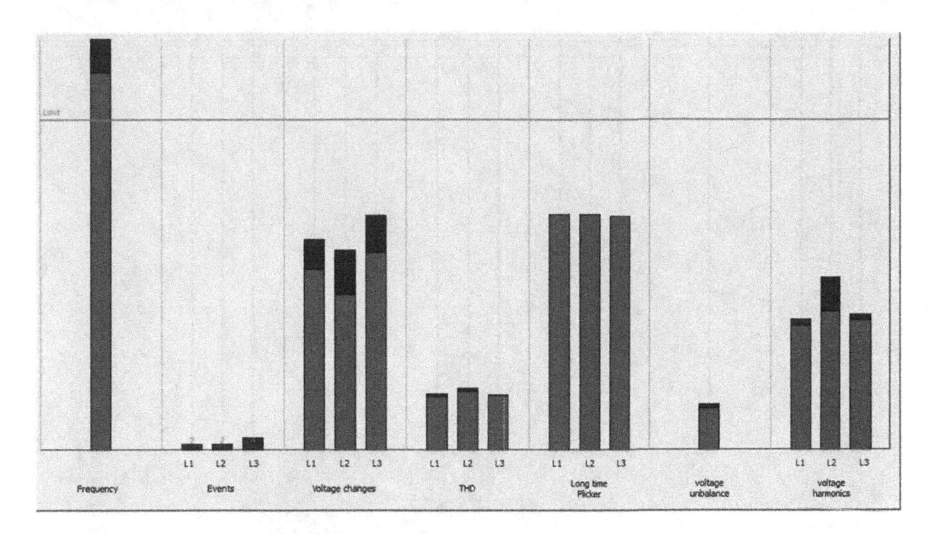

Fig. 6 Overview of interface of diesel generator in system

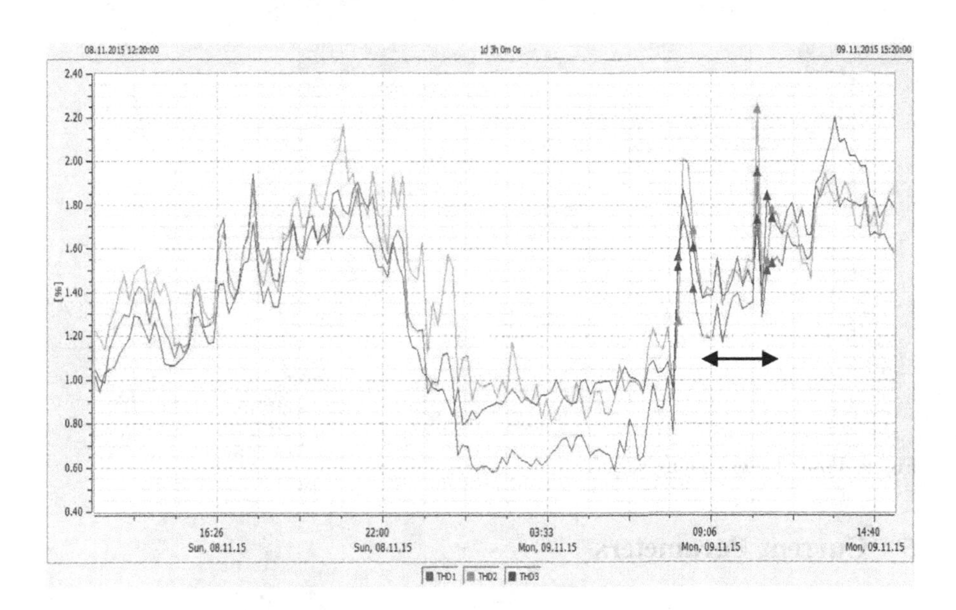

Fig. 7 VTHD during DG operation

and in negative due to diesel generator supply. The oscilloscopic events of diesel generator interference at main power shutdown are shown in Figs. 11, 12, and 13.

The sudden interfacing of diesel generators leads to frequency deviations in existing system and that will cause power quality issues.

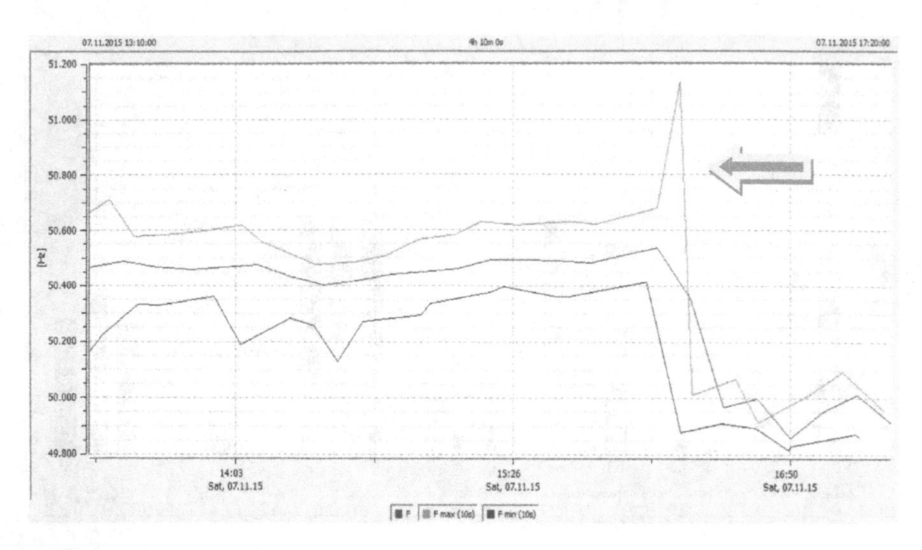

Fig. 8 Frequency crossed its standard limit

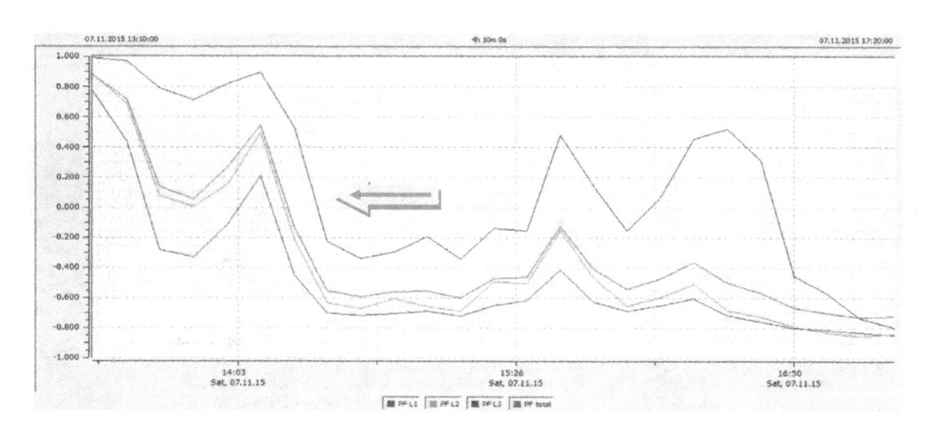

Fig. 9 Power factor coming negative

5 Current Parameters

The utilizing of nonlinear loads such as Switch mode power supplies, variable speed drives, UPS, electronic ballasts and other power electronic controlled loads is raised, along with this the harmonics in currents also raised. It is important to record the harmonic distortion in particular load to design and install a filter that suits to reduce the distortion within limits of standards. The mean values of THDs of all feeders are shown in Table 4.

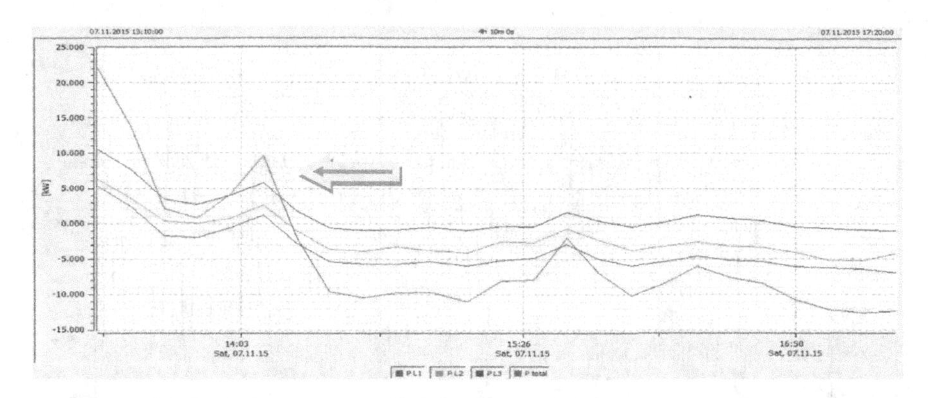

Fig. 10 Real power coming negative

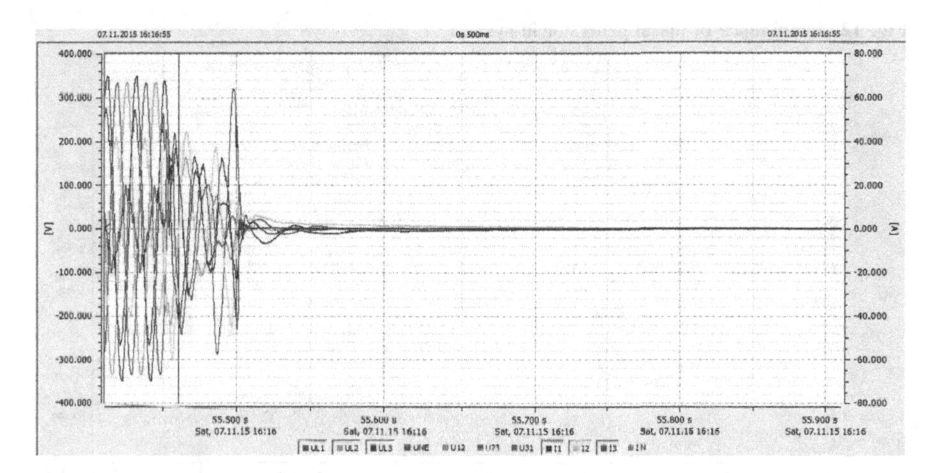

Fig. 11 Main power shutdown in system

6 Other Power Quality Parameters

6.1 Frequency

The frequency is one important parameter in the power quality, the effects of frequency deviations in other electrical equipments like filters, capacitor banks, machines like motors and transformers, etc., are detailed in [5]. It is necessary to find the variations of frequency in power supply to determine the effect on the loads. The readings of frequency variations are tabulated and shown in Table 5.

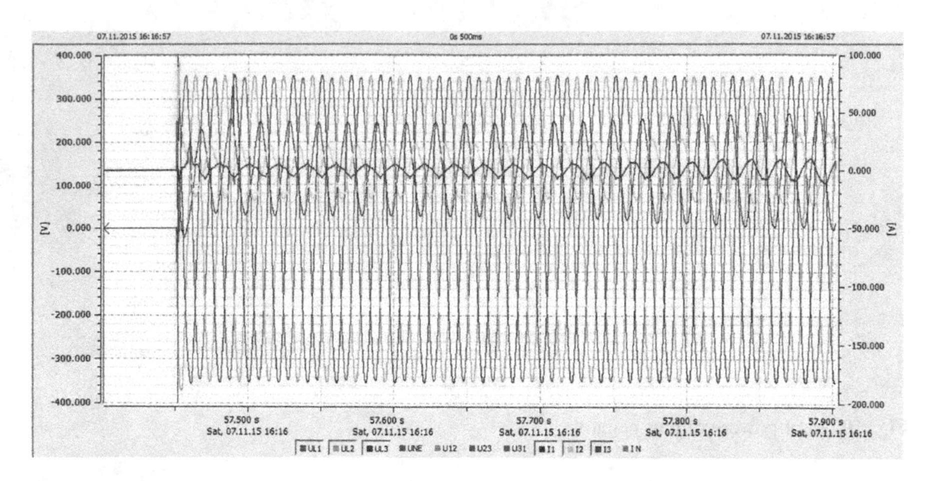

Fig. 12 Interference of diesel generator in system

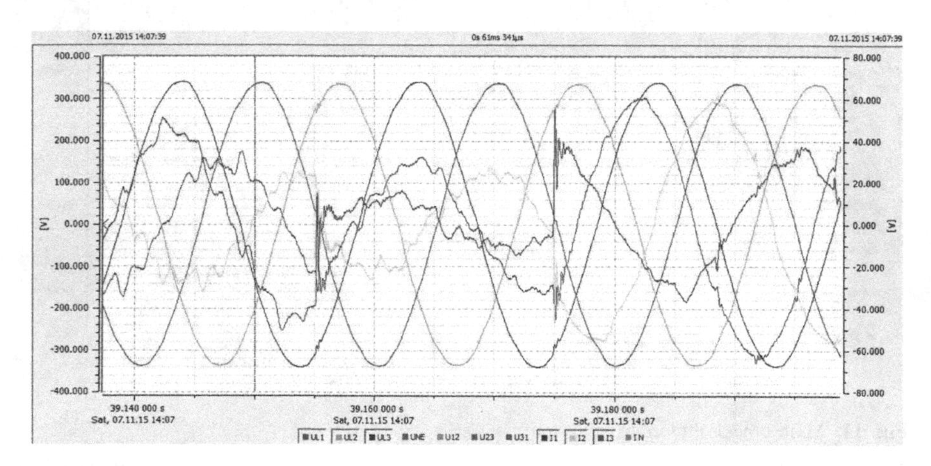

Fig. 13 Current shifting to above 90° phase

6.2 *Flicker*

Voltage flickers called fast varying voltages in short duration are caused by starting of nonlinear loads like arc furnaces, welding machines, induction motors, and also of generators. According to IEC 61000-3-3 standard the limits of long-term and short-term flicker is given as

- For short-time flicker in observed interval of 10 min the limit must be in 1.0
- For long-time flicker in observed interval of 2 h duration, the limit must be 0.65.

Table 4 Current harmonics of various feeders

S.no.	Name of load	Percentage THD			
		Line 1	Line 2	Line 3	Line *N*
1	SJT lighting	21.113	45.906	133.074	267.150
2	SJT pump	79.219	170.831	55.990	349.444
3	SJT UPS	27.762	24.364	19.454	37.217
4	E-annex lighting	5.782	5.582	7.601	88.190
5	EF pump	13.352	25.276	17.747	413.465
6	F-lighting	13.111	13.760	16.490	140.164
7	F-power feeder 1	14.885	14.886	30.790	169.435
8	F-power feeder 2	18.041	16.659	16.746	763.202
9	SJT STP	3.995	5.396	4.569	135.295

Table 5 Frequency of all load feeders

S.no.	Name of load	Frequency (Hz)		
		Min	Mean	Max
1	SJT lighting load (solar)	49.742	50.017	51.022
2	SJT power & UPS	49.781	50.008	50.395
3	SJT pump	49.811	50.016	50.260
4	E-annex block lighting	49.750	49.952	50.114
5	F-lighting	49.786	50.010	50.276
6	F-power and AC feeder 1	49.727	49.995	50.250
7	F-power and AC feeder 2	49.764	49.984	50.250
8	E&F pump	49.877	50.052	50.218
9	STP	49.867	50.048	50.203

It is observed that the short and long time flicker crossed the limits in lighting feeder according to standards given. The obtained readings of short time flicker and long time flicker are shown in Tables 6 and 7.

6.3 Power Parameters

The power readings of all feeder parameters including apparent power shown in Table 8 and real power and reactive power is shown in Table 9. The high power consumption load is SJT power and UPS feeder. The power factor of all loads is given in Table 10. Here the pump load has poor power factor which to be considered.

Table 6 Short-time flicker of all load feeders

Name of the load	Short time flicker (PST)		
	Line 1	Line 2	Line 3
SJT lighting load (solar)	1.286	1.313	1.324
SJT power & UPS	0.221	0.275	0.287
SJT pump	0.185	0.215	0.209
E-annex block lighting	0.259	0.271	0.275
F-lighting	0.292	0.295	0.287
F-power and AC feeder 1	0.591	0.563	0.601
F-power and AC feeder 2	0.525	0.608	0.555
E&F pump	0.243	0.285	0.279
Sewage treatment plant (STP)	0.186	0.227	0.240

Table 7 Long-time flicker of all feeders

Name of the load	Long time flicker (PLT)		
	Line 1	Line 2	Line 3
SJT lighting load (solar)	4.953	4.956	4.973
SJT power & UPS	0.247	0.362	0.331
SJT pump	0.222	0.236	0.218
E-annex block lighting	0.291	0.305	0.305
F-lighting	0.196	0.240	0.248
F-power and AC feeder 1	0.626	0.570	0.617
F-power and AC feeder 2	0.583	0.625	0.561
E&F pump	0.268	0.326	0.315
Sewage treatment plant (STP)	0.226	0.307	0.281

Table 8 Apparent power readings of all feeders

S.no.	Apparent power readings (KVA)			
	Name of the load	Minimum	Average	Maximum
1	SJT lighting load (solar)	1.431	13.971	46.726
2	SJT power & UPS	32.935	63.607	182.145
3	SJT pump	0.2646	14.442	60.392
4	E-annex block lighting	13.208	20.999	37.390
5	F-lighting	44.005	70.105	124.559
6	F-power and AC feeder 1	30.183	54.356	85.838
7	F-power and AC feeder 2	42.980	76.916	129.652
8	E&F pump	3.242	31.959	41.143
9	Sewage treatment plant (STP)	37.785	50.981	66.2665

Table 9 Real and reactive power readings of all feeders

Load number	Power parameters					
	Real (KW)			Reactive (KVAr)		
	Min	Avg	Max	Min	Avg	Max
1	−35.4	−6.50	41.68	0.8169	9.202	28.92
2	31.12	61.14	179.8	6.008	16.90	38.48
3	0.138	12.39	48.45	0.225	7.348	36.24
4	3.607	6.807	11.67	8.706	14.58	25.78
5	35.22	55.81	102.1	25.93	42.12	73.34
6	27.86	51.81	82.31	4.64	15.84	35.06
7	35.18	73.40	129.6	10.59	21.89	49.53
8	2.214	26.67	34.55	2.365	17.56	22.42
9	37.66	50.81	66.19	23.17	40.73	74.96

Table 10 Power factor of all load feeders

S.no.	Name of the load	Power factor		
		Min	Mean	Max
1	SJT lighting load (solar)	−0.889	−0.542	0.946
2	SJT power & UPS	0.858	0.955	0.990
3	SJT pump	0.509	0.757	0.948
4	E-annex block lighting	0.670	0.723	0.787
5	F-lighting	0.720	0.804	0.910
6	F-power and AC feeder 1	0.792	0.954	0.994
7	F-power and AC feeder 2	0.743	0.951	0.995
8	E&F pump	0.668	0.815	0.853
9	STP	0.991	0.997	0.999

7 Discussion on All Readings

- The most common issue in all readings is voltage harmonics, including with other PQ events.
- The harmonics in all feeders are analysed including even harmonics, odd harmonics, and triple-*n* harmonics (3*n*).
- Voltage unbalance is the main reason for formation of triple-*n* harmonics in the power system.
- The odd harmonics occurred in the system due to nonlinear loads connected to the system.
- The flicker in voltage is in standard limit in all feeders.
- The even harmonics in feeders were generated due to induction motor loads in specific feeders like pumps and STP loads.

- Power factor in all loads is very poor due to inductive loads except in STP as it is almost near to unity. The necessary action must be implemented for reaching unity.
- The power electronic switches in converter will cause the distortion in voltages and currents and leads to generate inter-harmonics of high frequency and also reason for rapid voltage changes.
- PQ events in all readings are due to inductive loads in all feeders which causes dip, rapid voltage change, etc.
- THD in voltage and currents in UPS feeder is more mainly due to PWM inverter operation.
- The common issue in both solar lighting and in ups is harmonics and max voltage harmonics (200 ms) that due to mainly inverter operation in two systems that generated by switching operations, this leads to damage in load side.

8 Recommended Tips for Improving Quality of Power

- To avoid the effects of harmonics on sensitive loads, it is necessary to maintain harmonics in safe limit by installing filter at load side either in shunt or series.
- In interfacing with solar and diesel generation, certain necessary protecting devices are to be taken connected for avoiding unwanted issues appearing in quality of power.
- Desired FACTS devices based on obtained results should be connected to reduce the losses involving in power quality.
- Filters should be connected in interfacing of inverter in power system to reduce the harmonics appeared in system.

9 Conclusion

Our future outcome is to improve the quality of power since the actual need of power is significantly increasing daily, as well as the issues also increasing. So it is often difficult to take steps to maintain quality of power along with increasing of load demand. The collected data from the survey is to analyze the quality of power in academic and hostel blocks, from these obtained results, new devices have to be designed that persuade with improving the quality of power along with existing devices.

Acknowledgments I'm highly indebted to VIT University, Danfoss industries Private Ltd., Chennai and technical staff in power house, for their guidance and constant support as well as cooperation to collect the data for this project.

References

1. N. Karthikeyan, "Power Quality Survey on Technological Institute" IEEE Trans., power systems 2009.
2. Indian Electricity Grid code March 2002 website: nerldc.org/COMMERCIAL/iegc.pdf.
3. EPRI, "Impact of Nonlinear Loads on Motor Generator Sets," Electr. Power Res. Inst. (EPRI), Palo Alto, CA, Tech. Rep. 1006489, 2001.
4. Degan, R.E., MF. McGranaghan, S. Santoso, and H.W. Beaty, "Electrical power systems equality," McGraw-Hili companies, second edition, 2006.
5. Angelo Baggini, Handbook of Power Quality, John Wiley & Sons Publication, 2008.

Dynamic Resource Allocation Mechanism Using SLA in Cloud Computing

Jayashree Agarkhed and R. Ashalatha

Abstract Resource provisioning is greatly required to expand the performance in cloud. The hardware technique used for resource allocation methods can be central processing unit (CPU) scaling, which is the frequency of physical cores. The software technique used can be horizontal and vertical elasticity. The resource management affects the evaluation of a system by performance, functionality, and cost. The resource management in cloud environment should have enormous policies and decisions for adhering various objcctive optimization. Efficient resource management is the process of allocating the resources and handling the workload variations effectively. The computing resources have to be handled efficiently among the users of virtual machines. Service level agreement (SLA) is defined as the contract made between the providers and the customers of cloud for guaranteeing the quality of service (QoS) issues. This paper gives various policies defined in cloud computing related to SLA issues.

Keywords Cloud computing · Resource provisioning · Resource allocation · Service-level agreement · Quality of service

1 Introduction

Cloud computing can be referred as applications which are delivered as a service over Internet and hardware systems. It provides various services through systems software in the cloud data centers. While providing basic resources and services, cloud computing has been devoted to guarantee the quality of service (QoS). The software

J. Agarkhed (✉) · R. Ashalatha
Poojya Doddappa Appa College of Engineering, Kalaburagi 585102, Karnataka, India
e-mail: jayashreeptl@yahoo.com

R. Ashalatha
e-mail: ashalatha.dsce@gmail.com

S.S. Dash et al. (eds.), *Artificial Intelligence and Evolutionary Computations in Engineering Systems*, Advances in Intelligent Systems and Computing 517,
DOI 10.1007/978-981-10-3174-8_61

as a service (SaaS) providers comply with the requirements of QoS which are given in SLA contract made between the cloud providers and the users [1].

SLA contains the terms and conditions for any type of service maintained by the provider for user applications. SLAs are generally negotiated between context-aware application provider and cloud provider for QoS-related issues. They are used to specify the needs required for providing services in cloud computing [2]. The physical resources in cloud include CPU, memory, storage, workstations, network elements, and sensors. The logical resources are system abstractions which include operating system, energy, network throughput or bandwidth, information security, protocols, network loads, and delays.

The rest of the paper is organized as follows. Section 2 deals with related work. Section 3 gives resource allocation mechanism in cloud computing environment. Section 4 shows some important policies in cloud. Section 5 discusses various performance metrics for cloud system. Section 6 gives the comparative study and Sect. 7 ends with conclusion.

2 Related Work

Nemati et al. [3] have introduced two adaptive SLA-based elasticity management algorithms for virtualized IP multimedia system (IMS) environments which are mentioned as USA and BSA algorithms. Call setup delay and user priority are the two SLA attributes used for controlling CPU resources.

The resource management is greatly required in cloud computing which provides different policies for obtaining at the correct solution. The proposed policies rely on information about the system performance history and apply at runtime for their reduced computational time [4]. The SLA model provides enormous knowledge for negotiation of SLAs between providers and clients. The priority-based scheduling mechanism uses COSIM-CSP simulation framework [5].

Resource allocation methods are mainly used for the optimization of cloud resources. They are used when the clients have certain SLAs for meeting the cloud requirements. Different resource provisioning techniques based on SLA's are specified by large Cloud Service Providers (CSPs). Different CSPs have to be managed efficiently to maximize the total profit in the system [6].

The end-to-end framework design and its implementation provide provisioning of database tier. For guaranteeing SLA performance requirements, the applications are based on consumer-centric policies and issues. This framework provides flexible mechanisms for fine-granular SLA metrics at the application level for consumer applications [7].

The application partition offloading scheme for mobile users are provided by resource allocation algorithm. This method uses semi markov decision process. The asset assignment strategy utilizes arrangement cycle approach for accelerating application execution. The algorithm performs better than greedy admission control for throughput and QoS requirements to achieve high efficiency [8].

3 Resource Allocation Mechanism

The resource allocation policies are used for assigning resources to the requests made in the cloud environment. Various parameters of resource allocation include resource contention, fragmentation, scarcity, and over-provisioning factors. Cloud resource management policies can be categorized as admission control, capacity allocation, load balancing, energy optimization, and QoS guarantees. Implementation of resource management policies can be classified as control theory, machine learning, utility-based approach, and market-oriented mechanisms. Many auto-scaling methods provide performance metrics like CPU utilization for reducing the execution cost. But, the user requirements in hybrid clouds for variable resources utilization must maintain service-level agreements like deadline cost, cost-oriented policies, or performance-oriented policies. Using auto-scaling method, the users can make use of required resources for any type of applications thereby reducing the usage cost. The performance-oriented policy helps the users to meet the deadline requirements. SLA-oriented resource allocation methods are used for resource management system in cloud. The resource management issues such as SLAs are most importantly required for customer-driven, computational risk management, and autonomic resource management system [9] (Fig. 1).

Providing authentic QoS is a main area of cloud computing environment. SLAs are used for providing QoS characteristics such as throughput, response time, latency, etc. QoS-based resource allocation mechanisms are the important area to be covered in providing cloud security [10]. There exist many components in SLA data centers with respect to resource allocation scheme. Major entities involved in resource allocation scheme are as follows.

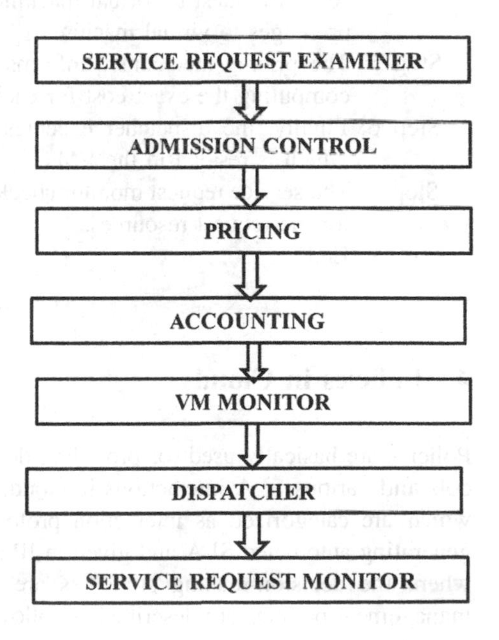

Fig. 1 SLA-oriented resource allocation

(a) Cloud Users/Brokers: This is the entity who submits the request for getting the service to the cloud.
(b) Resource Allocator: SLA resource allocator is the intermediate between the CSPs and the consumer.
(c) Virtual Machines: Multiple VMs can be started simultaneously in cloud which runs on multiple operating systems in a single system.
(d) Physical Machines: Each data center consists of multiple servers for servicing every incoming request coming into the cloud.

Pseudo code for SLA-based resource allocation policy is given in Algorithm 1.

Algorithm 1: SLA-oriented resource allocation

Step 1: Start
Step 2: Each service request is submitted by the user or the cloud broker.
Step 3: The service request examiner and admission control design admits the incoming request for QoS mechanisms for providing various services.
If (each availability of requested resource = true)
Status of request = Accept
else
Status of request = Reject
Step 4: The service request examiner and the admission control gets the information about the resource availability from virtual machine monitor (VMM) and service request monitor for allocating the resources to each cloud user.
For (each available resource in cloud)
Assign request to virtual machines (VMs) and allocate the resource privileges to virtual machine
Step 5: The pricing mechanism informs the correct usage of resources for computing the exact cost for each service to the user.
Step 6: Finally, the dispatcher executes the service for the given request which is resided in the VMs
Step 7: The service request monitor checks the overall progress of the service for the request resources.
Step 8: End

4 Policies in Cloud

Policies are basically used for providing domain knowledge for interaction protocols and various SLA interactions in cloud. There are two types of SLA policies which are categorized as interaction protocol IP policies which are defined by generating automated SLA and given in IP policy. Next one is the strategy policy where the decision-making strategies are specified in this policy [11]. Cloud management policies are described as follows:

A. *Resource Allocation Policy*: The resource management affects the evaluation of a system by performance, functionality, and cost. Auto-scaling method is used for unplanned workloads when there is pool of resources available and when there is a monitoring system for real-time resource allocation. The autonomic policies are of great demand for scaling of system, large population, and many user request as well as load fluctuation. Implementation of resource management policies can be classified as control theory, machine learning, utility-based approach, and market-oriented or economic mechanisms [12].

The resource allocation policy allocates the unused resources coming from various computing platforms to each of its users. For adapting to the dynamic environment, system has to consider current resource allocation strategies. Types of resource allocation policies are given below.

a. *Fairness policy*: This policy allocates the computing platform resources equally among all the users of cloud.

$$\text{fairness} = 1 - \frac{\sigma t}{\mu t} \tag{1}$$

b. *Greedy efficiency policy*: Greedy policy distributes computing platform resources to the user with maximum throughput.
c. *Fair efficiency policy*: This policy provides competitive efficiency along with user satisfaction. It distributes equal CPU cores to single user by taking into consideration higher throughput of the system [13].

B. *SLA Policy*: The SLA policy is used between the cloud providers and the users for controlling the services levels. It uses the QoS attribute values of various VMs and consolidates into one QoS index value [14]. The SLA proposal given by client to the providers is finalized between both the parties, then the QoS requirements and the economic conditions are kept in SLA [15].

The SLA algorithm maintains the priority of the tasks in task queue and handovers each task to their respective server cluster. Different QoS parameters are used for finding the response time of each task for effective resource utilization in the system. For every task queue maintained, length, deadline, and cost of each task are calculated. The server cluster can be categorized into high processing power, medium processing power or low processing power servers clusters based on SLA [16].

The service-level agreements consist of data storage, utilization of available bandwidth, and security standards. The problem of over-provisioning of resources arises between cloud consumers and cloud providers. The situation of over-provisioning of cloud resources leads to over utilization or underutilization of resources in cloud. Therefore, proper resource provisioning algorithms are required for resource provisioning and allocation of requests between clients and the providers [17]. The biggest challenge in cloud computing relies on allocation of virtual resources on the hardware in virtualized Infrastructure as a Service. The virtualization technique provides a set of virtual resources on top of physical resources.

The cloud computing resource management problems are based on static or dynamic problems. The static or offline requests for cloud resources are already known and are constant in nature. But, the dynamic requests are online and can change over time depending on the user requirements [18]. An SLA-based resource allocation optimization strategy boosts the aggregate benefit which is the aggregate cost picked up from serving the customers subtracted by the operation expense of the server cluster. A joint optimization framework consists of request dispatching, dynamic voltage, and frequency scaling for individual cores. Each core in a single cluster is modeled using a Continuous Time Markov Decision Process (CTMDP). A near optimal hierarchical solution consists of a central manager and distributed local agents. The near optimal resource allocation and consolidation algorithm consistently performs better than baseline algorithms [19]. Steps required for evaluating service-level agreements in cloud is given in Algorithm 2.

Algorithm 2: Evaluation of service level agreements

Step 1: Start
Step 2: Understand Roles and Responsibilities
Step 3: Evaluate Business Level Policies
Step 4: Understand Service and Deployment Model Differences
Step 5: Identify Critical Performance Objectives
Step 6: Evaluate Security and Privacy Requirements
Step 7: Identify Service Management Requirements
Step 8: Prepare for Service Failure Management
Step 9: Understand the Disaster Recovery Plan
Step 10: Develop an Effective Management Process
Step 11: Understand the Exit Process
Step 12: End

C. *Resource Management Allocation Policies and Mechanisms*

The principle guiding decision-making polices in cloud computing are mentioned below.

(i) Resource management allocation policies

- Admission control mechanisms
- Capacity allocation methods
- Load balancing strategies
- Energy optimization techniques
- QoS guarantees issues

(ii) Mechanisms based on control theory

- Feedback and feed forward mechanism
- Utility-based approaches
- Market-oriented mechanism
- Machine learning mechanism.

5 Performance Metrics

The performance metrics are used to compare different works under resource management techniques. Some of them are reliability, deployment ease, QoS, delay and control overhead. The virtual machines are assigned dynamically for meeting SLA's and offers auto-scaling features by cloud providers. Various QoS requirements used in SLA include throughput, delay, latency, etc. Resource provisioning is an important aspect of SLA.

The SLA metrics refers to the values which are measured for a given service. The SLA values must agree to the contract terms and the service to achieve the required business objective. Every SLA comprises of service performance metrics with its corresponding service-level objectives (SLO) [20]. The metrics are mentioned below.

a. *Mean time to failure* (*MTTF*): MTTF measures the mean time between two consecutive failures for the same service S_A. If we have N monitored events $t_{u,i}^{S_A}$ indicates the time from which the service is up and $t_{d,i+1}^{S_A}$ is the time at which the same service fails again, MTTF is defined as follows.

$$\mathrm{MTTF}(S_A) = \left(\sum_{i=1}^{N-1}\left(t_{d,i+1}^{S_A} - t_{u,i}^{S_A}\right)\right)\Big/N \tag{2}$$

b. *Mean time to repair* (*MTTR*): MTTR measures the mean time needed to repair a specific service S_A for N times. MTTR is given by

$$\mathrm{MTTR}(S_A) = \left(\sum_{i=1}^{N-1}\left(t_{u,i}^{S_A} - t_{d,i}^{S_A}\right)\right)\Big/N \tag{3}$$

c. *Mean time between failures* (*MTBF*): MTBF measures the mean time occurred between two consecutive failures of the same service. It can be simply considered as the sum between MTTF and MTTR. It is mentioned below.

Table 1 Types of policies in cloud SLA

SLA data policies	SLA business-level policies
a. Data preservation	a. Acceptable use policy
b. Data redundancy	b. Excess usage
c. Data location	d. Payment and penalty models
d. Verification of new data location	d. Governance/versioning
e. Data seizure	e. Subcontracted services
f. Data privacy	f. Licensed software
	g. Industry-specific standards

$$\mathrm{MTBF}(S_A) = \mathrm{MTTR}(S_A) + \mathrm{MTTF}(S_A) \tag{4}$$

d. *Response time*: having its own response time *rtS*$_A$(*t*)*i*, the overall response time RT is the average of the distinct response times as given below. Response time quantifies how long it takes to a service SA to answer client's interrogation. If we consider N interrogations from client, each

$$\mathrm{RT}(S_A) = \left(\sum_{i=1}^{N} rt^{S_A}(t)_i\right) \bigg/ N \tag{5}$$

e. *Availability*: Availability describes how long a service is available within its temporal window of analysis. Being TAW the length of this temporal window and TSD the administration downtime period, the availability *A* is expressed as

$$A(S_A) = (T_{\mathrm{TW}} - T_{\mathrm{SD}})/T_{\mathrm{TW}}. \tag{6}$$

6 Comparative Study

Table 1 gives the types of SLA policies in cloud computing environment. It includes critical data policies and critical business level policies in cloud SLA.

7 Conclusion

Efficient resource management is the process of allocating the resources and handling the workload variations effectively. The computing resources have to be handled efficiently among the users of virtual machines. SLA-related resource management is an important issue to be taken into consideration in cloud computing. Various research challenges based on QoS and throughput for achieving low cost and high performance is considered enormously.

References

1. Fox, A. Griffith, R. Joseph, A. Katz, R., Konwinski, A. Lee, G..& Stoica, I. (2009). Above the clouds: A Berkeley view of cloud computing. Dept. Electrical Eng. and Comput. Sciences, University of California, Berkeley, Rep. UCB/EECS, 28(13), 2009.
2. Boloor, K. Chirkova, R. Salo, T. & Viniotis, Y. (2011, July). Management of SOA-based context-aware applications hosted in a distributed cloud subject to percentile constraints. In Services Computing (SCC), 2011 IEEE International Conference on (pp. 88–95). IEEE.
3. Nemati, H. Singhvi, A. Kara, N. & El Barachi, M. (2014, December). Adaptive SLA-based elasticity management algorithms for a virtualized IP multimedia subsystem. In Globecom Workshops (GC Wkshps), 2014 (pp. 7–11). IEEE.
4. Cardellini, V. Casalicchio, E. Presti, F. L. & Silvestri, L. (2011, November). Sla-aware resource management for application service providers in the cloud. In Network Cloud Computing and Applications (NCCA), 2011 First International Symposium on (pp. 20–27). IEEE.
5. Peng, G. Zhao, J. Li, M. Hou, B. & Zhang, H. (2015, May). A SLA-based scheduling approach for multi-tenant cloud simulation. In Computer Supported Cooperative Work in Design (CSCWD), 2015 IEEE 19th International Conference on (pp. 600–605). IEEE.
6. Wang, Y. Lin, X. & Pedram, M. (2014, April). A game theoretic framework of sla-based resource allocation for competitive cloud service providers. In Green Technologies Conference (GreenTech), 2014 Sixth Annual IEEE (pp. 37–43). IEEE.
7. Zhao, L. Sakr, S. & Liu, A. (2015). A framework for consumer-centric SLA management of cloud-hosted databases. Services computing, IEEE Transactions on, 8(4), 534–549.
8. Liu, Y. & Lee, M. J. (2015, March). An adaptive resource allocation algorithm for partitioned services in mobile cloud computing. In Service-Oriented System Engineering (SOSE), 2015 IEEE Symposium on (pp. 209–215). IEEE.
9. Buyya, R. Garg, S. K. & Calheiros, R. N. (2011, December). SLA-oriented resource provisioning for cloud computing: Challenges, architecture, and solutions. In Cloud and Service Computing (CSC), 2011 International Conference on (pp. 1–10). IEEE.
10. Buyya, R, Yeo, C. S. & Venugopal, S. (2008, September). Market-oriented cloud computing: Vision, hype, and reality for delivering it services as computing utilities. In High Performance Computing and Communications, 2008. HPCC'08. 10th IEEE International Conference on (pp. 5–13). Ieee.
11. Chhetri, M. B. Vo, Q. B. & Kowalczyk, R. (2012, May). Policy-based automation of SLA establishment for cloud computing services. In Cluster, Cloud and Grid Computing (CCGrid), 2012 12th IEEE/ACM International Symposium on (pp. 164–171). IEEE.
12. Wang, Y. Chen, S. & Pedram, M. (2013, April). Service level agreement-based joint application environment assignment and resource allocation in cloud computing systems. In Green Technologies Conference, 2013 IEEE (pp. 167–174). IEEE.
13. Hwang, E. Kim, S. Yoo, T. K. Kim, J. S. Hwang, S. & Choi, Y. R. Resource Allocation Policies for Loosely Coupled Applications in Heterogeneous Computing Systems.
14. Nakamura, L. H. Estrella, J. C. Santana, R. H. Santana, M. J., & Reiff-Marganiec, S. (2014, November). A semantic approach for efficient and customized management of IaaS resources. In Network and Service Management (CNSM), 2014 10th International Conference on (pp. 360–363). IEEE.

15. Hamsanandhini, S. & Mohana, R. S. (2015, January). Maximizing the revenue with client classification in Cloud Computing market. In Computer Communication and Informatics (ICCCI), 2015 International Conference on (pp. 1–7). IEEE.
16. Rajeshwari, B. S. & Dakshayini, M. (2015, June). Optimized service level agreement based workload balancing strategy for cloud environment. In Advance Computing Conference (IACC), 2015 IEEE International (pp. 160–165). IEEE.
17. Patel, K. M, & Bohara, A. P. M. H. Self-adaptive Resource Allocation using Feedback and Ranking in Cloud Computing by Crawler.
18. Guzek, M. Bouvry, P. & Talbi, E. G. (2015). A Survey of Evolutionary Computation for Resource Management of Processing in Cloud Computing [Review Article]. Computational Intelligence Magazine, IEEE, 10(2), 53–67.
19. Wang, Y. Chen, S. Goudarzi, H. & Pedram, M. (2013, March). Resource allocation and consolidation in a multi-core server cluster using a Markov decision process model. In Quality Electronic Design (ISQED), 2013 14th International Symposium on (pp. 635–642). IEEE.
20. Longo, A. Zappatore, M. & Bochicchio, M. A. (2015, June). Service Level Aware-Contract Management. In Services Computing (SCC), 2015 IEEE International Conference on (pp. 499–506). IEEE.

A Look-up Table-Based Maximum Power Point Tracking for WECS

J. Antony Priya Varshini

Abstract This paper proposes a look-up table method for tracing the maximum power generated in the wind energy conversion system (WECS). The look-up table is used to develop the corresponding change in the duty ratio with the change in the current. Self-excited induction generator (SEIG) is the perfect choice for generating power from the wind. A diode bridge rectifier (DBR) is used to convert renewable energy into electrical output and a dc–dc converter maintains the grid voltage. This method of tracking the maximum power from the wind is determined using the dc grid current. The dc grid current is the control variable that tracks the maximum power in the system and it is not dependent on the machine and wind parameters. Simplicity in circuit and control algorithm is the key advantage for supplying maximum power from the WECS to the dc microgrid.

Keywords Wind energy conversion system (WECS) · Maximum power (MP) · DC microgrid · Look-up table · Maximum power point tracking (MPPT) · Self-excited induction generators (SEIG) · Diode bridge rectifier (DBR) · DC–DC converter

1 Introduction

Countries in the world fully recognize the urge to promote wide spread adoption of renewable energy into the energy sources of the country, with the intention of promoting environmental stewardship, social development, and sustained economic growth [1]. Renewable energy sources which can be called as non-conventional energy are sources that are continuously replenished by natural processes. A renewable energy system converts the energy from wind, falling water, sunlight, sea waves, geothermal heat, or biomass into the form we can use such as heat or

---

J. Antony Priya Varshini (✉)
Electrical and Electronics Department, Karunya University, Coimbatore, India
e-mail: antonypriyavarshini14@karunya.edu.in

S.S. Dash et al. (eds.), *Artificial Intelligence and Evolutionary Computations in Engineering Systems*, Advances in Intelligent Systems and Computing 517,
DOI 10.1007/978-981-10-3174-8_62

electricity. Various renewable energies are the solar energy, wind energy, bioenergy, hydroenergy, geothermal energy, wave energy, and tidal energy. Wind energy is the most readily available and free source of energy since prehistoric times. India receives solar energy in the region of 5–10 kWh/m^2 for 250–330 days in a year. This energy is sufficient to set up 20–40 MW wind power plant per square kilometer land area.

The development and deployment of wind technologies for more than two decades are the strong research infrastructure and a good manufacturing base for production of single vertical axis and horizontal axis. Wind turbine has been established in India ranking third in renewable energy production in world economy [2]. The renewable energy sources like wind energy conversion systems (WECS) produce wind energy that varies throughout the day and it is also dependent on the geographical location. For a specific wind velocity for WECS there is a peak point at which maximum power can be obtained.

The output power of the wind turbine depends on the speed and accuracy at which peak power points are tracked by the implementation of a maximum power point tracking (MPPT) control techniques [3–6]. The output from the WECS depends on the rotor speed that changes with the variation of wind speed. There is always an optimum rotor speed for WECS for a specific wind speed at which maximum power can be extracted out of the system. The location of the maximum power point (MPP) is unknown but can be determined, either through calculations or through search algorithm techniques.

Figure 1 shows the proposed MPPT technique for tracking the MP in the WECS. A wind-driven induction generator (IG) can operate in self-excited mode with the help of capacitor banks that are connected to a three-phase diode bridge rectifier (DBR) and then to a dc–dc converter to supply a dc microgrid [7]. The dc microgrid is being replaced by resistor. This project proposes look-up table-based MPPT. To develop look-up tables, detailed simulations are carried out. In the WECS, the dc grid current is the control variable to track maximum power. Two look-up tables have been developed that track the MP for the proposed MPPT technique in the WECS. The first look-up table has been developed to track the change in current and it relates the current with the duty ratio that is generated. The corresponding duty ratio is given to the second look-up table and it generates the corresponding constant to be given as an input for the generation of duty ratio by PWM.

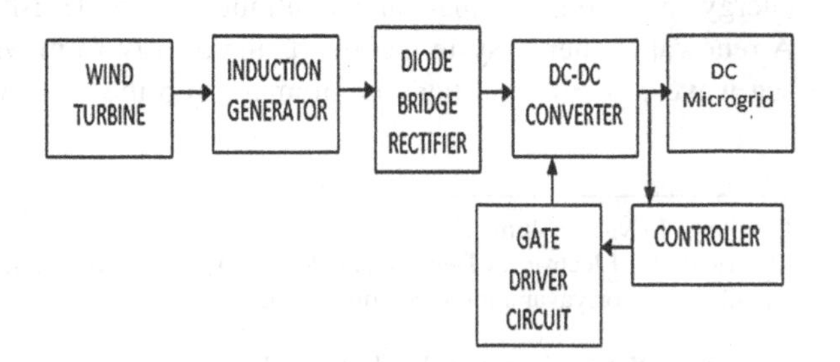

Fig. 1 WECS with proposed MPPT strategy for DC microgrid

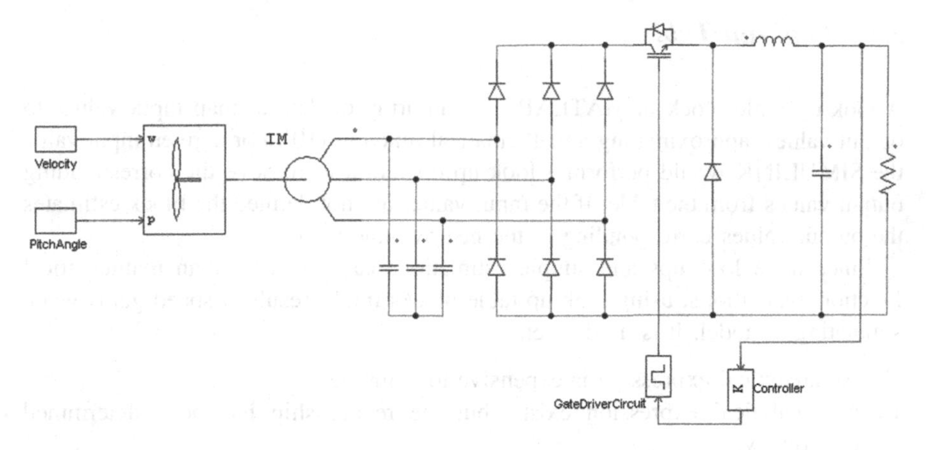

Fig. 2 Circuit diagram for the look-up table-based maximum power point tracking for WECS

The proposed algorithm for MPPT is not dependent of the machine and wind turbine characteristics [8]. Simplicity in circuit and control algorithm simplicity are the main advantages of the proposed configuration for providing power to the dc microgrid from the WECS.

The circuit diagram for the look-up table-based MPPT for WECS is shown in Fig. 2. The torque which is the mechanical input from the wind turbine is given to the induction generator which is operated in self-excited mode using capacitor banks at the stator. The induction generator is used to generate the ac voltage and the ac is converted into dc using the DBR. The DBR converts the ac into dc. The dc–dc converter bucks the dc to the given voltage to the dc microgrid. The output current of the dc–dc converter (dc grid current) is the control variable and is controlled to determine the maximum power from the WECS.

2 Look-up Table Strategy for WECS

Wind power generation involves energy extraction from the wind using a wind turbine generator. The wind energy extracted from renewable energy sources like wind energy conversion systems (WECS) varies throughout the day and it is also dependent on the geographical location. The output power of wind turbine depends upon the speed and accuracy at which peak power points are tracked by the implementation of a MPPT control techniques. The output power from WECS is a function of rotor speed that will change with respect to change in wind speed. The maximum power is determined either using complex calculation or by various algorithm techniques [9]. The control strategy used is a look-up table-based MPPT control strategy for determining the maximum power from the wind. Two look-up tables are used. The first look-up table provides the corresponding duty ratio for the corresponding change in current value. The second look-up table generates the ratio.

2.1 Look-up Table

A look-up table block in MATLAB uses an array of data to map input values to output values, approximating a mathematical function [10]. For a given input value the SIMULINK would perform a look-up operation to retrieve the corresponding output values from the table. If the input values are not defined the block estimates the output values corresponding to the nearby table values.

Since table look-ups and simple estimations can be faster than mathematical function evaluations, using look-up table blocks might result in speed gains when simulating a model. It is used when

I. An analytical expression is expensive to compute.
II. No analytical expression exists, but the relationship has been determined empirically.

SIMULINK provides a broad assortment of look-up table blocks, each geared for a particular type of application. The different types of look-up in MATLAB/SIMULINK are as follows:

1. 1-D Look-up Table
2. 2-D Look-up Table
3. Sine block
4. Cosine block
5. Direct Look-up Table (*n*-D)
6. Interpolation using Prelook-up
7. Look-up Table Dynamic
8. Prelook-up
9. *n*-D Look-up Table.

For determining the MPPT for the WECS the 1-D look-up table is used. The 1-D look-up table is a one-dimensional function which generates output corresponding to the input value.

2.2 Look-up Table Data

The 1-D look-up table is a one-dimensional function which generates one set of output for the corresponding input value. In this proposed MPPT control technique two look-up tables are used. The first look-up table has been developed to track the change in current and it relates the current with the corresponding duty ratio generated. This duty ratio is given to the second look-up table which generates the corresponding constant to be given as an input for the generation of duty ratio by pulse width modulation for the IGBT-based dc–dc converter. The look-up table is developed by detailed simulation. Thus using the look-up table the time consumed is less; thus the system is faster.

1. *Current–to-Duty Ratio*: The current-to-duty ratio look-up table is used to generate the corresponding value of duty ratio for the change in the current. To develop this look-up table, detailed simulations are carried out. The simulation has been done for various duty ratios and the current values are noted for the change in the duty ratio. The values of currents are taken as the input for the look-up table and their corresponding duty ratios are generated as output. The values of the currents are to be entered in the breakpoints and corresponding values of the duty ratios are entered in the table data in the 1-D look-up table in MATLAB. Thus for the corresponding change in the current its duty ratio is generated as listed in Table 1. Corresponding value of constant for the change is the corresponding duty ratio. The constant value is then compared with the triangular wave for the pulse width modulation for the generation of gate pulse for the dc–dc converter.
2. *Duty Ratio to Constant*: The duty ratio to constant look-up table is used to generate the corresponding value of constant for the change in the duty ratio which is the input from the first look-up table. To develop this look-up table, detailed simulations are carried out. The simulation has been done by considering the change in constant value that is related to the pulse width modulation technique in order to generate the duty ratio for the IGBT-based dc–dc converter and the corresponding duty ratio values are noted. The values of duty ratios from the first look-up table are taken as the input for the second look-up table and their corresponding constants are generated as output. The values of the duty ratios are to be entered in the breakpoints and the corresponding values of the constant are entered in the table data in the 1-D look-up table. Thus for the corresponding change in the duty ratio a constant is generated as listed in Table 2.

Table 1 Current-to-duty ratio

S. No.	Current	Duty ratio
1	1.661	0.15
2	2.322	0.25
3	2.653	0.35
4	2.813	0.45
5	2.855	0.50
6	2.885	0.55
7	2.922	0.65
8	2.953	0.75
9	2.972	0.85
10	2.67	0.95

Table 2 Duty ratio to constant

S. No.	Duty ratio	Constant
1	0.15	0.50
2	0.25	0.75
3	0.35	1.00
4	0.45	1.25
5	0.50	1.50
6	0.55	1.75
7	0.65	2.00
8	0.75	2.25
9	0.85	2.50
10	0.95	2.75

3 MPPT Strategy for Wind-Driven Induction Generator

The MPPT control strategy used for the WECS is a look-up table-based MPPT control strategy for determining the MP that is determined from the wind as shown in Fig. 3. There are two look-up tables that are used. The first look-up table provides the corresponding duty ratio for the corresponding change in current value. The second look-up table determines the corresponding value of constant for the change in the corresponding duty ratio [11–14]. The constant value is then compared with the triangular wave for the pulse width modulation for the generation of gate pulse for the dc–dc converter.

The power obtained from the wind turbine is expressed as

$$P_T = 0.5C_P\rho \mathrm{AV}^3 \quad (1)$$

where ρ is the air density in kg/m^3, A is the swept area of turbine in m^2, V is the wind velocity in m/s, and Cp is the maximum power coefficient which is the value dependent on the ratio between the turbine rotor's angular velocity (ω_r) and wind speed (V). This ratio is known as the tip speed ratio (TSR) as in [4]:

$$\lambda = \frac{\omega_r}{V}R \quad (2)$$

The DC grid is replaced with a resistor value by setting a constant dc current at the DBR, and the equivalent load at the generator terminal can be represented as a pure resistive load as given in [7] by

$$R = 3V_p^2 / P_e \quad (3)$$

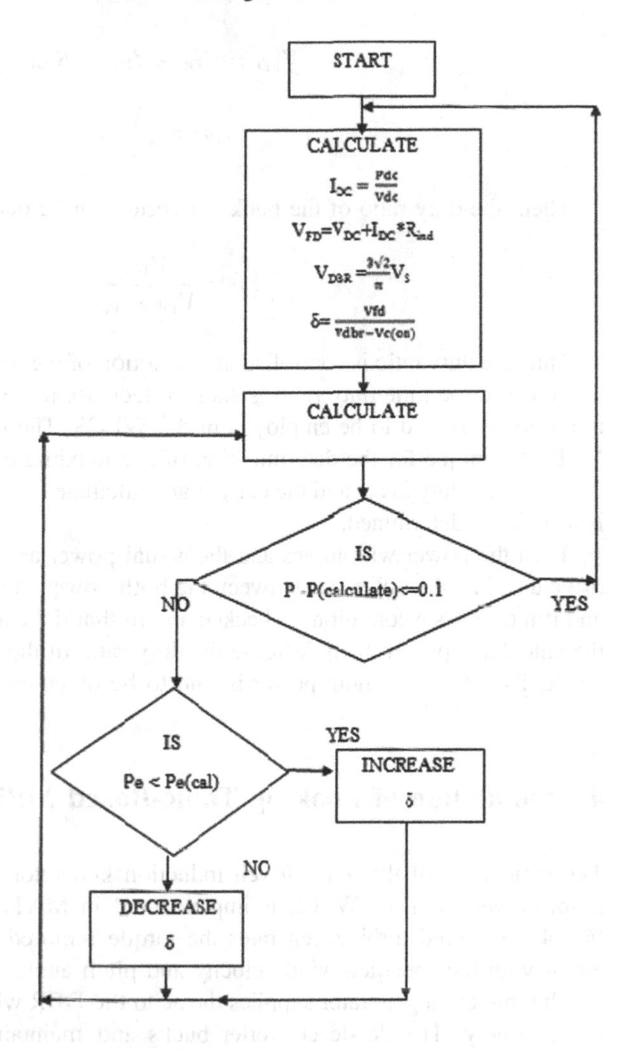

Fig. 3 MPPT implementation

The dc output voltage of an uncontrolled rectifier, V_{DBR}, is given as follows:

$$V_{DBR} = \frac{3\sqrt{2}}{\pi} V_S - V_D, \tag{4}$$

where $V_{D(ON)}$ is the on-state voltage across each diode in DBR and V_S is the line voltage at the input side of DBR. The delta-connected stator has been used so $V_s = V_p$ [7]. Further, considering the continuous current conduction at the direct current side of the 3-phase DBR, two diodes will conduct for any instant. This aspect has also been included for evaluating the dc voltage. For known value of dc grid voltage and inductor resistance, the value of freewheeling diode is as follows:

$$V_{\mathrm{FD}} = V_{\mathrm{DC}} + I_{\mathrm{DC}} * R_{\mathrm{ind}} \tag{5}$$

$$I_{\mathrm{DC}} = \frac{P_{\mathrm{DC}}}{V_{\mathrm{DC}}}. \tag{6}$$

Then, the duty ratio of the buck converter can be determined as

$$\delta = \frac{V_{\mathrm{FD}}}{V_{\mathrm{DBR}-V_{\mathrm{C}}}}. \tag{7}$$

Thus the duty ratio is controlled as a function of the dc grid current. If the current is increased then the duty ratio is said to decrease and vice versa. Thus the maximum power is said to be employed in the WECS. The flowchart demonstrates the MPPT technique for the determination of the maximum power in the WECS. The values of the duty cycle and the current are calculated and then the calculated power is said to be determined.

Then the power with losses and the actual power are said to be in a same range of values, i.e., the difference between the both power should not be greater than 0.1 and if it exceeds a condition is checked and in that if the actual power value exceeds the calculated power, then decrease the duty ratio of the buck converter or the vice versa. Thus the maximum power is said to be obtained in the system.

4 Simulation of Look-up Table-Based MPPT

The simulation of the wind-driven induction generator for determining the maximum power from the WECS is implemented in MATLAB R2014a as shown in Fig. 4. The wind turbine generates the torque required to run the induction generator with the specified wind velocity and pitch angle.

The induction generator supplies the ac to the DBR which rectifies the ac supply to dc supply. The dc–dc converter bucks and maintains the supply to the grid voltage. The grid current from the dc is the control variable. Two look-up tables are used to implement the proposed MPPT control strategy for the WECS. The dc grid current is given to the first look-up table which generates the corresponding duty ratio. This duty ratio is given to the second look-up table which generates the corresponding constant for the pulse width modulation to be given to the IGBT-based dc–dc converter and thus maximum power is obtained.

A. *MPPT Implementation*

The MPPT implementation is given by means of a look-up table method. In accordance with the change in the wind velocity the change in the maximum power is tracked. The corresponding change in the current is reflected in the change in duty ratio of the dc–dc converter.

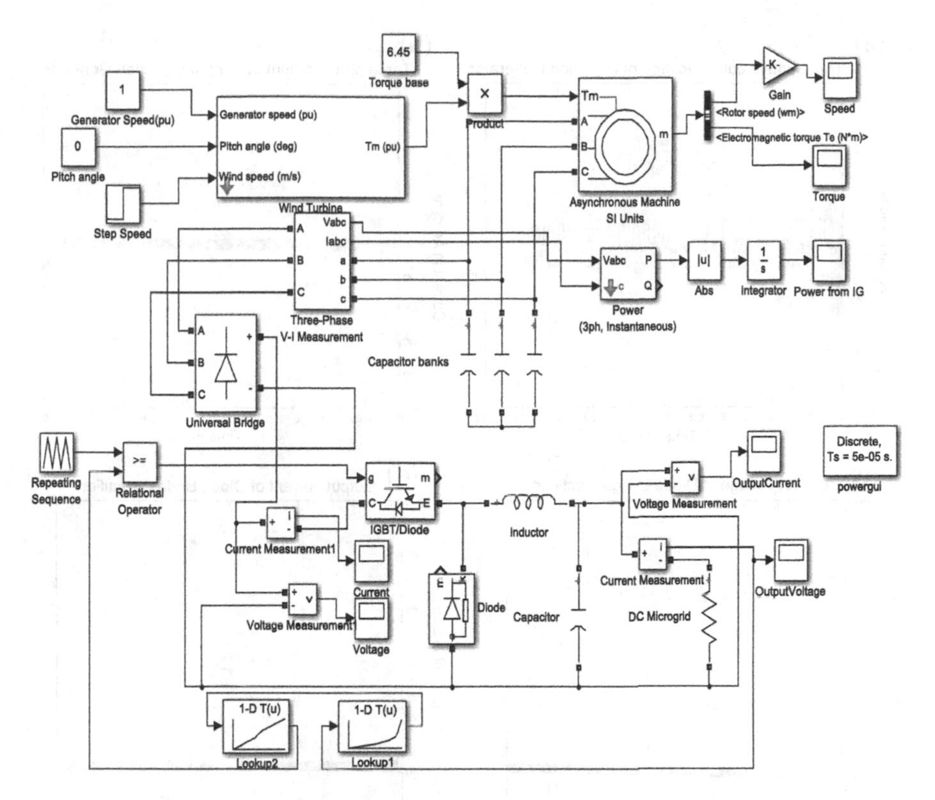

Fig. 4 Simulation diagram of wind-driven IG

The look-up table is used to develop the corresponding change in the current with a change in the duty ratio; if the current increases, then the duty ratio is decreased and vice versa. The first look-up table gives the value of the duty ratio by taking current as the input and the second look-up table gives the value of constant for generating the duty ratio by PWM by taking the duty ratio from the first look-up table as input. The system has been simulated for various load conditions and the corresponding changes in the current and power are noted. When the resistor (DC Microgrid) varies, the corresponding value of current also varies. When the resistor increases, the corresponding value of the current also increases in order to attain the maximum power with constant dc voltage. The experimental waveform of the wind-driven IG based on look-up table-based MPPT implementation is shown in Fig. 5.

B. *Maximum Power With Change In Wind Velocity*

The change in the wind velocity affects the maximum output from the converter. If we consider a step input for the wind velocity, it varies from 11 to 13 m/s with a step time of 5 s in a simulation period of 10 s which is shown in Fig. 6. When the velocity of the wind changes at time 5 s, there is a corresponding change in the

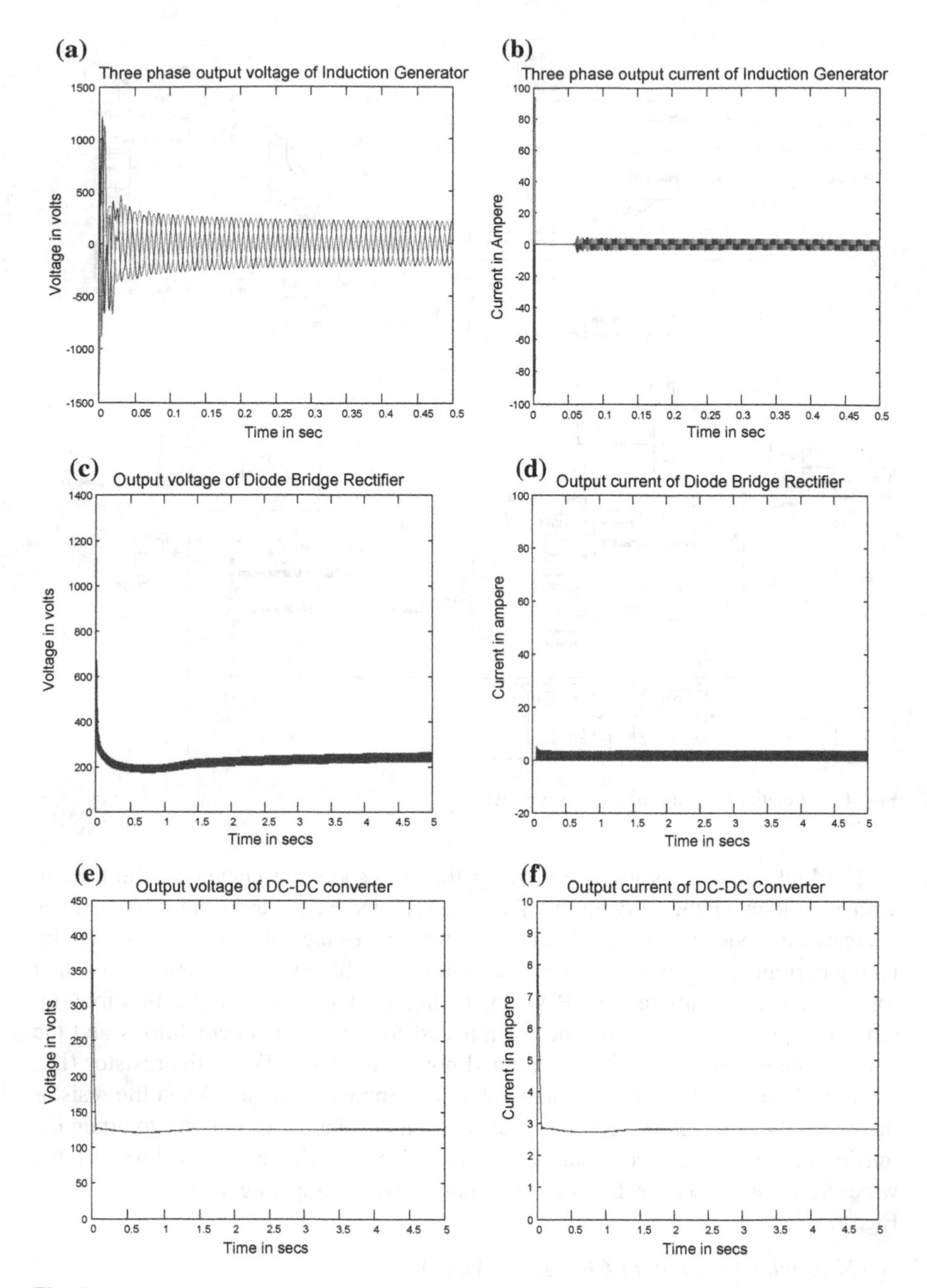

Fig. 5 Experimental waveform of look-up table-based WECS. **a** Three-phase output voltage of IG. **b** Three-phase output current of IG. **c** Output voltage of DBR. **d** Output current of DBR. **e** Output voltage of dc–dc converter. **f** Output current of dc–dc converter

output power of the dc–dc converter but takes time to reach the corresponding power value. Thus by varying the velocity of the wind the value of maximum power is obtained by the corresponding change in the output power. Thus in the proposed MPPT control strategy the look-up table-based technique has been implemented to attain the maximum power from the WECS. The change in the velocity of the wind affects the output power of the system.

The proposed MPPT control strategy for the wind-driven induction generator is simulated in MATLAB and the results are obtained. The results show that the proposed model is best suited for low-cost applications for determining the maximum power and with simple circuit algorithm. The time taken is also effectively reduced by the look-up tables.

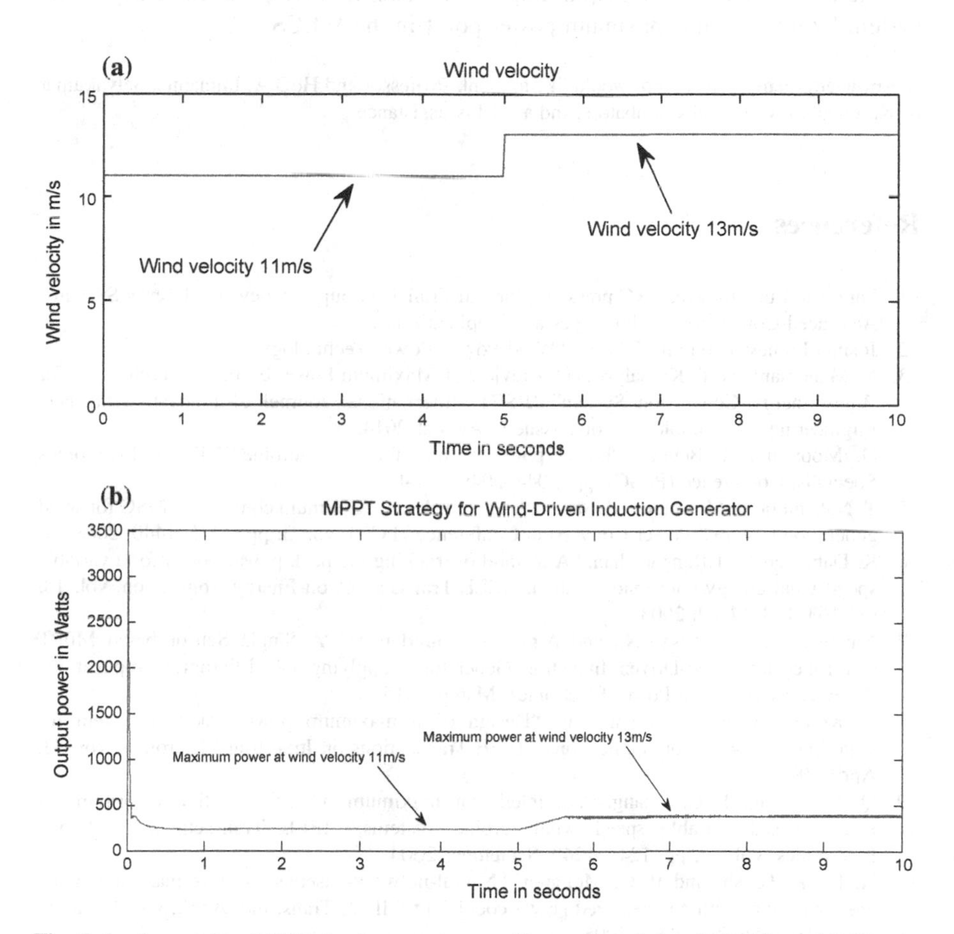

Fig. 6 Look-up table MPPT implementation. **a** Change in wind velocity from 11–13 m/s. **b** Change in maximum power from 11–13 m/s

5 Conclusion

The proposed MPPT algorithm and topology of the circuit for dc microgrid application is supplied for a small-scale WECS using look-up tables. The dc grid current is the control variable and the change in the current changes the duty ratio of the dc–dc converter to track the maximum power point. The logic of decreasing the duty ratio with the increasing value of the dc grid current for MPPT is done with the help of look-up tables which thereby simplify the complication of the algorithm. Without the need for the measurement of wind velocity or turbine rotor speed, i.e., with no mechanical sensors, the MPPT algorithm proposed is simple and effective to implement.

The look-up tables used help in time saving and improving the efficiency of the system by tracking the maximum power point in the WECS.

Acknowledgments The author would like to thank Professor and HoD A. Immanuel Selvakumar of Karunya University in Coimbatore, India for his assistance.

References

1. Fang Lin Luo Hongye, CRC press, Taylor and Francis Group, "Renewable Energy Systems: Advanced Conversion Technologies and Applications".
2. Joshua Earnest Economy Edition, "Wind Power Power Technology".
3. N. Manomani and P. Kausalyadevi "A review of Maximum Power Extraction Techniques for Wind Energy Conversion System" IJISET- International Journel of Innovative Science, Engineering & Technology, Vol 1 Issue ^, August 2014.
4. G. Moor and H. Beukes, "Power point trackers for wind turbines," Power Electronics Specialist Conference (PESC), pp. 2044–2049, 2004.
5. T. Nakamura, S. Morimoto, M. Sanada, and Y. Takeda, "Optimum control of PMSG for wind generation system," Power Conversion Conference (PCC), vol. 3, pp. 1435–1440, 2008.
6. R. Datta and V.-T.Ranganathan, "A method of tracking the peak power points for a variable speed wind energy conversion system," IEEE Transactions on Energy Conversion, vol. 18, pp. 163–168, March 2003.
7. Nayanar V., Kumaresan N., and Ammasai Gounden N. "A Single Sensor based MPPT Controller for Wind-Driven Induction Generators supplying DC Microgrid Application" IEEE Transactions on Power Electronics, March 2015.
8. E. Koutroulis and K. Kalaitzakis, "Design of a maximum power tracking system for wind-energy-conversion applications," IEEE Transactions on Industrial Electronics, vol. 53, April 2006.
9. Q. Wang and L.-C. Chang, "An intelligent maximum power extraction algorithm for inverter-based variable speed wind turbine systems," IEEE Transactions on Power Electronics, vol. 19, pp. 1242–1249, September 2004.
10. H. Li, K. L. Shi and P. G. McLaren, "Neural-network-based sensorless maximum wind energy capture with compensated power coefficient," IEEE Trans. Ind. Appl., vol. 41, no. 6, pp. 1548–1556, Nov./Dec. 2005.
11. Galdi V., Piccolo A. and Siano P., "Designing an adaptive fuzzy controller for maximum wind energy extraction," IEEE Transactions on Energy Conversion, vol. 2, pp. 559–569, 2008.

12. H. M. Boulouiha, A. Allali, A. Tahri, A. Draou, and M. Denai, "A simple maximum power point tracking based control strategy applied to a variable speed squirrel cage induction generator," *J. Renew. and Sustain. Energy*, 4, 053124, Oct. 2012.
13. S. Senthil Kumar, N. Kumaresan and M. Subbiah "Analysis and Control of Capacitor-Excited Induction Generators Connected to a Micro-Grid through Power Electronic Converters" *Accepted for publication in IET Gener. Transm. Distrib.* (doi:10.1049/iet-gtd.2014.0529).
14. S. Senthil Kumar, N. Kumaresan, N. Ammasai Gounden, and N. Rakesh, "Analysis and control of wind-driven self-excited induction generators connected to the grid through power converters," *Front. Energy*, vol. 6, no. 4, pp. 403–412, 2012.

Author Biography

Antony Priya Varshini J was born on March 23, 1992 in Tamil Nadu, India. She obtained her BE degree from SCAD Engineering College, Tirunelveli India (Anna University) in 2014 and the MTech degree in Power Electronics and Drives from Karunya University in 2016.

Interval Valued Hesitant Fuzzy Soft Sets and Its Application in Stock Market Analysis

T.R. Sooraj and B.K. Tripathy

Abstract Molodtsov introduced soft set theory in 1999 to handle uncertainty. It has been found that hybrid models are more useful than that of individual components. Yang et al. introduced the concept of interval valued fuzzy soft set (IVFSS) by combining the interval valued fuzzy sets (IVFS) and soft set model. In this paper we extend it by introducing interval valued hesitant fuzzy soft sets (IVHFSS) through the membership function approach introduced by Tripathy et al. in 2015. To illustrate the application of the new model, we provide a decision making algorithm and use it in stock market analysis.

Keywords Soft sets · Fuzzy sets · Fuzzy soft sets · IVFS · IVFSS · Decision making

1 Introduction

The notion of fuzzy sets introduced by Zadeh [1] in 1965 is one of the most fruitful models of uncertainty and has been extensively used in real life applications. In order to bring topological flavor into the models of uncertainty and associate family of subsets of a universe to parameters, Molodtsov [2] introduced the concept of soft sets in 1999. A soft set is a parameterized family of subsets. Many operations on soft sets were introduced by Maji et al. [3, 4]. Hybrid models are obtained by suitably combining individual models of uncertainty have been found to be more efficient than their components. Several such hybrid models exist in literature. Tripathy and Arun [5] defined soft sets through their characteristic functions. Similarly, defining membership function for FSSs will systematize many operations

T.R. Sooraj (✉) · B.K. Tripathy
School of Computer Science and Engineering, VIT University,
Vellore, Tamil Nadu, India
e-mail: soorajtr19@gmail.com

B.K. Tripathy
e-mail: tripathybk@vit.ac.in

S.S. Dash et al. (eds.), *Artificial Intelligence and Evolutionary Computations in Engineering Systems*, Advances in Intelligent Systems and Computing 517,
DOI 10.1007/978-981-10-3174-8_63

defined upon them as done in [6]. Many of soft set applications have been discussed by Molodtsov in [2]. An application of soft sets in decision making problems is discussed in [3]. Among several approaches, in [7], FSS and operations on it are defined. This study was further extended to the context of fuzzy soft sets by Tripathy et al. in [6], where they identified some drawbacks in [3] and took care of these drawbacks while introducing an algorithm for decision making. The concept of interval valued fuzzy sets is introduced by Zadeh. Later on, Yang et al. [8] introduced the concept of Interval valued hesitant fuzzy soft sets (IVHFSS) by combining the interval valued fuzzy sets (IVIFS) and soft sets. The hesitant fuzzy sets introduced by Torra [9] are an extension of fuzzy sets, where the membership of an element is a set of values. It is sometimes difficult to determine the membership of an element into a set and in some circumstances this difficulty is caused by a doubt between a few different values. Some operations on hesitant fuzzy sets are defined in [10]. He also discussed an application of decision making. The concept of hesitant fuzzy soft sets (HFSS) was introduced by Sunil et al. They also discussed an application of decision making. In this paper, we introduce interval valued hesitant fuzzy soft sets (IVHFSS). Applications of various hybrid models are discussed in [6, 11–20]. The major contribution in this paper is introducing a decision making algorithm which uses IVHFSS for decision making and we illustrate the suitability of this algorithm in real life situations. Also, it generalizes the algorithm introduced in [7] while keeping the authenticity intact. In this paper, we introduce the concept of IVHFSS with the help of membership function.

2 Definitions and Notions

In this section we present some of the definitions to be used in the paper. We assume that U is a universal set and E is a set of parameters defined over it.

Definition 2.1 A fuzzy set A is defined through a function μ_A called its membership function such that

$$\mu_A : U \to [0, 1].$$

Definition 2.2 A soft set over the soft universe (U, E) is denoted by (F, E), where

$$F : E \to P(U) \tag{1}$$

Here $P(U)$ is the power set of U.

Let (F, E) be a soft set over (U, E). Then in [6] it was defined as a parametric family of characteristic functions $\chi_{(F,E)} = \{\chi^a_{(F,E)} | a \in E\}$ of (F, E) as defined below.

Definition 2.3 For any $a \in E$, we define the characteristic function $\chi^a_{(F,E)} : U \to \{0, 1\}$ such that

$$\chi^{a}_{(F,E)}(x) = \begin{cases} 1, & \text{if } x \in F(a); \\ 0, & \text{otherwise.} \end{cases} \quad (2)$$

Definition 2.4 An interval valued fuzzy set A over U is given by a pair of functions $\mu_A : U \to D([0,1])$, where D ([0, 1]) is the set of all closed sub intervals of [0, 1], such that $\mu_A(x) = [\mu_A^-(x), \mu_A^+(x)] \subset [0,1]$.

Definition 2.5 A tuple (F, E) is said to be an interval valued fuzzy soft set over (U, E) by a family of membership functions $\mu_{(F,E)} = \left\{\mu^{a}_{(F,E)} | a \in E\right\}$ such that $\forall x \in U$, $\mu^{a}_{(F,E)}(x) \in D[0,1]$, we represent $\mu^{a}_{(F,E)}(x) = \left[\mu^{a-}_{(F,E)}(x), \mu^{a+}_{(F,E)}(x)\right]$.

Definition 2.6 A hesitant fuzzy set on U is defined in terms of a function that returns a subset of [0, 1] when applied to U, i.e.,

$$T = \{\langle x, h(x)\rangle | x \in U\} \quad (3)$$

where $h(x)$ is a set of values in [0, 1] that denote the possible membership degrees of the element $x \in U$ to T.

Definition 2.7 A pair (F, E) is called a hesitant fuzzy soft set if $F : E \to \mathrm{HF}(U)$, where HF(U) denotes the set of all hesitant fuzzy subsets of U.

Definition 2.8 An interval valued hesitant fuzzy set, H on U is defined in terms of its membership function $\mu_H : U \to P(D[0,1])$ where $P(D[0,1])$ denotes the set of all subsets of the set of closed sub intervals of [0, 1].

3 Interval Valued Hesitant Fuzzy Soft Sets (IVHFSS)

In this section, we introduce the notion of IVHFSS.

Definition 3.1 A pair (F, E) is called an interval valued hesitant fuzzy soft set if $F : E \to \mathrm{IVHF}(U)$, where IVHF($U$) denotes the set of all interval valued hesitant fuzzy subsets of U.

An IVHFSS H on U is defined in terms of its membership function $\mu_H : E \to P(\mathrm{IVHFS})$, such that $\forall a \in E$ and $\forall x \in U$, $\mu^{a}_{H}(x) \in P([0,1])$.

Given three IVHFEs in an IVHFSS is represented by h, h_1, and h_2. Then we can define union and intersection operations as follows.

Definition 3.2 Let (*F*, *E*) and (*G*, *E*) be two IVHFSSs over a common universe (*U*, *E*). Then their union is the IVHFSS (*H*, *E*) such that $\forall a \in E$ and $\forall x \in U$, we have

$$\begin{aligned} h_1^a \cup h_2^a &= \{\max(\alpha_1, \alpha_2) | \alpha_1 \in h_1^a, \alpha_2 \in h_2^a\} \\ &= \left\{ \left([\max(\mu_{\alpha_1}^{e-}, \mu_{\alpha_2}^{e-}), \max(\mu_{\alpha_1}^{e+}, \mu_{\alpha_2}^{e+})], |\alpha_1 \in h_1^a, \alpha_2 \in h_2^a \right) \right\} \end{aligned} \tag{4}$$

where $\alpha_1 \in h_1^a, \alpha_2 \in h_2^a$. h_1^a *and* h_2^a denote the hesitant fuzzy sets.

Definition 3.3 Let (*F*, *E*) and (*G*, *E*) be two IVHFSSs over a common universe (*U*, *E*). Then their intersection is the IVHFSS (*H*, *E*) such that $\forall a \in E$ and $\forall x \in U$, we have

$$\begin{aligned} h_1^e \cap h_2^e &= \{\min(\alpha_1, \alpha_2) | \alpha_1 \in h_1, \alpha_2 \in h_2\} \\ &= \left\{ \left([\min(\mu_{\alpha_1}^{e-}, \mu_{\alpha_2}^{e-}), \min(\mu_{\alpha_1}^{e+}, \mu_{\alpha_2}^{e+})], |\alpha_1 \in h_1, \alpha_2 \in h_2 \right) \right\} \end{aligned} \tag{5}$$

Definition 3.4 The complement of IVHFSS (*F*, *E*), represented as $(F, E)^c$, is defined as

$$h^c = \left\{ \left(1 - \mu_{(F,E)}^{a-} \right) - \left(1 - \mu_{(F,E)}^{a+} \right) \right\} \tag{6}$$

Definition 3.5 An IVHFSS (*F*, *E*) is said to be a null IVHFSS if and only if it satisfies

$$\mu_{(F,E)}^a(x) = [0, 0] \tag{7}$$

Definition 3.6 An IVHFSS (*F*, *E*) is said to be an absolute IVHFSS if and only if it satisfies

$$\mu_{(F,E)}^a(x) = [1, 1] \tag{8}$$

4 Application of IVHFSS in Decision Making

Some of the applications of soft sets are discussed in [20]. Later on many researchers discussed application of hybrid soft set models in decision making problems. In [11], Tripathy et al. classified the parameters into to positive and negative parameters. Here we discuss an application of DM in IVHFSSs.

Algorithm

1. Input the parameter data table. The parameter data table contains all the information about the parameters.

2. Input the IVHFSS table
3. Repeat the following steps for pessimistic, optimistic and neutral cases

   a. Construct the priority table. It is obtained by multiplying the parameter values with the corresponding priority values.
   b. Construct the comparison table by finding the entries as differences of each row sum in priority table. Also rank the objects as per the score obtained.

4. Construct the decision table using the normalized score Eq. (9).

$$\text{Normalized Score} = \frac{2 * \left(|c| * |k| * |j| - \sum_{i=1, x \in K}^{i=|j} RC_{ix}\right)}{|k| * |j| * |c| * (|c| - 1)} \tag{9}$$

Here, 'c' is the number of objects, 'k' is the number of cases, 'j' is the number of judges.

4.1. Assign rankings to each candidate based upon the score obtained.
4.2. If there is more than one having same score than who has more score in a higher ranked parameter will get higher rank and the process will continue until each entry has a distinct rank.

4.1 Application in Stock Market Analysis

Consider the case of a stock market. Suppose a person wants to buy a stock from the available stocks. Before he selects a stock, he needs to consider the parameters of the stock. If he selects a stock which is underperforming, then it will be a big loses to him. So, the person should be careful of choosing the stocks. This can be done with the help of the parameters. Some of the parameters to be considered are Dividend, one year low and up, Exchange, Industry type, Base price etc. Here, we can take the parameter "Stock price" as a negative parameter. If the base price of stock is more, then the interest of the person to that purchase that stock will also decrease.

Let U be the universal set of stocks and E be the parameter sets. Then $U = \{s_1, s_2, s_3, s_4, s_5, s_6\}$ be the stocks and E = {Dividend, one year low and up, Exchange, Base price, Industry type, Industry owner}. For simplicity, we can take $E = \{e_1, e_2, e_3, e_4, e_5, e_6\}$. Here, the parameter "Base price" is the negative parameter because as

Table 1 Parameter data table

Parameter	e_1	e_2	e_3	e_4	e_5	e_6
Priority	0.2	0.4	0.2	−0.3	0.1	0
Parameter rank	3	1	3	2	5	6

Table 2 IVHFSS

U	e_1	e_2	e_3	e_4	e_5	e_6
C_1	0.1–0.5 0.1–0.3 0.2–0.5	0.2–0.3 0.2–0.4 0.1–0.3	0.2–0.3 0.3–0.5	0.8–0.8 0.8–0.9	0.4–0.5 0.5–0.7 0.4–0.7	0.6–0.9 0.5–0.6 0.5–0.7
C_2	0.8–1.0	0.4–0.8 0.5–0.7	0.7–0.9 0.6–0.9 0.6–0.8	0.2–0.5 0.2–0.4 0.3–0.4	0.7–1.0	0.5–0.5 0.5–0.6
C_3	0.1–0.5 0.1–0.3 0.2–0.5	0.5–0.8	0.7–0.9 0.7–0.8	0.7–0.8	0.8–1.0	0.5–0.7 0.6–0.7
C_4	0.5–0.9 0.6–0.8	0.6–0.8 0.6–0.7	0.5–0.9	0.8–1.0	0.5–0.9 0.6–0.8	0.7–0.8
C_5	0.1–0.2	0.1–0.4	0.9–1	0.3–0.6	0.1–0.5	0.8–1.0
C_6	0.9–1.0	0.7–0.9 0.7–0.8	0.6–0.7 0.5–0.7	0.1–0.3	0.2–0.4 0.3–0.4	0.3–0.7

Table 3 Pessimistic values

U	e_1	e_2	e_3	e_4	e_5	e_6
c_1	0.133333	0.166667	0.25	0.8	0.433333	0.5333333
c_2	0.8	0.45	0.633333	0.233333	0.7	0.5
c_3	0.133333	0.5	0.7	0.7	0.8	0.55
c_4	0.55	0.6	0.5	0.8	0.55	0.7
c_5	0.1	0.1	0.9	0.3	0.1	0.8
c_6	0.1	0.1	0.9	0.3	0.1	0.8

the price of stock increases, the customer's interest of selecting that stock will decrease.

The parameter data table contains all the information about the parameters and it is shown in Table 1. Based on the absolute value, we can rank the parameters.

Consider the interval valued hesitant fuzzy soft set (*F*, *E*), "Selection of stocks". This is given in the Table 2.

In the case of IVHFSS, we have to consider pessimistic case, optimistic and neutral cases. The pessimistic values are obtained by taking the average of the lower values in the hesitant elements. Optimistic values are obtained by taking the average of the upper interval values in the hesitant elements. Neutral values are obtained by taking the average of pessimistic and optimistic values. Tables 3, 4, and 5 show the pessimistic, optimistic and neutral values.

Since the parameter "Industry owner" is having zero priority, we can eliminate that parameter from further calculations. Then, we have to find priority table for each cases. Priority table is obtained by multiplying the priority values with the corresponding parameter values. The priority tables for pessimistic, optimistic and neutral cases are shown in Tables 6, 7, and 8.

Table 4 Optimistic values

U	e_1	e_2	e_3	e_4	e_5	e_6
c_1	0.433333	0.333333	0.4	0.85	0.633333	0.7333333
c_2	1	0.75	0.866667	0.433333	1	0.55
c_3	0.433333	0.8	0.85	0.8	1	0.7
c_4	0.85	0.75	0.9	1	0.85	0.8
c_5	0.2	0.4	1	0.6	0.5	1
c_6	1	0.85	0.7	0.3	0.4	0.7

Table 5 Neutral values

U	e_1	e_2	e_3	e_4	e_5	e_6
c_1	0.283333	0.25	0.325	0.825	0.533333	0.633333
c_2	0.9	0.6	0.75	0.333333	0.85	0.525
c_3	0.283333	0.65	0.775	0.75	0.9	0.625
c_4	0.7	0.675	0.7	0.9	0.7	0.75
c_5	0.15	0.25	0.95	0.45	0.3	0.9
c_6	0.55	0.475	0.8	0.3	0.25	0.75

Table 6 Priority table for pessimistic case

U	e_1	e_2	e_3	e_4	e_5	Row sum
c_1	0.026667	0.066667	0.05	−0.24	0.043333	−0.05333
c_2	0.16	0.18	0.126667	−0.07	0.07	0.466667
c_3	0.026667	0.2	0.14	−0.21	0.08	0.236667
c_4	0.11	0.24	0.1	−0.24	0.055	0.265
c_5	0.02	0.04	0.18	−0.09	0.01	0.16
c_6	0.02	0.04	0.18	−0.09	0.01	0.16

Table 7 Priority table for optimistic case

U	e_1	e_2	e_3	e_4	e_5	Row sum
c_1	0.086667	0.133333	0.08	−0.255	0.063333	0.108333
c_2	0.2	0.3	0.173333	−0.13	0.1	0.643333
c_3	0.086667	0.32	0.17	−0.24	0.1	0.436667
c_4	0.17	0.3	0.18	−0.3	0.085	0.435
c_5	0.04	0.16	0.2	−0.18	0.05	0.27
c_6	0.2	0.34	0.14	−0.09	0.04	0.63

Table 8 Priority table for neutral case

	e_1	e_2	e_3	e_4	e_5	Row sum
c_1	0.056667	0.1	0.065	−0.2475	0.053333	0.0275
c_2	0.18	0.24	0.15	−0.1	0.085	0.555
c_3	0.056667	0.26	0.155	−0.225	0.09	0.336667
c_4	0.14	0.27	0.14	−0.27	0.07	0.35
c_5	0.03	0.1	0.19	−0.135	0.03	0.215
c_6	0.11	0.19	0.16	−0.09	0.025	0.395

Table 9 Comparison table for pessimistic case

	c_1	c_2	c_3	c_4	c_5	c_6	Row sum	Rank
c_1	0	−0.52	−0.29	−0.31833	−0.21333	−0.21333	−1.555	6
c_2	0.52	0	0.23	0.201667	0.306667	0.306667	1.565	1
c_3	0.29	−0.23	0	−0.02833	0.076667	0.076667	0.185	3
c_4	0.31833	−0.20167	0.02833	0	0.105	0.105	0.354993	2
c_5	0.21333	−0.30667	−0.07667	−0.105	0	0	−0.275	4
c_6	0.21333	−0.30667	−0.07667	−0.105	0	0	−0.275	4

Table 10 Comparison table for optimistic case

	c_1	c_2	c_3	c_4	c_5	c_6	Row sum	Rank
c_1	0	−0.535	−0.32833	−0.32667	−0.16167	−0.52167	−1.87333	6
c_2	0.535	0	0.206667	0.208333	0.373333	0.013333	1.336667	1
c_3	0.32833	−0.20667	0	0.001667	0.166667	−0.19333	0.096663	3
c_4	0.32833	−0.20833	−0.00167	0	0.165	−0.195	0.08833	4
c_5	0.16167	−0.37333	−0.00167	−0.165	0	−0.36	−0.73833	5
c_6	0.52167	−0.01333	0.19333	0.195	0.36	0	1.256667	2

Table 11 Comparison table for neutral case

	c_1	c_2	c_3	c_4	c_5	c_6	Row sum	Rank
c_1	0	−0.5275	−0.30917	−0.3225	−0.1875	−0.3675	−1.71417	6
c_2	0.5275	0	0.218333	0.205	0.34	0.16	1.450833	1
c_3	0.30917	−0.21833	0	−0.01333	0.121667	−0.05833	0.140837	4
c_4	0.3225	−0.205	0.01333	0	0.135	−0.045	0.22083	3
c_5	0.1875	−0.34	−0.12167	−0.135	0	−0.18	−0.58917	5
c_6	0.3675	−0.16	0.05833	0.045	0.18	0	0.49083	2

Comparison tables are constructed by taking the entries as differences of each row sum in priority table values. The comparison tables for pessimistic, optimistic and neutral cases are shown in Tables 9, 10, and 11.

Table 12 Decision table

U	Pessimistic	Optimistic	Neutral	Normalized score	Rank
c_1	6	6	6	0.266667	6
c_2	1	1	1	0.377778	1
c_3	3	3	4	0.311111	4
c_4	2	4	3	0.333333	3
c_5	4	5	5	0.288889	5
c_6	4	2	2	0.355556	2

Then, construct the decision table with the help of the normalized score equation. Then rank the objects as per the score obtained. It is shown in Table 12.

From this table, we can see that the stock s_2 is the best stock to purchase.

5 Experimental Verification

The algorithm was coded by using Python in a laptop having Intel core i3 processor, 2 GB RAM, 320 GB HDD and 1.7 GHz clock speed. A synthetic data table having 300 candidates and 5 judges with 20 parameters was taken as input. The results are encouraging and we have not produced the tables because of space constraints and the complexity of the computation.

6 Conclusions

In this paper, we introduced IVHFSS with the help of membership functions. We also defined some of the operations of IVHFSS like union, intersection etc. The earlier decision making algorithms proposed by different researches through soft set model were faulty as discussed in [11]. In this paper, we proposed an algorithm which uses the concept of negative parameters [6]. Also, an application of this algorithm in solving a real life problem is demonstrated.

References

1. L.A. Zadeh, "Fuzzy sets", Information and Control, vol. 8, 1965, pp. 338–353.
2. D. Molodtsov, "Soft Set Theory—First Results", Computers and Mathematics with Applications, vol. 37, 1999, pp. 19–31.
3. P.K. Maji, R. Biswas, and A.R. Roy, "An Application of Soft Sets in a Decision Making Problem", Computers and Mathematics with Applications, vol 44, 2002, pp. 1007–1083.

4. P.K. Maji, R. Biswas, and A.R. Roy, "Soft Set Theory", Computers and Mathematics with Applications, vol. 45, 2003, pp. 555–562.
5. B.K. Tripathy and K.R. Arun, "A New Approach to Soft Sets, Soft Multisets and Their Properties", International Journal of Reasoning-based Intelligent Systems, Vol. 7, no. 3/4, 2015, pp. 244–253.
6. B.K. Tripathy, Sooraj, T.R., R.K. Mohanty, A new approach to fuzzy soft set and its application in decision making, Advances in intelligent systems and computing, vol. 411, 2016, pp. 305–313.
7. P.K. Maji, R. Biswas, and A.R. Roy, "Fuzzy Soft Sets", Journal of Fuzzy Mathematics, vol. 9 (3), 2001, pp. 589–602.
8. X.B. Yang, T.Y. Lin, J.Y. Yang, Y. Li and D. Yu, "Combination of interval-valued fuzzy set and soft set", Computers & Mathematics with Applications 58 (3)(2009) 521–527.
9. Torra V., Hesitant fuzzy sets and decision, International Journal of Intelligent Systems 25 (6), 2010, pp. 395–407.
10. Torra V., Narukawa Y., On hesitant fuzzy sets and decision, The 18th IEEE international conference on Fuzzy Systems, Jeju Island, Korea, 2009, pp. 1378–1382.
11. B.K. Tripathy, Sooraj T.R., R.K. Mohanty and K.R. Arun, "A New Approach to Intuitionistic Fuzzy Soft Set Theory and its Application in Decision Making", Advances in Intelligent Systems and Computing, 439, 2016, pp. 93–100.
12. Sooraj T.R., R.K. Mohanty and B.K. Tripathy, "Fuzzy Soft Set Theory and its Application in Group Decision Making", Advances in Intelligent Systems and Computing, 452, 2016, pp. 171–178.
13. B.K. Tripathy, R.K. Mohanty, Sooraj T.R. "A New Approach to Intuitionistic Fuzzy Soft Set Theory and its Application in Decision Making", Proceedings of ICICT 2015.
14. B.K. Tripathy, R.K. Mohanty, T.R. Sooraj, "On Intuitionistic Fuzzy Soft Sets and Their Application in Decision Making", Lecture Notes in Electrical Engineering, 396, 2016, pp. 67–73.
15. B.K. Tripathy, R.K. Mohanty, Sooraj, T.R., "On Intuitionistic Fuzzy Soft Set and its Application in Group Decision Making", In: Proceedings of the International Conference on Emerging Trends in Engineering, Technology and Science (ICETETS), Pudukkottai, 2016, pp. 1–5.
16. B.K. Tripathy, R.K. Mohanty, Sooraj T.R., A. Tripathy, "A modified representation of IFSS and its usage in GDM", Smart Innovation, Systems and Technologies, 50, pp. 365–375.
17. R.K. Mohanty, Sooraj, T.R., B.K. Tripathy, "An application of IVIFSS in medical diagnosis decision making", International Journal of Applied Engineering Research, vol. 10, pp. 85–93.
18. B.K. Tripathy, Sooraj, T.R., R.K. Mohanty, "A New Approach to Interval-valued Fuzzy Soft Sets and its Application in Decision Making", Advances in Intelligent Systems and Computing, 509, 2017, pp. 3–10.
19. R.K. Mohanty, Sooraj, T.R., B.K. Tripathy, "IVIFS and Decision making", Advances in Intelligent Systems and Computing, 468, 2017, pp. 319–330.
20. Tripathy, B.K., Sooraj, T.R., Mohanty,RK., "Advances Decision Making using Hybrid Soft set Models", International Journal of Pharmacy and Technology, 8(3), pp. 17694–17721.

A Novel Control Strategies for Improving the Performance of Reduced Switch Multilevel Inverter

K. Venkataramanan, B. Shanthi and T.S. Sivakumaran

Abstract Multilevel Inverter extensively used in medium and high voltage applications, because of their lower harmonic distortion, electromagnetic interference, and higher DC bus utilization. Still some demerits presence in MLIs is more number of power semiconductor devices. Due to this, complexity arises in switching strategies and voltage imbalance occurs across each level. In addition to the above, harmonic spectrum of MLIs depends on the switching strategy. In order to improve the performance of MLIs, novel hybrid switching strategies were developed and implemented in seven-level reduced switch inverter. This paper analysis the potential of above strategies interns of performance parameters like THD, V_{RMS}, and DF are compared with sub-harmonic pulse width modulation strategy. Simulations were carried out using MATLAB/SIMULINK.

Keywords Symmetrical · VFHBC · THD · V_{RMS} · DF

1 Introduction

In recent years, the advancement in power semiconductor devices in particularly MLI offers handing high voltage and power rating with minimum harmonic distortions. The conventional two-level inverter is quite hard to connect directly in higher voltage levels. A MLI is a system that synthesis an expected voltage from multistep, we can get better power quality with least harmonic distortions. The multistep produce higher power quality waveform, thereby voltage and

---

K. Venkataramanan (✉)
Department of E&I, Annamalai University, Chidambaram, Tamilnadu, India
e-mail: raceramana@gmail.com

B. Shanthi
Centralized Instrumentation and Service Laboratory, Annamalai University, Chidambaram, Tamilnadu, India

T.S. Sivakumaran
Arunai College of Engineering, Tiruvannamalai, Tamilnadu, India

S.S. Dash et al. (eds.), *Artificial Intelligence and Evolutionary Computations in Engineering Systems*, Advances in Intelligent Systems and Computing 517,
DOI 10.1007/978-981-10-3174-8_64

electromagnetic compatibility drastically reduced. Various topologies have been available for MLI: DCMLI, FCMLI, and CHBMLI; out of these CHBMLI are most trusted in industrial needs due to its simple construction and easy mode of controlling nature. However, complications are developed whenever move to higher number of levels, so researchers are keen to reduce a number of devices. Even though harmonic distortion is limited by higher levels, further reduction can be achieved through modulation strategies. From various strategies, carrier-based pulse width modulation is most fascinating, feasible, and flexible in controlling the firing angle of power semiconductor devices. Loh et al. [1] described the error reduction during phase synchronization in further sampling the error, which eliminates common mode voltage. To achieve optimum harmonic, phase shift the sampling instances is very necessary. Selective harmonic elimination becomes widely acceptable, which is alternative to PWM strategy. There are several solutions to control the triplen harmonics discussed in literature [2]. Ben-brahim and Tadakuma [3] proposed adding a bias to the reference voltage and switching patterns to improve the output with reduced switching losses. Lu and Corzine [4] made detailed analysis of CHMLI-based real and reactive power compensations by introducing advanced control techniques. Jeevananthan et al. [5] proposes inverted sine carrier based modulation strategies for PWM inverter. Jang-Hwan et al. [6] introduced carrier-based novel switching sequence for voltage source inverter. The developed new strategy used to modify the carriers or references directly to eliminate lower order harmonics [7]. Aghdam et al. [8] analyzed various multicarrier by applying in asymmetric multilevel inverter. The combination of fundamental switching with pulse width modulation makes a hybrid switching scheme for effective control of the multilevel inverter which prescribed in [9]. Sun [10] made a study over different cells operations when in different frequency condition to ensure the low switching losses. Seyezhai [11] proposed hybrid modulation technique for asymmetric inverter. Bahr Eldin et al. [12] introduced multicarrier PWM technique for converters. In literature [13], alternative switching sequence developed to reduce the total harmonic distortions, normally in space vector modulation the pivot vector utilized only one time but in this method it employs twice within the sub cycle. Dordevic et al. [14] made a comprehensive comparison if carrier-based modulation with space vector modulation for voltage source inverter.

2 Symmetrical Seven-Level Inverter

The chosen topology is a symmetrical topology since all input DC sources are equal. Figure 1 shows a configuration of single-phase symmetrical seven-level inverter with reduced switch count. The inverter is unique when compared to various topologies; it consists of three independent DC sources with same values and eight active switching elements with two diodes. Four active switches and two diodes are used for level generation and remaining four switches, which utilized for

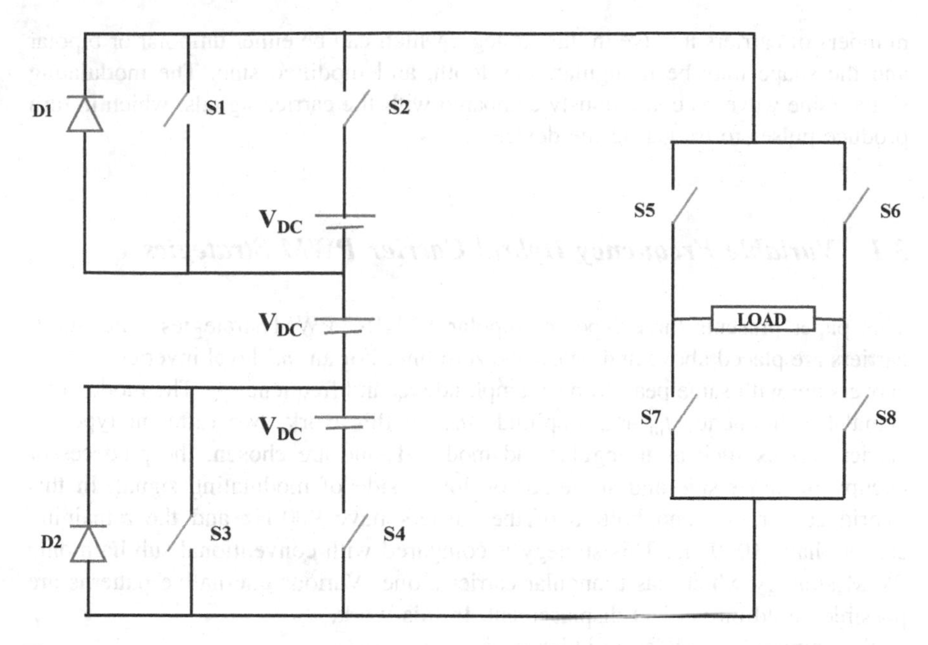

Fig. 1 Symmetrical seven-level inverter

Table 1 Switching sequence

Switches/levels	S1	S2	S3	S4	S5	S6	S7	S8
$3V_{DC}$	0	1	0	1	1	0	0	1
$2V_{DC}$	0	1	0	0	1	0	0	1
$1V_{DC}$	0	0	0	0	1	0	0	1
0	0	0	0	0	1	1	0	0
$-1V_{DC}$	0	0	0	0	0	1	1	0
$-2V_{DC}$	0	1	0	0	0	1	1	0
$-3V_{DC}$	0	1	0	1	0	1	1	0

polarity reversal. Moreover, further improvement to a high level is also easy. The switching patterns for seven-level inverter are given in Table 1.

The diodes concern only one diode by pass the conduction signal to get $\pm 2V_{DC}$ levels and both the diodes into the action for producing $\pm V_{DC}$ levels.

3 Modulation Strategies

Among the various strategies, carrier-based PWM is quite often used owing to their simplicity and flexibility in controlling action. Multicarrier-based PWM techniques emphasis lower harmonic distortion with high RMS output voltage. Multiple numbers of carriers involve in this strategy, which can be either unipolar or bipolar and the shape may be triangular, saw tooth, and modified sine. The modulating signal (sine wave) is continuously compared with the carrier signals, which in turn produce pulses to triggering the device.

3.1 Variable Frequency Hybrid Carrier PWM Strategies

This paper presents three types of bipolar VFHBC PWM strategies. The multi-carriers are placed above and below the zero line. For an 'm' level inverter $(m - 1)$ carriers are with same peak-to-peak amplitude a_c and frequency f_c. The modulating signal has frequency f_m and amplitude a_m. In this work, two different types of carrier signals such as triangular and modified sine are chosen, the predecessor occupy in upper side and successor on lower side of modulating signal. In this hybrid carrier, top and bottom of the carriers have 900 Hz and the remaining carriers have 1000 Hz. This strategy is compared with conventional sub-harmonic PWM strategy which has triangular carrier alone. Various alternative patterns are possible based on vertical displacement. In this work,

the amplitude modulation index m_a is

$$m_a = \frac{2A_m}{(m-1)A_c} \tag{1}$$

In further, the frequency modulation index is

$$m_f = \frac{f_c}{f_m} \tag{2}$$

3.1.1 Variable Frequency Hybrid Carrier-A (VFHBC-A)

The triangular and modified sine carriers of same amplitude are disposed such that bands they occupy are contiguous.

3.1.2 Variable Frequency Hybrid Carrier-B (VFHBC-B)

In this strategy, the triangular carriers and modified sine carriers have same amplitude which falls in the above and below the zero reference, respectively, are same in phase but they are 180° phase shifted with respect to each other.

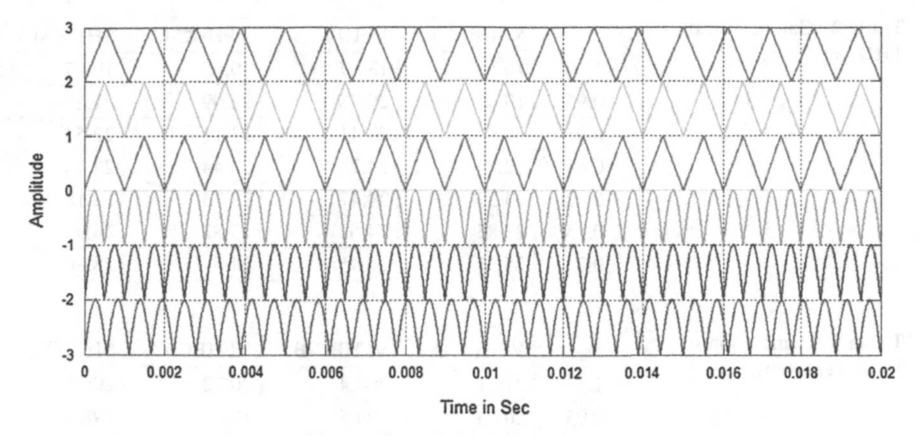

Fig. 2 Carrier arrangement for VFHBC-A

3.1.3 Variable Frequency Hybrid Carrier-C (VFHBC-C)

This strategy provides 180° displacement of all carriers with each other alternately.

The triangular and modified sine carrier's arrangement for VFHBC-A only displayed as sample in Fig. 2. The frequency modulation index of outer most triangular carrier is 18 whereas inner triangular carrier is 20 and outer most modified sine carrier is 36, remaining inner modified sine carrier as 40.

4 Simulation Results

The single-phase symmetrical seven-level inverter is modeled in MATLAB/SIMULINK using power system block set. Gate signals for this inverter using VFHBC strategies are simulated. Simulations were performed for different values of modulation index ranging from 0.7 to 1 and the corresponding %THD are observed using the FFT block and displayed in Table 2. Table 3 shows the fundamental RMS output voltage of inverter. Table 4 displays the corresponding percentage and distortion factor (DF) of the output voltage, respectively. The simulated output voltage with above strategies are displayed for only one sample value of $m_a = 0.8$. Figure 3 show the output voltage generated by VFHBC-A strategy and its FFT plot in Fig. 4. Figures 5 and 6 display the output voltage generated by VFHBC-B strategy and its FFT plot, respectively. Figures 7 and 8 show the output voltage generated by VFHBC-C strategy and its FFT plot, respectively. Figures 9 and 10 show the output voltage generated by sub-harmonic pulse width modulation strategy and its FFT plot, respectively. Figures 11 and 12 provide the graphical comparison of %THD and V_{RMS} with various m_a.

In common 18th harmonic energy is dominant in all VFHBC-A, B, C strategies. In addition to this 16th and 17th harmonic energy are dominant in VFHBC-A and

Table 2 Comparison of % THD versus m_a

m_a	VFHBC-A	VFHBC-B	VFHBC-C	SHPWM
1	16.88	18.37	19.65	19.77
0.95	19.05	20.78	22.09	21.25
0.9	21.27	22.41	23.17	22.54
0.85	22.73	24.21	23.44	23.15
0.8	23.53	24.47	23.73	24.18
0.75	23.85	24.6	23.82	26.93
0.7	25.51	26.45	24.5	28.80

Table 3 Comparison of V_{RMS} versus m_a

m_a	VFHBC-A	VFHBC-B	VFHBC-C	SHPWM
1	216.1	204.4	207.2	206.8
0.95	207.3	194.5	197	198.6
0.9	197.4	184.8	187.4	198.6
0.85	187.6	174.2	178	181.6
0.8	176.3	164.1	168.3	175.5
0.75	166.8	154.6	159.1	163.3
0.7	154.8	143.5	150.3	149.7

Table 4 Comparison of % DF versus m_a

m_a	VFHBC-A	VFHBC-B	VFHBC-C	SHPWM
1	0.0065	0.0048	0.0057	0.0059
0.95	0.0050	0.0051	0.0073	0.0042
0.9	0.0033	0.0047	0.0073	0.0026
0.85	0.0016	0.0027	0.0045	0.0017
0.8	0.0019	0.0018	0.0032	0.0012
0.75	0.0022	0.0025	0.0023	0.0029
0.7	0.0033	0.0024	0.0026	0.0041

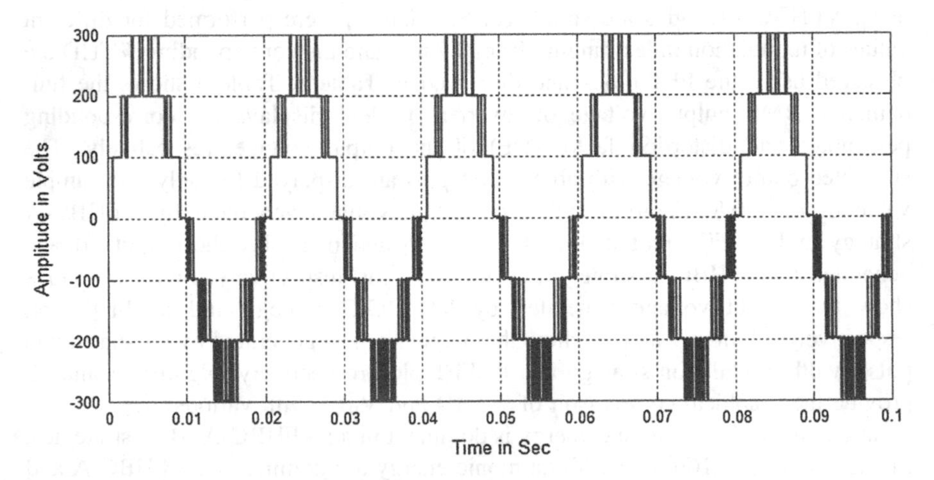

Fig. 3 Output voltage generated by VFHBC-A strategy

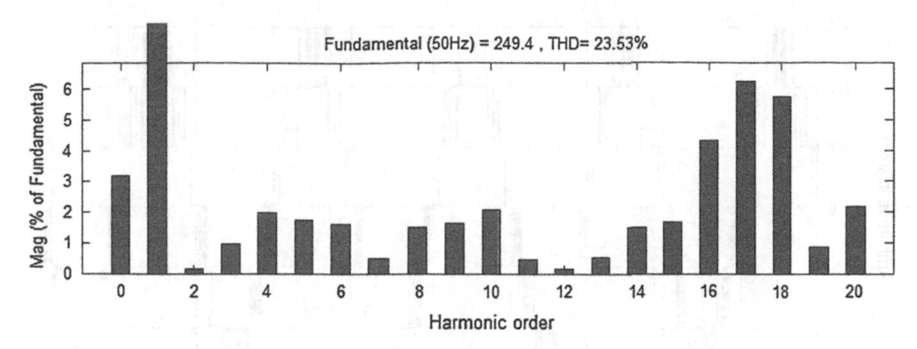

Fig. 4 FFT plot of output voltage generated by VFHBC-A strategy

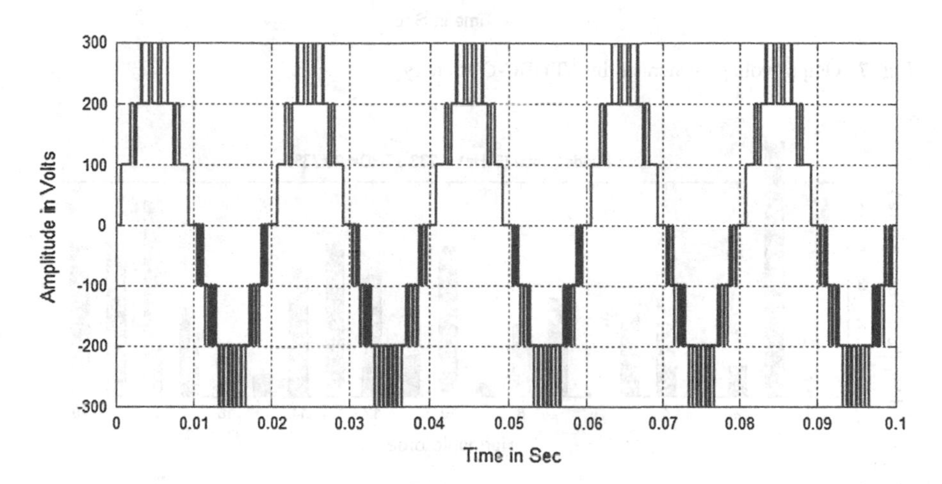

Fig. 5 Output voltage generated by VFHBC-B strategy

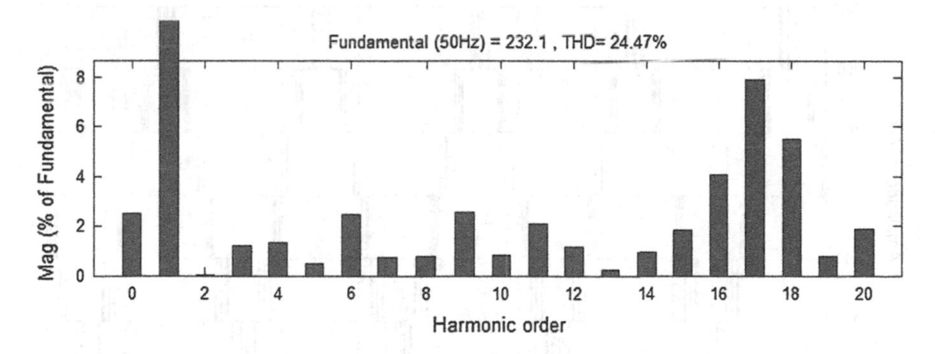

Fig. 6 FFT plot of output voltage generated by VFHBC-B strategy

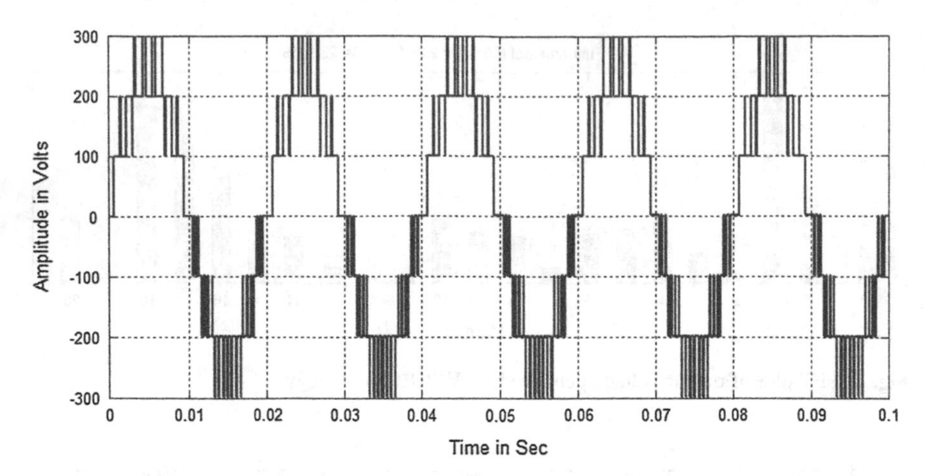

Fig. 7 Output voltage generated by VFHBC-C strategy

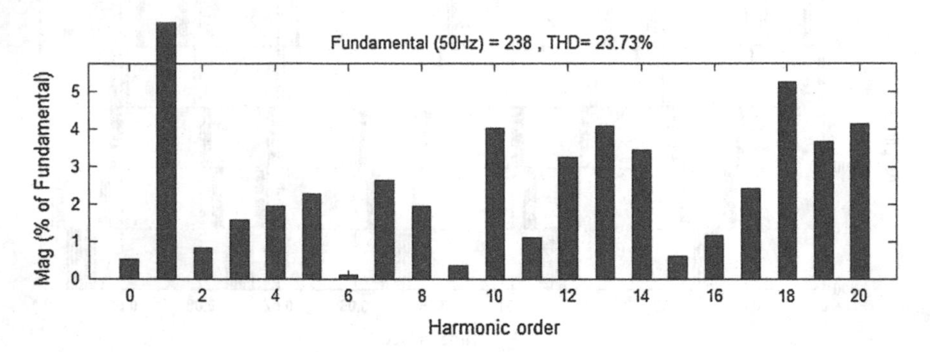

Fig. 8 FFT plot of output voltage generated by VFHBC-C strategy

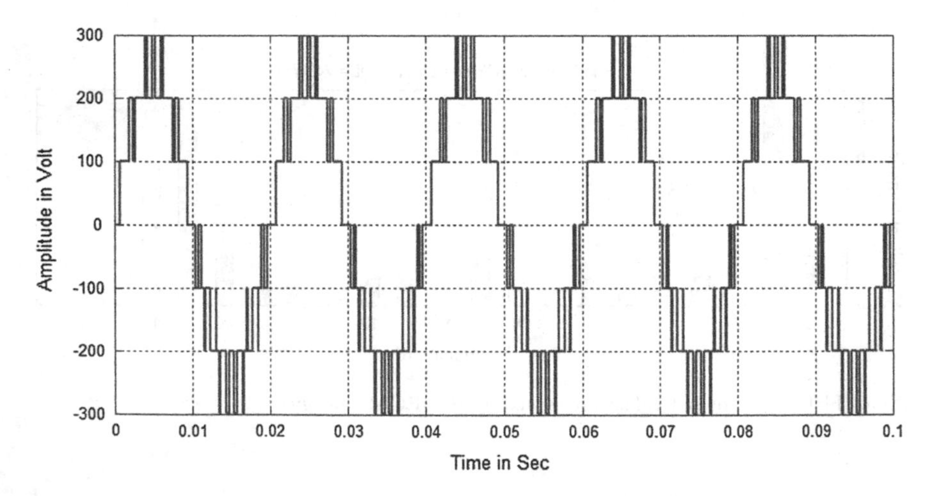

Fig. 9 Output voltage generated by SHPWM strategy

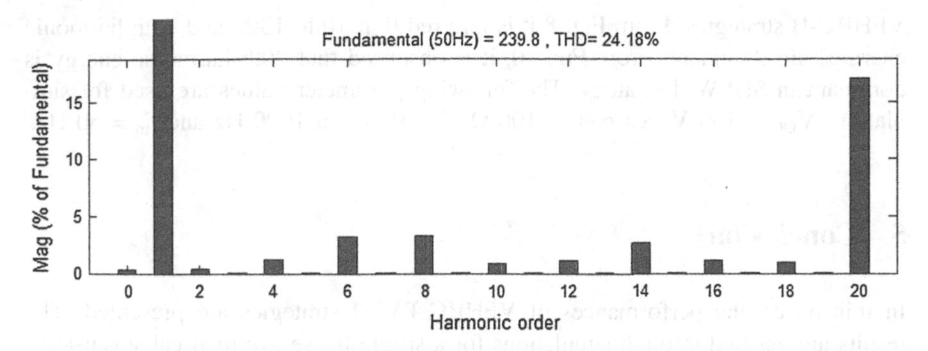

Fig. 10 FFT plot of output voltage generated by SHPWM strategy

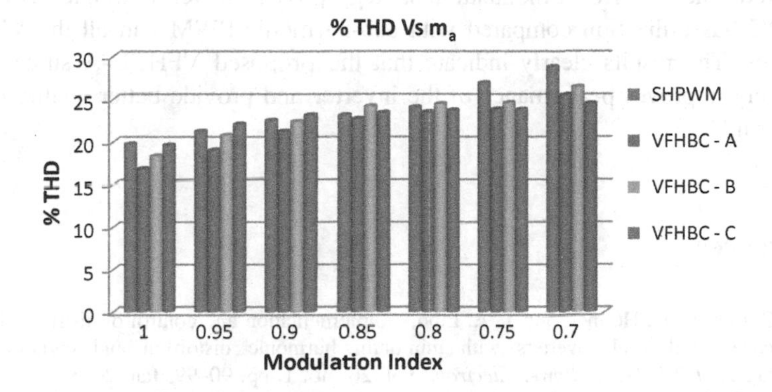

Fig. 11 %THD versus m_a for all strategies

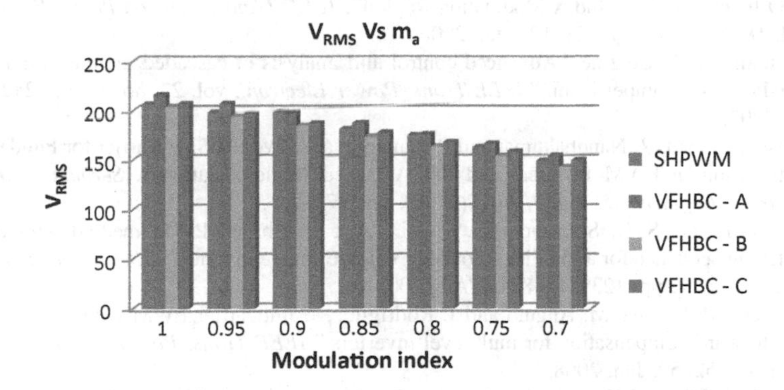

Fig. 12 V_{RMS} versus m_a for all strategies

VFHBC-B strategies. From Fig. 8 it is inferred that 10th, 13th, and 19th harmonic energies are dominant. From Fig. 10, it is observed that 20th harmonic energy is dominant in SHPWM strategy. The following parameter values are used for simulation: V_{DC} = 100 V, R (load) = 100 Ω, f_c = 900 and 1000 Hz and f_m = 50 Hz.

5 Conclusion

In this paper, the performances of VFHBC PWM strategies are presented. The results are verified through simulations for a single-phase symmetrical seven-level inverter in MATLAB/SIMULINK. The output quantities such as %THD, fundamental V_{RMS} and %DF are measured with all PWM schemes and the results are compared. The VFHBC-A modulation strategy provide lower harmonics as well as high DC bus utilization compared with sub-harmonic PWM and all the VFHBC strategies. The results clearly indicate that the proposed VFHBC-A strategy can effectively improves performance of the inverter and provide better quality output waveform.

References

1. P. C. Loh, D. G. Holmes, and T. A. Lipo, "Implementation and control of distributed PWM cascaded multilevel inverters with minimum harmonic distortion and common-mode voltages," *IEEE Trans. Power Electron.*, vol. 20, no. 1, pp. 90–99, Jan. 2005.
2. J. R. Wells, B. M. Nee, P. L. Chapman and P. T. Krein, "Selective harmonic control: A general problem formulation and selected solutions", *IEEE Trans. Power Electron.*, vol. 20, no. 6, pp. 1337–1345, Nov. 2005.
3. L. Ben-brahim and S. Tadakuma "Novel Multilevel Carrier Based PWM Control Method for GTO Inverter in Low Index Modulation Region", *IEEE Transactions on Power Electronics,* Vol. 18, No. 1, 42(1), 121–127, Jan 2006.
4. S. Lu and K. A. Corzine, "Advanced control and analysis of cascaded multilevel converters based on *P–Q* compensation," *IEEE Trans. Power Electron.*, vol. 22, no. 4, pp. 1242–1252, Jul. 2007.
5. S. Jeevananthan, R. Nandhakumar and P. Dananjayan,"Inverted Sine Carrier for Fundamental Fortification in PWM Inverters and FPGA Based Implementations", *Serbian Journal of Electrical Engineering, Vol. 4, No. 2(2007), pp. 171–187.*
6. K. Jang-Hwan, S.-K. Sul, and P.N. Enjeti, "A carrier-based PWM method with optimal switching sequence for a multilevel four-leg voltage source inverter," *IEEE Trans. Ind. Appl.*, Vol. 44, No. 4, pp. 1239–1248, Jul./Aug.2008.
7. S. Kouro, P. Lezana, M. Angulo and J. Rodriguez, "Multicarrier PWM with DC-link ripple feed forward compensation for multilevel inverters," *IEEE Trans. Power Electron.*, vol. 23, no. 1, pp. 52–59, Jan. 2008.
8. M.G.H. Aghdam, S.H. Fathi, B. Gharehpetian "Analysis of multicarrier PWM methods for asymmetric multilevel inverter", in *Proc. 3rd IEEE Conference on Industrial Electronics and Applications,* ICIEA'08, pp. 2057–2062, 2008.

9. Govindaraju C. and Baskaran K. "Optimized hybrid phase disposition PWM control method for multilevel inverter", *International Journal of Recent Trends in Engineering*, Vol. 1, No. 3, pp. 129–134, 2009.
10. X. Sun, "Hybrid Control Strategy for A Novel Asymmetrical Multilevel Inverter," *IEEE Transactions on Industrial Electronics pp. 5–8, 2010.*
11. *R. Seyezhai "Investigation of Performance Parameters For Asymmetric Multilevel Inverter" Using Hybrid Modulation Technique. International Journal of Engineering Science and Technology (IJEST) (2011)., 3(12), 8430–8443.*
12. Bahr Eldin, S Mohammed, KS RamaRao, "A New Multicarrier Based PWM for Multilevel Converter", *IEEE Applied Power Electronics Colloquium (IAPEC),* 63–69, 2011.
13. Soumitra Das and G. Narayanan, "Novel Switching Sequences for a Space-Vector-Modulated Three-Level Inverter," *IEEE Transactions on Industrial Electronics,* Vol. 59, No. 3, March 2012.
14. O. Dordevic, M. Jones and E. Levi, "A Comparison of Carrier-Based and Space Vector PWM Techniques for Three-Level Five-Phase Voltage Source Inverters," *IEEE Transactions on Industrial Informatics*, vol. 9, no. 2, pp. 609–619, 2013.

Implementation of Connected Dominating Set in Social Networks Using Mention Anomaly

S. Nivetha and V. Ceronmani Sharmila

Abstract Social network sites (SNS) are open platform that allows users to create an account in public sites. Social network site contains more active users, who share the posts, messages, or valuable information in real time. Even though social network provides timely updating still it is hard, because the current event mentions are overwhelmed by other mentions, which are totally irrelevant with the current affairs. The proposed methods mainly concentrate on text mentions in social networks and avoid the current affairs of the social aspect. Here two main methodologies are proposed that will improve the web content mining in social network sites. GBMAD (Geographically based mention anomaly detection), is a one of the best methods to manage dynamic links (i.e., Mentions: shares, likes, re-tweets, comments, etc.). GBMAD detects current events over the particular geographical location and it determines the importance of mentions over the crowd. The main advantage is it dynamically works in the real-time data rather than defined data. The next method is connected dominating set is used for identifying the closely related structure by using graph theory techniques. In the proposed work, the anomaly users are identified using Mention anomaly detection. In the future work, the implementation of Sybil attacks in the network will be identified and prevent from unsolicited friend requests in social networks using graph-based theory.

Keywords Social network · Connected dominating set · Anomaly detection · Geographical heuristic · Greedy algorithm

---

S. Nivetha (✉) · V. Ceronmani Sharmila
Department of Information Technology, Hindustan Institute of Technology and Science, Chennai, India
e-mail: nivetha2k5@gmail.com

V. Ceronmani Sharmila
e-mail: csharmila@hindustanuniv.ac.in

S.S. Dash et al. (eds.), *Artificial Intelligence and Evolutionary Computations in Engineering Systems*, Advances in Intelligent Systems and Computing 517,
DOI 10.1007/978-981-10-3174-8_65

1 Introduction

1.1 Social Network Sites

Social network sites become the most important and powerful medium to share information faster and easier throughout the globe. Social network sites work based on SaaS. Using SNS an individual can build a secure and closed network which can also be connected with the entire public network. The individuals who share the information can also view the data shared by others in the group. The privacy and security of social network sites are controlled by the user so the nature of the social network profile differs from one to another.

Social network sites are a public discourse, which was the most powerful tool in the 20th century to make an individual to connect with entire world easily. SNS is not a pure networking, SNS allows relation between unknown persons. Using social network sites, public can share ideas, interest, activities, and a lot more. Maintenance of social network sites is made by the site owners. The information collected by the social network is allowed to access by the other users on the same social network site and share with other sites. SNS allow users to share their digital data among the other users like friends and followers. It is always advised to avoid posting more sensitive and personal data on social network sites to avoid security and privacy issues. This ensures the privacy of the users among the social network. The most widely used social network sites as on date is Facebook and Twitter which hold billions of active user's data in dynamic networks. Social networking has become one of our modern lifestyle, which can easily spread the view of the common man toward the society, politics, sharing knowledge, and help to update every second. SNS helps to share their view in any format like text, image, video, or any other format.

1.2 Data Mining

Data Mining is a systematic search thoroughly on data, basically it contains a huge volume of data that contains information, so it also called as big data. Where searching of data is done with consistent patterns or by the systematic relation within the variables. Then to certify, find the detected patterns and that are applied into another fragment of data. The main purpose of data mining is predictive and prediction. The predictive data mining is the general data mining type that is used for most of the business applications. The steps of data mining having three different stages are: the initial exploration, pattern recognition, or model building with validation and/or verification, and deployment.

Data mining is the study of "Knowledge Discovery in Databases" method, or KDD. Data mining is also one of the fields in computer science, which are used to find or compute patterns for big data. It has several interdisciplinary, such as machine learning, database; cloud computing, artificial intelligence, and so on. The main objective of data mining technique is to analyze the data set and gathering information and pattern it for reuse. The KDD term is a misnomer, because it is not dealing with data alone, it also gains the knowledge and drawn out the patterns from the huge volume of data.

In the field of big data, data mining is a job, which is either in the way of fully automated and/or semi-automated. It helps to remove previously unknown, interesting patterns such as category of data records (cluster analysis), different records (anomaly detection), and dependencies (association rule mining). Here the patterns are used like an abstract for large volume of data, which are used for later analysis in machine learning and predictive analytics. For example, the knowledge mining procedure may recognize different categories in the data, it is used for the prediction that supporting to make clear decisions about the data preparation or data collection. The reporting and result translation are additional steps of the KDD process. The knowledge discovery in databases process is commonly defined with three stages: preprocessing, data mining, and results validation.

1.2.1 Preprocessing

A target set of data must be massed before using the data mining algorithms. The data mining discovers the patterns that are present in the data set. The expected data set is large sufficient to handle within the time limit. In data mining, the preprocessing is necessary to evaluate the multiple variables in data sets. The final data set should be clean, that is, the data set eliminates noise in the missing data.

1.2.2 Data Mining

Data mining having six tasks that imply in every data set. They are (1) anomaly detection (Outlier/change/deviation detection)—The identification of uncommon data records, that might be interesting or data errors that need further investigation. (2) Association rule learning (Dependency modeling)—Searches for relationships between variables. For instance, a supermarket might collect data on customer purchasing habits. Using association rule learning, the supermarket can decide which products are often bought together and use this information for marketing purposes. This is occasionally mentioned to as market basket analysis. (3) Clustering—is the function of finding groups and structures in the data that are in some way or another "similar," without using known structures in the data.

(4) Classification—is the task of generalizing recognized structure to apply to new data. For instance, an e-mail program might attempt to divide an e-mail as "legitimate" or as "spam." (5) Regression—attempts to discover a function which models the data with the least error. (6) Summarization—providing a more compact representation of the data set, incorporating visualization and report generation.

1.2.3 Result Validation

While in the process of data mining the data may corrupt accidently, when it happens to gain the significant results, but future prediction is impossible and it will not duplicate new sample on the data set. A straightforward form of this problem in machine learning is notable as over fitting, but the same problem can arise at contrary phases of the task and thus a train/test split—when applicable at all—may not be enough to stop this from happening.

1.3 Anomaly-Based Detection

Anomaly-based detection is a way to define the behavior of networks. In the data mining, the anomaly detection is the way of detecting an event, data, or observations which are not adapted to any anticipated pattern in a dataset. An anomaly is referred as noise, deviations, outliers, and exceptions, while the intrusion detection occurs in the network, which shows unexpected burst happen at the time analyzing frequently used data then rare data.

Anomaly detection is not having any standard principle for analyzing any data set. There are three categories of anomaly detection in data mining; they are supervised, unsupervised and semi-supervised. Using a supervised anomaly detection, the anomalous data are removed from the data set. It gives the result of accuracy, which are statistically more important.

The extensive obstacle in detecting anomalies is to predefine the set of rules for an analyzing. The performance depends on how to test the given data set. Some predefined rules are not valued by the testers. Even though it is not a valuable in testers its most need while handling difficult data set. The deep data set knowledge is about to know the behavior of the data in each and every condition. If the predefined rules are controlled the data, then it makes anomaly detection is simple things. If any sudden changes in the data set behavior, then the anomaly detection happened. But once the rules are defined and protocol is made then anomaly detection systems works well. The main advantage of anomaly detection is signature based engines.

Nowadays, the network anomaly detection is widely used in research and commercial industries, for network monitoring and network security. The most interested data in a network are captured to reduce heavy flow of data in the predefined capacity and flow of a network. In network security, the interest lies in defining known or unknown anomalous patterns of an attack or a virus.

1.4 Geographical Greedy Heuristic

A simple greedy algorithm can be progressed by gaining information from the profile of user's. Then the geographically greedy algorithm detects user as nodes for establishing a network depends on the geographical location of social network users. From the public profile, the country and the city information where filtered. The geographical greedy algorithm first looks at the geographic location information as a target node. In other words the non-geographical users if may not post the information regarding their location while posting posts on social networks that situation algorithm consider location as it from user profile. The geographical greedy algorithm collects all the user information from the targeted node and the chosen target location by more often used in the set. The dataset which has all information from the profile of users, like city, country, etc., is considered in the algorithm. In a social network, while posting an information user may enter just a city name as location, then the country name has been gathered from dataset by using an algorithm.

In the implementation of greedy algorithm in defining social network, it collects location information of the node and their neighbor nodes. The neighboring node also exists in the same geographical location, then select that node in the set for future establishment. Otherwise, check any other neighbor node with the same geographical location either as from the same city or country. If suppose the users having different geographical location, then go with maximum number of followers from the same location is selected.

1.5 Connected Dominating Set

A connected dominating set (CDS) S is a fragment of a graph $G = (V, E)$ in that S forms a dominating set when the nodes are connected with one another. If these dominating nodes are connected to form a graph, then the set is called connected dominating set. Then the CDS graph is another fragment of connected nodes S. So that each node in G is either in set S or adjacent to that set S. From the concept of spanning tree in a graph G, set S act as non-leaf nodes in that tree. In the set S, the nodes that are acting as dominating set are called as dominators and the nodes that are adjacent to a dominator are called dominatees (Fig. 1).

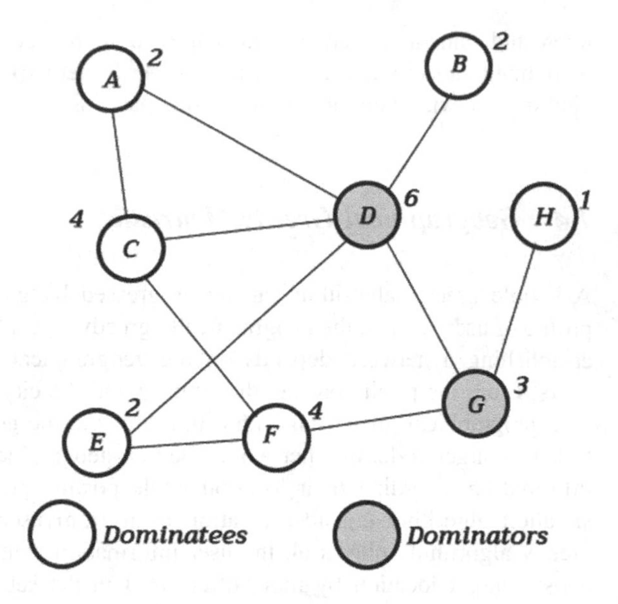

Fig. 1 Model of connected dominating set graph

2 Related Works

Toshimitsu Takahashi et al. [1] proposed a probability model of the mentioning behavior of a social network user and detected the emergence of a new topic from the anomalies, which was measured through the model. The proposed model was applied to four real data sets collected from Twitter. This approach gave better results when compared with text-anomaly-based approach.

Yuan Cheng et al. [2], proposed a novel user-to-user relationship-based access control (UURAC) model for online social network (OSN) systems that utilized regular expression notation for policy specialization. The system was validated by implementing a prototype system and the performance of the algorithm was evaluated.

Qiaozhu Mei and Cheng Xiang Zhai [3] proposed a text mining algorithm for studying a particular temporal text mining (TTM) task. The proposed approach was evaluated on two different domains such as news articles and literature, and the results proved that interesting evolutionary theme patterns are discovered effectively.

Zhi Yang et al. [4], proposed a vote trust model, which is a scalable defense system that leverages the user activities. This system models the friend invitation interactions as directed signed graphs and uses two key mechanisms for detecting Sybils. The results demonstrated that vote trust detected large-scale collusion among Sybils.

Donghyun Kim et al. [5], proposed a dominating set to identify an effective leader group of social network. The problem was formulated for finding a minimum sized leader group satisfying the three essential requirements such as influenceability, partisanship, and bipartisanship. The performance ratio achieved was $O(\ln \Delta)$, where Δ is the maximum node degree.

Ceronmani Sharmila and George [6] proposed four CDS algorithms for mobile ad hoc networks. The strategy chosen was used as a factor for constructing the

CDS. A strategic node can be a source, DNS, web proxy, internet provider, gateway or doorway, beacon, security key provider, repeater, rescuer, router, etc. The efficiency of the algorithm is evaluated using the parameters CDS node size, CDS edge size, and CDS circuit size. Simulation results show that the proposed algorithm gives the freedom to have any node as the starting node, without any significant change in the number of nodes and number of edges.

3 Proposed Method

3.1 Network Creation

Here first need to create an account in social networks. If the user does not have any account in social networks like Facebook, Twitter, etc., it means he/she needs to create the account and must register into the network. Once registered in the network, then it used to view the node (user/friends) information's in the network. If he/she is already member he/she may login into the network. View the information's like User id, location, friend list, followers list, etc., For analyzing the information's in the network.

3.2 Heuristic Preprocessing

After the network creation has been completed, here heuristic process is implemented. Categorization of users based on mentions or by geographic location. By identification of the individual users in the network apply the connected dominating set greedy algorithm for finding the maximal node identification.

Data preprocessing explains different types of processing methodologies that are carried out in the experimental dataset. It's mostly used as a primary level data mining process, data preprocessing qualifies the data into an appropriate form that will be more effectively adapted for the user. The different types of tools and techniques are used for preprocessing, including: (a) sampling—collection of a model data subset from a huge amount of data; (b) transformation—manage raw data to yield an input; (c) demising—dismiss unwanted noise from data; (d) normalization—gathers data for more desired effect; and (e) access—draw out mentioned data that is appropriate in some particular context (Fig. 2).

3.3 Fine-Grained Results

After the heuristic processing is completed, need to generate fine-grained results. From the algorithm results, we have to identify the maximal bounded node (User)

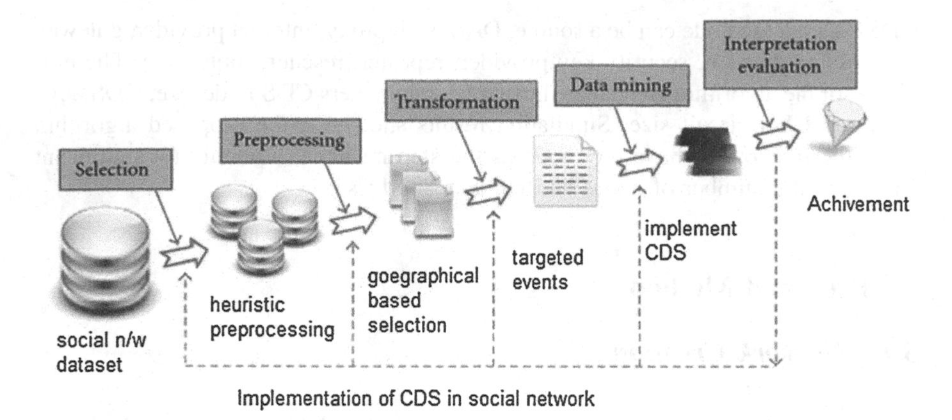

Fig. 2 Process for implementing of CDS in a social network Graph

in the network and filter fine-grained results. Here, the filter has been done by using some fixed parameters such as user's geographical location, area of interest, nativity, gender and so on. The geographical-based greedy heuristic algorithm are implemented in network for expansion. From the profile of users, the geographical location has been traced. Using geographical location and profile, the city and the country of the users are separately listed. By implementing algorithms in a defined network, first, it identifies at geographical location details of the user from their profile. If the user is not mentioned their location in profile by using preprocessing, it will remove by heuristic processes. At the same time if a user posts a message and did not mention their location, then it automatically considers the location as they entered in their profile. Before implementing CDS, the chosen nodes are fine-grained by analyzing the list of friends and followers in the network (Fig. 3).

If the targeted nodes are using the targeted location frequently, then consider that node as one of the CDS nodes in the network. From the dataset, the different cities and countries of users are identified and plotted as a separate table if one or more cities having same name then they are plotted under the same list of tables. If a user mention only a city in their post-algorithm identifies country by perused database.

3.4 Message Broadcasting

After identifying the geographical-based users, it has to identify the users who have the maximum number of followers in the network and broadcast the messages to that particular user. Through this multiple users can receive the message and identified by multiple users. Through this goal are achieved easily. A connected dominating set (CDS) act as a strength of the network it improves the routing methodology of the network. A CDS method minimizes the communication overhead, improve the bandwidth utilization, efficient energy and it also improves

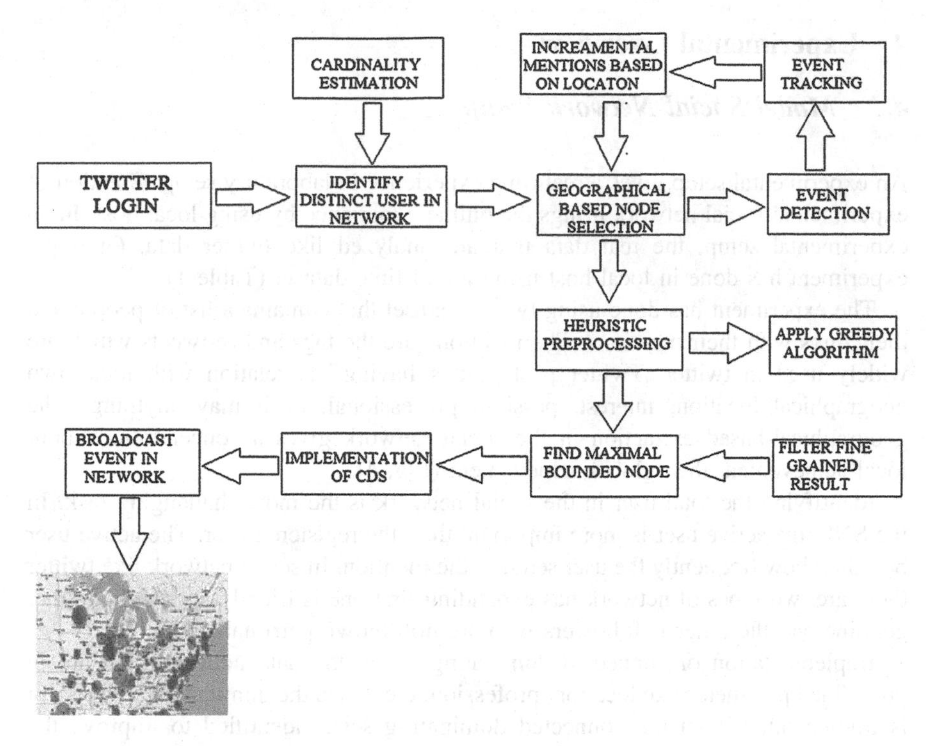

Fig. 3 System flow diagram

network effectiveness of a social network. The CDS nodes are called dominator (backbone node), else are called dominatees (non-backbone node). Using CDS, routing method is easy and it can adapt to network topology changes quickly.

Algorithm:

1. Select n number of nodes to populate set S

2. Another set C contain geographic player with the higher metric

3. For each node p in S and each node p1 not in S

   a. If C contains u

   b. Then set u as dominating node

4. If v is also a dominating node

5. Connect nodes u and v to form Connected Dominating Set

6. Select another node repeat from step 3

4 Experimental Overview

4.1 Model Social Network Setup

An experimental setup gives a real-time experience in laboratory setup. Creation of experimental social network brings the online experience by using local host. In an experimental setup, the real data that are analyzed like twitter data. Only the experiment has done in local host using a real-time dataset (Table 1).

The experiment has done using twitter dataset that contains a list of people and their links with their mentions. The mentions are the tags and re-tweets which are widely used in twitter. Twitter post that is having the relation with their own geographical location, interest, passion, professional, or it may anything. The geographical-based extraction in the social network gives a concentration about local achievement then global achievement (Fig. 4).

Identifying the total user in the social network is the most challenging task. In the SNS, the active user is more important than the registered user. The active user based on how frequently the user sending the mention. In social network like twitter there are two types of network has expanding first one is friend list, which is more genuine and the other is followers who are not known personally (Fig. 5).

Implementation of connected dominating set in the data nodes, which are in predefined parameter like location, profession, etc. Once the dominating set concept is implemented, then the connected dominating set is identified to improve the

Table 1 Fine-grained result of geographical-based model

For location in …			For other countries		
Aadhav	5922	322	Aahesh	533	377
Aadhesh	5651	392	Aahilyan	36,438	1941
Aadhi	5728	1531	Aahithyan	177	147
Aadhisan	4333	15	Aajayan	599	5634
Aadhithya	2500	593	Aajeevan	6042	177
Aadilen	216,920	11	Aajish	526	314
Aadish	85,112	67	Aakas	32,957	253
Aadithyavarman		522	Baabu	30,363	2585
Aadityaa	8883	524	Baalen	2621	849
Aadylen	42,894	12,768	Baarath	41,495	1901
Aageethan	4446	2895	Baarathy	5905	486
Aaharan	27,314	27,028	Babishan	540,683	253,891
Aahilyan	36,438	1941	Babu	756	511
Aajayan	599	5634	Babusankar	3498	244
Aajeevan	6042	177	Badmapriyan	10,181	6156
Aakas	32,957	253	Bageerathan	113,507	207
Babu	756	511	Baharani	100,935	103
Baheer	141,821	398	Bahilan	1,784,697	1945

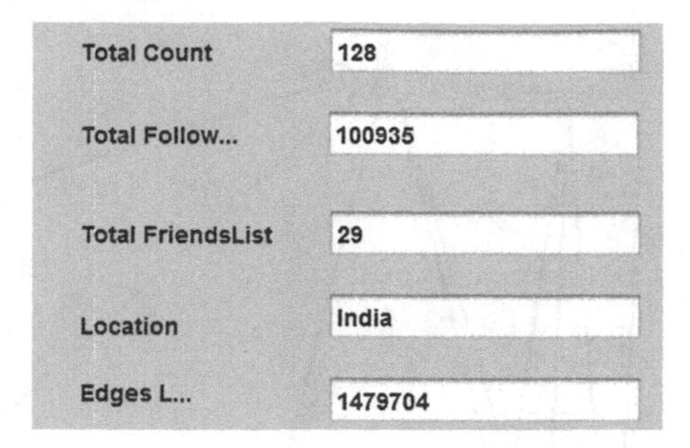

Fig. 4 Identification of user in the network

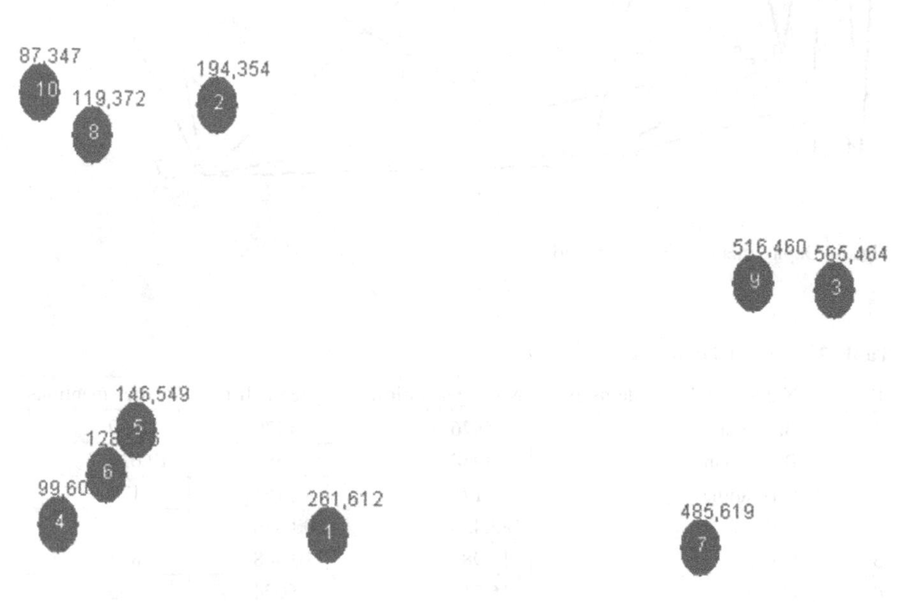

Fig. 5 Plotted social network based on location

performance of dynamic links. Deployment of CDS in social networks is to find the maximum number of linked users (Fig. 6).

The target in the network is to identify the maximum number of friends and followers from the sample twitter dataset. The nodes/user links in the twitter are analyzed to find out the maximum node identification, i.e., friends and followers of the users. Implementation of CDS in the network reduces routing time. Like that, in the social network like twitter identifying maximum friends and followers to reach maximum reachability of the post in the defined network. In a sample twitter dataset applying the concept of finding a maximum link user from the sample set.

List of CDS implied users in the predefined sample twitter dataset. The following table contains a list of users, who acting as a CDS node in the network with their node generation, friends list, and their twitter mentions. In a dynamic network, like social network, the node generation and the mentions are directly proportional

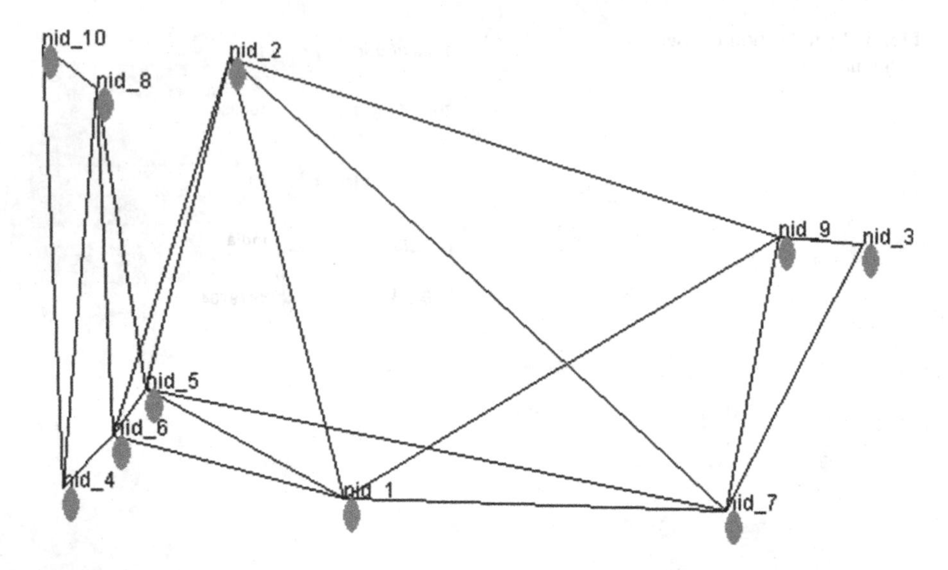

Fig. 6 Deployment of social network

Table 2 List of CDS nodes

Row	Name of CDS node users	Node generation	Friends list	Twitter mentions
1	Baheerathi	5876	1479	569
2	Baheethan	4958	569	19,616
3	Gaayantha	174	151	174
4	Gajaani	183,430	200,736	34
5	Gajanya	11,985	10,368	8833
6	Gajeeva	16,719	6024	793
7	Gana	21,266	23,212	183,430
8	Gananthini	5565	471	878
9	Ganga	367,128	146,125	88,333
10	Gangalakshmi	333,717	48,590	2068
11	Gangashri	879	1990	11,985
12	Gangesh	156,169	69,998	188
13	Gauratha	323	408	16,719
14	Gaurika	178	158	3060
15	Gaveetha	336	1	35,426
16	Saadhana	1944	2	21,266
17	Saakisri	21,963	18,988	48,188
18	Saakshi	12,586	518	14,926

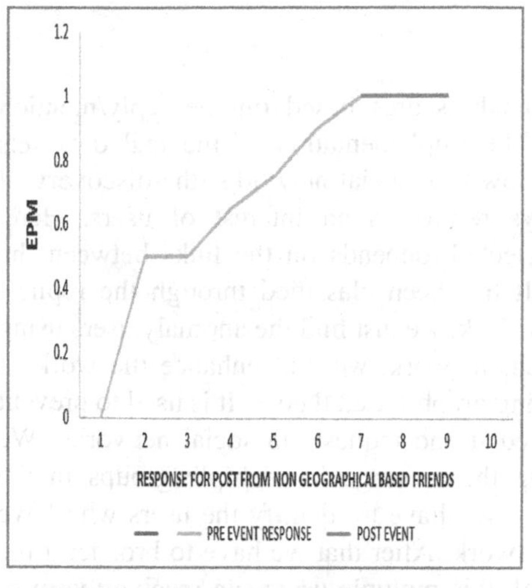

Graph based on non-geographical mentions

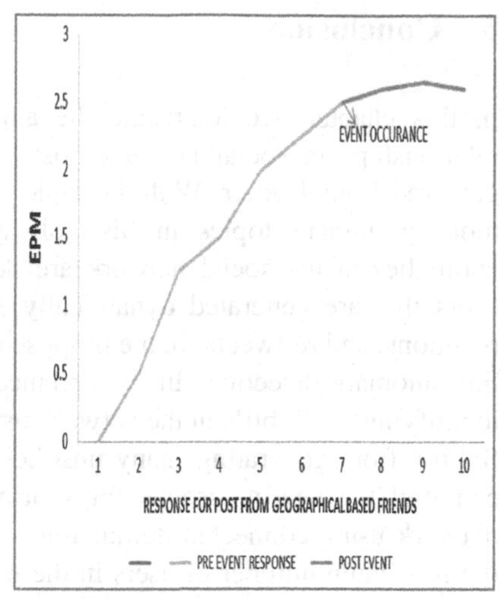

Graph based on geographical mentions

Fig. 7 Graph's for different types of munitions

to one another, but in some case even a small node can create most of mentions in a particular peak time. It depends on the users how actively using their social network (Table 2).

4.2 Result Analysis

Analyzing the above experimental method with various parameters shows how the real-time data works under the situation. The twitter data set is analyzed and the result are plotted depending upon the values which are resulted. The parameters used for plotting the graph is that number of mentions based on geographical friends versus event-post-mentions EPM. In that mentions have two types of values that are/pre-event value and post-event value (Fig. 7).

The mentions of both geographical and non-geographical-based friends are analyzed very carefully and plotted the graph based on the output. The value of mentions is taken as an exponential growth for both positive and negative mentions. The pre-event posts have exponential growth than post-event posts. By event success, growth always depends on the geographical location based friends mention rather than non-geographically based friends mention. The comparison of CDS network with the existing network shows that the CDS network taking lesser time when compared with existing networks without CDS. The above bar chart that have the systems like CDS network and existing network versus the time taken from the network to achieve the desire broadcasting of messages.

5 Conclusion

In this chapter, we aggregate the anomaly scores based on the reply/mention relationships in social network posts. The implementation of the real data sets gathered from Twitter. With the rapid growth of social networks, the discovery of more prominent topics in SNS always reinforces an interest of users. Here, anomalies in the social network are detected, depends on the links between the users that are generated dynamically. It has been classified through the replies, mentions, and re-tweets. In the proposed work, we just find the anomaly users using link anomaly detection. In the enhancement work, we will enhance the work of identifying the Sybil's in the network using graph-based theory. It is used to prevent Sybil's from generating many unsolicited friend requests in social networks. We can find the nearby groups for sharing the message in multiple groups in the network using connected dominating set. We have to identify the users who have the maximum number of users in the network. After that we have to broadcast the messages to that particular user. Through this multiple users can receive the message and identified by multiple users and we can easily achieve our goals.

References

1. Takahashi. T, et al. "Discovering Emerging Topics in Social Streams via Link-Anomaly Detection", IEEE Transactions on Knowledge and Data Engineering, 2014, 26(1), pp. 120–130.
2. Yuan Cheng, Jaehong Park and Ravi Sandhu, "An Access Control Model for Online Social Networks Using User-to-User Relationships", IEEE Transactions on Dependable and Secure Computing, 2015. (in press).
3. Q. Mei and C. Zhai, "Discovering Evolutionary Theme Patterns from Text: An Exploration of Temporal Text Mining," Proc. 11th ACM SIGKDD Int'l Conf. Knowledge Discovery in Data Mining, 2005, pp. 198–207.
4. Zhi Yang, et al. "VoteTrust: Leveraging Friend Invitation Graph to Defend Social Network Sybils", IEEE Transactions on Dependable and Secure Computing, 2015. (in press).
5. Kim, Donghyun, et al. "A Dominating Set Based Approach to identify Effective Leader Group of Social Network", Computing and Combinatorics. Springer Berlin Heidelberg, 2013, pp. 841–848.
6. V. Ceronmani Sharmila, A. George, "Construction of Strategic Connected Dominating Set for Mobile Ad Hoc Networks," Journal of Computer Science, Science Publications, 2014, 10(2), pp. 285–295.

Control of Induction Motor Using Artificial Neural Network

Abhishek Kumar, Rohit Singh, Chandan Singh Mahodi and Sarat Kumar Sahoo

Abstract The main objective of this paper is to design a controller for control of an Induction motor. In this paper, we have proposed v/f control of induction motor using artificial neural network, the network is trained using back propagation algorithm and Levenberg–Marquardt learning is used faster computation. The main approach is to keep voltage and frequency ratio constant to obtain constant flux over the entire range of operation and thus to have precise control of the machine. The effectiveness of the controller is demonstrated using MATLAB/Simulink simulation.

Keywords Induction motor · v/f control · Artificial neural network · Back propagation learning · Levenberg–marquardt algorithm

Abbreviations

ANN Artificial Neural Network
BP Back Propagation
PI Proportional-Integral
NN Neural Network
IM Induction motor
VSI Voltage Source Inverter
PWM Pulse Width Modulation
EMF Electromagnetic Force
RPM Rotation Per Minute

A. Kumar (✉) · R. Singh · C. Singh Mahodi · S. Kumar Sahoo
VIT University, Vellore, India
e-mail: abhishek_kumar.2013@vit.ac.in

R. Singh
e-mail: rohitsingh3594@gmail.com

C. Singh Mahodi
e-mail: csmahori@gmail.com

S. Kumar Sahoo
e-mail: sksahoo@vit.ac.in

S.S. Dash et al. (eds.), *Artificial Intelligence and Evolutionary Computations in Engineering Systems*, Advances in Intelligent Systems and Computing 517,
DOI 10.1007/978-981-10-3174-8_66

1 Introduction

Induction motors are the most preferred motors for any industrial or domestic purpose applications. For last few years, it was being used in the only applications which require a constant speed because of the expensive or inefficient speed control of the conventional methods. However, advancement in power electronic devices and converter technologies in the past few decades has improved the efficient speed control by varying the supply frequency, which made possible the various forms of adjustable speed induction motor drives. Along with that there has been a quite recognizable improvement in control methods and artificial intelligence which incorporates neural network, fuzzy logic, and genetic algorithm [1–4]. One of the major changes occurred in the context of induction motor control was the invention of field-oriented control (FOC). In this method, determining correctly the orientation of the rotor flux vector is mandatory, rather it leads to the poor response of the drive, but due to its complexity, it is not that feasible [1, 4–6]. Although, direct torque control (DTC) has a very good torque response without any complex orientation transformation and controls the inner loop current but still it possesses some drawbacks, like torque and flux ripple [7–9]. In this paper, scalar control is implemented where the motor is fed with variable frequency signals generated by the PWM control from an inverter where the v/f ratio is maintained constant so that constant torque can be retained over the entire operating range [3, 10]. There are a number of ways to implement scalar control. One common linear control strategy is proportional-integral (PI) control which is very easy to be implemented; although it cannot lead to good tracking and regulating performance simultaneously, also its control performance parameters sensitivity is more toward the load disturbances and system parameters variations [11]. Such controllers with fixed parameters cannot provide these requirements until the use of some unrealistically high gains. That's why the conventional constant gain controller used in the variable speed induction motor drives considered poor in practical applications due to the irregularities in the drive such as load disturbance, variations in mechanical parameter and unmodeled dynamics. For last few years, artificial neural network intelligent and fuzzy logic controllers have been gaining a great importance and proving their numerous advantages in many respects. That also makes artificial intelligent controller (AIC) the best controller for induction motor control. Usually, it becomes difficult to develop an accurate system mathematical model because of the unavoidable and obscure variations in the parameter. One of the major advantages of ANN-based technique is that there is no need of any mathematical modeling of the motor under consideration subsequently the time required for drive development can be substantially reduced. ANN possesses the ability to learn the desired mapping between the input and output signal for a system. The multilayer perceptron neural network is used for function approximation. The back propagation algorithm (including its variations) is the principal procedure for training multilayer perceptrons [1, 4–6, 12].

2 Induction Motor Drive

The drive system discussed in this paper consists of three-phase voltage source inverter (VSI), PWM generator, three-phase induction motor (IM), and artificial neural network controller (ANN controller) for the v/f control (Figs. 1 and 2).

The induction motor or asynchronous motor is well known for its commercial usage as it occupy less space than DC motor drives for any given rating and do not need maintenance although the fact that its control circuitry is costly, but cost is a changing factor, controller cost will reduce as it has reduced over period of time. The equivalent circuit of an Induction motor referring to the stator side of the Induction motor is shown in Fig. 3.

where

R_s = Resistance of the stator conductors
R_r = Resistance of the rotor conductor referred to the stator side
X_s = Leakage reactance of the stator conductor
X_m = Magnetizing reactance of the stator conductor
X_r = Leakage reactance referred to the rotor side of the stator side

Induction motor runs at a speed always less than the synchronous speed. This difference in speed is generally referred as the slip speed

$$\omega_{sl} = \omega_n - \omega_r \tag{1}$$

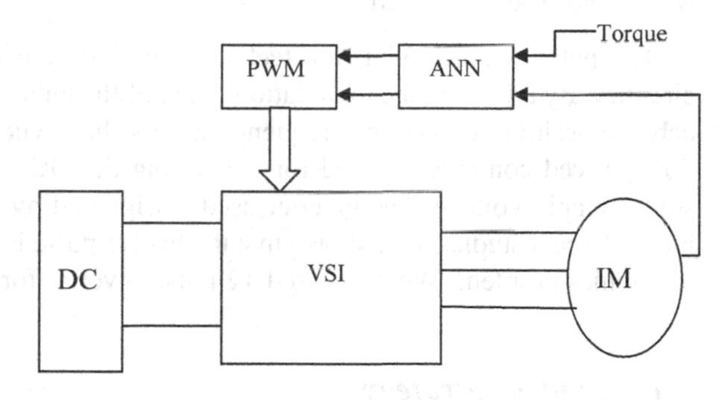

Fig. 1 Block diagram of open loop control of the Induction motor drive

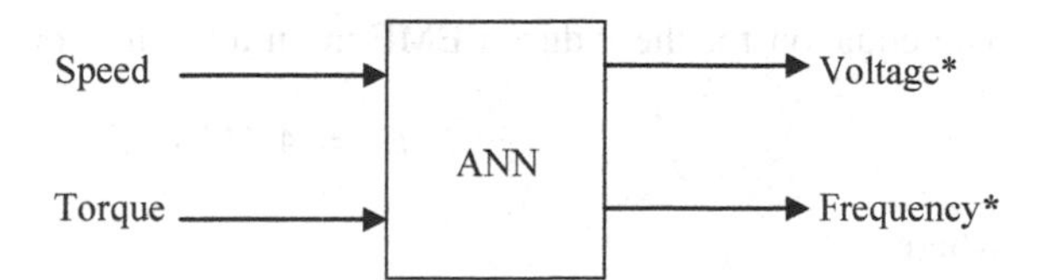

Fig. 2 Neural network to calculate the Voltage and frequency for various torque and rotor speed

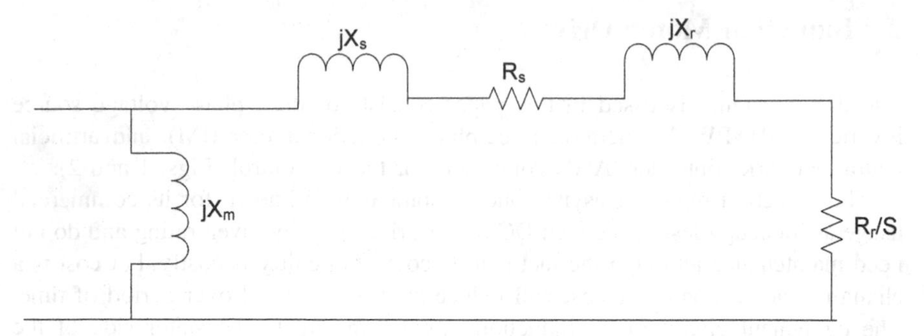

Fig. 3 Most simplified equivalent circuit of an Induction motor

where

ω_{sl} = Slip Speed in radian/second
ω_n = Synchronous speed in radian/second
ω_r = Rotor speed in radian/second

And the slip of the induction motor is given by the formula

$$\text{Slip} = \frac{N_n - N_r}{N_n} \quad (2)$$

where

N_n = Synchronous speed in RPM
N_r = Rotor speed in RPM

The pulse width modulation technique is widely used in sophisticated control circuitry. By adjusting the modulation index of the pulse width modulation block we achieve desired voltage and frequency across the asynchronous motor. Whereas a voltage feed converter is used for converting the DC voltage to three-phase sinusoidal supply voltage. The inverter used is triggered by a 12 pulse PWM generator block. Recent studies have shown that the twelve pulse inverter is less affected by the harmonics problem. We have used 12 pulse inverter for our IM drive.

2.1 Control Strategy

The equation for the induced EMF in an asynchronous motor is given by

$$E_{\text{in}} = 4.44N\psi_m f \quad (3)$$

where

E_{in} = Induced EMF
E_s = Terminal Voltage across the Induction motor

ψ_m = Flux developed by the stator current
f = Supply frequency

On neglecting the stator resistance drop and stator inductance drop in an asynchronous motor the supply voltage is equal to the induced EMF in the machine, and hence we have

$$E_s = 4.44N\psi_m f \tag{4}$$

Now for changing the speed of the machine we can either change the frequency of the machine or the supply but both of the methods results in the reduction of air gap flux moreover increasing voltage could also result in insulation failure which is not at all advisable. So for maintaining the air gap flux constant for a certain load voltage and frequency both need to be decreased such that their ratio remains constant as which could be clearly seen from Eq. (4). Therefore, for a maximum stator current that is driven for a machine the flux will remain constant.

$$\psi_m = \frac{E_s}{f}$$

The formula of the rotor speed is given by:

$$N_r = \frac{120f(1-S)}{P} \tag{5}$$

where

S = Slip
P = Number of poles

The torque developed by an asynchronous motor can be approximately written as

$$T_d = \frac{E_s^2 S}{\omega_n R_r} \tag{6}$$

where

T_d = Torque developed in the Induction motor

Now by maintaining the v/f (i.e., $v = E_s$) ratio constant we can maintain a constant developed torque for the asynchronous motor at variable slip.

3 Neural Network Controller

Artificial Intelligence techniques such as particle swarm optimization (PSO), fuzzy logic (FL), artificial neural network (ANN), and genetic algorithm (GA) have recently been implemented in many areas of electric drives for and control for better

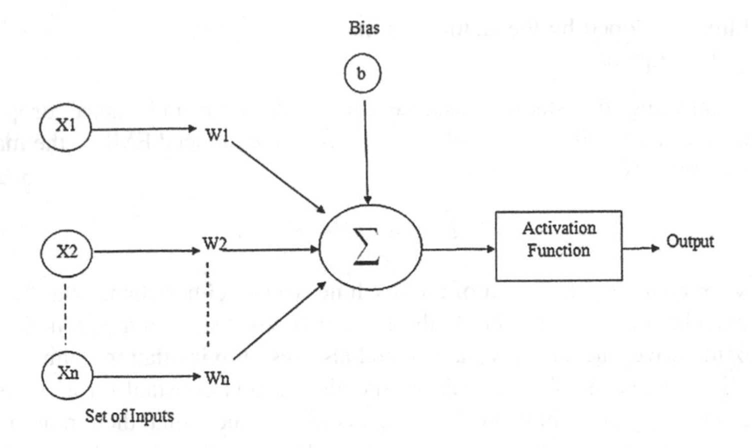

Fig. 4 ANN basic model

and accurate control of the electric machines and inverters. The main objective of artificial Intelligence techniques is to implement human brain in controller for better control of the system. Such system with embedded computational intelligence is often termed as an intelligent system.

ANN has been put in use since the early 1990s. Since then it has continuously attracted many researchers and scientist and has found its application in many areas. It is based on biological neuron model (Fig. 4).

It comprises of a number of artificial neurons that is interconnected together. Fundamentally, it has an Op-amp summer like structure. The artificial neuron is also called processing element. In neural network, all input signals flow through a weight (gain) to the processing unit of the networks. These weights can be positive or negative based on the neurons. These synaptic weights (W_{ik}) signifies the strength between the neuron k and neuron *i*. The summer accumulates all the input-weighted signal and weighted bias signal b, and then pass it through the transfer function also known as activation units. Activation units can be linear on nonlinear depending upon the neural network. Mainly three types of activation function are used (Table 1).

(i) Tan-sigmoidal
(ii) Log-sigmoidal
(iii) Threshold model
(iv) Linear model

3.1 Backpropagations Learning

The interconnections of artificial neurons results in neural network (NN). The Main aim of NN is to behave as a human brain in a certain domain to solve many real life

Table 1 Activation function

Name	Formula
Tan-sigmoidal	$F(x) = \tan h(x)$
Log-sigmoidal	$F(x) = (1 + \exp(-ax))^{-1}$
Threshold model	$F(x) = 1; x > 0$
Linear model	$F(x) = x$; for all x

problems. Structure of biological neural network is not well known; so many researchers have designed their own model to replicate the biological neuron system with high accuracy. Some of the commonly used models are listed below:

(1) Perceptron Model
(2) Adaline and Madaline
(3) Backpropagation (BP) Model
(4) Self-organizing Maps
(5) Boltzmann Model

We have used multilayered feedforward Backpropagation model to train our ANN controllers. It is mainly used in power electronics and motor control. BP-Model is a multilayered and feedforward model. It is most commonly used because it is one of the simple and the most general method to for supervised training. Basically, it is worked by approximating the nonlinear relationship between input and output by adjusting the synaptic weight.

Backpropagation model is divided in two parts: feedforward and backpropagations. At first, input signal is been applied and its response is been propagates, layer by layer through the network until an output is produced. Then the network output is compared with the expected output and the error is been computed. These errors are again feedback the system to adjust the weight of each layer as such to give the desired output (Fig. 5).

Mainly gradient descent (delta rule), Levenberg–Marquardt and Gauss–Newton method are used for minimizing the error function. In this paper, we have chosen Levenberg–Marquardt for the training of the dataset. The main reason behind choosing Levenberg–Marquardt algorithm is its fast convergence rate and stability. It is more robust and stable as compare to Gauss–Newton method (Fig. 6).

For our NN model, forward computation is been explained in the following steps:

- Using the given formula, calculate net values, slopes, and outputs for all neurons from layer = 1:

$$\text{net}_j^1 = \sum_{i=1}^{2} I_i w_{j,i}^1 + w_{j,0}^1 \tag{7}$$

$$x_j^1 = f_j^1\left(\text{net}_j^1\right) \tag{8}$$

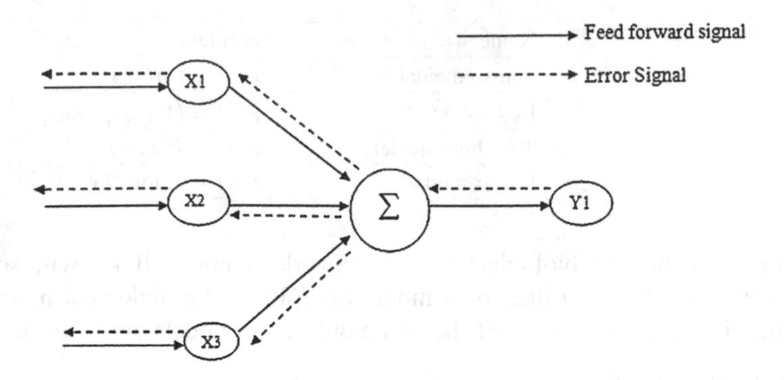

Fig. 5 Basic model of back propagation algorithm

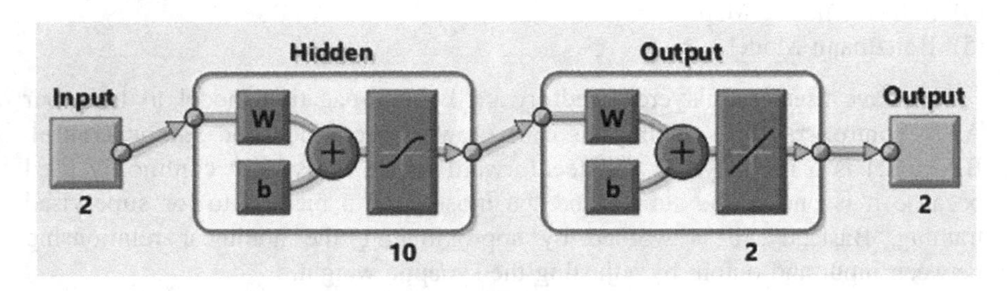

Fig. 6 Neural network model for control of induction motor

$$S_j^1 = \frac{\partial f_j^1}{\partial \text{net}_j^1} \tag{9}$$

where

I_i = Input the network 1
net_j^1 = output of the summer in layer 1
x_j = output of layer 1
S_j^1 = slopes of the activation unit in layer1
j = index of the neuron in first layer

- Using the output of layer 1 calculate the net values, slope and the output of layer 2 and 3 (i.e., hidden layer and output layer)

After computing the output value and slope of the activation units for all three layer, node matrix x and slope matrix s can be formed. Modeling of matrix x and s represent the end of the forward computation. Here, for a given output j, we can use the result obtained from the forward computation in backward computation j

- Calculate the error at output neuron j and initialize δ as the slope of output of neuron j:

$$e_j = d_j - o_j \tag{10}$$

$$\delta^3_{j,j} = s^3_j$$

$$\delta^3_{j,k} = 0$$

where

d_j = Target output at neuron j
o_j = Output at neuron j obtained in the forward computation
$\delta^3_{j,j}$ = Back propagation between neuron j and neuron j(Self Back propagation)
$\delta^3_{j,k}$ = Back propagation between neuron j and neuron k

- Back propagate δ from the inputs of nth layer to the output of n – 1st layer

$$\delta^{n-1}_{j,k} = w^3_{j,k}\delta^3_{j,j}$$

k = Index of neuron in (n − 1)th layer
- Back propagate δ from the output of nth layer to the input of n – 1st layer

$$\delta^n_{j,k} = \delta^{n-1}_{j,k} s^{n-1}_k$$

k = Index of neuron in (n − 1)th layer
For backpropagation process of other outputs, the last three steps are repeated After performing backpropagation computation and feedforward computations, the whole δ matrix and y matrix can be obtain for the controlling of an induction motor.
- After obtaining both the matrix, elements of jacobian matrix can be obtained using

$$\frac{\partial e_{p,m}}{\partial w_{j,i}} = -\delta_{m,j} x_{j,i} \tag{11}$$

where

$e_{p,m}$ = Mean Error function between predicted and measured output

4 Simulations and Analysis

4.1 Conventional Controller

The conventional control technique involves the use of PI (proportional integral) controller for the control of the machine. It includes the calculation and minimization of error. The speed from the induction motor is compared with the reference speed and then the values and the error is calculated and then it then as the input signal. The proportional, integral, and the derivative terms are calculated and added to find the output of the PI controller so as to minimize the error, but the problem with the PI controller is that it is not robust. Although it cannot lead to good tracking and regulating performance simultaneously, also its control performance parameters sensitivity is more toward the load disturbances and system parameters variations (Fig. 7).

Such controllers with fixed parameters cannot provide these requirements until the use of some unrealistically high gains. That is why the conventional controller used in the variable speed induction motor drives is considered to be poor in practical applications due to the abnormal behavior in the drive such as load unbalancing, variations in various parameters and unmodeled dynamics.

4.2 Neural Network Controller for 3ph Induction Motor

We have designed a controller for 3ph induction motor using ANN. In our ANN, we have taken electromagnetic torque (N-m) and speed (rad sec^{-1}) as the input and have taken voltage (V) and frequency (Hz) as the output (Fig. 8).

Based on the output of the controller, PWM pulse is been generated and send to the gate of the three-phase inverter. We have compared our result with the conventional Induction motor drive (PI Controller) for different loads and the NN controller is able to control the induction motor efficiently.

The simulation results of both ANN and conventional controller are shown in Figs. 9 and 10.

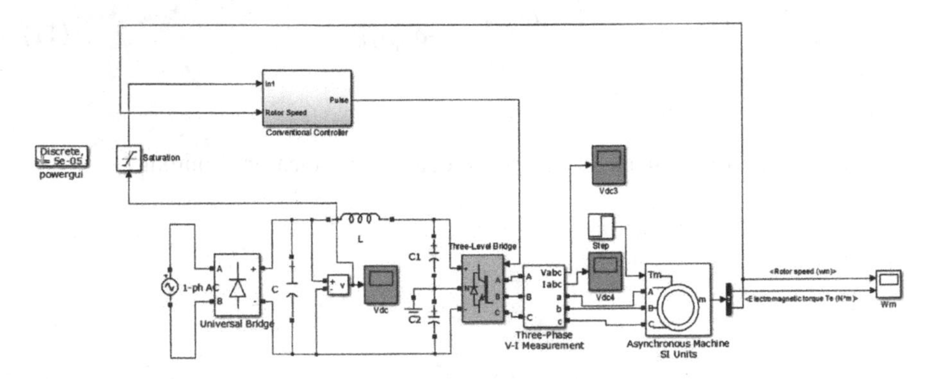

Fig. 7 Conventional controller of induction motor

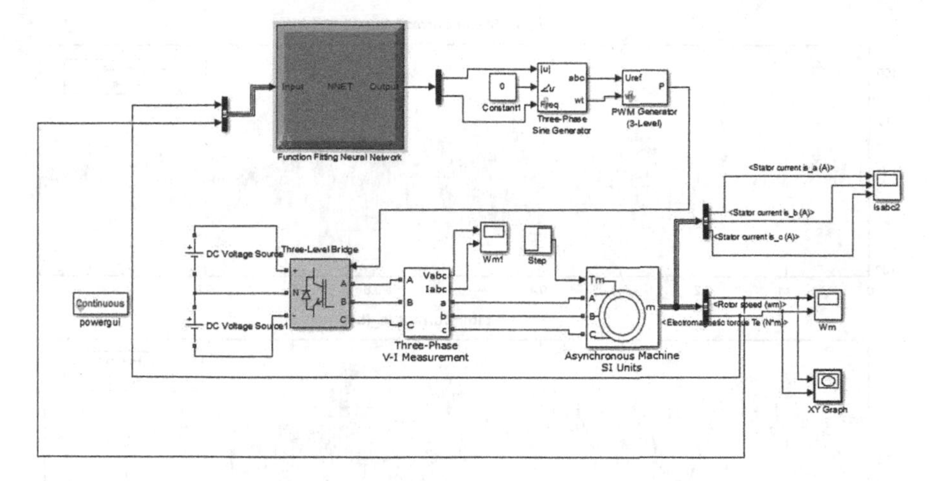

Fig. 8 ANN controller for induction motor

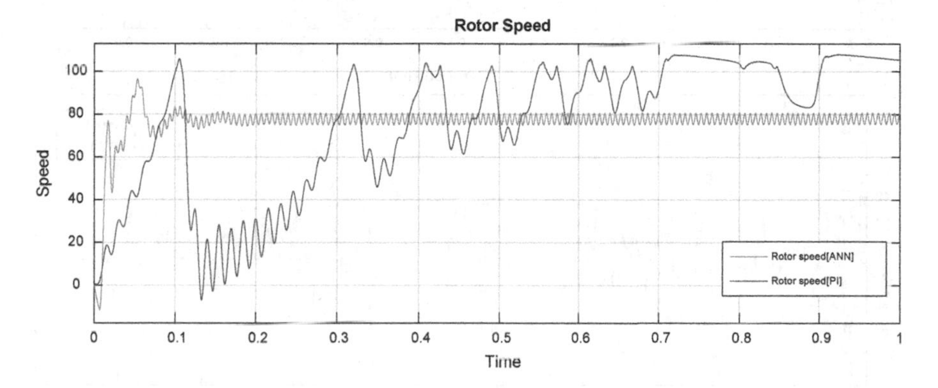

Fig. 9 Speed versus time curve comparison between conventional and ANN controller for IM

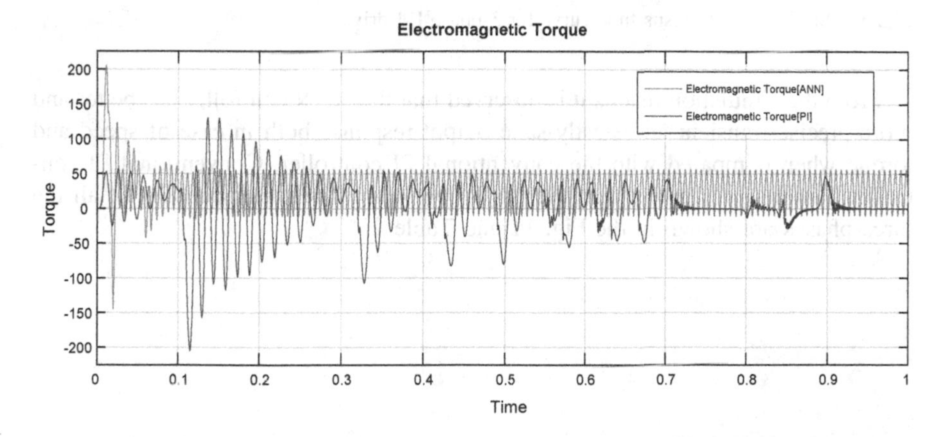

Fig. 10 Toque versus time curve comparison between conventional and ANN controller for IM

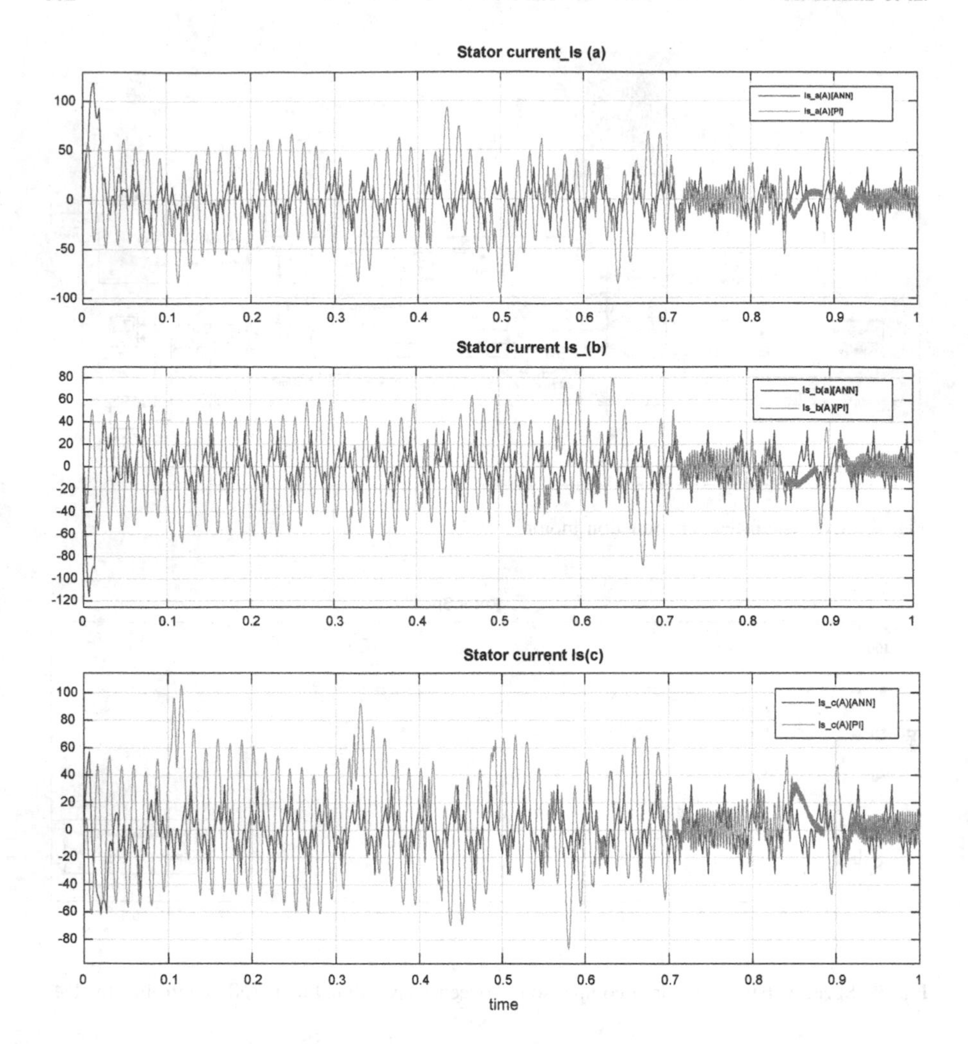

Fig. 11 Stator current versus time curve for 3-phase IM drive

From the simulation results it is observed that the ANN controller has better and more precise transient and steady-state output response both in case of speed and torque when compared with the conventional PI controller. Conventional PI controller is more error prone than the ANN controller. The stator currents of all the three-phases are shown in the Fig. 11 and Table 2.

Table 2 Dataset used for training of neural network controller

Torque	Speed	Frequency	Voltage
25	26.02456335	10	80
25	41.72456335	15	120
25	57.42456335	20	160
25	73.12456335	25	200
25	88.82456335	30	240
25	104.5245634	35	280
25	120.2245634	40	320
25	135.9245634	45	360
25	151.6245634	50	400
20	27.09965053	10	80
20	42.79965053	15	120
20	58.49965053	20	160
20	74.19965053	25	200
20	89.89965053	30	240
20	105.5996505	35	280
20	121.2996505	40	320
20	136.9996505	45	360
20	152.6996505	50	400
15	28.1747379	10	80
15	43.8747379	15	120
15	59.5747379	20	160
15	75.2747379	25	200
15	90.9747379	30	240
15	106.6747379	35	280
15	122.3747379	40	320
15	138.0747379	45	360
15	153.7747379	50	400

5 Conclusions

ANN-based controller for the Induction motor drive is presented in this paper. MATLAB/Simulink models for Induction motor drive with ANN controller and conventional PI controller is compared and observed that ANN has better performance with minimum ripples in the output than the conventional controllers. Thus, the use of ANN controller has properly addressed the problem related to the cost as well as complexity.

References

1. S. Rafa, A. Larabi, L. Barazane, M. Manceur, N. Essounbouli, A. Hamzaoui.: Fuzzy vector control of induction motor. In: Networking, Sensing and Control (ICNSC), 10th IEEE International Conference, pp. 815–820 (2013).
2. M. S. M. Aras, S. N. B. S. Salim, E. C. S. Hoo, I. A. b. W. A. Razak, M. H. b. Hairi.: Comparison of Fuzzy Control Rules Using MATLAB Toolbox and Simulink for DC Induction Motor-Speed Control. In: Soft Computing and Pattern Recognition (SOCPAR), IEEE Conference, pp. 711–715 (2009).
3. N. Venkataramana Naik, S.P. Singh.: A novel type-2 fuzzy logic control of induction motor drive using Scalar Control. In: Power Electronics (IICPE), IEEE 5th India International Conference, pp. 1–6 (2012).
4. A. K. P. Toh, E. P. Nowicki, F. Ashrafzadeh.: A flux estimator for field oriented control of an induction motor using an artificial neural network. In: Industry Applications Society Annual Meeting, pp. 585–592 (1994).
5. B. Karanayil, M. F. Rahman, C. Grantham.: Online Stator and Rotor Resistance Estimation Scheme Using Artificial Neural Networks for Vector Controlled Speed Sensorless Induction Motor Drive. In: IEEE Transactions on Industrial Electronics, vol. 54, no. 1, pp. 167–176 (2007).
6. A. Ba-Razzouk, A. Cheriti, G. Olivier, P. Sicard.: Field oriented control of induction motor using neural networks decouplers. In: Industrial Electronics, Control, and Instrumentation, pp. 1428–1433 (1995).
7. V. Ambrozic, G. S. Buja, R. Menis.: Band-constrained technique for direct torque control of induction motor. In: IEEE Transactions on Industrial Electronics, vol. 51, no. 4, pp. 776–784 (2004).
8. C. Attaianese, S. Meo, A. Perfetto.: A voltage feeding algorithm for direct torque control of induction motor drives using state feedback. In: Industrial Electronics Society. Proceedings of the 24th Annual Conference of the IEEE, vol. 2, pp. 586–590 (1998).
9. Z. Sorchini, P. T. Krein.: Formal Derivation of Direct Torque Control for Induction Machines. In: IEEE Transactions on Power Electronics, vol. 21, no. 5, pp. 1428–1436 (2006).
10. P. M. Menghal, A. J. Laxmi..: Scalar control of an induction motor using artificial intelligent controller. In: Power, Automation and Communication (INPAC), IEEE Conference, pp. 60–65 (2014).
11. S. Pati, M. Patnaik, A. Panda.: Comparative performance analysis of fuzzy PI, PD and PID controllers used in a scalar controlled induction motor drive Circuit. In: Power and Computing Technologies (ICCPCT), International IEEE Conference, pp. 910–915 (2014).
12. A. M. A. Amin, A. A. El-Samahy.: Real-time tracking control of squirrel cage induction motor using neural network. In: Industrial Electronics Society (IECON), Proceedings of the 24th Annual Conference of the IEEE, vol. 2, pp. 877–882 (1998).

A Novel Image Intelligent System Architecture for Fire Proof Robot

B. Madhevan, Sakkaravarthi Ramanathan and Durgesh Kumar Jha

Abstract The aim of this research is to design and analyze a fireproof firefighting robot that can enter into fire environment and navigate itself through the fire and send information about the fire behavior. This robot would help the fire rescue team to better understand the fire behavior and trapped person location and thus would be a critical advantage in term of time saving and rescue teams risk for their own life. In this paper, an image processing system and communication architecture for firefighting robot based on GSM technology and microcontroller is designed. Camera connected to microcontroller using serial cable will capture the image data and store it on a storage device. The processed image are sending to the predefined mobile number using GPRS technology. The encoder is used to improve the efficiency of compressed image. MATLAB software is used for the image processing which uses Fuzzy Coded Means to complete this process.

Keywords Firefighting mobile robot · Intelligent system · GSM · Image processing · System architecture

1 Introduction

Natural disaster mainly involves fire, takes so many lives, because victims do not get proper attention on the time. Life of innocents can be saved if proper care reached to them within time. The scope of the current work is to design and develop a fireproof—firefighting-surveillance robot that can traverse into fire environment and send information about the fire behaviour and the location of people trapped in these hazardous areas. When a fire breaks out in an industrial environment or a

B. Madhevan (✉) · D.K. Jha
School of Mechanical and Building Sciences, VIT University, Chennai, India
e-mail: madhevan.b@vit.ac.in

S. Ramanathan
School of Computing Science and Engineering, VIT University, Chennai, India
e-mail: sakkaravarthi.r@vit.ac.in

S.S. Dash et al. (eds.), *Artificial Intelligence and Evolutionary Computations in Engineering Systems*, Advances in Intelligent Systems and Computing 517,
DOI 10.1007/978-981-10-3174-8_67

Table 1 Fire accidents in Tamilnadu

Year	Total fire accidents	Property lost (Cr Rs) in fire calls	Lives lost in rescue calls	Lives lost in
2013	25,109	42.55	75	1586
2012	32,273	27.02	87	1722
2011	22,219	27.59	84	1878
2010	18,311	24.60	75	1773
2009	21,840	29.51	127	1438

house fire environment, it takes approximately 30–40 min for the fire rescue team to reach the spot. Thus the developed robot would help the fire rescue team to better understand the fire behaviour and trapped person in these hazardous areas. The proposed model is an amateur attempt at creating an autonomous fire proof rescue machine, aiding humans in fire and hazardous situations. When fire breaks out in a building or any closed spaces, it is quite risky for humans to traverse through this hazardous situations, as one may get trapped in such closed spaces. For the year 2009–2013, a brief information regarding the fire accidents in Tamil Nadu-India is given in Table 1.

This can be achieved by using autonomous firefighting robot which has a cameramount on it. The camera is protected with transparent glass ceramic which can sustain temperature up to 1292° Fahrenheit. The first step of image processing system is capturing the live image. Once it is done captured image is sent to the predefined phone number at receiving side. Simulation of proposed circuit diagram is done in PROTEUS software and acquired images are analyzed in MATLAB software. Hardware parts include mainly microcontroller, GSM module, and camera. Microcontroller (atmega328) is used as a controller and also work like a mediate system between camera and GSM module. The camera is used to get the information of victims trapped under fire and it sends the captured information to the microcontroller. GSM connected with microcontroller gathers information from it and sends to the mobile phone using GPRS techniques. Thus, information from inside firehouse is communicated to outside without much consumption of time. While capturing the image inside fire, the main problem with the camera is the lighting system, but that can be overcome by increasing shutter speed of the camera. Apart from this introductory section, this paper is presented in the following manner. Section 2 provides a more detailed study on literature study. Section 3 provides a more detailed study of robotics module. Section 4 explains about the system design of firefighting robot. Section 5 explains briefly about communication architecture. Methodology is explained in Sect. 6. Results and discussions are presented in Sect. 7. Finally, conclusions are given followed by references.

2 Literature Survey

Robotics have found its way into many applications in our day to day life and it has been of great help to mankind. The firefighting robots discussed here is used to reduce the loss of human lives and reaction times under emergency situations. Standard reaction times are sometimes too long and can cause catastrophic effects, similarly, under severe circumstances, the lives of the firefighter are put at risk. To prevent such unfortunate events firefighting robots are designed. In [1], GSM was used for sending information and display the amount of information on LCD display. Wireless sensor network and GSM technology were used to capture and send the images to a pre-defined phone number, as described in [2]. Wireless monitoring system using S3C2440 microcontroller based on arm 9 core has been discussed in [3]. In [4], a surveillance system has been designed which uses obstacle detection sensor to detect the victim, capture image using jpeg camera store it in USB and then send it as an SMS. Face detection system has been used to detect the face and compare it with the stored image if it doesn't match then it sends information, it also sends the location over GPS discussed in [5]. In [6], the wireless video monitoring system based on GPRS and ARM processor has been designed, which was used to monitor the video continuously and send it to a client. In [7], image filtering was done using TPSO or TPFF based fuzzy filtering. Colour image segmentation using Fuzzy C-Means has been described in [8–10]. Image filtering using interval-valued fuzzy sets has been proposed in [11]. Video monitoring and remote video surveillance using S3C240 and GPRS has been discussed in [12–15]. In [16] video monitoring has been done using GSM in embedded Linux. Video surveillance using S3C2440 and GSM has been discussed in [17]. GSM technology and wireless sensor network have been proposed in [18–20]. Authors in [21, 22] explain about the mathematical modeling for a mobile robot traversing in an unknown environment with obstacles.

3 Robotics Module for Firefighting Robot

3.1 *Material for Body and Insulation*

The robot has to move inside a fire environment consisting of very high temperatures, ranging from 1000 to 2000 °C. In order to make the robot approximately fire proof it has to be shielded using proper material as shown in Table 2. The body of the robot can be made of Aluminium, Carbon Steel, Iron, Copper [1] as shown in Table 3.

Table 2 Material selection-I

Material	Density Kg/m^3	Thermal conductivity (W/m K)
Fiber glass <250 °C	14–100	0.04 at 20 °C
Rock wool <700 °C	30–200	0.04 at 20 °C
Ceramic fiber paper	200	0.08 at 600 °C
Polyurethane foam	62	0.095 at 350 °C

Table 3 Material selection -II

Material	Tensile strength (MPa)	Compressive strength (MPa)	Thermal conductivity 25 °C (W/m °K)	Melting point (°C)	Density g/m^3
Aluminum	70–700	NA	230	660.3	2.7
Carbon steel	500	500	43	1425–540	8.05
Iron	170	550	75	1538	7.9
Copper	70–220	117–220	390	1085	8.9

4 System Design

4.1 System Architecture

System architecture includes various subsystems as shown in Fig. 1 and function of each subsystem is briefly described below.

4.2 Mechanical Design

Using solid works software, a 3D model of the robot is generated as shown in Fig. 2.

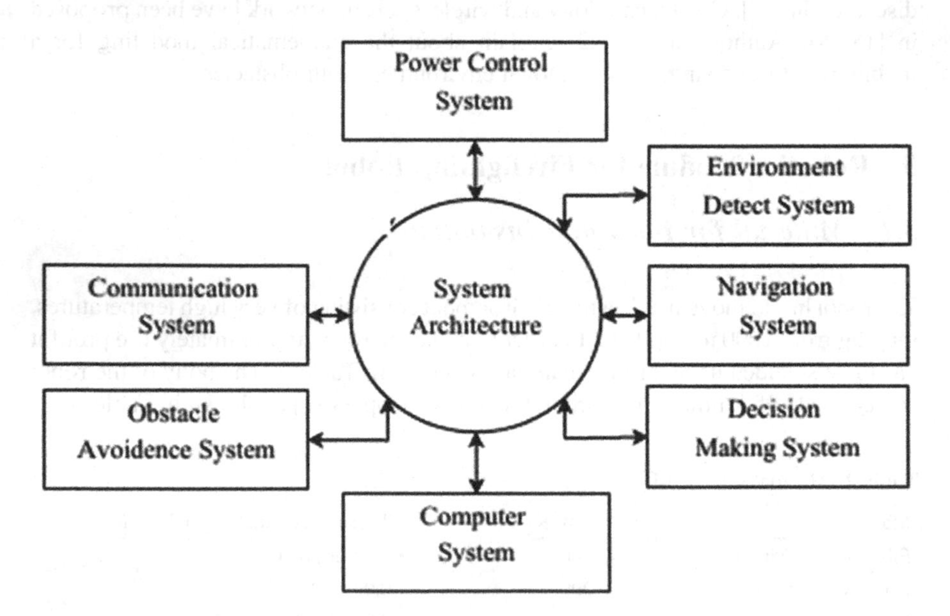

Fig. 1 System architecture of firefighting robot

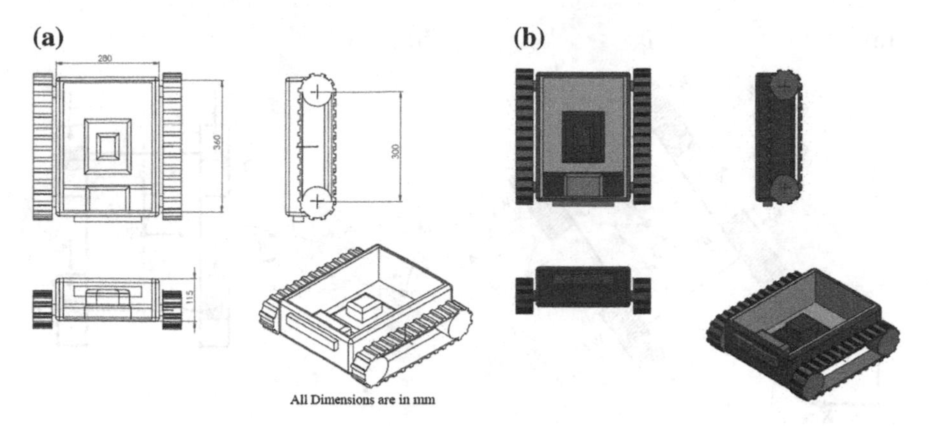

Fig. 2 Mechanical design

4.3 Kinematic and Dynamic Model

The robot's mass center is at the body frame's geometric center. The two adjacent wheel rotate with equal speeds. All the four wheels keep constant ground contact. Kinematic and dynamic model are shown in Fig. 3a, b respectively.

- Relationship equation for Control inputs and speed of wheels:

$$v_x = \frac{(\omega_l * r + \omega_r * r)}{2} = \frac{v_l + v_r}{2} \tag{1}$$

$$\omega_z = \frac{(-\omega_l * r + \omega_r * r)}{2y_0} \tag{2}$$

  l-left, *r*-right and y_0 is the instantaneous tread (instantaneous center of rotation) ICR value.
- Instantaneous radius of the path curvature:

$$R = \frac{v_g}{\omega_z} = \frac{v_l + v_r}{-v_l + v_r} * y_0 \tag{3}$$

  Hence, Straight line motion, R = ¥ and Rotational motion, $R = 0$

- Dynamic model equations are written below [2]:

$$F_{xfr} + F_{xrr} + F_{xfl} + Fxrl - R_x - \frac{m^* (v_g)^2}{R} * \sin(\beta) = 0 \tag{4}$$

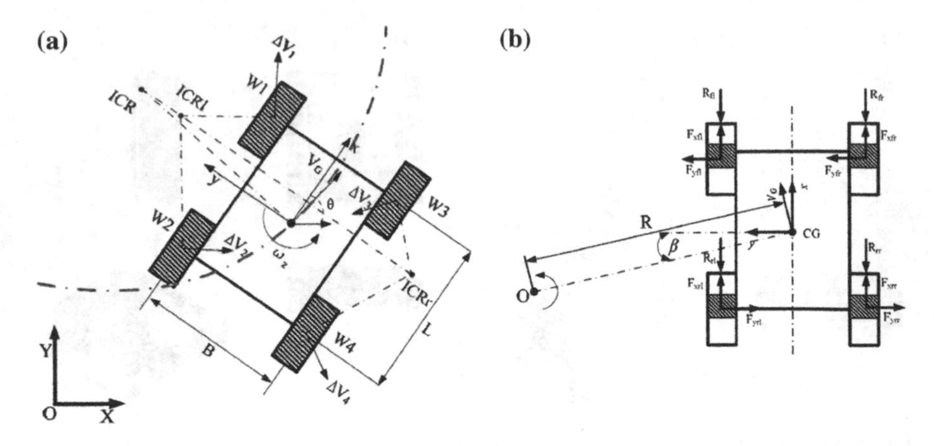

Fig. 3 Kinematics diagram for proposed robot with force reactions

$$F_{yfr} + F_{yrr} + F_{yfl} + F_{yrl} = \frac{m^* \left(v_g\right)^2}{R} {}^* \cos(\beta) \tag{5}$$

$$M_d - M_r = 0 \tag{6}$$

where, $F_{xfr}, F_{xrr} F_{xfl}, Fxrl$ are the forces in longitudinal direction. $F_{yfr}, F_{yrr}, F_{yfl}, F_{yrl}$ are the forces in lateral direction. v_g is the velocity of the vehicle, (β) is the angle between the velocity of the vehicle and x-axis of the local frame of reference. R_x is motion resistances on the four wheels acting externally. M_d and M_r are the drive and resistance moments respectively.

- Ground-wheel interaction, shear stress τ_{ss} and shear displacement j can have relationshipas written below:

$$\tau_{ss} = p^* \mu^* \left(1 - e^{\frac{-j}{k}}\right) \tag{7}$$

where, p is the pressure applied normally, m is the friction coefficient and K is the deformation shear modulus.

5 Communication Architecture

The system architecture is shown in Fig. 4. It consists of a microcontroller, camera, GSM module power supply and LCD display. The camera is interfaced with the microcontroller using a serial cable. GSM module is interfaced with a controller using this serial driver. This system will work such that when a device starts, the camera captures the video and using a serial cable it sends this signal to the microcontroller. GSM will get the captured video in Jpeg format and send it to the other communication

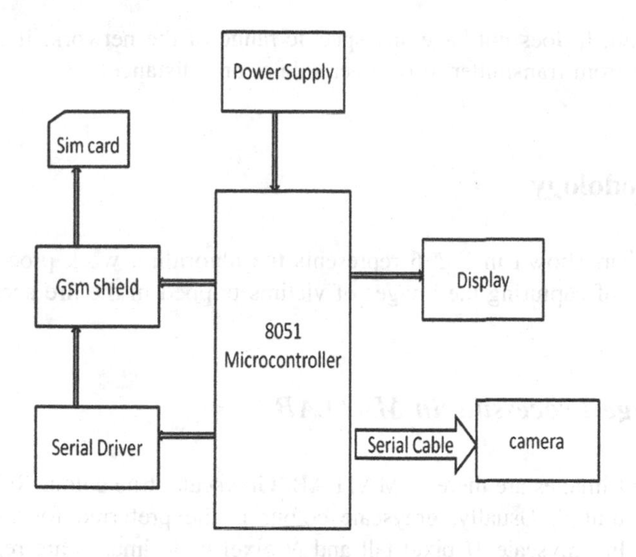

Fig. 4 Communication structure

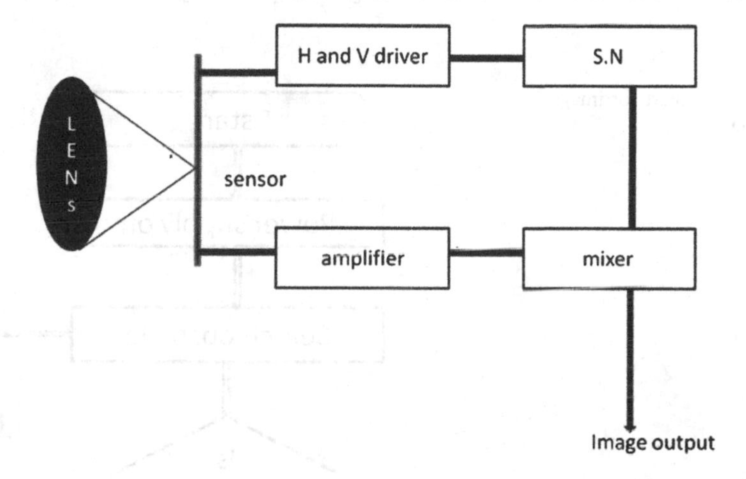

Fig. 5 Camera structure. Where, S.N = Synchronous Generator, Sensor = CMOS Sensor

device using GPRS. OV7670 CMOS camera is used to capture the video. It has CMOS (Complimentary Metal Oxide Semiconductor) sensor.

Internal architecture and function of CMOS camera are shown in following Fig. 5. Simcom sim 900 GSM (Global System for Mobile Communication) module is used for this research. It is dual band 900–1800 MHz with dimensions 24 * 24 * 3 mm. It is controlled via AT command. AT command is the command used to control the modem. General packet radio service is used for the global system for mobile communication. The advantage of using GPRS is since it works with sim card, It does not have any specific range of the network. It can send the information from transmitter to receiver side at any distance.

6 Methodology

The flow chart shown in Fig. 6 represents the algorithm, work process and systematic way of capturing the images of victims trapped in the fire accidents.

6.1 Image Processing in MATLAB

Five types of images are there in MATLAB. Grayscale, true colour RGB, indexed, binary and unit 8. Usually, grayscale colour is the preferred format for image processing. In grayscale M pixel tall and N pixel wide images are represented in $[M * N]$ matrix form. grayscale intensities are represented in [0,1] form where 0 is for black and 1 is for white. When the image is captured by the camera because of

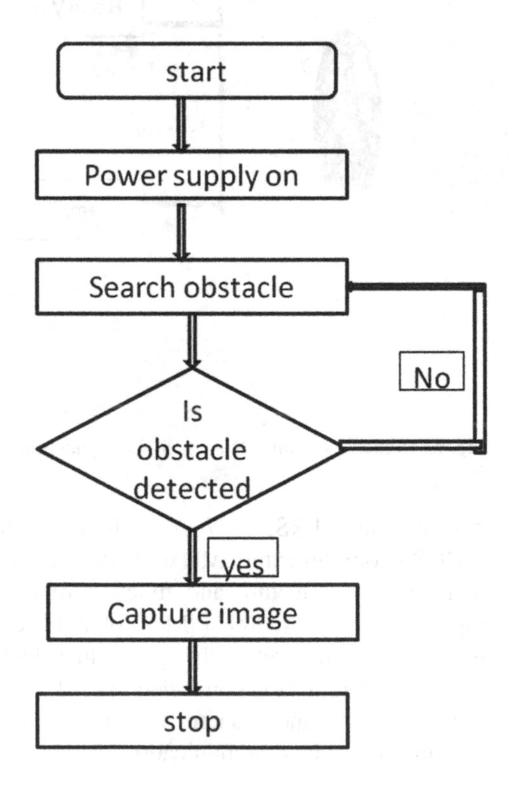

Fig. 6 Flow chart for image capturing

Fig. 7 Original image

an improper lighting system, fire and smoke captured image is not much clear. It is a very tedious task for rescuers to extract exact information from that image. Figure 7 shows the exact image captured by the camera which is not so clear hence it is necessary to do image processing.

6.2 Image Segmentation

Image segmentation process is done in order to get a clear picture of captured image. It divides an image into multiple parts and identify the relevant information or other hidden objects as shown in Figs. 8 and 9. Many methods are there for image segmentation like thresholding methods, colour-based segmentation, transform methods and texture methods. In thresholding method images are partitioned in foreground and background. Image thresholding can be performed in two ways one is image thresholding using multi-level thresholding and another is using set level. Thresholding technique is one of the most used segmentation technique in the MATLAB. The Imread command is used to read the image. Now the image is displayed and its contrast is set by using Matlab toolbox. It has two image display function imshow and imtool. Imshow is used for fundamental image display function. Imtool performs some image processing task.

Fig. 8 Segmented image

Fig. 9 Segmented image for different intensity

7 Results and Discussion

Circuit for communication architecture is designed in scheme it circuit. Proper microcontroller, GSM module, and camera is selected for designing the circuit. Testing of the designed circuit is done in the same software. Image processing and

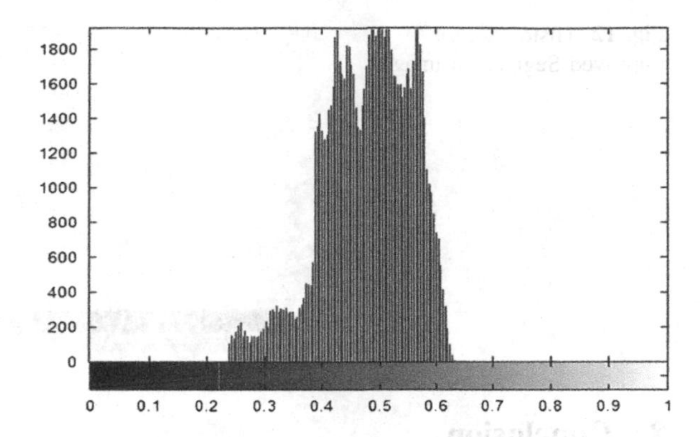

Fig. 10 Histogram of original image

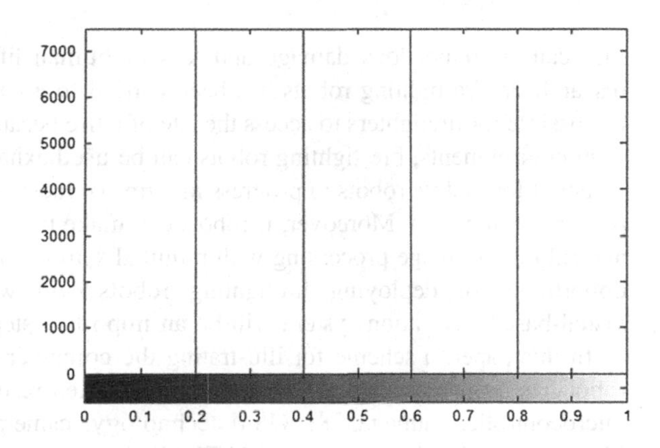

Fig. 11 Histogram of processed image

image segmentation part of the system is done in the MATLAB using FCM (Fuzzy C-Means) code. The FCM algorithm is implemented to a black and white image to cluster the grayscale intensities to five variations, which intern gives distinct shapes in the image which otherwise was blurred due to various circumstances. It is a clustering method which allows each data point to belong to multiple clusters. It is a clustering method which allows each data point to belong to multiple clusters. Figure 10 shows the histogram of the original image (pixel versus intensity). In below figure, it is noticed that intensity range is very narrow. Intensity range can be made wider by using another segmentation. Fig. 11 shows the histogram for processed image. It is noted that intensity range of processed image is wider than the original image. It covers the potential range of 0–1 over full area hence the contrast of processed image is more. To export image data from the MATLAB workspace to a graphics file in one of the supported graphics file formats, imwrite function is used. When using imwrite, MATLAB variable name and the name of the file is specified. Figure 12 shows the histogram for improved segmented image, which covers the potential range of 0–1 over a full area as well as its intensity is reaching to peak.

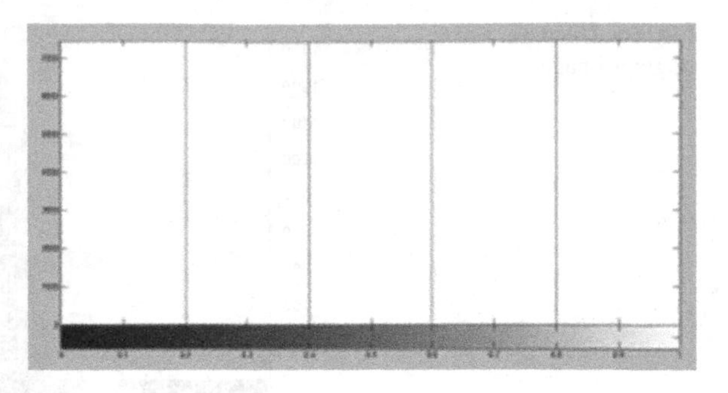

Fig. 12 Histogram for Improved Segmented Image

8 Conclusion

Fire causes tremendous damage and loss of human life and property. Recently, research on fire-fighting robots has been carried to a large extent. It is sometimes impossible for firefighters to access the site of a fire because of drastic conditions. In such environments, fire-fighting robots can be used exhaustively. Visual sensing is essential for mobile robots to progress in terms of increased robustness and reduced power consumption. Moreover, if robots can make use of computationally efficient algorithms for image processing with minimal setup with zero calibration, then the opportunity of deploying firefighting robots will widely increase. A novel visual-based navigation system will be an important step in this direction.

In this paper, a scheme for illustrating the computer architecture of firefighting robot has been presented along with the simulated result. This scheme combines microcontroller (atmega328), GSM technology, camera and dynamic algorithm. Image processing is done using MATLAB for the image captured by the camera. Histogram (pixel versus intensity) of the original image and segmented image is obtained and compared. Fuzzy Coded Means (FCM) is used for the image segmentation.

References

1. Karthik, P., Deepika, S., Haritha, N., Punitha, S.: A Prevention and Automation of Public Distribution System using RFID and Facial Recognition Camera. IOSR Journal of Engineering (IOSRJEN), 4 (2014) 09–1.
2. Mousami S Vanjale, Asmita Deshpande, Krunal Pawale, PoojaBhagwa,: Automatic Security System Based on Wireless Sensor Network and GSM Technology. International Journal of Engineering Research Technology (IJERT). 3(4) (2014) 467–470.
3. Sridhar, V., Nagendra, G., Zubair, M.: Mobile Robots for Search and Rescue, Design of Wireless Monitor System Based On S3C2440 and GPRS. February 2013.
4. Sunil, M.P.: Design of an Event Based Surveillance System Using Advanced Technology. International Journal of Engineering Science and Innovative Technology (IJESIT), 2(6) (2013) 451–459.

5. Shaik Meeravali, Vamshidhar Reddy, A. Sudharshan Reddy, S.: An Inexpensive Security Authentication System Based on a Novel Face-Recognition Structure. International Journal of Engineering Trends and Technology (IJETT), 4(9) (2013) 4094–4097.
6. Dhonde Vilas, R., Joshi, S. G.,: Remote Video Monitoring System Based On ARM and GPRS. International Journal of Innovative Research in Science, Engineering and Technology, 3(4) (2014) 463–470.
7. Hsien-Hsin Chou, Ling-Yuan Hsu, and Hwai-Tsu Hu,:Turbulent-PSO-Based Fuzzy Image Filter With No-Reference Measures for High-Density Impulse Noise. IEEE Transactions on Cybernetics, 43(1) (2013) 296–307.
8. Rafika Harrabi, Ezzedine Ben Braiek,:Color Image Segmentation Using a Modified Fuzzy C-Means Technique and Different Color Spaces. 1st International Conference on Advanced Technologies for Signal and Image Processing - ATSIp 2014 March 17–19, 2014.
9. Eng, H. L., Ma, K. K.: Noise Adaptive soft-switching Median Filter, IEEE 10(2) (2001) 242–251.
10. Cavus, N.: The Evaluation of Learning Management Systems Using an Artificial Intelligence Fuzzy Logic Algorithm. Adv. Eng. Softw., 41(2) (2010) 248–254.
11. Bigand, A., Colot, O.: Fuzzy Filter Based on Interval Valued Fuzzy Sets for Image Filtering, Fuzzy Sets Syst., 161(1) (2010) 96–117.
12. An Inexpensive Security Authentication System Based on a Novel Face-Recognition Structure, International Journal of Engineering Trends and Technology (IJETT) Volume 4 Issue 9- Sep 2013.
13. Remote Video surveillance System Based on S3C2440 and GPRS Volume 2, Issue 6, June 2014 International Journal of Advance Research in Computer Science and Management Studies.
14. Huiping Huang, Shide Xiao, Xiangy in Meng and Ying Xiong,: A Remote Home Security System Based on Wireless Sensor Network and GSM Technology. Second IEEE International Conference on Networks Security, Vol. 21, 24–25 April 2010.
15. Huiping Huang, Shide Xiao, Xiangy in Meng and Ying Xiong,: A Remote Home Security System Based on Wireless Sensor Network and GSM Technology. Second IEEE International Conference on Networks Security, Vol. 21, 24–25 April 2010.
16. Wang, Y. T., Chen Y.C.: Multiple-Obstacle Avoidance in Role Assignment of Formation control. I. J. Robotics and Automation, 27 (2) (2012) 177–184.
17. Amoozgar, Mohammad Hadi, Sadati, SeyedHossein, Alipour, Khalil,: Trajectory Tracking of wheeled Mobile Robots Using a kinematical Fuzzy Controller. I. J. Robotics and Automation, 27(1) (2012).
18. Yan Zhuang, Ke Wang, Wei Wang, and Huosheng Hu,: A Hybrid Sensing Approach to Mobile Robot Localization in Complex Indoor Environments. I. J. Robotics and Automation, 27(2) (2012).
19. Hui, Z., Zhang, Shan Y. Xiong, and Yue Liu,: Real-Time Path Planning of Multiple Mobile Robots in a Dynamic Certain Environment. I. J. Robotics and Automation, 28(1) (2013).
20. Madhevan, B., Sreekumar, M.: Tracking Algorithm Using Leader Follower Approach for Multi Robots. Procedia Engineering, (64) (2013) 1426–1435.
21. Madhevan, B., Sreekumar, M.: A Systematic Implementation of Role Assignment in Multi Robots Using Leader Follower Approach: Analytical and Experimental Evaluation. 13th International Conference on Control, Automation, Robotics and Vision (ICARCV), Singapore, December 2014.
22. Madhevan, B., Sreekumar, M.: Implementation of Role Assignment and Fault Tree Analysis for Multi Robot Interaction. I. J. Robotics and Automation, 2013.

Hardware Implementation of DSP-Based Sensorless Vector-Controlled Induction Motor Drive

Rajendrasinh Jadeja, Ashish K. Yadav, Tapankumar Trivedi, Siddharthsingh K. Chauhan and Vinod Patel

Abstract Due to low-cost and high-reliability, induction motors have discovered significant applications for variable speed applications. The vector control method is used to achieve better dynamic response and use of induction motor for a wide range of speed variations. However, it uses shaft encoder for rotor speed estimation that suffers from added cost, measurement noise, and maintenance associated with them. To overcome these shortfalls, speed is estimated with the help of measured voltages and currents in sensorless vector-controlled drive. The rotor speed estimation method is based on a direct synthesis of induction motor state equations. Simulation of the drive is carried out using PSIM software. In order to validate the method, laboratory prototype is developed using DSP TMS320C2811 for 10 H.P. and 15 H.P. squirrel cage induction motor.

Keywords Digital signal processor · Direct synthesis · State estimation · Sensor less vector control

1 Introduction

The development in the power electronics field and variable speed applications of AC machines have acquired importance. The variable speed induction motor drives fundamentally utilize voltage source inverters (VSI). On the other hand, induction motors have advantages in terms of robustness, high torque-to-weight ratio, reliable operation, operation ability in a hazardous environment, and cost-effectiveness. Field-oriented control (FOC) provides dynamic control of induction motor-based adjustable speed drives (ASDs) by independently controlling torque and flux [1].

R. Jadeja (✉) · A.K. Yadav · T. Trivedi · S.K. Chauhan
Marwadi Education Foundation's Group of Institutions, Rajkot, Gujarat, India
e-mail: rajendrasinh.jadeja@marwadieducation.edu.in

V. Patel
AMTECH Electronics India Ltd., Gandhinagar, Gujarat, India

S.S. Dash et al. (eds.), *Artificial Intelligence and Evolutionary Computations in Engineering Systems*, Advances in Intelligent Systems and Computing 517,
DOI 10.1007/978-981-10-3174-8_68

For speed measurement, shaft encoders are required by FOC drives [1]. As these sensors have their own shortfalls, sensorless vector control is implemented wherein the speed is estimated and not measured. This method without sensors has benefits of higher reliability and lower cost.

Speed estimation is an issue specifically compelling with impelling induction motor drive where the mechanical speed of the rotor is by and large not quite the same as the velocity of the spinning magnetic field [4].

In this work, direct synthesis of state equations method is used for robust and accurate rotor speed estimation. DSP is used since it gives better performance out of the simple modern controllers for industrial applications [1]. The motor parameter variations cause the disturbances in the system responses [2]. The PI controller gains require regular tuning with the parameter variation.

2 Sensorless Vector Control

In vector control, three-phase stator current are resolved into torque producing component i_{qs} and flux producing component i_{ds} which are orthogonal to each other in synchronously rotating reference frame. Conventionally, shaft-mounted speed encoder is used for measurement of speed. For control over an entire range of speed including starting conditions, a speed signal is required in indirect vector control. For sensorless drives, the rotor speed is estimated by measuring stator currents [2–11].

Figures 1 and 2 show the phasor diagram of vector control with an increase of torque component and flux component of the current, respectively. For increasing in the torque component of the current, mutual flux is equal to the rotor flux on the *d* axis reference current. Whereas the i_{ds} components is reduced by weakening of flux [2].

1. Induction Motor
2. 3 Phase Bridge Rectifier and Inverter
3. SVPWM block
4. Park Inverse Transformation Block
5. PI Controller Block
6. Clarke Forward Transformation
7. Park Forward Transformation Block
8. Estimator block (Direct synthesis from state equation)

Figure 3 shows a basic block diagram of vector control [2] whereas Fig. 4 shows the basic sensorless vector control by using direct synthesis from the state equations. Motor currents are sensed from the motor input. Using Clarke transformation, sensed motor current is converted in terms of α, β reference frame and then converted into *dq* reference frame using the Park transformation. This *d* and *q* currents are compared with reference *d* and *q* currents. Estimation block calculates the rotor

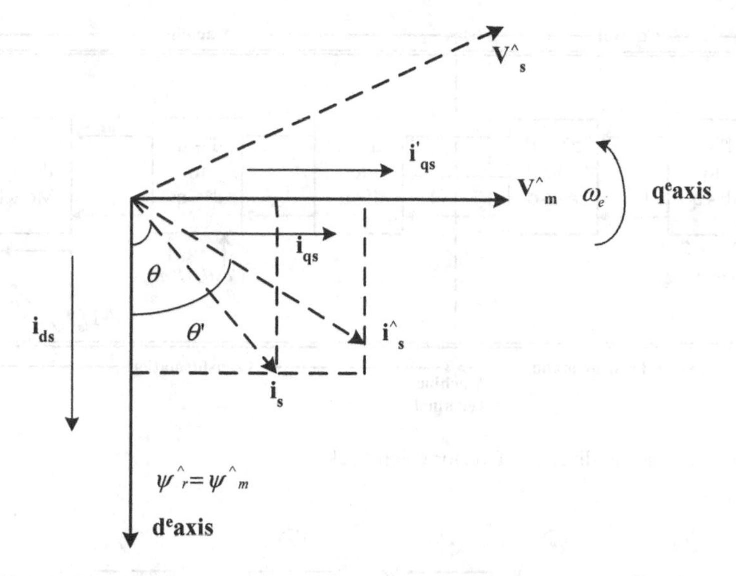

Fig. 1 Phasor diagram of vector control with an increase of torque component of current [2]

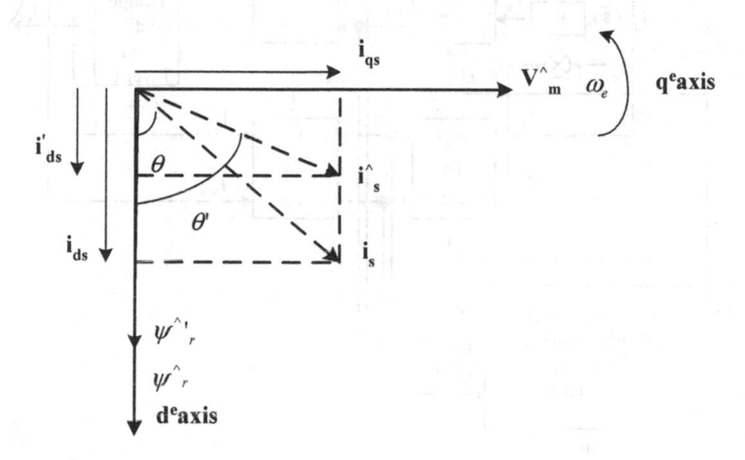

Fig. 2 Phasor diagram of vector control with an increase of flux component of current [2]

angle θ_e and rotor speed that is estimation speed by using α and β voltages and currents [3].

2.1 Variables

1. i_a, i_b, i_c—Three-phase currents of the motor
2. i_α, i_β—Currents from Clarke transformation

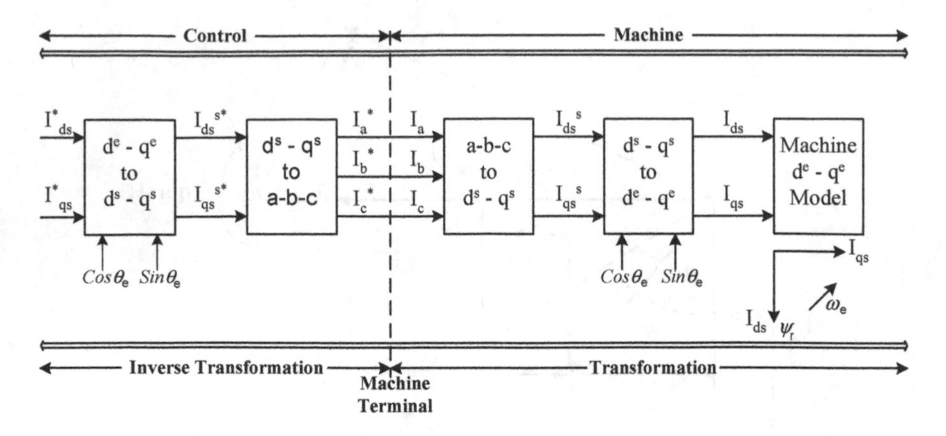

Fig. 3 Basic principle diagram of vector control [2]

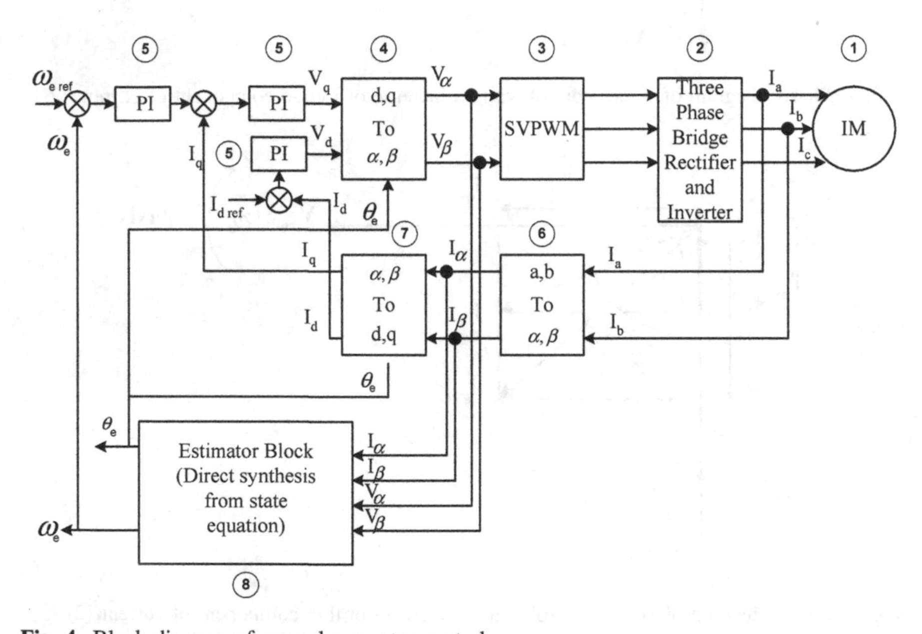

Fig. 4 Block diagram of sensorless vector control

3. i_d, i_q—D and Q axis currents
4. v_α, v_β—Voltages from Park transformation
5. v_d, v_q—D and Q axis Voltages

6. PI—Proportional Integral
7. $\omega_{\rm ref}$—Reference Speed
8. θ_e—Rotor angle
9. $i_{\rm qs}$, $i_{\rm ds}$—q and d axis stator currents
10. L_m—Magnetizing Inductance
11. ω_r—Rotor Speed
12. $\Psi_{\rm ds}$—d axis flux from stator side
13. $\Psi_{\rm qs}$—q axis flux from stator side
14. $\Psi_{\rm dr}$—d axis flux from rotor side
15. $\Psi_{\rm qr}$—q axis flux from rotor side
16. Ψ_m—Mutual flux
17. Ψ_r—Rotor flux
18. R_s—Stator resistance
19. L_r—Rotor inductance
20. L_s—Stator inductance
21. $\frac{d\theta_e}{d_t}\omega_e$—Synchronous speed
22. τ_r—Rotor time constant

3 Direct Synthesis from State Equations

The dynamic d_s-q_s frame state equations of an induction motor are used to estimate speed directly based on machine parameters [2]. Evidently, this method is highly machine parameter sensitive [2]. In this method, only the machine terminal currents are sensed and rotor fluxes calculated. The actual system can be modeled with a high degree of accuracy in this from state equations. The voltages in α, β reference frame is given by the following equation:

$$v_\alpha = v_d * \cos(\theta_e) + v_q * \sin(\theta_e) \tag{1}$$

$$v_\beta = -v_q * \cos(\theta_e) + v_d * \sin(\theta_e) \tag{2}$$

The current in α, β reference frame is calculated directly from phase currents as

$$i_\alpha = i_a \tag{3}$$

$$i_\beta = \frac{1}{3} * (i_a + (2 * i_b)) \tag{4}$$

Using current of α, β reference frame, the currents in d, q reference frame are given by

$$i_d = i_\alpha * \cos(\theta_e) + i_\beta * \sin(\theta_e) \tag{5}$$

$$i_q = -i_\alpha * \sin(\theta_e) + i_\beta * \cos(\theta_e) \tag{6}$$

Now, stator flux linkages in *d-q* reference frame are given as

$$\Psi_{\text{ds}} = \int (v_\alpha - i_{ds} * R_s)\text{d}t \tag{7}$$

$$\Psi_{\text{qs}} = \int (v_\beta - i_{qs} * R_s)\text{d}t \tag{8}$$

The rotor flux linkages calculate based on the voltage model is given by following equations:

$$\Psi_{\text{dr}} = \frac{L_r}{L_m} * \Psi_{\text{ds}} - \left[\frac{L_s * L_r - L_m^2}{L_m} * i_\alpha\right] \tag{9}$$

$$\Psi_{\text{qr}} = \frac{L_r}{L_m} * \Psi_{\text{qs}} - \left[\frac{L_s * L_r - L_m^2}{L_m} * i_\beta\right] \tag{10}$$

Now, rotor angle (θ_e) is calculated from the above Eqs. 9 and 10.

$$\theta_e = \tan^{-1}\frac{\Psi_{\text{qr}}}{\Psi_{\text{dr}}} \tag{11}$$

The synchronous speed is easily calculated by derivative of rotor angle which is given in below equation:

$$\frac{\text{d}\theta_e}{\text{d}t} = \frac{\Psi_{\text{dr}} * \Psi_{\text{qr}} - \Psi_{\text{qr}} * \Psi_{\text{dr}}}{\Psi_r^2} \tag{12}$$

Now the rotor speed is calculated by subtracting slip speed in above Eq. 12.

$$\omega_r = \frac{\text{d}\theta_e}{\text{d}t} - \left[\frac{L_m}{\tau_r} * \frac{(\Psi_{\text{dr}} * i_{\text{qs}} - \Psi_{\text{qr}} * i_{\text{ds}})}{\Psi_r^2}\right] \tag{13}$$

4 Experimental Setup

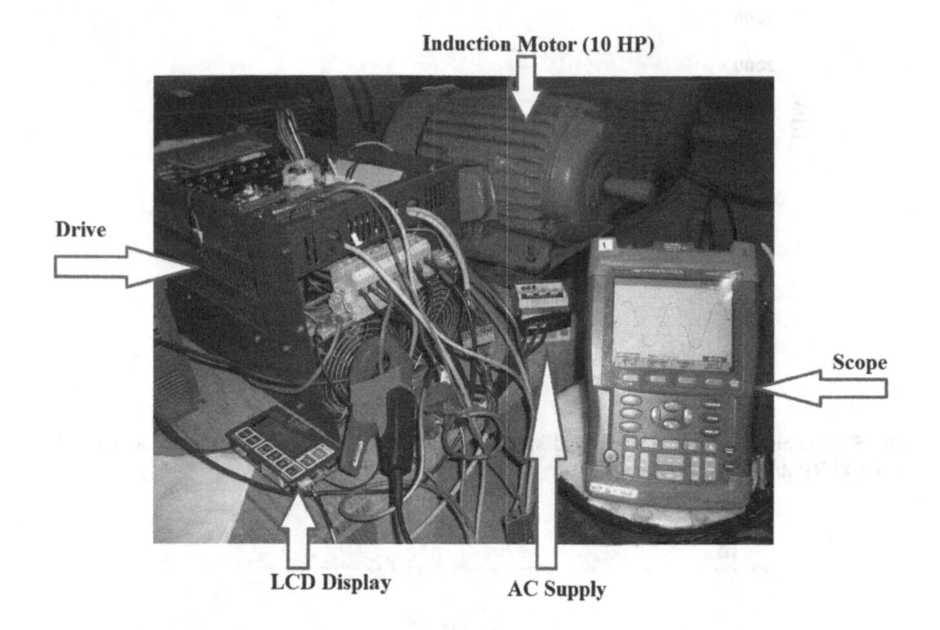

The picture shows the experimental setup of sensorless vector control of induction motor drive. Drive rating is 7.5 kW and at the output of the drive 10 HP motor is connected. The input to the system is three-phase 415 V, 50 Hz supply. The entire control of the system is done using DSP TMS320C2811, and the experimental results of induction motor drive are taken. Using current sensors the phase currents are measured and it is used for calculating rotor speed. LCD display is used to load motor parameters. The waveforms are taken using fluke 199C scope meter.

5 Simulation and Experimental Results

In Fig. 5, the actual speed of the motor (10 HP, 3000 rpm) and estimated speed that is calculating by direct synthesis from the state equations are same. This result is shown in the no load condition.

In Fig. 6, it shows the current waveform for phase A; that is at no load condition.

In Fig. 7, it shows the d and q axis currents which are in DC quantity. This waveform is observed at no load condition.

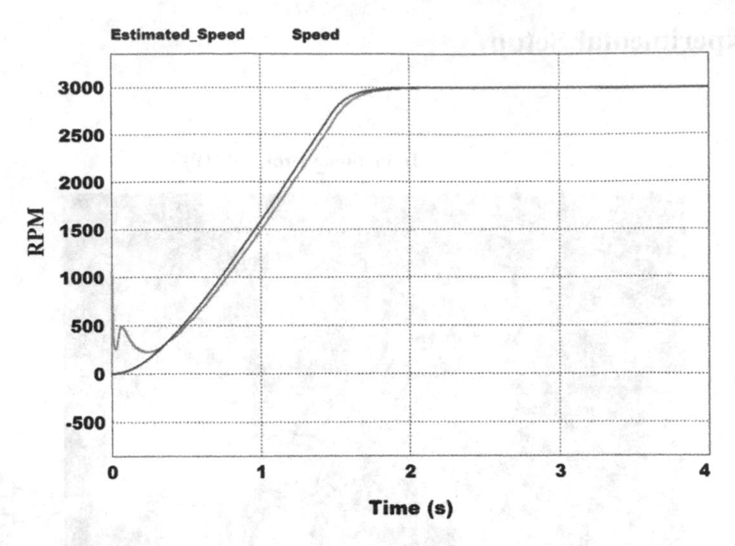

Fig. 5 Combined estimated speed and speed (actual speed) (3000 RPM), (Scale: $X = 1$ s/Div., $Y = 500$ RPM/Div.)

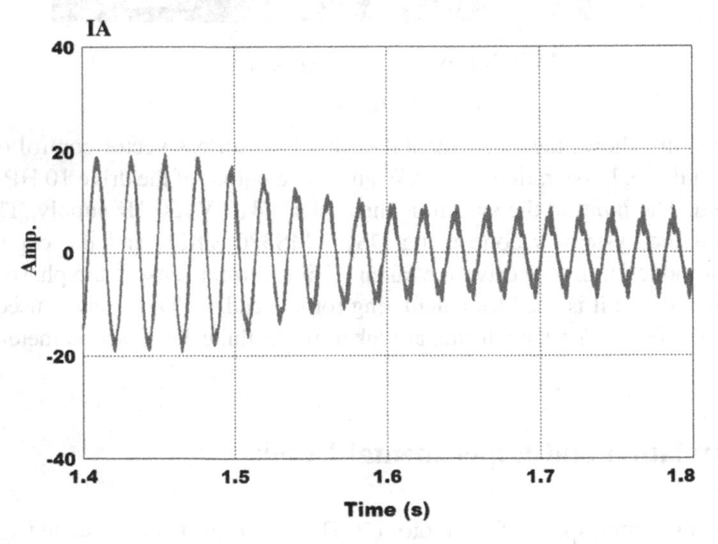

Fig. 6 Phase Current (IA) for 50 Hz, (At no load condition) (Scale: $X = 0.1$ s/Div., $Y = 20$ A/Div.)

In Fig. 8, it shows the torque waveform with no load condition. And observed the I_q current and torque waveform is same.

In Fig. 9, it shows the hardware results of actual and estimated speed, for variation in frequency at 20 Hz (1200 rpm) from the rated frequency 50 Hz

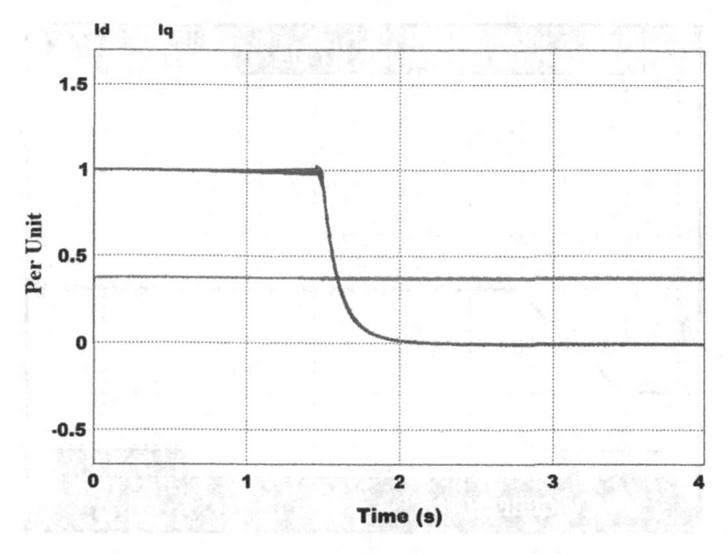

Fig. 7 *d* and *q* axis Currents I_d and I_q, (Scale: X = 1 s/Div., Y = 0.5/Div.)

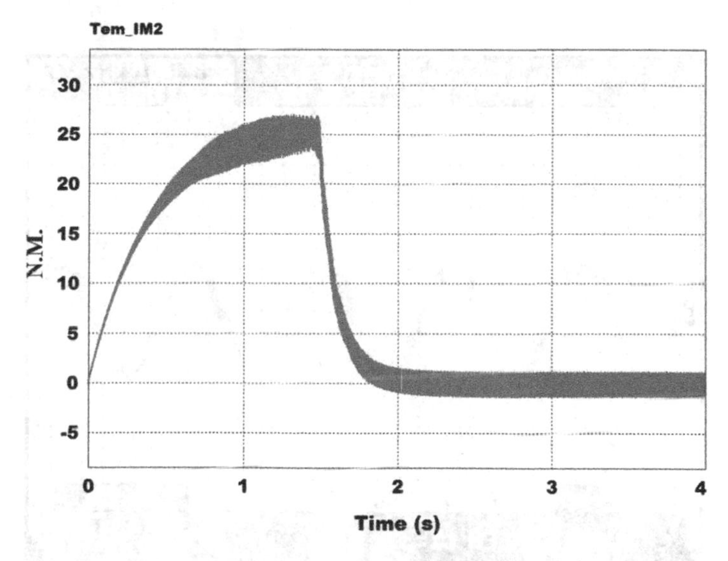

Fig. 8 Torque (Tem_IM2) in N.M., (At no load condition) (Scale: X = 1 s/Div., Y = 5 N.M./Div.)

(3000 rpm). In Fig. 9, A indicates estimated speed and B indicates actual speed. It is observed from the results that estimator and controller logic works properly as actual speed is tracking the estimated speed.

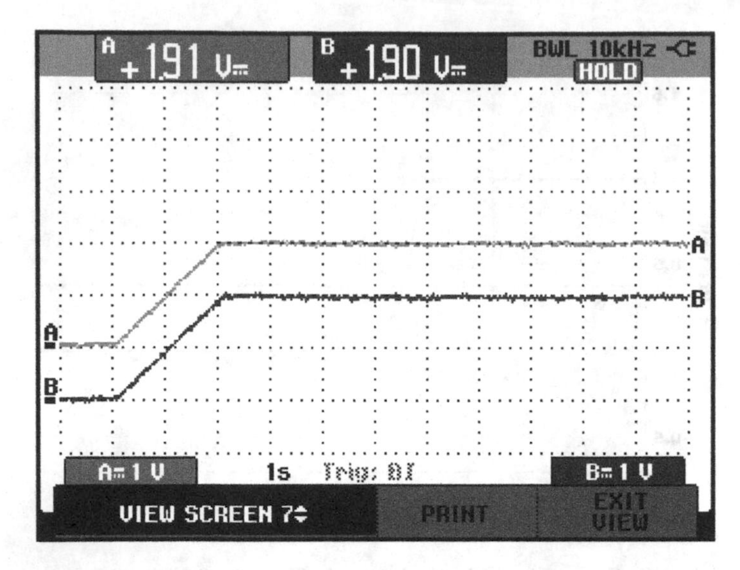

Fig. 9 Experimental speed waveforms at ~20 Hz (1200 RPM) (Scale: X = 1 s. (10 Hz)/Div., Y = 1 V (600 RPM)/Div.)

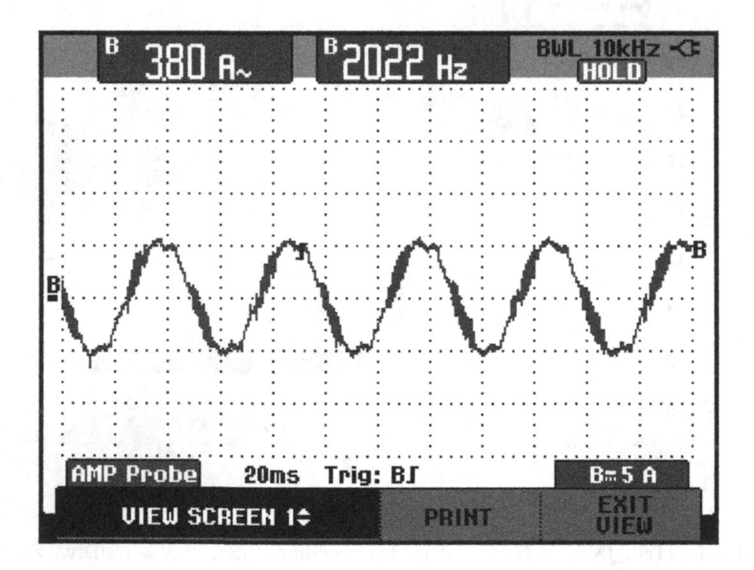

Fig. 10 Experimental current waveform at ~20 Hz, (Scale: X = 20 ms./Div., Y = 5 Amp./Div.)

In Fig. 10, it shows the hardware results of phase current at 20 Hz from rated frequency (50 Hz) where 3.80 A current is flowing in the motor (10 HP) at no load condition.

In Fig. 11, it shows experimental result of the torque in N.M., at no load condition and rated frequency (50 Hz) its look like I_q current. Hence, it is observed

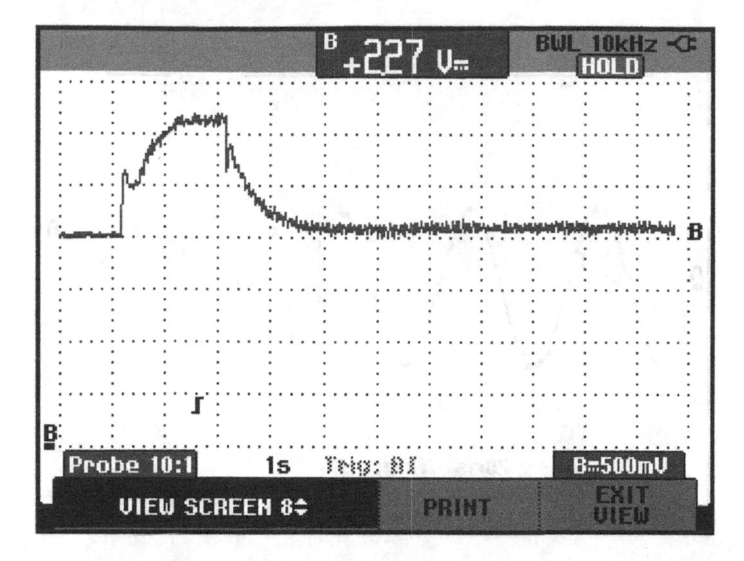

Fig. 11 Experimental torque waveform at no load condition (Scale: $X = 1$ s./Div., $Y = 500$ mV (10 N.M.)/Div.)

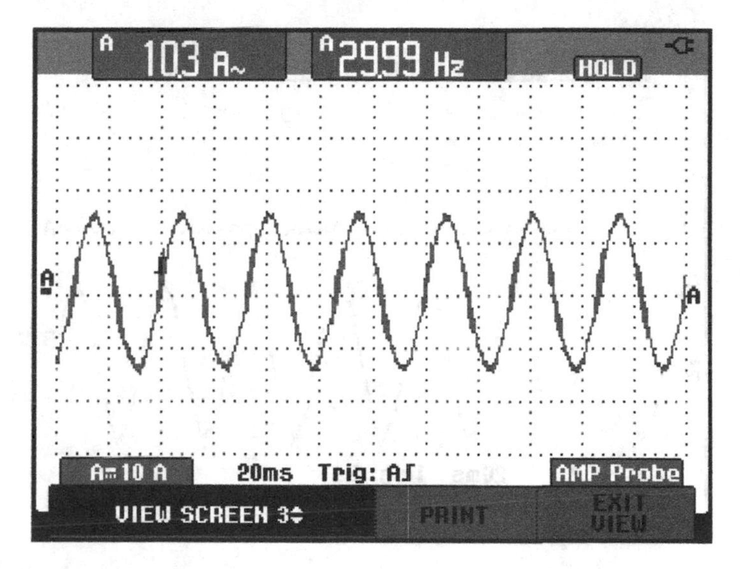

Fig. 12 Experimental current waveform at ~30 Hz (15 HP) (Scale: $X = 20$ ms./Div., $Y = 10$ Amp./Div.)

from Figs. 8 and 9; experimental results of estimate and actual speed are in line with the simulation results.

In Fig. 12, it shows the experimental current waveform at ~30 Hz, at no load condition, this current waveform is for 15 HP induction motor.

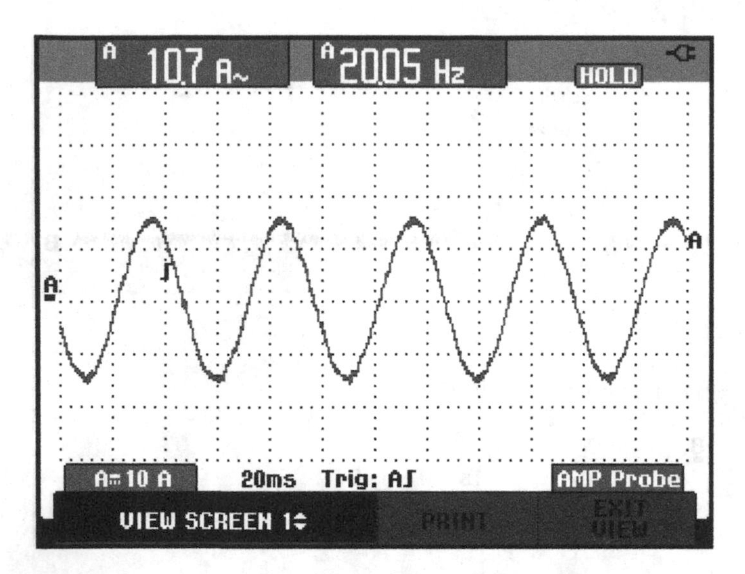

Fig. 13 Experimental current waveform at ~20 Hz (15 HP) (Scale: $X = 20$ ms./Div., $Y = 10$ Amp./Div.)

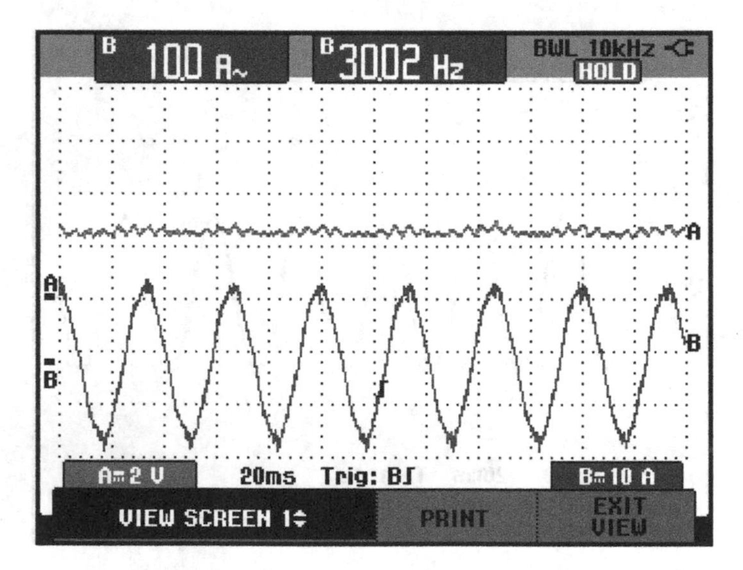

Fig. 14 Experimental current and torque waveform at ~30 Hz (15 HP) (For A (Torque): Scale: $X = 20$ ms./Div., $Y = 2$ V (10 N.M.)/Div.) (For B (Current): Scale: $X = 20$ ms./Div., $Y = 10$ Amp./Div.)

In Fig. 13, it shows the Experimental Current waveform at ~20 Hz, at no load condition, this current waveform is for 15 HP induction motor.

In Fig. 14, it shows the hardware results of current and torque, at no load condition for 30 Hz Frequency current is 10 A.

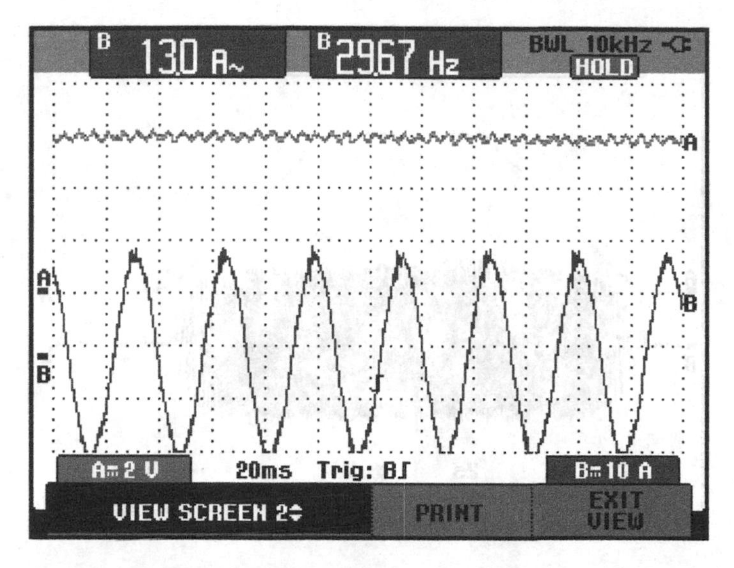

Fig. 15 Experimental current and torque waveform (50% load) at ~30 Hz (15 HP) (For A (Torque): Scale: X = 20 ms./Div., Y = 2 V (10 N.M.)/Div.) (For B (Current): Scale: X = 20 ms./Div., Y = 10 Amp./Div.)

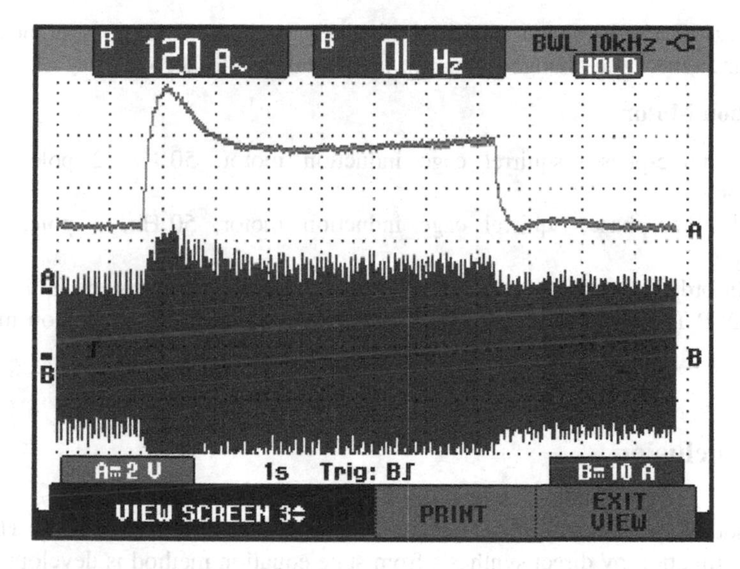

Fig. 16 Experimental current and torque waveform at ~30 Hz (15 HP) (For A (Torque): Scale: X = 1 s./Div., Y = 2 V (10 N.M.)/Div.) (For B (Current): Scale: X = 1 s./Div., Y = 10 Amp./Div.)

In Fig. 15, it shows the hardware results of current and torque, at 50% load condition for 30 Hz Frequency current is increasing (13 A).

In Fig. 16, it shows the hardware results of current and torque, given step in load change in currents has been seen.

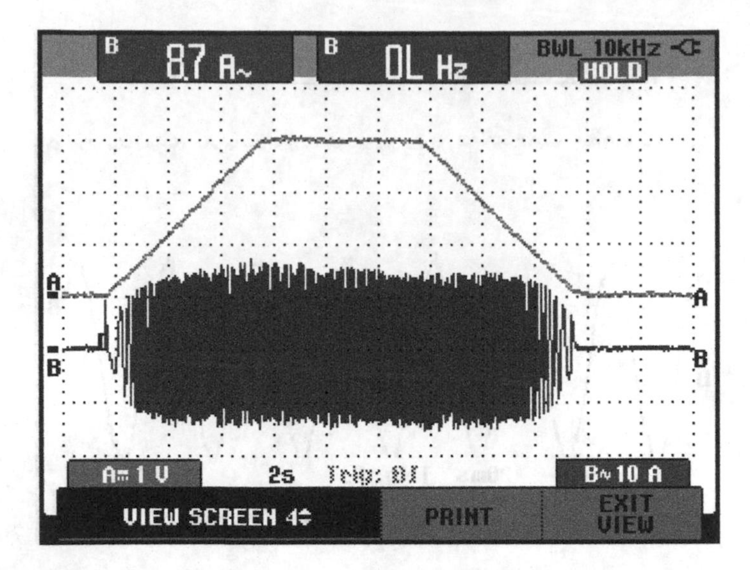

Fig. 17 Experimental current and speed waveform at ~30 Hz (15 HP) (For A (Speed): Scale: X = 2 s./Div., Y = 1 V (300 R.P.M.)/Div.) (For B (Current): Scale: X = 2 s./Div., Y = 10 Amp./Div.)

In Fig. 17, experimental results are obtained for step change in reference speed command where the change in current is also taken into account.

Induction Motor

10 HP: three-phase, squirrel cage induction motor, 50 Hz, 2 pole, 415 V, 3000 rpm.
15 HP: three-phase, squirrel cage induction motor, 50 Hz, 4 pole, 415 V, 1500 rpm.
DSP, In order to validate the developed model laboratory prototype is developed using DSP TMS320C2811 for 10 HP and 15 HP squirrel cage induction motor.

6 Conclusion

A sensorless vector-controlled induction motor drive for 10 HP and 15 HP with speed estimation by direct synthesis from state equation method is developed. This work presents simulation analysis of the proposed system using PSIM for various operating frequency. The results show that the estimated speed and actual speed are same. Laboratory prototype for the proposed drive is developed, and experimental results are presented. Prototype Experimental results are similar to the simulation results and show the satisfactory operation of developed sensorless vector control of induction motor drive.

Appendix

Parameter of 10 HP Induction motor:

$R_s = 0.7215\ \Omega$, $R_r = 0.4357\ \Omega$, $L_{ls} = 0.003351$ H, $L_{lr} = 0.003351$ H, $L_m = 0.1694$ H, $L_s = 0.172751$ H, $L_r = 0.172751$ H, Moment of Inertia (MI) = 0.0980 J.

Parameter of 15 HP Induction motor:

$R_s = 0.4748\ \Omega$, $R_r = 0.1900\ \Omega$, $L_{ls} = 0.0028745$ H, $L_{lr} = 0.0028745$ H, $L_m = 0.06838$ H,
$L_s = 0.0712545$ H, $L_r = 0.0712545$ H, Moment of Inertia (MI) = 0.229 J.

References

1. Raju Yanamshetti, Sachin S. Bharatkar, Debashis Chatterjee, and A.K. Ganguli, 'A simple DSP based speed sensorless field oriented control of induction motor. International Journal of Modelling and Simulation of Systems', Vol. 1, no. 4 (2010): 213–218.
2. Bimal Bose, '*Modern Power Electronics and AC Drives*' Prentice Hall, (2001).
3. Paul C. Krause, Oleg Wasynczuk, Scott D, '*Analysis of Electric Machinery and Drive Systems*', 2002.
4. J. Holtz, "Sensorless control of induction motor drives," in Proceedings of the IEEE, vol. 90, no. 8, pp. 1359–1394, Aug 2002.
5. J. Holtz and Juntao Quan, "Sensorless vector control of induction motors at very low speed using a nonlinear inverter model and parameter identification," in IEEE Transactions on Industry Applications, vol. 38, no. 4, pp. 1087–1095, Jul/Aug 2002.
6. E. D. Mitronikas and A. N. Safacas, "An improved sensorless vector-control method for an induction motor drive," in IEEE Transactions on Industrial Electronics, vol. 52, no. 6, pp. 1660–1668, Dec. 2005.
7. C. Caruana, G. M. Asher and M. Sumner, "Performance of HF signal injection techniques for zero-low-frequency vector control of induction Machines under sensorless conditions," in IEEE Transactions on Industrial Electronics, vol. 53, no. 1, pp. 225–238, Feb. 2006.
8. R. Raute et al., "A review of sensorless control in induction machines using hf injection, test vectors and PWM harmonics," 2011 Symposium on Sensorless Control for Electrical Drives, Birmingham, 2011, pp. 47–55.
9. N. C. Park and S. H. Kim, "Simple sensorless algorithm for interior permanent magnet synchronous motors based on high-frequency voltage injection method," in IET Electric Power Applications, vol. 8, no. 2, pp. 68–75, February 2014.
10. I. Benlaloui, S. Drid, L. Chrifi-Alaoui and M. Ouriagli, "Implementation of a New MRAS Speed Sensorless Vector Control of Induction Machine," in IEEE Transactions on Energy Conversion, vol. 30, no. 2, pp. 588–595, June 2015.
11. Texas Instruments, 'Digital Motor Control Software Library', October 2003.

Author Biographies

Rajendrasinh Jadeja is Professor and Dean-Faculty of Engineering at Marwadi Education Foundation's Group of Institutions. His area of interest includes control of power electronic converters, AC drives, Pulse Width Modulated power electronic converters, and Power Quality Conditioning Equipment. He is an active member of Global Engineering's Dean Council (GEDC). He is also a member of American Society for Engineering Education and American Education Research Association.

Ashish Yadav is PG Scholar at Marwadi Education Foundation's Group of Institutions. He completed his B.E. from Marwadi Education Foundation's Group of Institutions. His area of interest include Pulse Width Modulation techniques and control of drives.

Tapankumar Trivedi is Asst. Professor at Marwadi Education Foundation's Group of Institutions. He completed B.E. from Vishwakarma Government Engineering College, Chandkheda and M. Tech. from Indian Institute of Technology (IIT), Roorkee. His areas of interest include Power Quality Improvement devices, control of power electronic converters, and Multilevel Inverter-fed IM Drives.

Siddharthsingh Chauhan is Associate Professor at Institute of Technology, Nirma University, Ahmedabad. His area of interest includes Advanced Current Controllers, Digital Signal Processor Based Power Electronic Systems and Power Electronics applications to Power Systems. Dr. Siddharthsingh Chauhan is Reviewer for various journals—IEEE Transactions on Industrial Electronics, IEEE Transactions on Power Electronics, IET Power Electronics, Electric Power Components and Systems—Taylor & Francis, Elsevier Journals, etc.

Vinod Patel is currently working with R&D department at Amtech electronics (I) ltd, as AGM (R&D). He had received his B. E. (Instrumentation and control) from L. D. College of Engineering, Ahmadabad, India in 1994. He had varied experience of about 22 years in the field of power electronics, electrical drives and DSP based control of Industrial automation products. He had worked with development of specialized power supplies for plasma applications, converter/inverter for wind power generation, traction drives, Active Harmonic Filters, specialized high frequency magnetic, different types of IGBT drivers, DSP control board development. He is currently involved with development of specialized power supplies, Vector control and sensor less control of AC/DC drives and wind generators, multilevel inverters, solar inverters. He has guided more than 85 PG students for their project work in different area.

Artificial Intelligence-Based Technique for Intrusion Detection in Wireless Sensor Networks

Gauri Kalnoor and Jayashree Agarkhed

Abstract Network with large number of sensor nodes distributed spatially is termed as Wireless Sensor Network (WSN). The tiny devices called as sensor nodes are cheap, consume less power, and the capabilities of computation is limited. The most challenging issue for WSN is protecting the network from misbehavior of intruders or adversaries. One of the major techniques used to prevent from any type of attack in the sensor network is artificial intelligence system (AIS). Intrusion Detection System (IDS) is considered to be the second line of defense, as sensor nodes are first defense line. WSNs are highly vulnerable to intrusions and different types of attacks. In most critical applications of WSN, the human intervention or some physical devices are not sufficient for protecting the network from strong adversaries and attacks. Thus, artificial intelligence techniques are used for intrusion detection and prevention of sensor networks.

Keywords Artificial intelligence · IDS · WSN · Attack · Computational intelligence

1 Introduction

Wireless Sensor Network (WSN) is a collection of tiny devices spatially distributed sensor nodes. These nodes are independent and hence operate individually and do not require the authority centrally. Some of the limitations of sensor nodes in design aspects and functionality are mainly in terms of processing, storage, and communication [1]. WSNs are growing in vast areas of applications and considered to be the most emerging technology. It is growing consistently in many fields in our daily life. Its distributed, fast growing, design limitations, and wireless nature result in

G. Kalnoor (✉) · J. Agarkhed
P.D.A College of Engineering, Kalaburagi 585102, Karnataka, India
e-mail: kalnoor.gauri@gmail.com

J. Agarkhed
e-mail: jayashreeptl@yahoo.com

S.S. Dash et al. (eds.), *Artificial Intelligence and Evolutionary Computations in Engineering Systems*, Advances in Intelligent Systems and Computing 517,
DOI 10.1007/978-981-10-3174-8_69

frequent threats and attacks creating harm to the network. The best and high accurate security mechanism needs to be provided for preventing potential attacks and threats to take place [2, 3].

Most of the security mechanisms that are designed for protecting WSNs are packet or data encryption, authentication required to stop or restrict the adversaries attack on the packet, and malicious access by malicious users. Other type of security mechanisms are secure exchange of key, secure routing, and so on. Frequent analysis and timely responses are required for security of WSN, defending from different attacks. Detection of misbehavior act in the network is done by artificial intelligence system (AIS). The computational ability of sensor networks are more vulnerable to misbehaviors and malfunctions. Hence, a system which is computationally friendly is needed to protect WSN from intruders. An autonomous built-in mechanism is provided to the system to identify behavior of the user that can potentially damage the network.

Based on human immune system (HIS), Intrusion Detection System (IDS) is designed using artificial immune system that has the capability of protecting human from foreign germs. Artificial Neural Networks (ANN)-based IDS, Genetic algorithm-based IDSs are few of the artificial intelligence mechanisms. To improve the efficiency and reliability of the network, Artificial Intelligence techniques helps reducing alerts, incidents prioritizing, reducing or eliminating false alarms, and also increases self-learning [4] of response for an incident. It automates detection of an intruder by IDS and intervention of human is reduced. Artificial intelligence for IDS is classified into three types: Multi-agent based Computational Intelligence (MCI), Computational Intelligence (CI), and Traditional Artificial Intelligence (TAI).

In the next section, the survey of various authors with their proposed work is discussed based on the AI techniques.

2 Related Work

Artificial Intelligence techniques are used to improve performance of the system. Wide variety of research has been done in security of WSN. The works proposed by many authors are discussed in this section.

The ability to undertake a particular action against or many user's misbehavior is the major criteria of WSN. Thus the best solutions that can facilitate these criteria are Artificial Immune System. Shamshirband et al. [5], have discussed the principle of HIS used by Artificial Intelligence techniques. In Shamshirband et al. [6], the proposed system explains about the HIS having the ability to efficiently detect potential harmful foreign agents. The main aim of AIS is to identify the node's behavior if it has negative impact on the sensor network.

In Kaliyamurthie and Suresh [7], the basic design and its challenges are discussed by the authors. Most efficient set of genes that are suitable for deciding the node's behavior in the network is explained. The network's performance based on node's viewpoint is the important measure considered in Patel et al. [8].

The requirement for the system to perform efficiently is that the computation should be robust and easy against deception.

In Mavee et al. [9], the author specifies the misbehavior action in WSN can take place in three different forms: Dropping of a packet, data structures modification that was set for routing and packets modification. Fictitious node creation and skewing the topology of the network are also possible forms of misbehavior that occurs in WSN. One of the reasons for the sensor nodes to execute any kind of misbehavior is the desire to consume less battery life and make the WSN nonfunctional.

Based on probabilistic modeling, the authors in Kumar and Reddy [10] proposed the broadcasting mechanism called as knowledge acquisition in dynamic environment. In Fang et al. [11], the proposed system called multi-agent based IDPS and non CI methods are discussed. This system detects automatically and neutralizes the snippers of the enemy. In Hassan [12] and Jongsuebsuk et al. [13], the authors have discussed and derived intrusion detection probability of linear intruder that is detected by k-set of sensors in the range of maximum intrusion distance that is allowable in WSN.

In Chaudhary et al. [14], the author have characterized the detection probability rate in both the single-sensing and multiple-sensing scenarios of detection. A sleep-scheduling algorithm is discussed that guarantees minimum response time for detecting an intruder and also minimum consumption of power. The authors in Benaicha et al. [15] have discussed different types of mobility models that are designed for sensor nodes that have mobility nature moving in a random manner. Sensor mobility improves the network performance by sensor sensing coverage and also time for detection is also reduced using these mobility models.

In the next section, we will discuss the problem statement and the proposed system for intrusion detection in WSN.

3 IDS Based on Artificial Intelligence Technique

In this section, the techniques used in designing IDS for intrusion detection is discussed and explained.

3.1 Artificial Immune System

The security threats that occur in WSN are the major challenging issues. An anomaly-based IDS is designed using artificial immune technique for preventing such security threats against WSN. Immune-based mechanism [16] is used in many applications particularly for computing classification of data, optimization of the system, and detection of an intruder. IDS-based artificial immune system is designed and a general architecture using such technique is shown in Fig. 1.

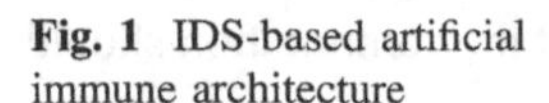

Fig. 1 IDS-based artificial immune architecture

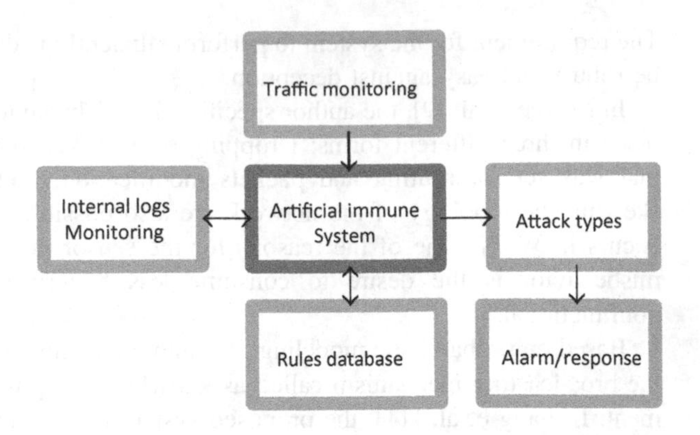

In WSN, the immune-based system can be applied easily compared to other types of network architecture. One of the recently introduced security threats or attack types in WSN is Interest Cache Poisoning (ICP) attack. This type of attack occurs at network layer and has the capability of routing packets disruption. Direct diffusion [17] and Dentritic Cell Algorithm (DCA) are applied to every sensor node for detecting misbehavior nodes and their respective antigens. The direct diffusion maintains two tables: data cache and interest cache. It handles two types of routing packets over WSN: data packets and interest packets.

To detect an ICP attack, DCA algorithm is applied to monitor two types of signals called as danger signals and safe signals [18]. An Intrusion prevention and prediction technique is also designed using immune systems. Also, another technique known as adaptive detection of threat is designed using immune systems. It is implemented at MAC layer, in which a gene is identified for detection. A natural selection algorithm is equipped in every node of WSN to monitor the behavior of traffic in the surrounding neighbor nodes. WSN demands a light weight IDS as it has constraints on resources such as energy, processing, and memory.

The other technique for artificial intelligence is designed considering the resource constraints of WSN.

3.2 Artificial Neuron Technique

This technique has the capability to learn and understand by method of training used in complex trends identification. It has two ways of detection an intruder [19]: feedforward and the feedback mechanism. The signals move from input to output considering only one direction for data flow. This is called feedforward. In feedback, the signals move in both the directions. In different areas, specially, in the field of pattern recognition and intrusion detection, the artificial neural method can be applied.

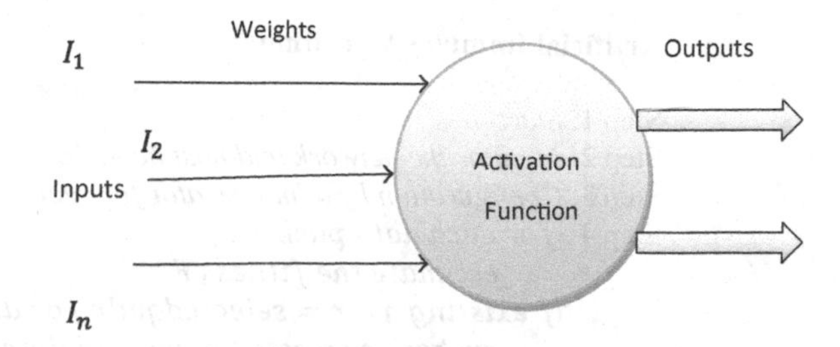

Fig. 2 Artificial neuron-based IDS architecture

The rule-based IDS has many limitations and it can be overcome by using artificial neural technique based on intrusion detection [20]. It uses both normal and abnormal type of data sets, which is more effective if it is trained properly. The general architecture based on IDS using this technique is shown in Fig. 2.

The newly introduced attack called as exhaustion of energy attack is detected using the artificial neuron mechanism. It is mainly designed for WSN when clustering is applied. Energy harvesting takes place in all the sensor nodes consisting of three layers: input, hidden, and output layer. Discriminating the events to be both normal and abnormal is the main aim that is trained to such systems. In channel access attack, the detection rate is found to be very low and is not suitable to be detected by this mechanism.

An unsupervised neural network is used for intelligent response mobile robot system for detection of intruder. Changes related to time are detected using this model. The investigation is done by the robot traveling in the infected area, after the intruder is detected.

Due to the deployment of sensor nodes in WSN is in harsh and unattended area, the attacker can manage to compromise few nodes which may further result in fault data forwarding to the associated sink of the network. Hence, a mechanism called as malicious node detection is used to increase accuracy in detection.

In the next section, the algorithms designed are discussed based on the proposed mechanisms and types of attacks that occur in WSN.

4 IDS-Based Algorithm Using Artificial Intelligence

In this section, the algorithm used for deetction of intruder using the techniques in artificial intelligence is discussed. Some of the primary assumptions made in the algorithm are:

1. The activities of the system are observable and
2. The normal and intrusive activities have distinct evidence.

Types of intruders considered are both internal and external intruders.

Artificial Immune Algorithm

Step 1: *Start*
Step 2: *Monitor the network and host behavior.*
Step 3: *Configuration by administrator for security.*
Step 4: $for\ each\ data\ packet\ P,$
$estimate\ the\ fitness\ F$
$if\ existing\ pair = selected\ pair\ for\ detecting$
$exchange\ selected\ pairs\ and\ detect$
$else$
$generate\ randome\ strings\ and\ verify$
Step 5: $for\ each\ self\ learning$
$if\ self\ strings = generated\ random\ strings$
$Reject$
$else$
$set\ detector$
Step 6: *Calculate the cost of the path discovered with Detection rate and FAR.*
Step 7: *Drop packets, if an intruder is detected.*
Step 8: *End.*

In the proposed algorithm, a random-generate and test process is used. A self-learning process determines if the random bit strings are generated by the sensor nodes, sent to each node of WSN and later self-strings which is already assigned to the nodes during initialization are matched. If a match is found true, then IDS is not activated and the detection process is rejected. Otherwise, detector is set and an alarm is sent to all the neighboring nodes. This process of self-learning is shown in Fig. 3.

At network layer, a four-layered architecture is designed for protecting the WSN from misbehaviors and abuses by adversaries. The two phases of the proposed architecture includes learning phase and detection phase. In the learning phase, there are two possibilities: learning and detection is implemented separately for each neighboring node or the learning and detection can be implemented at a single

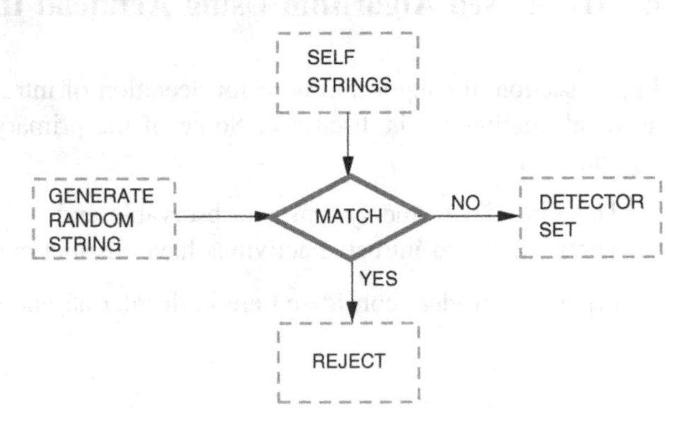

Fig. 3 Flow diagram for self-learning process

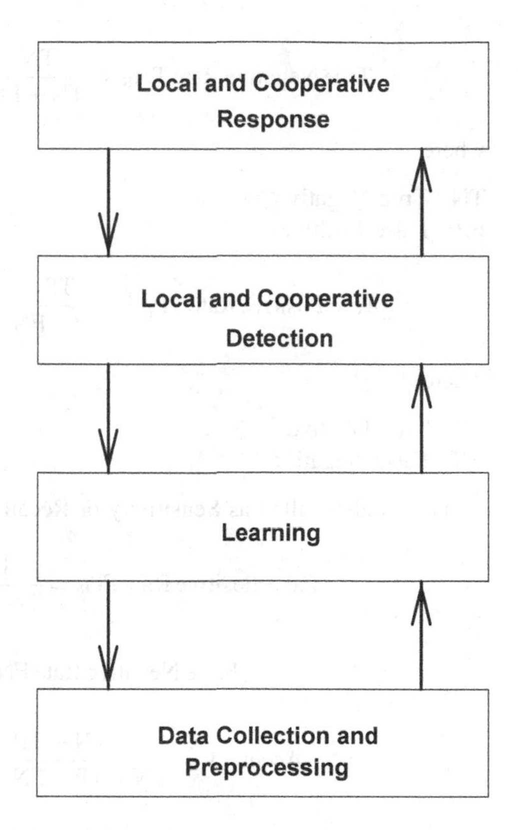

Fig. 4 A four layered architecture of IDS

instance for all the surrounded neighboring nodes in the predicted area of attack. This layered architecture is shown in Fig. 4.

A local response is given by each node which is cooperative and the local detection is performed. At lower level, the data is collected for preprocessing. This preprocessed data is learnt by each node in the learning phase. Later, at higher level the detection is performed locally with a local response sent at higher level.

5 Performance Results and Discussions

The performance of WSN is analyzed and the results obtained are discussed.

5.1 Performance Analysis

In this section, we discuss some of the measures for quantifying the performance of WSN when the proposed model is used for detecting intruder. The performance is calculated using the following Eqs. (1–6):

$$\text{True Negative Rate TN}_\text{R} = \frac{\text{TN}}{\text{TN} + \text{FP}} = \frac{\text{Total no. of True alerts}}{\text{No. of Alerts}} \tag{1}$$

where,

TN True Negative
FP False Positive

$$\text{True Positive Rate TP}_\text{R} = \frac{\text{TP}}{\text{TP} + \text{FN}} = \frac{\text{No. of attacks detected}}{\text{No. of observable attacks}} \tag{2}$$

where

TP True Positive
FN False Negative

This is also called as Sensitivity or Recall (R).

$$\text{False Positive Rate FP}_\text{R} = \frac{\text{FP}}{\text{TN} + \text{FP}} = 1 - \frac{\text{TN}}{\text{TN} + \text{FP}} \tag{3}$$

$$\text{False Negative Rate FN}_\text{R} = \frac{\text{FN}}{\text{TP} + \text{FN}} \tag{4}$$

$$\text{Accuracy} = \frac{\text{TN} + \text{TP}}{\text{TN} + \text{TP} + \text{FN} + \text{FP}} = \left[1 - \frac{n}{N}\right] 100 \tag{5}$$

$$\text{False Alarm Rate FA}_\text{R} = \left[\frac{n}{N}\right] 100 \tag{6}$$

where n is no. of n samples and N is the initial size of the command set.

5.2 *Results*

The artificial intelligence-based techniques are used for IDS in WSN and the results observed after applying different techniques for different types of attacks are discussed in Table 1.

In Table 1, the backpropagation mechanism used for AI has high accuracy rate for the flooding type of attack. But using genetic algorithm mechanism, the DoS attack type has very high accuracy rate when compared to all other mechanisms. In Table 2, the results of detection rate (DR) and false alarm rate (FAR) are shown for different methods used in WSN.

The results computed based on DR and FAR using different techniques are explained in Table 2. High detection rate and very low FAR is found for each method applied, so that the performance of WSN is increased [21].

Table 1 Performance results

AI mechanisms	Attack type	Accuracy (%)
DCA	ICP	Less than 70
Adaptive immune	Anomalies in network layer	Above 75
Co-evolutionary	DOS (denial of service)	50
Back propagation	Flooding	95
Unsupervised artificial neuron	Timely changes	88
Feedforward	Malicious nodes	98.74
Genetic algorithm	DoS	99
Artificial neuron	Energy exhaustion attack, routing attacks	96

Table 2 Results

AI method	Objective	Performance (DR/FAR) (%)
Combination of data mining and fuzzy rules	Identifying misused behavior	DR increased to 50
Fuzzy logic with double sliding window detection	Identifying abnormal behavior	DR = 99.97 FAR = 0.05
Genetic programming with grammatical evolution	Identify route disruption attacks	DR = 99.41 FAR = 1.23
Dynamic Bayesian model	Identify intrusion and improve performance of IDS	DR = 99.08 FAR = 1.9

Figure 5 shows the screenshot of the detection of intruders in WSN which includes n sensor nodes and an IDS designed for this purpose. AI mechanism is used for detecting the attacks as shown in the screenshot.

6 Conclusion

The development of information technology has high positive impact on WSN and causes critical issues that are difficult and time consuming to manage. The proposed system for attack detection for high order of traffic behavior is discussed so that the correlation with traffic and time consumption is maintained. Different mechanisms of artificial intelligence are designed such that the detection rate is increased and FAR is decreased accordingly. Based on the types of attacks, the mechanisms are proposed and the simulation results are discussed to show that the performance of the network increases. Thus, AI techniques are used to discover essence of intelligence and also intelligent machines can be developed. It is also applied for finding solutions or methods for complex problems by using some intelligent methods and proper decision making process. A sensor network is protected using such techniques and detecting or preventing intruders from damaging the network.

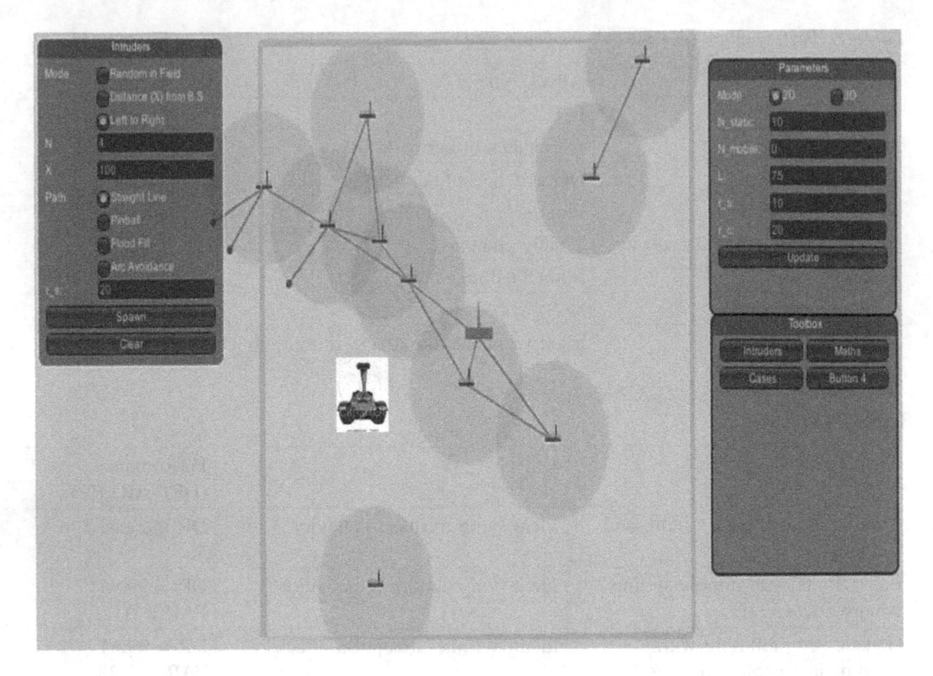

Fig. 5 Screenshot of detection in WSN

References

1. Tyugu, E. (2011, June). Artificial intelligence in cyber defense. In Cyber Conflict (ICCC), 2011 3rd International Conference on (pp. 1–11). IEEE.
2. Patel, A., Taghavi, M., Bakhtiyari, K., & Júnior, J. C. (2012). Taxonomy and proposed architecture of intrusion detection and prevention systems for cloud computing. In Cyberspace Safety and Security (pp. 441–458). Springer Berlin Heidelberg.
3. Çakır, H., & Sert, E. (2010). Bilişim Suçları ve Delillendirme Süreci. 2. *Uluslararası Terörizm ve Sınıraşan Suçlar Sempozyumu Bildirisi*, 143.
4. Artificial Intelligence, Wikipedia, http://en.wikipedia.org/wiki/Artificial_intelligence, (24/11/2014).
5. Shamshirband, S., Patel, A., Anuar, N. B., Kiah, M. L. M., & Abraham, A. (2014). Cooperative game theoretic approach using fuzzy Q-learning for detecting and preventing intrusions in wireless sensor networks. *Engineering Applications of Artificial Intelligence*, *32*, 228–241.
6. Shamshirband, S., Anuar, N. B., Kiah, M. L. M., & Patel, A. (2013). An appraisal and design of a multi-agent system based cooperative wireless intrusion detection computational intelligence technique. *Engineering Applications of Artificial Intelligence*, *26*(9), 2105–2127.
7. Kaliyamurthie, K. P., & Suresh, R. M. (2012). Artificial Intelligence Technique Applied to Intrusion Detection. *International Journal of Computer Science and Network Security (IJCSNS)*, *12*(3), 87.
8. Patel, A., Taghavi, M., Bakhtiyari, K., & Júnior, J. C. (2013). An intrusion detection and prevention system in cloud computing: A systematic review. *Journal of Network and Computer Applications*, *36*(1), 25–41.

9. Mavee, S. M., & Ehlers, E. M. (2012, June). A Multi-agent Immunologically-inspired Model for Critical Information Infrastructure Protection–An Immunologically-inspired Conceptual Model for Security on the Power Grid. In *Trust, Security and Privacy in Computing and Communications (TrustCom), 2012 IEEE 11th International Conference on* (pp. 1089–1096). IEEE.
10. Kumar, G. V., & Reddy, D. K. (2014, January). An agent based intrusion detection system for wireless network with artificial immune system (AIS) and negative clone selection. In *Electronic Systems, Signal Processing and Computing Technologies (ICESC), 2014 International Conference on* (pp. 429–433). IEEE.
11. Fang, X., Koceja, N., Zhan, J., Dozier, G., & Dipankar, D. (2012, June). An artificial immune system for phishing detection. In *Evolutionary Computation (CEC), 2012 IEEE Congress on* (pp. 1–7). IEEE.
12. Hassan, M. M. M. (2013). Network intrusion detection system using genetic algorithm and fuzzy logic. *International Journal of Innovative Research in Computer and Communication Engineering, 1*(7).
13. Jongsuebsuk, P., Wattanapongsakorn, N., & Charnsripinyo, C. (2013, May). Real-time intrusion detection with fuzzy genetic algorithm. In *Electrical Engineering/Electronics, Computer, Telecommunications and Information Technology (ECTI-CON), 2013 10th International Conference on* (pp. 1–6). IEEE.
14. Chaudhary, A., Tiwari, V. N., & Kumar, A. (2014, February). Design an anomaly based fuzzy intrusion detection system for packet dropping attack in mobile ad hoc networks. In *Advance Computing Conference (IACC), 2014 IEEE International* (pp. 256–261). IEEE.
15. Benaicha, S. E., Saoudi, L., Guermeche, B., Eddine, S., & Lounis, O. (2014, August). Intrusion detection system using genetic algorithm. In *Science and Information Conference (SAI), 2014* (pp. 564–568). IEEE.
16. Padmadas, M., Krishnan, N., Kanchana, J., & Karthikeyan, M. (2013, December). Layered approach for intrusion detection systems based genetic algorithm. In *Computational Intelligence and Computing Research (ICCIC), 2013 IEEE International Conference on* (pp. 1–4). IEEE.
17. Wattanapongsakorn, N., Srakaew, S., Wonghirunsombat, E., Sribavonmongkol, C., Junhom, T., Jongsubsook, P., & Charnsripinyo, C. (2012, June). A practical network-based intrusion detection and prevention system. In *Trust, Security and Privacy in Computing and Communications (TrustCom), 2012 IEEE 11th International Conference on* (pp. 209–214). IEEE.
18. Aziz, A. S. A., Salama, M. A., ella Hassanien, A., & Hanafi, S. E. O. (2012). Artificial immune system inspired intrusion detection system using genetic algorithm. *Informatica, 36* (4).
19. Barani, F. (2014, February). A hybrid approach for dynamic intrusion detection in ad hoc networks using genetic algorithm and artificial immune system. In *Intelligent Systems (ICIS), 2014 Iranian Conference on* (pp. 1–6). IEEE.
20. Sharma, S., Kumar, S., & Kaur, M. (2014). Recent trend in Intrusion detection using Fuzzy-Genetic algorithm. *International Journal of Advanced Research in Computer and Communication Engineering, 3*(5).
21. Patel, A., Qassim, Q., Shukor, Z., Nogueira, J., Júnior, J., Wills, C., & Federal, P. (2011). Autonomic agent-based self-managed intrusion detection and prevention system. In *Proceedings of the South African Information Security Multi-Conference (SAISMC 2010)* (pp. 223–234).

Multiconstrained QoS Multicast Routing for Wireless Mesh Network Using Discrete Particle Swarm Optimization

R. Murugeswari and D. Devaraj

Abstract With the growing demand of multimedia applications, quality of service (QoS) assured that multicast routing is an important issue in wireless mesh networks. Finding a multicast tree which satisfies the multiple constraints is a NP-Complete problem. In this paper, we propose a discrete particle swarm optimization (DPSO) approach for finding the minimum tree cost from a given source to a set of destination with delay and bandwidth constraint. The concept of relative position base indexing (RPI) is used to convert a continuous space to discrete space for multicast routing. The simulation is carried out in NS-2.26 and the comparison is made with respect to tree cost under two scenarios, namely varying node mobility and increasing number of nodes. The results demonstrate that DPSO has better speed, performance, and efficiency than multicast routing based on genetic algorithm.

Keywords Wireless mesh network · Discrete particle swarm optimization · Relative position base indexing

1 Introduction

Wireless mesh networks (WMNs) have been emerging for the past years as an important thing for providing lot of broadband Internet access services. WMNs are dynamically self-organized, self-configured and the nodes in the networks are automatically creating and maintaining mesh connectivity among themselves.

R. Murugeswari (✉)
Department of Computer Science and Engineering, Kalasalingam University, Krishnankoil 626126, Tamilnadu, India
e-mail: murugesananth@gmail.com

D. Devaraj
Department of Electrical and Electronics Engineering, Kalasalingam University, Krishnankoil 626126, Tamilnadu, India
e-mail: deva230@yahoo.com

S.S. Dash et al. (eds.), *Artificial Intelligence and Evolutionary Computations in Engineering Systems*, Advances in Intelligent Systems and Computing 517,
DOI 10.1007/978-981-10-3174-8_70

These features bring many advantages to WMN such as ease of deployment, low installation cost, and reliability. In WMN, nodes act as both relays, forwarding traffic to or from other mesh nodes. A WMN [1] is composed of wireless mesh routers (MRs) and mesh clients (MCs). Mesh routers provide connectivity between mesh clients which form a static mesh networking infrastructure called a wireless mesh backbone. The end user devices such as mesh clients with wireless access capability may change their locations frequently. One or more MRs in WMN performs as the Internet gateways and it enables to integrate WMN with existing wireless networks.

As a result of broadcasting nature of wireless communications and the community-oriented nature of WMNs, group communications based on multicasting [2] are expected to be a common communication paradigm in WMNs. Multicast [3–5] is the delivery of information simultaneously from a source to multiple destinations and it reduces the channel bandwidth, delivery delay, and router processing time. The messages are delivered over every link of the network only once, generate copies only when the links to the destination split. It delivers underlying network support for collaborative multimedia applications such as video conference, e-learning, content distribution, and distance education.

Multicast applications in WMN require guaranteed quality of service (QoS), and it is difficult to design QoS in multicast routing schemes for WMNs. The goal of QoS routing is to find the best path which meets the constraints of some QoS metrics in the network, the metrics include bandwidth, delay, delay jitter, packet loss rate, etc. NP-complete problem for finding feasible paths with multiconstraints cannot be solved precisely in polynomial time. To reduce its complexity, a heuristic or approximate algorithm must be designed. The evolutionary technique takes advantage to solve these problems. Many algorithms based on ant colony optimization, genetic algorithm, and intelligent computational methods (i.e., genetic algorithm, simulated annealing, and tabu search) to find an optimal multicast tree have been proposed [6–8].

Jun Zhuo et al. [9] described an improved discrete particle swarm optimization (IDPSO) approach for QoS Multicast routing problem which simultaneously considers QoS parameter as Maximum delay, cost of the tree, average delay. These algorithms include the mutation and crossover operations in genetic algorithm. In the first part of velocity updation, the mutation operation is included. The crossover operator is used for the second and the third part of velocity updation. Experimental results demonstrate that discrete PSO algorithm can not only produce cheaper routes with lower average delay than KPP and Shortest Path Tree (SPT) in several cases, but also find a set of nondominated multicast routes to distribute on the Pareto front uniformly. The multiobjective routing algorithm for wireless mesh network using discrete particle swarm optimization proposed by Muhua Zhong and Yuzhong Chen [10]. This algorithm simultaneously minimizes three objectives, namely delay, packet success rate, and available bandwidth of the path. The available bandwidth of the path is calculated using contention graph. It considers only the intra-flow interference in WMN. Simulation results show that discrete PSO-based QoS routing is efficient than enumeration method and GA algorithm.

Zongwu Ke et al. [7] proposed a multiconstrained QoS-based multicast routing algorithm using Genetic Algorithm. The constraints are bandwidth, delay, and cost. The edge set is used to represent the multicast tree. Mesh client sent the multicast requirement to its nearby mesh router, and the mesh router searches an optimal route, then send the route to the source node. The Simulation results have shown that GA solve the problem caused by random selection of base GA. Ant colony-based multiconstraints QoS aware service selection algorithm was proposed by Neeraj Kumar et al. [6]. Ants are launched from the source node and it is based upon the guided search evaluation (GSE) criterion. The best path is chosen based on the cost effective (CE) metric which is a combination of various parameters such as hop count, service availability, and jitter. The results show that proposed algorithm is more effective than improved ant colony QoS routing algorithm (IAQR) algorithm with regard to convergence, end-to end delay, and service availability. A simulation-based performance comparison of Shortest Path Trees (SPT), Minimum Steiner Tree (MST), and Minimum number of transmission (MNT) trees in WMNs presented by Uyen Trang Nguyen [11]. The simulation results show that SPTs are better performance to multicast flows than MCTs and also support for dynamic join and leaves. Hui Cheng et al. [8] described the quality of service multicast routing and channel assignment problem using three intelligent computational methods such as genetic algorithm, simulated annealing, and tabu search. This method consists of problem formulation, solution representation, fitness function, and the channel assignment algorithm. To search the minimum-interference, multicast trees satisfy the end-to-end delay constraint of WMN and it is achieved by intelligent computational method. It also optimizes the usage of the scarce radio network resource in WMNs. Simulation results show that the proposed method attained better performance than shortest path trees and level channel assignment(LCA) multicast algorithm regarding total channel conflict and tree cost.

Meie Shen et al. [12] developed a bi-velocity discrete particle swarm optimization (BVDPSO) approach to optimize multicast routing problem in communication networks. A novel bi-velocity strategy is used to represent the possibilities of each dimension between 1 and 0, where 1 represents a node which is selected to construct the multicast tree, whereas 0 stands for otherwise. This algorithm updates the velocity and position based on the original PSO. The simulation results demonstrate that BVDPSO provides better solutions with higher accuracy than the heuristic methods and with faster convergence speed than the GA and previous PSO-based methods. A set-based PSO vehicle routing problem with time windows (S-PSO-VRPTW) is proposed by Yue-Jiao Gong [13] to solve the VRPTW. The discrete search space is an arc set of the complete graph that is defined by the nodes in the VRPTW and the subset of arcs represents the candidate solution. In this algorithm, operators are defined on the set rather than the arithmetic operators in the original PSO algorithm. Each particles position represents a set of delivery routes and defined as a subset of arcs. Experimental results show the effectiveness and efficiency of the proposed algorithm which provides better results than Solomons instances, two GAs, and a hybrid ACO algorithm.

A novel multicast routing in mobile ad hoc network based on PSO algorithms was proposed by Alireza Sajedi Nasab [14]. The constraints are delay and battery energy of the node. The particle is encoded using extended sequence and topology-based encoding. The spanning tree is created using prims algorithm. After creating trees, preorder traversal algorithm gives a sequence number specifing the index of nodes in the tree. The simulation results show that PSO algorithm has better performance than GA. Yannis Marinakis [15] introduces a new hybrid algorithmic approach based on particle swarm optimization (PSO). PSO algorithm is suitable for continuous optimization problems. To convert continuous to discrete values the relative position-based indexing method is used. A number of various alternatives PSO are tested and the proposed algorithm performs better for solving all benchmark instances.

The rest of the paper is described as follows. In Sect. 2, the Mathematical problem formulation of WMN is given. The proposed Discrete PSO for solving the Multicast routing problem is discussed in Sect. 3. In Sect. 4, the simulation details and result of the algorithm are presented and analyzed. We summarize the conclusion and future work which is given in last section.

2 Problem Formulation

The Wireless Mesh Network can be modeled as an undirected graph $G = (V, E)$, where V is the set of nodes and E is the set of all edges representing the connection between nodes. Let $d(v_0, v_1)$ be the distance between two nodes v_0 and v_1. An edge $(v_0, v_1) \in E$ between two nodes exists if $d(v_0, v_1) \in R$, where R is the transmission range. We define the set MT as the set of possible multicast trees in G. In the Multicast communication, the source node is $S \in V$ and the set of destination is $U = (u_1, u_{2,}, u_n) \in V - S$, where the node u_i is the group member of U and n refers to the number of members.

The multicast routing problem is stated as, Given a graph G (V, E) find a multicast tree MT $\in G$ of an optimal multicast route that minimizes the cost of multicast tree MT satisfy the QoS parameters delay and bandwidth. Mathematically this is stated as

$$\text{Minimize} \cos t_{p(S,U)} = \sum_{(i,j) \in p(S,U)} c_{ij}$$

Subject to

$$\text{Delay}_{p(S,U)} = \sum_{(i,j) \in p(S,U)} d_{ij} \leq D$$

$$\text{Bandwidth}_{p(S,U)} = \min_{(i,j)\in p(S,U)} \{b_{ij}\} \geq B$$

where d_{ij} is the delay of link (i, j), c_{ij} is the cost of link (i, j) used as hop count, b_{ij} is the bandwidth of link (i, j), $p(S, U)$ is the path from source node S to destination node U, D is the maximum end-to-end delay constraint and the minimum bandwidth constraint of multicast tree as B.

3 Particle Swarm Optimization: A Brief Overview

Kennedy and Eberhart [16] proposed the particle swarm optimization (PSO) algorithm. It is a population-based stochastic optimization technique to simulate the social behavior of social organisms such as bird flocking and fish schooling. Individuals in the population are referred to as particles that are associated with velocities and the individuals move around a multidimensional search space seeking an optimal or good solution. In PSO algorithm, initially a set of particle is created randomly where each particle represents a possible solution. Every particle consists of current position, velocity, *p*best and *g*best. In this technique every particle move toward the direction of particle which is close to the solution and after some time all particles ends with a particular solution which is an optimal solution. After each iteration, *p*best and *g*best are updated for every particle if better or more fitness function is found. This process continues iteratively until either the required result is converged or it reaches the user defined maximum number of iteration.

The position of each particle is represented by X_{ij} and its performance is evaluated by the predefined fitness function ($f(X_{ij})$). The velocity of each particle (V_{ij}) represents the changes that will be made in the particle to move from one position to another. In each iteration, every particle adjusts its velocity V_{ij} and position X_{ij} using the following equation:

$$V_{ij}(t+1) = wV_{ij}(t) + C_1\,\text{rand1}(p\text{best}_{ij} - X_{ij}(t)) + C_2\text{rand2}(g\text{best}_j - X_{ij}(t)) \quad (1)$$

$$X_{ij}(t+1) = X_{ij}(t) + V_{ij}(t+1), \quad (2)$$

where w is inertia weight and it is used to control the impact of previous histories of velocities on the current velocity, c_1 and c_2 are acceleration coefficients, r_1 and r_2 are two independent-generated random numbers in the interval [0, 1]. $p\text{best}_{ij}$ is the best solution this particle has reached; $g\text{best}_i$ is the global best solution of all the particles until now and t represents the index of the generation.

The design method and the algorithm process of PSO is easy to understand and simple to implement. PSO algorithms have a global search ability, fast convergence speed, strong robustness, etc. In recent years, PSO algorithms have been widely used in several applications [17–20].

4 Discrete Particle Swarm Optimization for QoS Multicast Routing

The original PSO is simple and efficient, but it is mostly used to find solutions in a continuous space. The multicast routing problem is a discrete problem, standard PSO is not appropriate for this problem and some modification must be done. In order to represent a set of solutions, the particle is encoded to an integer sequence in which each number stands for a path from the source node to a destination node.

4.1 Encoding

In our multicast routing problem, the method of integer encoding is used to represent the particle. Each particle represents the sequence of nodes specifically a path from source to destination node. For example, a particle $P = [v_1, v_2, v_3, v_t]$, where v_1 denotes source node, v_t denotes destination node.

4.2 Particle Initialization

The Multicast tree can be generated using RandomWalkRSMT [21]. Consider S as the Multicast source node and D is the set of receivers. This algorithm starts with the source node S and it moves randomly to choose one of its neighbors within the transmission range. This process is repeated until all the destination nodes are found and finally return the multicast trees.

Algorithm 1. RandomWalkRSMT Algorithm

1. MT= φ.
2. D=φ.
3. S=V0.
4. D=Set of receivers.
5. Visited []=S.
6. While D not equal to D do
7. V1 =choose a random neighbor of V0
8. If V1 is not marked then
9. MT= MT U V0 , V1
10. Visited [ ] = visited [ ] U V1;
11. If (V1= D) then
12. D= D U V1.

13. End if
14. End if
15. V1=V0;
16. End while;
17. Return MT;

After obtaining the input multicast tree using RandomWalkRSMT algorithm, we compute all possible paths from source node to every destination nodes by applying Depth First Search (DFS) recursively.

Consider a particle with six nodes from the possible paths. Here 1 is the source node and 6 is the destination node.

$$X_{ij} = \begin{bmatrix} 1 & 3 & 2 & 4 & 5 & 6 \\ 1 & 3 & 5 & 4 & 2 & 6 \\ 1 & 4 & 5 & 2 & 3 & 6 \end{bmatrix}$$

At the starting of algorithms initial paths are the current personal best paths of particles defined by

$$Pbest = \begin{bmatrix} 1 & 3 & 2 & 4 & 5 & 6 \\ 1 & 3 & 5 & 4 & 2 & 6 \\ 1 & 4 & 5 & 2 & 3 & 6 \end{bmatrix}$$

And the global best is the particles which have smallest fitness function (Tree cost) among all particles defined by

$$G_{\text{best}} = [1 \quad 4 \quad 5 \quad 2 \quad 3 \quad 6]$$

After initialization of particles, updation took place. These particles are in discrete form, so updation cannot take place as normal continuous value. To address this problem, in this paper we have used relative position-based indexing (RPI) [22]. This approach transforms every element of the particle into a float interval [0, 1], compute the velocities and the position of all particles, and then convert back the particles position into the integer domain.

The transformation from integer to floating-point values is achieved by dividing every element of the vector by the biggest one of them. Thus, the position vector of the previous particle has become

$$X_{ij} = \begin{bmatrix} 1 & 0.6 & 0.4 & 0.8 & 1 & 6 \\ 1 & 0.6 & 1 & 0.8 & 0.4 & 6 \\ 1 & 0.8 & 1 & 0.4 & 0.6 & 6 \end{bmatrix}$$

After applying Eq. (1) the velocity of the particle becomes

$$V_{ij} = \begin{bmatrix} 1 & 0.2 & 0.6 & 0.4 & 0.4 & 6 \\ 1 & 0.2 & 0 & 0.4 & 0.2 & 6 \\ 1 & 0 & 0 & 0 & 0 & 6 \end{bmatrix}$$

Using Eq. (2) the position of the particle is

$$X_{ij} = \begin{bmatrix} 1 & 0.8 & 1.0 & 1.2 & 1.4 & 6 \\ 1 & 0.8 & 1 & 1.2 & 0.6 & 6 \\ 1 & 0.8 & 1 & 0.4 & 0.6 & 6 \end{bmatrix}$$

The conversion of real particle into integer domain is performed by replacing the smallest floating-point value by the smallest integer value, and subsequently replaces the value by the next integer value until all the elements have been converted. The final particle position is

$$V_{ij} = \begin{bmatrix} 1 & 2 & 3 & 4 & 5 & 6 \\ 1 & 3 & 4 & 5 & 2 & 6 \\ 1 & 4 & 5 & 2 & 3 & 6 \end{bmatrix}$$

This process can be repeated until the maximum number of generation is reached. This approach always yields a feasible solution, except when two or more floating values are same.

4.3 Fitness Function

PSO generally maximizes the fitness function. Hence the minimization objective is converted into maximization objective as

$$F = \left\{ \begin{array}{ll} 1/\cos t(\mathrm{MT}) & \mathrm{bandwidth}(\mathrm{MT}) \geq B \wedge \mathrm{delay}(\mathrm{MT}) \leq D \\ 0 & \mathrm{bandwidth}(\mathrm{MT}) < B \vee \mathrm{delay}(\mathrm{MT}) > D \end{array} \right\}.$$

4.4 Particle Repair

When the conversion between two operational domains (real to integer) is applied infeasible solution may be created. Three unique strategies have been described by Godfrey and Donald [22] to repair the replicated values. In this paper, back mutation method is used. Back mutation indexes repair the rear of the replicated array with the values selected randomly from the missing array value. The steps are as follows:

1. To find all repetitive values in the infeasible solution.
2. An array of missing value is generated based on the replicated position.
3. An insertion array is randomly generated based on the missing values. At Last, the feasible solution is generated.

4.5 Discrete PSO Algorithm

The random multicast tree can be generated using RandomwalkMST. After generating the tree, we can get the entire possible path from source to a set of destination using depth first search algorithm. Calculate the fitness value of each path. Using relative position-based indexing, we can convert integer to floating path conversion and then change the velocity and particle position using Eqs. (1) and (2) and finally it converts the floating path to integer using RPI method. This process is repeated until we can obtain the maximum number of generation. The pseudo code of the algorithm is given below.

Algorithm 2. Discrete PSO Algorithm

1. Initialize the Multicast tree using RandomwalkMT
2. Select all possible particle using DFS
3. Initialize the velocity and position of each particle.
4. Calculate the fitness value of each particle
5. Compute the best solution of each particle
6. Find the best particle of the entire swarm
7. Do while the maximum number of iterations has not be reached
8. Convert particles positions into continuous form using RPI
9. Compute the velocity of each particle
10. Calculate the new position of each particle
11. Modify the best solution of each particle
12. Find the best particle of the whole swarm
13. Convert particles positions into integer form using RPI
14. End do
15. Return the best particle (best solution)

5 Experimental Setup

The Simulation study is performed in a WMN with 30 nodes. The source node is 1 and the destination nodes are 5, 10, 12, 15, 21. The mobility speed is considered in the range [0, 25] m/s. The transmission ranges of all nodes are fixed as 250 m. The nodes are allowed to move randomly as random waypoint mobility model. Only one channel on each wireless link is considered. We have selected NS-2.24 for the simulation and the protocol used is Multicast Adhoc On demand Distance Vector (MAODV) protocol. Each experiment is executed for 500 s of simulated time. The terrain size for simulation is 1000 m × 1000 m for all experiments. The total number of population is 20 and the maximum number of generation is 100. The value of c_1, c_2, w, rand1 and rand2 is 1. The proposed DPSO algorithm is compared with Genetic Algorithm, where the crossover probability is taken as 0.8 and mutation probability 0.1. The tree cost is obtained by calculating the average result of 20 runs. The average value with standard deviation is plotted as error graphs with 95% confidence interval.

5.1 Performance Analysis

In this section, we have described the simulation results as the performance evaluation of our algorithm. The same sender, receivers and the same network configuration were used for all the algorithms (GA and DPSO). All receivers joined a multicast group at the starting and it stayed until the whole group terminated. Figure 1 compares the running time between DPSO and GA algorithm with different number of nodes. When we increase the number of nodes, DPSO algorithm finds optimal multicast tree in less time than GA. Hence the running time of DPSO algorithm is lesser than GA algorithm.

Figure 2 shows the tree cost values in various generations which are presented. Tree cost found by DPSO algorithm in each generation is lower than GA algorithm. DPSO has no crossover and mutation operator and converged as fast, this leads automatically to reduction in execution time.

Figure 3 indicates the tree cost of DPSO and GA algorithm with increase in the number of nodes when the maximum speed of mesh client is 5 m/s. The tree cost is

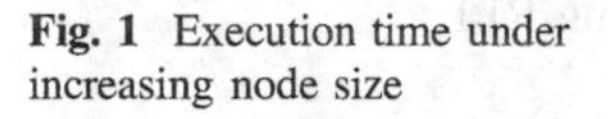

Fig. 1 Execution time under increasing node size

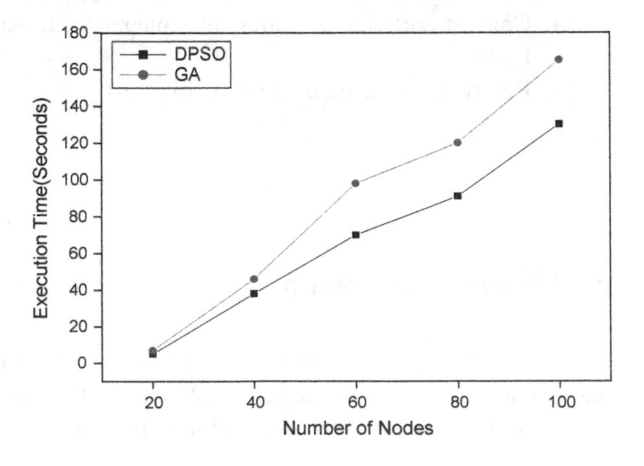

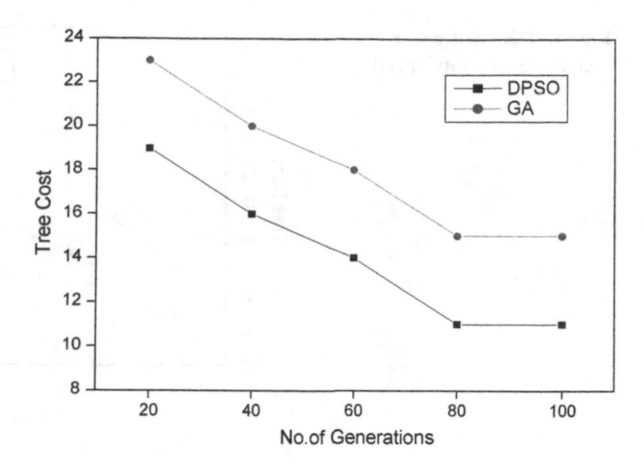

Fig. 2 Tree cost under increasing no. of generation

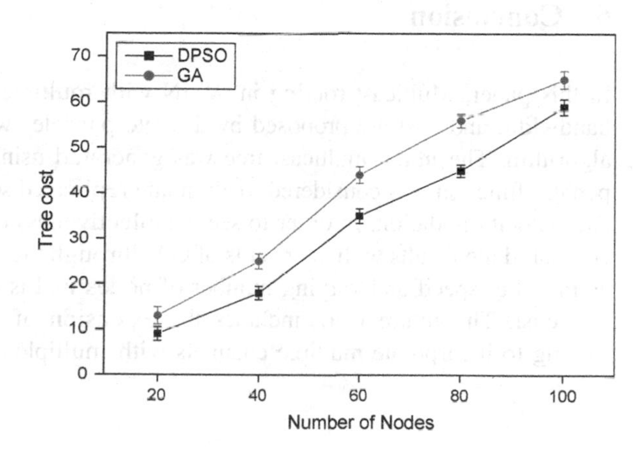

Fig. 3 Tree cost under increasing node size

increased with the increasing number of nodes. It is obvious that DPSO perform better than GA.

Figure 4 shows the tree cost for 30 node network by varying the node mobility. Compared to GA, DPSO performs well in case of low node mobility as well as in the high node mobility. To increase the mobility speed, the paths between communication end points will break frequently. Using back mutation method, the redundancy of nodes in the path can be avoided. Hence the tree cost of DPSO has lower than GA.

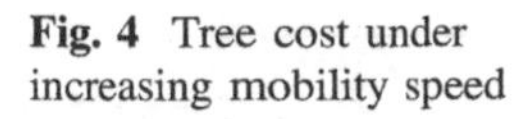

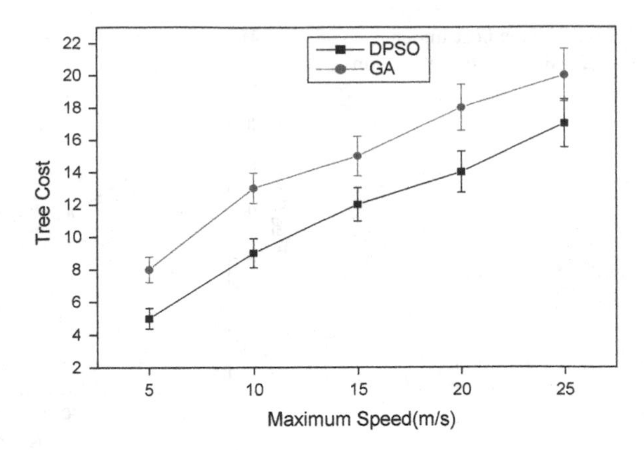

Fig. 4 Tree cost under increasing mobility speed

6 Conclusion

In this paper, Multicast routing in WMN with multiple constraints namely delay, bandwidth and cost are proposed by discrete particle swarm optimization (DPSO) algorithm. The initial multicast tree was generated using RandomwalkRSMT and penalty function was considered to eliminate replicated solutions during the particle and velocity updation. In order to see the effectiveness of the DPSO algorithm, we compared the results with the results of GA, through the two types of scenarios such as mobility speed and varying number of nodes and is observed in minimizes the tree cost. The future work includes the extension of DPSO based on multicast routing to incorporate multiple channels with multiple radio metrics.

References

1. I.F. Akyildiz, X. Wang, W. Wang, Wireless mesh networks: a survey, Computer Networks, 47(4), pp. 445–487, 2005.
2. R. Morales, S. Monnet, I. Gupta, G. Antoniu, Move: design and evaluation of a malleable overlay for group-based applications, IEEE Transactions on Network and Service Management, 4(2), pp. 107–116, 2007.
3. B. Wang, J. Hou, A survey on multicast routing and its QoS xtensions: problems, algorithms, and protocols, IEEE Networks, 14, pp. 22–36, 2000.
4. X. Wang, J. Cao, H. Cheng, M. Huang, QoS multicast routing for multimedia group communications using intelligent computational methods, Computer Communication, 29, pp. 2217–2229, 2006.
5. C. Oliveira, P. Pardalos, A survey of combinatorial optimization problems in multicast routing, Computer Operational Research, 32, pp. 1953–1981, 2005.
6. Neeraj Kumar, Rahat Iqbal, Naveen Chilamkurti, Anne James, An ant based multi constraints QoS aware service selection algorithm in Wireless Mesh Networks, Simulation Modelling Practice and Theory, 19, pp. 1933–1945, 2011.
7. Zongwu Ke, Layuan Li, Qiang Sun and Nianshen Chen, QoS Multicast Routing Algorithm for Wireless Mesh Networks, Eighth ACIS International Conference on Software

Engineering, Artificial Intelligence, Networking, and Parallel/Distributed Computing, pp. 835–840, 2007.
8. Hui Chenga, Shengxiang Yangb, Joint QoS multicast routing and channel assignment in multi-radio multichannel wireless mesh networks using intelligent computational methods, Applied Soft Computing, 11, pp. 1953–1964, 2011.
9. Jun Zhuo, Yu-zhong Chen, Yi-ping Chen, A Novel multiobjective optimization algorithm based on Discrete PSO for QoS Multicast routing in Wireless Mesh Networks, International Symposium on Intelligent Signal Processing and Communication Systems (ISPACS 2010), pp. 1–4, 2010.
10. Muhua Zhong and Yuzhong Chen, Discrete PSO Algorithm for QoS Routing optimization in Wireless Mesh Network, American Journal of Engineering and Technology Research, 11(9), pp. 136–141, 2011.
11. Uyen Trang Nguyen, On multicast routing in wireless mesh networks, Computer Communications, 31, pp. 1385–1399, 2008.
12. Meie Shen, Zhi-Hui Zhan, Wei-Neng Chen, Yue-Jiao Gong, Jun Zhang and Yun Li, Bi-Velocity Discrete Particle Swarm Optimization and its application to Multicast Routing Problem in Communication Networks, IEEE Transactions on Industrial Electronics, 61(12), pp. 7141–7151, 2014.
13. Yue-Jiao Gong, Jun Zhang, Ou Liu, Rui-Zhang Huang, Henry Shu-Hung Chung and Yu-Hui Shi, Optimizing the Vehicle Routing Problem With Time Windows: A Discrete Particle Swarm Optimization Approach, IEEE Transactions on Systems, Man and Cybernetics PART C: Applications and Reviews, 42(2), pp 254–267, 2012.
14. Alireza Sajedi Nasab, Vali Derhamia, Leyli Mohammad Khanlib, Ali MohammadZareha Bidoki, Energy-aware Multicast routing in manet based on particle swarm Optimization, Procedia Technology, 1, pp. 434–438, 2012.
15. Yannis Marinakis, Georgia-Roumbini Iordanidou, Magdalene Marinaki,Particle warm Optimization for the Vehicle Routing Problem with Stochastic Demands, Applied Soft Computing, 13, pp. 1693–1704, 2013.
16. J. Kennedy, R.C. Eberhart, Particle swarm optimization, Proceedings of the IEEE International Conference on Neural Networks, pp. 1942–1948, 1995.
17. R.V. Kulkarni and G.K. Venayagamoorthy, Bio-inspired algorithms for autonomous deployment and localization of sensor nodes, IEEE Transactions on Systems, Man and Cybernetics-Part C: Applications and Reviews, 40(6), pp. 663–675, 2010.
18. P. Kanakasabapathy and K.S. Swarup, Evolutionary tristate PSO for strategic bidding of pumped-storage hydroelectric plant, IEEE Transactions on Systems, Man and Cybernetics-Part C:Applications and Reviews, 40(4), pp. 460–471, 2010.
19. R.V. Kulkarni and G.K. Venayagamoorthy, Particle swarm optimization in wireless-sensor networks: A brief survey, IEEE Transactions on Systems, Man and Cybernetics-Part C: Applications and Reviews, 41(2), pp. 262–267, 2011.
20. Q. Zhu, L.M. Qian, Y.C. Li and S.J. Zhu, An improved particle swarm optimization algorithm for vehicle routing problem with time windows, IEEE Congress Evolutionary Computation, pp. 1386–1390, 2006.
21. Andrei Broder, Generating random spanning trees, IEEE 30th Annual Symposium on Foundations of Computer Science, pp. 442–447, 1989.
22. Godfrey C, Donald D, Differential evolution: a handbook For global permutation based combinatorial optimization, First edition, Springer, Germany, 2009.

Author Index

S.S. Dash et al. (eds.), *Artificial Intelligence and Evolutionary Computations in Engineering Systems*, Advances in Intelligent Systems and Computing 517,
DOI 10.1007/978-981-10-3174-8

Printed in the United States
By Bookmasters